JOURNAL OF CHROMATOGRAPHY LIBRARY - volume 67

monolithic materials

preparation, properties and applications

JOURNAL OF CHROMATOGRAPHY LIBRARY - volume 67

monolithic materials
preparation, properties and applications

edited by

František Švec
Department of Chemistry, University of California,
Berkeley, USA

Tatiana B. Tennikova
Institute of Macromolecular Compounds,
Russian Academy of Sciences,
St. Petersburg, Russia

Zdeněk Deyl
Institute of Physiology, Academy of Sciences
of the Czech Republic,
Prague, Czech Republic

2003

ELSEVIER
Amsterdam - Boston - London - New York - Oxford - Paris
San Diego - San Francisco - Singapore - Tokyo

ELSEVIER SCIENCE B.V.
Sara Burgerhartstraat 25
P.O. Box 211, 1000 AE Amsterdam, The Netherlands

First edition 2003

Library of Congress Cataloging in Publication Data
A catalog record from the Library of Congress has been applied for.

British Library Catologuing in Publication Data
A catalogue record from the British Library has been applied for.

ISBN: 0-444-50879-1
ISSN: 0301-4770 (Series)

♾ The paper used in this publication meets the requirements of ANSI/NISO Z39.48-1992 (Permanence of Paper).
Printed in The Netherlands.

Preface

Although the word *monolith*, relating to a large single block of stone, often shaped into a pillar ("monolithos"), was introduced by ancient Greeks, it first emerged in chromatography only within the last decade. At first, many practitioners did not believe that materials such as we will describe might attract the attention of the chromatographic community or, even less, be really useful. However, monolithic separation media were developed rapidly and their application soon became a reality thanks to the lucky coincidence of, (i), the considerable efforts of enthusiastic scientists who, despite hostile reactions of some reviewers, invoked the interest of separation scientists for these novel media by publishing their results, reporting in symposia, and giving seminars, and, (ii), the attention of the industry which was eager to achieve quantum-leap innovations in stationary phases for chromatography to enable both rapid separations and high throughput at low cost. Today, the situation is completely different. For example, special sessions concerning monolithic columns are becoming almost regular parts of prestigious scientific meetings. This has been demonstrated recently at the International Symposium HPLC 2002 in Montreal.

Another significant contribution to the popularity of monolithic materials has resulted from developments in the emerging field of separation science, capillary electrochromatography (CEC). The *in situ* preparation of the monoliths, without need for retaining frits, provides an almost ideal approach to the very efficient yet easy fabrication of CEC columns. Indeed, a number of research teams led by renowned scientists from all round the world, such as Stellan Hjertén, Czaba Horváth, Milos Novotny, Gerard Rozing, Richard Zare, Hanfa Zou, and many others — including the editors and authors of this volume — have recognized the potential of monoliths in CEC and quickly developed their own continuing, "monolithic programs". An even brighter future is foreseen for monolith in the field of microfluidic analytical systems.

Despite of their short history, Alois Jungbauer has recently called the monoliths "the fourth generation of chromatographic sorbents". Several monolithic devices for chromatographic separations have already been commercialized by companies such as BIA Separations (Ljubljana, Slovenia), BioRad Laboratories (Richmond, CA), Isco (Lincoln, NE), and Merck (Darmstadt, Germany).

The general problem typical for all monographs concerned with rapidly developing fields, including this one, is that they can never embrace everything. Although the vast majority of monoliths and their applications are directed towards bioseparations, which also represent a significant part of this book, we have also included chapters that demonstrate that monoliths can serve well in other important areas such as catalysis, solid-phase chemistry, optics, and microfluidics.

This book is a snapshot of the situation at the specific time at which the chapters were written. Obviously, the number of new papers relating to monoliths and published during the time of preparation of this book has increased significantly. Numerous new papers concerning monoliths are published every week. Since the Editors needed about six months to collect all of the contributions, some of them cover their areas up to December 2001 while others up to May 2002.

Similarly, it is difficult to cover all aspects of the subject to the equal extent. Each author, and therefore each chapter, is different. Some chapters are written more encyclopaedically, trying to cover all the material, while others provide a distilled survey of concepts. In contrast to some authors who present yet-unpublished results, others have focused largely on their own work from the recent past. Some areas are reviewed very deeply, with a number of experimental details, others demonstrate the breadth by covering a very wide field in a limited number of pages. However, all authors of this book are the best qualified experts in their fields and used their experience and knowledge to show what they consider most important in the areas on which they write.

All contributors certainly deserve our thanks and appreciation for their job well done. Obviously, this book would never come to reality were it not for the excellent chapters they have prepared. We also wish to thank Mr. Gilles Jonker and Ms. Reina Bolt from Elsevier Science Publishers who encouraged us to start, and Dr. Ivan Mikšík for his excellent desk-editing of all manuscripts and converting them into perfect camera-ready pages.

This book is, to the best of our knowledge, the first monograph concerning monoliths. Our aim was to achieve a maybe-less-than-perfect book and to publish it quickly, rather than to endlessly perfect something that would never see the printer's press. Since nothing in this world holds forever, and the field of monoliths is no exception, a new edition of this book might be desirable in the future. Therefore, we will certainly appreciate a feedback from the readers that may help us to improve our later work.

Berkeley, Prague, and St. Petersburg   František Švec
June 2002   Zdeněk Deyl
           Tatiana B. Tennikova

List of Contributors

Robert W. ALLINGTON
ISCO Inc., 4700 Superior Street, Lincoln, NE 608504-1328, USA

Miloš BARUT
BIA Separations d.o.o., Teslova 30, SI-1000 Ljubljana, Slovenia

Michael R. BUCHMEISER
Institute of Analytical Chemistry and Radiochemistry, University of Innsbruck, Innrain 52a, A-6020 Innsbruck, Austria

Jeremiah BWATWA
Department of Biomedical Engineering, Purdue University, West Lafayette, IN 47907-1295, USA

Neil R. CAMERON
Department of Chemistry, University of Durham, South Road, Durham, DH1 3LE, U.K.

Zdeněk DEYL
Institute of Physiology, Academy of Sciences of the Czech Republic, Videnská 1083, 142 20 Prague 4, Czech Republic

Christer ERICSON
Genomics Institute of the Novartis Research Foundation, 10675 John Jay Hopkins Drive, San Diego, CA 92121 USA

Mark R. ETZEL
University of Wisconsin, Department of Chemical Engineering, Madison, WI 53706, USA

Yolanda FINTSCHENKO
Sandia National Laboratories, Livermore, CA 94551, USA

Jean M. J. FRÉCHET
Department of Chemistry, University of California, Berkeley, CA 94720-1460, USA

Per-Erik GUSTAVSSON
Department of Pure and Applied Biochemistry, Center for Chemistry and Chemical Engineering, Lund University, P.O. Box 124, SE-221 00 Lund, Sweden

Rainer HAHN
   Institute for Applied Microbiology, University of Agricultural Sciences, Muthgasse 18, A-1190 Vienna, Austria

Ernest F. HASSELBRINK
   University of Michigan, Ann Arbor, MI 48109-2121, USA

Ken HOSOYA
   Department of Polymer Science and Engineering, Kyoto Institute of Technology, Matsugasaki, Sakyo-ku, Kyoto 606-8585, Japan

Christian G. HUBER
   Institute of Analytical Chemistry and Radiochemistry, Leopold-Franzens-University, Innrain 52 a, A-6020 Innsbruck, Austria

Tohru IKEGAMI
   Department of Polymer Science and Engineering, Kyoto Institute of Technology, Matsugasaki, Sakyo-ku, Kyoto 606-8585, Japan

Tao JIANG
   ISCO Inc., 4700 Superior Street, Lincoln, NE 608504-1328, USA

Djuro JOSIĆ
   Octapharma Pharmazeutika ProduktionsgesmbH, Research & Development, Oberlaaer Strasse 235, A-1100 Vienna, Austria

Alois JUNGBAUER
   Institute for Applied Microbiology, University of Agricultural Sciences, Muthgasse 18, A-1190 Vienna, Austria

Craig KEIM
   Department of Agricultural and Biological Engineering, Purdue University, West Lafayette, IN 47907-1295, USA

Brian J. KIRBY
   Sandia National Laboratories, Livermore, CA 94551, USA

Hiroshi KOBAYASHI
   Department of Polymer Science and Engineering, Kyoto Institute of Technology, Matsugasaki, Sakyo-ku, Kyoto 606-8585, Japan

Christine LADISCH
   Department of Consumer Sciences and Retailing, Purdue University, West Lafayette, IN 47907-1295, USA

Michael LADISCH
   Department of Biomedical Engineering, Purdue University, West Lafayette, IN 47907-1295, USA

Michael LÄMMERHOFER
   Christian Doppler Laboratory for Molecular Recognition Materials, Institute of Analytical Chemistry, University of Vienna, Währingerstrasse 38, A-1090 Vienna, Austria

IX

Per-Olof LARSSON
Department of Pure and Applied Biochemistry, Center for Chemistry and Chemical Engineering, Lund University, P.O. Box 124, SE-221 00 Lund, Sweden

Milton L. LEE
Department of Chemistry and Biochemistry, Brigham Young University, Provo, UT 84602-5700, USA

Chenghong LI
Department of Consumer Sciences and Retailing, Purdue University, West Lafayette, IN 47907-1295, USA

Wolfgang LINDNER
Christian Doppler Laboratory for Molecular Recognition Materials, Institute of Analytical Chemistry, University of Vienna, Währingerstrasse 38, A-1090 Vienna, Austria

Audrius MARUŠKA
Vytautas Magnus University, Department of Chemistry, Vileikos 8, LT-3035 Kaunas, Lithuania

Vera G. MATA
Laboratory of Separation and Reaction Engineering, Department of Chemical Engineering, Faculty of Engineering, University of Porto, Porto, Portugal

Mojca MERHAR
BIA Separations d.o.o., Teslova 30, SI-1000 Ljubljana, Slovenia

Igor MIHELIČ
Faculty for Chemistry and Chemical Technology, University of Ljubljana, Hajdrihova 19, 1000 Ljubljana, Slovenia

Ivan MIKŠÍK
Institute of Physiology, Academy of Sciences of the Czech Republic, Videnská 1083, 142 20 Prague 4, Czech Republic

Masanori MOTOKAWA
Department of Polymer Science and Engineering, Kyoto Institute of Technology, Matsugasaki, Sakyo-ku, Kyoto 606-8585, Japan

Herbert OBERACHER
Institute of Analytical Chemistry and Radiochemistry, Leopold-Franzens-University, Innrain 52 a, A-6020 Innsbruck, Austria

Luís PAIS
Laboratory of Separation and Reaction Engineering, Department of Chemical Engineering, Faculty of Engineering, University of Porto, Porto, Portugal

Eric C. PETERS
Genomics Institute of the Novartis Research Foundation, 10675 John Jay Hopkins Drive, San Diego, CA 92121 USA

X

Karin PFLEGERL
 Institute for Applied Microbiology, University of Agricultural Sciences, Muthgasse 18, A-1190 Vienna, Austria

Galina A. PLATONOVA
 Institute of Macromolecular Compounds, Russian Academy of Sciences, Bolshoy pr. 31, 199 004 St. Petersburg, Russia

Aleš PODGORNIK
 BIA Separations d.o.o., Teslova 30, SI-1000 Ljubljana, Slovenia

Alírio E. RODRIGUES
 Laboratory of Separation and Reaction Engineering, Department of Chemical Engineering, Faculty of Engineering, University of Porto, Porto, Portugal

Börje SELLERGREN
 University of Dortmund, INFU, Otto Hahn Strasse 6, 44221 Dortmund, Germany

Timothy J. SHEPODD
 Sandia National Laboratories, Livermore, CA 94551, USA

Anup K. SINGH
 Sandia National Laboratories, Livermore, CA 94551, USA

Aleš ŠTRANCAR
 BIA Separations d.o.o., Teslova 30, SI-1000 Ljubljana, Slovenia

František ŠVEC
 University of California, Department of Chemistry, Berkeley, CA 94720-1460, USA

David SÝKORA
 Department of Analytical Chemistry, Institute of Chemical Technology, Technická 5, 166 28 Prague, Czech Republic

Nobuo TANAKA
 Department of Polymer Science and Engineering, Kyoto Institute of Technology, Matsugasaki, Sakyo-ku, Kyoto 606-8585, Japan

Qinglin TANG
 Analytical Research and Development, Pharmaceutical Research Institute, Bristol-Myers-Squibb Company, Deepwater, NJ 08023, USA

Tatiana B. TENNIKOVA
 Institute of Macromolecular Compounds, Russian Academy of Sciences, Bolshoy pr. 31, 199 004 St. Petersburg, Russia

Shaofeng XIE
 ISCO Inc., 4700 Superior Street, Lincoln, NE 608504-1328, USA

Peidong YANG
 Department of Chemistry, University of California, Berkeley, CA 94720, USA

Yiqi YANG
> *Department of Textiles Clothing and Design, University of Nebraska, Lincoln, Lincoln, NE 68583-0802, USA*

Michal ZABKA
> *Laboratory of Separation and Reaction Engineering, Department of Chemical Engineering, Faculty of Engineering, University of Porto, Porto, Portugal*

Contents

XIV

Chapter 9   Monolithic Columns Prepared from Particles 197
Qinglin TANG and Milton L. LEE

Chapter 10   Layered Stacks 213
Mark R. ETZEL

Chapter 11   Biotextiles — Monoliths with Rolled Geometrics 235
Jeremiah BWATWA, Yiqi YANG, Chenghong LI, Craig KEIM,
Christine LADISCH and Michael LADISCH

Chapter 12   Polymerized High Internal Phase Emulsion Monoliths 255
Neil R. CAMERON

F. Švec, T.B. Tennikova and Z. Deyl (Editors)
Monolithic Materials
Journal of Chromatography Library, Vol. 67

Chapter 1

Historical Review

František ŠVEC[1] and Tatiana B. TENNIKOVA[2]

[1]*University of California, Department of Chemistry, Berkeley, CA 94720-1460, USA*
[2]*Institute of Macromolecular Compounds, Russian Academy of Sciences, Bolshoy pr. 31, 199 004 St. Petersburg, Russia*

CONTENTS

1.1   DEFINITIONS

A number of alternative names have been coined in the literature concerned with the new generation of materials covered by this book. For example, Hjertén introduced term "continuous polymer bed" [1] for his compressed polyacrylamide gel. In the late 1980s, we started to call our macroporous poly(glycidyl methacrylate-*co*--ethylene dimethacrylate) sheets prepared by a bulk polymerization process "macroporous polymer membranes" [2], whereas materials having similar macroporous structure and cylindrical shape that emerged in the early 1990s have been called "continuous polymer rods" [3], "porous silica rods" [4], or "continuous column support" [5].

The first appearance of the term "monolith" to describe a single piece of functionalized cellulose sponge used for the protein separation emerged in 1993 [6]. Similarly, expression "monolithic" related to rigid macroporous polymers prepared by bulk polymerization in a closed mold has also been introduced and rapidly became a standard [7]. Obviously, these single words are handier than the early multi-word expressions. The popularity of these terms is now supported by trade names of commercial processes and products that include in one way or another the word monolith. For example, BIA Separations (Ljubljana, Slovenia) is using their CIM® (Convective Interaction Media) disks for separation processes they call "high-performance *monolithic* chromatography" since 1998. Recently, Merck K.G. (Darmstadt, Germany) commercialized their silica rod columns under the trade name Chromolith® in which the expression *monolith* is also hidden. Based on these developments, the Editors of this book took the liberty to stick with the terms "monolith" and "monolithic" and minimize the use of the alternative expressions, thus unifying all the materials and methods under a single banner. It is worth noting that extruded ceramic materials with well-defined cellular structure widely used as supports for catalysts in automobiles are also called monoliths [8].

The word *monolith* originates from Greek μονολιτηοσ that combines two words μονο (single) and λιτηοσ (stone). While some areas of current chemistry such as ion exchange are finding their roots in The Bible [9], monoliths may be considered even older since the first exemplars have already been created during the early existence of this planet. Remember the well-known Ayers Rock in Australia, perhaps the largest monolith on the Earth. Their natural appearance has also led to the definition of monolith. For example, the Webster's College Dictionary [10], defines the noun *monolith* as (i) a large block of stone, (ii) something, such as a column, made from one large block, and (iii) something suggestive of a large block of stone. Although the word *column* in the second definition sounds familiar to those who practice chromatography, these definitions relate to the topic of this book only vaguely. In contrast, adjective *monolithic* explained as (i) constituting a monolith, (ii) massive, solid, and uniform, and (iii) constituting or acting as a single, often rigid, uniform whole, refers much closer to what we will call monoliths on the following pages. Obviously even this definition does not cover all the possible "clones" of monolithic technologies as they developed until these days. For example, the stacks of thin membranes such as Sartobind membrane absorbers (Sartorius, UK) or the discontinued MemSep chromatography cartridges (Milipore) used for the separations are difficult to accommodate under this roof. In contrast, homogeneous gels developed by Hjertén in Uppsala belong in this category. However and as always, one size never fits all.

1.2 DEVELOPMENT OF MONOLITHIC MATERIALS

1.2.1 Early attempts

Although many review articles concerning monolithic polymers in general and monolithic devices for chromatography in particular refer to the prior state-of-the-art, they most often do not describe the specifics of the pioneering work in this field. It is even likely that references to this work are often just copied from one paper to another and only a few authors may really know in detail what these "ancient" communications actually describe. Therefore we decided to uncover more details of these early approaches to monolithic materials and show examples of both their preparation and application.

1.2.1.1 Hydrogels

Historically, Kubín, Špaček, and Chromeček in the Institute of Macromolecular Chemistry in Prague appear to be the first who published the preparation of a continuous polymer matrix and its use for the chromatographic separation in size-exclusion mode. [11]. Their motivation was to find a synthetic alternative to crosslinked polysaccharides that were very popular at that time as separation media for "gel-filtration". These authors polymerized 100 mL of a 22% aqueous solution of 2-hydroxyethyl methacrylate containing 0.2% ethylene dimethacrylate in a 25 mm i.d. glass tube using a redox free-radical initiating system. After 24 h reaction at room temperature, the spongy elastic gel was removed from the tube and literally cooked in water to extract the unreacted components. The highly swollen 22 cm long soft rod was then drawn by vacuum in a column and used for the separation of water-soluble polymers. Simultaneously, another extracted monolith was disintegrated in a blender, the irregular particles dried, sieved, fraction 60–200 μm collected, re-swollen in water, and packed in a column for comparative experiments. Using a hydrostatic pressure of 0.5 m water, the packed 23.5 mm i.d. column enabled a flow rate of 25 mL/h. In contrast, the permeation through the monolithic gel was only 4 mL/h (0.07 mL/min). The operational ability of this monolithic column has been demonstrated on the separation of polyvinylpyrrolidone and sodium chloride in water (desalination). The flow rate was too small to make the column useful for any actual separations and the longitudinal diffusion considerably contributed to a low efficiency of 552 plates/m that equaled a height equivalent to theoretical plate (HETP) of 1.81 mm. The column packed with the hydrogel particles afforded slightly better HETP of 0.88 mm but the selectivity was also rather poor. Unfortunately, the limited range of experimental methods available at that time did not allow any better characterization of this first monolithic material.

The failure of this early approach most likely results from both unsuitable porous structure of the gel and column clogging by the action of hydrostatic pressure used to drive the flow. Although some additional optimization could perhaps slightly improve the separation properties of these materials, this first attempt was not very successful and has been forgotten for a number of years. Only very recently Hjertén revisited the homogeneous swollen gels, used them in capillary electrochromatographic mode with electroosmosis as the flow driving force and achieved excellent separations [12].

1.2.1.2 Foams

In the early 1970s, two groups experimented independently with polyurethane foams prepared in situ within the confines of large chromatographic columns [13,14]. The Monsanto group [13] adopted the original Albert Zlatkis' idea of shaping polymer foams and inserting them into a column. They extended the concept and prepared open pore polyurethane foam in situ. The authors claim the following advantages of their material [13]:

> *"1. Glass and metal columns of various configurations, lengths, and diameters (including capillary columns) can readily be filled because of the low viscosity of the precursor reagent.*
>
> *2. The polyurethane structure adheres tightly to the interior walls of columns of most types of materials of construction, thus preventing channeling along the support-column wall interface and providing "built in" continual baffles.*
>
> *3. The porosity, density, surface area, and flow characteristics can be controlled by varying reaction conditions.*
>
> *4. The material can be used as a gas-solid, gas-liquid, liquid-liquid, and thin layer chromatographic support.*
>
> *5. Stationary phases can be added either by incorporation with the reactants or by a solution method after the support is formed.*
>
> *6. Compounds with relatively low vapor pressures can be analyzed at low column temperatures."*

Although the first communication falls short on details concerning the preparation of these monolithic columns, most likely to avoid interference with their patent application prosecuted at the same time [15], it clearly demonstrates that a decent gas chromatographic separations of various hydrocarbon mixtures could be achieved. Fig. 1.1 shows the separation of C_6–C_9 alkanes as an example. The second publication that

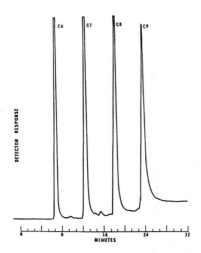

Fig. 1.1. GC separation of C₆–C₉ alkanes on open pore polyurethane column containing 10 wt.% of Dow Corning Silicone Fluids 550 included in the polymerization mixture. (Reprinted with permission from [13]. Copyright 1974 Preston Publ.). Conditions: Glass column 50 × 0.6 cm i.d., temperature gradient 50–140°C in 5 min.

followed in 1973 has already fully revealed the specifics of the technology [16]. The polyurethane foams were produced from a mixture containing 4,4'-diphenyl-methanediisocyanate and a polyol, which was defined as a reaction product of diethylenetriamine with propylene oxide, in various proportions dissolved in an isodensity solvent composed of 60:40 toluene–carbon tetrachloride mixture. This mixture was drawn in the column tube, sealed, and polymerized at room temperature. The polymerization reaction was initiated by the tertiary amine functionalities of the polyol component and continued for 18 h upon continuous tumbling that prevented settling of the polymer and eliminated channeling. Once the reaction was completed, the solvent was removed from the pores of the monolith by pressure of nitrogen and the column conditioned at 100°C for 24 h.

The crosslinked polyurethane formed during this process was not soluble in the polymerization mixture and separated from the system as a new phase in a shape of interconnected spherical units. Except for the size of the globules, this structure shown in Fig. 1.2 is quite similar to that of the typical current macroporous polymer mono-liths. The size of the globules could be controlled in a broad range of 1–10 μm by varying the dilution of the reaction mixture. These globules were nonporous as can be inferred from a low specific surface area of only 0.4 m²/g found for the foam [13]. Obviously, the larger the globules, the larger the pores, and the better the permeability

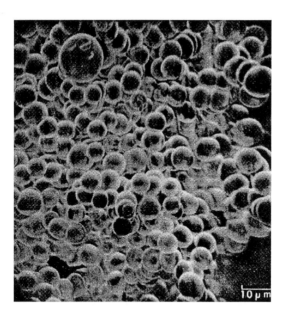

Fig. 1.2. Photomicrograph of open pore polyurethane structure (Reprinted with permission from [13]. Copyright 1970 Preston Publ.)

of the monolithic foam. However, the best chromatographic properties such as a column efficiency of 800 plates/m determined for decane and an average resolution of 5.50 for the separation of C_{12}, C_{13}, and C_{14} alkanes were only achieved with an optimized material that had a bulk density of 0.178 g/mL. The polyurethane was stable at temperatures up to about 200°C, a limit that seriously restricted its application in GC [16].

Since the surface of the "bare" polyurethane is rather polar, it could be used directly for the GC separations of volatile compounds in the gas-solid mode. In addition, the surface polarity could also be readily changed using a variety of liquid phases anchored to the surface. This was achieved either by admixing the liquid stationary phase to the polymerization mixture or by more common coating of the surface of the monolithic polyurethane. The coated GC columns operating in the gas-liquid mode had both higher efficiency and sample capacity [16].

The polyurethane monoliths were also used in liquid chromatography [17]. Tests demonstrated that the columns resisted pressures of up to 10 MPa at which the monolith was detached from the wall and compacted. The back pressure of only 0.2 MPa at a flow rate of 4 mL/min through a 100 × 0.2 cm i.d. column determined for the isopropanol/heptane mobile phase indicated a good permeability of the monolith

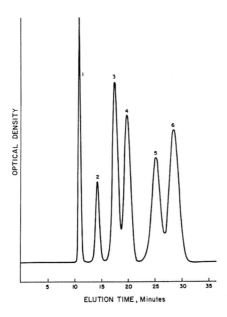

Fig. 1.3. Separation of dichloroanilines by liquid chromatography on open pore polyurethane column. (Reprinted with permission from [17]. Copyright 1974 Preston Publ.). Conditions: Polyurethane OH:NCO ratio 2:1, column 100 × 0.2 cm i.d., solvent 20% 2-propanol in heptane, flow rate 0.26 mL/min, inlet pressure 0.34 MPa. Peaks: benzene (1), 2,6-dichloroaniline (2), 2,4-dichloroaniline (3), 2,3-dichloroaniline (4), 3,5-dichloroaniline (5), and 3,4-dichloroaniline (6).

for liquids. Except for dimethylformamide, the polyurethane did not swell appreciably in any common LC solvent.

The proportion of diisocyanate and polyol affected the chromatographic performance as it controlled both internal structure and polarity of the surface. An excess of the polyol has been required to achieve the best result. Fig. 1.3 shows the separation of dichloroanilines in a normal-phase mode. The column efficiency calculated for 2,6-dichloroaniline was 1720 plates/m.

The German group [14] that independently explored the polyurethane foams at about the same time demonstrated two techniques: (i) production of a foam in situ and (ii) packing columns with the powdered foam. They expanded the arsenal of polymers and, in addition to polyurethane, also used other foamed polymers such as styrene copolymers, natural rubber, and polyethylene. All these columns were used in gas-solid GC. Unfortunately, almost no experimental details and only a limited discussion are presented in their report.

References pp. 14-15

By today's standards, the performance of the monolithic polyurethane columns was poor. However, it was comparable with the packed columns of the early 1970s. In fact, one of the reports [16] indicates that these monolithic columns were commercially available from Analabs Inc. located in North Haven, CT. It is now difficult to speculate why these columns were not more widely accepted. Perhaps their preparation was too complicated compared to a simple packing of columns with particulate solids. Their thermal stability was also much lower that that of inorganic packings typically used in GC. Maybe, their fixed surface chemistry was less suitable for LC in a wide variety of separation modes. The stability of the urethane bonds of the polymer under long-term exposure to solvents and analytes has never been tested. No word concerning column-to-column reproducibility either. Most likely though, it was too difficult for these novel columns to compete at that time with the just establishing technologies relying on packed inorganic particles.

However, similar to renewed interest in hydrogel-based columns, the polyurethane monoliths have also been recently partly revitalized since a monolithic polyurethane foam was used as a template for the preparation of large blocks of macro/microporous zeolites with the potential application in catalysis [18].

1.2.2 Recent developments

After the initial efforts described in the previous sections, the interest in monolithic separation media faded for almost two decades. In contrast to the monoliths described above, technical details of monolithic technologies developed recently are thoroughly presented in the following chapters of this book. Therefore, using our own memory and detailed description presented by Professor Hjertén [19], we will focus more on the reasons why and how the three most often used technologies were introduced rather than on the detailed description of the procedures and applications.

1.2.2.1 Compressed beds

Professor Hjertén at the University of Uppsala in Sweden knew from his early experiments that beds of soft beads packed in chromatographic columns could be deformed to make the distances between the beads smaller thus achieving an enhanced resolution [20]. The major difficulty was to design such beds that simultaneously fulfilled two contradictory requirements in order to obtain the desired high resolution: a relatively low backpressure upon a strong compression of the beads. His team designed such beds already in the 1960s and separated monomers and dimers of albumin. However, similar to the experiments described by Kubín [11], the flow rate was low and, therefore, they never published the results.

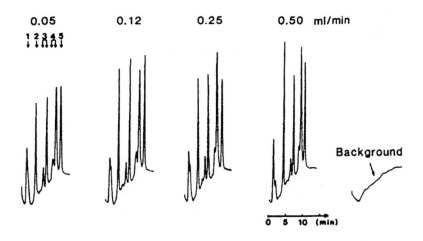

Fig. 1.4. HPLC of model proteins in cation-exchange mode using compressed continuous gel at different indicated flow rates. (Reprinted with permission from [1]. Copyright 1989 Elsevier) Conditions: gel plug 3 × 0.6 cm, linear gradient from 0.01 mol/L sodium phosphate buffer pH 6.4 to 0.25 mol/L sodium chloride in the buffer, gradient volume 5 mL. Peaks: alcohol dehydrogenase (1), horse skeletal myoglobin (2), whale myoglobin (3), ribonuclease A (4), and cytochrome c (5).

Continuing in these efforts, the workers in Uppsala observed again in the late 1980s the counterintuitive increase in the resolution of peaks in columns packed with crosslinked non-porous agarose following the increase in the flow rate. Since a higher pressure has been used to achieve the desired flow rate and the beads were soft, the chromatographic bed was compressed under these conditions. The interstitial volumes between the beads became smaller and also changed the flow profile through the bed. Surprisingly, the coincidence of these effects was advantageous and significantly improved the separation [21]. Since Hjertén also had an extensive experience with crosslinked polyacrylamide gels and observed changes in their appearance upon a considerable increase in crosslinking, he integrated both these observations and designed a chromatographic medium using an entirely new concept first published in 1989 [1]. The simplest implementation involved polymerization of a dilute solution of 240 mg N,N'-methylenebisacrylamide and 0.005 mL acrylic acid in 10 mL phosphate buffer that formed bed of a soft polymer. In an alternative approach, the gel was cut to small pieces, modified if desired, and its dispersion poured in a tube. The following compression of the bed to only about 17 % of its original length resulted in a column enabling an excellent chromatographic separation of proteins in ion-exchange mode as demonstrated in Fig. 1.4. Professor Hjertén comments the success of his approach in the following way [19]:

> *"However, it is not an easy task to design a synthesis method that affords a bed with the desired properties, which is illustrated very well by the following quotation from a letter I received shortly after our first paper on the continuous beds was published: "These papers are of great interest to me: For years I have been considering 'continuous' beds, but Prof. XX and Dr. YY warned me that the high resistance and slow flow-rate would be difficult to overcome." My answer, based on the experiments described below, is, certainly, also illuminating. "As your colleagues have pointed out, it is not easy to synthesize continuous beds with low flow resistance. I believe that many researchers have tried to prepare such beds, but the difficulties to achieve both high flow rates and high resolution are not so easy to overcome."*

After their additional development, these compressed beds with ion-exchange functionalities became commercially available from BioRad Laboratories (Richmond, CA) under the trade name UNO®. Further significant refinements and applications of this technology occurred once it approached the field of capillary electrochromatography in the mid 1990s [22].

1.2.2.2 Monolithic disks

In the mid 1980's, Professor Belenkii and his group at the Institute of Macromolecular Compounds in St. Petersburg, which was then a part of the Academy of Sciences of USSR, studied very thoroughly gradient methods for the chromatographic separation of proteins using stationary phases with a variety of chemistries and column geometries. Their theoretical considerations published later [23,24] have led to a conclusion that only a certain and often a rather short distance is required for the separation of proteins in the gradient elution mode. This later resulted in the concept of thin separation beds. However, it was very difficult to create such short beds from particulate sorbents due to irregularities in packing density and excessive channeling. Therefore, novel approaches were needed.

The Prague Institute of Macromolecular Chemistry studied the preparation of macroporous polymer beads with different chemistries since the early 1970 [25,26]. One of the projects in 1980s concerned highly selective sorbents for metal ions using chelating functionalities attached to the surface of porous beads. Since objects in the bead shape were not well suited for the measurements of mass transfer through the pores, macroporous poly(glycidyl methacrylate-*co*-ethylene dimethacrylate) were also prepared in format of a thin layer by bulk polymerization in a closed flat mold. Although this study has never been published, it happened to be an excellent starting

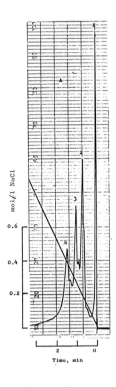

Fig. 1.5. Authentic chart demonstrating separation of myoglobin (1), conalbumin (2), ovalbumin (3), and soybean trypsin inhibitor (4). Conditions: Disk poly(glycidyl methacrylate-co-ethylene dimethacrylate) modified with diethylamine, 2 mm thick, 20 mm diameter, mobile phase 0.01 mol/L TRIS buffer pH 7.6, 2 min gradient from 0 to 0.6 mol/L sodium chloride in the buffer.

point for a successful cooperation between both Institutes that has led to the successful development of chromatographic media in a novel shape – flat disks.

In 1987, both authors of this chapter joint their forces and worked together in the laboratories of the Institute in Prague. Their collaborative work considerably accelerated the development of the new thin disk technology for the separation of proteins. Fig. 1.5. shows an authentic charts of the very early ion-exchange chromatographic separation of four proteins in gradient mode using monolithic disk placed in a simple home-made cartridge. The first communication concerning the new separation method, which we called at that time *high-performance membrane chromatography* (HPMC), has been submitted to Journal of Chromatography already in 1988. Unfortunately, the reviewers did not recognize the potential of the new approach, it took many months until the paper was rejected, and its publication considerably delayed. The manuscript that was then accepted in Journal of Liquid Chromatography and published in 1990 [2] stirred up a considerable interest documented by more than 300 cards obtained by the authors requesting sending the preprint of this paper. Interestingly enough, Hjertén had very similar experience with publication of his pioneering work introducing the compressed beds [20].

In contrast to some parts of the scientific community, the industry has quickly recognized the potential of the separation media in this atypical shape. Thus, Knauer GmbH in West Berlin (Germany) quickly acquired the rights to produce and market the HPMC technology. The license was later transferred to the spin-off company Säulentechnik GmbH where the group led by Josic and Reusch continued in the product development [27]. The first series of Quick-Disks® was introduced at the market in 1991. Due to a number of unfavorable circumstances, Säulentechnik discontinued the development and marketing. However, soon thereafter, a small company BIA (recently renamed to BIA Separations, d.o.o.) located in Ljubljana (Slovenia) also noticed the advantages of the monolithic disk technology and got rapidly involved. Using this technology, BIA Separations has successfully developed a series of products called CIM® Disks in a short period of time and commercialized them in 1998. Their current success relies on a balanced combination of research, development, and marketing.

1.2.2.3 Monoliths in cylindrical shapes

The development of rigid monolithic columns begun at Cornell University in the early 1990s [3]. Although the chemical nature and the macroporous structure of these monoliths is similar to that of the disks, in contrast to the disks that are produced from pre-cast blocs, placed in a recyclable cartridge, and then used for the separations, cylindrical monolithic columns that have a different aspect ratio are prepared directly in a column tube by in situ polymerization and used immediately as the separation device.

A completely new and important issue of this research was to achieve filling of the entire tube cross section with the monolithic polymer and avoid channeling at the column wall-monolith interface. Series of experiments clearly demonstrated that the mechanism of bulk polymerization in an unstirred tube was quite different from that of typical suspension polymerization. By the choice of suitable reaction conditions at which the polymerization proceed slowly, the nuclei and their clusters could sediment to the bottom of the upright standing mold where they formed growing layer of porous structure. Since the monolith grew from the bottom up, the free space created by radial shrinkage of the polymer accompanying the polymerization process was filled with the liquid polymerization mixture remaining at the top of the already created monolith thus ensuring success of this approach [28].

Another unknown territory was the flow through the long body of the macroporous monolithic column. Many authorities in the chromatographic field we discussed this issue with were very skeptical in this respect. This is not dissimilar to the experience of S. Hjertén. It was also unclear at the beginning whether the viscous porogenic

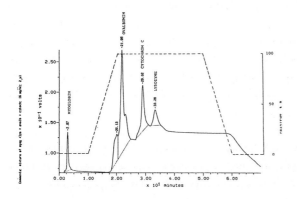

Fig. 1.6. Authentic chart demonstrating separation of myoglobin (1), ovalbumin (2), cytochrome C (3), and lysozyme (4) using monolithic 30 × 8 mm i.d. column. Conditions: poly(glycidyl methacrylate-co-ethylene dimethacrylate) monolith modified with diethylamine, mobile phase 0.01 mol/L TRIS buffer pH 7.6 for 10 min, followed by a 10 min gradient from 0 to 1 mol/L sodium chloride, UV detection at 218 nm.

liquid could be removed from the pores of a five or more centimeters long rod using pressures tolerable by typical chromatographic equipment. Surprisingly, the flow through occurred to be only a small problem thanks to the well-designed porous structure that was very different from that of classical macroporous beads. The porogens could also be easily removed from the pores using pressures of only 5–15 MPa depending on the length of the columns. The first monolithic column was successfully used for the separation of proteins in the ion exchange mode. The original chart of this separation is shown in Fig. 1.6. These monolithic columns have been recently commercialized by ISCO Inc. under the trade name Swift (Lincoln, Nebraska).

1.3 WHAT FOLLOWED

After the publications of the original communications describing the modern monoliths that have their roots in the 1980s, the field of monolithic materials grew steadily and several new approaches emerged. Most noticeable is the development of inorganic monoliths in Japan. In 1991, Professors K. Nakanishi and N. Soga at the Kyoto University reported for the first time formation of bulk silica gels from sol mixtures containing organic polymers that have a morphology including micrometer range pores [29]. Knowing the development of organic monolithic separation media at Cornell, Professor Nobuo Tanaka at the Kyoto Institute of Technology quickly realized the potential of these monoliths with well-controlled properties. He modified their surface with octadecyl functionalities, encased the monolith in heat shrinking Teflon tubing and placed it in radially compressed cartridge. This column was then successfully used for chromatographic separation of polypeptides and the results presented in a Japanese journal already in 1993 [30]. However, it took another three years before this research was accepted and published in Analytical Chemistry [4]. Recently commercialized and very successful Chromolith® columns are based on this technology.

The monolithic tubes developed short time ago for large-scale separations are another interesting contribution to the field of monoliths [31]. These columns are used in the radial flow mode. The major advantage of this shape is the easy dissipation of the heat of polymerization during the preparation and high loading capacities.

Real explosion of monolithic materials followed the renewed interest in capillary electrochromatography (CEC) in the second half of 1990s. A wide variety of monolithic approaches has been developed and successfully applied for the efficient separations in CEC mode [32].

As at the present, the monoliths are no longer used in chromatographic separations only. They also found their way to a variety of other fields such as enzyme immobilization and bioengineering, solid phase organic synthesis, flow injection analysis, and diagnostics. Most of these developments are detailed on the following pages of this book.

1.4 ACKNOWLEDGMENT

Support of this work by grant of the National Institute of General Medical Sciences, National Institutes of Health (GM–48364) is gratefully acknowledged. The authors also wish to thank Professor Stellan Hjertén for sharing his memories with us that reminisced the early times of the development of monolithic materials in Sweden.

1.5 REFERENCES

1   S. Hjertén, J.L. Liao, R. Zhang, J. Chromatogr. 473 (1989) 273.
2   T.B. Tennikova, F. Svec, B.G. Belenkii, J. Liquid Chromatogr. 13 (1990) 63.
3   F. Svec, J.M.J. Fréchet, Anal. Chem. 54 (1992) 820.
4   H. Minakuchi, K. Nakanishi, N. Soga, N. Ishizuka, N. Tanaka, Anal. Chem. 68 (1996) 3498.
5   S.M. Fields, Anal. Chem. 68 (1996) 2709.
6   R. Noel, A. Sanderson, L. Spark, in: J.F. Kennedy, G.O. Phillips, P.A. Williams (Eds), Cellulosics: Materials for Selective Separations and Other Technologies, Horwood, New York, 1993, pp. 17-24.
7   C. Viklund, F. Svec, J.M.J. Fréchet, K. Irgum, Chem. Mater. 8 (1996) 744.
8   J.L. Williams, Catal. Today 69 (2001) 3.

9   F. Helfferich, Ion Exchange, McGraw-Hill, New York, 1962.

10   Webster's College Dictionary, Random House Inc., New York 1991, p. 877.

11   M. Kubín, P. Špaček, R. Chromeček, Coll. Czechosl. Chem. Commun. 32 (1967) 3881.

12   A. Vegvari, A. Foldesi, C. Hetenyi, O. Kocnegarova, M.G. Schmid, V. Kudirkaite, S. Hjertén, Electrophoresis 21 (2000) 3116.

13   W.D. Ross, R.T. Jefferson, J. Chrom. Sci. 8 (1970) 386.

14   H. Schnecko, O. Bieber, Chromatographia 4 (1971) 109.

15   I.O. Salyer, R.T. Jefferson, W.D. Ross (1971). US pat.3580843.

16   F.D. Hileman, R. E. Sievers, G. G. Hess, W. D. Ross, Anal. Chem. 45 (1973) 1126.

17   T.R. Lynn, D.R. Rushneck, A.R. Cooper, J. Chromatogr. Sci. 12 (1974) 76.

18   Y.J. Lee, J.S. Lee, Y.S. Park, K.B. Yoon, Adv. Mater. 13 (2001) 1259.

19   S. Hjertén, Ind. Eng. Chem. Res. 38 (1999) 1205.

20   S. Hjertén, Personal Communications.

21   S. Hjertén, Y. Konguan, J.L. Liao, Macromol. Chem., Macromol. Symp. 17 (1988) 349.

22   S. Hjertén, D. Eaker, K. Elenbring, C. Ericson, K. Kubo, J.L. Liao, C.M. Zeng, P.A. Lindström, C. Lindh, A. Palm, T. Srichiayo, L. Valcheva, R. Zhang, Jpn. J. Electrophor. 39 (1995) 105.

23   V.G. Mal'tsev, D.G. Nasledov, S.A. Trushin, T.B. Tennikova, L.V. Vinogradova, I.N. Volokitina, V.N. Zgonnik, B.G. Belenkii, J. High Resolut. Chromatogr. 13 (1990) 185.

24   B.G. Belenkii, A.M. Podkladenko, O.I. Kurenbin, V.G. Mal'tsev, D.G. Nasledov, S.A. Trushin, J. Chromatogr. 645 (1993) 1.

25   J. Čoupek, M. Křiváková, S. Pokorný, J. Polym. Sci., Polym. Symp. 42 (1973) 185.

26   F. Svec, J. Hradil, J. Čoupek, J. Kálal, Angew. Macromol. Chem. 48 (1975) 135.

27   D. Josic, J. Reusch, K. Loster, O. Baum, W. Reutter, J. Chromatogr. 590 (1992) 59.

28   F. Svec, J.M.J. Fréchet, Chem. Mater. 7 (1995) 707.

29   K. Nakanishi, N. Soga, J. Am. Ceram. Soc. 74 (1991) 2518.

30   N. Tanaka, N. Ishizuka, K. Hosoya, K. Kimata, H. Minakuchi, K. Nakanishi, N. Soga, Kuromatogurafi 14 (1993) 50.

31   A. Podgornik, M. Barut, A. Štrancar, D. Josic, T. Koloini, Anal. Chem. 72 (2000) 5693.

32   Z. Deyl, F. Svec, Capillary Electrochromatography, Elsevier, Amsterdam, 2001.

Preparation of Different Types of Monolithic Materials

F. Švec, T.B. Tennikova and Z. Deyl (Editors)
Monolithic Materials
Journal of Chromatography Library, Vol. 67

Chapter 2

Rigid Macroporous Organic Polymer Monoliths Prepared by Free Radical Polymerization

František ŠVEC and Jean M. J. FRÉCHET

Department of Chemistry, University of California, Berkeley, CA 94720-1460, USA

.

CONTENTS

2.1 INTRODUCTION

Since their inception in the late 1980s, a vast variety of useful porous monolithic materials have been described and their applications demonstrated. These monoliths were primarily used in chromatographic applications and emerged to address some of the well-known drawbacks of particulate sorbents such as the micrometer-sized beads used in packed high-performance liquid chromatography (HPLC) columns. Despite the immense popularity of bead-based separation media, the slow diffusional mass transfer of macromolecular solutes into stagnant pool of the mobile phase present in their pores and the large void volume between packed particles remain significant issues in applications such as the rapid and efficient separations of macromolecules [1]. Considerably improved mass transfer characteristics were observed with partly perfused beads [2,3], particles consisting of a rigid porous silica matrix with pores filled with a soft hydrogel [4,5], and also with non-porous beads made of silica [6,7] or of synthetic polymers [8,9]. Decreasing the size of the packings can also facilitate mass transfer, though this also decreases interstitial volumes and therefore very small beads can only be used in relatively short columns to avoid unreasonably high back pressures. The problems of interparticular volume and slow mass transfer do not exist in membranes and monolithic systems for which all of the fluids flow through their pores [10]. A broad variety of approaches to these systems including monolithic disks [11-13], compressed gels [14], stacks of membranes [10], and rolled fabrics [15] are described in the other chapters of this book.

In the early 1990s, we introduced *rigid* macroporous monoliths, usually in cylindrical shapes, and obtained by a very simple "molding" process [16,17]. This initial research was later followed by other groups [18-24] and, in addition to their original application in HPLC, the rigid monolithic rods have found a large number of applications in areas such as gas chromatography [25], capillary electrochromatography [26], microfluidics [27-30], catalysis [31,32], solid phase extraction [33], and supported organic reactions [34-36]. These applications are covered in other chapters.

2.2 PREPARATION OF MONOLITHS

The preparation of monolithic macroporous organic polymer rods using a "molding" process is simple and straightforward. A simplified scheme of this process is shown in Fig. 2.1. The mold, typically a tube, is sealed at one end, filled with a

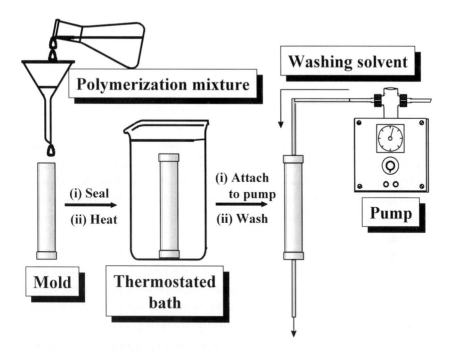

Fig. 2.1. Scheme of preparation of macroporous monolith using a simple "molding" process.

polymerization mixture, and then sealed at the other end. The polymerization is then triggered by heating in a bath or by UV irradiation. The latter approach also enables the formation of monoliths in specific location within transparent containers using irradiation through a mask in a lithography-like process. Once polymerization is completed, the seals are removed, the tube is provided with fittings, attached to a pump, and a solvent is pumped through the monolith to remove the porogens and any other soluble compounds that might remain in the polymer. A broad variety of molds of different sizes ranging from microfluidic channels to large bore columns several centimeters in diameter have been used. Mold materials included stainless steel, silicon, synthetic polymers (polyether-ether-ketone, poly(methyl methacrylate), polypropylene, cyclic olefin copolymer, polycarbonate), fused silica, quartz, and glass and mold shapes varied from tubular (rigid tubes, flexible capillaries) or channel-like (microfluidic chips) the even large area flat sheets [16,19-24,28,37-40].

While the preparation of cylindrical monoliths with a homogeneous porous structure in capillaries and tubes with a diameter of up to about 10–25 mm is readily achieved in a single polymerization step, larger size monoliths are somewhat more difficult to prepare due to slow dissipation of the heat of polymerization. However, two methods [39,42] that circumvent this problem have been demonstrated recently and are described elsewhere in this book .

2.3 MACROPOROUS POLYMERS

The most valued feature of porous monoliths is their well-defined macroporous structure. It is therefore appropriate to review here the concepts that are key to the formation and control of macroporous structures. Particulate macroporous polymers emerged in the late 1950s as a result of the search for polymeric matrices suitable for the manufacture of ion-exchange resins with better osmotic shock resistance and faster kinetics. The history of these inventions has been recently reviewed [43]. In contrast to the polymers that require solvent swelling to become porous, macroporous polymers are characterized by a permanent porous structure formed during their preparation that persists even in the dry state. Their internal structure consists of numerous interconnected cavities (pores) of different sizes, and their structural rigidity is secured through extensive crosslinking. These polymers are typically produced as spherical beads by a suspension polymerization process that was invented in Germany in the early 1910s [44-47]. Macroporous polymers are finding numerous applications as both commodity and specialty materials. While the former category includes ion-exchangers and adsorbents, supports for solid phase synthesis, polymeric reagents, polymer-supported catalysts, and chromatographic packings fit well into the latter [48]. Although the vast majority of current macroporous beads are based on styrene-divinylbenzene copolymers, other monomers including acrylates, methacrylates, vinylpyridines, vinylpyrrolidone, and vinyl acetate have also been utilized [48]. Numerous reviews of the preparation of macroporous polymers by suspension polymerization have appeared [49-52].

2.3.1 Mechanism of pore formation in beads

To achieve the desired porosity, the polymerization mixture should contain both a crosslinking monomer and an inert diluent, the porogen [49-52]. Porogens can be solvating or non-solvating solvents for the polymer, or they can be soluble polymers, or mixtures of soluble polymers and solvents.

The "classical" mechanism of pore formation during polymerization depends on the type of porogen used to create the porous structure. For example, during a typical polymerization in the presence of a precipitant (non-solvating solvent), the following mechanism applies [49,50,52]. The organic phase contains both monovinyl and divinyl monomers, initiator and porogenic solvent and the polymerization starts when the mixture is heated to create initiating radicals within the organic solution. The polymers thus formed become insoluble and precipitate in the reaction medium as a result of both their crosslinking and their lack of solubility in the porogen. In this implementation, the monomers are better solvating agents for the polymer than the porogen. Therefore, the precipitated insoluble gel-like species (nuclei) are swollen with the monomers that are still present in the surrounding liquid. The polymerization then continues both in solution and within the swollen nuclei. Polymerization within the latter is kinetically preferred because the local concentration of monomers is higher in the individual swollen nuclei than in the surrounding solution. Branched or even crosslinked polymer molecules formed in the solution are captured by the growing nuclei and further increase their size. The crosslinked character of the nuclei prevents their interpenetration and loss of individuality through coalescence. The nuclei, enlarged by the continuing polymerization, associate in clusters held together by polymer chains that crosslink the neighboring nuclei. The clusters remain dispersed within the liquid phase rich in the inert solvent (porogen) and continue to grow. In the latter stages of polymerization, the clusters are large enough to come into contact with some of their neighbors, thereby forming an interconnected matrix within the polymerizing system. The interconnected matrix becomes reinforced by both interglobular crosslinking and the capture of chains still polymerizing in solution, which leads the to the formation of the final porous polymer body. The fraction of voids, or macropores, within this final porous polymer is, at the end of the polymerization, close to the volume fraction of the porogenic solvent in the initial polymerization mixture because the porogen remains trapped in the voids of the porous crosslinked polymer.

2.3.2 Pore formation during bulk polymerization

The classical mechanism of the pore formation during the heterogeneous polymerization of monomers in the presence of thermodynamically poor solvents used as porogens does not allow a prediction of the porous properties that should result. The current knowledge of factors that control the pore size of bulk macroporous polymers remains mostly empirical as it is largely derived from the extensive published data concerning macroporous beads [49-52]. In addition, the knowledge acquired for the preparation of porous beads does not appear to be directly applicable to macroporous monoliths. For example, the pore size distribution of the molded monolith shown in Fig. 2.2 is quite different from that observed for "classical" macroporous beads prepared from the same polymerization mixture using both identical reaction time and temperature [38]. The former clearly contain very large pores or channels not commonly found in beads. This raises the question of why the pore size distribution of beads obtained by suspension polymerization is so much different from that of the monoliths? Could this difference be a direct result of the polymerization technique?

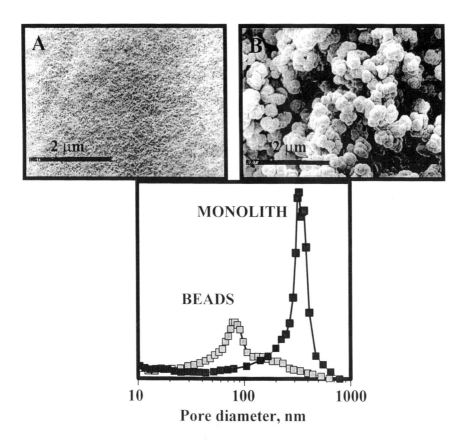

Fig. 2.2. Morphology and differential pore size distribution curves of poly(glycidyl methacrylate-*co*-ethylene dimethacrylate) beads and monolith prepared from identical polymerization mixtures (adapted from [53]). Polymerization mixture: glycidyl methacrylate 24%, ethylene dimethacrylate 16%, porogenic solvent cyclohexanol 48%, dodecanol 12%, AIBN 1% (with respect to monomers), temperature 70°C, time 12 h.

To find answers to these questions, a detailed comparative study of both these processes has been carried out [53].

2.3.2.1 Effect of the tension at the liquid-liquid interface

An analysis of the system used for suspension polymerization reveals that the interfacial tension between the aqueous and organic phases plays a very important role in droplet formation upon mixing. This role includes the control of both the size and the spherical shape of the droplets [45,46]. Another largely ignored effect of the interfacial tension, which is particularly important when designing the suspension polymerization process for the preparation of macroporous beads, is related to the shrinkage that occurs during any vinyl polymerization. Assuming that the process is free of coalescence, the size of the original droplets decreases during suspension polymerization as a result of the unavoidable volume shrinkage. Referring back to the mechanism of heterogeneous polymerization that leads to macroporous polymers, it is known that at the early stage of the suspension process, the nuclei are randomly dispersed within the rotating droplets of the polymerization mixture. Since the interfacial tension exerts a constant pressure on the surface of the droplets, their spherical shape is retained, but the combined effects of the interfacial tension and the shrinkage push the nuclei closer to one another. This process may even lead to the formation of a dense shell covering the outer surface of the beads [54]. Since the monomers are diluted with the porogen, the overall volume shrinkage is only about 6% as opposed to the value of approximately 15% expected for undiluted monomers [55]. Therefore, the interfacial tension is a likely contributor to the denser assembly of globules found in beads. However, this can hardly be the only reason for the dramatic differences observed between the pore size distributions of beads and monoliths.

2.3.2.2 Pore formation during polymerization in a mold

The external conditions accompanying the polymerization within a mold are quite different from those found in the suspension polymerization process. First, only one phase, the organic mixture, is present in the mold. Therefore, the interfacial tension between the aqueous and organic phases characteristic of the suspension process is absent. Moreover, in contrast to the droplets that revolve in the aqueous phase as a result of stirring, the contents of the tubular mold remain stationary during the polymerization.

The basic mechanism of the polymerization in the presence of a porogenic solvent outlined above, including the precipitation of nuclei and shrinkage, is general and remains operative regardless of the polymerization technique. However, the solid nuclei and their clusters have a higher density than the polymerization mixture. Therefore, in the absence of mixing and if the overall rate of polymerization is relatively slow, the solid can sediment and accumulate at the bottom of the mold. This is confirmed in experiments in which the mold is opened at an early stage of the process revealing the presence of a short, loosely packed rod submerged in the still liquid polymerization mixture above it. Further confirmation is obtained by visual observation of polymerization process carried out in a transparent mold. Thefore, the unstirred nuclei and their clusters fall to the bottom of the mold where they form a very loose, highly porous, and poorly organized structure early on during the polymerization.

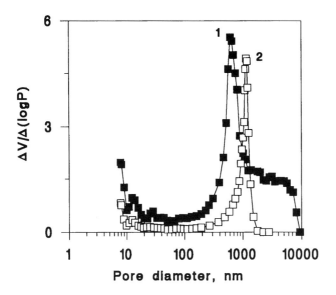

Fig. 2.3. Differential pore size distribution curves of the poly(glycidyl methacrylate-*co*-ethylene dimethacrylate) rods after 1 h (1) and 14 h (2) of polymerization in a 8 mm stainless steel tube at a temperature of 55°C (Reprinted with permission from [53], copyright 1995 American Chemical Society). Polymerization mixture: glycidyl methacrylate 24%, ethylene dimethacrylate 16%, porogenic solvent cyclohexanol 54%, dodecanol 6%, AIBN 1% (with respect to monomers).

This was confirmed by the presence of very large pores, as seen in the pore size distribution curve of the monolith isolated after only 60 min of polymerization (Fig. 2.3) and as confirmed by the large measured pore volume of 3.8 mL/g. The very high specific surface area of 523.9 m^2/g that is observed at this stage indicates that the nuclei retain a great deal of their individuality. However, as the polymerization proceeds further, they come into contact with each other and the loose structure becomes more cohesive as nuclei become interconnected as a result of continued formation of new polymer chains. Therefore, even the monolith obtained after 1 h of polymerization at 55°C retains its shape and does not fall apart upon removal from the tubular mold. This solid polymer has a very high porosity of 80% and therefore it is rather fragile.

The polymerization process continues both in the monomer solution above the growing rod, and in the mixture that permeates the swollen very large pores of the early rod resulting in the formation of numerous new nuclei. The following observations support the suggested mechanism: (i) the disappearance of the very large pores (as measured by mercury porosimetry) in the latter stages of the polymerization within a mold is a linear function of conversion, and (ii) the reduced individuality of the nuclei resulting from polymerization within the swollen nuclei and also by capture of the dissolved polymers. This leads to a decrease in both pore volume and surface areas for monoliths examined during the latter stages of the polymerization (Fig. 2.3).

In contrast to the monoliths, the growing nuclei in the droplets of the suspension polymerization mixture dispersed in the aqueous phase do not lose their individuality that early. The droplets revolve and the nuclei within the droplet move randomly as a result of the stirring and the centrifugal force lets. The nuclei are therefore able to keep their individuality longer, grow separately and "pack" better within the bead (droplet). As a result, the voids between the globules that constitute a single bead are smaller. Therefore, the dynamics of the system appears to be the main cause for the difference in the pore size distribution between the monoliths and "classical" porous beads.

2.3.3 Absence of radial shrinkage within the mold

The above discussion described the mechanism of pore formation during a polymerization within a mold but it still does not explain completely why no radial shrinkage of the polymer rod within the mold is observed. This is also likely the result of the absence of interfacial tension compressing the polymer as it is formed, the lack of mixing during polymerization, and, in some specific cases, the interaction of the monolithic material with the wall of the mold.

As was mentioned earlier, the overall volume shrinkage during the preparation of a typical monolith is only about 6%. Clearly, some of this shrinkage occurs during the early nucleation and growth within the "free-flowing" swollen nuclei. This decreases the overall volume of the polymerization mixture within the mold but does not affect the monolith that has not yet been formed. Any shrinkage in the size of the polymer monolith would only be expected to occur in the latter stages of the polymerization when the monolith already exists. At that time, the matrix is already heavily crosslinked and can hardly change its overall size. Any remaining shrinkage would therefore occur within the monolith itself, leading to the formation of larger or more numerous pores rather than leading to a reduction in its external dimensions.

However, even if some radial shrinking did occur, the "space" created would be filled by some of the polymerization mixture that is still present on top of the growing monolith up to a conversion of almost 70%. Therefore, radial shrinkage, if any takes place, is not deleterious to the function of the molded monoliths as no void is created between the monolith and the walls.

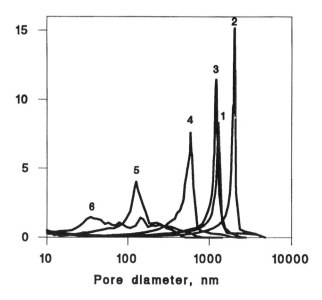

Fig. 2.4. Differential pore size distribution curves of the poly(glycidyl methacrylate-*co*-ethylene dimethacrylate) monoliths prepared by polymerization in a 8 mm stainless steel tube at a temperature of 55 (1), 60 (2), 65 (3), 70 (4), 80 (5) and 90°C (6) (Reprinted with permission from [56], copyright 1995 American Chemical Society). Polymerization mixture: glycidyl methacrylate 24%, ethylene dimethacrylate 16%, porogenic solvent cyclohexanol 54%, dodecanol 6%, AIBN 1% (with respect to monomers).

2.3.4 Effect of the polymerization rate on pore formation

The reaction rate for a free-radical polymerization is not a simple function of temperature as the process consists of several consecutive reactions. Typically, the initiation step has the highest activation energy. Therefore, it is the initiation rate that is most temperature dependent. For example, the half-life of azobisisobutyronitrile is 37 h at a temperature of 55°C but only 6 h at 70°C. Fig. 2.4. clearly document the effect of the polymerization temperature on the final pore size distribution of monoliths. The lower the temperature, the larger the pores. This is quite consistent with the suggested mechanism of pore formation. The decomposition rate of the initiator, the number of growing radicals, as well as the overall polymerization rate are higher at elevated temperature. Therefore, at higher temperatures, more nuclei are formed at once and they all compete to swell with the remaining monomers. The nucleation rate is faster than the swelling and the supply of monomers becomes exhausted after a shorter period of time. Since the number of nuclei that grow to the globular size is large but their sizes remain relatively small, the interstitial voids between smaller globules are also smaller. Thus, the size of pores is inversely related to temperature for both beads and monoliths. However, since the dynamic effects are not affected by temperature, the differences in porosity profiles between the two shapes persist.

If the polymerization is carried out in a mold at a lower temperature, the reaction rate is considerably slower. The number of nuclei is smaller and therefore they have more time to swell. As a result, their growth is perpetuated by the polymerization that proceeds within and they are able to sediment because space is available and the overall process is slow. This results in larger size globules and clusters. Therefore, the voids between clusters are also larger and the pore size distribution profile is shifted towards larger pores.

2.4 Control of pore size

2.4.1 Monoliths prepared using thermally initiated polymerization

Many applications of porous materials such as catalysis, adsorption, ion exchange, chromatography, solid phase synthesis, etc., rely on intimate contacts with a surface that supports the active sites. In order to obtain a large surface area, a large number of smaller pores should be incorporated into the polymer. The most substantial contributions to the overall surface area arises first from the micropores, with diameters smaller than 2 nm, and then from the mesopores with sizes in the range 2 to 50 nm. Large macropores only make an insignificant contribution to the overall surface area. However, these pores are essential to allow liquid to flow through the material at a reasonably low pressure. This pressure, in turn, depends on the overall porosity profile of the material. Therefore, the pore size distribution of the monolith is extremely important and it should be adjusted properly to fit each type of application. Key variables such as temperature, composition of the pore-forming solvent mixture, and content of crosslinking monomer enable the tuning of the average pore size within a broad range spanning several orders of magnitude from tens of nanometers to tens of micrometers.

2.4.1.1 Polymerization temperature

The polymerization temperature, through its effects on the kinetics of polymerization described above, is a particularly effective means of control, enabling the preparation of macroporous polymers with different pore size distributions from a single polymerization mixture. The effect of temperature can be readily explained in terms of the nucleation rates, and the shift in pore size distribution induced by changes in the polymerization temperature can be accounted for by the difference in the *number of nuclei* that result from these changes [56,57]. For example, while the sharp maximum

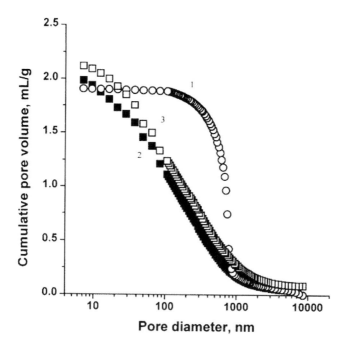

Fig. 2.5. Integral pore size distribution profiles of porous polymer monoliths prepared by a typical polymerization at 70°C (1), and in the presence (2) and the absence (3) of 2,2,6,6-tetramethyl-1-piperidyloxy (TEMPO) at 130°C. Reaction conditions: polymerization mixture, styrene, 20 wt %; divinylbenzene, 20 wt %; 1-dodecanol, 60 wt %; benzoyl peroxide (BPO) 0.5 wt % (with respect to monomers), TEMPO 1.2 molar excess with respect to BPO, reaction time 8, 24, and 46 h, respectively.

of the pore size distribution profile for monoliths prepared at a temperature of 70°C is close to 1000 nm, a very broad pore size distribution curve shown in Fig. 2.5 spanning from 10 to 1000 nm with no distinct maximum is typical for monolith prepared from the same polymerization mixture at 130°C [58].

2.4.1.2 Porogenic solvents

The choice of pore-forming solvent is another tool that may be used for the control of porous properties without affecting the chemical composition of the final polymer. Phase separation of crosslinked nuclei is a prerequisite for the formation of the macroporous morphology (*vide supra*). The polymer phase separates from the solution during polymerization as a result of its limited solubility in the polymerization mixture as well as the intrinsic insolubility that results from crosslinking.

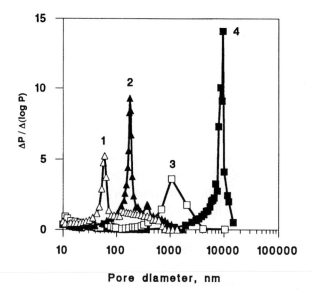

Fig. 2.6. Effect of toluene in the porogenic solvent on differential pore size distribution curves of molded poly(styrene-*co*-divinylbenzene) monoliths (Reprinted with permission from [57], copyright 1996 American Chemical Society). Conditions: polymerization time 24 h, temperature 80°C, polymerization mixture: styrene 20%, divinylbenzene 20%, cyclohexanol and toluene contents in mixtures 60+0 (1), 50+10 (2), 45+15 (3), and 40+20% (4).

The effect of the thermodynamic quality of the porogenic solvent system on the properties of the porous polymers has been documented for several polymerization systems. For example, porous poly(glycidyl methacrylate-*co*-ethylene dimethacrylate) monoliths are often prepared using cyclohexanol-dodecanol mixtures as porogens. A higher content of dodecanol leads to monoliths with larger pores. This results from an earlier onset of phase separation (nucleation) in the system that contains more dodecanol since this porogenic component is a more potent precipitant than toluene. The actual pore size distribution profiles Fig. 2.6 demonstrate that the addition of even a relatively small percentage of dodecanol to the polymerization mixture results in a dramatic decrease in pore sizes.

The shift in distribution towards smaller pore sizes by adding a better solvent results from the shift in timing of the phase separation that occurs during polymerization. In contrast, the addition of a poorer solvent to the polymerizing system results in an earlier phase separation of the polymer. The new phase preferentially swells with the monomers because these are thermodynamically much better solvents for the polymer than the porogenic solvent. As a result, the local concentration of monomers in the swollen nuclei is higher than that in the surrounding solution and the polymerization reaction proceeds mainly in these swollen nuclei rather than in the solution. Those newly formed nuclei obtained in the solution are likely to be adsorbed by the large pre-globules formed earlier by coalescence of many nuclei and further increase their size. Overall, the globules that are formed in such a system are larger and, consequently, the voids between them (pores) are larger as well. As the solvent quality improves, the good solvent competes with monomers in the solvation of nuclei, the local monomer concentration within the swollen nuclei is lower and the globules grow less and are smaller. As a result, porous polymers formed in good solvating solvents have smaller pores. Clearly, the porogenic solvent controls the porous properties of the monolith through the *solvation of the polymer chains* in the reaction medium during the early stages of the polymerization [57,59].

The pore size is also affected by the nature of the monomers used for the preparation of the monoliths. For example, polymerization mixtures containing 24% 2-hydroxyethyl methacrylate, 16% ethylene dimethacrylate, 54% cyclohexanol, and 6 % dodecanol did not afford materials with large pores while monoliths prepared from similar mixtures containing styrene or even glycidyl methacrylate were porous as expected [60]. A recent study has revealed that in order to obtain sufficiently large pores in monoliths containing 2-hydroxyethyl methacrylate, a porogenic mixture with a much higher content of dodecanol must be used than is needed for monoliths based on glycidyl methacrylate [61]. For example, GMA monoliths with a pore size of 1,000 nm are readily obtained by thermal polymerization at 60°C using a polymerization mixture containing 20% dodecanol and 40% cyclohexanol. In contrast, a much higher percentage of the less polar dodecanol (50%) is required for the preparation of 2-hydroxyethyl methacrylate monoliths with a similar pore size. An alternative approach involves the use of porogenic mixtures containing higher aliphatic alcohol such as tetradecanol with only a small amount of co-porogen [62].

Supercritical carbon dioxide was recently added to the broad family of porogenic solvents used for the preparation of porous monoliths using thermally initiated free radical polymerization [63]. This type of porogen is attractive since it is non-toxic, non-flammable, and inexpensive. In addition, the porogenic properties of this "solvent" can be fine tuned by varying the pressure in the mold. Once the polymerization is completed, the porogen is simply allowed to evaporate with no need for washing and no residual solvents in the monolith. Using ethylene dimethacrylate and trimethylolpropane trimethacrylate as monomers, broad range of materials was prepared featuring a typical macroporous structure and pore sizes ranging from 20 to 8,000 nm.

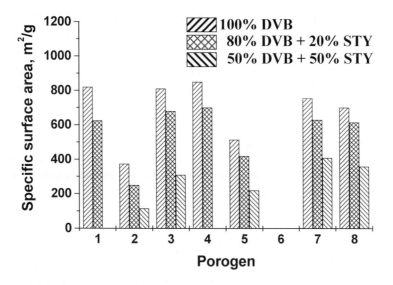

Fig. 2.7. BET surface areas for monoliths prepared from divinylbenzene (DVB) and styrene (ST) using various porogenic solvents (Reprinted with permission from [21], copyright 2001 American Chemical Society). Conditions: polymerization time 24 h, polymerization mixture: monomers 50%, porogenic solvent: tetrahydrofuran (1), acetonitrile (2), toluene (3), chlorobenzene (4), hexane (5), methanol (6), dimethylformamide (7), methyl *t*-butyl ether (8) 50%, AIBN 1% (with respect to monomers), temperature 80°C for porogens 2,3,4, and 7, 65°C for other.

2.4.1.3 *Percentage of crosslinking monomer*

In contrast to temperature, increasing the proportion of the crosslinking agent present in the monomer mixture affects not only the porous properties but also the chemical composition of the final monoliths. The average pore size decreases as a result of early formation of highly crosslinked globules with a reduced tendency to coalesce. Experimental results suggest that, in this case, the pore size distribution is controlled by the limited *swelling of crosslinked nuclei* [57]. However, the polymerization of mixtures with a high content of crosslinker is useful for the preparation of monoliths with very large surface areas. The effect of crosslinking on the preparation of poly(styrene-*co*-divinylbenzene) monoliths from three different mixtures using a variety of single solvent porogens is illustrated in Fig. 2.7. Clearly, the highest surface area, exceeding 800 m^2/g, is achieved for the polymerization of "pure" high grade divinylbenzene (80%) and the lowest for its 1:1 mixture with styrene regardless of the porogen [59].

Obviously, the pore size of monoliths with surface areas in the range of hundreds of square meters is very small and these materials are not permeable to liquids at reasonable pressures. However, they may serve as supports for the immobilization of catalysts active in reactions carried out in the gas or vapor phase. Since many catalytically active metallomonomers are air and water sensitive, preparation of porous beads from these monomers by means of classical suspension polymerization would not be possible. In contrast, the intrinsic advantage of the single phase preparation of monolithic materials enables the use of any monomer and, in addition, the broad variety of solvents that can be chosen as porogens may also help to achieve the desired solubility of the active monomer in the polymerization mixture [59].

2.4.2   Monoliths prepared using thermally initiated polymerization in the presence of stable free radicals

Living free-radical polymerization has recently attracted considerable attention since it enables the preparation of polymers with well-controlled compositions and molecular architectures, previously the exclusive domain of ionic polymerizations, using very robust conditions akin to those of a simple free radical polymerization [64-73]. Due to its potential ability to create materials with specific properties, living free radical polymerization has also been used for the preparation of porous polymer monoliths. In one of the implementations, the monolith is prepared first using bulk polymerization in the presence of a stable free radical such as 2,2,5,5-tetramethyl-1-pyperidinyloxy (TEMPO) [58]. The terminal nitroxide moieties that remain in the monolith can be used in a subsequent grafting step involving growth of the functional polymer within the pores of the previously formed parent monolith. This makes the stable free-radical assisted two-step polymerization process a versatile tool, since it allows the preparation of functionalized porous materials with a large variety of surface chemistries that originate from only a single type of parent monolith.

Unfortunately, this polymerization leads to monoliths with a less permeable porous structure shown in Fig. 2.5 as a result of the rather high reaction temperature of 130°C typically required to obtain high conversions using TEMPO as a stable free radical [58]. In contrast, the use of "low" temperature mediators such as 2,2,5-trimethyl-3-(1-phenylethoxy)-4-phenyl-3-azahexane [74] for the preparation of porous monoliths substantially simplifies the control of porous properties and polymers with a pore size of 50–1100 nm can be prepared [75].

Viklund *et al.* have utilized yet another type of stable free radical to prepare macroporous poly(styrene-*co*-divinylbenzene) monoliths. They used 3-carboxy-2,2,5,5-tetramethylpyrrolidinyl-1-oxy (carboxy-PROXYL) and 4-carboxy-2,2,6,6-tetramethyl-1-piperidinyloxy (carboxy-TEMPO), respectively, as the mediators and poly(ethylene glycol) with 1-decanol as the binary porogenic solvent [76]. These polymerizations were found to be faster and led to higher degrees of monomer conversions in a shorter period of time than analogous reactions using TEMPO. The use

TABLE 2.1

EFFECT OF COMPOSITION OF 50:50 wt.% METHANOL–COPOROGEN MIXTURES USED AS THE POROGENIC SOLVENT ON THE PORE SIZE OF POLY(GLYCIDYL METHACRYLATE-*CO*-ETHYLENE DIMETHACRYLATE) MONOLITHS [77].[a]

Coporogen	Boiling point, °C	Median pore size nm	Pore volume mL/g	Spec. surface area, m²/g
Ethanol	78	4700	1.8	0.6
Tetrahydrofuran	66	64	1.2	105.8
Acetonitrile	81	41	0.8	110.5
Chloroform	61	52	1.0	76.2
Ethyl acetate	77	53	1.4	63.5
Hexane	68	5220	1.5	0.43

[a]Polymerization conditions: ethylene dimethacrylate 0.96 g, glycidyl methacrylate 1.421 g, azobisisobutyronitrile 24 mg, total porogenic solvent 3.6 g, UV initiation at 365 nm for 16 h, room temperature.

of mediators with carboxylic functionality also accelerates the reaction kinetics and improved the permeability of the prepared monoliths. Modification of the composition of porogenic mixture enables control of the porosity profiles of the monolithic polymers over a wide range even allowing the preparation of monolithic HPLC column useful for size exclusion chromatography.

Grafting of these preformed monoliths containing "dormant" radicals is then achieved by filling their pores with a monomer solution and heating to the desired temperature to activate the capped radicals. For example, the functionalization of poly(styrene-*co*-divinylbenzene) monolith with chloromethylstyrene or 2-vinylpyridine leads to materials with up to 3.6 mmol/g of functionalities [75].

2.4.3 Monoliths prepared using photopolymerization

Only a limited number of porogenic solvents have been used for the preparation of large porous polymer monoliths using thermally initiated polymerizations. Specifically, mixtures of solvents with rather high boiling points such as dodecanol with toluene or cyclohexanol proved to be suitable for the preparation of numerous porous polymers from both styrenic and methacrylate monomers at temperatures as high as 130°C. In contrast, UV initiated polymerizations are typically carried out at room temperature. Therefore, solvents with much lower boiling points such as the lower alcohols and hydrocarbons and their mixtures shown in Table 2.1 can safely be used

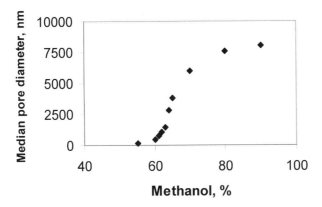

Fig. 2.8. Effect of composition of methanol-ethyl acetate mixture used as the porogenic solvent on the median pore size of poly(glycidyl methacrylate-*co*-ethylene dimethacrylate) monoliths (Reprinted with permission from [77], copyright 2002 Wiley-VCH). Polymerization conditions: ethylene dimethacrylate 0.96 g, glycidyl methacrylate 1.42 g, AIBN 24 mg, methanol+ethyl acetate 3.6 g, UV initiation at 365 nm for 16 h, room temperature.

even in less perfectly sealed molds such as channels of microfluidic devices. For example, Fig. 2.8 shows the very broad range of pore sizes that can be obtained for poly(glycidyl methacrylate-*co*-ethylene dimethacrylate) monoliths using mixtures of methanol and ethyl acetate [77].

The morphology of the monoliths is closely related to their porous properties, and, as was the case for the thermally initiated preparation, it again reflects the quality of the porogenic solvent, the percentage of crosslinking monomer, and the ratio between monomer and porogen. The synergism of these reaction condition variables was verified using multivariate analysis [78].

2.4.4 Monoliths prepared using γ-radiation

The use of γ-rays to initiate polymerization and produce porous polymer monoliths is rare and only one example can be found in the literature [79]. In a typical experiment, a 30% solution of glycidyl methacrylate and diethylene glycol dimethacrylate in a porogenic solvent such acetone, ethyl acetate, *t*-butanol, 1,4-dioxane, and methanol was placed in a Teflon tube, sealed in a plastic bag filled with nitrogen, and irradiated in a ^{60}Co γ-source at different temperatures and dose rates. The total dose typically used was in a range of 1–50 kGy. As expected, the porous properties of these monoliths depended on the choice of porogen, the percentages of the two monomers, their overall concentration in the solution, and also on both the dose of radiation and its

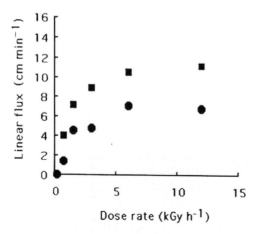

Fig. 2.9. Effect of radiation dose on the permeability of the monoliths for acetonitrile (Reprinted with permission from ref. [79], copyright 2001 Elsevier). Column 30×5 mm i.d., polymerization mixture: 30 (■) and 40% (●) diethylene glycol dimethacrylate in methanol. Permeability is calculated from the time needed for 5 ml of acetonitrile to pass trough the monolith at a pressure of 0.1 MPa.

rate. Obviously, these variables are almost identical to those used in the other polymerizations described above with the radiation dose relating to the concentration of free radical initiator. For example, Fig. 2.9 demonstrates that the pore size in monoliths prepared using γ-radiation initiated polymerization measured by permeability for flow depends on the dose and reaches its maximum for doses exceeding 5 kGy/h.

2.4.5 Mesoporous monoliths

Although the vast majority of rigid monolithic polymer posses large pores, other materials with mesopores in a range of 2–50 nm are likely to find applications in areas such as shape or size selective catalysis, optical devices, dielectrics, etc. The undeniable success achieved with mesoporous inorganic materials is believed to be a powerful driving force for the development of their organic counterparts.

Monolithic molecularly imprinted materials have been prepared and subsequently crushed and sieved to obtain irregular particles. Many of these monoliths that are described elsewhere in this book clearly fit in the category of mesoporous materials [80]. Steinke *et al.* have prepared a series of monolithic polymers from pure ethylene dimethacrylate or trimethylolpropane trimethacrylate in the presence of a variety of porogens to afford materials with mean pore size diameters of 4–8 nm, well within the mesopore range [81,82].

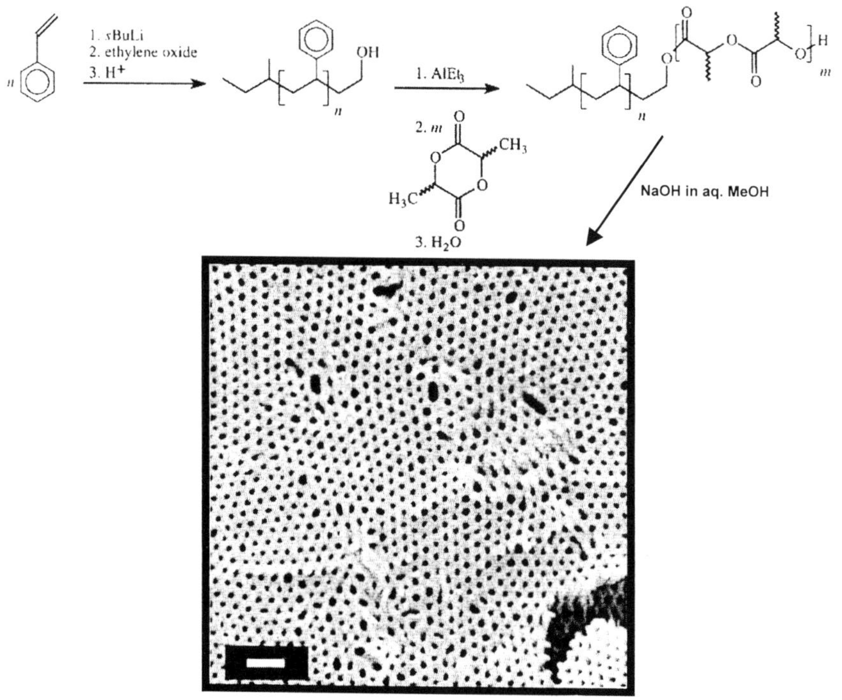

Fig. 2.10. Preparation and scanning electron micrograph of the fractured surface of mesoporous degraded poly(styrene-block-lactic acid) monolith (adapted from [21]). The white scale bar in the lower left corner of the micrograph represents 100 nm.

Recently, Hillmyer developed an interesting approach to the synthesis of well-defined mesoporous monoliths using poly(styrene-block-lactic acid) as a precursors [59]. The starting block copolymer was prepared by anionic polymerization of styrene terminated with ethylene oxide, followed by activation of the terminal hydroxyl groups with triethylaluminum. This polymer was then used to initiate the ring-opening polymerization of lactic acid affording the desired diblock copolymer. Application of sinusoidal shear onto molten copolymer led to the parallel alignment of cylindrical poly(lactic acid) domains. Finally, the materials was treated with a solution of sodium hydroxide at 65°C to hydrolyze and etched away the poly(lactic acid) chains leaving behind uniform 16 nm large pores. The preparation and morphology of the monolith are shown in Fig. 2.10.

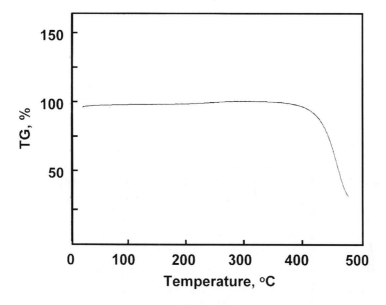

Fig. 2.11. Thermal gravimetric curve of the poly(divinylbenzene) monolith in air at a heating rate of 20°C/min (Reprinted with permission from ref. [25], copyright 2000 Wiley-VCH).

2.5 Thermal stability

Most rigid polymer monoliths are designed for use at room temperature thus making considerations of their thermal stability less important. However, specific applications in catalysis reactions or gas chromatography (GC) may require the use of elevated temperatures to increase the rate of the processes in which they are involved. For example, elution times in GC become shorter at higher temperatures thus decreasing the time required for analysis.

Typically, gradients spanning a wide range of temperatures are used in GC to achieve good resolution and high efficiency. Therefore, the thermal stability of a porous monolith used as a separation medium in gas-solid GC is an important characteristic for this application. Thermogravimetric analysis (TGA) shown in Fig. 2.11 indicates that the porous poly(divinylbenzene) monolith we used in our GC studies did not undergo any significant thermal degradation until a temperature of 380°C was reached [25]. This excellent thermal stability enables the monolith to operate routinely at temperatures up to 300°C, and even up to 350°C for short periods of time without deterioration of properties.

In contrast, methacrylate-based monoliths exhibit lower thermal stability. The first loss of mass accompanying their degradation can be observed at about 215°C. Although this temperature is lower that that found for poly(divinylbenzene), it is still sufficient for numerous applications including GC. This temperature of onset of decomposition correlates well with those found earlier for macroporous beads with similar chemical structures [83,84].

2.6 HYDRODYNAMIC PROPERTIES

For practical reasons, the pressure needed to drive a liquid through any system should be as low as possible. Because in the vast majority of current applications liquids flows through the monoliths, the first concern is their permeability to those liquids, which depends on the size of their pores. Therefore, a monolith with relatively small pores of the size found in typical macroporous beads is likely to be physically damaged by the extremely high pressures required for flow. Obviously, lower flow resistance can be achieved with materials that have a large number of broad channels. However, many applications also require a large surface area in order to achieve a high loading capacity. Since high surface area is generally a characteristic of porous material that contains smaller pores, a balance must be found between the divergent requirements of low flow resistance and large surface area. An ideal monolith should contain both large pores for convection and a connected network of shorter and smaller pores for high capacity [57].

Fig. 2.12 shows the flow resistance expressed in back pressure per unit of flow rate as a function of the flow rate. Typically, the pressure needed to sustain even a very modest flow rate is quite high for materials that have a mean pore diameter of less than about 500 nm, while high flow rates can be readily achieved at low pressures with materials that have pores larger than 1000 nm. Although the shape of pores within a monolith is very different from that of a tube, the Hagen-Poiseuille equation essentially holds for the cylindrical flow-through pores of poly(glycidyl methacrylate-*co*-ethylene dimethacrylate) and poly(styrene-*co*-divinylbenzene) monoliths regardless of their chemistry [57].

2.7 Surface chemistry

Obviously, the monolithic material may often serve its purpose only if provided with the surface chemistry that matches the desired application. For example, hydrophobic moieties are required for solid phase extraction, while ionizable groups must be present for separations in ion-exchange mode, and chiral functionalities are the

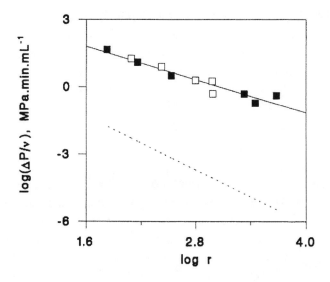

Fig. 2.12. Flow resistance (back pressure per unit of flow rate) of the molded poly(glycidyl methacrylate-*co*-ethylene dimethacrylate) (open points) and poly(styrene-*co*-divinylbenzene) (closed points) 100 mm × 8 mm monoliths as a function of pore diameter at the mode of pore size distribution curve (full line) and flow resistance calculated from Hagen-Poiseuille equation (dotted line) (Reprinted with permission from ref. [57], copyright 1996 American Chemical Society).

prerequisite for enantioselective separations. Several methods can be used to prepare monolithic columns with a wide variety of chemistries.

2.7.1 Preparation from functional monomers

The number of monomers that may be used directly in the preparation of polymer monoliths is much larger than that available for classical suspension polymerization because the mold used to prepare a monolith always contain only a single phase. Therefore, almost any monomer, including hydrophilic water-soluble and reactive monomers, which are not suitable for standard polymerization in aqueous suspensions, may be used to form a monolith. This significantly increases the variety of surface chemistries that can be obtained directly. Some of the monomers (**1**–**9**) and crosslinking agents (**10**–**13**) that have been used for the preparation of rigid porous monoliths are shown in the Fig. 2.13. These monomers incorporate a broad variety of functionalities varying from very hydrophilic (acrylamide **8**, 2-acrylamido-2-methyl-1-propanesulfonic acid **6**), reactive (glycidyl methacrylate **5**, chloromethylstyrene **2**, 2-vinyl-4,4-dimethylazlactone **7**), to latent nucleophiles (4-acetoxystyrene **3**), to hydrophobic (styrene **1**, butyl methacrylate **4**), zwitterionic **9**, and even chiral monomers

References pp. 47-50

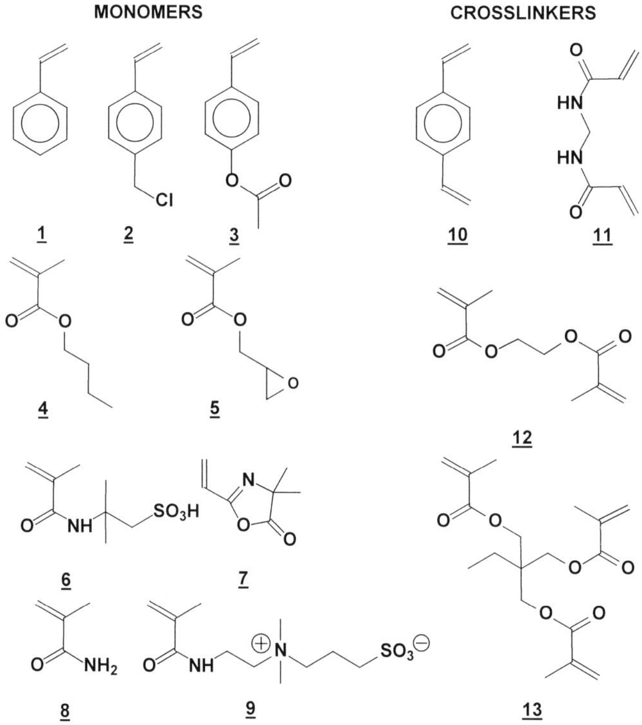

Fig. 2.13. Examples of monomers used for the preparation of porous monoliths.

[32,61,85-89]. A drawback of the direct polymerization approach is that the polymerization conditions optimized for one polymerizing system cannot be transferred directly to another without further experimentation. Therefore, the use of new monomer
mixtures always requires optimization of polymerization conditions in order to
achieve the desired properties, such as sufficient permeability, of the resulting monolith [85].

2.7.2 Modification of reactive monoliths

Chemical modification is another route that increases the number of available
chemistries, also allowing the preparation of monoliths with more exotic functionalities for which monomer precursors may not be readily available. These reactions are
easily performed using monoliths prepared from monomers with reactive group such
as **2** and **5**. For example, Fig. 2.14 shows the reaction of poly(chloromethylstyrene-*co*-
divinylbenzene) with ethylenediamine followed by γ-gluconolactone, a process that
completely changes the surface polarity from hydrophobic to highly hydrophilic [90].

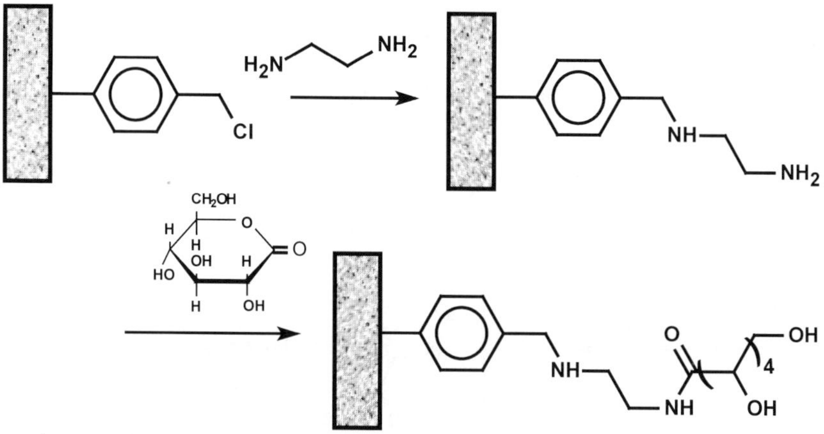

Fig. 2.14. Hydrophilization of hydrophobic surface of a poly(chloromethylstyrene-*co*-divinylbenzene) monolith [90].

A similar reaction of glycidyl methacrylate-based monoliths with diethylamine leads to useful ion-exchangers for bioseparations [16].

2.7.3 Grafting

The grafting of chains onto the pore surface of monoliths is also a very promising route for control of surface chemistry and is particularly well suited for microfluidic devices. As discussed above, the preparation of functionalized monoliths by direct copolymerization of functional monovinyl and divinyl monomers requires repeated optimizations of the polymerization conditions for each new pair of functional mono-mer and crosslinker. Since the functional monomer constitutes both the bulk and the active surface of the monolith, a substantial percentage of the functional units remains inaccessible since their reactive groups are buried within the highly crosslinked poly-mer matrix. The simple modification processes described in the previous section most often leads to the introduction of a single new functionality at each reactive site of the surface. In contrast, the grafting of rare or custom-prepared functional monomers on the surface of the large pores of a "generic" monolith enables a better utilization of these specialty monomers. In addition, the attachment of chains of functional poly-mers to the reactive sites at the surface of the pores affords multiple functionalities emanating from each individual surface site, thus dramatically increasing the density of surface group. Such materials, which intrinsically possess higher binding capaci-ties, are attractive for use in chromatography, ion-exchange, and adsorption.

References pp. 47-50

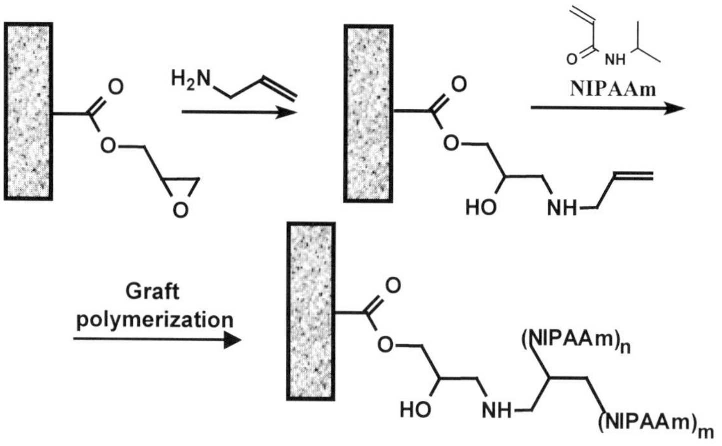

Fig. 2.15. Grafting of poly(glycidyl methacrylate-*co*-ethylene dimethacrylate) monolith with N-isopropylacrylamide [91].

The simplest grafting approach involves filling the pores of a monolith prepared from polymerization mixtures with a high percentage of crosslinking monomer (and presumably containing a large number of unreacted vinyl functionalities) with secondary polymerization mixture, consisting of a solution of a functional polymer and a free radical initiator. The secondary polymerization is triggered within the pores by applying heat or UV light, and the grafting is achieved through incorporation of surface vinyl group in the growing chains. This approach usually suffers from the formation of many non-grafted chains that must be subsequently removed from the pores. However, Viklund and Irgum have successfully used this technique to modify the surface of porous poly(trimethylolpropane trimethacrylate) monoliths with up to 1.5 mmol/g of functionalities derived from poly(N,N-dimethyl-N-methacryloxyethyl-N-(3-sulfopropyl) ammonium betaine) chains, and have subsequently used the grafted monolith for the separation of proteins [86].

Grafting can also provide the monolithic polymers with rather unexpected properties. For example, the two step grafting procedure summarized in Fig. 2.15, which involves the vinylization of the pore surface by reaction of the epoxide moiety with allyl amine, and a subsequent in situ radical polymerization of *N*-isopropylacrylamide isopropylacrylamide (NIPAAm) initiated by azobisisobutyronitrile within these pores leads to a composite that changes its properties, such as polarity and swelling, in response to changes in external temperature [91]. Using this composite, separation of proteins in hydrophobic interaction mode shown in Fig. 2.16 can be achieved by a simple isocratic elution by merely changing the column temperature.

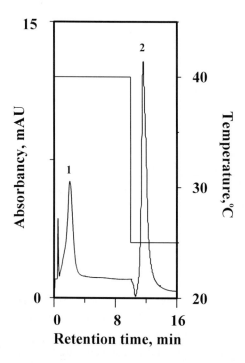

Fig. 2.16. Temperature controlled hydrophobic interaction chromatography of carbonic anhydrase (1) and soybean trypsin inhibitor (2) using porous poly(glycidyl methacrylate-*co*-ethylene dimethacrylate) monolith grafted with poly(*N*-isopropylacryl-amide-*co*-methylenebisacrylamide) (Reprinted with permission from ref.[91], copyright 1997 Wiley-VCH). Conditions: monolithic cylinder 10 × 10 mm i.d. in stainless steel cartridge, mobile phase 1.4 mol/L ammonium sulfate in 0.01 mol/L phosphate buffer (pH 7), flow rate 1 mL/min.

We have also demonstrated that the cerium (IV) initiated grafting of polymer chains onto the internal surface of porous media originally developed by Müller for beads [92] could also be used to graft poly(1-acrylamido-1-methyl-3-propanesulfonic acid) **6** onto the internal surface of hydrolyzed poly(glycidyl methacrylate-*co*-ethylene dimethacrylate) monoliths and considerably improve separation of proteins [87]. Since the free radicals that initiate grafting reside at the pore surface, the main advantage of this approach is the predominant growth of grafted chain and the absence of polymerization in solution.

Yet another example of grafting involves the reaction of an azoinitiator, 4,4'-azo-bis(4-cyanovaleric acid), with the surface functionalities of a poly(chloromethyl-styrene-*co*-divinylbenzene) monolith. The resulting surface-bound initiating moieties may then be used to initiate growth of polymers that may contain a variety of functional groups. A small percentage of a crosslinking monomer typically added to the

References pp. 47-50

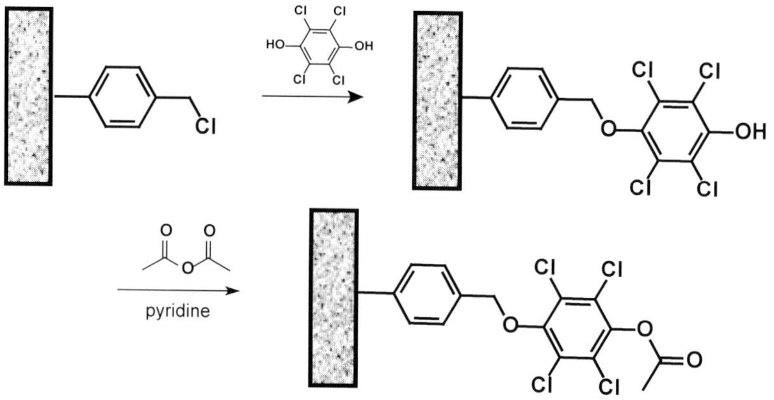

Fig. 2.17. Preparation of a solid phase acylation reagent using chemical modification of poly(chloromethylstyrene-*co*-divinylbenzene) monolith [93].

polymerization mixture to create a gel-like structure at the pore surface, thereby avoiding the loss of functional monomer by formation of ungrafted soluble chains. The amount of the crosslinker used also controls the swelling of the gel and thus the flow properties of the final monolith. This has been demonstrated with the grafting of 4,4-dimethyl-2-vinylazlactone in a monolith subsequently used as a nucleophile scavenger, and of 4-acetoxystyrene in a monolith that could be further modified for use as a polymeric acylating reagent (Fig. 2.17) [93].

Recently, we have developed another method for grafting porous polymer monoliths that uses UV light to initiate growth of a polymer from the surface. In this process, a derivative of benzophenone used as part of the polymerization mixture that fills the pores, abstracts hydrogen atoms from methylene groups exposed at the surface of the monolithic material and enables graft polymerization to be initiated from the surface We have now shown that this newest approach is particularly amenable to the grafting of polymers with desired chemistries at specific locations within the channels of a microfluidic system through the use of photolithographic techniques [94].

2.8 CONCLUSION

Although much remains to be done in the study of rigid macroporous monoliths, the fast pace of recent developments has opened new vistas for the preparation of supports and separation media with exactly tailored properties. Among the numerous advantages of the continuous polymer monoliths are their ease of preparation and ruggedness when used under demanding conditions, the versatility of their chemistries, and their excellent overall properties in a number of applications. The experimental work done so far confirms the great potential of these monolithic materials, which are expected to be particularly useful in a variety of micro- and nanoscale applications. Besides conventional chromatography of biological and synthetic molecules characterized by both very high efficiency and speed, these materials have also been used for electrochromatography in capillaries and microfluidic channels. The low flow resistance enables fast mass transport and contributes substantially to an increase in the activity of immobilized reagents and catalysts. As a result, high flow rates can be used to achieve high throughputs.

2.9 ACKNOWLEDGMENT

Thanks are due to our numerous co-workers listed in the references for their most valuable contributions to this research. Support of this work by a grant of the National Institute of General Medical Sciences, National Institutes of Health (GM-44885) is gratefully acknowledged. This work was also partly supported by the Division of Materials Sciences of the U.S. Department of Energy under Contract No. DE-AC03-76SF00098.

2.10 REFERENCES

1   K.K. Unger, Packings and Stationary Phases in Chromatographic Techniques, M. Dekker, New York, 1990.
2   F.E. Regnier, Nature 350 (1991) 643.
3   N. Afeyan, S.P. Fulton, F.E. Regnier, J. Chromatogr. 544 (1991) 267.
4   J. Horvath, E. Boschetti, L. Guerrier, N. Cooke, J. Chromatogr. A 679 (1994) 11.
5   E. Boschetti, J. Chromatogr. A 658 (1994) 207.
6   L.F. Colwell, R.A. Hartwick, J. Liquid Chromatogr. 10 (1987) 2721.
7   G. Jilge, B. Sebille, C. Vidalmadjar, R. Lemque, K.K. Unger, Chromatographia 32 (1993) 603.
8   W.C. Lee, J. Chromatogr. B 699 (1997) 29.
9   C.G. Huber, P.J. Oefner, G.K. Bonn, J. Chromatogr. 599 (1992) 113.
10   D.K. Roper, E.N. Lightfoot, J. Chromatogr. A 702 (1995) 3.
11   T.B. Tennikova, M. Bleha, F. Švec, T.V. Almazova, B.G. Belenkii, J. Chromatogr. 555 (1991) 97.
12   D. Josic, J. Reusch, K. Loster, O. Baum, W. Reutter, J. Chromatogr. 590 (1992) 59.
13   D. Josic, A. Štrancar, Ind. Eng. Chem. Res. 38 (1999) 333.
14   S. Hjertén, Ind. Eng. Chem. Res. 38 (1999) 1205.
15   K.H. Hamaker, S.L. Rau, R. Hendrickson, J. Liu, C.M. Ladish, M.R. Ladish, Ind. Eng. Chem. Res. 38 (1999) 865.
16   F. Švec, J.M.J. Fréchet, Anal. Chem. 54 (1992) 820.

17   F. Švec, J.M.J. Fréchet, Science 273 (1996) 205.

18   R.E. Moore, L. Licklider, D. Schumann, T.D. Lee, Anal. Chem. 70 (1998) 4879.

19   I. Gusev, X. Huang, C. Horváth, J. Chromatogr. A 855 (1999) 273.

20   B.H. Xiong, L.H. Zhang, Y.K. Zhang, H.F. Zou, J.D. Wang, J. High Resolut. Chromatogr. 23 (2000) 67.

21   B.P. Santora, M.R. Gagne, K.G. Moloy, N.S. Radu, Macromolecules 34 (2001) 658.

22   M.L. Zhang, Y. Sun, J. Chromatogr. A 912 (2001) 31.

23   P. Coufal, M. Čihák, J. Suchánková, E. Tesařová, Z. Bosáková, Chem. Listy 95 (2001) 509.

24   P. Coufal, M. Čihák, J. Suchánková, E. Tesařová, Z. Bosáková, K. Štulík, J. Chromatogr. A 946 (2002) 99.

25   D. Sýkora, E.C. Peters, F. Švec, J.M.J. Fréchet, Macromol. Mat. Engin. 275 (2000) 42.

26   F. Švec, in: Z. Deyl and F. Švec (Eds.), Capillary Electrochromatography, Elsevier, Amsterdam, 2001, p. 183-240.

27   C. Yu, F. Švec, J.M.J. Fréchet, Electrophoresis 21 (2000) 120.

28   T. Rohr, C. Yu, M.H. Davey, F. Švec, J.M.J. Frechet, Electrohoresis 22 (2001).

29   C. Yu, M.H. Davey, F. Švec, J.M.J. Fréchet, Anal. Chem. 73 (2001).

30   Y. Fintschenko, W.Y. Choi, S.M. Ngola, T.J. Shepodd, Fresenius' J. Anal. Chem. 371 (2001) 174.

31   M. Petro, F. Švec, J.M.J. Fréchet, Biotechnol. Bioeng. 49 (1996) 355.

32   S. Xie, F. Švec, J.M.J. Fréchet, Polym. Prepr. 38 (1997) 211.

33   S.F. Xie, F. Švec, J.M.J. Fréchet, Chem. Mater. 10 (1998) 4072.

34   J.A. Tripp, J.A. Stein, F. Švec, J.M.J. Fréchet, Org. Lett. 2 (2000) 195.

35   J.A. Tripp, F. Švec, J.M.J. Fréchet, J. Comb. Chem. 3 (2001) 216.

36   J.A. Tripp, F. Švec, J.M.J. Fréchet, J. Comb. Chem. 3 (2001) 216.

37   Q. Wang, F. Švec, J.M.J. Fréchet, Anal. Chem. 65 (1993) 2243.

38   F. Švec, J.M.J. Fréchet, J. Molec. Recogn. 9 (1996) 326.

39   E.C. Peters, F. Švec, J.M.J. Fréchet, Chem. Mater. 9 (1997) 1898.

40   E.C. Peters, F. Švec, J.M.J. Fréchet, Adv. Mater. 11 (1999) 1169.

41   S.M. Ngola, Y. Fintschenko, W.Y. Choi, T.J. Shepodd, Anal. Chem. 73 (2001) 849.

42   A. Podgornik, M. Barut, A. Štrancar, D. Josic, T. Koloini, Anal. Chem. 72 (2000) 5693.

43   I.M. Abrams, J.R. Millar, React. Funct. Polym. 35 (1997) 7.

44   R. Arshady, A. Ledwith, React. Polym. 1 (1983) 159.

45   B.W. Brooks, Macromol. Symp. 35/36 (1990) 121.

46   H.G. Yuan, G. Kalfas, W.H. Ray, J. Macromol. Sci., Chem. Phys. C31 (1991) 215.

47   P.J. Dowding, B. Vincent, Coll. Surf. A 161 (2000) 259.

48   R. Arshady, J. Chromatogr. 586 (1991) 181.

49   J. Seidl, J. Malínský, K. Dušek, W. Heitz, Adv. Polym. Sci. 5 (1967) 113.

50   A. Guyot, M. Bartholin, Progr. Polym. Sci. 8 (1982) 277.

51   P. Hodge, D.C. Sherrington, Syntheses and Separations Using Functional Polymers, Wiley, New York, 1989.

52   O. Okay, Progr. Polym. Sci. 25 (2000) 711.

53   F. Švec, J.M.J. Fréchet, Chem. Mater. 7 (1995) 707.

54   Z. Pelzbauer, J. Lukáš, F. Švec, J. Kálal, J. Chromatogr. 171 (1979) 101.

55   T. Takata, T. Endo, Progr. Polym. Sci. 18 (1993) 839.

56   F. Švec, J.M.J. Fréchet, Macromolecules 28 (1995) 7580.

57   C. Viklund, F. Švec, J.M.J. Fréchet, K. Irgum, Chem. Mater. 8 (1996) 744.

58   E.C. Peters, F. Švec, J.M.J. Fréchet, C. Viklund, K. Irgum, Macromolecules 32 (1999) 6377.

59   A.S. Zalusky, R. Olayo-Valles, C.J. Taylor, M.A. Hillmyer, J. Am. Chem. Soc. 123 (2001) 1519.

60   J. Hradil, M. Jelinková, M. Ilavský, F. Švec, Angew. Macromol. Chem. 185/186 (1991) 175.

61   M. Lämmerhofer, E.C. Peters, C. Yu, F. Švec, J.M.J. Fréchet, W. Lindner, Anal. Chem. 72 (2000) 4614.

62   S. Xie, F. Švec, J.M.J. Fréchet, Chem. Mater. 10 (1998) 4072.

63   A.I. Cooper, A.B. Holmes, Adv. Mater. 11 (1999) 1270.

64   B.K. Chong, T.T. Le, G. Moad, E. Rizzardo, S.H. Thang, Macromolecules 32 (1999) 2071.

65   R.A. Mayadunne, E. Rizzardo, J. Chiefari, J. Krstina, G. Moad, A. Postma, S.H. Thang, Macromolecules 33 (2000) 243.

66   M.K. Georges, R.N. Veregin, P.M. Kazmaier, G.K. Hamer, Macromolecules 26 (1993) 2987.

67   K.A. Moffat, G.K. Hamer, M.K. Georges, Macromolecules 32 (1999) 1004.

68   C.J. Hawker, Acc. Chem. Res. 30 (1997) 373.

69   C.J. Hawker, J. Am. Chem. Soc. 116 (1994) 11185.

70   T. Nishikawa, M. Kamigaito, M. Sawamoto, Macromolecules 32 (1999) 2204.

71   K. Matyjaszewski, M.L. Wei, J.H. Xia, N.E. Mcdermott, Macromolecules 30 (1997) 8161.

72   T.E. Patten, K. Matyjaszewski, Adv. Mater. 10 (1998) 901.

73   T. Fukuda, A. Goto, K. Ohno, Macromol. Rapid Comm. 21 (2000) 151.

74   D. Benoit, V. Chaplinski, R. Braslau, C.J. Hawker, J. Am. Chem. Soc. 121 (1999) 3904.

75   U. Meyer, F. Švec, J.M.J. Fréchet, C.J. Hawker, K. Irgum, Macromolecules 33 (2000) 7769.

76   C.Viklund, A. Nordsrtöm, K. Irgum, F. Švec, J.M.J. Fréchet, Macromolecules 34 (2001) 4361.

77   C. Yu, M. Xu, F. Švec, J.M.J. Fréchet, J. Polym. Sci. Polym. Chem. 40 (2002) 755.

78   C. Viklund, E. Ponten, B. Glad, K. Irgum, P. Horsted, F. Švec, Chem. Mater. 9 (1997) 463.

79   M. Grasselli, E. Smolko, P. Hargittai, A. Safrany, Nucl. Instr. Meth. Phys. Res. B 185 (2001) 254.

80   G. Wulff, Angew. Chem. 34 (1995) 1812.

81   J.H.G. Steinke, I.R. Dunkin, D.C. Sherrington, Macromolecules 29 (1996) 5826.

82   J.H.G. Steinke, I.R. Dunkin, D.C. Sherrington, Macromolecules 29 (1996) 407.

83   J. Lukáš, F. Švec, J. Kálal, J. Chromatogr. 153 (1978) 15.

84   J. Lukáš, F. Švec, E. Votavová, J. Kálal, J. Chromatogr. 153 (1978) 373.

85   Q. Wang, F. Švec, J.M.J. Fréchet, J. Chromatogr. A 669 (1994) 230.

86   C. Viklund, K. Irgum, Macromolecules 33 (2000) 2539.

87   C. Viklund, F. Švec, J.M.J. Fréchet, K. Irgum, Biotech. Progr. 13 (1997) 597.

88   S. Xie, F. Švec, J.M.J. Fréchet, J. Polym. Sci., Polym. Chem. 35 (1997) 1013.

89   E.C. Peters, K. Lewandowski, M. Petro, F. Švec, J.M.J. Fréchet, Anal. Commun. 35
      (1998) 83.
90   Q. Wang, F. Švec, J.M.J. Fréchet, Anal. Chem. 67 (1995) 670.
91   E.C. Peters, F. Švec, J.M.J. Fréchet, Adv. Mater. 9 (1997) 630.
92   W. Müller, J. Chromatogr. 510 (1990) 133.
93   J.A. Tripp, F. Švec, J.M.J. Fréchet, J. Comb. Chem. 3 (2001) 604.
94   T. Rohr, D.F. Ogletree, F. Švec, J.M.J. Fréchet, Chem. Mater., in press.

F. Švec, T.B. Tennikova and Z. Deyl (Editors)
Monolithic Materials
Journal of Chromatography Library, Vol. 67

Chapter 3

Short Monolithic Columns – Rigid Disks

Miloš BARUT, Aleš PODGORNIK, Mojca MERHAR, Aleš ŠTRANCAR

BIA Separations d.o.o., Teslova 30, SI-1000 Ljubljana, Slovenia

CONTENTS

3.1 INTRODUCTION

During the last fifteen years, a tremendous increase in the speed of chromatographic separations of large molecules having low mobility, such as proteins, DNA, and even viruses, can be noticed. These fast separations, with concomitant reduction in the contact time between the labile biomolecules and the surface of the stationary phase, have enabled a considerable increase in both the specific activity and yield of various therapeutics. Furthermore, rapid HPLC can also be applied to the on-line monitoring of biotechnology processes in real time. These benefits make HPLC one of the key analytical techniques requested by regulatory authorities [1-3].

The increase in speed of the HPLC separations has resulted mainly from improvements in both column design and hydrodynamics, as well as the mass-transfer characteristics of the stationary phases [4]. The latter originated from the use of; (i), non-porous beads made of silica [5,6] and synthetic polymers [7,8], (ii), gigaporous perfusive [9]- or gel-in-a-shell particles [10], and (iii), membranes [11-13] and monoliths [14-25]. These separation media allow much higher flow rates to be applied, thus enabling very fast separations without deterioration in resolution.

The speed of the separation also depends on the length of the column, that determines the residence time of the analyte molecules within the separation device. Since, in general, the residence times in shorter columns are shorter at the same flow velocity, the separation cycle is faster. However, such short columns must possess sufficient separation power to afford the desired resolution. Snyder and Stadalius [26] and Chen and Horváth [1] have shown that for proteins this could be achieved using gradient elution. Particle diameters of less than 1 μm and column lengths shorter than 1 cm were suggested to be optimal for the separation of these large biomolecules. However, it was difficult to efficiently pack very small particles in short columns. In addition, only small samples could be separated on these short columns and, therefore, this approach was not suitable for preparative work.

Resulting from these theoretical considerations, a new type of monolithic stationary phases has been introduced into the chromatographic scene in the late 1980s and the beginning of the 1990s. The current literature describes a large number of monolithic materials differing in both chemistry [14-25] and geometry [14,16,22,25,27].

3.2 SHORT MONOLITHIC COLUMNS

The experience with the preparation of macroporous polymers, combined with the realization that the separation of large molecules in the gradient mode is achieved in only a thin layer of the separation medium, has led to the development of a novel stationary phase in the shape of a flat "membrane", consisting of a polymer with internal structure similar to that of macroporous beads. The novel concept of short monolithic rigid disks suitable for the separation of proteins has been both theoretically elaborated [28-30] and practically implemented [16,17] at the beginning of the 1990s. In contrast to conventional columns that are typically several centimeters long and only a few millimeters wide, these polymer disks had completely different geometry. Specifically, the first disks had a diameter of 20 mm and thickness (length) of

Fig. 3.1. Examples of CIM Convective Interaction Media® rigid disks.

only 1 or 2 mm. The results that could be achieved with these disks in the HPLC separation of proteins, using the hydrophobic interaction chromatographic mode, were fully comparable to the separation on a capillary column, despite their application in a less-than-optimal cartridge [16]. The most striking features of this new separation device were a very low pressure-drop across the membrane (up to 80 times lower than for a conventional column) and very high loading capacity. Owing to the use of the continuous thin polymer layer and the high efficiency, the new chromatographic method has been termed, "high performance membrane chromatography" (HPMC). The technology and a variety of applications of monolithic disks have been summarized in several reviews [31-34]. Today, these rigid disks, and the short monolithic columns obtained by their stacking are commercially available [35].

The early work with the disks mostly concerned the optimization of the polymerization process and demonstration of the separations [36-39]. This, was followed by the design and optimization of suitable cartridges, carried out by the German group led by Josić et al. [18]. In the period 1994-1998, we carried out extensive studies aimed at improvements in the mechanical and chemical properties of the rigid disks, their batch-to-batch reproducibility, and optimization of the cartridge. These efforts resulted in the construction of short monolithic columns called CIM Convective Interaction Media® [40] that are presented in Fig. 3.1.

3.3 PREPARATION OF RIGID DISKS

Preparation of the rigid disks mostly involves three major steps: (i), polymerization of a monolithic material in a suitable format of sheets (membranes) or rods, (ii), cutting disks from them, and, (iii), their chemical modification. Once the preparation of rigid disks is completed, a short monolithic column is obtained by inserting them in an appropriate cartridge or other device.

TABLE 3.1

MONOMERS AND CROSSLINKERS USED FOR THE PREPARATION OF RIGID
MACROPOROUS DISKS

Monomer	Crosslinking agent	Reference
Glycidyl methacrylate	Ethylene dimethacrylate	[16,17,28,36]
Glycidyl methacrylate + styrene	Ethylene dimethacrylate	[52]
Glycidyl methacrylate + 2-hydroxyethyl methacrylate	Ethylene dimethacrylate	[37]
Glycidyl methacrylate + butyl methacrylate	Ethylene dimethacrylate	[37]
Glycidyl methacrylate + octyl methacrylate	Ethylene dimethacrylate	[28,37]
Glycidyl methacrylate + dodecyl methacrylate	Ethylene dimethacrylate	[28,37]
Glycidyl methacrylate + N-vinyl-pyrrolidone	Ethylene dimethacrylate	[41]
Styrene	Divinylbenzene	[28,47]
Styrene	Poly(ethylene glycol) (PEG)400 diacrylate	[47]
Styrene	Poly(ethylene glycol) (PEG)1000 diacrylate	[47]
Glycidyl methacrylate + glycidyl methacrylate-peptide conjugate	Ethylene dimethacrylate	[51]

3.3.1 Polymerization of monoliths

The rigid disks or macroporous polymeric monoliths are prepared by a free-radical copolymerization of a reactive monomer with a cross-linking monomer in the presence of a pore-forming solvent and an initiator. Selected examples of monomers and cross-linkers used for the preparation of rigid disks are shown in Table 3.1. The initial polymerization procedure has been, with respect to the polymerization mixture and conditions, similar to that used for the preparation of macroporous beads [41,42]. However, the conditions typical of suspension polymerization — and therefore the properties of the product — are rather different from those of the polymerization in bulk [43]. Therefore, the optimization of the latter process, in terms of the composition of the polymerization mixture and the polymerization conditions, had to be carried out.

TABLE 3.2

BASIC CHARACTERISTICS OF CHROMATOGRAPHIC "MEMBRANES" [17]
Abbreviations: GMA = glycidyl methacrylate; EDMA = ethylene dimethacrylate; G-5 = polymerization mixture containing 5 vol. % GMA and 95 vol. % EDMA; G-60 = polymerization mixture containing 60 vol. % GMA and 40 vol. % EDMA

Monolith	Monomer ratio, wt%		S_g [a] m^2/g	L [b] $l/h.m^2MPa$	r_p [c] nm
	GMA	EDMA			
G-5	5	95	250	–	5[d]
G-25	25	75	139	390	10[e]
G-40	40	60	80	1120	17[e]
G-50	50	50	60	1940	23[e]
G-60	60	40	43	–	25[d]

[a]Specific surface area

[b]Permeability for water

[c]Mean pore radius

[d]Mean pore radius measured on beads of identical composition

[e]Mean pore radius calculated from permeability

3.3.1.1 Parameters affecting free radical polymerization

In contrast to columns packed with particulate stationary phases, in which the mobile phase flows through the interparticle voids, the monolith completely fills the column and all the mobile phase flows through its pores. Therefore, the permeability of a monolith featuring pores with a size smaller than 100 nm, typical of macroporous beads, would be very low, and the resulting extremely high pressure drops might destroy the column. Therefore, one of the first parameters that had to be optimized was the pore size of the monolith [17,36,37]. This property is controlled by, (i), the percentage of monomers in the mixture, their type, and the monomer to crosslinker ratio [17]; (ii), the percentage of porogen in the mixture and its composition [43], and (iii), the polymerization temperature and time [43,44]. For example, Table 3.2 shows the effect of the monomer-to-cross-linker ratio on the specific surface area, permeability, and porosity of the rigid disks. The Table shows that these properties vary widely with the changes in the proportions of the monomers. An increase in the relative amount of a cross-linker leads to disks with higher specific surface area, indicating a

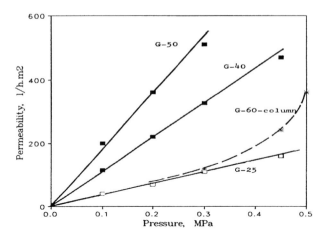

Fig. 3.2. Effect of pressure on flow through a 1 mm thick poly(glycidyl methacrylate-*co*-ethylene dimethacrylate) monoliths containing various percentages of glycidyl methacrylate. The results are compared to a 300 × 8 mm i.d. column packed with 5–7 μm copolymer beads containing 60% glycidyl methacrylate (Reprinted with permission from ref. [17]. Copyright 1991 Elsevier).

larger number of smaller pores. This also results in lower permeability of these disks. Tennikova *et al.* [17] monitored the pressure required to achieve a flow of 1 ml/min. This pressure was 0.75 MPa for a disk prepared from the polymerization mixture containing 75 % ethylene dimethacrylate, but only 0.1 MPa for mixtures with 50 % cross-linker. Figure 3.2 shows the flow through 1 mm thick disks prepared from various mixtures of glycidyl methacrylate and ethylene dimethacrylate as a function of hydrodynamic pressure. However, an increase in the percentage of cross-linker results in slower polymerization kinetics and in a decrease in surface density of the reactive epoxy groups [36]. Thus, a cross-linker concentration of 40-50% represents an optimum at which monoliths with both sufficiently large pores and surface area are obtained.

3.3.1.2 Morphology and characteristics of the rigid disks

The morphology of monolithic disks depends on a number of parameters. Their structure consists of interconnected microspheres (globules) that are aggregated in larger clusters. The pores in the macroporous polymers are irregular channels between clusters of these globules (macropores), between globules within each cluster (mesopores), and within the globules themselves (micropores) [45]. The globular

Fig. 3.3. Scanning electron micrograph of poly(glycidyl methacrylate-*co*-ethylene dimethacrylate) disk. Magnification 10,000x.

internal structure of a rigid porous polymer disk is shown in Fig. 3.3. Application of the knowledge of both polymerization kinetics and factors affecting the polymerization process, such as the composition of the polymerization mixture, the temperature, the size and shape of the mould, as well as the heat transfer properties of both mould and polymer, enables the preparation of monolithic disks with a well controlled optimized internal structure [46]. The polymerization is carried out in a way that affords material containing a sufficient volume of large pores with a diameter of about 1–3 μm. These pores are required to achieve high permeability for liquids. In addition to these flow-through channels, the disks also contain a network of highly connected pores smaller than 100 nm that contribute significantly to the surface area needed for the desired high binding capacity. Therefore, typical monolithic disks feature a bimodal pore-size distribution. An example of pore-size distribution of the monolithic disk is shown in Fig. 3.4.

The total porosity of the disks mainly depends on the relative amount of monomers and pore-producing solvents (porogens) in the polymerization mixture. The higher the amount of porogens in the mixture, the higher the porosity. The reason for this lies in the fact that the pore-producing solvents do not react during polymerization and remain entrapped within the monolithic structure. Once the polymerization is completed, the solvents are washed from the polymeric matrix, and the volume originally occupied by these solvents represents the pore volume. In theory, very high porosities

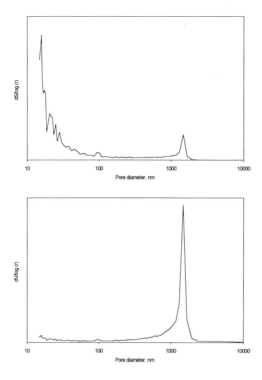

Fig. 3.4. Differential pore size distributions of CIM disk determined by mercury porosimetry.

can be obtained by simply increasing the percentage of these solvents in the polymerization mixture. However, the mechanical stability of these disks would be severely
reduced. Since the disks are used in HPLC-like applications, they must withstand the
unidirectional pressure exerted by the flow of the mobile phase. The highly porous
monolithic disks are also softer and more elastic. Thus, they could get compressed
upon applying the flow and the pore size would decrease. As a consequence, the
chromatographic characteristics might change during the separations, which is not
acceptable. The optimized structure of CIM monolithic disks exhibits a porosity of
approximately 62 % with the size of the largest pores approximately 1.5 μm. This
material swells in water and various organic solvents to an extent of less than 7 % and
resists pressures of up to 20 MPa.

3.3.2 Different approaches to the preparation of bare rigid disks

At the very beginning, the rigid disks were cut from a flat sheet of the porous monolith prepared by polymerization within a mould that consisted of two heated metal plates, whose surfaces were covered with a polypropylene film to prevent sticking of the formed monolith to the mould. The space between the plates was adjusted by inserting a silicone rubber gasket that controlled the thickness of the sheet [16,17,36]. Owing to the volume shrinkage of about 6 % during the polymerization, the actual thickness of the gasket had to be slightly larger than the desired thickness of the disks. The gasket had an opening through which the mould was filled with the monomer mixture. The polymerization was initiated by placing the mould into the water bath. After completion of the polymerization and cooling, the mould was disassembled and the polymer sheet recovered. The disks were then cut from this sheet.

Another approach to the preparation of rigid disks involved polymerization of the monolith within the confines of a tube made of polypropylene, stainless steel, or other inert materials. The tube was sealed on one side with a rubber septum, filled with the monomer mixture, purged with nitrogen to remove oxygen, sealed at the top side using another rubber septum, and placed in a thermostated water bath or oven in an upright position. After the polymerization was completed, the tubular mould was opened and the rod-shaped monolith removed from the tube either by applying pressure or with the help of compressed air. In a similar approach, Hird *et al.* [47] prepared the monolithic rod in Pyrex glass tube, which was then carefully broken. The disks were easily cut from the rod using a razor blade or lathe. This approach facilitates the preparation of disks varying in diameters and thickness. Švec and Fréchet [44] have demonstrated that the basic mechanism of the monolith creation by bulk polymerization in the presence of porogenic solvents is general and independent of the polymerization technique. The solid nuclei and their clusters have a higher density than the liquid polymerization mixture. Since mixing is absent and the overall rate of polymerization is slow, they sediment and accumulate at the bottom of the mould. This may result in the morphology varying along the length of the rod. Consequently, disks cut from the same rod might have different properties at different locations. In order to check this, we cut ten disks from a single rod polymerized in a polypropylene tubular mould, subjected them to the mercury porosimetry, and compared their porosity, specific surface area, cumulative pore volume, and the median pore diameter. The results summarized in Table 3.3 clearly demonstrate the excellent structural similarity of all of these disks.

The limitation of this technology is the diameter of the mould from which the heat of polymerization can still be dissipated efficiently without affecting the polymerization temperature and thus the morphology. Free-radical polymerization is an exother-

TABLE 3.3

STRUCTURAL CHARACTERISTICS OF RIGID DISKS CUT FROM A POLYMERIC ROD

Disk No.	Surface area m^2/g	Pore volume ml/g	Median pore radius nm
1 [a]	23.48	1501	473
2	23.48	1444	464
3	23.01	1403	446
4	22.91	1415	446
5	23.53	1382	430
6	23.15	1409	446
7	23.50	1410	446
8	23.35	1388	446
9	22.84	1399	430
10 [b]	24.18	1401	414

[a] Top of the rod

[b] Bottom of the rod

mic process and the released heat dissipates only when the inner diameter of the tube is small. Both Peters *et al.* [48] and Podgornik *et al.* [49] have demonstrated that this occurs in tubes with a diameter of up to 25 mm. In moulds with larger diameters, the chromatographic characteristics of the monolith change considerably.

All of the above approaches afford "bare" disks, which can readily be used in a variety of applications. However, it has been observed that these disks can be easily scratched and damaged by chipping away the edges or by breaking the monolith completely. These disks have an isotropic structure that makes them permeable in both axial and radial directions. Therefore, the rigid disks must be operated in a way that allows flow only in the axial direction. In order to increase their mechanical stability, and to prevent leakage through the side walls, it is advantageous to encase the disks by using a mechanically and chemically stable non-porous fitting ring [40,50].

3.3.3 Control of surface chemistry

3.3.3.1 Direct polymerization

Monolithic disks with some specific chemistries can be prepared easily by direct polymerization of suitable monomers. For example, monoliths with reactive epoxy groups are obtained by copolymerization of glycidyl methacrylate and ethylene dimethacrylate. Several attempts have also been reported in which three monomers were copolymerized to modulate the chemistry. The first study concerned the direct preparation of alkyl-substituted disks for hydrophobic interaction chromatography, by direct copolymerization of glycidyl methacrylate and ethylene dimethacrylate with an addition of other monomers such as 2-hydroxyethyl methacrylate, butyl methacrylate, octyl methacrylate, or dodecyl methacrylate, that did not require any subsequent chemical modifications [37]. The addition of the third monomer such as butyl methacrylate or 2-hydroxyethyl methacrylate led to monoliths with changed porous properties only when a rather high concentration, exceeding 20 vol.% of this monomer in the polymerization feed, was used. This was particularly true for monomers with hydrophobicity similar to that of glycidyl methacrylate. It has to be pointed out that other monomers such as octyl methacrylate or dodecyl methacrylate affected the morphology of the disk significantly and made it rather different from that of parent poly(glycidyl methacrylate-*co*-ethylene dimethacrylate). Obviously, the polymerization process had to be re-optimized taking into account parameters such as the reactivity of the monomers and their hydrophobic/hydrophilic character to achieve the desired porous properties. Table 3.1 presents monomers that were used for the direct preparation of functional monolithic disks.

Hahn *et al.* [51] recently prepared a custom-made monomer with an attached peptide ligand and copolymerized it with ethylene dimethacrylate to directly obtain disks useful for affinity chromatography or for coupling of an additional ligands to further increase their binding capacity.

3.3.3.2 Chemical modification

Another route for the preparation of polymer monoliths with a variety of chemistries uses the subsequent chemical modification of a "generic" monolith containing reactive groups. These "generic" disks have a highly optimized structure that results from polymerization under fine-tuned conditions. Chloromethyl groups originating from chloromethylstyrene or the epoxy groups of polymerized glycidyl methacrylate, are examples of the reactive chemistries mostly used. The CIM rigid disks that are prepared from the latter have a very high density of epoxy groups that exceeds 4

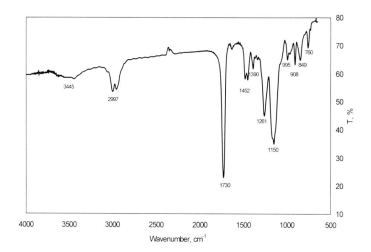

Fig. 3.5. IR spectrum of CIM Epoxy disks. Vibrational bands at 908 cm^{-1} and 850 cm^{-1} are characteristic for the epoxy groups.

mmol per gram of the dry support. The characteristic fingerprint of the chemical structure of these rigid disks is determined by IR spectroscopy. A typical IR spectrum of this material is shown in Fig. 3.5.

Several chemical reactions of epoxy functions, such as hydrolysis, sulfonation, carboxymethylation, and amination,, shown in Fig. 3.6 have been adopted from the studies carried out with macroporous beads [52,53] and used for the modification of the disks [54]. For example, reaction of poly(glycidyl methacrylate-*co*-ethylene dimethacrylate) monolith with amines, shown in Fig. 3.6, (Schemes **1** and **2**), leads to suitable weak- and strong anion exchangers, respectively. Reaction with a complex formed from oleum and 1,4-dioxan (Scheme **3**) affords a strong cation exchanger. Reaction of epoxy groups with amino groups of peptides or proteins such as protein A and protein G provides highly specific functionalities for affinity chromatography [55-57].

Besides the suitable surface chemistry, for optimal chromatographic properties of separation media such as the disks, one also requires a sufficient amount of functionalities (ligand density). A low group density results in low binding capacity for the molecules of interest. Figure 3.7 demonstrates the linear relationship between the group density and the resulting dynamic binding capacity for ion-exchange CIM disks. A high active group density is also beneficial for purification of real-life samples containing increased amounts of salts. For example, the weak anion-exchange CIM disks exhibited a relatively high dynamic binding capacity even for bovine

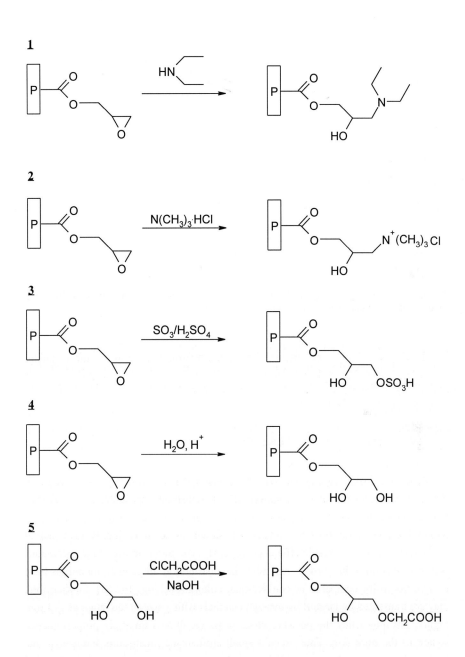

Fig. 3.6. Chemical reactions of epoxy groups of poly(glycidyl methacrylate-*co*-ethylene glycol dimethacrylate) disks with various reagents. **1** amination with diethylamine; **2** amination with trimethylamine hydrochloride; **3** sulfonation; **4** hydrolysis; **5** carboxymethylation.

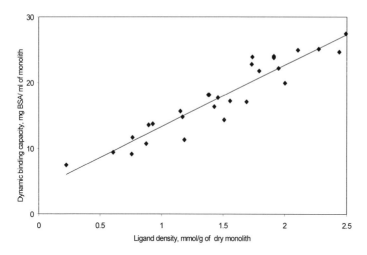

Fig. 3.7. Dynamic binding capacity of CIM DEAE disk for bovine serum albumin (BSA) as a function of ligand density. BSA concentration 1 mg/ml, flow rate 3 ml/min.

serum albumin dissolved in the buffer containing 0.3 mol/l NaCl [58], as shown in Fig. 3.8. Therefore, these disks can be used for both real-time on-line- and off-line monitoring of biotechnology processes since they avoid time-consuming sample pre-treatment such as dialysis [59].

3.4 HYDRODYNAMIC PROPERTIES

Monolithic disks are characterized by a linear dependency of back pressure on flow rate within a wide range [17,60,61]. This demonstrates the rigidity of these disks, that are not deformed or compressed even at high flow rates. The permeability of monolithic phases for flow has often been shown to be much higher than that of packed beds with comparable efficiencies [62,63]. The permeability of the continuous bed is a function of its porosity and pore size distribution. In general, monoliths have a higher porosity compared to conventional column packings [62,63]. Furthermore, they are transected by large flow-through channels with a size in the range of 1–3 μm. Figure. 3.9 shows that the pressure drop across the disk is inversely proportional to square of the pore size. Thus, even a small shift in the mean pore diameter to the lower values can result in a significantly reduced permeability. The intrinsically high permeability of rigid disks is supported by two additional parameters that help to achieve very low pressure drops: (i), the small thickness of the disk and, (ii), the large cross-sectional area that affords low linear velocities even at high flow rates.

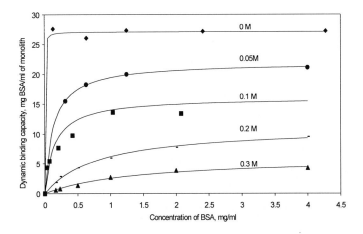

Fig. 3.8. Effect of NaCl concentration on adsorption isotherms of bovine serum albumin (BSA). (Reprinted with permission from ref. [58]. Copyright 2000 Wiley). Solid lines were calculated using the Langmuir equation. Conditions: Loading phase, 20 mmol/l Tris-HCl pH 7.4 with varying concentrations of sodium chloride; Elution, 1 mol/l sodium chloride in loading buffer; Stationary phase, CIM DEAE monolithic disk; Flow rate, 5 ml/min; Detection, 280 nm.

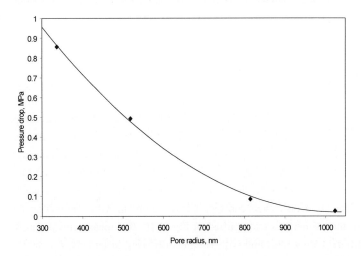

Fig. 3.9. Effect of the modal pore radius on the pressure drop in monolithic disks. Disk, 12 mm diameter and 3 mm thickness. The pressure drop was determined using deionized water at a flow rate of 4 ml/min.

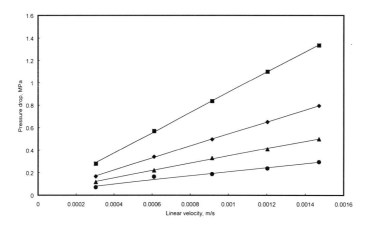

Fig. 3.10. Pressure drop in CIM disks as a function of linear flow velocity and the number of disks in the stack. Number of disks: 1 (●); 2 (▲); 3 (♦); 4 (■).

In general, the pressure drop of both columns packed with porous particles and monolithic beds can be described using the Darcy equation [64].

$$\frac{\Delta p}{L} = -\frac{\eta}{B_0} \cdot u$$

where Δp is the pressure drop [Pa], L is the column length [m], η is the viscosity of the mobile phase [Pa.s], B_0 is the specific permeability [m^2], and u is the linear velocity [m/s]. This equation predicts a linear relationship between the pressure drop, the linear velocity, and the thickness of the bed (column length). The pressure drop in a rigid disk can easily be determined by pumping a suitable mobile phase first through the cartridge containing the disk followed by pumping the same mobile phase at the same flow rate through the cartridge from which the disk was removed. The actual pressure drop of the rigid disk itself can then easily be obtained by subtracting the latter value from the former. As exemplified in Fig. 3.10, this measurement can be carried out at different flow rates and with columns differing in lengths, *i.e.*, with a cartridge containing a varying number of disks. The specific permeability of the rigid disks, B_0, can easily be calculated from the data of Fig. 3.10 using the above equation. The resulting value of B_0, equal to 3.10^{-14} m^2, is only slightly larger than the permeability of conventional columns packed with 5 μm particles and much less than the values reported for monoliths by other researchers [60,62,65]. This can readily be explained by the lower porosities of the poly(glycidyl methacrylate-*co*-ethylene dimethacrylate) monolithic disks compared to silica monoliths [62], and to monoliths obtained by ring-opening metathesis polymerization [63]. However, the back pressure of the monolithic disks remains very low. For example, a pressure drop of less than 2 MPa at a high flow rate of 8 ml/min was observed for a stack composed of four disks. The average residence time of the mobile phase within a single rigid disk at this flow rate was only 1.5 s. This enabled very fast separations at only a fraction of the pressure drops usually encountered in conventional HPLC.

3.5 OPTIMIZATION OF THE CARTRIDGE

The new stationary phases in the format of rigid disks were used for the HPLC separations of proteins. To achieve this, the disk had to be placed in a device that allowed connection to a common HPLC system that facilitated injection of the sample and pumping of the mobile phase without bypassing or leakage of the liquids and without damaging the wide and thin disk. For fast analytical separations, the sample distribution across the whole surface of the disk-face, and peak broadening, become important issues. Good sample distribution can be achieved by using a sufficiently large dead volume in front of the disk. Similarly, the use of an inlet capillary with a large radius leads to lower pressure in the system and better distribution of the sample. However, the separation performance of the disk in such a device is reduced owing to excessive peak broadening.

Tennikova *et al.* [16,17] carried out the separations of proteins in a specifically designed magnetically stirred vessel with the disk fixed at the bottom, thus addressing the difficulties concerning sample distribution. The vessel had an inlet capillary for the supply of the mobile phase and an outlet, in which the solution which had passed the disk was collected and than directed to a UV detector. In a typical experiment, the sample solution was added at the top of the disk and the binding buffer solution pumped through the vessel to remove all non-adsorbed proteins. The elution buffer was then pumped through the vessel and gradually diluted the binding buffer, thus creating an exponential gradient of the mobile-phase strength. Josić and Štrancar [31] realized that this method allowed application of fairly large amounts of dilute sample and its good distribution across the entire disk surface. Furthermore, they noticed that this approach is advantageous in preparative chromatography such as the separation of components from cell culture supernatants in which large volumes are applied. This method was also useful for affinity chromatography, and generally for all chromatographic separations where elution is achieved with a step-wise gradient. However, they also pointed to the complicated sample application and the poor reproducibility of the gradient as the main weaknesses. Therefore, this method was assumed to be less suitable for rapid chromatographic separations.

References pp. 73-75

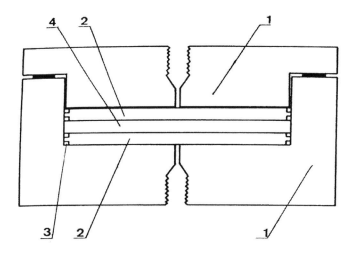

Fig. 3.11. Schematics of a HPMC cartridge. (Reprinted with permission from ref. [18]. Copyright 1992 Elsevier). 1, holder; 2, distribution plates; 3, O-rings; 4, porous polymer disk.

As a result, a new cartridge with a low dead-volume containing optimized flow distributors was designed and used [18]. The development of this new housing aimed at avoiding the main problems observed previously: (i), leakage and by-passing of the mobile phase; (ii), poor distribution of the mobile phase; and, (iii), limited mechanical stability of thin and wide rigid disks. The first series of problems were the consequence of the fact that the disks have been designed to be exchangeable. Since the disks may not fit exactly in the cavity of the cartridge, the mobile phase could leak through the sidewall or bypass the disk. This problem was partly eliminated in cartridge designs involving O-rings placed at both top and bottom of the disk, thus sealing its perimeter. However, this approach was less than ideal since the O-rings exerted pressure at the edges of the disk upon tightening the fittings of the cartridge. This has often led to cracks in the monolithic structure, and eventually to a complete breakage of the disk. A scheme of this cartridge used by Josić *et al.* [18] is shown in Fig. 3.11.

The second problem was the poor distribution over the entire surface of the disk because the jet of the liquid leaving the inlet tube with a maximum diameter of 1 mm could not be distributed evenly across the surface having a diameter of 20 mm or more. Experiments have shown that only a small portion of the disk, usually its center, was used for the separation. In addition, the center of the disk experienced a much higher pressure that reduced the life of the separation unit. This problem has been solved after extensive experimental work involving twenty different designs of plates that were placed at both the top and bottom of the disk [18]. Using a ferritin solution,

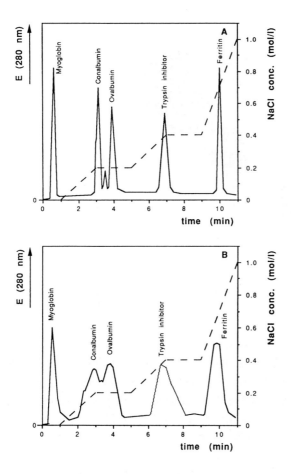

Fig. 3.12. Anion-exchange chromatography of myoglobin (1), conalbumin (2), ovalbumin (3), soybean trypsin inhibitor (4), and ferritin (5) using a CIM DEAE disk in a cartridge with an optimized solvent distribution: (A), compared to that obtained with a cartridge containing a less-than-optimal distribution plate, which afforded a large void volume, (B). (Reprinted with permission from ref. [18]. Copyright 1992 Elsevier). Chromatographic conditions: disk diameter 25 mm, thickness 2 mm; flow rate, 3 mL/min, pressure 0.1–0.4 MPa.

the distribution of the colored protein over the surface of the disk could be monitored and the distributor plate enabling the best distribution selected. The effect of the distributor on the separation of proteins is demonstrated in Fig. 3.12. Clearly, the separation is much better using the cartridge containing the optimized distribution plate.

Finally, the third problem, of exchanging the disks in the cartridge simply, allowing for easy column "packing" and "unpacking" has been addressed by Štrancar *et al.* [40]. They have developed a new cartridge consisting of a hollow cylinder, two

Fig. 3.13. Photograph of the optimized cartridge for CIM disks produced by BIA Separations.

retaining fittings with flow distributors, and two screw caps (Fig. 3.13). The flow distributors were made of thin PAT (Peek Alloyed with Teflon) frits with a thickness of 0.78 mm and a porosity of 5 μm. The rigid disks used with this type of cartridge were encased in a non-porous fitting ring that directed the flow of the mobile phase exclusively to the axial directions. The design of this cartridge also allowed for the stacking of up to four disks within the housing and enabled devices with various separation lengths. The major advantage of this approach compared to connection in series of several cartridges, each having a single disk is the lack of dispersion effects that would reduce the separation power. Consequently, the peak-widths of proteins separated using a single disk or a stack of two or three disks are very similar [40]. We have also demonstrated that the efficiency of the monolithic device containing one or more disks related to a unit length remained unchanged [66]. For example, Fig. 3.14 shows that HETP for stacks of monolithic disks with the overall length of 3-, 6-, 9-, and 12 mm, located in a single housing, remain practically constant over the entire range of lengths. An additional benefit of this approach is an increase in the dynamic binding capacity of the device containing multiple disks [40,66].

3.6 CONJOINT LIQUID CHROMATOGRAPHY (CLC)

But far the most advantageous feature of disk-stacking is the option to perform conjoint liquid chromatography (CLC) that we have introduced recently [40]. This

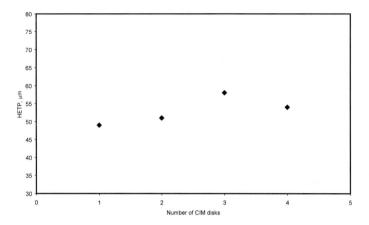

Fig. 3.14. Effect of the number of CIM disks placed in a single housing on the separation efficiency. HETP values were calculated from the width of a citric acid peak (Reprinted from ref. [66]. Copyright 2002 Marcel Dekker).

technique is based on a combination of different chromatographic modes in a single run by stacking several disks with different functionalities in the same housing. This operational mode, that was impossible until now, allows a single-step separation using different chromatographic modes to be achieved in a system characterized by a very low dead volume. Our approach is a much faster, more precise, and cheaper version of multidimensional chromatography, and does not require complex automation to switch numerous pumps and valves. Another advantage of CLC is its scaleability by applying the "tube-in-a-tube" technique described elsewhere in this book. The only limitation of the CLC is the incompatibility of certain mobile phase systems, which do not allow free combinations of all chromatographic modes. Since its inception, the use of CLC has been reported several times in the literature [40,55,67].

One can argue that the disk stacking may result in poor reproducibility. However, experimental work has proved that an analytical method involving disk-stacks affords reproducible results and can be fully validated. The determination of impurities in highly concentrated IgG products is an example of such a validated method which is routinely used in quality-control of IgG concentrates [68]. This clearly confirms the potential of conjoint liquid chromatography.

3.7 SCALE-UP AND MINIATURIZATION OF RIGID DISKS

The original rigid disks described by Tennikova *et al.* [16,17] had a diameter of 20–25 mm and a thickness of 1–2 mm. The bed volume of these disks was approximately 1 ml. The first commercially available Quick Disks™ produced by Säuelentechnik were available in three different diameters of 10, 25, and 50 mm, with thickness ranging from 1 to 7 mm. The 50 mm type was also the largest disk reported. The reasons why even larger disks were not used was the difficulty encountered in the preparation of large-diameter monoliths as well as the absence of an efficient distribution of the sample over such a wide and thin object [18].

Since the first commercial product was introduced, the technology for producing monoliths and membranes has developed substantially and large diameter membranes [69] and rigid disks [70] can be currently prepared. The efficient distribution across the entire the large surfaces has also been achieved [69]. These developments now enable construction of large units including wide disks for preparative purifications. However, handling thin disks with a large diameter is difficult, making them impractical. Therefore, the development of scaled-up units has been directed towards the design of tubes, presented elsewhere in this book, that operate in the radial flow mode.

In contrast, it appears much easier to miniaturize the disks. The smallest disks reported so far had a diameter and thickness of 2 mm, representing a volume of 6 µl. These microdisks were used for a solid-phase synthesis of peptides [71]. The small diameter rigid disks are also suitable for fast chromatographic screening in a 96-well-plate format and for direct solid phase synthesis [47].

3.8 CONCLUSION

The preparation of monolithic rigid disks is simple and consists in polymerization leading to the parent monoliths, optionally followed by chemical modification. Optimization of polymerization in terms of composition of the polymerization mixture, temperature, type and shape of the mould, was required to obtain rigid disks with the desired structure. This disk structure enables high permeability and provides for both high capacity and high resolution-power. Rigid disks are monolithic stationary phases. Therefore, they can easily be used in a variety of separation systems without a need for packing. As a result of the very low back-pressure, high resolution power, and wide range of surface chemistries, the disks are stationary phase of choice for separation and purification of large molecules. Since they enable extremely fast separations of biomacromolecules, they are also suitable high-throughput tools in genomic and proteomic studies. In addition, the combination of different chromatographic modes in a single run, by stacking several disks with different functionalities in the same housing or well using the conjoint liquid chromatography approach, opens new frontiers of sample preparation and purification.

3.9 REFERENCES

1   H. Chen, Cs. Horváth, J. Chromatogr. A 705 (1995) 3.
2   E. Boschetti, J. Chromatogr. A 658 (1994) 207.
3   S.K. Paliwal, T.K. Nadler, D.J.C. Wang, F.E. Regnier, Anal. Chem. 65 (1993) 3363.
4   J.J. Kirkland, J. Chromatogr. Sci. 38 (2000) 535.
5   L.F. Colwell, R.A. Hartwick, J. Liquid Chromatogr. 10 (1987) 2721.
6   G. Jilge, B. Sebille, C. Vidalmadjar, R. Lemque, K.K. Unger, Chromatographia 32 (1993) 603.
7   W.C. Lee, J. Chromatogr. B 699 (1997) 29.
8   C.G. Huber, P.J. Oefner, G.K. Bonn, J. Chromatogr. 599 (1992) 113.
9   N.B. Afeyan, N. F. Gordon, I. Mazsaroff, L. Varady, S.P. Fulton, Y.B. Yang, F.E. Regnier, J. Chromatogr. 519 (1990) 1.
10   J. Horvath, E. Boschetti, L. Guerrier, N. Cooke, J. Chromatogr. A 679 (1994) 11.
11   S. Brandt, R.A. Goffe, S. Kessler, J.L. O'Connor, S.E. Zale, Biotechnology 6 (1988) 779.
12   D.K. Roper, E.N. Lightfoot, J. Chromatogr. A 702 (1995) 3.
13   E. Klein, J. Membrane Sci. 179 (2000) 1.
14   S. Hjertén, J.-L. Liao, R. Zhang, J. Chromatogr. 473 (1989) 273.
15   C. Ericson, S. Hjertén, Anal. Chem. 71 (1999) 1621.
16   T.B. Tennikova, B.G. Belenkii, F. Švec, J. Liq. Chromatogr. 13(1) (1990) 63.
17   T.B. Tennikova, M. Bleha, F. Švec, T.V. Almazova, B.G. Belenkii, J. Chromatogr. 555 (1991) 97.
18   Dj. Josić, J. Reusch, K. Löster, O. Baum, W. Reutter, J. Chromatogr. 590 (1992) 59.
19   A. Štrancar, P. Koselj, H. Schwinn, Dj. Josić, Anal. Chem. 68 (1996) 3483.
20   F. Švec, J.M.J. Fréchet, Anal. Chem. 64 (1992) 820.
21   F. Švec, J.M.J. Fréchet, Ind. Eng. Chem. Res. 38 (1999) 34.
22   K. Nakanishi, N. Soga, J. Am. Ceram. Soc. 74 (1991) 2518.
23   H. Minakuchi, K. Nakanishi, N. Soga, N. Ishizuka, N. Tanaka, Anal. Chem. 68 (1996) 3498.
24   K. Cabrera, G. Wieland, D. Lubda, K. Nakanishi, N. Soga, H. Minakuchi, K.K. Unger, Trends Anal. Chem. 17 (1998) 50.
25   P.-E. Gustavsson, P.-O. Larsson, J. Chromatogr. A 925 (2001) 69.
26   L.R. Snyder, M.A. Stadalius, in: Cs. Horváth (Ed.), High-Performance Liquid Chromatography, Advances and Perspectives, Vol. 4, Academic Press, Orlando, 1986, p 195.
27   A. Štrancar, M. Barut, A. Podgornik, P. Koselj, H. Schwinn, P. Raspor, Dj. Josić, J. Chromatogr. A 760 (1997) 117.
28   T.B. Tennikova, F. Švec, J. Chromatogr. 646 (1993) 279.
29   B.G. Belenkii, A.M. Podkladenko, O.I. Kurenbin, V.G. Maltsev, D.G. Nasledov, S.A. Trushin, J. Chromatogr. 645 (1993) 1.
30   N.I. Dubinina, O.I. Kurenbin, T.B. Tennikova, J. Chromatogr. A 753 (1996) 217.
31   Dj. Josić, A. Štrancar, Ind. Eng. Chem. Res. 38 (1999) 333.

32   T.B. Tennikova, R. Freitag, J. High Resolut. Chromatogr. 23 (2000) 27.

33   Dj. Josić, A. Buchacher, A. Jungbauer, J. Chromatogr. B 752 (2001) 191.

34   A. Štrancar, A. Podgornik, M. Barut, R. Necina in: R. Freitag (Ed.), Adv. Biochem. Eng. Biotechnol. 76 (2002) 50.

35   www.biaseparations.com

36   F. Švec, M. Jelínková, E. Votavová, Angew. Makromol. Chem. 188 (1991) 167.

37   J. Hradil, M. Jelínková, M. Ilavský, F. Švec, Angew. Makromol. Chem. 185/186 (1991) 275.

38   H. Abou-Rebyeh, F. Körber, K. Schubert-Rehberg, J. Reusch, Dj. Josić, J. Chromatogr. 566 (1991) 341.

39   Dj. Josić, F. Bal, H. Schwinn, J. Chromatogr. 632 (1993) 1.

40   A. Štrancar, M. Barut, A. Podgornik, P. Koselj, Dj. Josić, A. Buchacher, LC-GC 11 (1998) 660.

41   J. Hradil, A.A. Azanova, M. Ilavský, F. Švec, Angew. Makromol. Chem. 205 (1992) 141.

42   F. Švec, J. Hradil, J. Čoupek, J. Kálal, Angew. Makromol. Chem. 48 (1975) 135.

43   C. Viklund, F. Švec, J.M.J. Fréchet, K. Irgum, Chem. Mater. 8 (1996) 744.

44   F. Švec, J.M.J. Fréchet, Chem. Mater. 6 (1995) 707.

45   F. Švec, J.M.J. Fréchet, J. Mol. Recognit. 9 (1996) 326.

46   I. Mihelič, A. Podgornik, M. Barut, M. Krajnc, T. Koloini, Temperature Distribution Modeling During Polymerization of a CIM Convective Interaction Media Monolithic Columns, 6th World Congress of Chemical Engineering, 2001, Melbourne, Australia.

47   N. Hird, I. Hughes, D. Hunter, M. G.J.T. Morrison, D.C. Sherrington, L. Stevenson, Tetrahedron 55 (1999) 9575.

48   E.C. Peters, F. Švec, J.M.J. Fréchet, Chem. Mater. 9 (1997) 1898.

49   A. Podgornik, M. Barut, A. Štrancar, Dj. Josić, T. Koloini, Anal. Chem. 72 (2000) 5693.

50   J.A. Tripp, F. Švec, J.M.J. Fréchet, J. Comb. Chem. 3 (2001) 216.

51   R. Hahn, A. Podgornik, M. Merhar, E. Schallaun, A. Jungbauer, Anal. Chem. 73 (2001) 5126.

52   V.V. Azanova, J. Hradil, F. Švec, Z. Pelzbauer, React. Polym. 12 (1990) 247.

53   J. Hradil, F. Švec, React. Polym. 13 (1990) 43.

54   V.V. Azanova, J. Hradil, G. Sytov, E.F. Panarin, F. Švec, React. Polym. 16 (1991) 1.

55   Dj. Josić, H. Schwinn, A. Štrancar, A. Podgornik, M. Barut, Y.-P. Lim, M. Vodopivec, J. Chromatogr. A 803 (1998) 61.

56   G.A. Platonova, G.A. Pankova, I.Y. Ilina, G. Vlasov, T.B. Tennikova, J. Chromatogr. A 852 (1999) 129.

57   K. Amatschek, R. Necina, R. Hahn, E. Schallaun, H. Schwinn, Dj. Josić, A. Jungbauer, J. High Resol. Chromatogr. 23 (2000) 47.

58   I. Mihelič, T. Koloini, A. Podgornik, A. Štrancar, J. High Resolut. Chromatogr. 23 (2000) 39.

59   H. Podgornik, A. Podgornik, A. Perdih, Anal. Biochem. 272 (1999) 43.

60   R. Hahn, A. Jungbauer, Anal. Chem. 72 (2000) 4853.

61   N.D. Ostryanina, O.V. Il'ina, T.B. Tennikova, J. Chromatogr. B 770 (2002) 35.

62   N. Tanaka, H. Nagayama, H. Kobayashi, T. Ikegami, K. Hosoya, N. Ishizuka, H. Minakuchi, K. Nakanishi, K. Cabrera, D. Lubda, J. High. Resolut. Chromatogr. 23 (2000) 111.

63   F. Sinner, M.R. Buchmeiser, Macromolecules 33 (2000) 5777.

64   S. Liu, J.H. Masliyah, Chem. Eng. Commun. 148-150 (1996) 653.

65   C.K. Ratnayake, C.S. Oh, M.P. Henry, J. High Resol. Chromatogr. 23 (2000) 81.

66   A. Podgornik, M. Barut, S. Jakša, J. Jančar, A. Štrancar, J. Liq. Chromatogr., submitted.

67   N.D. Ostryanina, G.P. Vlasov, T.B. Tennikova, J. Chromatogr. A 949 (2002) 163.

68   K. Branović, G. Lattner, M. Barut, A. Štrancar, Dj. Josić, A. Buchacher, J. Immunol. Methods, submitted.

69   M.A. Teeters, T.W. Root, E.N. Lightfoot, J. Chromatogr. A 944 (2002) 129.

70   M. Barut, personal comunication.

71   K. Pflegerl, A. Podgornik, E. Berger, A. Jungbauer, Biotech. Bioeng., accepted.

F. Švec, T.B. Tennikova and Z. Deyl (Editors)
Monolithic Materials
Journal of Chromatography Library, Vol. 67
© 2003 Elsevier Science B.V. All rights reserved.

Chapter 4

Tubes

Aleš PODGORNIK[1], Miloš BARUT[1], Igor MIHELIČ[2], Aleš ŠTRANCAR[1]

[1]*BIA Separations d.o.o., Teslova 30, 1000 Ljubljana, Slovenia*
[2]*Faculty for Chemistry and Chemical Technology, University of Ljubljana, Hajdrihova 19, 1000 Ljubljana, Slovenia*

CONTENTS

4.1 INTRODUCTION

Preparative chromatography is one of the most powerful methods in the production of larger quantities of highly purified substances. Among these, large biomolecules become more and more important. The purification can be carried out in one or, more commonly, several chromatographic steps using a variety of separation mechanisms such as size-exclusion, ion-exchange, hydrophobic interaction, or affinity chromatography, just to mention the most important ones. To perform the purification on the industrial scale, large chromatographic columns are preferred. They can have diameters of over 1 m and various lengths [1]. Historically for such columns, both the uniformity of packing of the stationary phase and of the sample distribution over the entire column were serious challenges. Also the supports for separation of large biomolecules were mostly soft and compressible. Matrix compression, owing to its weight reduced the flow through the column. A large bed-height combined with a high flow rate resulted in a high pressure drop, which further compressed the matrix and consequently changed the column characteristics. Having these problems in mind, Saxena [2] introduced radial chromatography. The main difference from the conventional, axial chromatography is that the mobile phase flows in a radial direction through the matrix bed, which typically has shape of a hollow cylinder. Mobile phase is distributed over the outer side of the tube and penetrates in a radial direction through the tube-shaped matrix bed. In this way, the matrix-bed height — better the tube thickness — is significantly smaller than in the case of axial columns. This reduces the pressure drop across the column while preserving the scalability of the process.

A similar chromatographic operation mode was introduced also for the scale-up of chromatographic membrane devices. Membranes are, by definition, very thin layers. Despite their hydrodynamic characteristics being superior to the conventional particulate supports, the total volume of a single membrane is very small. By increasing the cross-sectional area of such a device, the problem of uniform sample distribution is faced. Besides, the distribution of the mobile phase from a small inlet tube to a large-diameter separation device results in back-mixing, which inherently leads to higher dispersion and, consequently, reduced column performance. To overcome these problems the membrane can be wound to form a cylindrical shaped column – a spiral-wound membrane [3]. Since the number of turns is, at least, theoretically unlimited, the column of desired volume can be constructed. Again, a radial flow direction is applied. A detailed discussion can be found elsewhere in this book.

Monoliths represent a new generation of chromatographic phases. In contrast to other chromatographic supports they can be prepared in various shapes and various dimensions. They can be based on differing materials and therefore possess differing chemical and mechanical characteristics, from soft gels to rigid solids. Despite the many advantages of the monoliths over the conventional stationary phases described in details elsewhere in this book, there are very few reports in the literature concerning the preparation of large-volume monolithic columns. This is probably related to the difficulties of preparing such columns with a uniform structure. To understand the reasons, the mechanisms of formation of the monolithic structure should be elucidated. Owing to the difficulties with the preparation, one of the most promising designs is again a tube-shaped monolithic column geometry used with a radial flow direction.

4.2 PREPARATION OF MONOLITHS

Preparation of the polymeric monolith is commonly performed via free radical polymerization in bulk. Details are described elsewhere in this book. Here we briefly focus only on the formation of the monolith structure, which is essential for obtaining a column with the desired chromatographic characteristics. As an example, the preparation of a methacrylate-based monolith is described. A polymer mixture for preparation of this monolith consists of monomers — glycidyl methacrylate (GMA) and ethylene dimethacrylate (EDMA) — as well as pore-producing solvents ("porogens") cyclohexanol and dodecanol. At the polymerization temperature, the initiator which is also added to the polymerization mixture, starts to decompose, liberating free radicals that initiate the formation of the oligomeric nuclei. As the polymerization continues, these nuclei grow and become less and less soluble in the reaction mixture. Finally, they precipitate. The monomers are thermodynamically better solvating agents for the polymer than the porogens. Consequently, the precipitated nuclei are swollen with the monomers. Since the monomer concentration is higher in nuclei than in the surrounding solution, the polymerization in the nuclei is kinetically preferred. In the absence of mixing, insoluble nuclei sediment, owing to their higher density and accumulate in the bottom of the mold. Initially, they form a very loose structure, which is highly porous. During the course of polymerization the nuclei continue to grow and crosslink until the final structure is obtained. This mechanism has been described in detail by Švec and Fréchet [4].

The structure of the monolith depends highly on the polymerization temperature, as well as the initial composition of the polymerization mixture. The temperature actually controls the decomposition rate of the initiator and, therefore, also the number of nuclei formed in a given period of time. Since the amount of the monomers is limited, the lower the number of nuclei formed within a defined volume the larger their size. As they link together, the pores between them are also larger. In contrast, at higher polymerization temperatures, at which the initiator decomposition is much faster, the number of growing nuclei is considerably higher and, therefore, the pores between smaller nuclei are also smaller. Obviously, the polymerization temperature is a powerful tool to control the monolith structure [4].

Unfortunately, polymerization is typically an exothermic process. Therefore, the heat of polymerization is released during the course of the preparation of monoliths. In contrast to suspension polymerization of the particulate polymers, monoliths are typically prepared in the absence of mixing, which would destroy their structure. Since the mold size is most often in the range of millimeters or even centimeters, the generated heat might not be dissipated fast enough, leading to its rapid accumulation. The heat released per unit of reaction mixture volume is constant, therefore the larger

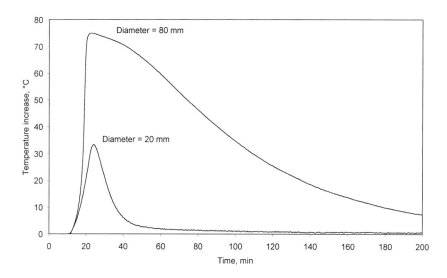

Fig. 4.1. Temperature increase during polymerization of glycidyl methacrylate-ethylene dimethacrylate monoliths. The temperature profile was measured in the center of the polymerization mixture placed in stainless steel molds with diameters of 20 or 80 mm.

the mold dimension, the higher the increase in the temperature inside the polymerization mixture during polymerization. The temperature measurements shown in Fig. 4.1 confirm this fact. The maximum temperature increase during the polymerization may exceed 70°C in a mold of 80 mm diameter while the increase inside a 20 mm mold is 32°C. Peters *et al.* studied in detail the effect of the temperature increase during the polymerization [5]. As expected, pore-size-distribution measurements revealed significant differences between the middle and the outer part of the monolith. The pore-size distribution was clearly non-homogeneous. As demonstrated by Hahn and Jungbauer [6], defects in the monolith structure might result in bimodal break-through curves and severe fronting. Consequently, the performance of such a column is significantly reduced and this column cannot be used for chromatographic analysis. Even in the absence of cracks, the pore-size distribution has to be carefully adjusted to obtain optimal column performance [7]. Therefore, any variations in the structure are undesirable and should be largely avoided.

Obviously, the exothermic nature of the polymerization results in a significant increase in temperature that affects the pore-size distribution and limits the preparation of large volume monoliths. Since this increase depends on the mold dimensions, an upper size limit exists for each geometry, which should not be exceeded if one is to obtain a monolith with a uniform structure. Three different approaches to the larger monolithic columns have been proposed in the literature to overcome this problem and the details with regard to scale-up are presented in the following section.

4.3 SCALE-UP APPROACHES

Basically, all the monolithic chromatographic columns described in the literature can be classified into three categories according to their geometries, schematically shown in Fig. 4.2; rods, disks, and tubes.

Rods are the most widely used chromatographic columns. They are characterized by the column length being larger then its diameter. Besides conventional analytical columns, capillary columns and microchips fall in this category. The extensive literature a this field is reviewed throughout this book. A uniform structure for these microcolumns is not a problem since their diameter rarely exceeds millimeters. Heat released during polymerization can be dissipated efficiently and the increase in temperature is negligible. In contrast, analytical and larger columns may have significantly larger volumes than capillaries. These columns also differ in their length and may have a diameter in the range of up to few centimeters depending on the monolith chemistry. For example, methacrylate-based monolithic columns used in chromatography had a diameter of 8- [8] and 16 mm [9], silica monoliths 4.6- [10] and 25 mm [11], acrylamide 7 mm [12], and superporous agarose 16 mm [13], just to men-

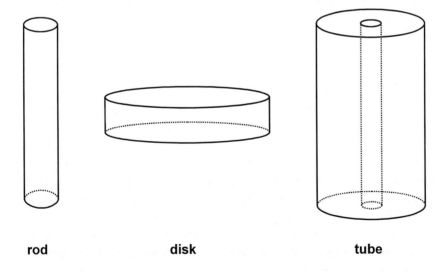

rod   **disk**   **tube**

Fig. 4.2. Different geometries of monoliths

TABLE 4.1

COMPARISON OF CHARACTERISTICS OF COLUMNS WITH DIFFERENT DESIGNS[a]

Column design	Column volume	Volume increase	Pressure drop	Average linear velocity
Rod	$V = r^2 \cdot \pi \cdot L$	L	$\Delta P = \dfrac{\eta \cdot F}{K} \cdot \dfrac{L}{r^2 \cdot \pi \cdot \varepsilon}$	$v = \dfrac{F}{r^2 \cdot \pi \cdot \varepsilon}$
Disk	$V = r^2 \cdot \pi \cdot L$	r^2	$\Delta P = \dfrac{\eta \cdot F}{K} \cdot \dfrac{L}{r^2 \cdot \pi \cdot \varepsilon}$	$v = \dfrac{F}{r^2 \cdot \pi \cdot \varepsilon}$
Tube	$V = \left(r^2 - r_{in}^2 \right) \cdot \pi \cdot L$	r_{in}^2 and/or L	$\Delta P = \dfrac{\eta \cdot F}{K} \cdot \dfrac{\ln\dfrac{r}{r_{in}}}{2 \cdot \pi \cdot L \cdot \varepsilon}$	$v = \dfrac{F \cdot \ln\dfrac{r}{r_{in}}}{2 \cdot \pi \cdot L \cdot \varepsilon \cdot \left(r - r_{in} \right)}$

[a] The equation for the pressure drop was derived from Darcy's law taking into account each particular geometry. V is the volume, r is the column outer radius, r_{in} is the tube inner radius, L is the column length, F is the flow-rate, and ε is the monolith porosity.

tion some of them. In all cases, this diameter appears to be small enough to avoid defects in the monolithic structure during the preparation, since excellent chromatographic properties have always been reported. However, the largest documented volume of those columns was 50 ml [11], which may not be sufficient for the majority of the preparative purifications. Details concerning the preparation of rod-shaped monoliths can be found in a different section of this book. Owing to the limited diameter, the overall volume of these columns can only be increased by increasing their length. Since the column volume increases linearly with its length (Table 4.1), very long columns would need to be constructed to achieve a significant volume increase. The longest methacrylate-based monolithic column reported had a length of 300 mm [14]. The pressure drop of this unit exceeded 10 MPa even at a flow rate of only 3.5 ml/min and would become the main obstacle in implementation of even longer columns. This seems not to be a problem for the monoliths with very high porosity, such as those based on silica [15]. Construction of a 1.4 m-long column with a total plate number of 108,000 has already been reported [16]. This enabled an excellent separation at a moderate pressure drop. However, such columns are difficult to handle because of their dimensions.

The second alternative is a disk shape. Several reports describe the application of methacrylate-based monolithic disk-shaped columns and this topic is discussed separately. These disks may have different sizes, starting from very small, having a volume of only 10 μl [17], up to those with a diameter of 50 mm and a thickness of 7 mm, representing an overall volume of 14 ml [18]. It is worth noting that this is also the largest thickness described so far for a disk. Compared to the rod-shaped columns, the volume increases with the square of the diameter (Table 4.1). Since this enlargement approach results in a large diameter column with a small thickness, problems related to the non-uniform sample distribution and back mixing, similar to those characterizing membranes, might occur. Another problem could be the mechanical stability of such a flat monolith because of its limited mechanical properties [19].

Both of the previous approaches have their intrinsic limitations, which cannot be easily overcome. Therefore, a third shape was introduced recently that enables the preparation of large methacrylate-based monoliths [20]. The basic idea behind this approach was to have more degrees of freedom in the preparation of the monolithic columns while preserving the limited allowable thickness. This can be achieved using monoliths in an annular-shaped tube. In this way, both the thickness of the tube and its inner diameter can be varied independently. While keeping the thickness of the monolith constant, the volume can be increased quadratically by increasing the outer column diameter and, in addition, linearly by variation of the total height (see Table 4.1). The largest column of this shape reported so far was a methacrylate-based monolith with a volume of 80 ml [21]. Recently, we also prepared a 800 ml methacrylate-based column. Radial methacrylate-based monolithic columns were used for the separation and purification of clotting factor VIII from human plasma [20], and for extremely fast flow-unaffected separation of myoglobin, conalbumin, and trypsin inhibitor and their purification on a gram scale [21]. Very recently, a superporous agarose radial column with the volume of 65 ml and bed thickness of 20 mm has also been developed [22]. This column exhibited constant performance (constant HETP value) at linear velocities in the range of 0.2–1.8 cm/min, and was successfully applied for purification of lactate dehydrogenase using a Cibacron Blue 3GA ligand. A similar column, with the immobilized enzyme β-galactosidase, was also used for the conversion of lactose to glucose and galactose.

The characteristics of all column designs are summarized in Table 4.1. To make the comparison even more clear, characteristic features were calculated for each geometry. Since a diameter of 10–25 mm was considered to be the upper limit for methacrylate-based monoliths [23] we set the maximum allowed linear dimension to 20 mm. For a rod shape this dimension is its diameter, for a disk shape it is the disk thickness, and for the annular shape this dimension is the thickness of the tube. Columns with two different volumes of 100 and 10,000 ml (10 l) are compared. The results of these calculations are presented in Table 4.2. Obviously, the situation for

TABLE 4.2

COMPARISON OF A PRESSURE DROP FOR THREE DIFFERENT GEOMETRIES AND
TWO DIFFERENT COLUMN VOLUMES

Column volume, ml	Geometry		Dimensions, cm	Relative pressure drop
100	Rod		$r = 1.0, L = 31.8$	254.4
100	Disk		$r = 4.0, L = 2.0$	1
100	Tube	$L = \dfrac{7.96}{r_{in} + 1}$	$r_{in} = 0.2, L = 6.6$	1.46
			$r_{in} = 1.0, L = 4.0$	1.10
			$r_{in} = 5.0, L = 1.3$	1.03
10 000	Rod		$r = 2.0, L = 3183.1$	$2.5 \cdot 10^6$
10 000	Disk		$r = 39.9, L = 2.0$	1
10 000	Tube	$L = \dfrac{796}{r_{in} + 1}$	$r_{in} = 1.0, L = 398$	1.10
			$r_{in} = 10.0, L = 72.4$	1
			$r_{in} = 20.0, L = 37.9$	0.5

The relative pressure drop was calculated for each pressure separately and can be compared
only for single column volume. Values were calculated using the equations from Table 4.1.
The flow rate was kept constant for all cases. Normalization was related to the pressure drop
of the disk.

columns of different volumes can be readily calculated using the Equations of Table
4.1.

Based on the results shown in Table 4.2, some useful information can be extracted.
Since two dimensions can be varied independently in the case of the annular geometry, namely the wall thickness and the tube diameter, and since the pressure-drop is a
function of both of these variables, different pressure drops can be obtained even for
columns of the same volume. This does not happen for other geometries, where the
pressure drop is defined already by the selection of the column volume. Disk-shaped
columns exhibit the lowest pressure drop among 100 ml columns and are closely
followed by tubes. The pressure drop for rod-shaped columns is about two orders of
magnitude higher. However, 10 ml columns with reasonable dimensions can be constructed using all geometries. In contrast, this is not possible for 10,000 ml columns.
The length of the rod-shaped column would be more than 3 m, which is quite impractical to handle. In addition, the pressure drop would be more than 10^6 times higher
then that of the disk-shaped column, although the dimensions of the latter are not

favorable either. The disk shape would require uniform distribution of the sample from a capillary to the disk having a diameter of 80 cm. As has been discussed above, the mechanical stability of such a column might also be an issue. Thus, the only currently acceptable geometry for this scale is the annulus. By adjusting its radius, an easily handed column can be readily constructed. Furthermore, the pressure drop decreases with the increase in the radius and may become even lower than that for the disk. It can be concluded from this simple calculation that the use of the tubular geometry appears to be the most promising scaling approach. It is worth noting that this conclusion is valid only for separations of large molecules and for elution in gradient mode, for which the resolution power is not related to the column length [24-27].

4.4 PREPARATION OF MONOLITHIC TUBULAR COLUMNS

Tubular monolithic columns can be prepared in three slightly different ways. The smaller monolithic tube columns can be prepared from polymerized rods by mechanically removing the inner part of the monolith after the polymerization is completed. This removes the part where the temperature-increase during polymerization is the highest and the deviation of the porous structure most pronounced. This approach is exemplified by the 22 ml methacrylate-based monolithic tube used for purification of plasma proteins [20] and a 65 ml superporous agarose monolith [22]. This technique can be applied only to rods with a certain diameter, since the middle part with the deformed porous structure, which should be removed, is becoming larger and larger and, consequently, the thickness of the remaining tube-wall is small.

The second option is to polymerize the monolith in an annular mold. The temperature on both the outer and inner sides of the mold is kept constant during polymerization and more of the surface is available for handling the heat dissipation. The thickness of the monolithic tube can then be larger than with the previous procedure. However, another alternative method must be used for even larger monoliths.

While the upper thickness of the monolith is limited, the selection of the tube diameter is flexible. Therefore, several tubes with different diameters can be prepared. The diameters are selected so that the inner diameter of the larger tube matches the outer diameter of the smaller tube. Thus, several tube-shaped monoliths can be inserted one into another. This approach is schematically shown in Fig. 4.3 and has been described by us in detail [21]. The separation characteristics of this 80 ml column are flow-independent for flow-rates of up to 250 ml/min. The possible voids between the individual assembled tubes can be filled by additional polymer. Namely, the voids are filled with a polymerization mixture and the polymerization is allowed to proceed. Since the thickness of these voids is very small, no increase in the temperature during

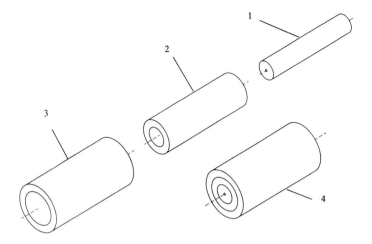

Fig. 4.3. Construction of a large-volume monolithic tube with a specific volume. The monolithic tube 4 consists of three monolithic annuli 1, 2, and 3 (Reprinted with permission from ref. [21], copyright 2000, American Chemical Society).

polymerization occurs and a single monolith with a uniform structure is obtained [28]. This so-called tube-in-tube approach has another advantage. Each of the several thinner tube-shaped monoliths that are assembled to form a large monolith may bear different active groups. This can be achieved either during their preparation using a suitable polymerization mixture containing monomers carrying desired functionalities or by chemical modification after the polymerization is completed. Using this approach, conjoint liquid chromatography, that has already been successfully demonstrated on small monolithic columns [29], can be transferred to the larger scale.

In contrast to the smaller units, which are routinely used for analytical purposes, large columns are mainly intended for purification and their capacity is very important. Using the design proposed above, the required column capacity for each active group can be set differently, depending on the amount of the target compound in the product to be purified. To achieve this, the volume of each tubular segment bearing the desired active functionalities can be adjusted accordingly, by varying the wall thickness or, when possible, also the position within the assembled column. If the required volume for the particular chemistry exceeds the allowable monolith thickness, two or more tubes with the same chemistry can be used to achieve the desired value. Therefore, this approach also enables very fine-tuning of the column characteristics on a large scale to meet the specific needs and, in addition, can substitute several columns with a single one.

4.5 MATHEMATICAL MODELING OF PREPARATION OF
METHACRYLATE-BASED MONOLITHS

As already mentioned, the preparation of large monoliths is limited by their tolerable thickness, which further depends on the preparation procedure and the monomer chemistry. Despite the different possible approaches for overcoming this problem, demonstrated in previous sections, the basic question remains unanswered. How does one determine the upper allowable thickness? As already discussed, the most important parameter that affects the morphology of a monolith is the temperature increase inside the polymerizing mixture. The problem of the allowable monolith thickness can therefore be formulated in terms of the determination of the maximum temperature increase with respect to the geometry and dimensions of the mold.

4.5.1 Rod

To estimate the temperature increase during this polymerization, the differential heat-balance equation 4.1 incorporating the term of heat generation has to be used:

$$\frac{\partial T}{\partial t} = \frac{\alpha}{r} \frac{\partial}{\partial r}\left(r \frac{\partial T}{\partial r}\right) + \frac{S}{\rho\, c_p} \tag{4.1}$$

where T is the temperature (K), t the time (s), r the radius (m), α the thermal diffusivity (m^2/s), ρ the density (g/cm^3), c_p the specific heat capacity (J/gK), and S is the released heat-flow per unit volume (W/cm^3). Equation 4.1 describes the changes of the temperature with time as a function of the radius and the heat release owing to the polymerization. For the solution of this equation, one initial- and two boundary conditions are required.
Initial condition:

$$T = T_0 \qquad\qquad 0 \le r \le r_0 \qquad\qquad t = 0 \tag{4.2}$$

Boundary conditions:

$$\frac{dT}{dr}\bigg|_{r=0} = 0 \qquad\qquad r = 0 \qquad\qquad t \ge 0 \tag{4.3}$$

$$T = T_\infty \qquad\qquad r = r_0 \qquad\qquad t \ge 0 \tag{4.4}$$

where r_0 is the outer radius (m), T_∞ the polymerization temperature (K), and T_0 the initial temperature (K).

References pp. 100-101

The experimental justification for these conditions has been demonstrated by Mihelič *et al.* [30]. Solution of Equation 4.1 requires estimation of the heat flow, *S*, during the polymerization. It depends on the reaction kinetics and the heat of polymerization. Using DSC measurements and applying the Ozawa-Flynn-Wall method [31], the polymerization of methacrylate-based monoliths was found to be a reaction with first-order kinetics for which the following equation applies:

$$\frac{dx}{dt} = (1 - x)\, A\, \exp[-E_{a,app}\, /RT] \qquad (4.5)$$

where *x* is the extent of reaction (–), *A* is the pre-exponential factor (s^{-1}), *R* is the gas constant (J/mol.K), and $E_{a,app}$ is the apparent activation energy (J/mol). Integration of Equation 4.5 leads to an expression for the extent of reaction as a function of time and temperature:

$$x(t,T) = 1 - \exp[-A \int_0^t \exp[-E_{a,app}/RT]\, dt] \qquad (4.6)$$

Without going into the details, which can be found elsewhere [30,32], Equation 4.1 can then be rewritten in the following form:

$$\frac{\partial T}{\partial t} = \frac{\alpha(x)}{r}\frac{\partial}{\partial r}\left(r\,\frac{\partial T}{\partial r}\right) + \frac{1}{c_p(x)}\frac{\partial}{\partial t}\left(\Delta H_r\left(1 - \exp[-A\int_0^t \exp[-E_{a,app}/RT]\, dt]\right)\right) \qquad (4.7)$$

The thermal diffusivity and the specific heat capacity in Equation 4.7 are not regarded as constants because the differences in physical properties between the liquid reagent mixture and the final two-phase product are too large. In this mathematical model they are both considered as linear functions of the extent of reaction.

$$\alpha(x) = \alpha_0\,(1 - x) + x\,\alpha_r \qquad (4.8)$$

$$c_p(x) = c_{p,0}\,(1 - x) + x\,c_{p,f} \qquad (4.9)$$

The subscripts *0* and *f* denote the initial and final values of the thermal diffusivity and the specific heat capacity.

One very important effect was observed during determination of the initial thermal diffusivity. We found that it cannot be considered constant but rather it should be treated as an effective diffusivity. This results from the convective mixing of the liquid phase at the beginning of polymerization. This mixing depends not only on the type of liquid, and the temperature difference, but also on the geometry of the mold in

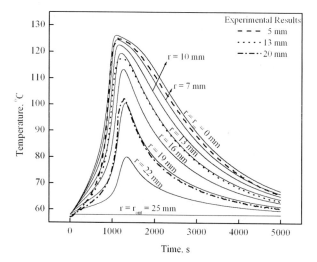

Fig. 4.4. Comparison of simulated and experimental temperature profiles inside a polymerizing mixture. Polymerization of the glycidyl methacrylate-ethylene dimethacrylate monolith was performed in a 35 mm diameter stainless steel mold. (Reprinted with permission from ref. [30], copyright 2002, Wiley InterScience).

which the polymerization is performed. Since the theoretical determination of effective thermal diffusivity is extremely difficult and not yet developed [33], the values must be determined experimentally for each particular system individually. Details concerning the estimation of required values and their experimental verification for our particular system can be found in the literature [30]. Using these findings, the validity of the mathematical model was verified by experimental measurements of the temperature profiles obtained during polymerization, and shown in Fig. 4.4. Three temperature sensors were placed into the polymerization mixture at different radial positions from the outer wall toward the center of the mold. The temperature was measured continuously during the polymerization. We concluded from the good agreement between experimental and simulated values that the mathematical model describes profiles of the temperature reasonably well and can be used for estimation of other parameters, which would otherwise require extensive experimental work or whose experimental determination would even be impossible.

The prediction of the maximum temperature increase inside the polymerization mixture during polymerization, as a function of the mold diameter and initial temperature, is the first direct outcome of the model. Having in mind that the maximum temperature should not exceed a certain value (which depends on the preparation technique and the monomer chemistry), the maximum allowable thickness of the rod

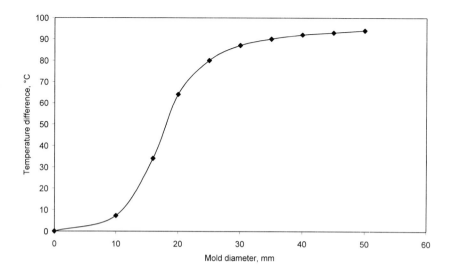

Fig. 4.5. Maximum temperature increases inside the polymerization mixture placed in molds of different diameters during the polymerization of glycidyl methacrylate-ethylene dimethacrylate monolith. Temperature values were calculated using our mathematical model.

to obtain a monolith with a uniform structure can be estimated. The mathematical prediction of the temperature increase is shown in Fig. 4.5. The curve shape is somewhat similar to the chromatographic break-through curve. The temperature increase is only moderate for the small-volume molds. This explains the uniformity of structure observed for the monolithic rods with small diameters. With the increase in mold diameter, a steep increase of the maximal temperature is observed, which appears to be proportional to the mold diameter. The temperature increase reaches a value of around 60°C. In molds of even larger diameter, further temperature increase is moderate. This is because the total heat release per unit of volume for polymerization mixture is constant, and this limits the temperature increase, which can reach an upper value of about 100°C. However, even a much smaller temperature increase affects unacceptably the monolithic structure [5] and the leveling-off has little overall effect.

Other interesting information which can be obtained from the model is the time required to complete the polymerization. Results of the simulation are presented in Fig. 4.6. While the polymerization in the center of the mold is completed in less then 20 min, the reaction in the areas adjacent to the mold wall continues for several hours. Since polymerization is completed only after all the polymerization mixture in the mold has reacted, the simulated conversion provides a very useful guide for determination of the time required.

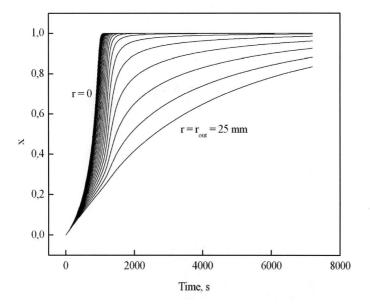

Fig. 4.6. Conversion of polymerization as a function of time and position inside the polymerization mixture. The simulation was performed for a mold with a diameter of 50 mm. (Reprinted with permission from ref. [30], copyright 2002, Wiley InterScience).

4.5.2 Tube

Modeling of the polymerization of the monolithic tube can be done in almost an identical way to that for the rod. Equations 4.1, 4.5, 4.6, and 4.7 apply again with the following initial and boundary conditions:

Initial condition:

$$T = T_\infty \qquad\qquad r_{in} \leq r \leq r_{out} \qquad\qquad t = 0 \qquad\qquad (4.10)$$

Boundary conditions:

$$T = T_\infty \qquad\qquad r = r_{in} \qquad\qquad t \geq 0 \qquad\qquad (4.11)$$

$$T = T_\infty \qquad\qquad r = r_{out} \qquad\qquad t \geq 0 \qquad\qquad (4.12)$$

As already discussed for the rod, the initial thermal diffusivity should also be determined since it depends on the mold geometry. While the maximum temperature-increase in the rod depends only on the mold diameter, both the inner diameter and its

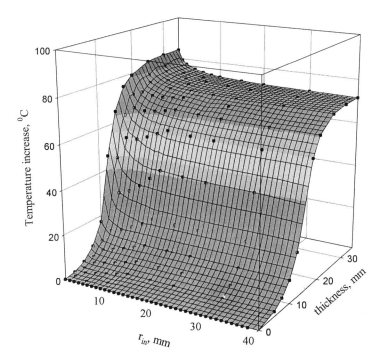

Fig. 4.7. Simulation of the maximum temperature increase inside a tube during polymerization as a function of inner radius of the annulus (r_{in}) and its thickness ($r–r_{in}$). Dots represent values calculated using the mathematical model.

thickness affect the temperature within the tubes. Results are shown in Fig. 4.7. For very small inner radii, the maximum temperature increase is the highest and exceeds 80°C. The maximum temperature increase becomes almost constant for monoliths with inner radii larger than 10 mm and no effect of further increase of the hole diameter can be observed. In contrast, the temperature-increase depends significantly on the wall thickness. While it is rather moderate for thicknesses of less than 10 mm, the temperature increase, which prevents the formation of a monolith with a uniform structure, can be observed in thicker tubes. As expected, the temperature increases with the tube thickness in similar way to that observed for the rod. The key difference in this implementation is that the maximal temperature is not reached in the middle of the polymerization mixture but is shifted towards the inner wall (see Fig. 4.8). This is more pronounced for small inner radii, while it is almost negligible in molds with larger ones. In fact, with the increase of the inner radius, the geometry is approaching that of a sheet. This is also confirmed by Fig. 4.7.

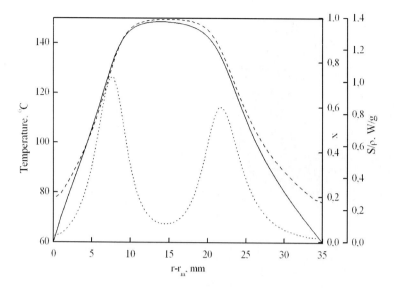

Fig. 4.8. Simulation of conversion, heat release, and temperature profiles inside an annular mold. Polymerization mixture was placed in the void between two concentric stainless steel tubes. The wall thickness was 35 mm and the inner diameter 5 mm. Profiles represent the situation calculated at 650 s after initiation of polymerization. (Reprinted with permission from ref. [30], copyright 2002, Wiley InterScience).

Differences between the rod- and tube geometry are also discernible from the conversion profile as well as from the heat flux. At the beginning, the temperature, conversion, and generated heat profiles are parabolic. Since the reaction is faster, and therefore completed sooner, in the center of the polymerization mixture, the heat distribution profile has a bimodal shape (Fig. 4.8). In the later stages, the polymerization continues in the areas close to the mold wall until is completed.

4.6 HYDRODYNAMIC CHARACTERISTICS OF TUBE MONOLITHS

The tube design was developed to avoid difficulties with the production of large monoliths having uniform porous structure. In addition, these large columns exhibit the lowest pressure drop for a given column volume (Table 4.2). The following analysis will demonstrate that monolithic columns with a radial flow may have efficiency superior to similar columns packed with beads. This advantage is particularly expressed for thicker columns.

In radial chromatography, the mobile phase is directed to the outer side of the tube and then penetrates through the bed to the center, in a radial direction. This path is

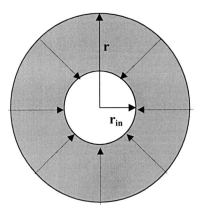

Fig. 4.9. Flow direction in the radial chromatographic column. Dashed arrows indicate the flow direction of the mobile phase.

schematically presented in Fig. 4.9. Since the area through which the mobile phase flows decreases from the outer- to the inner side of the annulus, this results in an increase in the flux and, consequently, in an increase in the linear velocity. This can easily be seen from the equation that enables the calculation of the linear velocity at certain positions within the tube:

$$v = \frac{F}{2 \cdot \pi \cdot r \cdot L \cdot \varepsilon}$$

(4.13)

where v is the linear velocity (m/s).

When the tube diameter is large and its thickness small, the change in linear velocity is also small and can be neglected. If, on the other hand, the inner radius is small and the thickness of the tube significant, significant changes in linear velocity through the column, shown in Fig. 4.10, are encountered. For example, the linear-velocity change is moderate for a tube column with the radius of the inner hole exceeding 3 cm and the thickness 3 cm. In contrast, the increase in linear velocity can even exceed one order of magnitude for very small inner radii. The largest monolithic tube on the market (CIM Convective Interaction Media® monolithic column produced by BIA Separations d.o.o.) currently has a hole diameter of 3 mm, a thickness of 16 mm, and a volume of 80 ml. The maximum allowed flow rate is 250 ml/min [34]. At this flow rate, the linear velocity at the outer surface is 255 cm/h, while in the inner part it reaches the value of 2972 cm/h, representing a twelve-fold increase.

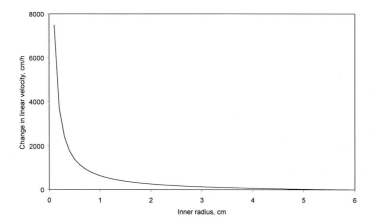

Fig. 4.10. Change in linear velocity through the annular column as a function of column dimensions. The calculation was performed for a monolithic column with a height of 5 cm and an outer radius of 6 cm at a flow rate of 200 ml/min. The inner radius was changed while the outer radius was kept constant. The change of linear velocity was calculated from the difference between linear velocities at the inner and outer surface of the tube.

It is generally known for columns packed with particulate stationary phases that the column efficiency depends on linear flow velocity according to the van Deemter equation. Typically, the efficiency of the column linearly decreases at high linear flow velocities. In contrast, the efficiency of monolithic columns for large molecules is flow independent. This is true both for the separation of biomolecules and the dynamic binding capacity [26,35-37]. Therefore, and despite the considerable increase in the linear velocity inside the column, the separation characteristics should not change. This has been demonstrated for the 80 ml volume methacrylate-based column having a hole diameter of only 1.5 mm [21]. Although the linear velocity was increased by a factor of 23 and the linear velocity in the center exceeded 4,000 cm/h, both resolution and dynamic binding capacity did not change (Fig. 4.11). These results are very important since they indicate that compact radial monolithic columns having a high resolution power, flow-independent separation characteristics, and a low pressure-drop can be constructed. Therefore, consideration of the tube geometry is an approach of choice for the construction of large monolithic columns.

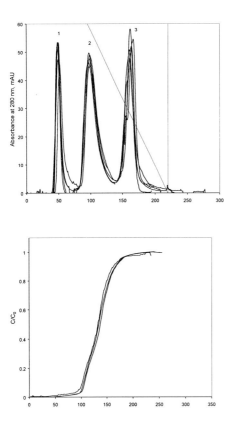

Fig. 4.11. Flow-unaffected resolution and dynamic binding capacity of an 80 ml glycidyl methacrylate-ethylene dimethacrylate monolithic tube column. Separations were performed at flow rates ranging from 40 to 240 ml/min and binding-capacity experiments at flow rates of 50–100 ml/min. Separation conditions: Mobile phase; buffer A, 20 mmol/l Tris-HCl buffer, pH 7.4; buffer B, 1 mol/l NaCl in buffer A; Gradient, 0–100 % buffer B in 200 ml; Sample, 2 mg/ml myoglobin (peak 1), 6 mg/ml conalbumin (peak 2), and 8 mg/ml soybean trypsin inhibitor (peak 3) dissolved in buffer A; Injection volume, 1000 µl; Detection, UV at 280 nm. Capacity determination: Sample, 10 mg/ml BSA in a 20 mmol Tris-HCl buffer, pH 7.4; Detection, UV at 280 nm. (Reprinted with permission from ref. [21], copyright 2000, American Chemical Society).

4.7 TRANSFER OF GRADIENT-CHROMATOGRAPHIC METHODS AMONG DIFFERENTLY SIZED MONOLITHIC COLUMNS

As already discussed, monoliths significantly reduce the time required for the separation of large biomolecules with a low diffusivity such as proteins and DNA. Separation is typically achieved in the gradient-elution mode, which is required to elute large molecules with a very narrow elution window [38]. The column length plays a less significant role in gradient elution [23-26]. Therefore, the method-transfer between differently sized columns is not necessarily based on the constant column length. For gradient-separation Yamamoto [39] introduced a dimensionless parameter, O, defined as:

$$O = \frac{L \cdot I_a}{G \cdot HETP_{LGE}} \tag{4.14}$$

where L represents the column length (m), G is the gradient slope normalized with respect to the column void-volume (M), I_a is the dimensional constant having a value of unity (1), and $HETP_{LGE}$ is the plate height in linear gradient-elution mode (m).

In order to obtain equal resolution for two columns, the dimensionless parameters O should also be equal [39]. Since for the same stationary phase, the HETP, should be equal, this criterion becomes:

$$\frac{L_1}{L_2} = \frac{G_1}{G_2} \tag{4.15}$$

Knowing the optimal gradient for a small column, a gradient time for a larger column can be calculated using Eq. 4.15 and the definition of G. The following equation is obtained:

$$t_{g,large} = t_{g,small} \cdot \left(\frac{V_{large}}{V_{small}}\right) \cdot \left(\frac{F_{small}}{F_{large}}\right) \cdot \left(\frac{L_{small}}{L_{large}}\right) \tag{4.16}$$

where t_g is the gradient time (s), V is the total column volume (m^3), F is the flow rate (m^3/s), and L is the column length (m). The porosities of the two columns are assumed equal.

This approach has been applied to compare the separations of standard test proteins on different commercially available methacrylate-based monolithic columns; axial CIM$^®$ disks with the thickness of 3, 6, 9, and 12 mm as well as CIM$^®$ tube columns with volumes of 8- and 80 ml. Axial columns having different thicknesses were prepared by placing several CIM$^®$ monolithic disks into a single housing [35]. Such a column behaves like a single monolithic column [40,41]. Since the method proposed by Yamamoto [39] has been developed only for axial columns, its application for radial columns is not obvious. In fact, the efficiency of the radial column could also be affected by the changes in linear velocity. As already discussed, the efficiency of monoliths does not depend on the linear velocity. Therefore, it is reasonable to assume that application of the criterion of equation 4.15 is meaningful also for monolithic columns with radial flow. The separation of myoglobin, conalbumin, and

TABLE 4.3

GRADIENT SLOPES FOR DIFFERENT MONOLITHIC COLUMNS CALCULATED
ACCORDING TO EQUATION 4.16

Column type	Column volume, ml	Column length[a], cm	Flow rate CV/min	Gradient time[b] s
Axial	0.34	0.3	10	78
Axial	0.68	0.6	10	39
Axial	1.02	0.9	5	52
Axial	1.36	1.2	2	97
Radial	8	0.65	2	180
Radial	80	1.6	2	73

[a] For operation in radial mode, column length actually represents a tube thickness.

[b] Concentration of NaCl was linearly increased from 0- to 0.7 mol/l during the specified gradient time.

trypsin inhibitor was performed to verify this hypothesis. The flow rate was set to two column-volumes (CV) per min for large columns and to 5 or 10 CV/min for small columns. These flow rates were used to perform separation of proteins in a reasonably short period of time on small columns also. Since the column efficiency defined in HETP does not depend on the mobile-phase linear velocity, these changes should not influence the column efficiency. The gradient was calculated according to Equation 4.16 and the results are shown in Table 4.3.

Chromatograms obtained after applying the methods described in Table 4.3 are presented in Fig. 4.12. All of the chromatograms are very similar, taking into account the fact that the smallest column has a volume of only 0.34 ml while the largest is 80 ml, over two powere of ten more. The flow rates ranged from 3.4 ml/min up to 160 ml/min. This required use of different HPLC systems to perform these experiments. As a consequence, extra column-band-spreading was also different, thus contributing to small differences among chromatograms. The amounts of injected proteins were significantly different. While 20 µl of protein solution with concentration of 4 mg/ml was injected into the axial columns, 500 µl of solution with a protein concentration of 24 mg/ml (150 times more) was applied on the 80 ml column. Based on these facts, and the similarities of the chromatograms, we can safely conclude that the criterion of Equation 4.15 can also be used for transfer of chromatographic methods from axial- to radial chromatographic monolithic columns as well as among radial monolithic

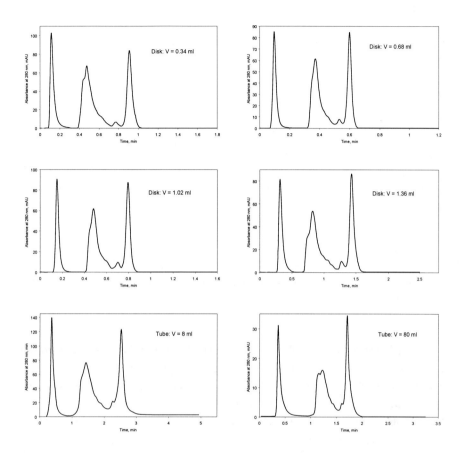

Fig. 4.12. Separation of myoglobin, conalbumin, and soybean trypsin inhibitor using six different CIM monolithic columns (BIA Separations) with different volumes, and axial- as well as radial-flow directions. Column dimensions are shown in Table 4.3. Conditions: Mobile phase; buffer A, 20 mmol/l Tris-HCl buffer, pH 7.4; buffer B, 1 mol/l NaCl in buffer A. For gradients see Table 4.3. Detection: UV at 280 nm.

columns of different volumes. These results also demonstrate that the characteristics of both large and small monolithic tube columns are equal. These conclusions are very important for the application of monolithic columns on industrial scale as well as for rapid and accurate method transfer.

4.8 CONCLUSIONS

This Chapter describes a practical solution for overcoming the problems related to the heat generation during polymerization of large-volume monoliths. This has been

References pp. 100-101

one of the most serious problems during the development of these units and its solution is based on a mathematical modeling of the polymerization process together with a rational design of both the monolithic polymer and appropriate housings.

Our analysis shows that tubular monoliths of defined dimensions can be prepared and merged to form one large monolith that is characterized by very good hydrodynamic and mass-transfer characteristics. Such performance can accelerate the application of these new promising chromatographic stationary phases in industrial separation and purification processes that afford higher yields as a result of lower product degradation. In addition, faster purification steps enabled by unparalleled hydrodynamic properties also increase the process throughput. Therefore, we may expect reports in the near future concerning the industrial downstream processes based on monolithic technologies.

4.9 REFERENCES

1   J.-C. Janson, T. Petterson in: G. Ganestos, P.E. Barker (Eds.) Preparative and Production Scale Chromatography. Marcel Dekker, New York, 1993, p. 559.

2   V. Saxena, US Pat. 4 627 918, (1986).

3   S.H. Huang, S. Roy, K. Hou, G. Tsao, Biotechnol. Prog. 4 (1988) 159.

4   F. Švec, J.M.J Frechet, Chem. Mater. 7 (1995) 707.

5   E.C. Peters, F. Švec, J.M.J. Fréchet, Chem. Mater. 9 (1997) 1898.

6   R. Hahn, A. Jungbauer, J. Chromatogr. A 908 (2001) 179.

7   M.B. Tennikov, N.V. Gazdina, T.B. Tennikova, F. Švec, J. Chromatogr. A 798 (1998) 55.

8   F. Švec, J.M.J. Fréchet, Anal. Chem. 64 (1992) 820.

9   F. Švec, J.M.J Fréchet, Biotechnol. Bioeng. 48 (1995) 476.

10   K. Cabrera, D. Lubda, H.-M. Eggenweiler, H. Minakuchi, K. Nakanishi, J. High Resolut. Chromatogr. 23 (2000) 93.

11   M. Schulte, D. Lubda, J. Dingenen, J. High Resolut. Chromatogr. 23 (2000) 100.

12   T.L. Tisch, R. Frost, J-L. Liao, W.-K. Lam, A. Remy, E. Scheinpflug, C. Siebert, H. Song, A. Stapleton, J. Chromatogr. A 816 (1998) 3.

13   P.-E. Gustavsson, P.-O. Larsson, J. Chromatogr. A 832 (1999) 29.

14   F. Švec, J.M.J. Fréchet, J. Chromatogr. A 702 (1995) 89.

15   H. Minakuchi, K. Nakanishi, N. Soga, N. Ishizuka, N. Tanaka, J. Chromatogr. A 762 (1997) 135.

16   K. Cabrera, D. Lubda, E. Dicks, H. Minakuchi, K. Nakanishi, N. Tanaka, Proceedings of HPLC Kyoto, September 11-14, 2001, Kyoto, Japan, p. 59.

17   K. Pflegerl, A. Podgornik, E. Berger, A. Jungbauer, Biotechnol. Bioeng. (2002) in press.

18   Dj. Josić, J. Reusch, K. Loster, O. Baum, W. Reutter, J. Chromatogr. 590 (1992) 59.

19   D. Horák, Z. Pelzbauer, M. Bleha, M. Ilavský, F. Švec, Angew. Makromol. Chem.185 (1991) 275.

20   A. Štrancar, M. Barut, A. Podgornik, P. Koselj, H. Schwinn, P. Raspor, Dj. Josić, J. Chromatogr. A 760 (1997) 117.

21 A. Podgornik, M. Barut, A. Štrancar, Dj. Josić, T. Koloini, Anal. Chem. 72 (2000) 5693.

22 P.-E. Gustavsson, P.-O. Larsson, J. Chromatogr. A 925 (2001) 69.

23 F. Švec, J.M. J. Fréchet, Ind. Eng. Chem. Res. 38 (1999) 34.

24 G. Vanecek, F. Regnier, Anal. Biochem. 109 (1980) 345.

25 T.B. Tennikova, F. Švec, J. Chromatogr. 646 (1993) 279.

26 M. Merhar, A. Podgornik, M. Barut, S. Jakša, M. Žigon, A. Štrancar, J. Liq. Chrom. Rel. Technol. 24 (2001) 2429.

27 A. Štrancar, A. Podgornik, M. Barut, R. Necina in: R. Freitag, (Editor), Advances in Biochemical Engineering/Biotechnology; Modern Advances Chromatography, Springer-Verlag, Heidelberg, 2002, in press.

28 A. Podgornik, M. Barut, A. Štrancar, Dj. Josić, WO9944053A2 (1999).

29 Dj. Josić, H. Schwinn, A. Štrancar, A. Podgornik, M. Barut, Y.-P. Lim, M. Vodopivec, J. Chromatogr. A 803 (1998) 61.

30 I. Mihelič, T. Koloini, A. Podgornik, J. Appl. Polym. Sci. Submitted for publication.

31 I. Mihelič, M. Krajnc, T. Koloini, A. Podgornik, Ind. Eng. Chem. Res. 40 (2001) 3495.

32 I. Mihelič, T. Koloini, A. Podgornik, M. Barut, A. Štrancar, Acta Chim. Slov. 48 (2001) 551.

33 F.M. White, Heat Transfer. Addison-Wesley, London, 1984.

34 http://www.biaseparations.com/Library/manuals/80_ml_Manual.pdf

35 A. Štrancar, M. Barut, A. Podgornik, P. Koselj, Dj. Josić, A. Buchacher, LC-GC 11 (1998) 660.

36 G. Iberer, R. Hahn, A. Jungbauer, LC-GC 17 (1999) 998.

37 I. Mihelič, T. Koloini, A. Podgornik, A. Štrancar, J. High Resol. Chromatogr. 23 (2000) 39.

38 R. Freitag, Cs. Horvath, in: A. Fiecher, (Ed.), Advances in Biochemical Engineering/ Biotechnology, v. 53, Springer, Berlin, 1995, p. 17.

39 S. Yamamoto, Biotechnol. Bioeng. 48 (1995) 444.

40 A. Podgornik, M. Barut, J. Jančar, A. Štrancar, J. Chromatogr. A 848 (1999) 51.

41 M. Vodopivec, A. Podgornik, M. Berovič, A. Štrancar, J. Chromatogr. Sci. 38 (2000) 489.

F. Švec, T.B. Tennikova and Z. Deyl (Editors)
Monolithic Materials
Journal of Chromatography Library, Vol. 67

Chapter 5

Rigid Polymers Prepared by Ring-Opening Metathesis Polymerization

Michael R. BUCHMEISER

Institute of Analytical Chemistry and Radiochemistry, University of Innsbruck, Innrain 52a, A-6020 Innsbruck, Austria

CONTENTS

5.1 INTRODUCTION

Monolithic separation media evolved as a successful "joint-venture" between material and separation science. Based on theoretical reflections, the common idea was to produce a support with a high degree of continuity that should meet the requirements for fast, yet highly efficient separations [1,2]. Despite the fact that the first experiments into this direction have been already carried out in the 1960s and 1970s [3,4], it took some 20 years to reach the final goal [5-8] and to open the new technology to the area of high-speed-high performance separations. Now, these supports, usually referred to as monolithic columns, continuous beds or rigid rods [4] are successfully used in standard HPLC and micro-separation techniques [9-15], capillary electrochromatography as well as in solid phase extraction (SPE) [16], focusing on medium and high-molecular mass biopolymers [17], and, quite recently, even on low-molecular mass analytes [18-21].

Generally speaking, the term "monolith" applies to any single-body structure containing interconnected repeating cells or channels. Such materials may either be metallic or prepared from inorganic mixtures, e. g. by a sintering process to form ceramics, or from organic compounds, e. g. by a crosslinking polymerization. In this chapter, the term "monolith" or "rigid rod" shall comprise crosslinked, organic materials which are characterized by a defined porosity and which support interactions/reactions between this solid and the surrounding liquid phase. Besides advantages such as lower backpressure and enhanced mass transfer [22,23], the ease of fabrication as well as the many possibilities in structural alteration should also be mentioned. In the following, the concept of monolithic media prepared by a transition metal-based polymerization is presented.

5.2 SYNTHESIS OF MONOLITHIC MEDIA

5.2.1 The traditional approach

Until now, a considerable variety of functionalized and non-functionalized monolithic materials based on either organic or inorganic polymers are available. While inorganic monoliths are usually based on silica and may conveniently be prepared *via* sol-gel techniques [18-21], organic continuous beds mostly consisting of polymethacrylates or poly(styrene-*co*-divinylbenzene) [5,7,8,24,25] are prepared almost exclusively by free radical polymerization. A profound insight into the technology of both sol-gel and free radical polymerization-based monoliths may be found in this book. Despite the comparably poor control over free radical polymerization-based systems, the porosity and microstructure of monolithic materials has successfully been varied [5]. In addition to the free radical process, we confirmed the general applicability of a transition metal-based polymerization technique to the synthesis of high-performance monolithic separation media.

5.2.2 Monolithic materials prepared by ring-opening metathesis polymerization (ROMP)

5.2.2.1 ROMP: Basics, scope and limitations

Transition metal-catalyzed polymerizations have generally gained significant interest due to the rapidly growing armor of well-defined, selective and active catalytic systems. In particular, polyinsertions such as that of Ziegler-Natta type as well as metathesis-based polymerizations have now reached a high level of control over polymerization kinetics, polymer structure, and properties. Thus, well-defined metathesis-active systems such as the highly selective and active Schrock catalysts of the general formula $Mo(N-2,6-R_2-C_6H_3)(CHCMe_2Ph)(OCR')_2$ or the more robust Grubbs-type initiators of the general formula $Cl_2Ru(CHPh)(PR''_3)_2$ offer access to basically every polymer architecture one might imagine [26]. Metathesis-based polymerization techniques in general and ring-opening metathesis polymerization (ROMP) in particular appear highly attractive for the synthesis of complex cross-linked architectures. One advantage of ROMP is its compatibility with functional monomers. This and the controlled "living" [27-31] polymerization mechanism allow a highly flexible yet reproducible polymerization setup. In course of our investigations to use ROMP [32] for the synthesis of functional high-performance materials, we have already combined this polymerization method with grafting and precipitation techniques for the preparation of functionalized separation media [33-44] and catalytic supports [45-49]. Due to the broad applicability of ROMP and the good definition of the resulting materials, we have also investigated to which extent this transition-metal catalyzed polymerization can be used for the synthesis of monolithic polymers [41]. We found that this may be accomplished by generating a continuous matrix by ring-opening metathesis copolymerization of suitable monomers with a crosslinker in the presence of porogenic solvents within a device (column).

5.2.2.2 Polymerization setup: Initiators and monomers

The choice of the suitable initiator represents a crucial step in creating a well-defined polymerization system in terms of initiation efficiency and control over propagation. Only if both quantitative and fast initiation occurs, the entire system may be designed on a *stoichiometric base*. This is important, since for control of microstruc-

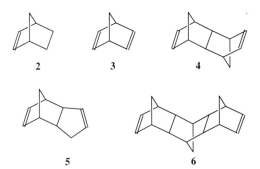

Fig. 5.1. Structures of monomers and crosslinkers. NBE (2), NBDE (3), DMNH-6 (4), DCPD (5), and 1,4a,5,8,8a,9,9a,10,10a-decahydro-1,4,5,8,9,10-trimethanoanthracene (6).

ture, the composition of the entire polymerization mixture needs to be varied using extremely small increments. The catalyst needs to be carefully selected from both chemical and practical points of view. Generally, Schrock and Grubbs systems, both highly active in the ROMP of strained functionalized olefins, can be used. Since the preparation and in particular derivatization of ROMP-based rigid rods requires some handling that may not be performed in a strict inert atmosphere, we focused on the less oxygen-sensitive ruthenium-based Grubbs-type initiators of the general formula $Cl_2(PR_3)_2Ru(=CHPh)$ (**1**), where R can be phenyl or cyclohexyl (Cy), rather than on the molybdenum-based Schrock-type initiators. Since $Cl_2(PPh_3)_2Ru(=CHPh)$ was not sufficiently reactive, we used $Cl_2(PCy_3)_2Ru(=CHPh)$. Among the possible combinations of monomers and crosslinkers such as norbornene (NBE), norbornadiene (NBDE), dicyclopentadiene (DCPD), 1,4,4a,5,8,8a-hexahydro-1,4,5,8-*exo-endo*-dimethanonaphthalene (DMN-H6), and 1,4a,5,8,8a,9,9a,10,10a-decahydro-1,4,5,8,9,10-trimethanoanthracene shown in Fig. 5.1, the copolymerization of NBE with DMN-H6 in the presence of two porogenic solvents, 2-propanol and toluene, with $Cl_2(PCy_3)_2Ru(=CHPh)$ was found to work best. Other monomers and crosslinkers such as norbornadiene and 1,4a,5,8,8a,9,9a,10,10a-decahydro-1,4,5,8,9,10-trimethanoanthracene are excessively reactive or afford unfavorable microstructures, while DCPD did not exhibit the desired crosslinking properties. This is in accordance with previous studies, where crosslinking of DCPD was found to occur only when induced thermally at high monomer concentrations [50-52].

5.2.3 Microstructure of metathesis-based rigid rods

In order to understand the synthesis of monolithic supports and the effects of polymerization parameters, a brief description of the general construction of a mono-lith in terms of microstructure, backbone and relevant abbreviations is given in Fig. 5.2. As can be seen in this Figure, monoliths consist of microstructure-forming micro-globules, which are characterized by a certain diameter (d_p) and microporosity (ε_p). Sum of volume fractions of both micropores and voids (intermicroglobular porosity) is the total porosity (ε_t). This value, representing a percentage of pores in the mono-lith, together with the pore size distribution that can be calculated from inverse size exclusion chromatography (ISEC) data [53], directly translates into a total pore vol-ume V_p, expressed in mL/g, and also allows calculation of the specific surface area σ expressed in m^2/g. In order to design monolithic supports for different separation tasks, we searched for tools that enable varying structural parameters in a controlled and reproducible way. Each component of the polymerization mixture, i. e. NBE, DMN-H6, solvents, and initiator as well as the temperature are variables that can be used to a certain extent for this purpose. The relative ratios of all components, i. e. NBE, DMN-H6, porogens and catalyst, allow varying the microstructure of the mono-lithic material. Structural data such as microglobule diameter (d_p) can be obtained from electron micrographs (ELMI) while ISEC allows the determination of all poros-

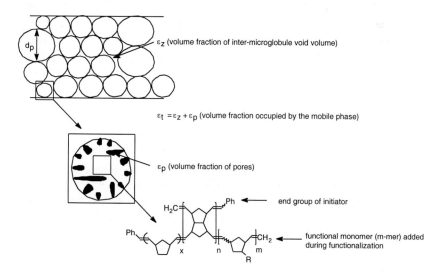

Fig. 5.2. Illustration of the physical meanings of d_p, ε_z, ε_p, ε_t and schematic drawing of the backbone structure. R = functional group.

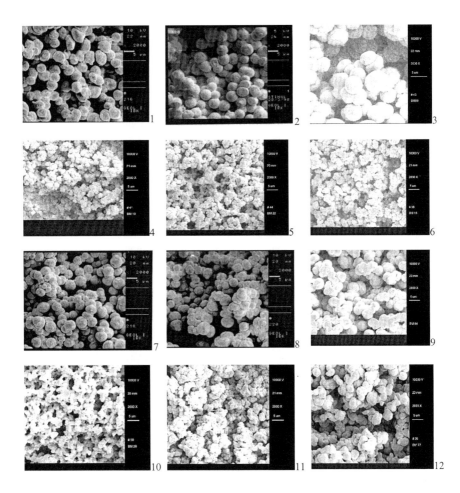

Fig. 5.3. Microstructure of metathesis-based monoliths varying in intermicroglubular void volume and volume fraction of pores.

ity data (ε_z, ε_p, ε_t) as well as specific surface area (σ) [53]. Alternatively, mercury intrusion porosimetry or nitrogen adsorption/desorption can also be used, since they represent alternatives for the analysis of porous systems. Mercury intrusion porosimetry is capable of providing data concerning the most relevant macropores (> 1000 Å). However, this method requires the drying of the materials, a process that may result in significant changes in the pore size. Since our monoliths mainly contain macropores, these can also be observed by electron microscopy (Fig. 5.3). Table 5.1 summarizes the structural variations that can be achieved. As shown, the volume

TABLE 5.1

CHARACTERISTIC PROPERTIES OF MONOLITHS PREPARED BY ROMP.

NBE wt-%	DMN-H6 wt-%	toluene wt-%	2-PrOH wt-%	**1** wt-%	Tp °C	$\sigma^{a)}$ m²/g	ε_p %	ε_z %	ε_t %	σ_p g/cm³	V_p mL	d_p μm
15	15	10	60	0.4	0	76	43	37	80	0.27	0.31	2±1
20	20	10	50	0.4	0	62	43	33	76	0.31	0.31	4±1
25	25	10	40	0.4	0	85	48	15	63	0.40	0.34	2±1
25	25	10	40	1	0	86	48	14	63	0.40	0.34	4±1
30	30	10	30	0.4	0	191	50	5	54	0.46	0.35	8±2
30	30	10	30	1	0	96	50	2	53	0.46	0.36	6±2
35	35	10	20	0.4	0	b)	b)	b)	b)	b)	b)	14±6
15	15	20	50	0.4	0	110	39	49	89	0.25	0.28	3±1
20	20	20	40	0.4	0	74	44	21	65	0.36	0.31	4±1
25	25	20	30	0.4	0	91	47	15	62	0.42	0.33	4±1
30	30	20	20	0.4	0	93	65	5	69	0.50	0.46	4±1

Continued on the next page

TABLE 5.1 (continued)

NBE wt-%	DMN-H6 wt-%	toluene wt-%	2-PrOH wt-%	$\underline{1}$ wt-%	Tp °C	$\sigma^{a)}$ m²/g	ε_p %	ε_z %	ε_t %	σ_p g/cm³	V_p mL	d_p μm
35	35	20	10	0.4	0	b)	b)	b)	b)	b)	b)	30±10
0	50	10	40	0.4	0	88	44	25	69	0.32	0.31	2±1
15	35	10	40	0.4	0	76	45	26	71	0.35	0.32	4±1
25	25	10	40	0.4	0	85	48	15	63	0.40	0.34	2±1
35	15	10	40	0.4	0	100	45	10	56	0.41	0.32	3±1
25	25	10	40	0.01	0	b)	b)	b)	b)	b)	b)	6±4
25	25	10	40	0.1	0	83	49	20	69	0.36	0.35	2±1
25	25	10	40	0.4	0	85	48	15	63	0.40	0.34	2±1
25	25	10	40	1	0	75	49	12	61	0.41	0.35	3±1
25	25	10	40	0.4	-30	97	50	13	63	0.39	0.35	8±2
25	25	10	40	0.4	-20	98	45	17	62	0.38	0.32	6±2
25	25	10	40	0.4	-10	85	47	10	58	0.39	0.33	4±2
25	25	10	40	0.4	0	85	48	14	63	0.40	0.34	2±1

15	15	60	0	12	9	64	73	0.27	0.06	2±0.5
20	10	50	0	13	13	42	55	0.38	0.09	2±0.5
25	10	40	0	11	15	19	34	0.46	0.11	2±0.5
10	10	50	0	20	13	46	58	0.30	0.09	2±1
30	10	50	0	10	12	35	47	0.38	0.08	n.a.
25	10	40	20	6	9	31	41	0.47	0.06	3±1
25	10	40	40	6	8	33	41	0.42	0.06	2.5±2
25	10	40	80	7	10	34	44	0.41	0.07	4.5±1

Abbreviations: NBE = norborn-2-ene, DMN-H6 = 1,4,4a,5,8,8a-hexahydro-1,4,5,8-exo,endo-dimethanonaphthalene, **1** = polymerization catalyst $Cl_2(PR_3)_2Ru(=CHPh)$, T_p = polymerization temperature, σ = specific surface area, ε_p = volume fraction of pores, ε_z = volume fraction of inter-microglobular void volume, ε_t = volume fraction occupied by mobile phase, ρ_p = apparent density, V_p = pore volume, d_p = microglobule diameter. Porous properties calculated according to ref. [53].

[a] Based on PS standards (2610 < M_p < 1 250 000).

[b] Not analyzed by ISEC due to high back pressure

fraction of the interglobular void volume (ε_z) and total porosity (ε_t) may be varied within a range of 0–50 % and 50–80 %, respectively. Fig. 5.3 illustrates some of the microstructures that were generated.

5.2.3.1 Effect of NBE:DMN-H6 ratio

Variations in NBE to DMN-H6 ratio do not result in any significant change in d_p, ε_z, and σ over a relatively wide concentration range. The monoliths typically contain microglobules with a diameter of 2–4 μm and surface areas in the range of 80–100 m^2/g. The NBE to DMN-H6 ratio does not affect the homogeneity of the microstructure. Monoliths with a NBE to DMN-H6 ratio of over 1.5 are non-porous, film-like structures, indicating the absence of any phase separation during the polymerization process. In contrast, monoliths consisting of pure DMN-H6 have been prepared successfully [43].

5.2.3.2 Effect of solvents

Based on the existing knowledge of pore-formation in monolithic materials [7,16,54,55], different mixtures of macro- and microporogens [5] were tested for their ability to afford the desired, well-defined microstructures. Macropore-forming properties of methanol, 2-propanol, cyclohexanol, 1-decanol, 1-dodecanol were investigated, while dichloroethane, dichloromethane and toluene were used as microporogens. Since tetrahydrofuran (THF) and dimethylsulfoxide (DMSO) are coordinating solvents that may reduce reactivity of the initiator, their use was avoided. Methanol turned out to be a poor solvent for NBE and the use of more carbon-rich alcohols such as cyclohexanol, 1-decanol, 1-dodecanol only resulted in the formation of gel-like structures. 2-Propanol was found to possess good macropore-forming properties. Toluene, dichloromethane and dichloroethane are capable of forming the desired microstructures in combination with 2-propanol. The choice between toluene or methylene chloride depends on the requirements of subsequent derivatization reactions (*vide supra*).

5.2.3.3 Effect of initiator concentration

Concentration of initiator represents an important issue in the preparation of monoliths. For example, any uncontrolled highly exothermic reactions must be strictly avoided. The total amount of initiator directly determines the number of growing nuclei that affect phase separation and microglobule size. However, for a desired *in situ* derivatization (*vide infra*), higher initiator concentrations are more favorable. In order to determine the number of active sites accessible for subsequent derivatization, "capping" with ethylvinyl ether was used to remove the living termini. Surprisingly, quantification of Ru by inductively-coupled plasma optical emission spectroscopy (ICP-OES) carried out with the effluent after the cleavage reaction revealed that more than 98% of the initiator is located on the surface of the microglobules [56]. This is in accordance with a micelle-based mechanism of microglobules formation, in which the catalytically active sites are located at the interphase between the solid and liquid phase. Besides some less significant effects on the microglobule shape, the initiator added in concentration within a range of 0.1–1 % does not affect size of both pores and microglobules. For experimental reasons, 0.4 % of the catalyst **1** were used for the preparation of monoliths and this percentage was also found sufficient for derivatization.

The role of free phosphine on the polymerization process has also been studied. Grubbs' catalyst-initiated metathesis polymerizations proceed *via* a dissociative mechanism [57], which means that dissociation of phosphine is required to start polymerization [57-59]. For the present catalytic system, rebinding of phosphine competes with olefin coordination under the reaction conditions. As a consequence, the presence of additional phosphine has a dramatic effect on the microstructure and hence on any subsequent application. For example, the presence of small amounts of PPh_3 in the range of 20–80 μg/g lead to a reduction in fraction of pores ε_p and pore volume V_p while the volume fraction of the void volume ε_z increases (Table 5.1). This also results in a decrease in the specific surface area σ by a factor of about 2. The presence of phosphine also affects the mean microglobule diameter that increases from 2 to 4.5 μm. These observations correlate with the reduced overall polymerization rate, which changes from diffusion to propagation controlled. Consequently, microglobules can grow in a controlled way to a larger size.

5.2.4 Porosity and permeability

Porosity [60] and pore size distribution of each monolithic column can be determined by ISEC in THF using polystyrene (PS) standards using the method first reported by Halász *et al.* [53]. Although somewhat controversial [61,62], this method presents a suitable way to perform such measurements as it operates under conditions similar to those used in actual HPLC separations. Furthermore, the choice of PS-standards allows the determination of the specific surface area and porosity relevant for chromatographic separations [63]. The flow resistance of monolithic columns that depends on porous properties must not exceed the upper back pressure limits of the equipment at flow rates desired for the use in the chromatographic separation. Some porous polymeric stationary phases may in contact with organic solvents swell and loose their mechanical strength. These polymers can then be deformed under the pressure gradient normally encountered in HPLC columns. Monoliths prepared using ROMP were evaluated by measurement of the pressure drop across the column using different solvents and a wide range of flow rates. The linearity of this plot confirms that the monolith is not compressed even at high flow velocities exceeding 5 mm/s.

5.2.5 Stability

Both chemical and thermal stability of monolithic materials is also very important. Differential scanning calorimetry coupled to thermogravimetry which was further coupled to mass spectrometry (DSC-TGA-MS) was used to obtain information concerning the stability. This measurement may also reveal any potential oxidation processes, which may affect surface chemistry during time. Three major breaks may be observed in TGA in a temperature range of T = 130–160°C, 305–310°C, and 454–471°C [64]. While the last two illustrate the destruction of the organic material at temperatures above 300 and 400°C, respectively, the first loss in weight observed at 130–160°C provides an useful insight into the chemistry. By comparing TGA data with those of pure poly-NBE and poly-DMN-H6, the first break in DSC is attributed to the loss of poly-NBE domains within the monolith. Complementary, this may be directly correlated with an enhanced poly-DMN-H6 content. Besides these structure-stability relationships, these investigations also provided an additional insight into the oxidation stability of these monoliths. Although the materials were stored in air at ambient temperature, the loss in weight was found to be less than 4% that roughly corresponds to 1% in oxidized carbon. Therefore oxidation is believed to play a minor role. This is also supported by the high reproducibility of separations. In contrast, the "capping"-process, i. e. the reaction that cleaves off the metal from the polymer chain, appears to have a considerble effect on the stability. By using different capping agents, the stability of the monolithic material may vary over several orders of magnitude [65].

5.2.6 Shrinkage

Shrinkage is unavoidable in vinyl polymerizations [66] and may lead to both longitudinal and radial contraction of monolithic polymers. Despite the low shrinkage of less that 5% typical for poly-NBE-based systems prepared by ROMP, all monoliths formed by in-column polymerization of NBE and DMN-H6 using catalyst **1** in the presence of any of the examined porogens exhibited longitudinal and radial shrinkage. The latter usually leads to a detachment of the continuous bed from the column wall and cannot be completely avoided even by changing the porogen to monomer ratio or by varying the crosslinker to monomer ratio. Consequently, covalent fixation of the polymer to the inner wall of the mold via anchor groups is necessary. In glass molds, this may be achieved by silanization of the activated inner wall using bicyclo[2.2.1]hept-2-ene-5-trichlorosilane that affords ROMP-active norbornene anchor groups. These groups then copolymerize during the polymer formation. While the use of anchor groups eliminates radial shrinkage, longitudinal shrinkage persists. However, this can be compensated for by use of an adequate polymerization setup [41,43] in which this longitudinal shrinkage occurs outside the confines of the mold and completely filled HPLC columns are obtained.

5.2.7 Metal removal and metal content

When dealing with transition metal catalyzed polymerizations, the efficiency of removing the metal from the monolith after polymerization needs to be addressed. The fact that ruthenium-initiated polymerizations may conveniently be capped with ethylvinyl ether may be demonstrated by ICP-OES investigations on the Ru-content of the final rods. These investigations revealed Ru-concentrations <10 μg/g, corresponding to a metal removal >99.8%.

5.2.8 Functionalization

While *functionalized* organic monolithic columns are well-defined in terms of microstructure characterized by the size and form of microglobules, specific surface area, as well as pore volume [6-8,16,54,67,68] and afford impressive separation results, their preparation was originally somewhat limited [5]. One synthetic protocol entailed the copolymerization of the corresponding functional monomers. Despite its simplicity, two problems have to be addressed. First, a part of the functional monomer is located in the interior of the material and is not available for interactions required for the separation process. An alternative approach that avoids this problem is the copolymerization of monomers possessing reactive groups such as epoxide or azlactone groups available for post-derivatization. Monoliths prepared from chloromethylstyrene, glycidyl methacrylate or acrylamides, are well suited for "simple" functionalizations such as the generation of amino-, alcohol-, phenol-, sulfonic acid- or carboxylate groups [24,69-72]. More sophisticated approach that involves the 2,2,6,6-tetramethylpiperidin-1-yloxy- (TEMPO-) or 2,2,5-trimethyl-3-(1-phenylethoxy)-4-phenyl-3-azahexane mediated surface grafting has been demonstrated recently [73-75].

Functionalization can be conveniently achieved in transition metal-based polymerizations. For example, in the present ruthenium-catalyzed polymerizations [29,30,76,77], the "living" character and the high tolerance of the catalytic system towards different functional monomers makes the ROMP approach very attractive. Since living polymerization systems are typically not immortal [27,78] and in view of the stability data for ruthenium-based catalysts such as 1 [79], optimum conditions for

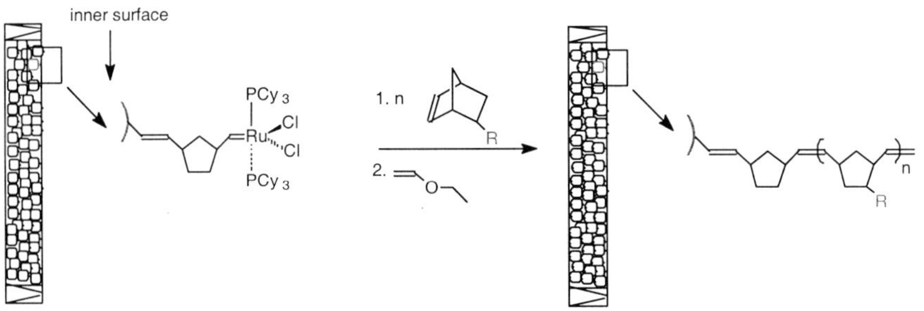

Fig. 5.4. Schematics of functionalization of monoliths prepared by ROMP.

grafting had to be elaborated in order to reduce loss of initiator activity to a minimum. First, the minimum time of 1 h needed to form the polymeric backbone was determined [43]. Using the high percentage of the catalyst covalently bound to the surface that exceeds 85% (*vide supra*), the new functional monomer is grafted onto the monolith surface by simply passing its solution through the mold (Fig. 5.4.). Since no crosslinking may occur, tentacle-like polymer chains attached to the surface are formed. The degree of this graft polymerization of functional monomers varies within almost two orders of magnitude, depending on their ROMP activity.

This approach offers multiple advantages. First, the structure of the "parent" monolith is not affected by the functional monomer and can be optimized regardless of the functional monomer used later. Second, solvents other than the porogens toluene and methanol (e. g. methylene chloride and DMF) may be used for the "*in situ*" derivatizations. This solves problems related to the solubility of the functional monomer in the porogens. For example, functional β-cyclodextrins are only soluble in DMF. A large variety of functional monomers may thus be grafted onto the monoliths. Restrictions resulting from the reduced activity of the catalyst towards some functional monomer can be solved by a careful monomer design. The versatility of this concept is demonstrated by the large variety of functional monomers shown in Table 5.2 that were grafted from the surface of monolith. For convenience, all monomers include norborn-2-ene or 7-oxanorborn-2-ene functionalities. The amount of functional monomer grafted onto the monolith was determined both in a *qualitative* way using FT-IR spectroscopy and *quantitatively* by acid-base titration and elemental analysis, respectively.

TABLE 5.2

CAPACITIES OF FUNCTIONALIZED ROMP MONOLITHS

Monolith	Functional monomer	Capacity [mmol/g]	Monolith	Functional monomer	Capacity [mmol/g]
I	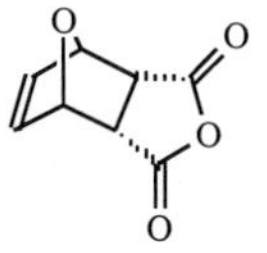	0.2[a]	V	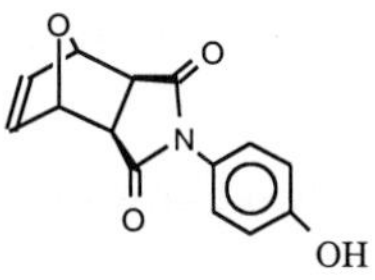	0.06[b]
II	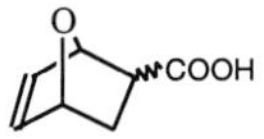	0.14[a]	VI	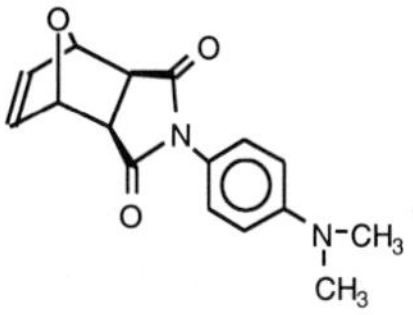	0.26[b]
III	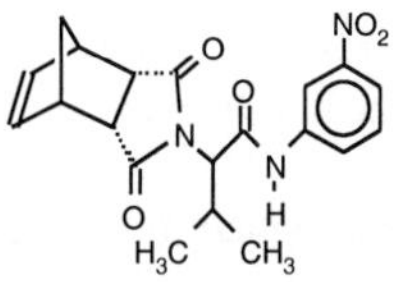	0.03[b]	VII	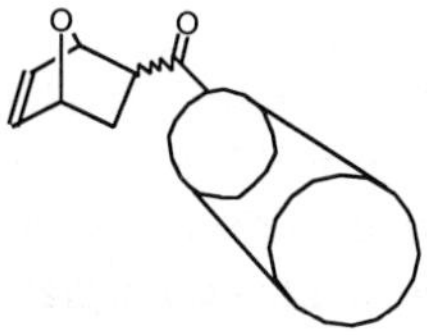	[c]
IV	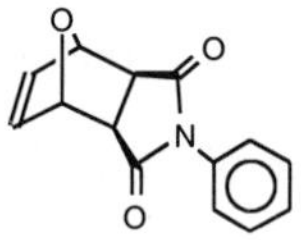	0.22[b]	VIII	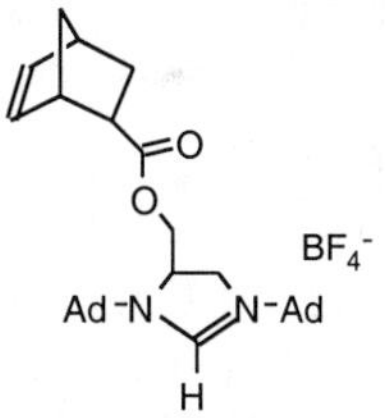	0.6

[a])Determined by titration

[b])Determined by elemental analysis (% nitrogen)

[c])Estimated from comparison with loading capacities and structural data of surface-grafted materials [40].

5.3 CONCLUSION

Metathesis-based monoliths described in this chapter add another dimension to the family of non-conventional separation media. They involve a non-aromatic poly(cyclopentadiylvinylen) backbone and enable very simple surface modification that provides the surface with a broad variety of functionalities including chiral selectors and chelating ligands. The microstructure of these monoliths may also be easily tailored. Consequently, these systems possess vast potential in the synthesis of novel separation media including those suitable for applications in the microscale separations such as μ-LC and CEC. Therefore they represent a valuable alternative that makes the existing armor of rigid monoliths even more comprehensive.

5.4 ACKNOWLEDGEMENT

Financial support provided by the Austrian Science Fund (FWF Vienna) and the Österreichische Nationalbank is acknowledged.

5.5 REFERENCES

1   N.B. Afeyan, N.F. Gordon, I. Mazsaroff, L. Varady, S.P. Fulton, Y.B. Yang, F.E. Regnier, J. Chromatogr. 519 (1990) 1.
2   N.B. Afeyan, S.P. Fulton, F.E. Regnier, J. Chromatogr. 544 (1991) 267.
3   M. Kubín, P. Špaček, R. Chromeček, Collect. Czech. Chem. Commun. 32 (1967) 3881.
4   L.C. Hansen, R.E. Sievers, J. Chromatogr. 99 (1974) 123.
5   E.C. Peters, F. Svec, J.M.J. Fréchet, Adv. Mater. 11 (1999) 1169.
6   F. Svec, J.M.J. Fréchet, Science 273 (1996) 205.
7   C. Viklund, F. Svec, J.M.J. Fréchet, K. Irgum, Chem. Mater. 8 (1996) 744.
8   C. Viklund, E. Pontén, B. Glad, K. Irgum, P. Hörstedt, F. Svec, Chem. Mater. 9 (1997) 463.
9   K. Hosoya, H. Ohta, K. Yoshizoka, K. Kimatas, T. Ikegami, N. Tanaka, J. Chromatogr. A 853 (1999) 11.
10 A. Maruska, C. Ericson, A. Végvári, S. Hjertén, J. Chromatogr. A 837 (1999) 25.
11 I. Gusev, X. Huang, C. Horváth, J. Chromatogr. A 855 (1999) 273.
12 Q. Tang, B. Xin, M.L. Lee, J. Chromatogr. A 837 (1999) 35.
13 R. Asiaie, X. Huang, D. Farnan, C. Horváth, J. Chromatogr. A 806 (1998) 251.
14 E.C. Peters, M. Petro, F. Svec, J.M.J. Fréchet, Anal. Chem. 69 (1997) 3646.
15 E.C. Peters, M. Petro, F. Svec, J.M.J. Fréchet, Anal. Chem. 70 (1998) 2288.
16 S. Xie, F. Svec, J.M.J. Fréchet, Chem. Mater. 10 (1998) 4072.
17 J.A. Gerstner, R. Hamilton, S.M. Cramer, J. Chromatogr. 596 (1992) 173.
18 N. Tanaka, H. Nagayama, H. Kobayashi, T. Ikegami, K. Hosoya, N. Ishizuka, H. Minakuchi, K. Nakanishi, K. Cabrera, D. Lubda, J. High Resolut. Chromatogr. 23 (2000) 111.
19 F. Rabel, K. Cabrera, D. Lubda, Int. Lab. 01/02 (2001) 23.
20 K. Cabrera, K. Sinz, D. Cunningham, Int. Lab. News 02 (2001) 12.
21 K. Cabrera, D. Lubda, H.-M. Eggenweiler, H. Minakuchi, K. Nakanishi, J. High Resolut. Chromatogr. 23 (2000) 93.
22 A.E. Rodrigues, J. Chromatogr. B 699 (1997) 47.
23 Y. Xu, A.I. Liapis, J. Chromatogr. A 724 (1996) 13.

24   D. Sykora, F. Svec, J.M.J. Fréchet, J. Chromatogr. A 852 (1999) 297.

25   Q.C. Wang, F. Svec, J.M.J. Fréchet, Anal. Chem. 65 (1993) 2243.

26   M.R. Buchmeiser, Chem. Rev. 100 (2000) 1565.

27   K. Matyjaszewski, Macromolecules 26 (1993) 1787.

28   T. Otsu, J. Polym. Sci. A: Polym. Chem. 38 (2000) 2121.

29   S. Penczek, P. Kubisa, R. Szymanski, Makromol. Chem. Rapid. Commun. 12 (1991) 77.

30   M. Szwarc, J. Polym. Sci., Polym. Chem. 36 (1998) ix.

31   T.R. Darling, T.P. Davis, M. Fryd, A.A. Gridnev, D.M. Haddleton, S.D. Ittel, R.R. Matheson Jr., G. Moad, E. Rizzardo, J. Polym. Sci., Polym. Chem. 38 (2000) 1706.

32   R.R. Schrock, in D. J. Brunelle (Ed.), Ring-Opening Polymerization, Hanser, Munich (1993), p. 129.

33   M.R. Buchmeiser, N. Atzl, G.K. Bonn, Int. Pat. Appl. AT404 099 (181296), PCT /AT97/00278

34   M.R. Buchmeiser, N. Atzl, G.K. Bonn, J. Am. Chem. Soc. 119 (1997) 9166.

35   F. Sinner, M.R. Buchmeiser, R. Tessadri, M. Mupa, K. Wurst, G.K. Bonn, J. Am. Chem. Soc. 120 (1998) 2790.

36   M.R. Buchmeiser, M. Mupa, G. Seeber, G.K. Bonn, Chem. Mater. 11 (1999) 1533.

37   M.R. Buchmeiser, R. Tessadri, G. Seeber, G.K. Bonn, Anal. Chem. 70 (1998) 2130.

38   M.R. Buchmeiser, F. Sinner, PCT A 960/99 (310599), PCT/EP00/04 768, WO 00/73782 A1.

39   M.R. Buchmeiser, F.M. Sinner, PCT A 604/99 (070499), PCT/EP00/02 846, WO 00/61288

40   M.R. Buchmeiser, F. Sinner, M. Mupa, K. Wurst, Macromolecules 33 (2000) 32.

41   F. Sinner, M.R. Buchmeiser, Angew. Chem. 112 (2000) 1491.

42   M.R. Buchmeiser, G. Seeber, R. Tessadri, Anal. Chem. 72 (2000) 2595.

43   F. Sinner, M.R. Buchmeiser, Macromolecules 33 (2000) 5777.

44   M.R. Buchmeiser, F. Sinner, M. Mupa, Makromol. Symp., 163 (2000) 25.

45   M.R. Buchmeiser, K. Wurst, J. Am. Chem. Soc. 121 (1999) 11101.

46   J. Silberg, T. Schareina, R. Kempe, K. Wurst, M.R. Buchmeiser, J. Organomet. Chem. 622 (2000) 6.

47   R. Kröll, C. Eschbaumer, U.S. Schubert, M.R. Buchmeiser, K. Wurst, Macromol. Chem. Phys. 202 (2001) 645.

48   U.S. Schubert, C.H. Weidl, C. Eschbaumer, R. Kröll, M.R. Buchmeiser, Polym. Mater. Sci. Eng. 84 (2001) 514.

49   M.R. Buchmeiser, R. Kröll, K. Wurst, T. Schareina, R. Kempe, C. Eschbaumer, U.S. Schubert, Makromol. Symp., 164 (2001) 187.

50   R.A. Fisher, R.H. Grubbs, Macromol. Symp. 63 (1992) 271.

51   T.A. Davidson, K.B. Wagner, J. Mol. Catal. A Chem. 133 (1998) 67.

52   T.A. Davidson, K.B. Wagener, D.B. Priddy, Macromolecules 29 (1996) 786.

53   I. Halász, K. Martin, Angew. Chem. 90 (1978) 954.

54   E.C. Peters, F. Svec, J.M.J. Fréchet, Chem. Mater. 9 (1997) 1898.

55   A.I. Cooper, A.B. Holmes, Adv. Mater. 11 (1999) 1270.

56   M. Mayr, B. Mayr, M.R. Buchmeiser, Angew. Chem. 113 (2001) 3957.

57   M.S. Sanford, M. Ulman, R.H. Grubbs, J. Am. Chem. Soc. 123 (2001) 749.

58   M.S. Sanford, L.M. Henling, M.W. Day, R.H. Grubbs, Angew. Chem. 112 (2000) 3593.

59   M.S. Sanford, J.A. Love, R.H. Grubbs, J. Am. Chem. Soc. 123 (2001) 6543.

60   Y.-F. Maa, C. Horváth, J. Chromatogr. A 445 (1988) 71.

61   J.H. Knox, H.P. Scott, J. Chromatogr. 316 (1984) 311.

62   J.H. Knox, H.J. Ritchie, J. Chromatogr. 387 (1987) 65.

63   B. Mayr, M.R. Buchmeiser, J. Chromatogr. A 907 (2001) 73.

64   B. Mayr, R. Tessadri, E. Post, M.R. Buchmeiser, Anal. Chem. 73 (2001) 4071.

65   M. Mayr, M.R. Buchmeiser, unpublished results (2002).

66   T. Takata, T. Endo, Progr. Polym. Sci. 18 (1993) 839.

67   F. Svec, J.M.J. Fréchet, Macromolecules 28 (1995) 7580.

68   S. Xie, F. Svec, J.M.J. Fréchet, J. Polym. Sci., Polym. Chem. 35 (1997) 1013.

69   A. Podgornik, M. Barut, J. Jancar, A. Strancar, J. Chromatogr. A 848 (1999) 51.

70   F. Svec, J.M.J. Fréchet, J. Chromatogr. A 702 (1995) 89.

71   M.B. Tennikov, N.V. Gazdina, T.B. Tennikova, F. Svec, J. Chromatogr. A 798 (1998) 55.

72   Q.C. Wang, F. Svec, J.M.J. Fréchet, Anal. Chem. 67 (1995) 670.

73   E.C. Peters, F. Svec, J.M.J. Fréchet, C. Viklund, K. Irgum, Macromolecules 32 (1999) 6377.

74   U. Meyer, F. Svec, J.M.J. Fréchet, C.J. Hawker, K. Irgum, Macromolecules 33 (2000) 7769.

75   C. Viklund, A. Nordström, K. Irgum, F. Svec, J.M.J. Fréchet, Macromolecules 34 (2001) 4361.

76   A.F. Johnson, M.A. Mohsin, Z.G. Meszena, P. Graves-Morris, J. Macromol. Sci.-Rev. Macromol. Chem. Phys. C39 (1999) 527.

77   O.W. Webster, Science 251 (1991) 887.

78   M. Szwarc, Makromol. Chem. Rapid Commun. 13 (1992) 141.

79   M. Ulman, R.H. Grubbs, J. Org. Chem. 64 (1999) 7202.

F. Švec, T.B. Tennikova and Z. Deyl (Editors)
Monolithic Materials
Journal of Chromatography Library, Vol. 67
2003 Published by Elsevier Science B.V.

Chapter 6

Monolithic Polysaccharide Materials

Per-Erik GUSTAVSSON and Per-Olof LARSSON

Department of Pure and Applied Biochemistry, Center for Chemistry and Chemical Engineering, Lund University, P.O. Box 124, SE-221 00 Lund, Sweden

CONTENTS

6.1 INTRODUCTION

Continuous beds – monoliths – were developed for separations of biomolecules around 1990 and have since become accepted as an alternative format to the established beaded supports. They are manufactured from most well known base materials, such as silica, polystyrene, polymethacrylates, cellulose, agarose, and polyacrylamide. Characterising features of these continuous beds are that they consist of one single piece of material in which is distributed a continuous network of wide, flow carrying pores. Besides the flow carrying pores, also narrow diffusion pores may exist. The proportion of "diffusive" pores is very much related to the base material used.

The advantages of the monolithic format compared to beaded supports are related to the production of the separation material as well as its performance. The preparation procedure is simplified because the monolith can be cast directly in the chromatography column avoiding the time consuming steps of sieving and packing usually associated with beaded supports. The chromatographic performance of monolithic supports is often remarkably good, which can be explained in terms of improved mass transport by convective flow in the flow-through pores. Furthermore, the percentage of active matrix in a continuous bed can be easily varied between 30 and 90 %, opening opportunities for high flow rates with very porous structures or high loading with denser structures.

This chapter will focus on the preparation and characterisation of monoliths made of polysaccharide materials. The literature primarily deals with agarose [1], and cellulose [2-4]. The use of both stacks of cellulose membranes [2], and rolled cellulose sheets [3] for chromatographic applications is covered elsewhere in this book. The use of other types of polysaccharides, such as dextran, alginate, k-carragenan, and starch for the preparation of monoliths should certainly be possible but has so far not been pursued.

6.2 MONOLITHS BASED ON AGAROSE

Polysaccharide materials for chromatography have been used successfully for a long time and include for example supports made of agarose, cellulose, and cross-linked dextran. The success of these materials relies primarily on their hydrophilicity/protein compatibility and their regenerability, *i.e.* they can be repeatedly treated with strong alkali, *e.g.* 0.5 mol/L NaOH, a preferred sanitation agent in industry. Their drawbacks include a low mechanical strength, which restricts the choice of particle size and flow rates. However, the mechanical strength can be improved considerably by chemical crosslinking. Beaded polysaccharide materials are manufactured by two-phase suspension processes [5].

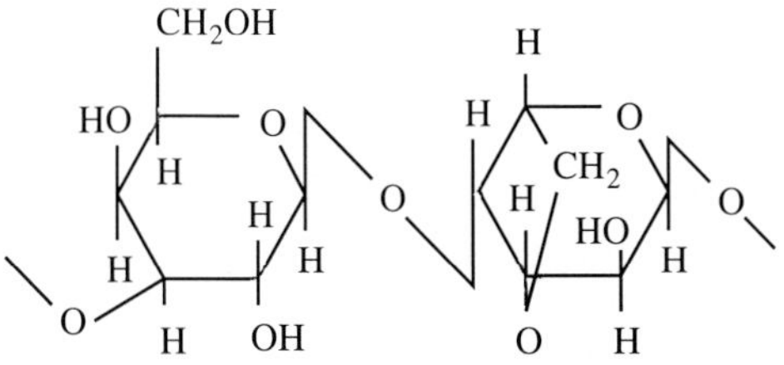

Fig. 6.1. The repeating unit of agarose consisting of D-galactose and 3,6-anhydro-L-galactose.

6.2.1 Properties of agarose

Agarose is a natural polymer prepared from seaweed (red algae) and consists of the D-galactose and 3,6-anhydro-L-galactose repeating units shown in Fig. 6.1. Agarose can be dissolved in boiling water and a gel is formed after cooling this solution below 45°C as a result of extensive hydrogen-bonding between the agarose chains. The gelling temperature may vary due to monomer composition (methoxyl content) and concentration of the solution and may also be altered by chemical derivatization of the polymer such as hydroxyethyl derivatization or addition of destructuring salts [6]. The pore size of agarose gels depends on the agarose content. Beads with a 6 % agarose content are frequently used and have an average pore size of approximately 30 nm, whereas 4 and 2 % agarose beads have a pore size of 70 and 150 nm, respectively [7]. Agarose with even larger diffusion pores are used in gel electrophoresis to allow the passage of very large molecules such as DNA. Prominent examples of commercial beaded materials for chromatography include Sepharose FF, Superose (Amersham Biosciences), Ultrogel A (BioSepra/Sigma-Aldrich), Bio-gel A (BioRad Labs) and Thruput (Sterogene).

6.2.2 Preparation of agarose monoliths

6.2.2.1 Principle of preparation

The concept of preparation of continuous agarose beds is explained in Fig. 6.2. Typically, an aqueous agarose solution (60°C) is mixed with a water-immiscible organic solvent containing a surface-active agent. The mixture is stirred vigorously for several minutes, forming a thick, white emulsion. The emulsion is poured into a mold such as a chromatography column. The emulsion actually consists of two continuous phases, an aqueous agarose phase and a water-immiscible organic phase. After a short time the mold is cooled and its content solidifies into a stiff rod. This rod consists of a

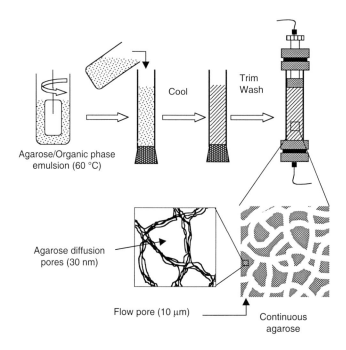

Fig. 6.2. Preparation of a continuous agarose bed suitable for chromatography. Experimental details can be found in ref. [1]. Note that the resulting agarose monolith has 10 μm wide flow-through pores as well as a system of much narrower 30 nm diffusive pores.

single piece of agarose gel, transected by a continuous flow pore system filled with the organic phase. Subsequently, the ends of the rod are trimmed, flow adapters are attached, and washing solutions are pumped through the column to remove the water-immiscible organic phase.

Figure 6.2 describes the preparation of a rodlike agarose monolith suitable for chromatography by allowing the agarose – organic phase emulsion to solidify in a tube. By using other molds for the solidification, a number of other useful monolith formats may be prepared, such as sheets, membranes, fibers, radial beds, and composites [1]. A few examples of these formats are shown in Fig. 6.3.

6.2.2.2 Experimental techniques

Although the basic concept for the preparation of superporous agarose monoliths indicated in Fig. 6.2 is straightforward, a number of experimental details has to be considered. Five main steps are involved: preparation of agarose solution, preparation

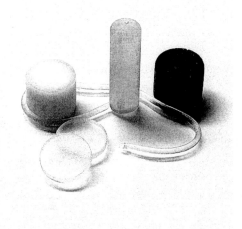

Fig. 6.3. Different self-supported continuous agarose formats. From left to right: Monolithic agarose – hydroxyapatite composite, membranes, rod, rod derivatized with Cibacron blue, and fiber. (Reprinted with permission from ref. [1], copyright 1999 Elsevier Sciences B. V.).

of pore-forming organic phase, preparation of emulsion, casting of monolith, and washing.

The first step is the preparation of aqueous agarose solution. Typically, agarose powder is suspended in water in a concentration of 4–8 %, and placed in a suitable container such as glass bottle or test tube, depending on the required amount of agarose solution. The suspension is conveniently heated in a microwave oven to 95–100°C and kept at that temperature for about 2 min. During heating the container must be occasionally shaken/stirred to keep the agarose powder well suspended and to level out non-uniform heating. The bottle/test tube must not be sealed during the heating since dangerous pressure may build up inside. Thus, a very loosely attached stopper should be used. The agarose solution is then placed in a water bath kept at 60°C until used (typically within one day).

The major issue in the above procedure is to keep the agarose powder well suspended during the heating. The powder starts to hydrate and transform into a viscous solution at temperatures higher than about 70°C. If the agarose is allowed to partially settle before this temperature is reached, the bottom part of the container may contain a concentrated and extremely viscous agarose solution that is very difficult to mix/equilibrate with the rest of the solution. An alternative procedure that avoids this problem is to heat the agarose suspension in an oil bath under constant stirring using a propeller stirrer.

Yet, another way is to prepare a stable stock agarose suspension that can be heated without stirring. For this, two grams of agarose powder are added to 200 ml distilled water in a bottle and heated in a microwave oven to 95–100°C with occasional shaking. The hot agarose solution is poured into 800 ml distilled water tempered to room temperature in a 1 L flask, forming a semiviscous agarose solution/suspension. Then, 58 g of agarose powder is added in portions with vigorous shaking. Finally, 100 mg of sodium azide is added this mixture to prevent microbial growth. This agarose suspension (6 % w/v) can be stored for long period of time and suitable portions withdrawn and conveniently heated without danger of settling of the agarose powder during the heating process.

The second step, i.e. the preparation of the organic phase (flow pore-forming phase) is brought about by mixing a water-immiscible solvent with the surface-active agent. A number of combinations are possible. The preferred one, at least for small-scale preparations, is a mixture of cyclohexane and Tween 80. Typically, Tween 80 is added to cyclohexane and the mixture is thermostatted at 60°C in a water bath. The concentration of Tween 80 may be varied but is most often kept at 6 %. Since cyclohexane is a highly flammable liquid and used at a temperature close to its boiling point (boiling point for pure cyclohexane is 81°C, boiling point for cyclohexane-water azeotrope is 70°C), work in a well-ventilated hood and use of explosion-proof electrical or pneumatic stirrers is recommended.

The third step is the formation of emulsion. The organic phase is vigorously mixed before it is added to the agarose solution. The amount of organic phase used directly determines the flow-through pore volume in the final monolith. Typical volumetric ratio is 1/3 organic phase mixed with 2/3 agarose solution. The agarose/organic phase mixture is immediately emulsified at 60°C by stirring at 1000 rpm for 5 min with a large blade overhead stirrer. A viscous white emulsion is soon formed. A suitable stirrer should have a rather large blade that sweeps a large part of the vessel, since the emulsion is very viscous. This diminishes convective flow and material distribution in the vessel. Due to the high viscosity, the use of a smaller stirrer cannot be compensated for by a higher stirring speed, since this may lead to a heterogeneous emulsion. Fig. 6.4 shows suitable vessels for the preparation of 30 ml and 300 ml of emulsion. The emulsion, which contains two continuous phases, the agarose phase and the organic, flow pore-forming phase, is unstable and gradually changes its properties. Thus the time span between emulsion formation and casting must be timed to allow consistent results.

The fourth step is the actual casting of the monolith. The agarose/organic phase emulsion is poured into a mold such as a glass tube pre-warmed to 60°C fitted with a rubber plug at the bottom end as shown in Fig. 6.2. After a short time, typically 1 min, during which time the emulsion matures, the system is cooled by transferring the glass

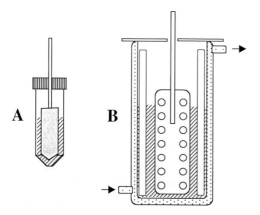

Fig. 6.4. Vessels suitable for formation of emulsion. Vessel with a stainless steel stirrer for preparing up to 30 ml of emulsion, constructed from a 50 ml disposable polystyrene plastic test tube (A). Thermostated glass reactor containing stainless steel baffles and stainless steel stirrer used for the preparation of 300 ml of emulsion with a total volume of 600 ml (B).

column to a water bath kept at 5°C. The agarose solution solidifies when its temperatures falls below about 45°C and a stiff superporous agarose monolith is obtained.

In the fifth and final step, the agarose monolith is removed from the glass tube, trimmed to an appropriate length and then reinserted. The tube is fitted with flow adapters and the organic phase in this column is washed out typically with 5 volumes of water, 5 volumes of 50 % aqueous ethanol, and another 5 volumes of water.

6.2.3 Factors influencing the properties of monolithic agarose

A number of factors affects the properties of monolithic agarose and some of them have already been briefly commented upon.

6.2.3.1 Diffusive pore size

The agarose content in its solution determines the size of the diffusive pores. For example, 4 % agarose affords a monolith with an average pore size of about 70 nm, 6 % 30 nm and 8 % 20 nm. Admittedly, control of size of diffusive pores is usually less important. The standard material, 6 % agarose, contains pores that allow most proteins to diffuse rather unhindered. More importantly the agarose concentration greatly influences the mechanical strength of the monolith.

6.2.3.2 Flow-through pore volume (porosity)

The volume of the flow-through pores is determined by the volume of the water-immiscible organic phase used in the emulsion. Therefore, it is usually easy to vary the percentage of these pores in the final monolith in the range 20–60 % and in some cases even broader.

6.2.3.3 Flow-through pore size

The flow-through pore size is controlled by both input of kinetic energy in the system during the preparation of the emulsion and concentration of the surface-active agent. A higher stirring speed gives narrower pores. Obviously, also the shape of the stirrer and baffles in the vessel affect the energy transfer in the system and the pore size. The effect of the surface-active agent is less transparent. We found that within certain limits, a higher percentage of surface-active agent affords monoliths with larger pores.

Another factor of critical importance for the pore size is the time span between the emulsion formation and the agarose gelling. The bi-continuous phase system formed during the emulsification process is inherently unstable. As soon as the stirring is switched off, the bi-continuous phase structure starts to degrade. The kinetics for this degradation is dependent on many factors. Initially this is manifested in an increase in the size of the flow-through pores and decrease in their number. Delayed cooling/solidification of an unstirred emulsion enables the preparation of monoliths with larger pores.

The size of the pores is obviously a very important factor controlling the flow properties of the monolith. This is an important issue considering that agarose gels have a limited mechanical strength. The flow resistance in a monolith is according to the Kozeny–Carman equation inversely proportional to the square of the pore diameter [8]. Thus, it is prudent to keep the pore size under control. As a rule of thumb, it is difficult to use monoliths with pore sizes smaller than 5 μm made of non-stabilized agarose with a bed length exceeding a few centimeters if reasonable flow rates are attempted. The literature describes agarose monoliths with pores often as large as 50 μm [1,9,10].

Monoliths with small pores are also desirable from another point of view. At a fixed total volume of the flow-through pores determined by the volume of organic phase used for the preparation, a decrease in the pore size increases the number of flow-through pores and, consequently, diminishes the distance between them. A shorter distance between the flow-through pores is desirable since it provides shorter diffusion path, resulting in faster mass transport, which is beneficial for improved chromatographic performance of the monolith. The choice of pore size is usually a compromise between both mechanical and mass transport requirements.

The fact that a longer period of time between formation of the emulsion and gelling increases the flow-through pore size imposes restrictions on the size of agarose monoliths. The emulsion is cooled from the outside and consequently the outer parts gel first and the inner parts later. This generates larger pores in the center of the resulting monolith. This effect is potentially very serious since it may lead to an uneven flow distribution with a considerably faster flow through the center of the monolith. Typically, monoliths wider than 3 cm prepared using standard techniques exhibit an appreciable distortion of the flow profile and are unsuitable for chromatography. However, a smart solution to this problem is to use the monoliths not in the traditional axial but rather in the radial direction. In this approach the asymmetric pore size distribution does not lead to uneven flow profiles. In fact, the slightly larger pores near the center of such radial-flow columns are beneficial at high flow rates.

Due to the many factors influencing flow-through pore size, it is difficult to predict the resulting pore properties with a certain type of equipment run under a certain set of conditions. However, after a few test runs with concomitant checking of the pore size in a microscope as described later, suitable conditions can readily be established.

6.2.3.4 Mechanical strength

As touched upon above, the mechanical strength of monolithic agarose may become a limiting factor for applications at high flow rates using deep beds. Once the pressure drop across the monolithic bed is too high as a result of a high flow rate, the structure compresses gradually. This compression is fully reversible once the flow rate is decreased again. Since the compression of the bed decreases primarily the size of the flow-through pores, an additional increase in the pressure drop can be observed leading to further compression of the bed. The pore structure completely collapses after exceeding a certain flow rate and the flow stops in much the same way as observed for a bed packed with agarose beads.

Several design features may be used to improve the mechanical strength. For example, higher agarose concentration and the use of chemical crosslinking of the agarose chains may help significantly. Even without a specific crosslinking, considerable improvement of mechanical strength is usually achieved upon chemical derivatization introducing ligands of various kinds such as ion exchange and affinity.

The mechanical strength is also improved if the total volume of flow-through pores is kept low. A high percentage of these pores afford a monolith with a "spongy" feel. However, a decrease in the pore volume by necessity also limits the pore cross-section area, which in turn leads to a higher flow resistance. This in fact counteracts the positive effects of lowering the pore volume.

6.2.4 Characterization of monolithic agarose

Characterization of properties of a new monolith is important to reproducibly achieve the desired function. Some of these characterization methods are standard techniques of determining the chromatographic performance, while others are specifically tuned to reveal the basic properties of the monolith, such as pore volume, pore size and its distribution, spatial distribution of pores, presence of dead end pores, etc.

6.2.4.1 Flow-through pore volume

The flow-through pore volume can be determined using size-exclusion experiments with molecules that do not penetrate the agarose matrix such as 0.5 μm latex particles or Blue Dextran (M_W 2,000,000). The elution volume for these molecules is a direct measure of the flow-through pore volume of the continuous agarose bed. Practical tests [1,11], have shown that the measured flow pore volume usually agrees very well with the designed pore volume represented by the volume of added organic phase in the emulsion.

6.2.4.2 Flow-through pore size, size distribution, and spatial distribution

Fundamental properties of the monolith such as the pore size, and spatial distribution of the pores can be rapidly determined by observing thin slices of the monolith under a microscope. The technique is very useful when developing new recipes. Thus, a few milliliters of an emulsion under preparation are withdrawn and solidified in a test tube or on a glass surface. The piece of gel is sliced by a microtome or a razor blade. It is not necessary to obtain extremely thin slices. The slices are then inspected and photographed using a microscope. A useful technique to improve the contrast between the flow-through pores and the agarose matrix is to allow the monolith surface to dry slightly. This partial drying leads to a preferential loss of water from the pores, which make them considerably more visible when illuminated from above and at an angle with no cover glass attached. Another simple trick to improve the contrast is a brief washing of the slices with ethanol. An example of this observation is given in Fig. 6.5. The pores are the dark areas, typically 50 μm large. The light areas constitute the agarose phase containing diffusive pores with a size of 30 nm that are not visible.

A complementary method to improve the visibility is filling the pores with a suspension containing colored latex particles that do not enter the agarose phase. A

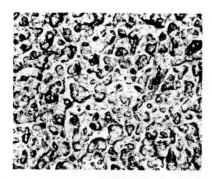

Fig. 6.5. Optical micrograph of a thin gel slice of continuous agarose bed. The contrast between the flow pores and the agarose has been enhanced by partial drying of the surface.

powerful technique, although not tested yet, would be to use fluorescent dyed latex particles and study the slice of a monolith with confocal microscopy. Related techniques are known to provide valuable structural information with superporous agarose beads [12-14].

We have also described a more elaborate way of determining pore size as well as the three-dimensional structure of the flow-through pore network [15]. In this technique, a replica of the pore structure is made and studied by scanning electron microscopy (SEM). Figure 6.6 shows the micrograph. A solution of ethylene dimethacrylate in toluene containing a free-radical initiator (azobisisobutyronitrile) was pumped through a continuous agarose bed column. The column inlet and outlet were then closed and polymerization was initiated using ultraviolet light (366 nm). After the polymerization was completed, the continuous bed was removed from the tube and the agarose phase was melted away in a boiling water bath. The obtained replica was then studied by SEM. The structure presented in Fig. 6.6 shows a fairly evenly distributed network of flow-through pores with diameters of between 25 and 75 μm. In some other monoliths the network was clearly less randomly distributed. Instead, the flow-through pores had a preferred orientation. Similar observations had been made under the light microscope, although not with such a clarity. A study of a set of gels prepared under different conditions demonstrated that the non-random pore distribution was due to mechanical stratification of the emulsion before the solidification of the agarose occurred. By careful treatment of the emulsion, the stratified pore system could be avoided. On the other hand, a stratified pore system could possibly be advantageous in some applications.

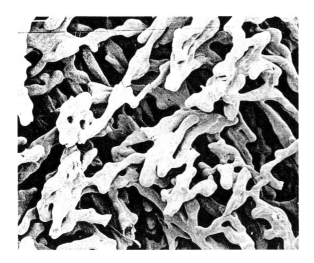

Fig. 6.6. SEM micrograph of pore structure replica of an agarose monolith. (Reprinted with permission from ref. [15], copyright 1996 Elsevier Sciences B. V.).

6.2.4.3 Chromatographic efficiency

The chromatographic efficiency of the continuous agarose beds can be determined by size-exclusion experiments. Pulse injection experiments with both a low-molecular-weight tracer such as sodium azide and medium-sized protein *e.g.* bovine serum albumin are routinely used in our laboratory to characterize continuous agarose beds. These results can be compared with the chromatographic efficiency of columns packed with standard agarose beads of different particle sizes. These comparisons reveal that the chromatographic efficiency of the continuous agarose beds is equal to the chromatographic efficiency of columns packed with standard agarose beads, having a diameter roughly equal to the distance between the flow-through pores of the continuous agarose matrix as determined by microscopy. Fig. 6.7 shows the HETP values obtained by pulse injections of sodium azide in a radial flow column as a function of the flow velocity. Also shown are theoretical HETP curves in packed columns for standard agarose beads with a particle size similar to the diffusion distance between the pores in the continuous radial column. The observed HETP-minimum is reasonably close to twice the distance between the flow-through pores, which could be expected for a well-behaved column.

Due to the transparency of continuous agarose beds, a convenient quick visual test of the homogeneity of the beds can be made by injecting colored latex samples to check that no gross pore distribution inhomogeneity exists. Similarly, by injecting a

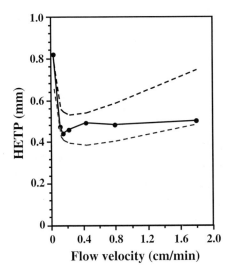

Fig. 6.7. Effect of flow velocity on efficiency of the monolithic agarose column expressed as HETP (curve 1). Column: 65 ml radial flow agarose monolith. Injection: 2 ml pulses of sodium azide solution (2 mg/ml). The dashed curves represent theoretical curves for 220 μm (curve 2) and 160 μm (curve 3) homogeneous agarose beads. (Reprinted with permission from ref. [11], copyright 2001 Elsevier Sciences B. V.).

colored low-molecular-weight sample an overall flow profile check can be obtained. By these experiments a quick check can be made to determine possible reasons for poorly performing beds.

6.2.5 Derivatization of monolithic agarose

Continuous agarose can be provided with functional chemistries equal to those known for agarose in bead form [16]. However, the protocols developed for agarose beads must be transformed to *in situ* activation/coupling conditions, which require pumping of reagent solutions through the continuous agarose bed column. This is usually successful. Difficulties may include the fact that derivatization often leads to slight shrinkage of the agarose gel. An obvious remedy is to cast the original gel slightly oversized. For small gel beds an axial compression of the gel with an aid of flow adaptors typically affords good result. Another potential difficulty may be associated with *in situ* derivatization using very reactive reagents. In such cases, preferential derivatization of the inlet portion of the continuous bed may result, while the outlet portion is poorly converted. This can be avoided by lowering the reaction rate by *e.g.* reacting at a low temperature and circulation of the reagents through the bed as

TABLE 6.1

OVERVIEW OF SURFACE CHEMISTRIES USED WITH SUPERPOROUS AGAROSE
MONOLITHS

Activation chemistry	Ligand	Bed dimensions (length × diameter, mm)	Application	Ref
Chlorotriazine	Cibacron Blue + various textile dyes	Various	Affinity chromatography	[11]
CNBr	NAD$^+$-derivative	60 × 16	Affinity chromatography	[1]
Glycidol/periodate-oxidation	Lactase	60 × 25 (radial column)	Bioreactor	[11]
CNBr	Acetylcholine esterase	15 × 5	Biosensor (electrochemical)	[10]
Tresylchloride	Glucose oxidase	15 × 5	Biosensor	[9]
CNBr	Antibody	15 × 5	Analyte trapping	[9]

fast as possible to ensure that all parts of the bed are in contact with the same concentration of the active reagent. Table 6.1 gives an overview of chemistries used.

Coupling of antibodies to a continuous agarose bed using the cyanogen bromide activation is an example of application [9]. The method was adapted from the activation procedure run at room temperature developed by March *et al.* [17], and modified for *in situ* conditions. The protocol was designed for a continuous agarose bed with dimensions of 15 × 5 mm but can be scaled to suit any bed dimension by following ordinary scale-up/down rules such as by maintaining constant both contact time and agarose/reagent ratio.

6.2.6 Composite agarose monoliths

In composites, two or more materials are combined to afford special advantages to the resulting product. An early example of this approach was the Ultrogel AcA media developed for size-exclusion chromatography of biomolecules [18]. Here, the good size-exclusion selectivity of the soft polyacrylamide gel could be exploited by incorporating the gel in the more mechanically stable agarose matrix. Another example is the Streamline adsorbents developed for expanded bed chromatography [19]. The

TABLE 6.2

AGAROSE MONOLITH COMPOSITES

Composite component in addition to agarose monolith	Application	Main advantages	Dimensions (length × diameter, mm)	Ref
Yeast cells	Catalytic reactor	Column usage possible	5 × 16, 10 × 10	[20]
Hydroxyapatite	Chromatography	Avoids column clogging	14 × 16	[1]
Ion-exchange beads	Chromatography	Application of dirty feed stock possible	5 × 16	[20]
1–2 μm graphite powder	Chromatography	Cheap reversed-phase adsorbent	5 × 16	[20]
Reticulated vitreous carbon	Electrochemical biosensor	Increased surface area/ Improved mechanical stability	15 × 5	[10]

Streamline adsorbents consist of quartz particles incorporated into agarose beads to increase their density.

By addition of a filling material to the agarose solution before making the agarose/organic phase emulsion, a range of composite agarose monoliths for different applications can be prepared (Table 6.2). The properties of the filling material such as its particle size and hydrophobicity determine the maximum percentage of the agarose phase that can be substituted with the filler. Mostly, up to 50 % of the agarose phase can be replaced while maintaining the stability of the agarose monolith. The preparation method is exemplified by the incorporation of hydroxyapatite particles in an agarose monolith [1].

A different way of making composite agarose monoliths is to cast the agarose/organic phase emulsion into a larger structure such as reticulated vitreous carbon (RVC). This preparation method was used for the preparation of a biosensor where RVC, apart from improving the mechanical stability of the bed, also provided an electrically conductive network through the bed [10]. This monolithic composite was then subjected to a cyanogen bromide activation followed by coupling of acetylcholinesterase. By using the electrode arrangement shown in Fig. 6.8, this composite monolith was used for the on-line determination of the pesticide paraoxon.

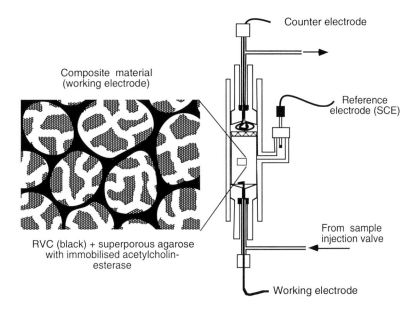

Fig. 6.8. Device for FIA determination of paraoxon that includes the continuous agarose/ reticulated vitreous carbon composite. Pore size 25 µm.

6.2.7 Special formats of monolithic agarose

6.2.7.1 Membranes

Monolithic agarose can also be manufactured in thick membrane/disc format. This format makes it easy to attain separations characterized by high flow velocities and short residence times at minimal pressure drops. Monolithic agarose membranes have potential as a cheap adsorbent for analysis in various chromatographic modes. A typical simple membrane-casting device that we assembled is shown in Fig. 6.9. It consists of two thick glass plates separated by a U-shaped gasket made of 5 mm thick silicon rubber. The rubber serves both as a seal and spacer, determining the thickness of the final membrane. The assembly is kept together by strong paper clamps and is pre-warmed to 60°C in a water bath. The agarose/organic phase emulsion is sub-sequently filled in the mold. After a pre-determined time the mold is transferred to a cold water bath to solidify the agarose. The casting device is then dismantled and circular pieces of the milky white membrane are punched from the sheet by a sharp steel puncher with the appropriate diameter. The membrane discs can be inserted in

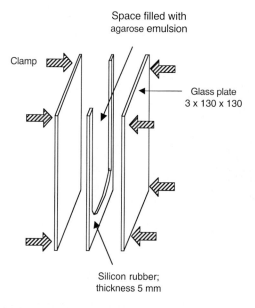

Space filled with
agarose emulsion

Clamp

Glass plate
3 x 130 x 130

Silicon rubber;
thickness 5 mm

Fig. 6.9. Casting device for agarose membranes. (Reprinted with permission from ref. [1], copyright 1999 Elsevier Sciences B. V.).

membrane holders and the organic phase removed by pumping water, 50 % ethanol, and water through the discs.

Another apparatus has also been constructed for continuous casting of monolithic agarose membranes. This device is schematically shown in Fig. 6.10. A hot agarose/organic phase emulsion is introduced in the container located at the top of the apparatus, where a turbine keeps the emulsion agitated. The emulsion is then drawn into the casting channel of the device by two endless Teflon sheets. The Teflon sheets are in contact with a thermostatted aluminum structure, which at the top is kept at 60°C and at the lower part at 10°C by aid of circulating water. When the agarose emulsion enters the zone of low temperature, it gels. At the bottom of the apparatus a continuous agarose membrane is delivered at a rate of 10 cm/min, which is determined by the speed of the Teflon sheets.

6.2.7.2 Monoliths in radial columns

The preparation of rods with a diameter exceeding 3 cm imposes problems due to the temperature gradient formed upon cooling the emulsion. As a result, the central part of the rod solidifies later than the outer parts, affording larger pores in the center. The resulting uneven flow profile leads to a decrease in chromatographic efficiency of

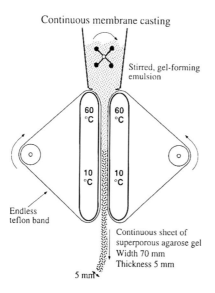

Fig. 6.10. Apparatus for continuous production/casting of monolithic agarose membranes.

the monolithic agarose. To circumvent this problem, large diameter rods should be operated in the radial direction. In this implementation, the asymmetric pore size distribution does not lead to uneven flow profiles. Figure 6.11 shows two types of radial column constructed for the application of monolithic agarose [11]. The radial column of Fig. 6.11a is constructed from two octagonal glass plates with a central hole functioning as end-pieces and a standard chromatography solvent filter from sintered stainless steel which acts as an outlet flow distributor. This simplified con-

Fig. 6.11. Radial columns for larger scale applications of monolithic agarose.

struct was used submerged in a beaker with a centripetal (inward) flow accomplished by connecting a pump or a vacuum to the center outlet. The radial column of Fig. 6.11b was designed to work in both centrifugal and centripetal flow directions. This column consisted of Plexiglas cylinder, Plexiglas top and bottom plate, and a stainless steel flow distributor.

The self-supported structure of the monolithic agarose greatly simplifies the packing procedure of the radial columns. A radial bed can be removed from the radial column B and inserted in the column A, and vice versa, ready for a new chromatographic run in a few minutes. The column in Fig. 6.11a also demonstrates another advantage of the self-supported structure of monolithic agarose. It enables a simple low cost construction of separation devices, which cannot be achieved with particle-based adsorbents. The flow properties of the radial beds are also excellent, allowing flow rates of over 100 ml/min to be used at a pressure drop less than 0.1 MPa.

6.2.7.3 Electrophoresis using the flow-through pores for internal cooling

A special format of monolithic agarose was briefly investigated for use in electrophoresis [1]. The idea was to use the flow-through pores to dissipate the Joule heat created during electrophoresis, and make it possible to use larger blocks of agarose for preparative electrophoretic separations. The heat normally produced in a larger gel block would lead to a severe temperature build-up with resulting loss of resolution or even breakdown of the gel structure. While this is not a problem for thin gels, it had a significant impact on the development of a large-scale use of this high-resolution technique [21]. The conceptual idea as well as one of the prototype designs is shown in Fig. 6.12. Initial tests showed indeed that this idea is viable since a substantial lowering of the internal gel temperature was observed. However, problems with protein recovery were observed, probably due to protein absorption at the agarose – organic solvent interface.

6.3 MONOLITHS BASED ON CELLULOSE

Cellulose found in the cell wall of plants consists of glucose molecules linked by 1,4 β-glucosidic bonds. Three different types of cellulose are currently available for chromatography, fibrous, microgranular, and regenerated. Fibrous cellulose consists of a highly inhomogeneous structure with essentially non-porous microcrystalline regions intertwined with less dense amorphous regions. This structure results in a material with little porosity. By acid treatment of cellulose followed by chemical cross-linking, microgranular cellulose have been developed featuring increased porosity and mechanical stability. Regenerated cellulose is prepared by first dissolving

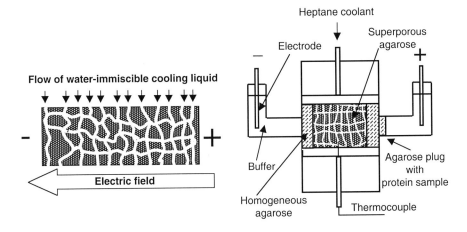

Fig. 6.12. Monolithic agarose for electrophoresis with internal cooling. The left part of the figure shows the concept of the internal cooling: A water-immiscible solvent is pumped through the electrophoretic bed carrying away the Joule heat formed during electrophoresis. The right panel shows a prototype design, where the superporous separation gel is surrounded by a layer of normal homogeneous agarose to prevent the escape of the heptane coolant pumped through the superporous separation gel

natural cellulose in a solvent, followed by a solidification step, leading, for example to beads with a high porosity suitable for separation of biomolecules. Cellulose has numerous hydroxyl groups available for coupling chemistries typical of all polysaccharide-based materials [16,22].

Cellulose membranes were used for chromatographic separations of biomolecules [23,24]. They are available in different formats such as layered stacks [2,25,26], and rolled layers [3,27]. These formats are described in detail elsewhere in this book.

Cellulose-based monoliths have been commercialized by Sepragen (San Leandro, CA) under the trade name SepraSorb [28]. These products are made of regenerated cellulose and fabricated in sheets. The monoliths have a continuous pore structure, with a pore diameter of 50–300 μm. The monoliths are available for ion-exchange chromatography, derivatized with both weak and strong ion-exchange functionalities. The large flow-through pore size of these monolith enables operation at high flow rates exceeding 100 ml/min at a very low back pressure. These monoliths can be used for capture steps with crude particle-containing feed streams without prior clarification.

6.4 ACKNOWLEDGEMENT

This work was supported by the Swedish Centre for BioSeparation (CBioSep).

6.5 REFERENCES

1   P.-E. Gustavsson, P.-O. Larsson, J. Chromatogr. A, 832 (1999) 29.

2   J.E. Kochan, Y.-J. Wu, M.R. Etzel, Ind. Eng. Chem. Res., 35 (1996) 1150.

3   J.F. Kennedy, M. Paterson, Polym. Int., 32 (1993) 71.

4   R. Noel, A. Sanderson, L. Spark, in: J.F. Kennedy, G.O. Phillips, P.A. Williams (Eds.) Cellulosics: Materials for Selective Separations and Other Technologies, Horwood, New York, 1993, p. 17-24.

5   R. Arshady, J. Chromatogr., 586 (1991) 181.

6   A.S. Medin, Studies on Structure and Properties of Agarose, Doctoral thesis, Department of Biochemistry, Uppsala University, Sweden, 1995.

7   P. Serwer, S.J. Hayes, Anal. Biochem., 158 (1986) 72.

8   G. Sofer, L. Hagel, Handbook of Process Chromatography, Academic Press, San Diego, 1997, p. 263.

9   M.P. Nandakumar, E. Pålsson, P.-E. Gustavsson, P.-O. Larsson, B. Mattiasson, Bioseparation, 9 (2000) 193.

10 M. Khayyami, M.T. Pérez Pita, N. Pena Garcia, G. Johansson, B. Danielsson, P.-O. Larsson, Talanta 45 (1998) 557.

11 P.-E. Gustavsson, P.-O. Larsson, J. Chromatogr. A, 925 (2001) 69.

12 P.-E. Gustavsson, A. Axelsson, P.-O. Larsson, J. Chromatogr. A, 795 (1998) 199.

13 E. Pålsson, A.-L. Smeds, A. Petersson, P.-O. Larsson, J. Chromatogr. A, 840 (1999) 39.

14 Anders Ljunglöf, Amersham Biosciences, personal communication.

15 P.-E. Gustavsson, P.-O. Larsson, J. Chromatogr. A, 734 (1996) 231.

16 G.T. Hermanson, A.K. Mallia, P.K. Smith, Immobilized Affinity Ligand Techniques, Academic Press, London, 1992.

17 S.C. March, I. Parikh, P. Cuatrecasas, Anal. Biochem., 60 (1974) 149.

18 J. Uriel, J. Berges, E. Boschetti, R. Tixier, C. R. Acad. Sc. (Paris) Série D, 273 (1971) 2358.

19 Expanded Bed Adsorption, Principles and Methods, Amersham Pharmacia Biotech, ISBN 91-630-5519-8.

20 P.-E. Gustavsson, P.-O. Larsson, WO 00/12618, 2000.

21 W. Thormann, in: J.-C. Janson, L. Rydén (Eds.), Protein Purification, Principles, High-Resolution Methods and Applications, Wiley, New York, 1998, pp. 651-678.

22 R. Arshady, J. Chromatogr., 586 (1991) 199.

23 D.K. Roper, E.N. Lightfoot, J. Chromatogr. A, 702 (1995) 3.

24 C. Charcosset, J. Chem. Technol. Biotechnol., 71 (1998) 95.

25 X. Santarelli, F. Domergue, G. Clofent-Sanchez, M. Dabadie, R. Grissely, C. Cassagne, J. Chromatogr. B, 706 (1998) 13.

26 D. Zhou, H. Zou, J. Ni, L. Yang, L. Jia, Q. Zhang, Y. Zhang, Anal. Chem., 71 (1999) 115.

27 K. Hamaker, S.-L. Rau, R. Hendrickson, J. Liu, C.M. Ladisch, M.R. Ladisch, Ind. Eng. Chem. Res., 38 (1999) 865.

28 http://www.sepragen.com

F. Švec, T.B. Tennikova and Z. Deyl (Editors)
Monolithic Materials
Journal of Chromatography Library, Vol. 67
© 2003 Elsevier Science B.V. All rights reserved.

Chapter 7

Monolithic Continuous Beds Prepared from Water-Soluble Acrylamide-Based Monomers

Audrius MARUŠKA

Vytautas Magnus University, Department of Chemistry, Vileikos 8, LT-3035 Kaunas, Lithuania

CONTENTS

7.1 INTRODUCTION

Non-particulate beds and monoliths are slowly becoming a mature class of chromatographic separation media covering a wide range of different formats and separation modes. More than a decade has passed since the pioneering studies concerning monoliths were carried out and the procedures describing the concepts were published in the literature [1]. During this period of time, these novel technologies have been evaluated for their suitability in terms of emerging needs and requirements. The great potential of the monoliths was recognized and demonstrated in the early studies [1,2]. Numerous publications, growing interest, and several monolithic chromatographic separation devices commercialised under the trade names, Uno (Bio-Rad Laboratories, Richmond, California, USA), CIM (Bia Separations, Ljubljana, Slovenia), Chromolith (Merck, Darmstadt, Germany), and Swift (Isco, Lincoln, Nebraska, USA) indicate the attractiveness of this technology. In early 2002, a Sciencedirect online database computer search for "Monoliths" resulted in 391 hits, "Silica rods" in 48 and "Continuous beds" in 116 hits when searched in combination with "chromatography". The monoliths primarily used as chromatographic media for the separations of biological compounds are also called, "the fourth generation of biochromatographic stationary phases" following the polysaccharides, highly crosslinked-, and monodisperse materials [3].

In terms of morphology, a chromatographic system involving the *monolithic packing* is characterized by *a continuum* of both *mobile* and *stationary* phases. Figure 7.1 shows an optical micrograph of such monolithic material. As is also shown in the vastly simplified scheme of this figure, flow-through channels, which are permeable for the mobile phase, transverse the monolith. Their size is rather large and exceeds that of typical particulate packings [4], reaching the area of so-called "flow-through macropores" through which the flow of the mobile phase occurs as a result of the pressure gradient [5]. In addition, the monolithic polymer matrix may also contain micropores and mesopores. In contrast, conventional particulate packing materials feature the continuous mobile phase; however, the stationary phase consists of a disperse material. A continuum of the mobile phase is an imperative for any chromatographic system no matter what packing is used. Collapse of this continuum leads to impermeability of the chromatographic device. The continuum of the stationary phase represents an important feature of the monolithic packing materials.

Non-particulate packing materials are mostly termed continuous beds or monoliths [6-12]. In terms of formation principles, the polymer-based monolithic packing materials may be divided into two groups: (i), monoliths prepared from polar and/or amphiphilic water-soluble monomers in the presence of a salt and, (ii), monoliths synthesized using monomers which are soluble in organic solvents, in the presence of

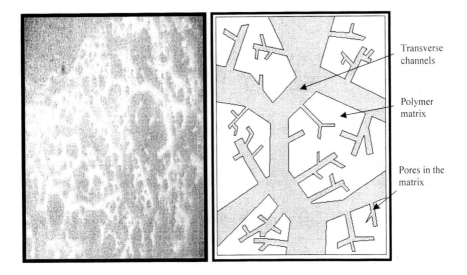

Fig. 7.1. Optical micrograph of a typical continuous bed and schematic presentation of the internal structure.

porogens. This rough classification is based on differences in the phase-separation required to form a continuous solid polymer matrix comprising the channel network that enables flow of a liquid through, and refers mainly to the preparation procedure not to the monomers utilized therein. It should be pointed out, that many of known monomers could be used in both approaches.

7.2 COMPRESSED CONTINUOUS BEDS

7.2.1 Preparation

One of the essential advantages of the monolithic materials is their manufacture *in situ*, enabling molds of any shape and format to be filled with a polymerization mixture or polymer solution and solidified to afford a chromatographic separation device which can be a cylindrical column, a capillary, or a groove in a microfluidic chip. Since the procedure is based on filling the device with a liquid precursor of the stationary phase, instead of packing the chromatographic column with a dispersed solid material, typical problems concerning packing quality, uniformity, and reproducibility are eliminated. Low pressure, or suction, applied without the need of specialized hardware is sufficient to fill the monomer solution, even in a narrow capillary, as opposed to the complex, high pressure packing devices required for packing columns with the particulate stationary phases.

Polymerization of a continuous bed occurs in the aqueous solution of divinyl monomer (cross-linker), monovinyl monomer and initiator, that is filled in a closed mold. After the polymerization is completed, the end seals are replaced with the column end fittings. Depending on the desired operational pressure, Plexiglas columns or disposable plastic syringes can be used as columns for low-pressure chromatographic applications, or stainless-steel columns that withstand even high pressures in the HPLC mode [13]. The column inlet fitting is adjustable and can be positioned to touch the top of the monolith after the bed has been washed and compressed by the flow of the mobile phase.

The axial compression of a chromatographic bed enables the reduction in the dead volume of the column packed with particulate materials of sufficient elasticity [14]. The compression results in shorter chromatographic runs and equilibration times, and improves efficiency and resolution, owing to narrowing interparticle channels and restricting resistance to the mass-transfer. This approach has been found to significantly improve the chromatographic characteristics of soft- or semi-rigid spherical or irregularly shaped particulate packings.

Axial compression has also been used to finalize the *in situ* preparation of the non-particulate chromatographic beds in order to achieve, (i), an increase in efficiency and resolution by minimizing the dead volume and narrowing the channels in the bed and, (ii), to match the monolith size with the desired column dimensions and exclude voids that might appear during the formation of the bed *in situ* as a result of the volume shrinkage typical of polymerization.

In Hjertén's pioneering work [1], the continuous bed was formed in a 30 cm long, 6 mm i.d. column, and subsequently compressed using the constant flow of the mobile phase. The flow rate used for the bed compression was always somewhat higher than that applied in the subsequent chromatographic experiments. The mechanical properties of the continuous bed had to match specific requirements. For example, the material had to be sufficiently flexible, not fragile, and to withstand the axial compression, while retaining the continuum of the polymeric matrix. Most importantly, the compressed bed has to remain permeable for flow-through. The hydraulic compression process is schematically shown in Fig. 7.2. First, the bed formed *in situ* shrinks during the polymerization and a void is formed between the column inner surface and the polymer. The flow of the mobile phase is then applied. Since the hydrodynamic resistance of the polymeric material is larger, the mobile phase passes through the empty space along the column wall at a higher velocity than through the porous polymer. Simultaneously, the longitudinal force that is applied on the top of the polymer rod, and is proportional to the cross-section and the pressure drop along the bed, pushes the polymer to the retaining (outlet) frit. The bottom layer of the column outlet experiences the maximum longitudinal strain (force per surface unit).

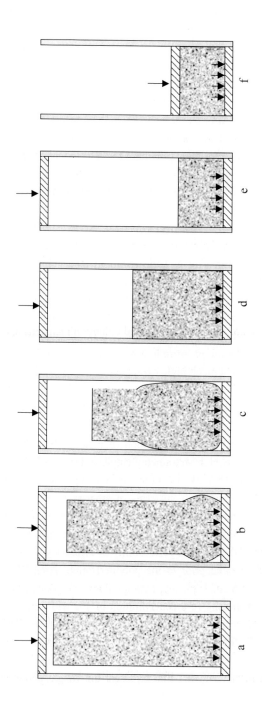

Fig. 7.2. Schematics of the preparation of a continuous chromatographic column by compressing the bed.

This strain tends to relax by changing the dimensions of the monolith in the direction perpendicular to the applied pressure. The relationship between the change in relative length along the two perpendicular directions is given by the Poisson ratio, with typical values of 0.3–0.5 for solids. This leads to an increase in the diameter in the vicinity of the outlet frit and the complete filling of the cross section. Continuing pumping of the mobile phase leads to changes in the aspect ratio of the entire mono-lith and disappearance of the voids between the column wall and the polymer. The hydraulic pressure of the flow through the bed becomes higher and further deforms the bed in the axial direction. This compression may lead to a bed length that is 5–10-fold smaller than that of the original polymer [15]. The extent of axial deforma-tion depends on the elasticity of the polymeric material. The hydrodynamic resistance of the bed increases with the compression, owing to narrowing of the flow-through channels in it. The final degree of deformation depends on both the permeability and mechanical properties of the compressed bed. This deformation is mostly permanent, although some relaxation can be observed after the hydrostatic pressure is released. The friction between the column wall and the monolith also helps to prevent the relaxation. However, only after using an inlet adapter, the column dimension is fixed and the bed remains compressed. Mechanical compression by means of an adapter sliding into the column can also be utilized instead of the hydraulic compression [13,16]. Compressing long beds with a small diameter is not recommended since the polymer rod may deform under the pressure in a way shown in Fig. 7.3. The friction of the folded rod would require a much higher flow rate to compress the bed and may lead to a total collapse of the channel structure.

The continuous bed for cation-exchange chromatography was originally formed from N,N'-methylene-*bis*-acrylamide cross-linker and an ionic monomer – acrylic acid – using common free radical polymerization in aqueous buffer, initiated by ammonium peroxydisulfate–N,N,N',N'-tetramethylethylenediamine (TEMED) system at ambient temperature [1]. The use of an entirely aqueous medium for the polymeri-zation is an advantage of this approach, in terms of environmental protection, that might be relevant in the potential scale-up of these columns for preparative separa-tions. Addition of salts to the polymerization mixture helps to optimize the phase separation process to achieve the best bed permeability-to-efficiency ratio. The mono-mer solution has to be de-aerated prior to the polymerization. After this, both the initiator and accelerator are added. Typically, both initiating components are added as a 5–10 % aqueous solution to achieve a concentration of 1% with respect to mono-mers [17]. This enables one to prepare the solid bed in approximately five minutes. This time is also sufficient to fill the column with the polymerizing solution and seal it prior to the phase separation. The overall composition of the polymerization mixture is often defined in terms of %C and %T that represent the weight percent of cross-

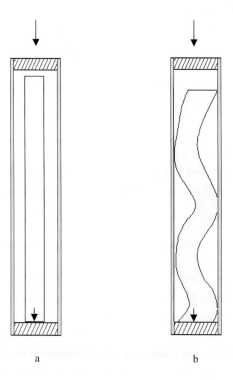

Fig. 7.3. Compression of the continuous bed in the narrow-bore column: a - shrunken polymer rod, b - compressed bed.

linker and the weight percent of total monomers (monovinyl plus cross-linker) in 100 mL of the aqueous solution, respectively.

7.2.2 Composition and characterization

The flexibility of the procedure leading to the continuous bed allows preparation of a broad range of the stationary phases from different monomers, and use of the compressed continuous beds for different chromatographic separations. For example, a compressed continuous bed for high-performance chromatography in the cation-exchange mode was prepared from a copolymer of acrylic acid and N,N'-methylene-*bis*-acrylamide [1]. Similarly, anion-exchange and hydrophobic-interaction chromatography columns were obtained by copolymerization of N,N'-methylene-*bis*-acrylamide with N-allyldimethylamine and butyl acrylate, respectively [17].

The porous properties of the compressed polymeric matrix were investigated by means of inverse-size-exclusion chromatography. The calibration curves indicated that the smaller pores in the continuous beds used for both anion exchange and hydrophobic interaction chromatography had an exclusion limit of 20,000 g/mol. The porosity of the bulk anion exchange polymer was approximately 8 vol. %, while this value for the monolith designed for hydrophobic interaction chromatography was 29 vol. %. The low exclusion limit of the continuous polymer matrix does not permit penetration of proteins into the small pores. Therefore, it is not surprising, that the resolution does not change with an increase in the flow rate of the mobile phase and in some cases even increases.

The mechanical and chemical stability of the continuous beds were improved upon substitution of the originally used crosslinker, N,N′-methylene-*bis*-acrylamide, with piperazine diacrylamide [18]. The resulting beds were completely stable in the pH range of 1–11 for 3 weeks while beds formed using N,N′-methylene-*bis*-acrylamide were partialy hydrolysed at pH 9 or 5.5 during a period of only 10–15 min. The stability of acrylamide-based continuous beds has been estimated from the uptake of hemoglobin, since this protein binds to carboxylic acid groups which are formed during the hydrolysis [16].

Detailed studies concerning the preparation of the continuous beds revealed that the hydrodynamic resistance of the bed depends on the ratio of the crosslinker to monovinyl monomer, such as piperazine diacrylamide/methacrylamide [16]. The smallest hydrodynamic resistance was achieved at a crosslinker concentration of 55.9%. The salt concentration in the polymerization medium had similar effect. The best permeability was found for the beds formed from monomer solution containing 60 mg/ml ammonium sulfate. The hydrodynamic resistance of the continuous bed also changes with the total monomer concentration and reaches a minimum value at %T = 15.

Compressible continuous beds containing reactive epoxide groups were synthesized from mixtures containing allyl glycidyl ether as a monomer. Poly(ethyleneimine) was later attached to this monolith, resulting a weak anion exchanger that was used for the purification of yeast enzyme concentrate and for chromatofocusing of human hemoglobin [18]. Epoxy-activated compressible continuous beds were also modified with alkyl ligands *via* an attachment of 1-octadecanol or hydrophilic compounds such as glucose, poly(vinyl alcohol), dextran, ethyl cellulose, and 1,3-(ditrimethylol)propane [16,19,20]. Hydrophilic functionalities were used for two reasons: (i), to increase hydrophilicity and reduce non-specific interactions during the separations of proteins and peptides or, (ii), to provide a hydroxyl-group-rich surface for coupling hydrophobic alkyl ligands such as 1,2-epoxyoctane and 1,2-epoxyoctadecane [19]. Epoxy-activated continuous beds were also used for covalent immobilization of proteins such as urease and applied for the determination of urea in clinical samples [20].

Piperazine diacrylamide was polymerized together with methacrylamide and 2-hydroxyethyl methacrylate to afford a continuous bed containing hydroxyl groups. Three triazine dyes, Cibacron Blue 3G-A, Pricon Red HE-3B, and Pricon Red H-3B, were then covalently attached to the continuous bed. First, the material of the polymerized continuous bed was suspended in a dye solution in 1 mol/l sodium hydrogencarbonate buffer (pH 9.5) and stirred for 48 h at room temperature. The suspension was then washed by decantation using a centrifuge, and finally packed into the Plexiglas column. The bed was then compressed, first by the flow, and additionally with the aid of the movable column adapter. Purification of dehydrogenases to a high degree was achieved in the dye-affinity chromatography mode using these compressed beds [21].

A detailed study of the preparation of continuous beds from piperazine diacrylamide, methacrylamide, and N-isopropylacrylamide for hydrophobic interaction chromatography of proteins has also been published [22]. The last monomer, containing the hydrophobic isopropyl functionality, is soluble in water. The hydrodynamic permeability of the bed, and the resolution of a mixture of chymotrypsinogen A and lysozyme, were improved with an increase in the N-isopropylacrylamide content.

A comprehensive study of the porosimetric and chromatographic properties was also carried out with the commercial UNO Q1 (quaternary amine) and UNO S1 (sulfonic acid) continuous ion-exchange beds [23]. An excellent batch-to-batch reproducibility was demonstrated on 23 production batches, with a RSD for retention of different compounds ranging from 0.7 to 1.3%.

7.3 NON-COMPRESSIBLE CAPILLARY FORMATS

Columns with an internal diameter of 2.1–4.6 mm are routinely used for conventional high-performance liquid chromatography. Typical flow rates used for these columns are in a range of 200–1000 µl/min, with sample volumes between 5 and 25 µl. In certain applications where both the sample concentration and its volume are limited, such as in biological analyses, more efficient and miniaturized analytical techniques are required. Capillary liquid chromatography (CLC) and capillary electrochromatography (CEC) using columns with a diameter of 10–300 µm and flow rates of 0.005–4 µl/min seem to be the solution. Sample volumes required for the separations in the capillary format can be then in the picoliter to nanoliter range. The capillary format also offers advantages in the coupling of liquid chromatography to mass spectrometry. Arrays of capillary columns or multichannel chips are potential means to a substantial increase in throughput.

7.3.1 Preparation *in situ*

The preparation of non-compressible beds differs from that used for their compressible counterparts described above, despite their possibly similar chemical composition. These monoliths are formed directly in their final size within fused silica capillaries without need for any compression. Both the synthesis and "packing" of the stationary phase are accomplished in a single step. The salt concentration in the polymerization mixture again plays an important role in the creation of the desired flow-through channels within the polymeric matrix. To retain the packing material in the fused silica capillary without supporting frits, the monolith is covalently attached to the capillary wall. This also stabilizes the continuous bed against both longitudinal and radial shrinkage and prevents creation of the undesired voids. To achieve this, the fused silica capillary is first pretreated with methacryloyloxypropyl trimethoxysilane ("Bind Silane") that reacts with silanol groups on the surface of the fused silica capillary *via* its methoxysilane groups, thereby attaching methacrylate ligands to the wall; these are able to form copolymers with the monomers in the aqueous mixture during the preparation of the continuous bed [17].

To improve the mechanical stability of the bed and to increase the phase ratio, the monomer concentration in this procedure is approximately twice as high as in compressible monoliths, reaching 20–30 %T. This procedure intrinsically does not require large quantities of monomers and other reagents since capillary columns can be produced from a few hundred microliters of the polymerization mixture. For example, only 0.058 mg of acrylic monomers and 0.106 mg of chiral selector (allyl-activated human serum abumin) were required to synthesize a 20 cm long, 50 μm i.d. capillary column for the separation of enantiomers [24]. Two hundred and fifty four columns of this diameter could be synthesized from 110 μl of the solution. A limiting factor in this method lies is the kinetics of polymerization. The phase separation and hardening of the polymeric material should occur within approximately 5 min after the addition of both initiator and accelerator. Slower polymerization often results in less opalescent, gel-like structures which have low hydrodynamic permeability, while fast polymerization and rapid phase separation may lead to problems if the time period is insufficient for filling the capillaries. Typically, 24 seconds are required to fill a 2 m long, 250 μm i.d. capillary with the polymerization solution using a water-jet pump. More capillaries can be filled simultaneously to finish before the polymerization sets on and the mixture gels. More efficient filling of the capillaries can be achieved with a device similar to that used for dynamic coating of capillary columns for gas chromatography, shown in Fig. 7.4, that uses a pressure of up to 0.5 MPa or a plastic syringe [12]. Active pressure is advantageous for filling viscous solutions containing polymer additives, or for narrow bore capillaries (50 μm or less). After the polymerization is

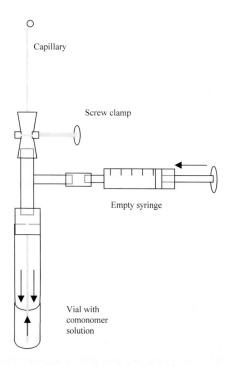

Fig. 7.4. Filling the capillary using pressure.

completed, the capillaries are cut to the desired length and washed with water. An inspection of the homogeneity of the continuous bed and the quality of the detection window is performed using an optical microscope at a 200x magnification. A homogeneous, opalescent continuous bed is seen if the synthesis was made properly. Most of the problems occur due to imperfect de-aeration of the monomer solution, that slows the polymerization and/or forms bubbles in the capillary. The capillary columns are then dipped in water or sealed with a silicone grease to avoid drying during their storage. Before their use for chromatographic experiments, these columns are conditioned with the mobile phase using an HPLC pump. The majority of these continuous beds may also be stored in a semi-dry state without losing their chromatographic performance.

7.3.2 Control of morphology

No matter whether the particulate or monolithic bed is packed into a chromatographic column, its efficiency and resolution depend on the structure of the packing — referred to as its morphology. In a monolithic column, the morphology primarily affects the eddy diffusion and resistance to the mass transfer in the flow-through channels. It also affects diffusion within the stationary phase, which depends on the wall thickness of the continuous beds. The overall structure of the continuous bed that includes both macropores and mesopores (*vide supra*) is formed simultaneously. Although the single-step manufacture of ready-to-use columns is a considerable advantage of the monolithic technologies, in terms of engineering, it requires additional efforts and methodologies to optimize the morphological characteristics of the chromatographic beds.

Equation 7.1 correlates the particle size, which defines the sizes of the interparticle voids, and the flow resistance defined as a pressure drop, ΔP, across the packed bed [25]

$$\Delta P = \frac{150\,\eta\,L\,F}{d_p^2\,d_c^2} \tag{7.1}$$

where η is the viscosity of the mobile phase, L is the bed length, F is the flow rate, d_p is the average diameter of the beads, and d_c is the column diameter. The flow properties of the continuous beds depend on the size of the flow-through channels in the polymeric matrix. A direct comparison of the flow properties of continuous and packed beds revealed that the hydrodynamic resistance of a monolithic column was close to that expected for a bed packed with 15 μm beads [26].

The relationship between resolution and flow velocity for the continuous beds was pointed out in the early studies [1,17]. The achievable flow velocity is a function of permeability, which in turn depends on the pore size, which is controlled by the composition of the polymerization mixture. We investigated the effect of ammonium sulfate concentration in this mixture on the flow rate achieved at a constant pressure through monoliths prepared using cross-linker together with either N-isopropylacrylamide or 2-hydroxyethyl methacrylate, and with a mixture of these two monomers (Fig. 7.5). The compositions of the specific polymerization mixtures are shown in Table 7.1. The flow rate for the nonpolar bed, I, increased exponentially with an increase in ammonium sulfate concentration in the polymerization mixture. An extremely permeable monolith is obtained in the presence of 30 mg/ml ammonium sulfate. A linear flow-velocity of 17 cm/min could be generated at 0.1 MPa. A bed with such high permeability may be used readily with a simple peristaltic pump or in a pipette-tip format. The more polar beds II and III exhibit different profiles of the hydrodynamic permeability, that depend less on the salt concentration and only decreased slightly at higher salt concentrations. This was not surprising, since hydrophobic interactions increase with the ionic strength. Therefore, larger flow-through channels were formed as a result of the early phase-separation process in systems includ-

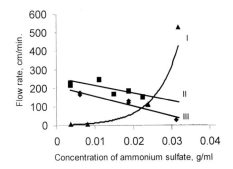

Fig. 7.5. Effect of ionic strength on the permeability of the monolithic continuous beds differing in polarity. Pressure, 3.4 MPa, capillary 100 μm i.d., length I - 13.4 (I) or 15.0 cm (II and III). For compositions of the monoliths see Table 7.1.

TABLE 7.1

POLYMERIZATION MIXTURES FOR THE PREPARATION OF MONOLITHS WITH DIFFERENT POLARITIES

Monolith	APS[a] μl	TEMED μl	PDA g	MA g	IPA g	HEMA ml	Buffer ml
I	20	20	0.0624	0.0376	0.05	–	1
II	20	20	0.0684	0.0124	0.076	0.0448	1
III	20	20	0.0752	–	–	0.09	1

[a] APS – ammonium peroxodisulfate; TEMED – N,N,N′,N′-tetramethylethylene diamine; PDA – piperazine diacrylamide; MA – methacrylamide; IPA – N-isopropylacrylamide; HEMA – 2-hydroxyethyl methacrylate

ing the hydrophobic monomer. Additives of other water soluble polymers such as poly(ethylene oxides) could also be used to promote the formation of the desired flow-through channels [27].

Figure 7.6 illustrates the effect of total monomer concentration on the morphology of the monolithic material and its performance, exemplified by the separation of small molecules. The higher the %T, the smaller is the size of the flow-through channels in the bed. This improved the separation ability of the column, although on account of the much higher back-pressure.

The proportion of cross-linker in the monomer mixture is typically kept at about 50% to afford rigid beds. A decrease in %C led to a loss of the mechanical strength of

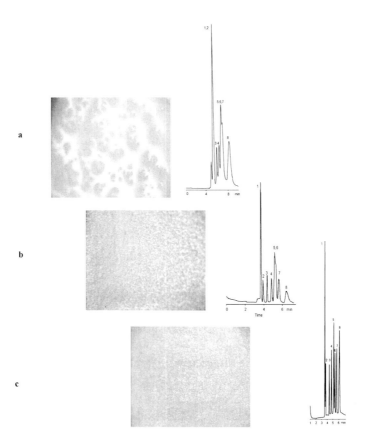

Fig. 7.6. Effect of total monomer concentration %T on morphology and resolution of monolithic continuous beds. % T, 16.8 (a); 23.8 (b); and 31.6 (c): mobile phase, 100% methanol. Peaks: pyridine (1), 4-hydroxymethylpyridine (2), 4-methoxyphenol (3), 2-naphthol (4), catechol (5), hydroquinone (6), resorcinol (7), 2,7-dihydroxynaphthalene (8). Capillary pressure-driven chromatography; capillary column, 175 × 0.1 mm i.d. (active length 125 mm). UV detection at 254 nm.

the polymer and formation of gel-like structures with extremely low permeability. Other parameters such as pH, temperature, and concentration of the initiator may also be used to control the morphology. Buffering the polymerization mixture is important for polymerizations of ionic monomers. Fast and reproducible free-radical polymerization initiated by the ammonium peroxodisulfate–TEMED system is achieved in a pH range of 5–7. The polymerization process is slower at higher pH values, even at high %C, and the product is a cloudy or transparent gel-like solid with both poor permeability and mechanical strength. These properties also deteriorate upon decreasing the cross-linking. For example, homogeneous gels with no flow-through channels were formed at %C=2. Interestingly, a good continuous bed could be synthesized easily from pure piperazine diacrylamide (%C=100). However, this material had only a limited use, owing to the lack of ligands or functional groups.

Although polymerizations of the continuous beds are most often performed at ambient temperatures (18–21°C), the hydrodynamic properties were considerably better when the reaction was carried out at higher temperatures (50-60°C).

7.3.3 Control of surface chemistry

Most of the current chromatographic applications are carried out in the reversed-phase mode [28]. Therefore, much of the experimental effort has been directed to-wards designing continuous beds for this specific separation mode. However, it is difficult to use highly hydrophobic monomers, owing to their limited solubility in the aqueous polymerization mixtures. One of the successful approaches described above involves synthesis of the continuous bed with epoxy functionalities used for sub-sequent attachment of the hydrophobic ligands. Alternatively, octadecyl ligands were attached *via* the reaction of 1,2-epoxyoctadecane with hydroxyl groups of the polymeric matrix [29].

Another option uses hydrophobic monomers such as stearyl methacrylate or butyl methacrylate emulsified in the aqueous polymerization mixture [30], or the addition of an organic solvent to increase the solubility of the non-polar components [27].

We have also used stearyl methacrylate, methacrylamide, vinylsulfonic acid, and piperidine diacrylamide in N,N-dimethylformamide or 1:1 2-propanol–N,N-dimethyl-formamide mixture to synthesize continuous beds for reversed-phase separations in the capillary format. These columns clearly exhibited attributes typical of reversed-phase chromatographic media. For example, their hydrophobicity, measured as the methylene group selectivity $\alpha_{(CH2)}$ and the change in free energy for transfer of a methylene group from the mobile phase to the stationary phase $\Delta\Delta G_{(CH2)}$ were in the range 1.48–1.86 and (–0.96) – (–1.517) kJ/mol, respectively, for the mobile phases consisting of 30–66% aqueous methanol. A typical reference material, LiChrospher 100 RP-18 silica beads, exhibits very similar values for both $\alpha_{(CH2)}$ (1.47–1.92) and $\Delta\Delta G_{(CH2)}$ (–0.942 to –1.589 kJ/mol) [31].

An elegant approach to reversed-phase continuous beds in capillary format was described by Ericson and Hjertén [32]. They synthesized the continuous beds from a solution prepared by dissolving piperazine diacrylamide and methacrylamide in phosphate buffer in which stearyl methacrylate in a small volume of dimethylform-amide was then dispersed by sonication, with no surfactants used. After the mixing of the piperazine diacrylamide and dimethyldiallylammonium chloride, the dispersion was pushed into the capillary and polymerized.

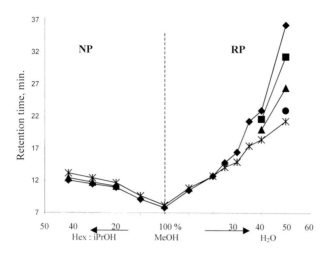

Fig. 7.7. Normal-phase (NP) and reversed-phase (RP) capillary electrochromatography separation of benzoic acid esters using a continuous bed with polymerized N-isopropyl-acrylamide. Capillary length 18.8 cm (effective 13.9 cm): mobile phase; NP, methanol (MeOH)–hexane (Hex) + 2-propanol (*i*-PrOH) mixture (85:15); RP, MeOH–H₂O; voltage 5 kV; UV detection at 254 nm; * - methyl benzoate, ● - ethyl benzoate, ▲ - propyl benzoate, ■ - butyl benzoate, ♦ - isoamyl benzoate.

A comprehensive evaluation of the acrylamide-based monolithic columns was published by Hoegger and Freitag [33]. Once again, piperazine diacrylamide, N,N-dimethylacrylamide, methacrylamide, butyl acrylate, and hexyl acrylate, dissolved either in aqueous buffer or its mixture with formamide, were used as respective components of the polymerization mixture. Porosimetric evaluation of the beds by means of mercury intrusion porosimetry and scanning electron microscopy revealed that the size of the flow-through channels could be reproducibly adjusted by selection of the ionic strength of the monomer solution.

Although seldom used, because of the requirement of less-polar mobile phases, which are incompatible with many classes of biologically important substances, the normal-phase chromatography is important for both analysis and preparative separations of a number of organic compounds including drugs and racemates. Continuous beds synthesized from piperazine diacrylamide, N-isopropylacrylamide, methacrylamide with either 2-hydroxyethyl methacrylate or vinylsulfonic acid were used for pressure-driven capillary HPLC using mobile phases consisting of hexane, ethanol, and methanol [34]. This system enabled separations of aromatic compounds in the normal-phase mode and was also used later in the CEC mode for the separations of benzoic acid esters. The effect of the mobile-phase composition on retention is shown in Fig. 7.7.

Continuous beds were also prepared by copolymerization of methacrylamide, piperazine diacrylamide, vinylsulfonic acid, and N-(2-hydroxy-3-allyloxypropyl)-L-4-hydroxyproline and used for chiral ligand-exchange chromatography [35,36]. A different approach was described by Kornyšova *et al.* [37] who prepared a monolith from N-(hydroxymethyl)acrylamide, allyl glycidyl ether, piperazine diacrylamide, and vinylsulfonic acid. The epoxy groups were then oxidized to an aldehyde and reacted with the chiral selector vancomycin. In contrast to the silica-beads-based counterpart, this monolith exhibited enantioselectivity even in the reversed-phase mode in the highly polar aqueous mobile phase. Racemic thalidomide was separated on this column in 10 min with a resolution R_S of 2.5 and an efficiency of 120,000 plates/m.

A monolithic column with covalently immobilized human serum albumin as the chiral selector was synthesized by Machtejevas *et al.* [24] using direct copolymerization of the allyl activated protein. The binding constant of free albumin is typically much higher than that of its immobilized counterpart, as a result of differences in accessibility of the active sites. To diminish the steric hindrance for the potential analytes, strongly interacting compounds such as acetylsalicylic acid and L-tryptophan were added to the protein solution first during its activation with allyl functionalities and then for synthesis of the monoliths, and removed only after the polymerization was completed [38]. Both additives exerted a positive effect on the resolution of kynurenine and tryptophan enantiomers, as demonstrated by the results obtained with the bed prepared in their absence.

7.3.4 Cyclodextrin derivatized continuous beds

The versatility of the preparation of monolithic materials enables the creation of a variety of stationary phases in a very simple manner. In developing new approaches, we have also made attempts to extend the range of the monoliths to those that are modified without using chemical reactions or additional steps as part of the synthetic path [39]. One of these approaches may include entrapment of compounds that could provide the bed with specific properties. However, this method is less favorable, owing to the continuous leakage (bleeding) of the entrapped compound, that changes the properties of the monolith [23]. This may be avoided if macrocyclic compounds are added to the polymerization mixture and threaded by the chains of the polymer to assemble the supramolecular structures called polyrotaxanes depicted in Fig. 7.8 [40-42].

First, we studied the generation of electro-osmotic flow in the continuous bed prepared from neutral acrylic monomers, a cross-linker, and an ionic derivative of β-cyclodextrin (β-CD) [39]. The effect of the hydrophobicity of the polymeric matrix, controlled by the ratio of N-isopropylacrylamide to methacrylamide, on the formation of polyrotaxane structures was also evaluated, assuming that template-directed thread-

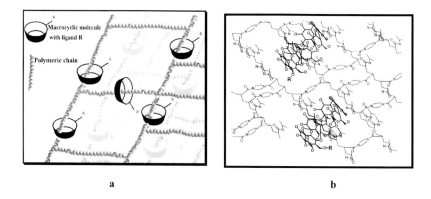

Fig. 7.8. Schematic representation (a) and computer simulation (b) of the polyrotaxane structure.

ing may occur and that this is mainly based on the interaction of the hydrophobic parts of the growing polymeric chain with the hydrophobic cavity of the cyclodextrin derivative. Cross-linking of these chains prevents dethreading of cyclodextrin units from the polymer chain and forms a three dimensional polyrotaxane structure. According to the magnitude of the electro-osmotic flow, formation of polyrotaxanes is favored in polymerization mixtures with higher hydrophobicity. The simulated polyrotaxane structure created from a polymerization mixture in which the ratio of the monomeric units and the cyclodextrin macrocycles is kept equal to that in the polyrotaxane structure is presented in Fig. 7.8b.

The capillary monolithic column prepared from mixtures containing 6-amino-β-CD were used to investigate their chiral recognition. The preliminary result shown in Fig. 7.9 is rather interesting, since the cavity of macrocycle in the rotaxane structure is filled with the polymer. Therefore, an inclusion complex cannot be formed. Since the enantioseparation has clearly been achieved, this result suggests that even the outer surface and the hydrophilic rims of the cyclodextrin may play important roles in enantiorecognition.

The monoliths discussed in this Section are very stable, since no change in the electro-osmotic flow was observed in the course of several weeks. This also largely excludes the idea that the β-CD moieties are only physically entrapped, since these would likely be washed out during that time, thus affecting the electro-osmotic flow [23].

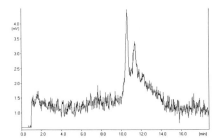

Fig. 7.9. CEC separation of metoprolol enantiomers using a continuous bed containing 6-amino-β-CD. Mobile phase, 1:1 methanol–water containing 0.05% acetic acid and 0.05% N,N,N′,N′-tetramethylethylenediamine, pH 5.8, conductivity adjusted to 165 μS/cm, voltage 5 kV, current 3 μA.

7.3.5 Monoliths with restricted-access

Drugs and their metabolites often need to be assayed in biological fluids such as plasma, serum, or urine. Typical RP-HPLC columns cannot be used for direct injection of biological fluids, because they contains protein that denature and irreversibly adsorb on the packing material. Therefore, column-switching techniques and restricted-access packing materials have been developed for direct injection and analysis of compounds in the biological fluids [43]. Restricted-access packings have a specific architecture [44,45]. They are covered at their outer surface with a hydrophilic semipermeable layer that prevents large molecules from reaching the pores inside the bead, owing to the size-exclusion, and the proteins are eluted in the void volume. In contrast, small solutes can penetrate the semipermeable interface layer and interact with the active sites within the pores.

We investigated the applicability of monolith-based restricted-access adsorbents in separations using the capillary format, and tested reversed-phase monoliths possessing different hydrophobicities owing to the incorporation of monomers with C8- or C18-substituents shielded with a hydrophilic polymer — shown schematically in Fig. 7.10 [43]. The quality of the hydrophilic shield proved to be of a great importance. If the hydrophobic surface is not sufficiently shielded, the proteins may adsorb and denature within the column. As a result, dramatic increases in the back-pressure and deterioration of separation efficiency after several injections of the protein-rich samples were observed.

The reversed-phase acrylamide-based monoliths were synthesized in a 100 μm i.d. fused-silica capillary or a 4.6 mm i. d. stainless steel chromatographic column, and shielded *in situ* with a hydrophilic polymer such as polyethylene glycol, dextran, and protein covalently attached to the surface. Figure 7.11 shows the separations of a

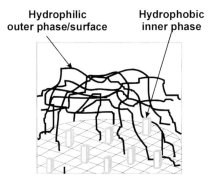

Fig. 7.10. Schematic representation of the restricted-access separation medium.

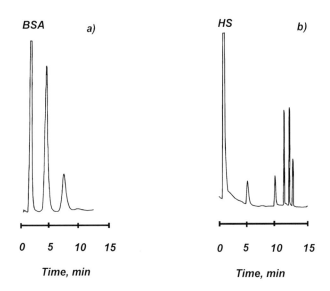

Fig. 7.11. HPLC, (a), and capillary LC, (b), separations using shielded reversed-phase monolithic continuous bed columns. HPLC column; 35 × 4.6 mm i.d.; mobile phase, A, 50 mmol/l sodium phosphate (pH 7.2), B, methanol; gradient from 20 to 80 % of B in A within 14 min; flow-rate 0.5 ml/min; detection UV 280 nm; sample 20 μl bovine serum albumin (BSA) solution containing methyl- and ethyl benzoate. Capillary LC column; 245 × 0.1 mm i.d. (active length 200 mm); mobile phase, A, 50 mmol/l sodium phosphate buffer, pH 7.2, B, methanol, gradient 5 to 80% B in A within 5 min; flow rate 0.6 μl/min; detection UV, 280 nm; sample, 15 nl human serum (HS) solution containing methyl-, ethyl-, propyl-, butyl-, and isoamyl benzoates.

TABLE 7.2

METHYLENE GROUP SELECTIVITY AND GIBBS FREE ENERGY FOR
MONOLITHIC COLUMNS USED IN REVERSE-PHASE CHROMATOGRAPHY

Monolith	Methanol content in mobile phase, % [a]	α_{CH2} [b]	$\Delta\Delta G_{(CH2)}$ [c] kJ/mol
RP	35	2.23	−1.96
	45	1.83	−1.47
	55	1.65	−1.22
RP-RAM	35	1.99	−1.67
	45	1.73	−1.33
	55	1.36	−0.75

[a] Mobile phase: methanol in 50 mmol/l sodium phosphate buffer pH 7.2

[b] Methylene group selectivity, $\alpha = k_2/k_1$

[c] Change in free energy for the transfer of a methylene group from mobile phase to stationary phase calculated using equation $\Delta\Delta G_{(CH2)} = -RT \ln \alpha_{(CH2)}$

mixture of benzoate esters dissolved in bovine serum albumin solution and human serum. Using the capillary format, a higher mass-sensitivity of 15 nl was found compared to the 20 µl required for HPLC. The column efficiency was also higher in the capillary column and the analysis time shorter, in spite of the 830-times lower flow-rate.

A linear relationship between retention factors and the number of methylene groups in the alkyl groups of alkyl benzoates confirmed a separation mechanism based on hydrophobic interaction. Interestingly, the shielding did not change considerably the hydrophobic affinity for small analytes (Table 7.2) and the excellent hydrodynamic permeability of these columns.

7.3.6 Homogeneous gels for CEC applications

Mass-transfer resistance in columns depends on the size of the flow-through channels. Although the monoliths mainly contain large pores, some zone broadening may result from the presence of limited amounts of micro- and mesopores, which affect the mass-transfer of small molecules. This is likely to be avoided by using macroscopically homogeneous media such as gels. Since the gel integrates the liquid with the solid polymer, and forms a uniform structure on the macromolecular level, it can be macroscopically regarded as an entirely homogeneous medium. Polymer gels are commonly used in electrophoresis as anti-turbulent or sieving media. Owing to their limited mechanical strength and small pores, liquids cannot be pumped through these gel-containing columns at velocities sufficient for chromatographic separations. In contrast, electro-osmotic flow can drive the mobile phase through ionized gels.

Fujimoto *et al.* prepared continuous gels using free-radical polymerization of the water-soluble monomers, N-isopropylacrylamide, 2-acrylamido-2-methylpropane-sulfonic acid, and N,N'-methylenebisacrylamide [46]. The total monomer concentration in these gels was only 6.9–10% and cross-linking 5.8–10%. Digital images of the Rhodamine B zone-front moving by electro-osmotic flow demonstrated a very flat flow profile [46]. The column efficiencies ranged from 107,000 to 160,000 plates/m.

An ionized gel was also prepared by copolymerization of acrylamide with N,N'--methylenebisacrylamide and 2-acrylamido-2-methylpropanesulfonic acid and used for the separation of both neutral- [47] and ionizable- [48] low-molecular-weight compounds. Molecular sieving was claimed to be the predominant separation mechanism. In contrast to packed columns and rigid monoliths, the homogeneous gels enable the "in-capillary" (in-gel) detection since they exhibit only a low UV absorbance.

Hjertén *et al.* [49] have proposed several approaches to homogeneous gel stationary phases for capillary electrochromatography. Since it was difficult to derivatize gel-forming polymers such as agarose in order to generate gels with both the desired chromatographic properties and pore-sizes large enough to enable the electro-osmotic flow, they developed a new method that included the replaceable low-melting-point agarose with an entrapped acrylate polymer bearing phenyl boronate and carboxylic acid groups. Boronate ligands had a dual duty: (i), they contributed to the generation of electro-osmotic flow and, (ii), interacted via complexation with compounds containing vicinal diol functionalities in the *cis*-configuration. This monolith separated the nucleosides cytidine, adenosine, uridine, and guanosine.

Poly(vinyl alcohol) gel was also formed *in situ* by filling the capillary with a 4% poly(vinyl alcohol) solution followed by the addition of borate ions using electrophoretic transport. The borates cross-linked the polymer and formed a gel that was used for the separation of parabenes [50].

Recently, we have synthesized hydrophobic alkyl-derivatized agarose-based homogeneous gels, which could be used for capillary electrochromatography in the reversed-phase mode. These gels tolerated mobile phases with high contents of the organic modifier, and even entirely organic mobile phases such as mixtures of methanol and acetonitrile. A CEC electrochromatogram obtained using a capillary filled with reversed-phase agarose gel is shown in Fig. 7.12. The column efficiency for the retained neutral compounds was close to 200,000 plates/m. Figure 7.13 shows another

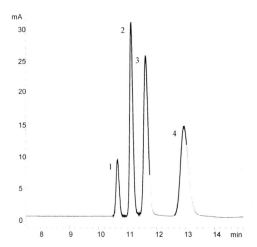

Fig. 7.12. CEC separation using an agarose-gel monolithic capillary column in reversed-phase mode. Capillary column; 320 × 0.1 mm i.d. (active length 230 mm); mobile phase 30% aqueous MeOH, 0.3% triethylamine acetate buffer pH 3.7. Peaks; acetone (1), methyl 4-hydroxybenzoate (2), ethyl 4-hydroxybenzoate (3), propyl 4-hydroxybenzoate (4).

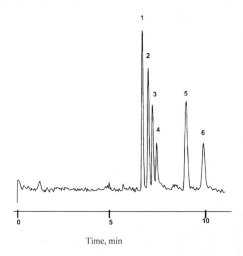

Fig. 7.13. CEC separation of a model mixture using homogeneous agarose gel with entrapped DEAE dextran. Stationary phase, 1% (w/v) of agarose gel with 20% (v/v) of DEAE dextran; mobile phase, 100 mmol/l Tris/boric acid buffer, pH 8, capillary, 20.3 cm (effective 16.7 cm) × 50 μm i.d.; injection; –1.5 kV, 15 s, voltage –5 kV, detection UV 254 nm.

example of capillary electrochromatography using homogeneous 1 % agarose gel with entrapped DEAE dextran, that afforded a constant electro-osmotic flow and a column efficiency of 500,000–800,000 plates/m [51].

7.4 MORPHOLOGY OF THE MONOLITHIC BEDS

7.4.1 Optical microscopy

We have demonstrated earlier that high-resolution optical microscopy together with other microscopic methods, is a very useful and convenient means for a simple and rapid evaluation of the morphology of monolithic beds and enables *in situ* inspection of their homogeneity [52,53]. Owing to its lower resolution, which enables one to see only features as small as 1–2 μm, optical microscopy cannot be used for investigation of the fine structural elements of the porous matrix. However, optical microscopy has the advantage that the sample preparation does not require any pre-treatment and, accordingly, fragile and thermally or chemically labile structures can be preserved. It is also easy to avoid drying of the sample and any consequent morphological changes of the matrix during its microscopic observation.

Figure 7.14 shows three-dimensional optical micrographs of β-CD modified monoliths formed from polymerization mixtures with various ionic strengths, obtained using Image-Pro Plus software and a programmable XYZ stage. The optical sectioning method was used to determine the location of each point on the surface. It is obvious from these micrographs that an increase in the ammonium sulfate concentration ranging from 0.016 to 0.08 g/ml results in an enhanced roughness of the surface, indicating the presence of larger channels in the polymeric matrix and better permeability of the monolith.

Fig. 7.14. Electron micrograph of typical dry acrylamide-based monolith at a magnification of 15,000x.

Fig. 7.15. Three-dimensional optical images of a continuous bed in different stages of the drying process. Wet monolith (a); 5 min drying (b); 10 min drying (c); imaged size 360 × 220 μm.

The permeability of the bed depends also on the density of the flow-through channels. Figure 7.15 shows optical micrographs of monoliths with different densities of the channels. Obviously, a smaller number of channels leads to lower permeability of the polymer matrix.

Scanning optical microscopy can also be used for the in-capillary inspection of the continuous beds. Figure 7.16a shows a bed that had shrunk during the polymerization, and formed a void between the monolith and the capillary wall. The shrinkage could even break the monolith and create an empty cavity. More detailed inspection of such a cavity, presented in Fig. 7.16c revealed that the inner wall of the capillary is coated with a layer of the polymer. This indicated that the void results from shrinkage and insufficient mechanical strength of the monolith rather than from the low quality of the wall-surface modification.

7.4.2 Scanning electron microscopy

The use of electron microscopy for morphological investigation of the monolithic materials is rather common but requires dry samples. However, drying may change

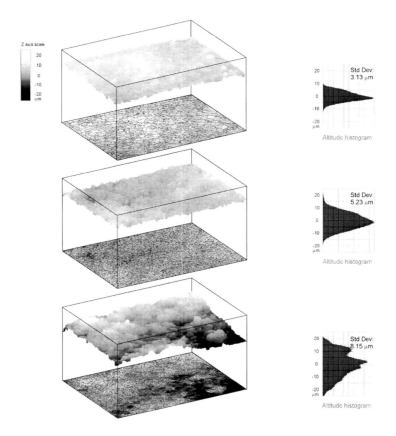

Fig. 7.16. Optical images of β-CD-modified continuous beds prepared by polymerization of mixtures varying in ionic strength. Concentration of ammonium sulfate in polymerization mixture 0.016 (a), 0.048 (b), and 0.08 g/ml (c).

dramatically the morphological features of the porous material. This is demonstrated on optical micrographs of Fig. 7.17. The roughness of the continuous bed surface is much lower (Fig. 7.17c) and the channels in the polymeric matrix more narrow when the bed is dried.

Electron micrographs showed that the acrylamide-based polymeric monolith consisted of channels and 2–4 μm thick "walls" between them (Fig. 7.18). The morphology is rough and randomly spread, without smooth walls and straight channels. The voids represent about 60% of the total bed volume. To avoid any changes in the morphological structure of the continuous beds, no metal coating was used in the sample preparation procedure for the electron microscopy. The samples were washed with water, methanol, and acetone, and then dried at 60°C.

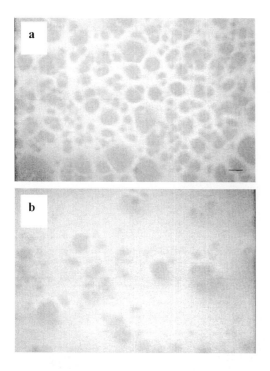

Fig. 7.17. Optical micrographs illustrating high (a), and low (b), density of the channels in the continuous beds.

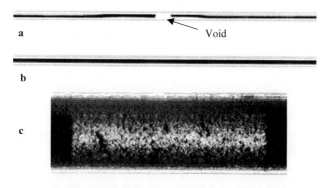

Fig. 7.18. Quality tests of capillary columns with *in-situ* polymerized continuous beds, using optical scanning microscopy. Low quality non-homogeneous bed, (a); high quality homogeneously polymerized continuous bed, (b); void in the capillary (a) at higher magnification (c).

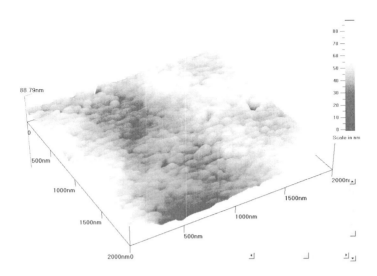

Fig. 7.19. AFM image of monolithic material based on acrylamide prepared from polymerization mixture containing a high concentration of ammonium sulfate.

7.4.3 Atomic force microscopy

We used intermittent contact atomic force microscopy (AFM) in contact mode for the phase imaging for investigating the β-CD modified continuous beds [39]. The AFM image of the continuous bed formed at a high ionic strength of 1.25 g/ml ammonium sulfate shown in Fig. 7.19, shows the globular structure of the surface, which comprises "fused" round-shaped polymeric units with a size of 65–115 nm and pores ranging from 23 to 60 nm. The volume fraction of these pores appears to be about 20%. Scanning of a larger area of 10 × 10 μm reveals 1.5–4 μm channels that are consistent with the optical microscopic observations.

7.5 CONCLUSIONS

The monolithic continuous beds formed from water-soluble acrylate monomers are versatile materials with applications that can be scaled down from typical HPLC columns to capillaries which are useful in CEC and microfluidic chip formats. Their preparation is simple, most often in a single step, and avoids the tedious column packing required for particulate separation media.

7.6 ACKNOWLEDGEMENTS

I would like to thank Professor Stellan Hjertén from Uppsala University, from whom I also learned the basics of the continuous beds, for inspirational discussions. I highly appreciate the work concerning monolithic packings done by my students and colleagues, specifically Dr. Olga Kornyšova, Dr. Vilma Kudirkaite, and Dr. Egidijus Machtejevas, and their help in the preparation of this chapter. Funding by the Volkswagen Foundation, grant No. I/75 926, and a grant from the Swedish Royal Academy of Sciences are also acknowledged.

7.7 REFERENCES

1   S. Hjertén, J.-L. Liao, R. Zhang, J. Chromatogr. 473 (1989) 273.
2   F. Svec, J.M.J. Fréchet, Anal. Chem. 64 (1992) 820.
3   G. Iberer, R. Hahn, A. Jungbauer, R.E. Majors, LC-GC Europe 13 (2) (2000) 88.
4   K.K. Unger, Packings and Stationary Phases in Chromatographic Techniques, Marcel Dekker, Inc., New York, 1990.
5   N.B. Afeyan, N.F. Gordon, I. Mazsaroff, L. Varady, S.P. Fulton, Y.B. Yang, F.E. Regnier, J. Chromatogr. 519 (1990) 1.
6   Y. Zhang, C.-M. Zeng, S. Hjertén, P. Lundahl, J. Chromatogr.A 749 (1996) 13.
7   F. Svec, E. C. Peters, D. Sýkora, J. M. J. Fréchet, J. Chromatogr. A 887 (2000) 3.
8   S.M. Fields, Anal. Chem. 68 (1996) 2709.
9   H. Minakuchi, K. Nakanishi, N. Soga, N. Ishizuka, N. Tanaka, Anal. Chem. 68 (1997) 3498.
10   I. Gusev, X. Huang, Cs. Horváth, J. Chromatogr. A 855 (1999) 273.
11   J. Matsui, T. Kato, T. Takeuchi, M. Suzuki, K. Yokoyama, E. Tamiya, I. Karube, Anal. Chem. 65 (1993) 2223.
12   J.-L. Liao, C.-M. Zeng, A. Palm, S. Hjertén, Anal. Biochem. 241 (1996) 195.
13   J. Mohammad, S. Hjertén, J. Biochem. Biophys. Methods 28 (1994) 321.
14   S. Hjertén in: M.T.W. Hearn (Ed.), HPLC of Proteins, Peptides and Polynucleotides: Contemporary Topics and Applications, VCH, Weinheim, 1991.
15   S. Hjertén, Y.-M. Li, J.-L. Liao, J. Mohammad, K. Nakazato, G. Pettersson, Nature 356 (1992) 810.
16   S. Hjertén, J. Mohammad, K. Nakazato, J. Chromatogr. 646 (1993) 121.
17   J.-L. Liao, R. Zhang, S. Hjertén, J. Chromatogr. 586 (1991) 21.
18   S. Hjertén, J. Mohammad, J.-L. Liao, Biotechnol. Appl. Biochem. 15 (1992) 247.
19   S. Hjertén, K. Nakazato, J. Mohammad, D. Eaker, Chromatographia 37 (1993) 287.
20   J. Mohammad, S. Hjertén, Biomed. Chromatogr. 8 (1994) 165.
21   J. Mohammad, A. Zeerak, S. Hjertén, Biomed. Chromatogr. 9 (1995) 80.
22   C.-M. Zeng, J.-L. Liao, K. Nakazato, S. Hjertén, J. Chromatogr. A 753 (1996) 227.
23   T.L. Tisch, R. Frost, J.-l. Liao, W.-K. Lam, A. Remy, E. Schienpflug, C. Siebert, H. Song, A. Stapleton, J. Chromatogr. A 816 (1998) 3.
24   E. Machtejevas, O. Kornyšova, V. Kudirkaite, S. Hjertén, A. Maruška, Chemine Technologija 19 (2001) 56.

25   L.R. Snyder, J.J. Kirkland, Introduction to Modern Liquid Chromatography, Wiley, New York, USA, 1979, p. 36.

26   S. Hjertén, Ind. Eng. Chem. Res. 38 (1999) 1205.

27   A. Palm, M.V. Novotny, Anal. Chem. 69 (1997) 4499.

28   R.E. Majors, LC-GC, 4 (1991) 686.

29   J.-L. Liao, Y.-M. Li, S. Hjertén, Anal. Biochem. 234 (1996) 27.

30   J.-L. Liao, N. Chen, C. Ericson, S. Hjertén, Anal. Chem. 68 (1996) 3468.

31   O. Kornyšova, S. Hjertén, A. Maruška, E. Markauskaite in: A. Šulčius (Ed.), Proceedings of the Conference on Inorganic Chemistry and Technology, Kaunas, Lithuania, 2000, p. 14.

32   C. Ericson, S. Hjertén, Anal. Chem. 71 (1999) 1621.

33   D. Hoegger, R. Freitag, J. Chromatogr. A 914 (2001) 211.

34   A. Maruška, C. Ericson, Á. Végvári, S. Hjertén, J. Chromatogr. A 837 (1999) 25.

35   M.G. Schmid, N. Grobuschek, C. Tuscher, G. Gübitz, Á. Végvári, E. Machtejevas, A. Maruška, S. Hjertén, Electrophoresis 21 (2000) 3141.

36   M.G. Schmid, N. Grobuschek, O. Lecnik, G. Gübitz, Á. Végvári, S. Hjertén, Electrophoresis 22 (2001) 2616.

37   O. Kornyšova, P.K. Owens, A. Maruška, Electrophoresis, 22 (2001) 3335.

38   E. Machtejevas, A. Maruška, Anal. Chem. 2002, submitted.

39   O. Kornyšova, E. Machtejevas, V. Kudirkaite, U. Pyell, A. Maruška, J. Biochem. Biophys. Meth. 50 (2002) 217.

40   A. Harada, Acta Polymer 49 (1998) 3.

41   H.W. Gibson, S. Liu, P. Lecavalier, C. Wu, Y.X. Shen, J. Am. Chem. Soc. 117 (1995) 852.

42   D. Callahan, H.L. Frisch, D. Klempner, Polym. Eng. Sci., 15 (1975) 70.

43   V. Kudirkaite, S. Butkute, E. Machtejevas, O. Kornyšova, D. Westerlund, A. Maruška, Chemine Technologija 19 (2001) 65.

44   K.A. Ramsteiner, J. Chromatogr. 456 (1988) 3.

45   D.J. Anderson, Anal. Chem. 65 (1993) 434R.

46   Ch. Fujimoto, Y. Fujise, E. Matsuzawa, Anal. Chem. 68 (1996) 2753.

47   Ch. Fujimoto, Anal. Chem. 67 (1995) 2050.

48   Ch. Fujimoto, J. Kino, H. Sawada, J. Chromatogr. A 716 (1995) 107.

49   S. Hjertén, Á. Végvári, T. Srichaiyo, H.-X, Zhang, C. Ericson, D. Eaker, J. Cap. Elec. 1/2 (1998) 13.

50   S. Hjertén, J. Chromatogr. 347 (1985) 195.

51   V. Kudirkaite, Doctoral Dissertation: Synthesis and Characterization of Polymer-based Adsorbents for High Performance Liquid Chromatography and Capillary Format Separations, Technologija, Kaunas, Lithuania, 2001.

52   A. Maruška, O. Kornyšova, S. Hjertén in: V. Snitka (Ed.), Proceedings of NEXUS PanEuropean Network of Excellence in Multifunctional Microsystems Microsystems Technology Activities in Baltic Region, Kaunas, Lithuania, 1999, p. 96.

53   O. Kornyšova, V. Snitka, V. Kudirkaite, E. Machtejevas, A. Maruška, Investigation of the Morphology of the Continuous Beds, Presented at Analysdagarna, Uppsala, Sweden June 14-17, 1999, p.246.

F. Švec, T.B. Tennikova and Z. Deyl (Editors)
Monolithic Materials
Journal of Chromatography Library, Vol. 67
2003 Published by Elsevier Science B.V.

Chapter 8

Monolithic Silica Columns for Capillary Liquid Chromatography

Nobuo TANAKA, Masanori MOTOKAWA, Hiroshi KOBAYASHI, Ken HOSOYA
and Tohru IKEGAMI

*Department of Polymer Science and Engineering, Kyoto Institute of Technology,
Matsugasaki, Sakyo-ku, Kyoto 606-8585, Japan*

CONTENTS

8.1 INTRODUCTION

One can obtain higher efficiency in a shorter separation time in HPLC by reducing the size of column packing materials. This approach has been taken for decades, and resulted in current columns packed with small particles providing typically 10,000–25,000 theoretical plates accompanied by a considerable pressure drop. The low permeability associated with small-sized packing materials is due to the small-sized interstitial voids in a column. High-speed operation of such a column is limited by the available pressure with current instrumentation of up to 350–400 kg/cm^2. A compromise has been necessary between the required column efficiency and the pressure drop in the column by adjusting the column length. It is desirable to obtain as high column efficiency in LC as in CE or GC, because HPLC can provide for the separations of a wider range of substances. High performance with a particle-packed column beyond the limit of common HPLC has been achieved recently by ultrahigh-pressure liquid chromatography (UHPLC) utilizing higher operational pressure, by capillary electrochromatography (CEC) utilizing electroosmotic flow, or by HPLC employing a monolithic column having a network structure instead of a particle-packed column.

8.2 THEORETICAL CONSIDERATIONS

Limit in performance of HPLC columns (and the reasons for the compromise) can be described as follows. Equations 8.1–8.9 define or express the properties of packing materials and columns [1-3] including the efficiency and flow resistance:

$$H = \sigma^2 / L = L / N \tag{8.1}$$

$$H = 1 / [(1 / C_e d_p) + (D_m / C_m d_p^2 u)] + C_d D_m / u + C_{sm} d_p^2 u / D_m$$
$$= A u^{1/3} + B / u + C u \tag{8.2}$$

$$h = A v^{1/3} + B / v + C v \tag{8.3}$$

$$h = H / d_p \tag{8.4}$$

$$v = u d_p / D_m = (u / d_p) / (D_m / d_p^2) \tag{8.5}$$

$$\varphi = \Delta P t_0 d_p^2 / \eta L^2 \tag{8.6}$$

$$K = u \eta L / \Delta P , u = L / t_0 \tag{8.7}$$

$$E = \Delta P t_0 / \eta N^2 = (\Delta P / N) (t_0 / N) (1 / \eta) \tag{8.8}$$

$$t_0 = h^2 \varphi N^2 \eta / \Delta P \tag{8.9}$$

where N stands for the number of theoretical plates, L is the column length, H is the height equivalent of a theoretical plate, σ^2 is the dispersion of a solute band, d_p is the particle size, u is the linear velocity of a mobile phase, D_m is the diffusion coefficient of a solute in the mobile phase, C_x is the coefficient for the contribution of each term, h is the reduced plate height, v is the reduced velocity, φ is the flow resistance parameter, ΔP is the column pressure drop, t_0 is the column dead time, η is the solvent viscosity, K is the column permeability, and E is the separation impedance.

Reduction of particle size can lead to higher column efficiency on the basis of the smaller contribution of eddy diffusion and mobile-phase mass transfer term (A-term) and the shorter diffusion path length (C-term) as indicated by eq. 8.2. If comparison is made at constant efficiency, one can achieve the reduction in t_0 with smaller d_p, while the pressure drop is inversely proportional to d_p^2 (eq. 8.6). This is why small particles in a range of 1.5–3 μm have been used in a short column for fast separations. There is a limit in overall performance ($N^2 / t_0 \Delta P$, cf. eq. 8.8) for a column packed with particles of certain size. The concept is known as Knox-Saleem optimum described by the basic chromatographic parameters, t_0, N, and ΔP in eq. 8.9 [2].

Generation of a higher N and a shorter separation time in LC has been achieved by using UHPLC [4,5] or CEC [6]. These approaches overcome high-pressure drop associated with a column having a tightly packed bed. Effort is being made now to solve practical problems of these techniques.

An increase in column performance can also be achieved by increasing column permeability (eq. 8.7). One possible way to attain higher permeability is to use a monolithic column, which is defined as a column consisting of single piece of solid that possesses interconnected skeletons and interconnected flow paths (through-pores) through these skeletons. A monolithic column can be made to have small-sized skeletons and large through-pores. In contrast to particle-packed columns, this structure can simultaneously reduce both flow resistance for a bed of similar skeleton-size and diffusion path length in the skeleton. Thus a support structure having a large (through-pore-size)/(skeleton-size) ratio can provide both high permeability and high column efficiency in HPLC. This is not possible with a particle-packed column with a (through-pore (interstitial void)-size)/(particle-size) ratio, commonly being in a range of 0.25–0.4 [7]. Of course, if the through-pores are too large, one cannot expect high column efficiency, due to a significant A-term contribution, i. e. the contribution of slow mass transfer in a mobile phase, as indicated by eq. 8.2 [8].

It has been reported that monolithic columns can be prepared by sintering or by embedding silica particles, after packing the particles into a capillary [9-13]. In principle, however, such columns prepared by the modification of a packed bed will not be able to escape from the limitations associated with a particle-packed column, unless higher permeability is provided using loosely packed particle beds. In CEC, lower performance was reported for these columns compared to particle-packed columns [14]. Recently, Minakuchi et al. and others reported that continuous porous-silica columns that feature high (through-pore-size)/ (skeleton-size) ratios can be prepared by a sol-gel method starting from alkoxysilanes in the presence of water-soluble organic polymers. As a result of the controllable sizes of silica skeletons and through-pores, the monolithic columns provided better overall performance, typically similar or higher column efficiency at a much lower pressure drop than conventional columns packed with particles [8,14-24].

8.3 PREPARATION OF MONOLITHIC SILICA COLUMNS

Monolithic silica columns can be prepared either in a mold such as a 6-9 mm i.d. glass test tube or in a fused silica capillary [8,14,16,20,21]. Nakanishi et al. reported the preparation of silica gel monoliths with a broad range of through-pores as well as with mesopores in a 10 nm range [25,26]. The preparation in a mold is accompanied by volume reduction of the whole silica structure. The diameters of products are ca. 4.6 or 7 mm when a mold of with 6 and 9 mm i.d., respectively, is used. The resulting silica monoliths must be then covered with PTFE tubing or with PEEK resin to fabricate a column for HPLC. Column length cannot be longer than 15 cm, since long, straight monoliths cannot be prepared. A PTFE-covered monolith (MS-PTFE) was used in an external pressurizing device, a Z-module (Waters). PEEK-covered columns (MS-PEEK) are commercially available, and can withstand inlet pressure of up to 12 MPa [20]. However, the fabrication of PEEK-covered columns is very difficult.

Compared to the preparation in a mold and subsequent column fabrication, the preparation of monolithic silica column in a fused silica capillary (MS-FS) is much easier, although the silica network structure has to be attached to the tube wall in order to prevent shrinkage of the skeletons. This method allows anyone to prepare a stable, high efficiency column. It used to be virtually impossible, if one started the preparation of beads from monomers, and then packed a column without considerable instrumentation.

Typical preparation conditions for MS-FS columns are as follows [8,27]: Tetramethoxysilane (TMOS, 4 mL) is added to a solution of poly(ethylene glycol) (PEG, 0.88 g, M_w=10,000) and urea (0.90 g) in 0.01 M acetic acid (10 mL) and stirred at 0°C for 45 min. Resulting homogeneous solution is charged into a fused–silica capillary tube, which has been treated with 1 mol/L NaOH solution at 40°C for 3 h in advance, and allowed to react at 40°C. Gelation occurs within 2 h and the gel is subsequently aged overnight at the same temperature. Then the monolithic silica column is treated at higher temperatures, for 3 h at 120°C to complete mesopore formation with ammonia generated by the hydrolysis of urea [26], followed by washing with water and methanol. After drying, heat-treatment is carried out at 330°C for 25 h, resulting in the decomposition of organic moieties in the capillary. After the preparation, each end (10–15 cm) of the capillary having large voids is cut off. A detection window (2 mm) is made by removing polyimide coating at a specified distance from capillary inlet to allow on-column detection through the silica monolith.

Surface modification of the monolithic silica is carried out on-column by continuous feeding of the solution of octadecyldimethyl-*N*,*N*-diethylaminosilane (2 mL) in 8 mL of toluene at 60°C for 3 h.

These monolithic columns showed reproducibility of about ±15% in terms of pressure drop and column efficiency. The columns can be used without retaining frits in CEC and in HPLC. A hybrid-type monolithic silica can be prepared similarly from a mixture of TMOS and MTMS [27]. The preparation is currently limited to a capillary of 250 μm or smaller diameter. Thus chromatographic evaluation or the use of the MS-FS columns must be carried out in a constant-pressure mode with a split flow after sample injection and on-column detection to avoid excessive band broadening [8,21,27], while it is possible to use an injector of 10–50 nL injection volume without flow splitting.

8.4 STRUCTURAL PROPERTIES OF MONOLITHIC SILICA COLUMNS

By controlling the composition of the starting mixture, or the concentrations of alkoxysilanes and PEG, one can control the size of silica skeletons and through-pores to produce monolithic silica columns having (through-pore-size)/(skeleton-size) ratios much greater than those found in a particle-packed column [14]. SEM photographs of the monolithic silica columns shown in Figures 8.1 and 8.2 reveal that both silica skeletons and through-pores are co-continuous and the domain size (a combined size of skeleton and through-pore, or a unit size of network structure after phase separation in the preparation mixture) decreases with the increase in PEG content in the starting mixture.

Figure 8.1 shows the SEM photographs of the fractured surfaces of the monolithic silica columns. The preparation of silica monolith from TMOS in a larger-size capillary was accompanied by the shrinkage of silica skeletons, leaving large voids along the wall of the 250 μm diameter capillary (Figure 8.1a). The attachment of the silica skeletons to the tube wall of a small-diameter capillary prevented the shrinkage of the whole network structure, while resulting in the large interstitial voids (through-pores). Figure 8.1(c-f) show the SEM photographs of MS(50)-A, -B, -C, and –D with domain sizes decreasing in this order (Table 8.1).

Figure 8.2 shows SEM photographs of monolithic silica columns prepared from a mixture of TMOS and MTMS. Figure 8.2a and 8.2b show the hybrid-type monolithic silica with different domain sizes in a 50 μm capillary, while Figure 8.2c and 8.2d show those prepared in a 100 and 200 μm capillaries, respectively. Monolithic silica columns were successfully prepared in the larger-sized capillary starting from a mixture of MTMS and TMOS, and they will be easier to operate.

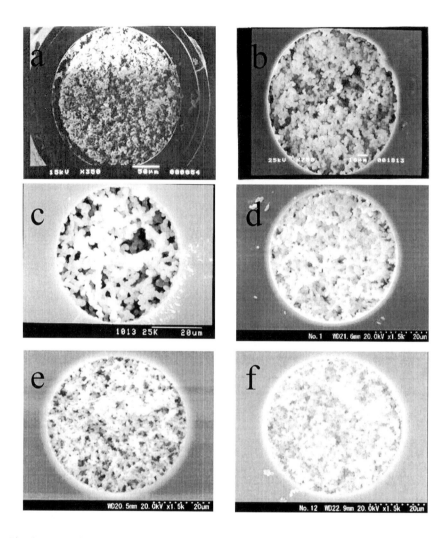

Fig. 8.1. Scanning electron micrographs of monolithic silica prepared from TMOS in fused silica capillaries of (a) 250 μm, (b) 100 μm, (c-f) 50 μm diameter that represent columns MS(50)-A, -B, -C, and -D shown in Table 8.1 (Reprinted with permission from refs. [8,27], copyright 2002 Elsevier).

It is also possible to prepare monolithic silica columns having various domain sizes between ca. 10 to 3 μm. In the preparation of MS(50)-A, B, C, and D from TMOS, the amount of PEG and reaction temperature were varied to reduce the domain size. With the increase in PEG content in the reaction mixture, silica monolith with a smaller domain size was formed as shown in Table 8.1 and Fig. 8.1. Similar

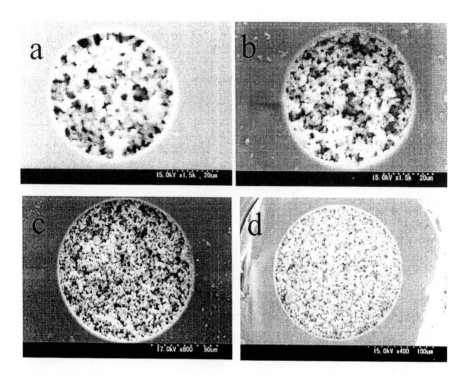

Fig. 8.2. Monolithic silica columns prepared from a mixture of MTMS and TMOS, (a) MS-H(50)-I and (b) MS-H(50)-II in a 50 μm capillary, (c) MS-H(100)-II in a 100 μm capillary, and (d) MS-H(200)-II in a 200 μm capillary (Reprinted with permission from ref. [27], copyright 2002 Elsevier).

results were obtained in the preparation from a mixture of TMOS and MTMS (Fig. 8.2). The relation between preparation conditions and resulting gel morphologies was recently described in detail [14].

Figure 8.3 shows the relation between the through-pore size and the skeleton size. The size of through-pores of monolithic silica in capillary was much larger than the skeleton size resulting in the (through-pore-size)/(skeleton-size) ratio of up to 2–4 while these ratios were reported to be only 0.25–0.4 for a column packed with particles [7], and found to be 1.2–1.5 for monolithic silica columns prepared in a mold enabling shrinkage of the whole network structures [17].

Table 8.2 shows that the porosity of a monolithic silica column is much greater than that of a column packed with particles. Major difference was seen in external porosities determined by size exclusion chromatography (SEC): 39 % for a particle-packed column, 60–70 % for MS-PTFE and MS-PEEK prepared in a mold, and even

TABLE 8.1

COMPOSITION OF THE PREPARATION MIXTURES AND PROPERTIES OF THE
RESULTING MONOLITHIC SILICA COLUMNS IN CAPILLARIES [8,27,28]

Column	PEG, g [a,b]	Urea, g	Skeleton size, μm [c]	Through-pore size, μm [c]
MS(50)-A	8.8	9.0	2.0	8.0
MS(50)-B	12.4	9.0	1.4	2.8
MS(50)-C	12.6	9.0	1.1	2.2
MS(50)-D	12.8	9.0	1.0	2.0
MS-H(50)-I	1.00	2.03	2.0	4.5
MS-H(50)-II	1.05	2.03	1.5	2.0

[a] Urea (9 g) and the specified amount of PEG added to 40 mL of TMOS and 100 ml of 0.01 mol/L aqueous acetic acid for MS(50)-A – D (27)

[b] Urea (2.03 g) and the specified amount of PEG added to 9 mL mixture of MTMS and TMOS at a 1/3 ratio and 20 ml of 0.01 mol/L acetic acid

[c] Measured from SEM photographs

higher for MS-FS prepared in a capillary. With the small amount of silica skeleton in MS-FS, total porosity exceeded 90 % of the column volume. The phase ratio in a column having C_{18} stationary phase decreased accordingly to less than 0.05 for MS-FS, compared to ca. 0.06 for MS-PTFE, and 0.19 for a column packed with the C_{18}-silica particles, as estimated from the porosity measurement before and after the bonding reaction as well as the comparison of retention factors (k values) [21]. The monolithic silica columns in capillary, however, possess much greater phase ratios than an open tube column or a column packed with 2 μm nonporous particles, affording higher sample loading capacity.

8.5 PERMEABILITY OF MONOLITHIC SILICA COLUMNS

The beneficial chromatographic features of monolithic silica columns arise from the large (through-pore-size)/(skeleton-size) ratios and high porosities, resulting in high permeability and a large number of theoretical plates per unit pressure drop. The pressure drop of MS-PTFE monoliths prepared in a mold accompanied by shrinkage was lower by a factor of 2–9 than that of a column packed with 5 μm ODS particles. MS-PEEK monolith also exhibited lower pressure drop compared to a column packed

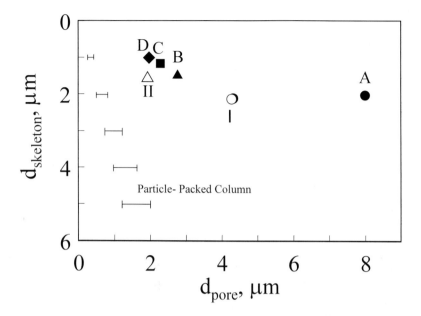

Fig. 8.3. Plots of the through-pore size against the skeleton size of monolithic silica prepared in a capillary. Columns: MS(50)-A (●), MS(50)-B (▲), MS(50)-C (■), MS(50)-D (□), MS-H(50)-I (O), MS-H(50)-II (△). Bars are shown to indicate the size of interstitial voids at 25-40 % of particle size in a column packed with particles.

with 5 μm particles. Typical column permeability or K values are $4 \cdot 10^{-13}$, $7 \cdot 10^{-14}$, and $4 \cdot 10^{-14}$ m^{-2} for MS-PTFE, MS-PEEK, and a column packed with 5 μm particles, respectively [20].

The permeability of a monolithic silica column in capillaries could be even higher than those of monolithic silica columns prepared in a mold. Fig. 8.4 compares the pressure drop of columns in both 80% aqueous methanol and acetonitrile. The pressure drop of MS(50)-A was much lower than that of MS-PTFE. The smaller flow resistance of the former monolith is due to the presence of much larger through-pores in an order of 8–10 μm. As shown in Table 8.3, the permeability of MS(50)-A, -B, -C and -D was $8 \cdot 10^{-14}$–$1.3 \cdot 10^{-12}$ m^2, that is about 2–30 times higher than that for a column packed with 5 μm particles, reflecting the high porosity and the large (through-pore-size)/(skeleton-size) ratios. Lower permeability was observed for MS-FS with smaller domain size just like for a column packed with smaller silica particles. The MS-FS column having the smallest domain size, or the skeleton size of 1 μm still exhibited higher permeability than a column packed with 5 μm particles. MS-H(50) columns prepared from a mixture of MTMS and TMOS showed slightly lower permeability than MS(50) columns at similar skeleton size. While skeleton sizes are

TABLE 8.2

POROSITIES OF COLUMNS

Column type	Particle-packed [b]	Monolithic silica column		
		MS-PTFE	MS-PEEK	MS(50)-A
Diameter, mm	4.6	7	4.6	0.05
Length, cm	15	8.3	10	25 [c]
Total porosity [a]	0.78	0. 86	(0.87)	0.92
Through-pore [a,d]	0.39	0.62	(0.69)	0.83
Mesopore [a]	0.40	0.24	(0.18)	0.09
Reference	17	17	20	28

[a] Porosity of monolithic silica column measured by size exclusion chromatography in THF using polystyrene standards and alkylbenzenes. The porosities in parentheses were obtained with C_{18}-bonded phase

[b] Develosil-C_{18} particles packed in a 4.6 mm i.d. and 10 cm long column

[c] Effective length between the inlet and the detection window. Total length 33.5 cm

[d] External porosity

uniform, the pore sizes representing inter-skeleton voids vary significantly, due to the agglomeration of skeletons.

8.6 CHROMATOGRAPHIC PERFORMANCE OF MONOLITHIC SILICA COLUMNS

MS-PTFE and MS-PEEK columns with a relatively small domain size of 2.5–3 μm afforded higher column efficiency at lower pressure drop than a column packed with 5 μm particles [20,29], especially at high linear velocities because of the smaller skeleton size. As in the case of a particle-packed column, higher column efficiency was obtained with a monolithic silica column having smaller domain size that resulted in higher pressure drop. Four monolithic silica columns (MS(50)-A, -B, -C, and -D in Table 8.1; 45 cm long) prepared from TMOS produced 36,000-57,000 theoretical plates for alkylbenzenes in 80 % acetonitrile at a linear velocity of ca. 1 mm/s, as shown in Figure 8.5. The efficiencies are similar to or greater than the efficiency of a 25 cm column packed with 3–5 μm silica-C_{18} particles, at much lower pressure drop.

A column of up to 150 cm long can be prepared in a capillary. This column generated a large number of theoretical plates utilizing high permeability. Figure 8.5e

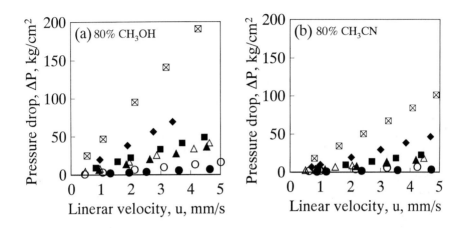

Fig. 8.4. Plots of column back pressure against linear velocity of mobile phase (Reprinted with permission from ref. [27], copyright 2002 Elsevier). Mobile phase: (a) 80% methanol, (b) 80% acetonitrile. The pressures were normalized to a column length of 15 cm. Columns: 5 μm silica-C_{18} particles, Mightysil RP18 (⊠). Monolithic silica columns in capillaries, MS(50)-A (●), MS(50)-B (▲), MS(50)-C (■), MS(50)-D (◆), MS-H(50)-I (○), MS-H(50)-II (△).

TABLE 8.3

PORE PROPERTIES AND PERMEABILITY OF COLUMNS

Column	Skeleton size, μm	Through-pore size, μm	Permeability, $K \times 10^{-14} \, m^2$
MS(50)-A	2.0	8.0	130
MS(50)-B	1.4	2.8	25
MS(50)-C	1.1	2.2	15
MS(50)-D	1.0	2.0	8
MS-H(50)-I	2.0	4.5	56
MS-H(50)-II	1.5	2.0	19
Mightysil	5.0	$(1.3-2.0)^a$	4

[a] Estimate at 25–40 % of particle size, 5 μm

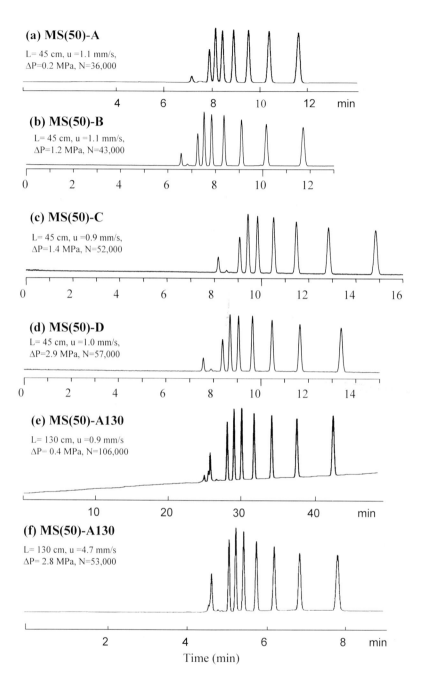

Fig. 8.5. Chromatograms of alkylbenzenes ($C_6H_5(CH_2)_nH$, n=0-6). Columns: (a) MS(50)-A, (b) MS(50)-B, (c) MS(50)-C, (d) MS(50)-D. Column size: 50 μm i.d. × 53.5 cm (effective length 45 cm). (e, f) MS(50)-A130, 50 μm i.d. × 138.5 cm (effective length 130 cm). Mobile phase: 80% acetonitrile in water. The linear velocity and the pressure drop are indicated.

shows a chromatogram of alkylbenzenes in 80 % acetonitrile at a linear velocity of ca. 1 mm/s obtained with a 130 cm long monolithic silica column, MS(50)-A130, prepared under the same conditions as MS(50)-A. Figure 8.5e shows that this 130 cm long column produced 106,000 theoretical plates for hexylbenzene at 0.9 mm/s linear velocity with a pressure drop of 0.37 MPa. At a higher linear velocity (4.7 mm/s), MS(50)-A130 column produced 50,000–60,000 theoretical plates at 2.8 MPa with t_0 of 4.5 min (Fig. 8.5f). Figure 8.5d and 8.5f show that MS-FS with large through-pores is advantageous for generating a large number of theoretical plates at similar pressure drop. Figure 8.5f suggests that it is possible to generate 100,000 theoretical plates with t_0 of 9 min and at ca. 6 MPa under similar conditions by using a 250 cm long column. Such a column would generate nearly 200,000 theoretical plates at a lower linear velocity.

Figure 8.6 shows chromatograms obtained with hybrid-type columns (MS-H) prepared from a mixture of TMOS and MTMS. MS-H columns were successfully prepared in a 50–200 μm capillary, to produce 20,000–30,000 theoretical plates/25 cm long column. MS-H(50)-II having both smaller through-pore and skeleton size afforded a higher number of theoretical plates and a higher pressure drop than MS-H(50)-I, while MS-H(200)-II prepared from the same mixture as MS-H(50)-II gave similar chromatogram. A 200 μm capillary column will be easier to work with than a 50 μm column, and MS-H columns will find practical utility with capillary HPLC instrumentation. A 50 cm long MS-H(100)-II column provided complete separation of 16 polynuclear aromatic hydrocarbons, priority pollutants designated by EPA, with ca. 50,000 theoretical plates in 30 min in 75% acetonitrile [28].

The retention factors of alkylbenzenes on MS-H columns in Fig. 8.6 are very similar to each other and slightly larger than those on MS(50) columns prepared from TMOS. This is presumably due to the higher phase ratio of MS-H columns based on the lower porosity and possibly due to the presence of methyl groups as well as C_{18} alkyl chains after chemical bonding. The separation factor for hexylbenzene and amylbenzene, $\alpha(CH_2)$, was 1.50–1.51 on MS-H and MS(50) columns, while these columns exhibited slightly different selectivity for planar and nonplanar compounds. MS(50) columns showed preferential retention of planar compounds compared to MS-H columns [28]. The reproducibility of preparation in terms of k values was about ±5% in each case.

8.7 KINETIC CHARACTERIZATION OF MONOLITHIC SILICA COLUMNS

MS-PEEK and MS-PTFE columns provided for lower plate height, specifically at high linear velocities compared to a column packed with 5 μm particles [20,29]. The

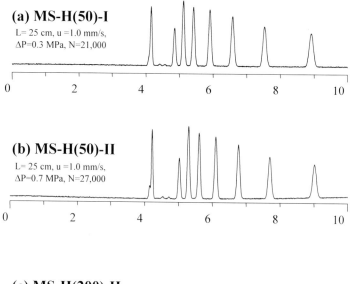

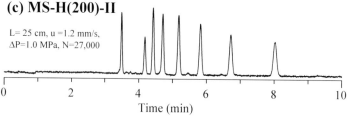

Fig. 8.6. Chromatograms of alkylbenzenes (C6H5(CH2)nH, n=0-6). Columns: (a) MS-H(50)-I 5 μm i.d. × 33.5 cm (effective length 25 cm), (b) MS-H(50)-II: 50 μm i.d. × 33.5 cm (effective length 25 cm), (c) MS-H(200)-II: 200 μm i.d. × 33.5 cm (effective length 25 cm). Mobile phase: 80 % acetonitrile in water.

small-sized silica skeletons resulted in a small increase in plate height accompanying an increase in linear velocity. It should be noted that these monolithic silica columns had lower pressure drop than the particle-packed column.

Figure 8.7 shows the van Deemter plots obtained for an alkylbenzene as a solute in both 80 % methanol and acetonitrile. The reproducibility of preparation in terms of plate height and pressure drop was within ±15% in most cases. The MS(50) and MS-H columns of smaller domain size exhibited a plate height minimum, while those with larger domain size did not run through a minimum at a linear velocity exceeding 0.5 mm/s. Figure 8.7 indicates that the MS(50) and MS-H columns have lower column efficiencies than a monolithic silica column of a similar skeleton size (1–2 μm) prepared in a mold. The plate height values obtained with these monolithic silica

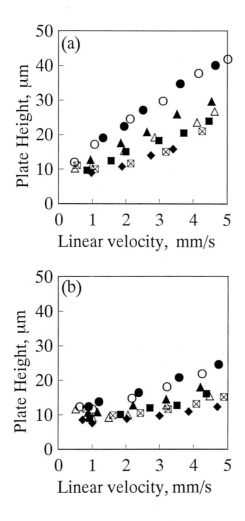

Fig. 8.7. Van Deemter plots obtained for C_{18} monolithic silica columns using hexylbenzene as a solute (Reprinted with permission from ref. [27], copyright 2002 Elsevier). Mobile phase: (a) 80 % methanol in water, (b) 80 % acetonitrile in water. For column symbols see Fig. 8.4.

columns were actually much larger than expected from the size of the silica skeletons that should result in small C-term in eq. 8.2. These results can be explained based on the contribution of the 2–8 μm large through-pores using the following equations.

$$H = A'u^n \tag{8.10}$$

$$H = 1 / [(1 / A_e) + (1 / A_m u)] + B / u + C u \qquad (8.11A)$$

$$H = 1/(1/A_e' + D_m / A_m' u) \qquad (8.11B)$$

When operated under CEC conditions in 80% acetonitrile–20% Tris-buffer, MS(50)-A column showed plate height of about 4 μm for hexylbenzene at 1 mm/s, and smaller plate height at higher linear velocities. Therefore B-term contribution in HPLC in 80% methanol that has higher viscosity than 80% acetonitrile should be much smaller at u=1.5–5 mm/s. C-term contribution to band broadening was also considered to be small for monolithic silica having 2 μm skeletons based on the results obtained with monolithic columns having a similar skeleton size prepared in a mold. Knox recently pointed out that A-term (eq. 8.2, 8.3) makes a major contribution to band broadening on a particle-packed column in HPLC, while C-term contribution is minor [31].

With small B-term and C-term contribution for MS(50)-A at a linear velocity range of 1.5–5 mm/s, eq. 8.10 characterizes the dependence of H on u. Knox approximated the dependence by eq. 8.10 with n = 0.3–0.4 for HPLC [31]. The plots in Figure 8.8a show a fair fit to eq. 8.10 at n = 0.5–0.6. The greater dependence of H on u for the MS-FS column suggests the higher contribution of mobile phase mass transfer (A_m term) to band-broadening relative to eddy diffusion (A_e term) of eq. 8.2.

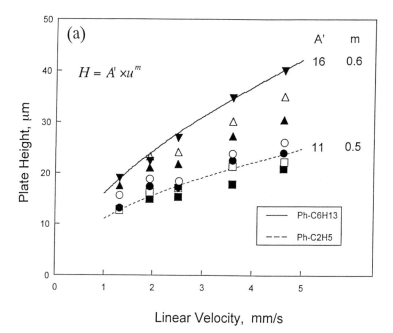

When the plots in Fig. 8.8 were fitted with eq. 8.11B (D_m was included to take into account the difference in molecular weight of the solute), a fair fit was observed with an A_e' value of 50–80 indicating that the eddy diffusion term is very large, and therefore the plate height is mainly controlled by the A_m' term. The results suggest the major contribution of A term to the plate height, especially of the A_m term that corresponds to the contribution of slow mobile phase mass transfer [8]. Large A_e' values suggest the irregularity of structure, or the presence of both large and small through-pores in a monolithic silica column, and/or the inefficient exchange of stream paths due to the characteristic structure of through-pores that are wide and relatively straight compared to interstitial voids of a particle-packed column. Dominating effect

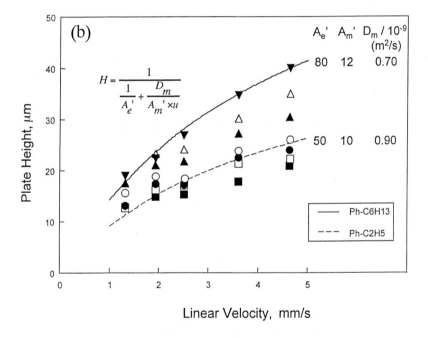

Fig. 8.8 (continued from previous page). (a) Use of the equation 10 for simulation of H against linear velocity plots in 80 % methanol–water for ethylbenzene and hexylbenzene as solutes (Reprinted with permission from ref. [8], copyright 2002 Elsevier). Column: MS(50)-A; Mobile phase: 80 % methanol in water; Samples: benzene (■), toluene (□), ethylbenzene (●), propylbenzene (○), butylbenzene (▲), amylbenzene (△), and hexylbenzene (▼). The curves were calculated for m=0.50 (ethylbenzene) and 0.60 (hexylbenzene) and the A' value was adjusted to fit the experimental points. (b) The simulation using equation 11B for H against linear velocity plots in 80 % methanol-water for ethylbenzene and hexylbenzene as solutes. Column: MS(50)-A; The curves were calculated for A_e'=50 (ethylbenzene) and 80 (hexylbenzene). The D_m values were calculated by using Wilkie-Chang equation [30]. The A_m' value was adjusted to fit the experimental points.

of slow mobile phase mass transfer is well known in open-tube liquid chromatography [2], where plate height shows nearly linear dependence on the linear velocity and strong dependence on retention factors. The latter is also seen in Fig. 8.8. Because the performance of the present monolithic silica in capillary seems to be dominated by A term in the velocity range studied, particularly large A_m resulting from the presence of the large through-pores together with a large A_e, much better performance can be expected in CEC, as shown below [32], and for monolithic silica columns having smaller through-pore size, as shown in Figs. 8.5-8.7. The adverse effect of the large through-pores, or significant band broadening caused by the slow mobile phase mass transfer in the presence of a large eddy diffusion term was minimized by the reduction in domain size of monolithic silica having large through-pores.

While the plate heights observed with MS(50)-D and MS-H(50)-II columns showing the highest efficiencies among the monolithic columns were comparable with that of a column packed with 5 μm particles, overall performance of the monolithic columns is much better than that of the particle-packed column. High performance of a MS-FS column is demonstrated through the smaller separation impedance or E value (eq. 8.8). The minimum E value obtained with MS(50)-A is close to 100, and that of MS-H(50)-I is about 300. These values are much smaller than minimum E values of 2000–3000, obtained with a column packed with 5 μm particles, as shown in Fig. 8.9. Generally, monolithic silica columns with the larger domain size provided for smaller E values, or better overall performance. In spite of the lower column efficiency observed with the monolithic silica column having large domain size shown in Figures 8.5-8.7, the generation of a very large number of theoretical plates will be easier with these columns by utilizing their high permeability. It is possible to generate 50,000 theoretical plates with a t_0 of less than 4 min at lower than 10 MPa with the monolithic columns by employing a longer column and a higher flow rate than indicated in Figs. 8.5 and 8.6. The monolithic silica columns with a smaller domain size will be useful for fast separations similar to a column packed with small particles.

Pressure-driven LC using long capillary-format monolithic columns should be examined further as a medium for high-efficiency separations, although CEC or UHPLC using a column packed with small silica particles may provide faster and more efficient separations. The monolithic columns can also be used in CEC or in UHPLC leading to even higher performance.

8.8 PROPERTIES OF MONOLITHIC SILICA COLUMNS IN CEC

Figure 8.10 shows the van Deemter plots obtained for an unmodified monolithic silica column (8 μm through-pores and 2 μm silica skeletons) by using thiourea as a solute, in 80% acetonitrile–20% 50 mmol/L TRIS-HCl buffer. Minimum plate height

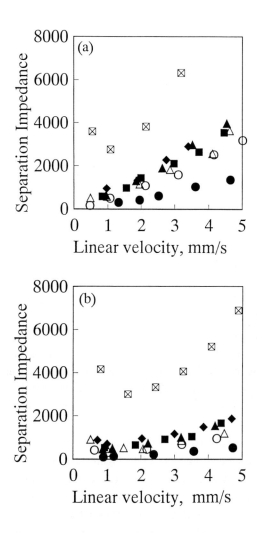

Fig. 8.9. Plots of separation impedance against linear velocity of mobile phase calculated for hexylbenzene as a solute. Mobile phase: (a) 80% methanol–water, (b) 80% acetonitrile–water. For column symbols see Fig. 8.4.

of about 3 μm was obtained for the unretained solute at $u = 1.5$ mm/s. Thus the capillary type monolithic column with 25 cm can potentially generate about 80,000 theoretical plates in CEC, although a 50 cm column produced only 100,000 plates under the electric field of 30 kV. This is due to the slow electroosmotic flow (EOF) in the monolithic silica column compared with a particle-packed column [32,33].

References pp. 195-196

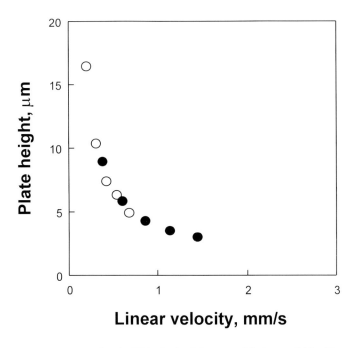

Fig. 8.10. Van Deemter plots in CEC obtained for unmodified monolithic silica columns with thiourea as a solute. Mobile phase: 80:20 acetonitrile–50 mmol Tris-HCl buffer pH 8; Column: MS(50)-A, effective length: (●) 25 cm, (O) 50 cm.

The use of low surface coverage C_{18} phase or a C_{18} phase prepared by using octadecyltrichlorosilane can to certain extent increase EOF. Figure 8.11 shows van Deemter plots under CEC conditions obtained with MS(50)-A-(I) having maximum surface coverage, MS(50)-A-(II) having very low surface coverage, and MS(50)-A-(T) having maximum surface coverage prepared by using octadecyltrichlorosilane. The three columns resulted in plots that seem to follow a common curve, although the retention factors k for these three columns were different: 0.7 for MS(50)-A-(I), 0.2 for MS(50)-A-(II), and 0.6 for MS(50)-A-(T). A plate height minimum of 4 μm was obtained, suggesting a potential column efficiency of about 60 000 plates for a 25 cm effective length. The EOF velocity, however, was not as high as expected from the availability of silanols, or the extent of silanol effect seen with a hydrogen-bonding solute in methanol-water mobile phase [34]. Disadvantages of using a low surface coverage material or polymeric C_{18} phase include the peak tailing for basic solutes and low chemical stability in a mobile phase with high pH values.

One possible way enabling fast CEC separations with monolithic silica columns is pressure-assisted operation [35]. Figure 8.12 shows chromatograms for the separation

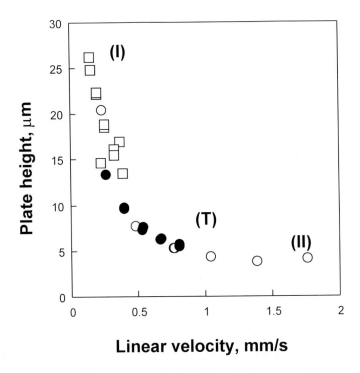

Fig. 8.11. Van Deemter plots obtained for C_{18} modified monolithic silica capillary columns in CEC mode with hexylbenzene as a solute. Mobile phase: 80:20 acetonitrile–50 mmol/L Tris-HCl buffer pH 8. ($\square$) MS-C_{18}(I), ($\bullet$) MS-C_{18}(T), and (O) MS-C_{18} (IV). Electric fields of up to 900 V/cm.

of alkyl phthalates in 80% acetonitrile–20% 5 mmol/L TRIS-HCl buffer. The chromatograms (a) and (b) in Fig. 8.12 show the separation of nine alkyl phthalates in 80% acetonitrile at 0.14 MPa (pressure-driven mode), the low-pressure mode of a Beckman P/ACE2200 instrument, and at 20 kV field strength (CEC mode), respectively. Figure 8.12(c) shows the chromatogram obtained by hybrid operation at 20 kV and 0.14 MPa. The linear velocity of the mobile phase in the hybrid operation was the sum of those achieved under the CEC conditions and the pressure-driven operation. The application of a pressure of 0.14 MPa shortened the separation time by 40% compared to operation in pure CEC mode. This is a unique advantage of the monolithic silica column prepared by a sol-gel method having a very high permeability of up to $1.2 \cdot 10^{-12} \, m^2$.

Plate heights were calculated to be 4.4 and 7.4 µm in the CEC mode, 7.5 and 13 µm in the pressure-assisted CEC mode, and 12 and 15 µm in the pressure-driven mode for dihexyl phthalate and di-2-ethylhexyl phthalate, respectively. Figure 8.12

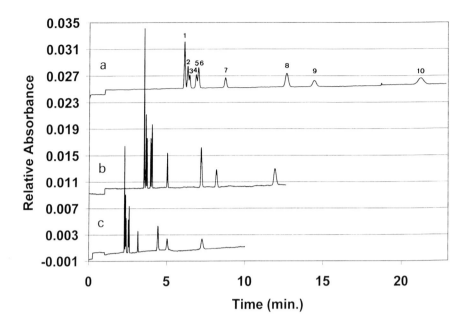

Fig. 8.12. Separation of alkyl phthalates on a monolithic silica column using pressure-driven flow at 0.14 MPa (a), electrically driven flow at 20 kV (b), and the hybrid operation using 20 kV and 0.14 MPa (c). Beckman P/ACE 2200 instrument; Temperature, 17°C; Mobile phase: 80:20 acetonitrile–50 mmol/L Tris-HCl buffer pH 8; Detection at 230 nm. Column: MS(50)-A-(III) 50 μm i.d., 30 cm effective length (37 cm total). Injection for 10 s by a nitrogen pressure of 3.4 kPa. Peaks: thiourea(1), dimethyl phthalate (2), diethyl phthalate (3), benzyl butyl phthalate (4), diisobutyl phthalate (5), dibutyl phthalate (6), dihexyl phthalate (7), di-2-ethylhexyl phthalate (8), dioctyl phthalate (9), dinonyl phthalate (10).

indicates that the column efficiency was highest in CEC, but pressure-assisted CEC also affords considerably higher column efficiency than pressure-driven separation in much shorter time. The high permeability is also helpful for rapid washing and equili-bration of columns under pressure. Thus, by utilizing the extremely high permeability of monolithic silica columns, it is possible to run pressure-assisted CEC at very low applied pressure. Since CEC can provide much higher efficiency by reducing the large A-term contribution associated with a monolithic silica column, and because the structure of monolithic silica columns without any retaining frits appears to be ideal for the use in CEC, it is desirable to prepare monolithic silica columns that can generate high EOF velocities for high efficiency- and high speed separations.

8.9 CONCLUSION

Monolithic silica columns prepared by a sol-gel method from TMOS or from a mixture of TMOS and MTMS, especially those prepared in a fused silica capillary, exhibit high permeability based on the large through-pores and small C-term contribution resulting from the small skeleton size. Both these features are favorable factors at high-speed separation compared to a column packed with particles. Therefore, fast separations can be achieved at high linear velocities, and also a large number of theoretical plates can be produced by using a long monolithic column using common HPLC equipment.

Reduction in through-pore size and skeleton size results in an increase in column efficiency. However, this is achieved at the expense of permeability. Hybrid-type silica monoliths prepared from a mixture of TMOS and MTMS in 50–200 μm capillaries will be practically useful in spite of the disadvantages including the individual preparation of each column and stationary phase, one at a time, and the limited capillary size. In a sense, the greatest merit of monolithic columns in the capillary format is that anyone can prepare a stable, highly efficient column without sophisticated instrumentation and examine various stationary phases or three-dimensional structures. Improved preparation method that enables the preparation in a larger-sized capillary, the increase in EOF, and the formation of a smaller domain size will further increase the practical utility.

8.10 REFERENCES

1   J.C. Giddings, Dynamics of chromatography, Part 1, Principles and Theory, Marcel Dekker, New York, 1965.
2   H. Poppe, J. Chromatogr. A, 778 (1997) 3.
3   P.A. Bristow, J.H. Knox, Chromatographia, 10 (1977) 279.
4   J.E. MacNair, K.C. Lewis and J.W. Jorgenson, Anal. Chem., 69 (1997) 983.
5   J.E. MacNair, K.D. Patel and J.W. Jorgenson, Anal. Chem., 71 (1999) 700.
6   M.M. Dittmann and G.P. Rozing, J. Chromatogr. A, 744 (1996) 63.
7   K.K. Unger, Porous Silica, Elsevier, Amsterdam, 1979, Ch. 5.
8   N. Ishizuka, H. Kobayashi, H. Minakuchi, K. Nakanishi, K. Hirao, K. Hosoya, T. Ikegami, N. Tanaka, J. Chromatogr. A, 960 (2002) 85.
9   M.T. Dulay, R.P. Kulkarni, R.N. Zare, Anal. Chem., 70 (1998) 5103.
10   C.K. Ratnayake, C.S. Oh, M.P. Henry, J. High Resol. Chromatgr., 23 (2000) 81.
11   R. Asiaie, X. Huang, D. Farnan, C. Horvath, J. Chromatogr. A, 806 (1998) 251.
12   Q. Tang, M.L. Lee, J. High Resol. Chromatogr., 23 (2000) 73.
13   T. Adam, K.K. Unger, M.M. Dittmann, G.P. Rozing, J. Chromatogr. A, 887 (2000) 327.
14   N. Tanaka, H. Kobayashi, K. Nakanishi, H. Minakuchi, N. Ishizuka, Anal. Chem., 73 (2001) 420A.

15   H. Minakuchi, K. Nakanishi, N. Soga, N. Ishizuka, N. Tanaka, Anal. Chem., 68 (1996) 3498

16   H. Minakuchi, K. Nakanishi, N. Soga, N. Ishizuka, N. Tanaka, J. Chromatogr. A, 762 (1997) 135.

17   H. Minakuchi, K. Nakanishi, N. Soga, N. Ishizuka, N. Tanaka, J. Chromatogr. A, 797 (1998) 121.

18   H. Minakuchi, N. Ishizuka, K. Nakanishi, N. Soga, N. Tanaka, J. Chromatogr. A, 828 (1998) 83.

19   K. Cabrera, G. Wieland, D. Lubda, K. Nakanishi, N. Soga, H. Minakuchi, K. K. Unger, Trends in Anal. Chem., 17 (1998) 50.

20   N. Tanaka, H. Nagayama, H. Kobayashi, T. Ikegumi, K. Hosoya, N. Ishizuka, H. Minakuchi, K. Nakanishi, K. Cabrera, D. Lubda, J. High Resol. Chromatogr., 23 (2000) 111.

21   N. Ishizuka, H. Minakuchi, K. Nakanishi, N. Soga, H. Nagayama, K. Hosoya, N. Tanaka, Anal. Chem., 72 (2000) 1275.

22   M.S. Fields, Anal. Chem., , 68 (1996) 2709.

23   C. Fujimoto, J. High Resol. Chromatogr., 23 (2000) 89.

24   J.D. Hayes, A. Malik, Anal. Chem., 72 (2000) 4090.

25   K. Nakanishi, J. Porous. Mater., 4 (1997) 67.

26   K. Nakanishi, H. Shikata, N. Ishizuka, N. Koheiya, N. Soga, J. High Resol. Chromatogr., 23 (2000) 106.

27   M. Motokawa, H. Kobayashi, N. Ishizuka, H. Minakuchi, K. Nakanishi, H. Jinnai, K. Hosoya, T. Ikegami, N. Tanaka, J. Chromatogr. A, 961 (2002) 53.

28   M. Motokawa, N. Tanaka, Unpublished results.

29   N. Tanaka, H. Kobayashi, N. Ishizuka, H. Minakuchi, K. Nakanishi, K. Hosoya, T. Ikegami, J. Chromatogr. A, In press.

30   C.R. Wilkie, P. Chang, Amer. Inst. Chem. Engr. J., 1 (1955) 264.

31   J.H. Knox, J. Chromatogr. A, 831 (1999) 3.

32   H. Kobayashi, C. Smith, K. Hosoya, T. Ikegami, N. Tanaka, Anal. Sci., 2002 (18) 89.

33   T.M. Zimina, R.M. Smith, P. Myers, J. Chromatogr., 758 (1997) 191.

34   K. Kimata, K. Iwaguchi, S. Onishi, K. Jinno, R. Eksteen, K. Hosoya, M. Araki, N. Tanaka, J. Chromatogr. Sci., 27 (1989) 721.

35   I.S. Krull, R.L. Stevenson, K. Mistry, M.E. Swartz, Capillary Electrochromatography and Pressurized Flow Capillary Electrochromatography, HNB Publishing, New York, 2000.

F. Švec, T.B. Tennikova and Z. Deyl (Editors)
Monolithic Materials
Journal of Chromatography Library, Vol. 67
© 2003 Elsevier Science B.V. All rights reserved.

Chapter 9

Monolithic Columns Prepared from Particles

Qinglin TANG[1] and Milton L. LEE[2]

[1] *Analytical Research and Development, Pharmaceutical Research Institute, Bristol-Myers-Squibb Company, Deepwater, NJ 08023*
[2] *Department of Chemistry and Biochemistry, Brigham Young University, Provo, UT 84602-5700*

CONTENTS

9.1 INTRODUCTION

Columns packed with particles, both porous and nonporous, have been widely used in high performance liquid chromatography (HPLC) for decades. Particulate packing materials have many advantages for HPLC applications. If porous, they have high surface area that leads to high sample capacity. Small particles generate high efficiency. Particles in large quantity can be manufactured reproducibly under strictly controlled conditions. Even if there are imperfections in the particles, the numerous trans-channel couplings in a packed column average laterally the impact of the imperfections by repeatedly mixing mobile phase streams at junctions between particles and spreading them across the column, which results in little fluctuation of column performance from batch-to-batch. Versatile particulate packing materials are commercially available for various applications.

Capillary electrochromatography (CEC) is the newest development in liquid chromatography [1-3]. Instead of using a high pressure pump, CEC employs high voltage across the column to drive the mobile phase through the column. Most publications about CEC describe the use of capillary columns of 50–200 μm i.d. packed with 1.5–10 μm. silica based particles. The main disadvantages of packed columns are the requirement for end-frits and the instability of the packed bed. End-frits, which are often difficult to make and which can deteriorate column performance, are needed to keep the particles in place in the column. Even if a column is tightly packed, voids are often observed in the column after use for a period of time.

Monolithic columns from particles are designed to inherit the versatility of well-developed particulate media but avoid the end-frits and instability of packed columns. The absence of end-frits reduces mobile phase flow resistance and allows easier interfacing to mass spectrometry. Also, the wall-supported monolith is expected to be stable and to suppress zone broadening caused by the so-called "wall effect". Various methods for preparing monoliths from particles have been reported. Generally, they can be classified into six types: particle-embedded, particle-sintered, particle-loaded, particle-entrapped, particle-bonded, and particle-crosslinked monoliths. The following sections describe in detail their preparation and properties.

9.2 PARTICLE-EMBEDDED MONOLITHIC COLUMNS

Tsuda *et al.* [4] first described a particle-embedded monolith in which silica gel was packed into a heavy-walled wide-bore glass tube that was then drawn into a capillary using a glass drawing machine. Under high temperature the glass tube melted and the particles were embedded in the wall, forming a particle-embedded monolith. Knox and Grant [5] reported a detailed procedure for fabricating particle-embedded capillary monoliths using particles with sizes of 3–50 μm. They first dried the silica gel overnight at 400°C and cooled it in a vacuum desiccator. A heavy-walled Pyrex glass tube of 1–2 mm i.d. sealed at one end with quartz wool was packed with the cooled silica gel by vibration and tapping while the tube was rotated. Approximately a 30-cm length of the tube was packed and the packing was topped with a glass wool plug. To remove water adsorbed during the packing process, the packed tube was evacuated using a vacuum pump while it was slowly passed through a furnace at 500°C. The tube was cooled under vacuum, positioned in the glass-drawing machine, and drawn to the required diameter. The drawing temperature was initially at

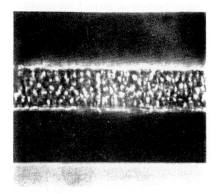

Fig. 9.1 Photograph of a particle-embedded monolithic capillary of 40 µm I.D. packed with 5 µm Hypersil silica gel (Reproduced with permission from ref. [5]. Copyright 1991 Friedr Vieweg & Sohn Verlagsgesellschaft mbH).

650°C, but ended at 560–600°C, depending on the final diameter of the capillary. After cooling to room temperature, a particle-embedded monolithic capillary as shown in Fig. 9.1 was obtained.

This particle-embedded monolithic capillary was stable under hydrodynamic flow if the capillary-to-particle diameter ratio was less than 10. It had a flow resistance factor as low as 120 compared to 500–1000 for slurry packed columns. An efficiency of 200,000 plates per meter in capillary electrochromatography (CEC) was obtained using 5-µm Hypersil silica-embedded in a 40 µm i.d. monolithic capillary derivatized with octadecyldimethylsilane (ODS). The monolithic capillary was transparent due to the thinness of the wall and, thus, could be used with fluorometric detection at excitation wavelengths higher than 240 nm.

There are several major problems associated with the preparation of particle-embedded monoliths. First, water must be evaporated during heating of the silica gel in the furnace of the drawing machine in order to obtain a good column. Second, a considerable lateral force is required to reorient the silica gel particles during drawing. Third, fused silica column preforms cannot be used because silica gel will fuse well below the melting point of the fused silica. Typically Pyrex glass results in fragile capillaries that are difficult to handle. Fourth, bonded-phase silica gels cannot be directly employed because the high drawing temperature can destroy the bonded phase. Any bonding of stationary phases must be carried out *in situ* after drawing, which can lead to irreproducibility in column-to-column performance. Fifth, particles are retained in the capillary because of the "keystone" effect from embedding the particles in the capillary inner wall. As a result, a hydrodynamically stable particle-embedded monolithic capillary with a capillary-to-particle diameter ratio greater than 10 cannot be prepared. Thus, the diameter of the particle-embedded monolithic column is limited.

9.3 PARTICLE-SINTERED MONOLITHIC COLUMNS

Asiaie *et al.* [6] described the preparation of particle-sintered monoliths in which the particles packed in a fused silica capillary were chemically sintered together and to the capillary inner wall. In this procedure, 6-μm ODS silica particles were packed into a fused silica capillary of 75 μm i.d. using a 50:50 v/v toluene–acetone slurry mixture. The packed capillary was rinsed with 0.1 mol/L NaHCO$_3$ for 5 min, followed by washing with water and acetone. The column, after purging with N$_2$ for at least 1 h, was heated at 120°C in a vacuum oven for 5 h to remove the physically adsorbed water, followed by 360°C in a GC oven for 10 h. NaHCO$_3$ at 360°C wetted and partially dissolved the surface of the ODS, forming a flux of low melting vitreous phase rich in sodium silicate that remained between the particles and between particle and capillary wall because of surface tension forces, forming a sintered monolith shown in Fig. 9.2. It was noticed that the reproducibility of preparation of sintered capillaries from ODS was higher than that from silica because of the adsorbed water, since silica has greater water-binding power than ODS. Water adsorbed to ODS was nearly eliminated during thermal treatment at 120°C, while water adsorbed to silica was just partially evaporated at that temperature. The residual water was released as a sudden burst of water vapor during the sintering step at 360°C, causing bed rupture and concomitant gapping. Due to the harsh experimental conditions (high temperature and basic NaHCO$_3$) in the sintering process, the stationary phase was partially destroyed and, thus, post deactivation and functionalization of the bed was necessary.

To avoid post-functionalization of the sintered monolith, Adam *et al.* [7] used water instead of NaHCO$_3$ and a movable heating coil to fabricate sintered monoliths. A 10% suspension of ODS in acetone was packed into fused silica tubing using a slurry packing method under a pressure of 60 MPa. The packed capillary was then filled with water. A heating coil driven by a processing motor moved along the length of the packed capillary with a constant velocity and at a constant temperature in the range of 300–400°C. As the heating coil moved along the packed capillary, the high temperature in the heated zone led to dissolution of silica to form polysilicic acids at the surface of the particles. After cooling, the solution in the interstitial voids became supersaturated with silica sol and re-deposition of silica occurred between particles and between particle and capillary wall, forming a monolith with particles retained in their original positions. The mild sintering conditions compared to those employed by Asiaie *et al.* maintained the integrity of the stationary phase and made post-function-

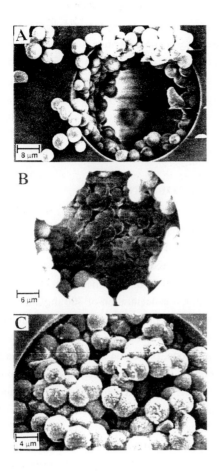

Fig. 9.2. Scanning electron micrographs of particle-sintered monolithic columns sintered at (a) 260°C, (b) 360°C without, and (c) 360°C with NaHCO3 flux (reproduced with permission from ref. [6]. Copyright 1998 Elsevier Science B.V.).

alization of the sintered monolith unnecessary. The quality of the sintered monolith was controlled by three operational parameters for a given particulate material: (i) the temperature of the heating coil, (ii) the speed by which the coil is moved along the packed column, and (iii) the number of cycles of this operation. Both too low temperature and number of cycles resulted in a monolith of low mechanical stability. Too high temperature or more cycles led to degradation of the surface of the particles and destruction of the bonded stationary phase. The best column efficiency for a column containing Hypersil-ODS1 was obtained after two cycles at a temperature of 300–400°C.

Particle-sintered capillary monoliths are mechanically strong and stable due to interparticle growth. They are also flexible because the polyimide coating on the fused silica capillary is not destroyed. The pore structure of the ODS in the sintered capillary remains essentially unchanged since irreversible destruction of the silica pore structures was observed only at temperatures above 600°C. An efficiency of 125,000 plates/m in CEC was obtained using a sintered 6 μm ODS column [6]. Under optimized operating conditions, no remarkable difference in retention factor, resolution, and efficiency were observed before and after sintering.

9.4 PARTICLE-LOADED MONOLITHIC COLUMNS

Dulay *et al.* [8] reported the preparation of particle-loaded continuous-bed columns, in which ODS was loaded in a sol-gel matrix that was subsequently covalently bonded to the capillary inner wall. In this procedure, the sol-gel solution was prepared by mixing 0.2 mL of tetraethoxylsilane (TEOS), 0.73 mL of ethanol, and 0.1 mL of 0.12 mol/L HCl. ODS particles were added at a concentration of 300 mg/mL to the sol-gel solution to create a suspension that was sonicated for several minutes and then introduced into 75 μm i.d. fused silica tubing of about 40 cm long by applying vacuum. The column was then coiled into a loop, placed on a hot plate, and heated above 100°C for about 24 h. During this period, TEOS hydrolyzed into tetrahydroxylsilane which then polycondensed into a silica sol-gel network filled with ethanol. Upon the evaporation of ethanol, the sol-gel network and the ODS particles were converted into a particle-loaded monolith shown in Fig. 9.3. As can be seen, the ODS particles were bonded evenly across the column and not merely cast along the capillary wall. Their loading into the sol-gel matrix was achieved either by physical trapping or by chemical bonding between the sol-gel matrix and the surface of the ODS through silanol groups. The sol-gel network was integrally fixed to the walls of the capillary through covalent bonding of silanol groups between the sol-gel matrix and the capillary inner wall. In contrast to a pure sol-gel monolith, the particle-loaded monolith did not display obvious shrinkage because the ODS particles helped to decrease the stress within the matrix. However, cracks around the ODS particles were noticed, possibly because of phase separation between the hydrophobic ODS surface and the hydrophilic sol-gel matrix.

Parameters controlling the success of the preparation were sol-gel matrix composition, ratio of particles to sol-gel, and suspension cast method. The sol-gel matrix solution was adjusted to a pH around 2, which enabled fast hydrolysis of TEOS, but slow polycondensation of silanol groups and, thus, slow solidification of the sol-gel matrix. The volumetric ratio of TEOS to solvent (ethanol) determined the porosity of the sol-gel matrix. The loading of ODS into the sol-gel matrix enhanced the mechani-

A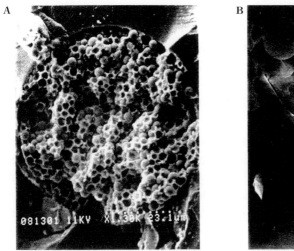

B

Fig. 9.3. Scanning electron micrographs of a particle-loaded monolithic column (reproduced with permission from ref. [8]. Copyright 1998 American Chemical Society).

cal strength of the monolith and provided the chromatographic selectivity. A good suspension of ODS particles in the sol-gel matrix solution is essential for obtaining a monolithic column with homogeneous distribution of ODS particles within the monolith. The optimum concentration was found to be 300 mg ODS per mL of the sol-gel solution. At lower concentration, such as less than 200 mg/mL the selectivity of the resulting column was poor. At higher concentration, e.g., 350 mg/mL, significant heterogeneity in the ODS packing density along the column led to numerous sections of the capillary column void of ODS particles. The suspension can be pushed or drawn into the tubing. The use of vacuum to fill the small bore capillary with the ODS/TEOS slurry suspension greatly improved the reproducibility of column preparation over pressurized syringe-filling methods.

Particle-loaded monolithic columns can sustain moderate pressures of up to 1.4 MPa. A 3 μm ODS-loaded monolithic column generated a specific permeability of 7.0 × 10^{-2} μm^2 [9,10] which is approximately 8 times greater than that of a tightly packed 3 μm ODS column. This high specific permeability is due to the low pressure method of preparation of the column in which 3 μm ODS particles are not as tightly packed into the tubing as they would be in a conventional packed LC column. Furthermore, large cracks around the ODS particles as shown in Fig. 9.3 and the highly porous sol-gel matrix allowed liquid to flow through the column at low pressure. An efficiency of 80,000 plates/m was obtained in CEC with a 3 μm ODS-loaded monolith. This moderate efficiency may be attributed to three factors. First, there may be inhomogeneous loading of the ODS column due to the different polarities and densities of the sol-gel solution and ODS particles. Second, some of the ODS particles can be too deeply embedded in the sol-gel matrix to play a role in the separation mechanism. Finally, cracks are often formed around the ODS particles.

9.5 PARTICLE-ENTRAPPED MONOLITHIC COLUMNS

Chirica and Remcho [11-13] published a method for preparing monolithic columns in which particles are entrapped in a monolithic structure by an entrapping matrix. Their procedure involved four steps: (i) the fused silica capillary was packed with ODS particles using a conventional solvent slurry packing method, and then dried by purging the column with nitrogen; (ii) the dried packed capillary was filled with entrapping matrix solution; (iii) the column was cured using different methods depending on the entrapping matrix; and (iv) the cured column was dried slowly at a specified temperature. Upon drying, a monolithic column with entrapped particles shown in Fig. 9.4 resulted [13].

The properties of particle-entrapped monoliths were determined by the homogeneity of the packed bed and the characteristics of the entrapment matrix, including porosity and surface chemistry. Because of the use of high pressure during column packing, the packed bed was tight and relatively uniform throughout the entire column. This ensured homogeneity throughout the column. Different entrapping matrices, such as silicate (Kasil 2130) [11], *tert*-butyl-triethoxylsilane [12], *n*-octyl-triethoxysilane [12], and methacrylate-based organic polymers [13] were investigated. Using various entrapping matrices, different packing materials can be entrapped in the monolith. All of the resultant particle-entrapped columns gave similar efficiencies which were nearly equal to those obtained for packed columns using the same particles. While the silicate matrix generated a monolith with serious peak tailing, other entrapping matrices gave satisfactory peak shapes and fine-tuned selectivity. Depending on the amount of entrapping matrix, the resultant particle-entrapped monoliths can have nearly the same or higher permeabilities than those observed for packed column beds containing the same packing materials. Retention factors for analytes on particle-entrapped columns using silicate entrapping matrices are significantly reduced compared to those for conventional packed columns due to a combination of masking of the stationary phase with the entrapping silicate and hydrolysis of the stationary phase during NH4OH rinse. In contrast, other entrapping matrices displayed no significant loss of retention.

The greatest disadvantage associated with particle-entrapped monoliths is that the preparation procedures are time-consuming because drying of the entrapped column must be performed slowly. Otherwise, surface tension forces will cause column bed

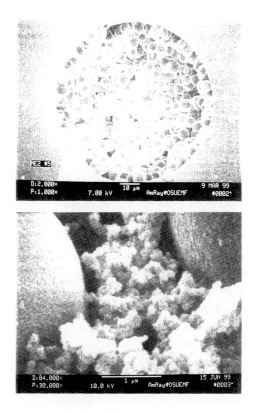

Fig. 9.4. Scaning electron micrographs of a particle-entrapped monolithic column (reproduced with permission from ref. [13]. Copyright 2000 American Chemical Society).

cracking and failure. Also the use of a retentive entrapping matrix adds to the retention of solutes, which makes it difficult to predict the overall retention behavior of analytes on the column.

9.6 SOL-GEL-BONDED MONOLITHIC COLUMNS

We developed a sol-gel-bonded monolith that employed a specifically designed sol-gel matrix and the use of supercritical fluid column packing and drying [14]. The preparation procedure involved four steps: (i) packing the column with a supercritical CO_2-slurry, (ii) filling the packed column with a sol solution, (iii) gelling and aging of the sol-filled packed column at room temperature, and (iv) drying the gelled, packed-column with supercritical carbon dioxide. A CO_2 slurry packing method developed in our laboratory [15] was used to pack the columns. In short, one end of a fused silica capillary was connected to a stainless steel vessel into which an approximate amount of packing material was placed. The other end of the capillary column was connected to a length of 25 μm i.d. restrictor tubing using a zero dead-volume union in which a stainless steel frit with 0.5 μm pores was placed. The stainless steel frit was used to retain the particles in the capillary, and the restrictor tubing was used to control the packing speed. The stainless steel vessel was connected to a syringe pump, and packing of the capillary column was carried out in an ultrasonic bath at room temperature by increasing the pressure from 5–35 MPa at 1 MPa/min. Finally the column was conditioned at 35 MPa for 30 min, and then depressurized overnight.

The sol solution was prepared by mixing appropriate amounts of tetramethoxysilane (TMOS), ethyltrimethoxysilane (ETMOS), methanol, trifluoroacetic acid (TFA) and water in a 0.6 mL polypropylene vial, followed by the addition of formamide. For example, to obtain a 9% (precursor/solution, v/v) sol solution, the following amounts were used: 20 μL TMOS, 20 μL ETMOS, 200 μL methanol, 15 μL TFA aqueous solution at pH 2, and 200 μL formamide. The apparent pH of the solution was approximately 5, measured using short-range pH test paper. The solution was vortexed for 5 min at room temperature using a vortex mixer and introduced into a packed capillary column to a desired length (observed with the naked eye or under a microscope with 10x magnification) using a 100-μL PEEK syringe mounted on a small syringe pump. The syringe was strong enough for filling 60 cm of capillary column packed with 5 μm ODS particles, or 45 cm of capillary column packed with 3 μm ODS particles, with the non-viscous sol solution at a speed of 1 μL/min.

Both ends of the sol-filled packed capillary were sealed using a silicone septum and the column was stored at room temperature for 24 h to achieve the conversion of the sol to a gel, and for aging of the resulting wet gel. After aging, the capillary was connected to an SFC pump using a zero dead-volume union. The column was flushed with supercritical CO_2 at 8 MPa inlet pressure and 40°C to replace the solvent in the column. Then, while supercritical CO_2 at 8 MPa was purged through the column, the column was dried by temperature programming from 40°C to 240°C at 1°C min^{-1}, and holding at 240°C for 1 h. Finally, the packing material at the end of the column that was not bonded by the sol-gel was flushed out, leaving a section of unpacked capillary.

A short length of the monolithic column was cut off and a gold coating of approximately 40 nm thickness was sputtered on it for observation using a scanning electron microscope. Typical scanning electron micrographs of the cross-section of a sol-gel bonded ODS monolithic column are shown in Fig. 9.5. As can be seen, the column was packed homogeneously without visible wall perturbations and the ODS particles were bonded to each other and to the capillary wall by the gel network, forming a sol-gel-bonded monolith.

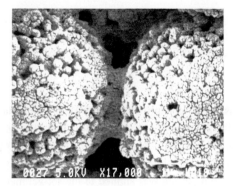

Fig. 9.5. Scanning electron micrographs showing cross-sectional views of the end of a sol-gel-bonded large-pore ODS monolithic column (reproduced with permission from ref. [20]. Copyright 1999 John Wiley & Sons).

The key for preparing a successful sol-gel-bonded monolith is the design of the sol-gel matrix and the drying process. Fig. 9.6 illustrates schematically the synthesis of the solgel bonded particles. At pH 2, both TMOS and ETMOS precursors quickly underwent hydrolysis in the solvent, forming tetrahydroxysilane and ethyltrihydroxysilane. After the addition of formamide, the solution pH was changed to approximately pH 5, where tetrahydroxysilane polycondensed much faster than ethyltrihydroxysilane. As a result, the sol solution was converted into a silica network end-capped with ethyl groups. The particles were bonded to each other and to the capillary inner wall through this inert sol-gel matrix, forming a wall-supported continuous-bed. In a classical sol-gel process for silica glasses, a single precursor, TMOS, is typically used [16]. However, a silica gel network fabricated from TMOS alone often shrinks and cracks during drying, and the silica gel network is active because of the exposed silanol groups on the gel surface. It was reported that a combination of alkyl derivatives of TMOS as co-precursors with TMOS resulted in a flexible gel network with a

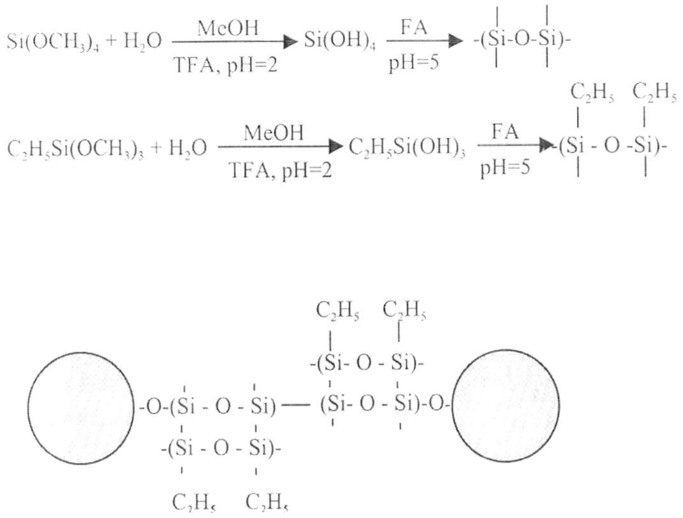

$$Si(OCH_3)_4 + H_2O \xrightarrow[\text{TFA, pH=2}]{\text{MeOH}} Si(OH)_4 \xrightarrow[\text{pH=5}]{\text{FA}} -(Si\text{-}O\text{-}Si)-$$

$$C_2H_5Si(OCH_3)_3 + H_2O \xrightarrow[\text{TFA, pH=2}]{\text{MeOH}} C_2H_5Si(OH)_3 \xrightarrow[\text{pH=5}]{\text{FA}} -(Si\text{-}O\text{-}Si)-$$

with C_2H_5, C_2H_5 substituents on the Si–O–Si network.

$$
\begin{array}{c}
C_2H_5 \quad C_2H_5 \\
| \qquad | \\
-(Si\text{-}O\text{-}Si)- \\
-O\text{-}(Si\text{-}O\text{-}Si) \longrightarrow (Si\text{-}O\text{-}Si)\text{-}O- \\
-(Si\text{-}O\text{-}Si)- \\
| \qquad | \\
C_2H_5 \quad C_2H_5
\end{array}
$$

Fig. 9.6. Synthetic scheme for sol-gel bonded particles (reproduced with permission from ref. [18]. Copyright 2000 Wiley-VCH Verlag GmbH).

more open pore structure [17], which could alleviate capillary stress during drying. The alkyl groups can also deactivate the gel network, thereby contributing little to the peak tailing of analytes. Even when the composition of the sol solution was well designed and controlled, extremely slow drying was required to minimize cracking of the column bed when conventional thermal drying was applied to the sol-gel matrix. This made column preparation rather time-consuming. Cracking of the column bed is believed to result from capillary forces generated during evaporation of the solvent. To avoid this, a supercritical fluid was used to dry the gel network since the surface tension of a supercritical fluid is zero [16]. In the fabrication of sol-gel-bonded mono-liths, supercritical CO_2 was pumped through the column to replace the solvent and residual reactants using a supercritical fluid chromatographic pump, and the column was then dried in supercritical CO_2 under elevated temperature. It was observed that column drying with supercritical CO_2 was fast and the resultant columns were crack-free.

Sol-gel-bonded monolithic columns from small-pore ODS [14,18,19], large-pore ODS [18,20,21], mixed-mode ODS/SCX [18,22], and C30-silica [23,24] have been successfully prepared. Column-to-column reproducibility was found to be less than 8% for three monolithic columns containing 9% sol-gel bonded 5-μm ODS [14]. An efficiency of 410,000 plates/m in CEC was obtained using a sol-gel-bonded 3 μm, 1500 Å ODS column, which is nearly double the efficiency from a sol-gel-bonded

3 µm, 80 Å ODS column [21]. This was attributed to EOF generation in the large-pores of the 3 µm, 1500 Å ODS.

9.7 PARTICLE-CROSSLINKED MONOLITHIC COLUMNS

We have also described a particle-crosslinked monolithic column in which polymer-encapsulated packing materials packed in polymer-coated fused silica tubing was crosslinked *in situ* using a free radical reaction [25]. A fused silica open tubular column with 0.25 µm film thickness of octyl-substituted polysiloxane was packed with polybutadiene (PBD)-encapsulated zirconia packing material using a supercritical CO_2 slurry packing method and placed in a GC oven and flushed with a free radical initiator, azo-*tert*-butane, in helium. The helium flow rate through the packed column was 0.5 cm/s. The oven temperature was elevated to 260°C at a rate of 0.5°C/min and held at 260°C for 5 h. Porous packing materials could be immobilized onto the PSB-Octyl capillary wall through free radical reactions [26] since about 50% of the vinyl groups on the PBD-encapsulated zirconia particles remained unreacted [27]. Under elevated temperature, azo-*tert*-butane decomposed to form free radicals and the PBD-encapsulated zirconia packing was crosslinked together and to the capillary wall through free radical initiated crosslinking reactions, forming a crosslinked-monolithic column shown in Fig. 9.7.

The experimental conditions, including the film thickness of the octylpolysiloxane coating in the open tubular column, the flow rate of the azo-*tert*-butane saturated helium through the column bed, and the time and temperature for the crosslinking reaction, determined the success of column preparation. When the octylpolysiloxane film thickness in the original capillary was 0.25 µm, the crosslinked packing material stayed in the column even when 40 MPa carbon dioxide at room temperature was applied at the column inlet. For achieving a mechanically strong monolithic column, the column must be purged for half an hour with azo-*tert*-butane vapors in helium at a flow rate of 2 mm/s before increasing the temperature. Using a crosslinking temperature of 260°C for 5 h, stable crosslinked-monolithic 3% PBD-encapsulated zirconia columns of 25–60 cm × 250 µm i.d. were obtained. The column efficiency, solute retention, and permeability of these monolithic columns were similar to those measured for columns freshly packed, but not crosslinked, with the same material under the same conditions.

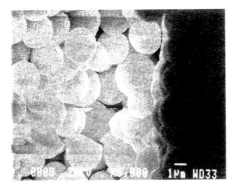

Fig. 9.7. Scanning electron micrograph of the end of a crosslinked monolithic column (reproduced with permission from ref. [25]. Copyright 1999 John Wiley & Sons).

9.8 CONCLUSIONS

Monolithic columns prepared from particles are designed to inherit the efficiency and selectivity of versatile LC packing materials. Monolithic columns from particles are expected to be stable without the need of end-frits. Various methods used for the preparation of monolithic columns from particles have been described, each with its own merits and shortcomings. Sol-gel-bonded monolithic columns employing an inert sol-gel matrix and supercritical CO_2 drying is currently the most promising method for obtaining high performance columns.

9.9 REFERENCES

1   V. Pretorius, B.J. Hopkins, J.D. Schieke, J. Chromatogr. 99 (1974) 23

2   J.W. Jorgenson, K.D. Lukacs, J. Chromatogr. 218 (1981)209

3   J.H. Knox, I.H. Grant, Chromatographia 24 (1987) 135

4   T. Tsuda, I. Tanaka, G. Nakagawa, Anal. Chem. 56 (1984) 1249.

5   J.H. Knox, I.H. Grant, Chromatographia 32 (1991) 317.

6   R. Asiaie, X. Huang, D. Farnan, C. Horvath, J. Chromatogr. A 806 (1998) 251.

7   T. Adams, K.K. Unger, M.M. Dittmann, G.P. Rozing, J. Chromatogr. A 887 (2000) 327.

8   M.T. Dulay, R.P. Kulkarni, R.N. Zare, Anal. Chem. 70 (1998) 5103.

9   C.K. Ratnayake, C.S. Oh, M.P. Henry, J. High Resol. Chromatogr. 23 (2000) 81.

10   C.K. Ratnayake, C.S. Oh, M.P. Henry, J. Chromatogr. A 887 (2000) 277.

11   G. Chirica, V.T. Remcho, Electrophoresis 20 (1999) 50.

12   G.S. Chirica, V.T. Remcho, Electrophoresis 21 (2000) 3093.

13   G.S. Chirica, V.T. Remcho, Anal. Chem. 72 (2000) 3605.

14   Q. Tang, B. Xin, M.L. Lee, J. Chromatogr. A 837 (1999) 35.

15   A. Malik, W. Li, M.L. Lee, J. Microcol. Sep. 5 (1993) 361.

16   C.J. Brinker, G.W. Scherer, Sol-gel Science: The Physics and Chemistry of Sol-gel Processing, Academic Press, San Diego, CA, USA, 1990.

17   J.D. Mackenzie, Hybrid Organic-Inorganic Composites, ACS Symposium Series 585, American Chemical Society, Washington, D.C., USA, 1995, p 227-236.

18   Q. Tang, M.L. Lee, J. High Resol. Chromatogr. 23 (2000) 73.

19   Q. Tang, N. Wu, M.L. Lee, J. Microcol. Sep. 12 (2000) 6.

20   Q. Tang, N. Wu, M.L. Lee, J. Microcol. Sep. 11 (1999) 550.

21   Q. Tang, N. Wu., B. Yue, M.L. Lee, Anal. Chem., in press.

22   Q. Tang, M.L. Lee, J. Chromatogr. A 887 (2000) 265.

23   L. Roed, E. Lundanes, T. Greibrokk, J. Microcol. Sep. 12 (2000) 561.

24   L. Roed, E. Lundanes, T. Greibrokk, J. Chromatogr. A 890 (2000) 347.

25   Q. Tang, Y. Shen, N. Wu, M.L. Lee, J. Microcol. Sep. 11 (1999) 415.

26   H. Yun, K.E. Markides, M.L. Lee, J. Microcol. Sep. 7 (1995) 153.

27   J. Li, P.W. Carr, Anal. Chem., 68 (1996) 2875.

F. Švec, T.B. Tennikova and Z. Deyl (Editors)
Monolithic Materials
Journal of Chromatography Library, Vol. 67

Chapter 10

Layered Stacks

Mark R. ETZEL

University of Wisconsin, Department of Chemical Engineering, Madison, WI 53706, USA

CONTENTS

10.1 INTRODUCTION

Traditionally, chromatographic separations utilize columns packed with beads. The size of the beads in part determines the balance between pressure drop and separation factor. Pressure drop across packed columns increases with decreasing size of the beads, while separation factors increase. Complicating phenomena in this balance are bead deformation under pressure and plugging of the column by particulates. As bead size decreases, and pressure drop increases, beads compress, resulting in a further increase in pressure drop and consequently an additional compression of the beads [1]. Eventually, the column may plug completely. At the same time, the void fraction in the packed column decreases, increasing filtration requirements for the feed solution. For an uncompressed bed of spheres, the voids between the beads have a size representing 15.5% of the bead diameter. Because filtration to about 25% of that size is required to prevent plugging of the packed bed by bridging particulates, filtration to about 4% of the bead diameter is required. For HPLC systems packed with 3 to 5 μm diameter beads, 0.1 to 0.2 μm filtration for the feed solution must precede the separation. For process chromatography systems, using 90 μm beads, 4 μm filtration is sufficient to avoid plugging of the column. If the beads are compressible, and all beads are to some extent, then finer filtration is required, because the void spaces are smaller. This also makes it difficult to scale up packed column chromatography systems.

In 1988, a revolutionary new approach was proposed to overcome the limitations of packed columns [2]. This approach involves using microporous membranes as chromatographic media. Because the membranes are thin (~ 0.1 mm), pressure drop limitations are not significant. Since solute transport in the membranes occurs primarily by convection, not diffusion, increased separation factors at increased flow rates are possible. Membrane compression is not significant, and increases in system capacity are obtained by simply enlarging membrane area. Therefore, scale up using membranes is less problematic than for packed columns.

The concept of membrane chromatography has evolved over the fifteen years since its introduction. Single thin sheets of membranes were found to be undesirable because of poor performance, and layered stacks of thin membranes are now the most common format. In this chapter, the preparation and characterization of layered stacks will be recounted, application niches and failures will be summarized, and future developments required for wider acceptance will be proposed.

10.2 POLYMERS AND METHODS OF CONSTRUCTION

Layered stacks are found in two different geometries: layers of membrane discs stacked into a circular holder, and layers of membranes wound around a porous core tube [3,4]. Both geometries start with the manufacture of microporous membrane sheets that are cut and fabricated into the final geometry. The membrane sheet is typically manufactured by starting with an existing microporous membrane sheet made previously for filtration applications. The internal surface of the micropores is then derivatized to include adsorptive sites suitable for separation in various modes such as ion exchange, immunoaffinity, hydrophobic interaction, reversed phase, as well as group specific (e.g. Protein A, Cibacron Blue), or immobilized metal affinity ligands. Past reviews have contained detailed descriptions of specific applications and chemistries, and separations methods development [5-10]. The focus of this section will be on the polymers used in construction of the leading commercialized membranes, modifications to the surface chemistry, and methods of fabrication. Some manufacturers and products are listed in Table 10.1.

10.2.1 Membrane materials

The most common commercially available membrane matrices for layered stacks consist of either poly(vinylidene fluoride) [11], crosslinked regenerated cellulose reinforced with a non-woven fabric [12,13], or polyethersulfone [14]. These materials were developed for microfiltration membranes and adapted for use in layered stacks. Other matrices prepared directly by copolymerization are described elsewhere in this book. Readers are also referred to the excellent review of Zeng and Ruckenstein [5] for the detailed polymer chemistry of membrane matrices.

10.2.2 Surface chemistry

Membrane matrices must be activated before adsorptive ligands can be coupled to the internal surface. The chemistry of activation is selected based on the reactivity of the membrane matrix, the shelf life of the activation chemistry, and other factors specific to a given application. For non-reactive membrane matrices such as polyethylene, polypropylene, polysulfone, nylon, and poly(vinylidene floride), reactive surface layers are created on the membrane by either coating or graft polymerization. Once the surface is activated, the desired ligand is covalently coupled to the surface. Sometimes, a small molecule spacer is inserted between the activated membrane surface and the ligand in order to prevent steric limitations to binding of the target solute to the membrane surface. Most activation and coupling chemistries are borrowed from methods developed for traditional chromatography [15].

Membrane matrices that are inert chemically, such as polyethersulfone and poly(vinylidene floride), can also be coated with hydroxyethyl cellulose that is covalently bound to the membrane surface activated with glycol diglycidylether via epoxy groups [16]. Alternatively, the membrane surface can be coated with a solution of reactive monomers that are polymerized in situ to form a crosslinked hydroxyalkyl acrylate or methacrylate surface [17].

An alternative to coating is graft polymerization. In radiation-induced graft polymerization, polyethylene or polypropylene membranes are grafted with different monomers to form a reactive surface [18]. Alternatively, amino acids such as glycine can be attached to the surface of poly(vinylidene floride) membranes by immersing the membranes in a boiling mixture of glycine and sodium hydroxide [19]. The grafted mem-

TABLE 10.1

EXAMPLE OF SOME COMMERCIALLY AVAILABLE MEMBRANE PRODUCTS

Manufacturer	Product	Surface chemistry*	Membrane matrix†	Membrane thickness	Pore size	Effective face area	Number of layers	Geometry
Millipore	Intercept	Q	PVDF	140 μm	0.65 μm	28 cm^2	8	25 mm disc
						1600 cm^2	8	160 mm disc
Pall	Mustang	Q, S	PES	140 μm	0.8 μm	16 cm^2	6	25 mm disc
						700 cm^2	80	37 mm disc
						7 m^2	80	368 mm disc
						0.07, 0.4, 2, 4, or 6 m^2	16	56 mm pleated cylinder
Sartorius	Sartobind	Q, S, C, D	RC	200 μm	3-5 μm	5 cm^2	1	25 mm disc
						15 cm^2	3	25 mm disc
						100 cm^2	5	50 mm disc
						120 cm^2	1	142 mm disc
						550 cm^2	1	293 mm disc
						600 cm^2	5	142 mm disc
						5500 cm^2	10	293 mm disc
						0.2, 0.5, 1, 2, 4, or 8 m^2	15, 30, or 60	96 mm cylinder

* Q - quaternary ammonium; S - sulfonic acid; C - carboxyl; D - diethylamine

† PVDF - polyvinylidene difluoride; PES - polyethersulfone; RC - regenerated cellulose

branes are hydrophilic and contain a reactive surface consisting of carboxylic acid groups.

As shown in Fig. 10.1, activation of the surface of regenerated cellulose membranes can occur via the hydroxyl groups using: (a) carbonyldiimidazole (CDI), (b) *N*-hydroxysuccinimide ester (NHS), (c) 2,2,2-trifluoroethanesufonyl chloride, (d) 1,4-butanediol diglycidyl ether, and (e) cyanogen bromide (CNBr). As shown in Fig. 10.1a, after CDI activation, the membrane surface contains acylimidazole groups that react readily with primary amines to form a stable carbamate linkage between the ligand and the membrane surface. Similarly, as shown in Fig. 10.1b, NHS activated matrices react with primary amines to form stable amide bonds. As shown in Fig. 10.1c tresyl chloride-activated matrices react with primary amines to form secondary amine linkages. Fig. 10.1d shows that one of the epoxide groups of diglycidyl ethers reacts with hydroxyl groups on the membrane surface to from a stable ether bond, and the other terminal epoxy group is available for reaction with primary amines to couple ligands to the surface. Perhaps the oldest activation chemistry is shown in Fig. 10.1e. CNBr reacts with hydroxyl groups to form cyanate esters and cyclic imidocarbonate moieties, which in turn react with primary amines to form isourea linkages. However, the isourea linkage is not stable, and ligands tend to leak off the matrix. Furthermore, isourea derivatives are weak anion exchangers, which can cause non-specific binding to the membrane. For these reasons, the alternative chemistries mentioned previously were developed to overcome the weaknesses of the CNBr method. However, CNBr activated matrices are still widely available and used extensively in research applications.

Hydrolysis of CDI and NHS activated supports results in reversion to the original hydroxyl group on the membrane matrix. Hydrolysis and ligand immobilization reactions compete during coupling. Therefore, non-aqueous and mixed aqueous-organic solvents should be used to eliminate hydrolytic side reactions and increase ligand density on the membrane surface [15,20].

Membranes other than regenerated cellulose can utilize similar activation chemistries. For example, surface carboxyl groups of glycine-grafted PVDF membranes react with CDI to yield the same reactive imidazole group as formed when using regenerated cellulose membranes, except an acylimidazole is formed from carboxyl-containing surfaces and an imidazole carbamate is formed from hydroxyl-containing surfaces. Similarly, surface hydroxyl groups of PVDF membranes coated with hydroxyethyl cellulose that is covalently bound to the membrane surface will react with CDI to form reactive imidazole carbamate groups.

Frequently, affinity ligands immobilized directly to the surface of membranes exhibit low binding capacity for large solutes [15,21]. Therefore, spacer arms are used as a bridge between the membrane surface and the affinity ligand. Spacer arms may

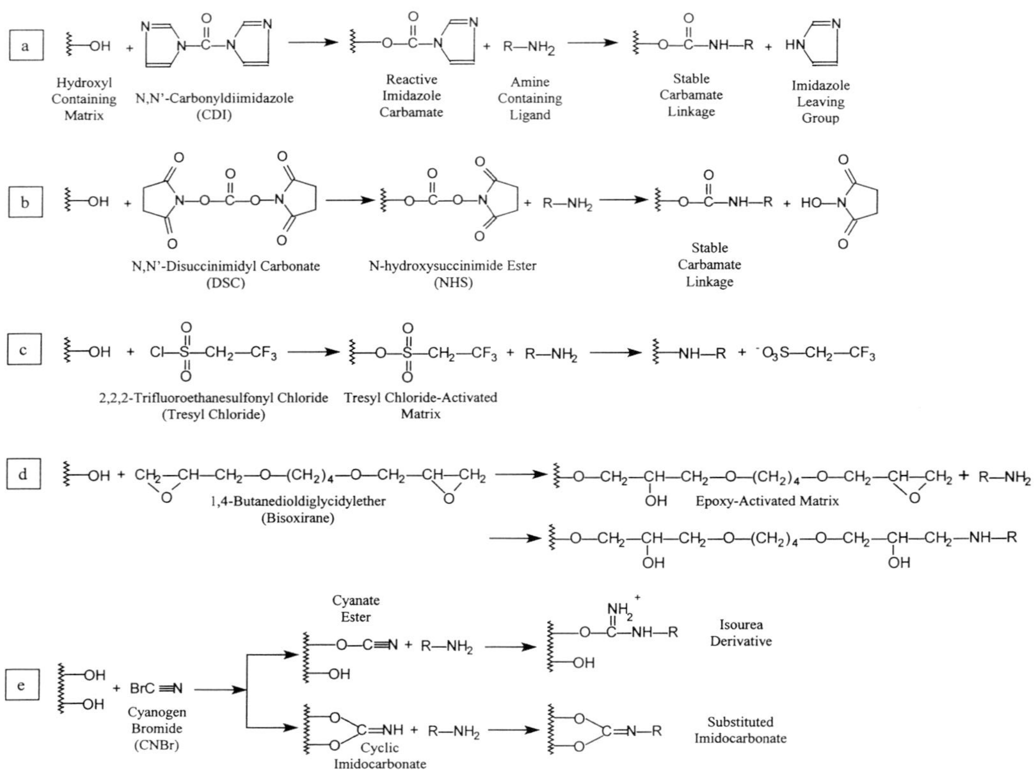

Fig. 10.1. Surface activation moieties and corresponding reactions.

consist of a difunctional linear molecule such as the small hydrophobic molecule of 1,6-diaminohexane or the high molecular weight hydrophilic amino-poly(ethylene glycol)-carboxylic acid (H_2N-PEG-COOH) with a molecular weight of 3,400. Often, binding sites on a target solute are buried or not close enough to the membrane surface to interact with a ligand immobilized directly on the membrane surface. A ligand that is attached to a spacer arm can protrude far enough from the membrane surface to reach the binding site. Furthermore, ligands immobilized directly to the membrane surface are immobile and inflexible and are deposited in a random fashion rather than in a face-centered-cubic (FCC) array. As a result, large solutes cannot approach monolayer coverage on the membrane surface and are bound to about 50% of the theoretical capacity [22]. Long flexible PEG spacer arms can impart mobility to the ligand allowing FCC packing of the solute, thus increasing binding capacity to the theoretical limit of monolayer coverage, while allowing free access of the ligand to the affinity binding site of the target solute.

10.2.3 Fabrication methods

Laboratory-scale separations are typically conducted using layers of membrane discs with a diameter of 25 or 47 mm stacked into a circular holder. Syringe filter housings are commonly employed to compact 3 or more layers into a normal-flow-mode device. The edges of the membranes are sealed under compression to prevent liquid from flowing around the edges of the stack and bypassing the membranes. Flow distribution channels are molded into the inlet and exit ends of the housing to mini-mize jetting of the feed solution through the center of the membrane stack. Ideally, flow would be distributed not uniformly across the face of the membrane stack, but rather such that the residence time is equal for all fluid paths through the membrane [23]. Fluid taking the shortest path through the center of the stack would have a lower velocity than fluid traveling through the longest path at the edges. Of course, actual flow paths fall short of this ideal, which leads to a reduction in separation perfor-mance.

For layers of membrane discs, scale-up to production-scale separations can be accomplished by increasing the diameter of the circular holder. Devices with diame-ters of 142 and 293 mm have successfully been used in larger-scale separations [24]. Increasing the diameter of the holder necessarily magnifies the problem of flow distri-bution, because the difference between the path length through the center and edges of the stack becomes larger with increasing holder diameter. A larger number of layers in the stack also increases capacity, but at the expense of higher pressure drop.

An alternative method of scale up is to wrap layers of membrane around a porous core tube [3,4]. In one device, 15 to 60 layers each 50 cm long were wound around a porous core tube of a diameter chosen to yield an outer diameter of 96 mm [25].

Membrane volumes up to 2 L were obtained in a single device. Breakthrough capacity (5 %) of these devices per unit volume of membrane increased logarithmically with the number of layers, which was not expected based on theory and was attributed to mal-distribution of flow within the system [26].

As described above, scale up of layered stacks by simply increasing the diameter of the circular holder, and the number of layers of discs and results in units with predictable performance is possible to a certain limit. Scale up to several liters of membrane volume will be problematic using increasingly larger circular holders containing more and more layers of discs, which leaves wrapped cylinders as an attractive alternative. However, improvements in the flow distribution of wrapped cylinders are required to increase the predictability of scale up. Ideally, all the streamlines in the wrapped cylinder will have the same residence time not equal velocities. Modeling approaches used for proper header design in columns [23] are likely to reveal the design required to reduce flow mal-distribution in wrapped cylinders.

10.3 SYSTEM PERFORMANCE – CHARACTERIZATION AND MODELING

The advantages of stacked layers of membranes over columns packed with beads lie in achieving fast binding of solute to the sites on the membrane surface and low trans-membrane pressure drop. However, certain phenomena compromise performance of layered stacks, and solutions to these technical challenges are needed to widen the range of applications beyond specific niche areas. These technical challenges are (1) minimizing dispersion in the flow system, (2) increasing uniformity in membrane thickness and pore size, (3) speeding the sorption kinetics, and (4) tailoring membrane designs for different size solutes. Deciphering the connections between these challenges and final system performance requires careful coupling of experimental characterization and mathematical models.

10.3.1 System dispersion

Reduction in separation performance can result from liquid mixing in the pump, tubing, fittings, membrane holder, layered stack, and detector system. For example, the unequal residence times of the liquid flowing through the layered stack result in reduced separation performance. These effects are termed system dispersion. The simplest model found to accurately describe system dispersion in layered stacks is the tandem combination of a continuously stirred tank reactor (CSTR) and an ideal plug flow reactor (PFR) as shown in Fig. 10.2 [27]. The system volume (V_{sys}) is

$$V_{sys} = V_{CSTR} + V_{PFR} \tag{10.1}$$

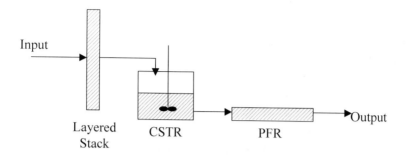

Fig. 10.2. Diagram of system dispersion model for layered-stack device.

where V_{CSTR} and V_{PFR} are the volumes of the individual parts of the model.
    The CSTR model is

$$dc_{out}/dt = (Q/V_{CSTR})(c_{in} - c_{out})$$
(10.2)

where c_{in} and c_{out} are the concentrations of the liquid flowing into and out of the CSTR, respectively, at a flow rate of Q, and where V_{CSTR} is the volume of the CSTR. In a CSTR, the contents are well stirred and uniform in composition throughout. Therefore, the liquid flowing out of the CSTR has the same composition as the liquid within the CSTR.

    The PFR model shifts the system response to longer times by a delay time t_{delay}. In a PFR, there is no axial diffusion or mixing, and the residence time is the same for all fluid elements. Because V_{PFR} includes the void volume of the layered stack V_{void}, the time of delay from the other parts of the flow system results from the remainder of the system volume $V_{PFR}-V_{void}$. The CSTR model accounts for mixing in the flow system. The first temporal moment method can be used to calculate V_{sys}, V_{CSTR}, and V_{PFR} from the measured response at the exit of the flow system to the introduction of a non-binding tracer at the entrance of the flow system [28,29].

    The importance of system dispersion in membrane stacks increases as the strength of binding of the solute to the matrix decreases. In analytical separations, typically a small aliquot of sample solution is injected into the mobile phase and carried into the layered stack. As the solutes in the sample are swept through the stack, they interact weakly and to different extents with the membrane surface and separate into bands. These bands broaden during travel through the device, and solutes interacting the least with the membrane surface occur first in the emerging liquid. This mode of operation, isocratic elution chromatography, requires buffers and operating conditions such that binding is weak, binding capacity is small, binding isotherms are linear, and the number of plates is large. If binding interactions were strong, then the sample would never be eluted from the membrane stack. If plate numbers were small, then the solutes would not separate. In this isocratic operation, the effects of system dispersion are magnified, because, for systems such as these, where the number of plates must be large, a small amount of mixing in the flow system can dramatically reduce the number of plates to a level that is smaller than that needed for the separation.

Conversely, a large aliquot of sample solution is loaded into the membrane stack for capture mode chromatography (frontal analysis), nearly saturating the membrane surface. In this mode, strong binding, high capacity, near monolayer surface packing (near the theoretical maximum), non-linear isotherms, and small plate numbers are desirable. For example, as few as 30 plates are sufficient to obtain a sharp break-through curve (BTC) and complete recovery of the target compound [30]. In this mode of operation, layered stacks are best as an alternative to column chroma-tography.

Conclusions concerning the importance of system dispersion, sorption kinetics, and other factors drawn from analytical chromatography may not be applied directly to capture chromatography. For example, a given extent of system dispersion in capture mode may not be a significant factor in determining system performance, while it may be very important in analytical chromatography. The following example illustrates the characterization and modeling of system dispersion for a system in capture mode [31] using a layered stack of three cross-linked regenerated cellulose membranes reinforced with a non-woven fabric and compressed in a 25 mm diameter syringe filter housing (Sartobind S15X, Sartorius, Edgewood, NY). The membranes were each 200 μm thick, had an average pore size of 5 μm, and contained sulfonic acid moieties on the surface. Bovine serum albumin (BSA) at a concentration of 0.2 g/L in 50 mmol/L sodium phosphate buffer, pH 7, was used as a non-binding tracer solute. At pH 7, BSA has a high negative charge (pI = 5.1), and does not bind to the sulfonic acid groups on the membrane. The flow rate was set at 10 mL/min (2030 column volumes/h), and the exit concentration was determined from absorbance at 280 nm (Fig. 10.3).

The system dispersion data were successfully fitted using the model shown in Fig. 10.3. The calculated parameter values were: V_{sys} = 1.5 mL, and V_{CSTR} = 0.4 mL. By difference, V_{PFR} is then 1.1 mL. The membrane volume was 0.295 mL and the void fraction was 0.8. Therefore, the membrane void volume was 0.24 mL, and the PFR volume from the other elements of the flow system was 0.86 mL. Thus, the impact of system dispersion for this system will be insignificant when loading more than about ten membrane volumes of feed solution. The loading volume is often one to two orders of magnitude larger than V_{CSTR}, making it unlikely that system dispersion will be significant for this device operating in capture mode.

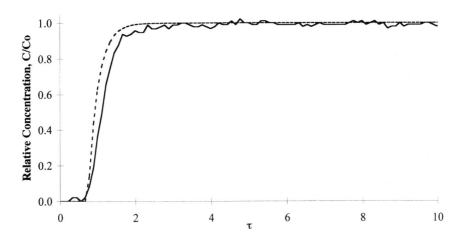

Fig. 10.3. Observed system dispersion curve (solid line) and predictions from the model (dotted line). The abscissa is the dimensionless time ($\tau = vt/L$), where v is the interstitial velocity, L is the stack thickness, and t is the elapsed time. The amount of feed solution loaded into the stack given in column volumes (CV) is $\varepsilon\tau$, where ε is the membrane void fraction ($\varepsilon = 0.8$). The flow rate was 2030 CV/h.

10.3.2 Membrane holder

Photographs of the 25-mm diameter membrane holder used in the previous example are shown in Fig. 10.4. The holder contains an inlet flow distributor that directs the feed solution across the face of the membrane stack, and an exit flow distributor that collects the flow from the back of the membrane stack. Eight radial channels move fluid directly away from the inlet port towards the edges of the stack. Fourteen concentric circular channels are interconnected with each radial channel to distribute flow. The inlet and exit ports contain an anti-jetting device that forces flow to move in only four of the eight radial channels and blocks most, but not all of the flow moving in the axial direction.

A test of the performance of this holder was conducted by loading 50 mL of 0.2 g/L porcine thyroglobulin in 50 mmol/L sodium citrate, pH 3.0, at 10 mL/min [31]. Unbound thyroglobulin was completely washed out using the citrate buffer. The membranes were cut out of the holder, stained using Coomassie blue, and destained using an aqueous solution containing 30 % methanol and 10 % acetic acid. The protein adsorption pattern is shown in Fig. 10.5 for the sides of the membranes facing upstream and downstream. No detectable flow mal-distribution can be detected. In other words, there is no evidence of a radial pattern in protein adsorption, because the

Fig. 10.4. Membrane holder for three 25 mm-diameter cation-exchange membranes (Sartobind S15X). Flow distributors at the inlet (top panel) and exit (bottom panel) of the stack contain anti-jetting devices, eight radial channels, and fourteen concentric circular channels.

holder enables a remarkably uniform protein adsorption pattern across the face of the membrane.

The downstream sides of the membranes contain more protein than the upstream sides. This was attributed to a higher concentration of sulfonic acid binding sites on the downstream side of the membranes, perhaps due to an asymmetric structure of the membrane. The first membrane in the stack (left) bound the most protein, the second membrane (middle) bound less protein, and the third membrane (right) bound the least amount of protein. This observation was expected, because protein was removed from the feed solution as it passed through the stack, leaving less protein available for adsorption on membranes in the lower layers.

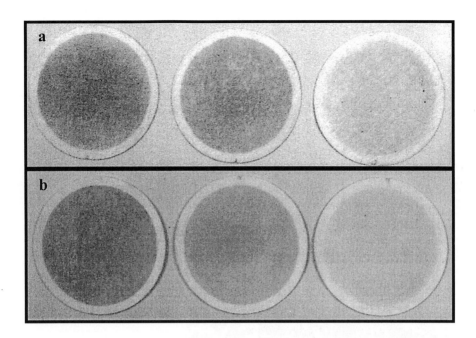

Fig. 10.5. Pattern of thyroglobulin adsorption on the three 25 mm diameter cation-exchange membranes inside the holder shown in Fig. 10.3. Sides facing upstream (a) bound protein slightly less than sides facing downstream (b). The first membrane to receive flow (left) bound most of the protein, followed by the second (middle), and the last membrane (right).

Scanning electron micrographs of the membrane surface shown in Figs. 10.6 and 10.7 demonstrate that the two faces of the membrane are asymmetric. The upstream side of the membrane (Fig. 6) has a more coarse pore structure, while the pore structure of the downstream side is finer. Micrographs of both sides show numerous fine pearl-like polymer deposits on a smooth web-like structure. In addition, thread-like strands ran through the microporous matrix. A cross-section of the membrane (Fig. 10.8) exhibits numerous 10 to 20 μm large polymer threads that are most likely fibers of the non-woven reinforcement fabric.

As mentioned above, the asymmetric structure of the membrane is likely the reason for increased binding of protein to the downstream side of the membrane. The opposite result would be expected for a symmetric membrane, because the upstream side would always be exposed to more concentrated protein solution. To test the asymmetry hypothesis, a single membrane disc was incubated in a Petri dish in 0.2 g/L thyroglobulin in 50 mmol/L sodium citrate, pH 3.0, washed, stained, and destained as before. Fig. 10.9 shows the result that supports the asymmetry hypothesis. Indeed, one side of the membrane adsorbed more protein than the other side.

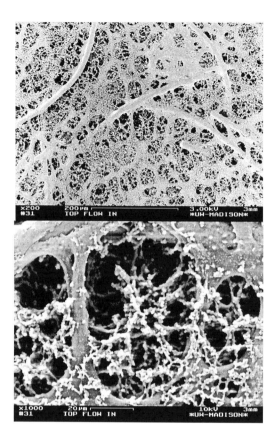

Fig. 10.6. Scanning electron micrographs of side of the membrane shown in Fig. 10.4 that faced upstream at magnifications of 200 (top) and 1,000 (bottom).

10.3.3 Variations in membrane characteristics

Membranes cannot be made perfect. No matter how controlled the manufacturing process is, variations in membrane thickness, void fraction, and concentration of binding sites will occur. The effects of thickness and void fraction variations can be severe. Thus, variations in thickness must be kept under 3% to have no effect on the performance of a single membrane layer, while membrane porosity should vary less than 1% to avoid the decline in performance [30]. Similarly, variations in the number of binding sites across a membrane, or between membranes, will reduce separation performance. However, layered stacks of membranes help to average out and reduce the effects of small variations in thickness, porosity, and ligand concentration characteristics of individual membranes. The example presented above clearly demonstrates

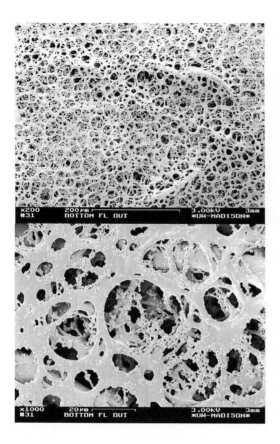

Fig. 10.7. Scanning electron micrographs of the side of the membrane shown in Fig. 10.4 that faced downstream at magnifications of 200 (top) and 1,000 (bottom).

that even a stack of only three membranes is entirely adequate to achieve excellent performance.

10.3.4 Sorption kinetics

One notable limitation to the success of layered stacks is the slow sorption kinetics of some solutes to bioaffinity ligands bound to layered stacks. For example, affinity systems consisting of solute~ligand combinations such as pepsin~pepstatin A, chymosin~pepstatin A [32], monoclonal antibody~Protein G [33], immunoglobulin G~Protein G [34], and immunoglobulin G~Protein A/G [35] display slow binding kinetics. Binding kinetics are sometimes so slow that the performance enhancements expected for layered stacks vs. beads cannot be observed. Rather than flow rate being

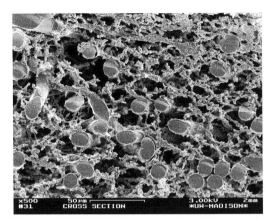

Fig. 10.8. Scanning electron micrograph of the cross section of the membrane shown in Fig. 10.4 at a magnification of 500.

limited by pressure drop, it is limited by slow adsorption-desorption kinetics. Thus, solute~ligand systems must be examined on a case-by-case basis to determine if the sorption kinetics are sufficiently fast for layered stacks to be an advantage. In general, sorption kinetics are faster for ion exchange systems than for affinity systems, and the advantages of layered stacks are more likely to be realized for the former systems.

10.3.5 Solute size

One of the advantages of layered stacks is the potential to eliminate limitations originating from mass transfer to the binding sites. For solute to be captured by the binding sites on the membrane surface, the residence time of the liquid in the membrane must exceed the time for the solute to diffuse to the binding site on the surface and bind. An order-of-magnitude conservative estimate for the time scale of diffusion t_D from the center of the pore to the wall is [28,36]

$$t_D = (d_p)^2/4D \tag{10.3}$$

where d_p is the diameter of the pore, and D is the diffusion coefficient of the solute. The residence time t_R in the membrane stack is

$$t_R = L/v \tag{10.4}$$

where L is the thickness of the stack and v is the interstitial velocity. In other words, mass transfer limitations are eliminated when $t_R/t_D \gg 1$. This situation does not occur

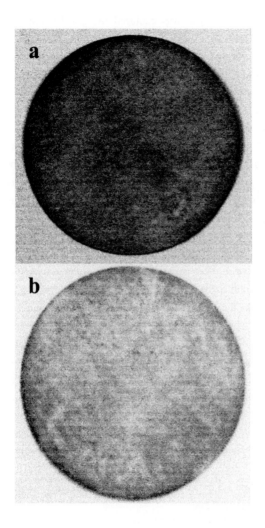

Fig. 10.9. Pattern of thyroglobulin adsorption after batch incubation for a single 25 mm diameter cation-exchange membranes that was similar to those in the holder shown in Fig. 10.3.

for thin membranes (small L) having large pores (large d_p) that are operated at high flow velocities (large v) in the separation of large solutes (small D). In general, membranes with a pore size of about 1 μm can be used to eliminate mass transfer limitations for large proteins when residence times are about one second [28]. However, for membranes having a pore size of 5 μm, residence times of 100 s or longer would be required to obtain sharp breakthrough curves for large proteins.

References pp. 233-234

For example, when bovine serum albumin was captured using a layered stack having a pore size of 150 µm, the breakthrough curves were broad and depended considerably on flow rate at residence times of 2–40 min [28]. Obviously, this pore size was too large to eliminate mass transfer limitations at high flow velocities. A pore size of about 9 µm was predicted to be required to eliminate mass transfer limitations for BSA at a residence time of about 1.5 min.

For a larger protein, such as thryoglobulin (THY) with a molecular weight of 660,000, residence times larger than 1.5 min and pore size smaller than 9 µm would be needed to observe sharp fronts. For example, the frontal experiment mentioned above including adsorption of 0.2 g/L THY in 50 mmol/L sodium citrate, pH 3.0 into the membrane stack at a flow rate of 10 mL/min, is shown in Fig. 10.10. As expected, the breakthrough was not sharp, because the 1.4 s residence time was too short for complete capture of a large protein such as THY within the 5 µm large pores. Even at a lower flow rate of 1 ml/min ($t_R = 14$ s), the curve was not sharp (see [31]). System dispersion was not a significant factor in these experiments as explained above (loading volume for BTC ≫ CSTR volume for SDC), and as seen by comparing the negligible time scale of the SDC to the BTC (Fig. 10.10). On the other hand, in a different report, BTCs were sharp when THY was captured onto a membrane stack having a pore size of 0.65 µm at a residence time of about 30 s [20]. This observation agrees well with our predictions as mentioned above [28].

In conclusion, the size of the solute to be separated directly determines the characteristics of the membrane stack such as pore size and thickness as well as the required operating conditions including flow rate to achieve complete capture of the solute. For large solutes, membrane stacks must have a small pore size to eliminate mass transfer limitations, and to realize the potential advantages of high flow rate and low pressure drop.

10.4 NICHE AND MISFIT APPLICATIONS

Layered stacks are a new technology designed to overcome the hydrodynamic and mass transfer limitations of traditional bead and column chromatography systems. Traditional column chromatography is ubiquitous in both analytical-scale and process-scale separations. Therefore, new technologies have to offer tremendous advantages to unseat well-accepted and successful existing technologies.

A typical misfit is the application of layered stacks to analytical separations. Slow affinity systems, and separation of peptides, or even small molecules are not suited for separations in layered stacks, and probably will not offer compelling advantages over bead-based chromatography systems. The capacity of layered stacks is generally lower than that of beads for these compounds.

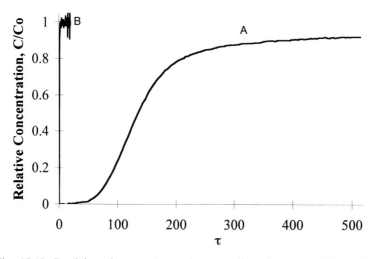

Fig. 10.10. Breakthrough curve (A) and system dispersion curve (B) for thyroglobulin adsorption onto the three 25 mm diameter cation-exchange membranes shown in Figs 10.3-7. Flow rate was 2030 CV/h.

In contrast, for separation of large proteins and very large solutes such as viruses and plasmid DNA, layered stacks offer compelling advantages. For example, the affinity sorption kinetics are relatively rapid in the capture of low-density lipoprotein (LDL) from blood plasma [21,37]. Thus, mass transfer limitations are the rate-controlling step in this case. Furthermore, LDL is too large (25 nm diameter) to enter pores of typical chromatographic beads, and binding occurs at the bead surface only, that is very small, representing only 0.7% of the surface area of membranes. Therefore, the greater surface area of membranes, and the elimination of mass transfer limitations constitute a niche application for this type of affinity membrane.

Similarly, beads are not suitable for the separation of large biomolecules such as viruses and plasmid DNA due to the small outer surface of beads, resulting in low capacity. This is another niche application for layered stacks that afford higher capacity and flow rate, and reduced pressure drop [20]. For example, anion-exchange layered stacks were successfully used for the capture and elimination of contaminants such as DNA and viruses from monoclonal antibody products [26].

Additional potential advantages of layered stacks include elimination of packing and re-packing that is typical of traditional process chromatography columns, elimination of channeling and bed compression, and elimination of cleaning and validation provided that the layered stacks are disposable. This alone will probably not create a niche for layered stacks. But, when combined with the other processing advantages mentioned above, these may sway more users to switch over from traditional bead-based chromatography systems.

References pp. 233-234

10.5 FUTURE DEVELOPMENTS

Applications in process chromatography are likely to emerge in the future as a favored application for layered stacks, especially for the capture of large biomolecules such as viruses and DNA. Viral clearance is an essential step in the manufacture of biopharmaceuticals. Traditionally, virus particles are removed by a sieving mechanism using nanofiltration membranes [38]. The development of humanized monoclonal antibodies have made them the fastest growing segment of therapeutic drugs. Because small pathogenic viruses such as the minute virus of mice (20 nm diameter) are similar in size to MAbs (10 nm diameter), it is difficult to achieve 99.99% reduction in viral load without losing a large amount of MAb in the process. Layered stacks containing anion exchange sites offer an attractive alternative to sieving membranes because the capacity of anion exchange membranes for large virus particles is greater than for small proteins [39].

This expectation can be understood using calculations of monolayer coverage of protein on the membrane surface [20] as indicated in Fig. 10.11. Assuming that a monolayer consists of a face-centered-cubic array on the surface, each protein occupies an area of $2 \sqrt{3} R_p^2$, where R_p is the geometric radius of the protein. For a spherical protein molecule of density 1 g/mL, $R_p = (3 M_W / 4 \pi N_A)^{1/3}$, where M_W is the protein molecular weight and N_A is Avogadro's constant. Each protein has a mass of $m_p = M_W / N_A$. The calculated monolayer coverage is 4.7 mg/m^2 for monoclonal antibodies with a molecular weight of 150,000, and 41 mg/m^2 for viruses having a molecular weight of 10^8. The internal surface area of a 0.65 µm pore size membrane is about 1.1 m^2/mL. Therefore, the calculated static capacity is 45 mg/mL for viruses and 5.2 mg/mL for monoclonal antibodies. In contrast, the capacity for viruses would be less than 1 mg/mL for chromatography beads, whereas their capacity for monoclonal antibodies is about 30 mg/mL. Thus, membranes have a higher capacity for the pathogen and a lower capacity for the target protein (MAb), and offer a compelling orthogonal technique to augment sieving membranes in viral clearance applications. In addition, membranes can be disposable, obviating the need for cleaning and validation.

Similarly, for other large biomolecules such as plasmid DNA for gene therapy, and viruses for vaccines, layered stacks are also assumed to offer advantages. However, large-scale systems must be developed for these and other process applications that maintain the performance of laboratory scale systems. Problems with flow mal-distribution, proper pore size and residence time combinations, developing surface chemistries with fast capture, and compete release during desorption, all must be solved before large scale applications will take off.

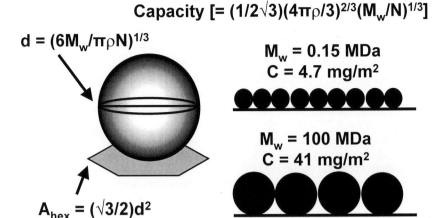

Fig. 10.11. Geometric model for the calculation of the effect of bimolecule diameter (d) on saturation capacity (C) for adsorptive membranes, where A_{hex} is the area of the bimolecule on the membrane surface when packed in a face-centered-cubic array, ρ is the bimolecule density (1 g/mL), M_w is the bimolecule molecular weight, and N is Avogadro's constant.

10.6 CONCLUSIONS

The successful development and application of layered stacks has helped to extend the limits of the chromatography industry. As with any revolutionary new technology, the rate of advancement is fast, and the target audience for the technology is evolving. Solid advances in the theoretical and experimental understanding of the performance of layered stacks has spawned new and improved designs, and re-focused the target applications towards new areas including gene therapy, viral clearance, and medical applications such as extracorporeal adsorption therapies. In the past, many applications of layered stacks were in analytical separations. However, in the future, layered stacks may have a greater impact in the biopharmaceutical process industries.

10.7 REFERENCES

1   J.J. Stickel, A. Fotopoulos, Biotechnol. Prog. 17 (2001) 744.

2   S. Brandt, R.A. Goffe, S.B. Kessler, J.L. O'Connor, S.E. Zale, Bio/Technology 6 (1988) 779.

3   M.S. Le, J.A. Sanderson, US Pat. 4,895,806 (1989).

4   D. Nussbaumer, K.T. Vinh, A. Abdul, W. Demmer, H.-H. Horl, A. Graus, G. Pradel, US Pat. 6,296,090 (2001).

5   X. Zeng, E. Ruckenstein, Biotechnol. Prog. 15 (1999)1003.

6   E. Klein, J. Membr. Sci. 179 (2000) 1.

7   C. Charcosset, J. Chem. Technol. Biotechnol. 71 (1998) 95.

8   D. Josic, A. Strancar, Ind. Eng. Chem. Res. 38 (1999) 333.

9   D.K. Roper, E.N. Lightfoot, J. Chromatogr. A 702 (1995) 3.

10   J. Thömmes, M.R. Kula, Biotechnol. Prog. 11 (1995) 357.

11   L.A. Blankstein, L. Dohrman, Am. Clin. Prod. Rev. 11 (1985) 33.

12   W. Demmer, H.-H. Horl, D. Dietmar, A.R. Weiss, E. Wunn, US Pat. 5,547,575 (1996).

13   H. Beer, W. Demmer, H.-H. Horl, D. Melzner, D. Nussbaumer, H.-W. Schmidt, E. Wunn, US Pat. 5,739,316 (1998).

14   T. Warner, P. Kostel, Genetic Eng. News 20 (2000) 1.

15   G.T. Hermanson, A.K. Mallia, P.K. Smith, Immobilized Affinity Ligand Techniques, Academic Press, San Diego, 1992.

16   E. Klein, E. Eichholz, D.H. Yeager, J. Membr. Sci. 90 (1994) 69.

17   M.J. Steuck, US Pat. 4,618,533 (1986).

18   I. Koguma, K. Sugita, K. Saito, T. Sugo, Biotechnol. Prog. 16 (2000) 456.

19   S. Sternberg, US Pat. 4,340,482 (1982).

20   H. Yang, C. Viera, J. Fischer, M.R. Etzel, Ind. Eng. Chem. Res. 41 (2002) 1597.

21   P.J. Soltys, M.R. Etzel, Biomaterials 21 (2000) 37.

22   X. Jin, J. Talbot, N.-H.L. Wang, Am. Inst. Chem. Eng. J. 40 (1994) 1685.

23   Q.S. Yuan, A. Rosenfeld, T.W. Root, D.J. Kingenberg, E.N. Lightfoot, J. Chromatogr. A 831 (1999) 149.

24   C.K. Chiu, M.R. Etzel, J. Food Sci. 62 (1997) 996.

25   W. Demmer, D. Nussbaumer, J. Chromatogr. A 852 (1999) 73.

26   H.L. Knudsen, R.L. Fahrner, Y. Xu, L.A. Norling, G.S. Blank, J. Chromatogr. A 907 (2001) 145.

27   H. Yang, M. Bitzer, M.R. Etzel, Ind. Eng. Chem. Res. 38 (1999) 4044.

28   F.T. Sarfert, M.R. Etzel, J. Chromatogr. A 764 (1997) 3.

29   E.N. Lightfoot, A.M. Lenhoff, R.L. Rodriguez, Chem. Eng. Sci. 37 (1982) 954.

30   S.-Y. Suen, M.R. Etzel, Chem. Eng. Sci. 47 (1992) 1355.

31   C.S. Rao, M.S. Thesis, University of Wisconsin, Madison, WI, 1998.

32   S.-Y. Suen, M.R. Etzel, J. Chromatogr. A 686 (1994) 179.

33   C. Viera, H. Yang, M.R. Etzel, Ind. Eng. Chem. Res. 39 (2000) 3356.

34   J.E. Kochan, Y.-J. Wu, M.R. Etzel, Ind. Eng. Chem. Res. 35 (1996) 1150.

35   O.P. Dancette, J.-L. Taboureau, E. Tournier, C. Charcosset, P. Blond, J. Chromatogr. A 723 (1999) 61.

36   I.A. Adisaputro, Y.-J. Wu, and M.R. Etzel, J. Liq. Chromatogr. 19 (1996) 1437.

37   P.J. Soltys, M.R. Etzel, Blood Purif. 16 (1998) 123.

38   H. Aranha, BioPharm 2 (2001) 32.

39   I. Schneider, Genetic Eng. News 22 (2002) 1.

F. Švec, T.B. Tennikova and Z. Deyl (Editors)
Monolithic Materials
Journal of Chromatography Library, Vol. 67
2003 Published by Elsevier Science B.V.

Chapter 11

Biotextiles — Monoliths with Rolled Geometrics

Jeremiah BWATWA[1,4], Yiqi YANG[5], Chenghong LI[3,4], Craig KEIM[2,4],
Christine LADISCH[3,4], and Michael LADISCH[1,2,4]

[1]*Department of Biomedical Engineering,*
[2]*Department of Agricultural and Biological Engineering,*
[3]*Department of Consumer Sciences and Retailing,*
[4]*Laboratory of Renewable Resources Engineering,*
Purdue University, West Lafayette, IN 47907-1295, USA

and

[5]*Department of Textiles Clothing and Design, University of Nebraska, Lincoln,*
Lincoln, NE 68583-0802, USA

CONTENTS

11.1 INTRODUCTION

Stationary phases that are formed from textiles are a continuous, interconnected fibrous matrix in the form of yarns and fabric. The fibers are assembled into yarns and the yarns are woven into fabric. Since individual fibers have exhibited poor flow properties when used as stationary phases, rolled fabric stationary phases have been developed. Rolled stationary phases enable a long bed length to be attained while retaining good flow properties [1,2]. This kind of stationary phase orients the fabric into a three-dimensional structure through contact between adjacent layers of fabric where the fabric [1,3,4] supports the fibers (Fig. 11.1(a)) [5] assembled into the yarns (Fig. 11.1(b)) [6], and the woven fabric (Fig. 11.1(c)) [6]. This is a type of monolithic material since there are no distinct or individual particles packed into the column. Further, since the material is a textile, and it is used to fractionate biomolecules, we have called this material a biotextile.

11.2 WOVEN MATRICES (TEXTILES) AS STATIONARY PHASES

11.2.1 Chemistry

11.2.1.1 Potential for derivatization

The use of fibers as chromatographic stationary phases provides a range of chemistries available for packed columns for industrial bioseparations. Cellulosic, polyphenylene, and poly(ethylene terephthalate) fibers separate proteins. Acrylic fibers exhibit hydrophobic and ionic interactions with some types of dye molecules [1,3,4]. Silica, polypropylene, polyamide, carbonaceous, and other textile fibers represent additional scaffolds upon which biospecific ligands could be grafted or ion-exchange capacity derivatized. The surface chemistries of fibers used in the textile industry can be both hydrophilic (cellulose) and hydrophobic (polyester, aramid, and acrylic). Fig. 11.2 shows some of the structures.

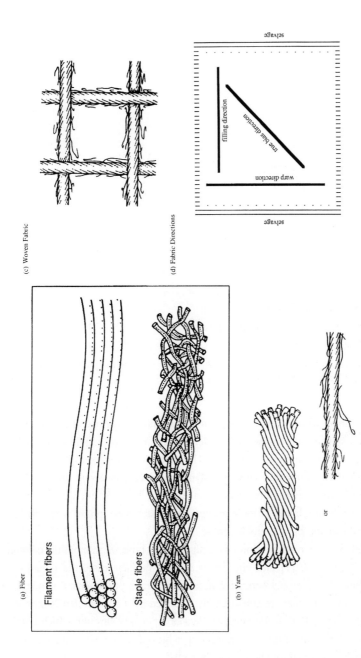

(c) Woven Fabric

(d) Fabric Directions

selvage

filling direction

true bias direction

warp direction

selvage

(a) Fiber

Filament fibers

Staple fibers

(b) Yarn

or

Fig. 11.1. Representation of fibers (a), yarn (b), woven fabric (c), and fabric directions and orientation (d), (Reprinted with permission from ref. [5,6,10], copyright 1999, American Chemical Society).

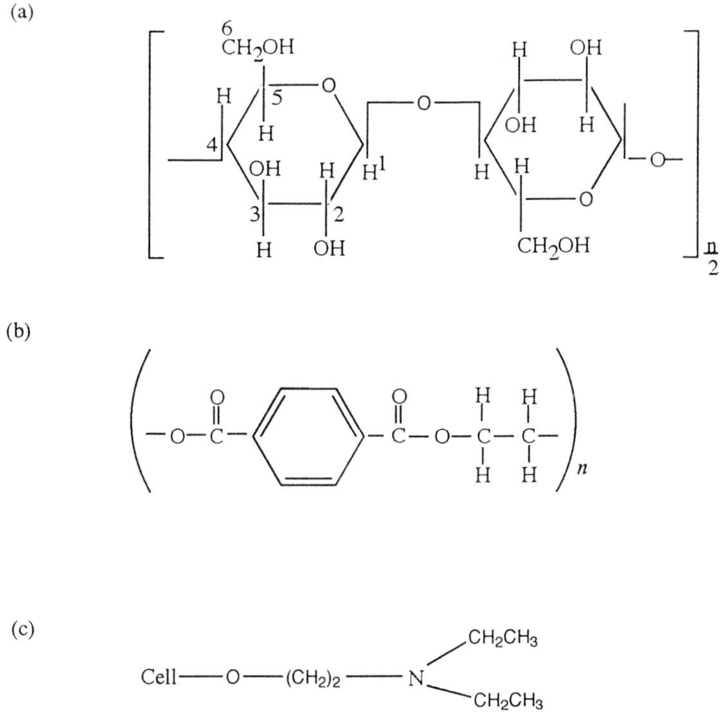

(a)

(b)

(c)

$$Cell\!-\!\!-\!O\!-\!\!-\!(CH_2)_2\!-\!\!-\!N\Big\langle \begin{array}{c}CH_2CH_3\\CH_2CH_3\end{array}$$

Fig. 11.2. Chemical structures of (a) cellulose (cotton), (b) poly(ethylene terephthalate) (polyester), and (c) DEAE cellulose.

Hydrophilic fibers with the appropriate porosities could prove useful as stationary phases for size exclusion chromatography. Derivatized forms of cellulose are viable for ion-exchange chromatography of proteins. Hydrophobic or reversed-phase interactions are possible with aromatic, vinyl copolymer, and carbonaceous fibers. These enable chromatographic separations to take place on the basis of interactions of the biomolecules with hydrophobic or ionic groups on the stationary phase when elution is carried out with an appropriate mobile phase.

This method of construction contrasts with another type of monolithic column developed by Svec *et al.* [7] and Hjertén *et al.* [8]. In their case, the column is prepared by filling it with a polymerization mixture containing porogens. Polymerization entraps the porogens that are subsequently removed by a solvent that is pumped into the column. Dissolution of the porogens leaves a porosity through which eluent flow occurs. Sharp peak and rapid separations are achieved since convective transport dominates diffusive factors throughout the column [9].

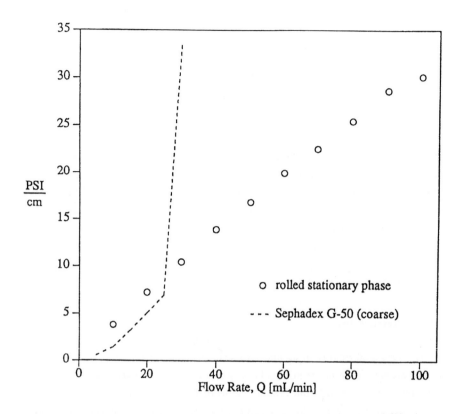

Fig. 11.3. Pressure drop for a 173 × 10 mm i.d. underivatized rolled stationary phase cotton/polyester column (O) and a 90 × 10 mm i.d. bed of Sephadex G-50 (---). (Reprinted with permission from ref. [14], Fig. 3, copyright 1999, by American Chemical Society). Mobile phase for rolled phase: deionized water; the mobile phase for the Sephadex column: 50 mmol/L Tris with 500 mmol/L NaCl at pH 8.

11.2.2 Forming cylindrical columns

11.2.2.1 Fabric preparation

All fabric is cut in a 45° diagonal (bias) direction [10] into swatches 20 cm wide and 60–100 cm long. Previous work on rolled stationary phases showed that columns rolled in the bias direction had a lower pressure drop and a smaller height equivalent of a theoretical plate (HETP) than did columns prepared from fabric rolled in the warp or filling direction [2]. The column is more stable than particulate packings such as Sephadex at higher flow rates (Fig. 11.3). Prior to rolling, the fabrics are scoured to remove antistatic coatings and any other substances that may have adsorbed to the fabric surface during its manufacture. This is accomplished by boiling the fabrics in a dilute detergent solution with sodium bicarbonate for one hour. The fabric is then removed and thoroughly rinsed with deionized (DI) water. All prepared fabrics are stored in DI water at 0–4°C until rolled and packed into a column.

11.2.2.2 Column rolling and packing

The fabric is immersed in DI water at room temperature for 3 h or more before packing in order to ensure pre-shrinkage. The wet fabric is then rolled into a cylinder and packed into a 10 mm (internal diameter) × 200 mm (length) glass tube as indicated in Fig. 11.4 in order to form a liquid chromatography column. The stationary phase is trimmed to length after packing so that one end of the rolled phase is flush with the column end fittings. Plungers are added on each end of the column in order to avoid dead volume [11, 12] (Fig. 11.4).

11.3 CHARACTERIZATION

11.3.1 Pressure drop

The structural rigidity of an underivatized rolled stationary phase is demonstrated by the pressure drop curve in Fig. 11.3. Stable operating pressures were achieved at flow rates up to 100 mL/min (18,000 cm/h) of DI water through a column with an inside diameter of 10 mm. At 25 mL/min, pressure drop of the Sephadex column [13] was 0.2 MPa/cm or about 3.7 MPa over the length of the bed. The particle Reynolds number corresponding to this flow rate was estimated to be about 8.7 [2] and is still considered to be in the laminar regime ($Re_p < 10$), although the decrease in the friction factor begins to level off in this range [15].

11.3.2 Peak broadening (plate height) according to modified van Deemter equation

The scanning electron micrographs of Fig. 11.5(a)-(b) show relatively large interfiber spaces (on the order of micrometers). Convection is likely to occur through both interyarn and interfiber channels, as illustrated schematically in Fig. 11.5(c). The interfiber channels have a small cross-sectional area while the interyarn channels are large. Just as flow in parallel pipes of equal lengths distribute according to cross-sectional areas, flow through these channels are envisioned to distribute in proportion to their cross-sectional areas. Consequently, pore or interfiber velocity, v_{pore}, is a small fraction of the interyarn velocity, and therefore of the chromatographic velocity,

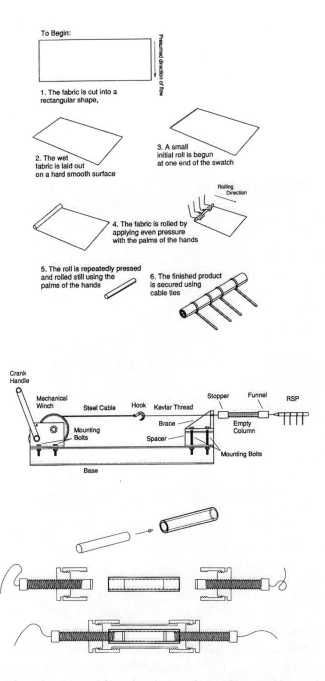

Fig. 11.4. Schematic representation of stationary phase rolling, packing, and the column apparatus used to minimize dead volume at inlet and outlet (Reprinted with permission from ref. [2], copyright 1998 American Chemical Society and American Institute of Chemical Engineers).

v_{chrom}. As the eluent flow rate increases, both the pore and the chromatographic velocity will increase proportionately while their ratio will remain constant, independent of flow rate. At higher flow rates, the pore velocity could become high enough to make convective transport through the interfiber space appreciable relative to diffusive transport. This would in turn affect the overall plate height expression. In order to estimate the contribution of convective intrafiber transport to band spreading, a simplifying assumption of plug flow is made. The effects of convection and diffusion can be added in parallel after Guttman and DiMarzio [16]. The convective contri-

(a)

(b)

(for plate c and legend see next page)

bution of plate height, H for piston flow is independent of velocity. Thus, the total plate height due to the "intraparticulate pore" space at high velocity v_{chrom} is given by equations 11.1 and 11.2:

$$\frac{1}{H_{pore,total}} = \frac{1}{H_{conv}} + \frac{1}{H_{diff}} = \frac{1}{D} + \frac{1}{C'v_{pore}} \tag{11.1}$$

$$H_{total} = A + \frac{B}{v_{chrom}} + \frac{DCv_{chrom}}{D + Cv_{chrom}} \tag{11.2}$$

The total plate height equals A + D when v_{chrom} is large.

11.3.3 Effect of characteristic dimensions on performance

Extensive modeling has been done on the effect of flow rate on pressure drop and plate height for rolled stationary phase columns, but there is relatively little information available on the effect of different fabric treatments on column separation performance. This drove the preliminary work by Keim *et al.* [17] which consisted of treating each fabric swatch with the cellulose enzyme Spezyme® CP, secreted by *Trichoderma longibrachiatum*. The enzyme treatment of cotton print cloth decreased

(c)

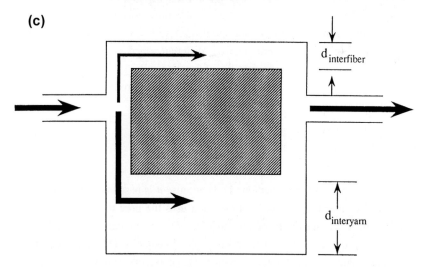

Fig. 11.5 (continued from previous page). Scanning electron micrographs of flat fabric (a), cross-section of rolled fabric (b), and schematic of the parallel flow patterns available, representing flow through the interyarn and interfiber channels (c). There are numerous interfiber channels for every interyarn channel, but for simplicity, the diagram only shows one channel of each kind. Scale bar indicated at bottom of micrographs equals 100 µm (Reprinted with permission from ref. [1], copyright 1992 Elsevier).

the water-accessible (total) pore volume by almost 20% (Table 11.1). Data of Table 11.1 originate from the simple logistic model explained in the next section.

Since enzyme treatment significantly affects the number of ionizable groups that can be attached to the cellulose surface, enzyme treatment may also affect resolution if the column is further derivatized with a ionizable group.

11.3.4 Models

One of the first models to estimate void fraction in rolled stationary phase columns is based on Lin *et al.* [18]. The model, which has four curve-fit parameters, is:

$$I = V_t - V_e = \frac{\alpha}{1 + e^{(\beta - \gamma x)}} \tag{11.3}$$

where I is the inaccessible pore volume, V_e is the elution volume of the probe of interest, V_t is the total accessible pore volume, α, β, and γ are logistic curve fit parameters, and x is the log of the pore diameter, D. Both V_t and V_e are divided by the mass of the rolled stationary phase to normalize the data on a weight basis.

Although the four-parameter logistic function fits the data fairly well, three of the parameters do not have any significant physical meaning and only two of the terms (V_t and α) can be estimated *à priori*. The curve-fitting procedure requires reasonable initial guesses for two parameters (β and γ) that cannot be easily estimated from the data. Thus, comparing the fit parameters from different rolled stationary phase columns does not provide much useful scientific information.

Another model (logistic) has been suggested that applies to rolled stationary phase columns is given by [17]:

$$V_e = \frac{\alpha}{1 + \beta\, e^{-\gamma D}} \tag{11.4}$$

Equation (11.4) differs from equation (11.3) because it is based on accessible void volume (instead of inaccessible void volume) and uses the pore diameter, D, in the exponential term in the denominator instead of $\log D$. The model also uses β as a pre-exponential term in the denominator instead of e^{β} in equation (11.3). For equation (11.4), α is approximately equal to the mass-normalized external void volume, V_0, and β can be calculated from equation:

$$\beta = -\left(\frac{V_0}{V_t}\right)^2 \tag{11.5}$$

TABLE 11.1

FITTED PARAMETERS FOR THE PORE SIZE DISTRIBUTION USING THE LOGISTIC FUNCTION MODEL GIVEN IN EQUATION (11.4). THE PREDICTED AND MEASURED WATER-ACCESSIBLE PORE VOLUME FOR EACH FABRIC IS SHOWN AND LABELED AS TOTAL PORE VOLUME [17].

| | α | V_0 | β | $-(V_0/V_t)^2$ | γ | r | Total pore volume, mL/g | | % Error |
							Predicted	Measured	
Cotton print cloth (CPC)- no enzyme	0.6491	0.647	-0.4602	-0.424	0.0265	0.995	0.351	0.347	1.3%
CPC with 4.1 GCU/g Spezyme	0.5643	0.567	-0.4501	-0.431	0.0249	0.995	0.303	0.297	2.0%
	0.5829	0.586	-0.4471	-0.442	0.0261	0.996	0.304	0.295	2.8%
CPC with 0.82 GCG/g Spezyme	0.6395	0.625	-0.4331	-0.431	0.0270	0.996	0.313	0.327	4.3%
	0.6052	0.592	-0.4329	-0.425	0.0254	0.995	0.304	0.316	3.9%

References p. 253

This logistic model has fewer curve-fit parameters than the model of Lin *et al.* [18]. Two of the three curve fit parameters can be easily estimated from the data.

11.4 APPLICATIONS

11.4.1 Separations

11.4.1.1 Desalting

Rolled stationary phase columns have many applications. One of these application is desalting. Hamaker *et al.* [2,12] showed that a DEAE derivatized 60/40 cotton/polyester blend fabric afforded size exclusion separations of bovine serum albumin (BSA) from sodium chloride which was added to the mobile phase to suppress the charges on the surface of the stationary phase. Li *et al.* [19], was able to show that complete size exclusion separations of BSA from NaCl and BSA from glucose were achieved in less than 8 minutes using cotton print cloth (Fig. 11.6).

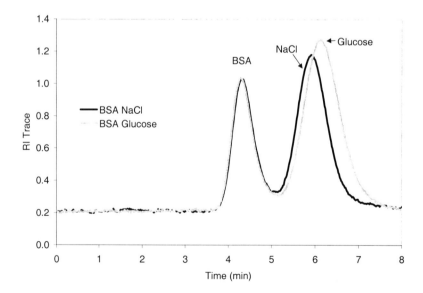

Fig. 11.6. Separation of BSA from NaCl, and from glucose using cotton print cloth rolled stationary phases (Reprinted with permission from ref. [19], copyright 2002 American Chemical Society and American Institute of Chemical Engineers).

11.4.1.2 Protein refolding

Another application for the rolled stationary phase involves refolding of a denatured protein, namely recombinant secretory leukocyte protease inhibitor (rSLPI) as described by Hamaker *et al.* [12]. rSLPI accumulates intracellularly in the form of inclusion bodies in an inactive form. After cell disruption, rSLPI is only partially soluble and to obtain full stability and activity, the protein must be solubilized and then refolded into the proper conformation. Previous work by Seely and Young [20] showed that the protein would refold at a 10-fold dilution factor. This was done because it was thought that if less denaturants were present, the protein would be more susceptible to refold correctly. Hamaker *et al.* [12] showed that the highest yield of refolded protein was obtained for a 3 mL fraction collected 3 min after the 2 mL sample had been injected (Fig. 11.7). This fraction contained 96% of the initial rSLPI injected into the column at a concentration of 1.28 mg/mL. This represents a 1.56x dilution of the protein, compared to a 10-fold dilution used by Seely and Young [20]. Hamaker *et al.* [12] also showed that this approach, if scaled up, would significantly reduce process volumes and therefore improve viability of downstream processing of large amounts of proteins.

11.4.1.3 Gradient chromatography

Rolled stationary phases also have the capability for gradient chromatography. Yang *et al.* [1] was able to separate a four protein mixture of BSA, anti-human IgG, insulin, and -galactosidase using a rolled stationary phase consisting of 95% Nomex (poly(*m*-phenylene isophthalamide)) / 5% Kevlar (poly(p-phenylene terephthalamide)) (Fig. 11.8). The effect of both ionic and hydrophobic interactions between these aramid fibers and proteins can be explained through Table 11.2 that summarizes retention characteristics for the four proteins in five different eluents.

Yang [3] was also able to separate the four-protein mixture stated above using a DEAE (anion exchanger) and sulfated (cation exchanger) cotton stationary phase using stepwise desorption conditions as shown in Tables 11.3 and 11.4.

11.4.1.4 Affinity chromatography of binding domains

Cellulosic affinity chromatography stationary phases must match binding preferences of the binding domain protein and give good flow properties. When using cotton cellulose, the material needs to be pretreated in some manner to remove adsorbed particles first. Otherwise, the cellulose-binding domain protein could cause release of cellulose particles, which in turn would migrate through the column to the retaining

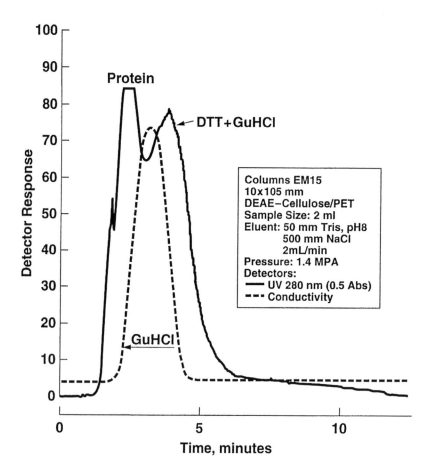

Fig. 11.7. Elution profile of rSLPI/denaturant mixture (Reprinted with permission from ref. [12], Fig. 4, copyright 1999, American Chemical Society). Conditions: flow rate 2 mL/min; buffer of 50 mmol/L Tris-HCl, pH 8.0, 500 mmol/L NaCl; protein concentration of 2 mg/mL dissolved in mobile phase, sample size 2 mL; UV detection at 280 nm.

frit at the outlet and plug it [9]. A suitable matrix is suggested by the work of Yang *et al.* [1] in which packed beds of woven cotton and ramie textiles have the desired porosity characteristics and mechanical flow stability when inserted in a chromatography column.

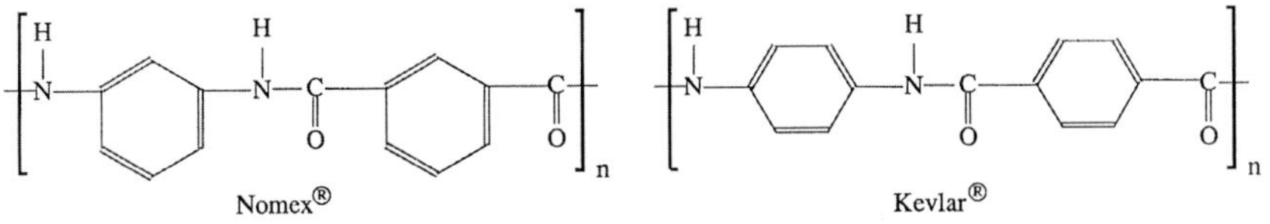

Fig. 11.8. Chemical structures of Nomex and Kevlar aramids.

TABLE 11.2

RETENTION CHARACTERISTICS OF ROLLED ARAMID STATIONARY PHASE [1]

Elution step	Protein [a]			
	IgG	BSA	Insulin	β-Gal
DI water (pH 5.5)	+	+	+	+
10 mmol/L Na$_2$B$_4$O$_7$ (pH 9.2)	–	+	+	+
1 mol/L (NH$_4$)$_2$SO$_4$ (pH 4.5)	–	–	+	+
80 mmol/L NaAc+80 mmol/L HAc (pH 4.7)	–	–	–	+
0.1 mol/L NaCl (pH 5.5)	–	–	–	–

[a] Retention of protein is marked with +, lack of retention with –

TABLE 11.3

EFFECT OF ELUENT PROPERTIES ON ADSORPTION OF PROTEINS ON ROLLED DEAE COTTON COLUMN [3]

Eluent	Protein [a]			
	IgG	BSA	Insulin	β-Galactosidase
DI water (pH 5.5)	+	+	+	+
25 mmol/L Trizma buffer (pH 7.2)	–	+	+	+
50 mmol/L Trizma buffer (pH 7.2)	–	–	–	+
2 mol/L NaCl	–	–	–	–

[a] Retention of protein is marked with +, lack of retention with –

TABLE 11.4

EFFECT OF ELUENT PROPERTIES ON ADSORPTION OF PROTEINS ON ROLLED
SULFATED COTTON COLUMN [3]

Eluent	Protein [a]			
	BSA	IgG	β-Galactosidase	Insulin
DI water (pH 5.5)	+	+	+	+
80 mmol/L NaAc-HAc (pH 4.7)	–	+	+	+
2.0 mmol/L NH$_4$Cl (pH 4.5)	–	–	+	+
5 mmol/L Na$_2$HPO$_4$KH$_2$PO$_4$ (pH 5.9)	–	–	–	+
10 mmol/L Na$_2$B$_4$O$_7$ (pH 9.2)	–	–	–	–

[a] Retention of protein is marked with +, lack of retention with –

11.4.2 Textile properties

There are many properties associated with certain textiles that make them great
candidates for separations. Napping, a mechanical textile finishing process that lifts
the fibers from yarns near the surface of the fabric to produce a hairy or fuzzy surface
[21] was shown to decrease the HETP of BSA and NaCl [19]. Plain weave fabrics
have a pronounced wale on the surface, which forms parallel diagonal ridges [22].
When twill fabrics are rolled into a stationary phase column, the curves between the
ridges may cause molecules to channel and thus result in a 400–500% increase in
HETP (Table 11.5).

Mercerization is a textile wet processing treatment that improves the porosity of
cotton fabrics in the textile industry [23,24]. Li *et al.* [19] showed that mercerization
improved porosity of cotton fabric columns by increasing the internal void fraction by
80% and decreasing the external void fraction by 36%. A side effect of mercerization
is that mercerized fabrics are not as compressible as unmercerized ones, which causes
an increase in HETP of BSA and NaCl by 170% and 270%, respectively. Even though
the unfavorable percent changes in HETP are larger than the favorable changes in
void fractions, the overall resolution was still improved by 35%. This indicates that
the void fractions have a greater effect on resolution than does the plate count. In
summation, mercerized, napped, plain weave fabrics with large yarn diameter, low
fabric count, and high compressibility are desired attributes for a good rolled chromatography column for carrying out size-exclusion chromatography. Of the ten fabrics

TABLE 11.5

PERCENT CHANGES IN THE COLUMN PERFORMANCE PARAMETERS BY NAP-
PING, TWILL, AND MERCERIZATION [19]

	Change in column performance, %		
	Napping [a]	Twill b	Mercerization [c]
H_{BSA}	− 49.9	+ 392.7	+ 171.8
H_{NaCl}	− 31.5	+ 500.8	+ 274.6
ε_p	+ 1	+ 35.9	+ 81.6
ε_b	+ 11.6	− 49.9	− 36.0
R_s	+ 28.2	− 100.0	+ 35.2

[a] Changes from unnapped fabric columns (sea island and cotton print cloth) to napped fabric columns (cotton flannel, cotton velveteen, and double napped cotton blanket) measured at 2 mL/min eluent flow rate.

[b] Changes from plain fabric columns (sea island cotton, cotton print cloth, cotton/polyester blends) to twill fabric columns (microdenier viscose rayon fabric) at 8 mL/min flow rate.

[c] Changes from cotton print cloth columns to mercerized cotton print cloth columns measured at 2 mL/min eluent flow rate.

tested, mercerized cotton flannel meets all the requirements and is the best candidate according to the value of resolution, as indicated in Fig. 11.9.

11.5 PERSPECTIVES AND FUTURE DIRECTIONS

Cellulose based stationary phases could be very useful at the process scale because cellulose is a natural, nontoxic, and renewable hydrophilic polymer. The work that has previously been done with the rolled stationary phase gives excellent preliminary results for its use at the process scale. One consideration on future applications involves work by Keim *et al.* [17]. Since enzyme treatment significantly affects the number of ionizable groups that can be attached to the cellulose surface, this treatment may affect separation resolution differently if the column is further derivatized with ionizable groups.

Li *et al.* [19] showed that cotton flannel met all the requirements for fabrics to make a good separation and is the best candidate based on resolution. This work also indicated that the mass of the stationary phase in the column affects HETP data for different fabric swatches inserted into a 10 mm × 200 mm glass tube. According to

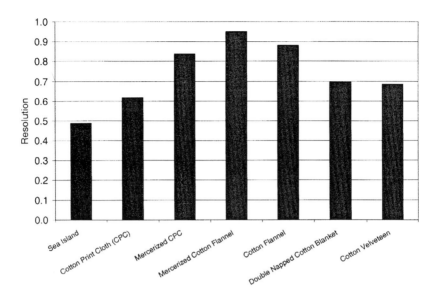

Fig. 11.9. Comparison of resolution in different types of cotton based rolled stationary phases packed in 10 mm × 200 mm liquid chromatography columns (Reprinted with permission from ref. [19], copyright 2002 American Chemical Society and American Institute of Chemical Engineers).

this work, 8.0 g in a 10 mm × 180 mm column gives the best resolution. Hence, further scale-up studies are needed.

11.6 CONCLUSIONS

We have demonstrated the reproducibility and utility of rolled continuous monolithic columns. Scale-up for industrial applications offers the promise of enabling rapid separations of large volumes of bioproducts, specifically for size exclusive separations. Other applications are envisioned for ion exchange and hydrophobic types of separations, as well. The biological origins of many textiles, and their application to separations of bioproducts, leads to the definition of a new type of stationary phase called biotextiles.

11.7 ACKNOWLEDGEMENTS

The material in this work was supported through the Purdue University Agricultural Research programs, and the University of Nebraska, Lincoln. We thank Tom

Huang, Winnie Chen, and Mark Brewer for their helpful comments during the preparation of this manuscript.

11.8 REFERENCES

1   Y. Yang, A. Velayudhan, C.M. Ladisch, M.R. Ladisch, J. Chromatogr. 598 (1992) 169.
2   K.H. Hamaker, J. Liu, R.J. Seely, C.M. Ladisch, M.R. Ladisch, Biotechnol. Prog. 14 (1998) 21.
3   Y. Yang, A. Velayudhan, C.M. Ladisch, M.R. Ladisch,in: G.T. Tsao, A. Feichter (Eds), Advances in Biochemical Engineering/Biotechnology; Springer-Verlag, Berlin, 1993; pp. 148-160.
4   Y. Yang, C.M. Ladisch, Textile Res. J. 63 (1993) 283.
5   V.H Elsasser, Textiles: Concepts and Principles, Delmar Publishers, New York 1997, p. 19.
6   K.L. Hatch, Textile Science, West Publishing Co., Minnesota, 1993, pp. 6, 266.
7   F. Svec, J.M.J. Fréchet, Anal. Chem. 64 (1992) 820.
8   S. Hjertén, L. Liao, R. Zhang, J. Chromatogr. 473 (1989) 273.
9   M.R. Ladisch, Bioseparations Engineering: Principles, Practice, and Economics. Wiley, New York, 2001, pp. 226-229, 693.
10   P.B. Hudson, A.C. Clapp, D. Knees, Joseph's Introductory Textile Science, Harcourt Brace Jovanovich College Publishers, Fort Worth, 1993, p. 19.
11   K.H. Hamaker, M.R. Ladisch, Sep. Purif. Methods 12 (1996) 184.
12   K.H. Hamaker, J. Liu, R.J. Seely, C.M. Ladisch, M.R. Ladisch, Biotechnol. Prog. 12 (1996) 184.
13   Pharmacia L.K.B. Biotechnology A.B. in Gel Filtration: Principles and Methods, 6th ed., 1991, Pharmacia, Uppsala, Sweden.
14   K.H. Hamaker, S-L Rau, R. Hendrickson, J. Liu, C.M. Ladisch, M.R. Ladisch, Ind. Eng. Chem. Res. 38 (1999) 865.
15   F.A.L. Dullien, Porous Media, Fluid Transport and Pore Structure, Academic Press, San Diego, 1992, pp. 240-254.
16   C.M. Guttman, E.A. DiMarzio, Macromolecules 3 (1970) 681.
17   C. Keim, C. Li, C.M. Ladisch, M.R. Ladisch, Biotechnol. Progr. (2002) In press.
18   J.K. Lin, M.R. Ladisch, J.A. Patterson, K. Noller, Biotech. Bioeng. 29 (1987) 976.
19   C. Li, C.M. Ladisch, Y. Yang, R. Hendrickson, C. Keim, N. Mosier, M.R. Ladisch, Biotechnol. Prog. (2002) In press.
20   R.J. Seely, M.D. Young, in: G. Georgiou, E. De Bernardez-Clark (Eds.); ACS Symposium Series No. 470; American Chemical Society, Washington, D.C., 1991; pp 206-216.
21   W.S. Perkins, Textile Coloration and Finishing, North Carolina Academic Press, North Carolina, 1996, chapter 7.
22   S.J. Kadolph, A.L. Langford, Textiles, 8th Ed., Merrill, New Jersey, 1998, pp 193-196.
23   R. Freytag, J.-J. Donzé, in: M. Lewin, S.B. Sello (Eds.) Handbook of Fiber Science and Technology: Volume I Chemical Processing of Fibers and Fabrics: Part A Fundamentals and Preparation, Marcel Dekker, New York, 1983; pp. 93 166.
24   T.L.Vigo, Textile Processing and Properties: Preparation, Dyeing, Finishing and Performance, Elsevier Publishers, New York, 1994, pp 37-42.

F. Švec, T.B. Tennikova and Z. Deyl (Editors)
Monolithic Materials
Journal of Chromatography Library, Vol. 67

Chapter 12

Polymerized High Internal Phase Emulsion Monoliths

Neil R. CAMERON

*Department of Chemistry, University of Durham, South Road, Durham, DH1 3LE,
U.K.*

CONTENTS

12.1 INTRODUCTION

In this chapter, the use of high internal phase emulsions (HIPEs) as templates to prepare porous monolithic polymers, known as polyHIPE (nomenclature devised by Unilever scientists), will be discussed. The materials themselves are highly porous and, usually, extremely permeable, due to a very high degree of interconnection between cavities throughout the matrix. The morphology shown in Fig. 12.1 is quite striking and is rather different from that of foams prepared by other techniques, such as gas blowing. At this point, it is worth defining two parameters that characterise the morphology and that will be referred to in this chapter; the large spherical cavities are termed "voids", whereas the circular holes connecting adjacent voids to one another are known as "interconnects". The term "cells" has been used previously for these cavities, however "voids" is now preferred to avoid confusion when (biological) cells are immobilised in polyHIPE materials.

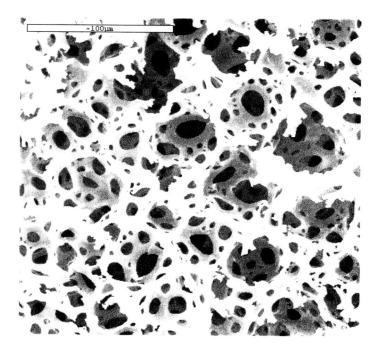

Fig. 12.1. SEM of a typical PolyHIPE material (reprinted with permission from [115], copyright 2000 Royal Society of Chemistry).

12.2 HIGH INTERNAL PHASE EMULSIONS

HIPEs have been known in the literature for a great many years [1]. Their defining characteristic is an internal (droplet) phase volume ratio of greater than 74.05 %, this number representing the maximum volume that can be occupied by uniform spheres when packed into a given space in the most efficient packing arrangement. In HIPEs, therefore, the droplets are either spherical and polydisperse, or are deformed into polyhedra. Like ordinary dilute emulsions, HIPEs can be obtained in either normal (oil-in-water, o/w) or inverse (water-in-oil, w/o) forms. In addition, the production of non-aqueous HIPEs of apolar-in-polar organic solvents has been described [2] and, more recently, Cooper and coworkers have reported the production of supercritical CO_2-in-water (c/w) HIPEs [3]. The preparation of such concentrated emulsions requires careful selection of surfactant, and it is generally accepted that this needs to be very insoluble in the droplet phase, to prevent emulsion inversion at high internal phase volume fractions. The study of HIPEs is a science in itself, and such emulsions have found many applications including safety fuels with low vapour pressures, systems for the recovery of oil and tar, detergents and in numerous cosmetic formulations. However, in this chapter the discussion is limited to the use of HIPEs to prepare monolithic porous polymers and the reader is directed elsewhere for discussions concerning the structure and properties of HIPEs [4] and their use in the preparation of particulate polymeric materials [5].

12.3 PREPARATION OF POLYHIPE MATERIALS

Conceptually, the preparation of porous materials by templating the structure of HIPEs involves polymerisation of the continuous phase around the droplets of the dispersed phase. Thus, the droplets create the voids in the foams and lead to the porous and permeable morphology. In principle, any curable liquid can be used to prepare the porous monolith, however in practice one is limited to liquids that are (i) either sufficiently hydrophilic or hydrophobic to form a stable emulsion and (ii) not adversely affected by water as one of the phases of the emulsion is usually aqueous. The procedure used to prepare polyHIPE materials consists of slowly adding the droplet phase solution to the monomeric continuous phase solution containing the surfactant, with constant mechanical agitation, to form the HIPE, followed by curing of the monomers in some manner (thermal, UV, redox, addition of a catalyst). The resulting porous monolith is washed exhaustively, usually in a Soxhlet apparatus, and then dried *in vacuo*. This process is depicted in Fig. 12.2.

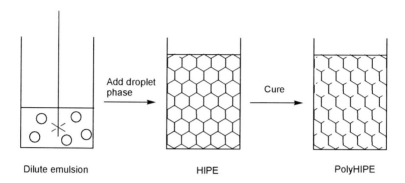

Fig. 12.2. Schematics of the preparation of HIPEs and polyHIPEs.

12.3.1 Chemistry

Both addition and step-growth polymerisation methods have been used to prepare polyHIPE materials. In this section, specific examples of each of these methods will be discussed and the properties as well as potential applications of the resulting monolithic materials will be described.

12.3.1.1 Addition Polymerisation

By far the most common method used to prepare polyHIPE monoliths is free-radical addition polymerisation. This is the case for a number of reasons: vinyl monomers can be sufficiently hydrophobic and can have a reasonably low vapour pressure at moderate temperatures to allow incorporation into a HIPE with an aqueous droplet phase, followed by thermal cure; foams are best prepared if the emulsion can be created in the absence of appreciable polymerisation, to allow efficient mixing of the phases and a homogeneous droplet size distribution – this is easily achievable if heating to e.g. 60°C is required to initiate polymerisation; monomers and initiators are usually tolerant of water; no small molecule by-products that require removal to allow high molar mass material to be obtained are produced.

12.3.1.1.1 Poly(styrene-co-divinylbenzene) PolyHIPE Materials

Foams derived from high internal phase emulsions have been known in the literature for some time. Seminal early work by Bartl and von Bonin [6-13] describes the preparation of highly porous materials using polymerisation in high internal phase emulsions, by thermally initiated radical polymerisation. Amongst the porous polymers described in that work were polystyrene, poly(methyl methacrylate), and copolymers with hydrophobic maleates, which were obtained with nominal porosities of up to 0.88. Subsequently, similar materials were described by Horie *et al.* [14,15], who prepared water-extended polyester resins by copolymerisation of styrene and an unsaturated polyester in a HIPE continuous phase.

However, after these initial studies, the area remained largely dormant until work carried out at Unilever was disclosed in the early 1980's. The system of styrene and divinylbenzene (S/DVB) was the focus of the original patent application [16]; this was stabilised with the nonionic surfactant sorbitan monooleate (Span 80) and polymerisation was initiated thermally by peroxodisulfate ($K_2S_2O_8$). Other monomers were also used to prepare porous materials, including butyl styrene, butyl methacrylate, allyl methacrylate, and monooctyl itaconate. The full patent [17] also describes the ability of such porous monolithic materials to absorb large amounts of liquids, up to 30 times of their own weight in some cases, by capillary action. An early idea for an application for such materials was as a component in highly absorbent substrates for spillage treatment [18,19].

Poly(styrene-*co*-divinylbenzene) polyHIPEs membranes have been prepared as matrices for polymer-dispersed liquid crystals (PDLCs) [20]. The membranes were prepared between indium-tin oxide (ITO) coated glass slides, which were sealed along two edges. The sandwiched membranes were filled with nematic liquid crystals by application of a vacuum. The resulting PDLC showed switchable transmission on application of a pulsed voltage. The use of PS/DVB polyHIPE monoliths as targets for inertially confined fusion (ICF) experiments has also been explored [21-25]. This involves filling the foams with a mixture of deuterium and tritium, then firing a high power laser at the resulting target causing fusion to occur. Composite foams (*vide infra*) were also developed for this purpose. More recently, Zhang *et al.* [26] have described the preparation of deuterated polystyrene polyHIPEs for the same application.

A further application investigated for poly(styrene-*co*-divinylbenzene) polyHIPE materials is as filter bodies. Gregory *et al.* [27] prepared polyHIPE discs, which were compressed before being tested for their filtration characteristics. Such filters were able to effectively remove 95% of polystyrene particles of diameter 1–5 μm from aqueous solution. Similarly, Walsh *et al.* [28-30] describe the use of PS/DVB materials as aerosol filters. The influence of water:monomer phase ratio and HIPE mixing time post-preparation on filtration characteristics were investigated. Optimised materials could remove all particles larger than 1 μm in diameter from gas flows. Similar work by Bhumgara [31] described the production of materials for cross flow microfiltration. In a slightly different vein, workers at Pharmacia [32] detail the preparation of poly(styrene-*co*-divinylbenzene) monolithic stationary phases for HPLC. Monoliths were prepared in moulds, placed in HPLC columns and kept in place either by apply-

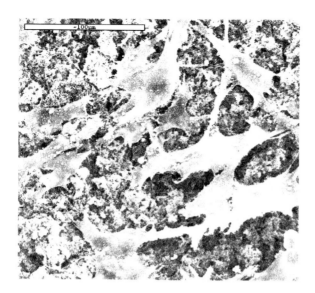

Fig. 12.3. Human skin fibroblasts growing on polyHIPE materials containing poly(ε-caprolactone) (reprinted with permission from [35], Copyright 2001 American Chemical Society).

ing air pressure between the column wall and a rubber lining, or by immobilising with an epoxy resin. The resulting monolithic columns were shown to be able to perform reversed-phase protein separations with plate heights down to 19 µm at 300 cm.h^{-1}. In addition, the pressure drop across such columns was observed to be very low. On a somewhat different note, but still related to the removal of particulates, Williams and Wrobleski suggest using PS/DVB polyHIPEs to entrap intact particles of cosmic dust for detailed study [33].

Biomedical applications for PS/DVB polyHIPE monoliths have also been pursued. Akay and coworkers have investigated the use of PS/DVB monoliths as substrates for cell and tissue growth [34]. Cell types such as chondrocytes, osteoblasts, and fibroblasts were grown on the substrate. Only the former were found to penetrate through the matrix over 21 days. Deposition of hydroxyapatite onto the internal surface of the material resulted in greater penetration and a higher number of attached cells. Busby *et al.* [35] have developed polyester-containing polyHIPE materials as substrates for tissue engineering. Poly(ε-caprolactone) macromonomers were incorporated into PolyHIPE materials, which, as seen in Fig. 12.3, supported the growth of human fibroblasts over a period of 7 days.

12.3.1.1.2 Elastomeric PolyHIPE Materials

Dyer and coworkers [36] at Procter and Gamble describe the production of rubbery polyHIPE materials by copolymerisation of conjugated polyenes, as a means of overcoming the difficulty of copolymerising monomers with high vapour pressures such as butadiene without using pressurised vessels. A more common route to the production of elastomeric emulsion derived foams, however, involves copolymerisation of a hydrophobic monomer such as 2-ethylhexyl acrylate (EHA). Again, this was first disclosed by Unilever [37,38]. A HIPE composed of an organic phase containing styrene, DVB, and EHA could be prepared in a similar fashion and using an identical aqueous phase composition as well as surfactant to those involving only styrene and DVB since EHA does not compromise emulsion stability. Materials with high levels of EHA (>60%) were rubbery at room temperature. The same patent also describes the production of similar materials from butyl acrylate. Analogous materials were later disclosed by researchers at Shell [39] and then Procter and Gamble [40,41]. The latter group was interested in developing materials with high adsorption capacities for bodily fluids, and accordingly demonstrated the ability to efficiently absorb relatively large volumes of synthetic urine with little leakage. Follow-up work relates the production of compressed materials of similar composition that expand and absorb aqueous fluids upon submersion [42]. Subsequently, they describe materials with very high porosities exceeding 99% obtained by using a more efficient surfactant system such as diglycerol monooleate [43,44]. The mechanical properties of the resulting materials were improved by addition of a second crosslinker, such as hexanediol diacrylate. A further use postulated for these EHA-containing monolithic materials is for insulation purposes (thermal, acoustic, or mechanical) [45,46]. The Procter and Gamble team have recently described another use for EHA-containing polyHIPE monolithic materials [47]. It was found that membranes of such foams could be used as efficient stain removers. Crucial to such an application was the incorporation of styrene, as a "toughening monomer" into the HIPE continuous phase, at an optimum level of 35 wt.%. Similar to styrene/DVB polyHIPE monoliths, EHA-derived materials have found use as filters for the removal of particulates. Chang *et al.* describe impregnating a porous substrate, such as a felt mat, with an EHA-containing HIPE and subsequent curing, to yield a high efficiency composite filter material [48]. These were found to operate at lower pressure drops than other high-efficiency filters.

We conducted a detailed study of the thermal properties of EHA-derived polyHIPE foams and those incorporating the similar monomer, 2-ethylhexyl methacrylate (EHMA) [49]. Both sets of materials were found to have a non-linear relationship between composition and T_g, which was explained qualitatively by considering the influence of EH(M)A units on chain flexibility and free volume together with a consideration of the likely comonomer unit diad sequencing predicted from reactivity ratio values.

The EHA-containing materials described so far were all prepared with sorbitan monooleate (Span 80) as surfactant. However it has been reported in the literature that this emulsifier may not be optimal for preparing such HIPEs. For example, Beshouri [50] describes the preparation of EHA-containing HIPEs from a mixture of two differ- ent sorbitan esters. The resulting emulsions were more stable than those prepared from single surfactants and could be obtained with higher internal phase volume fractions. Subsequent work by the same author [51,52] details the use of mixtures of sorbitan esters with a glycerol monofatty acid ester co-surfactant, leading to further optimisation of HIPE stability. The Procter & Gamble group [53] has also outlined alternative surfactants to Span 80. They found that stable HIPEs with uniform droplet sizes could be prepared using polyglycerol aliphatic ether surfactants. They claim that such surfactants are chemically less complex and have a more homogeneous composi- tion than those commonly employed, presumably sorbitan esters. Other investigations into optimised surfactant mixtures for preparing EHA-containing HIPEs have also been carried out. A mixture of a cationic surfactant, such as cetyltrimethylammonium chloride, with either a sorbitan ester [54] or an anionic surfactant (e.g. sodium do- decylbenzenesulfonate) [55], was found to result in a stable HIPE, which could be formed at a water/oil phase volume ratio of up to 60:1. Other advantages, particularly of the latter system, are that overall less surfactant was used (around 6 wt.% relative to the organic phase, compared to 20 wt.% typically employed with Span 80), and that the emulsions could be cured at higher temperatures, such as 100°C, without any noticeable emulsion collapse. Finally, workers at Dow [56] describe the production of stable HIPEs, and foams therefrom, using very low levels of surfactant. As little as 0.125 wt.%, relative to the oil continuous phase, of a polyoxyalkyl-based surfactant could be used to prepare stable HIPEs.

It is common knowledge in surfactant science that mixtures of emulsifiers can lead to more stable emulsions than single surfactants. Explanations for such behaviour are usually based on enhanced interactions between the components of surfactant mix- tures, leading to stronger interfacial films surrounding emulsion droplets. These are likely to be effective in reducing coalescence and Ostwald ripening, two processes leading to emulsion instability. It is conceivable that reasons such as these can explain the behaviour of the above multi-component systems; however, no explanations or supporting data were given as the work is reported in the patent literature.

Heterogeneous EHA-containing monoliths made from mixtures of two different precursor HIPEs have been reported [57-59]. Such materials may have, for instance, a bimodal distribution of void sizes as a result of employing HIPEs of different internal phase volume ratios. Heterogeneous foam formation simply involves mixing the two

HIPEs prior to curing. Presumably the high emulsion viscosity limits transport of aqueous phase liquid between droplets of each emulsion, thus producing the hetero-geneous structure. The Procter and Gamble group [59] also describe the production of striated materials by depositing alternate layers of HIPE in a mould.

Reducing the curing time required to produce monoliths from HIPEs is advanta-geous from the viewpoint of both a reduction in emulsion breakdown prior to gelation and the use of less energy in industrial processes. Brownscombe *et al.* [60,61] discuss prepolymerisation of the monomeric components of the HIPE oil phase as a strategy for achieving this. Such an approach can also be used to incorporate volatile mono-mers such as dienes. The same group [62] has also used a HIPE multi-step cure approach for the same purpose. Another strategy is to cure at elevated temperatures. As little as 4 min [63,64] were required to cure HIPEs of EHA, divinylbenzene, and hexanediol diacrylate at 182°C in a stainless steel pressure vessel. Similarly, Japanese workers [65] describe polymerising HIPEs at 97°C in stainless steel vessels for 7 min. In both these examples, the initiator was peroxodisulfate ($K_2S_2O_8$). However, alterna-tive initiators that can be decomposed by other means have also been employed, again with the objective of reducing curing time. The same Japanese group [66] has used redox initiator combinations, such as ascorbic acid and $FeSO_4$ with hydrogen perox-ide, to yield fully cured HIPEs in around one hour. UV-curing of HIPEs containing EHA has also been described [67]. A wide range of oil-soluble photoinitiators could be used, leading to solid foams following exposure to UV light. However, due to the opacity of the precursor emulsions, monoliths of limited width ($\leq$ 8 mm) could only be produced.

Some work has also been conducted on optimising process conditions in order to improve the homogeneity of polyHIPE monoliths. Desmarais *et al.* [68] claim that degassing the aqueous phase prior to HIPE formation leads to defect-free polyHIPE monoliths, when used in combination with a relatively high polymerisation tempera-ture of 80°C.

12.3.1.1.3 Other PolyHIPE Materials Prepared by Addition Polymerisation

Composite materials, in which an initially prepared porous monolith is impreg-nated with a second foam, have been described as targets for ICF experiments. Wil-liams *et al.* [21-25] prepared PS/DVB polyHIPE materials impregnated with silica aerogels, resorcinol-formaldehyde or phloroglucinol-formaldehyde foams, or polysty-rene, by filling under vacuum. The resulting composites had properties that were a combination of both base materials. For example, silica aerogel materials that were easily handleable and which could be machined without fracture were obtained. Poly-HIPE materials prepared from methacrylonitrile (MAN) have been used as precursors for monolithic porous carbons [69]. Such PMAN materials are made by templating

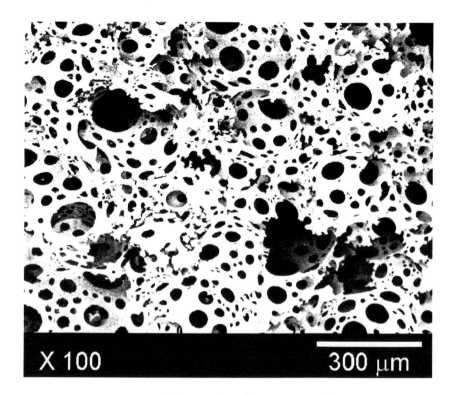

Fig. 12.4. Poly(acrylamide-*co*-N,N'-methylenebisacrylamide) polyHIPE material prepared using supercritical CO_2 as the droplet phase (reprinted with permission from [3], copyright 2001 Wiley-VCH).

"standard" (water-in-oil, w/o) HIPEs and are washed thoroughly prior to being oxidatively stabilised before carbonisation. Resorcinol-formaldehyde (RF) systems have also been prepared for the same purpose, using an o/w HIPE precursor. Finally, a recent development with potential "green chemistry" features describes the use of supercritical CO_2-in-water HIPEs stabilised by poly(vinyl alcohol) to prepare poly-HIPE materials shown in Fig. 12.4 from hydrophilic monomers such as acrylamide and 2-hydroxyethyl acrylate [3].

12.3.1.2 Step-Growth Polymerisation

The vast majority of polyHIPE materials have been prepared by addition polymerisation, however the patent literature describes the production of porous monoliths by step-growth polymerisation [70,71]. A range of different curing chemistries was used, including resorcinol-formaldehyde, urea-formaldehyde, polyvinyl formal, melamine-formaldehyde, and amine-cured epoxy and polyamide, to produce crosslinked, porous materials with void sizes in the range 0.5–100 μm. Such materials would be expected to have different thermal and mechanical properties compared to PS/DVB foams. However there have not been further reports in the literature of these or similar investigations.

12.3.1.3 Organic/Inorganic Composites

A number of organic/inorganic composite emulsion-derived foams are also known. Hubert-Pfalzgraf *et al.* [72,73] describe incorporation of up to 1.5 wt.% of Ti in the form of titania by formation of titanium complexes with polymerisable ligands and subsequent copolymerisation, in a HIPE continuous phase, with styrene and DVB. The resulting porous materials had morphologies similar to those of conventional styrene/divinylbenzene materials, and could potentially be used as reservoirs for liquid deuterium and tritium for inertially confined fusion (ICF) experiments. Workers from Dow [74] detail the addition of various filler materials such as clay particles, iron powder, hollow ceramic microspheres, and aluminium hydroxide to styrene/EHA/DVB foams, to prepare materials with enhanced flame retardancy or sound attenuation properties. It was found that an aluminium hydroxide impregnated material, unlike its unfilled counterpart, was difficult to ignite and quickly self-extinguished. Related work briefly describes the production of flame-retardant foams by the addition of a mixture of highly chlorinated paraffin wax and antimony oxide [75]. Silverstein *et al.* [76] prepared silica-organic polymer composite polyHIPE foams. A siloxymethacrylate monomer was copolymerised with styrene and DVB, and the trimethoxysilyl groups of the resulting monoliths were hydrolysed in-situ due to the acidity of the aqueous phase, arising from the decomposition of the initiator ($K_2S_2O_8$). The resulting materials were true inorganic/organic hybrids at the molecular level, since the silica formed on acidic crosslinking is covalently bound to the organic matrix.

12.3.1.4 Chemical Modification of PolyHIPEs

PolyHIPE monolithic materials have been modified chemically in a variety of ways, each of which falls under the headings of either covalent functionalisation or simple deposition.

12.3.1.4.1 Electrophilic Aromatic Substitution

Since the early 1980's, work on electrophilic aromatic substitution on polyHIPE materials has been performed. Sulfonation is by far the most common modification

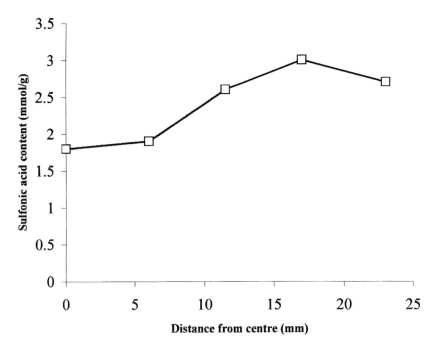

Fig. 12.5. Extent of sulfonation in different areas of polyHIPE monoliths after its reaction with lauroyl sulfate (data from [83]).

carried out. Unilever described sulfonation of poly(styrene-*co*-divinylbenzene) poly-HIPE discs with aggressive reagents such as sulfuric acid and oleum [77,78], yielding materials with degrees of sulfonation of up to 95 %. Schoo *et al.* [79] sulfonated polyHIPE monoliths to immobilise flavin to obtain bioreactors. This method affords reactors with higher activity than routes based on the modification of vinylbenzyl chloride residues or deposition (*vide infra*). Similarly, Akay *et al.* [80-82] sulfonated monoliths with sulfuric acid at 25°C, achieving a degree of modification of around 45 %. However, they did not examine the homogeneity of modification throughout the monolith. We found that a more hydrophobic sulfonating species, such as lauroyl sulfate in cyclohexane, yielded a lower level of surface functionalisation but a much more uniformly sulfonated monolith (Fig. 12.5) [83]. Such monoliths have been used as acid catalysts in the hydration of cyclohexene under two-phase conditions [84]. In addition to catalysing the reaction, the porous interconnected polyHIPE morphology acts as an efficient static mixer providing a large interfacial area between the two liquid phases.

Apart from sulfonation, nitration of poly(styrene-*co*-divinylbenzene) polyHIPE monoliths has been performed using ammonium nitrate and trifluoroacetic anhydride [85], giving a degree of nitration of 8.07 % which corresponds to a loading of nitro groups of 5.7 mmol.g^{-1} [86]. Electron spectroscopic imaging of the surface of the modified material indicated a distribution of nitrogen throughout the porous matrix. Similar to the sulfonation described above, increasing the hydrophobicity of the nitrating reagent by using tetrabutyl ammonium nitrate [83] resulted in a high and uniform degree of modification of large monoliths. This work also describes the uniform bromination of polyHIPE monoliths using tin tetrachloride/bromine in dichloromethane at room temperature.

12.3.1.4.2 Modification of Poly(4-Vinylbenzyl Chloride) Monoliths

PolyHIPE monoliths containing reactive benzyl chloride moieties are readily produced from the hydrophobic monomer 4-vinylbenzyl chloride (VBC), and can be modified easily using a variety of nucleophiles. Workers at Unilever [87-90] describe in the patent literature the modification of such polyVBC monoliths with a wide range of reagents, including amines, carboxylates, and alkoxides. Such modified monoliths were subsequently further elaborated: amines into quaternary ammonium groups, amine salts, amine oxides, and sulfonates; acetate esters into sulfates. Furthermore, carboxylated polyHIPE monoliths were produced by oxidation of benzyl chloride moieties. The ion exchange properties of the resulting materials were then investigated. As mentioned above, VBC polyHIPE was modified with flavin to yield bioreactor devices [79].

More recently, polyVBC monoliths have been surface-grafted with poly(4-vinylpyridine) for the sequestering of heavy metals from aqueous solution [91]. Benzyl chloride residues were modified with sodium thiosulfate, providing sites for the UV-initiated graft copolymerisation of 4-vinylpyridine. The resulting monoliths were able to remove efficiently, under flow-through conditions, Fe(II) and Pu(IV) with faster kinetics than poly(4-vinylpyridine) resin beads. Alexandratos *et al.* also describe the production of polyVBC polyHIPE monoliths for heavy metal chelation via a different route [92]. Benzyl chloride "handles" were in this case modified with either triisopropyl phosphite or sodium tetraisopropylmethylene diphosphonate. The monoliths so prepared were shown to scavenge efficiently various heavy metals such as Eu(III), Fe(III), Cu(II), and Pb(II) from aqueous solution in a batch process.

12.3.1.4.3 Modification of Poly(divinylbenzene) Monoliths

Crosslinked polymers prepared from mixtures containing high contents of divinylbenzene possess a certain level of unreacted residual double bonds. This may be up to 45 mol % of those present at the start of the curing process when 80% DVB is used

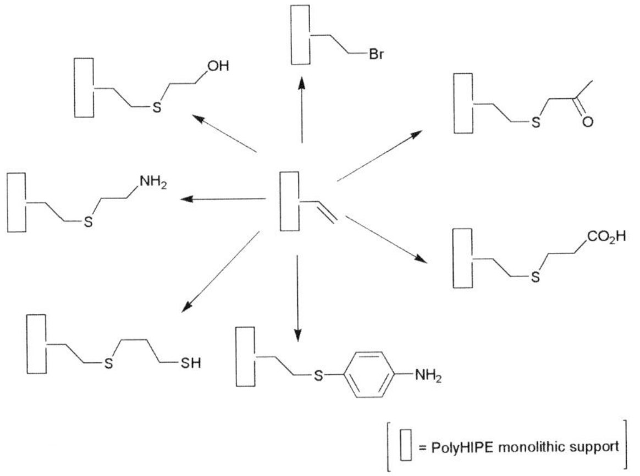

Fig. 12.6. Some chemical modifications of polyDVB polyHIPE monoliths (modified from [94]).

[93]. Such residual groups are amenable to further modifications to introduce useful functionality into the resulting polymers. This approach has been explored by Deleuze in Bordeaux [94-96]. PolyDVB polyHIPE monoliths were modified by hydroboration/oxidation, yielding primary alcohol residues, and by thiolation under radical conditions with a variety of thiols shown in Fig. 12.6. These modification reactions were performed using both batch and flow-through conditions [94]. In general, flow-through processes gave lower loadings [95,96]. Solid state NMR experiments [97] indicated that the thiol reagents were chemically bound to the monolith surface.

12.3.1.4.4 Modification of PolyHIPE Internal Surfaces by Deposition

Simple deposition by impregnating the porous matrix with a solution of the species to be deposited and subsequent drying is a facile method for modification of poly-HIPE monoliths, and has been pursued by several groups. For example, Schoo *et al.* [79] used this approach to deposit a flavin-containing polyelectrolyte onto the surface of PS/DVB polyHIPE monoliths, yielding oxidation biocatalysts. Park and Ruckenstein [98] deposited first a conducting polymer precursor, either pyrrole or thiophene, then an oxidant such as iron trichloride, onto the internal surface of poly(styrene-*co*-divinylbenzene) polyHIPE monoliths. Curing at room temperature followed by 50°C for 24 h afforded conducting porous composites with conductivities up to 4 S cm^{-2}. In subsequent work, Ruckenstein and Chen [99] employed a surface active pyrrole that became anchored to the foam surface by polymerisation. This enhanced the affinity of the porous host for compounds from the aqueous solutions subsequently imbibed.

Enzymes have also been immobilised on polyHIPE surfaces using a similar deposition approach. For example, lipases from *Candida rugera* and *C. cylindrecea* were deposited on poly(styrene-*co*-divinylbenzene) polyHIPE [100,101]. The resulting bioreactors were examined for their ability to hydrolyse tributyrin, olive oil, and triacyl glycerides and were found to have high activities and stabilities. Deposition has also been used as a means to increase the hydrophilicity of the monolith surface, thus enhancing water uptake. Drying hydrophobic monoliths prepared from emulsions with aqueous salt solution dispersed phases results in deposition of the electrolyte on the surface. This, together with the residual surfactant, renders the surface hydrophilic [102-104]. An alternative dopant is a water-soluble polymer, such as polyacrylic acid, which was deposited in a similar manner again to increase the hydrophilicity of the foam [105]. This work was carried out in order to prepare superabsorbent materials for consumer applications.

An electrodeposition approach has been used to prepare porous nickel electrodes. Sotiropoulos *et al.* [106] performed electroless plating of Ni from solutions inbibed into polyHIPE monoliths containing 2-ethylhexyl acrylate (EHA), however deposition was not uniform throughout the matrix. This situation could be improved by using an electrodeposition approach, yielding more uniformly coated materials of surface area similar to commercial Raney nickel catalysts. Further work from the same group [107] involved formation of the polyHIPE monolith around a Ni cathode. Subsequent Ni electrodeposition and polymer burnout yielded a porous Ni coating on the cathode. A later study [108] examined the electrochemical behaviour of such porous Ni electrodes. Current densities in the evolution of hydrogen were comparable to those obtained from sintered Ni electrodes. An alternative approach, in which electroplating takes place in the HIPE precursor during cure has also been described [109]. This yielded monoliths with deposited Ni on their internal surfaces, which had lower surface areas but higher current densities for hydrogen evolution than similar materials prepared by electroplating onto preformed polyHIPEs. Most recently, the same authors [110] have attempted to scale up their electroplating procedure under flow-through conditions. The resulting materials showed similar current densities in hydrogen evolution experiments to conventional Ni electrodes prepared by sintering. However, the preparation procedure of the porous monoliths is significantly easier and less energy intensive than that of the sintered materials.

12.3.2 Control of Morphology and Properties

12.3.2.1 Cellular Structure

PolyHIPE monoliths can be obtained as either closed or open cell materials. In the former, the thin monomer films surrounding adjacent emulsion droplets form intact polymer membranes, trapping the contents of the droplet within the foam. The aqueous phase therefore cannot be removed easily from the monolith, which is consequently of a relatively high density. The latter materials, however, have an open and interconnected structure in which each void is connected to some or all of its neighbours. As a result, these are highly permeable and exhibit a relatively low density, which is largely determined by the internal phase volume fraction of the HIPE precursor.

Williams and Wrobleski [111] undertook a study of the factors that controlled the cellular morphology of poly(styrene-*co*-divinylbenzene) polyHIPE foams. To their surprise, the authors found that the cellular structure was largely insensitive to the ratio of aqueous to oil phases but was affected to a very large degree by the amount of surfactant in the organic phase. Closed cell materials were formed at low (<5 wt.% relative to oil phase) surfactant concentration. Our subsequent work [112] shed some light on the cause of these unexpected results. By performing transmission electron microscopy (TEM) on sectioned, frozen samples of HIPE that had been cured for different periods of time, we found that the appearance of holes between adjacent cavities coincided with the gel-time. This suggests that the interconnecting holes are produced by shrinkage occurring during cure. The influence of surfactant concentration on this process is likely to relate to its effect on the thickness of the films separating emulsion droplets. At higher surfactant concentrations, the films become thinner, thus shrinkage during curing results in the production of interconnects between voids.

12.3.2.2. Void and Interconnect Size

A number of factors have been shown to affect the average size of polyHIPE voids and interconnects. Gregory *et al.* [113] investigated the influence of ionic strength of the aqueous phase on morphology and found that increasingly concentrated solutions of calcium chloride increased both the size and number of interconnecting holes. Similar observations were made independently by workers at Los Alamos National Laboratory [114], who also reported a drop in average void size with increasing either divinylbenzene content in organic phases composed of styrene and divinylbenzene or surfactant content. Further evidence for the strong effect of salt content on void and window size was provided by adding increasing amounts of potassium sulfate to HIPEs prepared with azobisisobutyronitrile rather than $K_2S_2O_8$ as initiator and observing up to a 100-fold drop in void size of the resulting foams.

It is likely that both of these effects i.e. the reduction in void size and the increase in interconnect size are related to emulsion stability. Larger droplet sizes are produced through lower emulsion stability, since the corresponding interfacial area is lower. Similarly, more stable emulsions have thinner films separating adjacent droplets so the foams will have larger interconnects. By increasing the surfactant concentration or making either the organic phase more hydrophobic or the aqueous phase more polar, the emulsion stability increases. Although there is no hard evidence to support this explanation, recent work by Barbetta *et al.* [115] has at least provided a strong indication that this reasoning is correct. It was noticed that foams prepared from DVB and 4-vinylbenzyl chloride exhibited a decrease in void size with increasing VBC content. Experiments involving monolayers of mixtures corresponding to the organic phase spread on a solution with composition identical to the aqueous phase showed that VBC was more strongly adsorbed at the air-water interface than DVB. This suggests that VBC displays co-surfactant behaviour and explains the reduction in void size observed. A similar effect was reported for monoliths derived from HIPEs containing chlorinated solvents in the organic phase [116].

12.3.2.3. Surface Area

PolyHIPE materials have relatively low surface areas of the order of ~5 m^2/g due to the void sizes being in the range of single to tens of micrometers. However, by replacing some of the monomer in the organic phase with a hydrophobic solvent, it is possible to increase surface areas [117]. The resulting monoliths have "dual porosity" – large voids characteristic of polyHIPE materials, and much smaller pores, which can be in the micro-, meso-, or macropore size range, resulting from phase separation of the polymer network from solution during polymerization. Using toluene in a 1:1 ratio relative to the monomer content as a porogen together with a high grade DVB (80%), materials with surface areas of 350 m^2/g were produced. More recently, polyHIPEs with surface areas of 550 m^2/g shown in Fig. 12.7 have been obtained using different porogenic solvents, in particular 2-chloroethylbenzene (CEB) [116]. The higher surface area is a result of the smaller agglomerate size produced during phase separation, which in turn is most likely due to CEB being a better solvent for polyDVB than e.g. toluene.

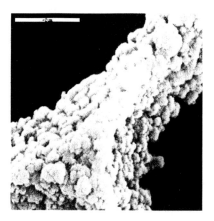

Fig. 12.7. High magnification SEM of a polyHIPE monolith prepared with porogen in the organic phase (reprinted with permission from [116], copyright 2000 Royal Society of Chemistry).

12.3.2.4. Thermal and Mechanical Stability

Poly(styrene-*co*-divinylbenzene) polyHIPE materials possess relatively low thermo-oxidative stability, which may limit their use in certain applications. The thermal behaviour can be enhanced, however, by including maleimide-based monomers. Increasing amounts of ethyl maleimide incorporated into S/DVB PolyHIPE foams increased T_g, as determined by dynamic mechanical thermal analysis (DMTA), to more than 200°C [118]. For comparison, the PS/DVB foam had a T_g of 129°C. Higher T_g values of up to 220°C could be obtained by using a bismaleimide crosslinker in place of DVB [119,120]. Increasing maleimide content also increases thermo-oxidative stability, as determined by thermogravimetry [121]. An extensive study of the effect of N-substituent size of alkyl maleimides indicated that smaller or bulkier moieties such as ethyl or cyclohexyl afford materials with the highest T_g values, while T_g was reduced by using N-propyl or N-butyl maleimide [122]. Independently and at around the same time, we reported producing polyHIPE monoliths by copolymerisation of maleimide-terminated poly(ether sulfone) with styrene, DVB, or a bis(vinyl ether) [123]. Non-aqueous HIPEs [2] were necessary as precursors to the porous materials, as the PES oligomers and styrene or DVB could only be co-solubilised in polar aprotic solvents such as dimethylformamide. The resulting porous monoliths were found to have much better thermo-oxidative stability than PS/DVB materials.

Another attempt to improve the mechanical properties of PS/DVB polyHIPE monoliths is described in the patent literature [124,125]. Mats consisting of thermally bonded bundles of different fibres were impregnated with S/DVB HIPEs and then cured. The composites so prepared were found to have much higher works of fracture, particularly in the dry state, than foams prepared without any added reinforcing material.

12.4 CONCLUSIONS

Templating the structure of high internal phase emulsions is an extremely useful way to prepare a wide variety of porous monolithic materials. Since the structure of the emulsion can be controlled via altering its physicochemical properties, the morphology of the resulting porous materials can be tailored to suit the desired application. Research concerning the production of polyHIPE materials is currently underway in a number of laboratories worldwide and one can expect the manufacturing of novel materials for a wide range of applications.

12.5 REFERENCES

1   K.J. Lissant (ed.), Emulsions and Emulsion Technology Part 1, Marcel Dekker Inc., New York, 1974.
2   N.R. Cameron, D.C. Sherrington, J. Chem. Soc., Faraday Trans. 92 (1996) 1543.
3   R. Butler, C.M. Davies, A.I. Cooper, Adv. Mater. 13 (2001) 1459.
4   N.R. Cameron, D.C. Sherrington, Adv. Polym. Sci. 126 (1996) 163.
5   E. Ruckenstein, Adv. Polym. Sci. 127 (1997) 1.
6   W. von Bonin, H. Bartl, DE 1137554, 1962.
7   W. von Bonin, H. Bartl, DE 1158268, 1963.
8   W. von Bonin, H. Bartl, DE 1160616, 1964.
9   W. von Bonin, H. Bartl, BE 623754, 1963.
10   W. von Bonin, H. Bartl, BE 622655, 1963.
11   H. Bartl, DE 1148382, 1963.
12   H. Bartl, W. Bonin, Makromol. Chem. 66 (1963) 151.
13   H. Bartl, W.v. Bonin, Makromol. Chem. 57 (1962) 74.
14   K. Horie, I. Mita, H. Kambe, J. Appl. Polym. Sci. 11 (1967) 57.
15   K. Horie, I. Mita, H. Kambe, J. Appl. Polym. Sci. 12 (1968) 13.
16   D. Barby, Z. Haq, Eur. Pat. Appl. EP 60138, 1982.
17   D. Barby, Z. Haq, US 4,522,953, 1985.
18   D. Barby, Z. Haq, Eur. Pat. Appl. EP 68830, 1983.
19   D. Barby, Z. Haq, U.S. 4,797,310, 1989.
20   I. Mason, S. Guy, D. McDonnel, N. Cameron, D. Sherrington, Mol. Cryst. Liq. Cryst. 263 (1995) 567.
21   A.M. Nyitray, J.M. Williams, J. Cell. Plastics 25 (1989) 217.

22   A.M. Nyitray, J.M. Williams, D. Onn, A. Witek, Mater. Res. Soc. Symp. Proc. 171 (1990) 99.

23   J.M. Williams, Jr., A.M. Nyitray, M.H. Wilkerson, U.S. 4966919, 1990.

24   J.M. Williams, A.M. Nyitray, M.H. Wilkerson, U.S. 5,037,859, 1991.

25   J.M. Williams, M.H. Wilkerson, Polymer 31 (1990) 2162.

26   L. Zhang, X. Luo, H.-q. Zhang, K. Du, Gongneng Gaofenzi Xuebao 14 (2001) 298.

27   P.G. Cummins, D.P. Gregory, Z. Haq, E.J. Staples, Eur. Pat. Appl. EP 240342, 1987

28   D.C. Walsh, J.I.T. Stenhouse, L.P. Kingsbury, E.J. Webster, J. Aerosol Sci. 27, Suppl. 1 (1996) S629.

29   D.C. Walsh, J.I.T. Stenhouse, L.P. Kingsbury, E.J. Webster, Proc. - World Filtr. Congr., 7th 2 (1996) 699.

30   D.C. Walsh, J.I.T. Stenhouse, L.P. Kingsbury, E.J. Webster, IChemE Res. Event, Eur. Conf. Young Res. Chem. Eng., 2nd 2 (1996) 1072.

31   Z. Bhumgara, Filtr. Sep. 32 (1995) 245.

32   K. Allmer, E. Berggren, E. Eriksson, A. Larsson, I. Porrvik, PCT Int. Appl. WO 9719347, 1997.

33   J.M. Williams, D.A. Wrobleski, J. Mater. Sci. 24 (1989) 4062.

34   G. Akay, S. Downes, V.J. Price, PCT Int. Appl. WO 0034454, 2000.

35   W. Busby, N.R. Cameron, C.A.B. Jahoda, Biomacromolecules 2 (2001) 154.

36   J.C. Dyer, B. Hird, P.K. Wong, S.M. Beshouri, PCT Int. Appl. WO 9621474, 1996.

37   C.J.C. Edwards, D.P. Gregory, M. Sharples, Eur. Pat. Appl. EP 239360, 1987.

38   C.J.C. Edwards, D.P. Gregory, M. Sharples, U.S. 4,788,225, 1988.

39   R.M. Bass, T.F. Brownscombe, U.S. 5,210,104, 1993.

40   T.A. DesMarais, K.J. Stone, H.A. Thompson, G.A. Young, G.D. LaVon, J.C. Dyer, U.S. 5,260,345, 1993.

41   T.A. DesMarais, K.J. Stone, H.A. Thompson, G.A. Young, G.D. LaVon, J.C. Dyer, U.S. 5,268,224, 1993.

42   J.C. Dyer, T.A. Desmarais, G.D. Lavon, K.J. Stone, G.W. Taylor, P. Seiden, S.A. Goldman, H.L. Retzsch, G.A. Young, PCT Int. Appl. WO 9413704, 1994.

43   T.A. Desmarais, K.J. Stone, J.C. Dyer, B. Hird, S.A. Goldman, M.R. Peace, P. Seiden, PCT Int. Appl. WO 9621680, 1996.

44   K.J. Stone, T.A. Desmarais, J.C. Dyer, B. Hird, G.D. Lavon, S.A. Goldman, M.R. Peace, P. Seiden, PCT Int. Appl. WO 9621681, 1996.

45   J.C. Dyer, T.A. Desmarais, PCT Int. Appl. WO 9640824, 1996.

46   J.C. Dyer, T.A. DesMarais, U.S. 5,633,291, 1997.

47   T.C. Roetker, T.A. Desmarais, B.A. Yeazell, PCT Int. Appl. WO 9946319, 1999.

48   B.T.A. Chang, P.K. Wong, T.V. Mai, PCT Int. Appl. WO 9737745, 1997.

49   N.R. Cameron, D.C. Sherrington, J. Mater. Chem. 7 (1997) 2209.

50   S.M. Beshouri, U.S. 5,200,433, 1993.

51   S.M. Beshouri, U.S. 5,334,621, 1994.

52   S.M. Beshouri, PCT Int. Appl. WO 9745457, 1997.

53   S.A. Goldman, J.J. Scheibel, U.S. 5,500,451, 1996.

54   R.P. Adamski, S.M. Beshouri, V.G. Chamupathi, PCT Int. Appl. WO 9745456, 1997.

55   R.M. Bass, T.F. Brownscombe, PCT Int. Appl. WO 9745479, 1997.

56   S.W. Mork, D.P. Green, G.D. Rose, U.S. 6147131, 2000.

57   T.F. Brownscombe, R.M. Bass, P.K. Wong, G.C. Blytas, W.P. Gergen, M. Mores, U.S. 5,646,193, 1997.

58   T.F. Brownscombe, R.M. Bass, P.K. Wong, G.C. Blytas, W.P. Gergen, M. Mores, PCT Int. Appl. WO 9719129, 1997.

59   T.M. Shiveley, T.A. DesMarais, J.C. Dyer, K.J. Stone, U.S. 5,856,366, 1999.

60   T.F. Brownscombe, R.M. Bass, L.S. Corley, U.S. 5,290,820, 1994.

61   T.F. Brownscombe, R.M. Bass, L.S. Corley, U.S. 5,358,974, 1994.

62   T.F. Brownscombe, W.P. Gergen, R.M. Bass, M. Mores, P.K. Wong, U.S. 5,189,070, 1993.

63   T.A. Desmarais, T.M. Shiveley, J.C. Dyer, B. Hird, S.T. Dick, PCT Int. Appl. WO 0050498, 2000.

64   T.A. Desmarais, T.M. Shiveley, J.C. Dyer, B. Hird, PCT Int. Appl. WO 0050502, 2000.

65   K. Nagasuna, H. Fujimaru, K. Kadonaga, K. Nogi, K. Sakamoto, M. Sasabe, K. Minami, PCT Int. Appl. WO 0127164, 2001.

66   K. Kadonaga, A. Mitsuhashi, H. Fujimaru, M. Sasabe, K. Takahashi, M. Izubayashi, PCT Int. Appl. WO 0136493, 2001.

67   K.L. Thunhorst, M.D. Gehlsen, R.E. Wright, E.W. Nelson, S.D. Koecher, D. Gold, PCT Int. Appl. WO 0121693, 2001.

68   T.A. Desmarais, T.M. Shiveley, J.C. Dyer, PCT Int. Appl. WO 0050501, 2000.

69   W.R. Even, D.P. Gregory, MRS Bull. 19 (1994) 29.

70   A.R. Elmes, K. Hammond, D.C. Sherrington, Eur. Pat. Appl. EP 289238, 1988

71   A.R. Elmes, K. Hammond, D.C. Sherrington, U.S. 4,985,468, 1991

72   N. Miele-Pajot, L.G. Hubert-Pfalzgraf, R. Papiernik, J. Vaissermann, R. Collier, J. Mater. Chem. 9 (1999) 3027.

73   L.G. Hubert-Pfalzgraf, L.C. Cauro-Gamet, R. Collier, Abstr. Pap. - Am. Chem. Soc. 221st (2001) INOR.

74   S.W. Mork, J.H. Solc, C.P. Park, PCT Int. Appl. WO 9909070, 1999.

75   Anon, Res. Discl. 439 (2000) 1892.

76   H. Tai, A. Sergienko, M.S. Silverstein, Polymer 42 (2001) 4473.

77   Z. Haq, Eur. Pat. Appl. EP 105634, 1984.

78   Z. Haq, US 4,536,521, 1985.

79   H.F.M. Schoo, G. Challa, B. Rowatt, D.C. Sherrington, React. Polym. 16 (1992) 125.

80   G. Akay, Z. Bhumgara, R.J. Wakeman, Chem. Eng. Res. Des. 73 (1995) 782.

81   Z.G. Bhumgara, R.J. Wakeman, G. Akay, Jubilee Res. Event, Two-Day Symp. 2 (1997) 1121.

82   R.J. Wakeman, Z.G. Bhumgara, G. Akay, Chem. Eng. J. 70 (1998) 133.

83   N.R. Cameron, D.C. Sherrington, I. Ando, H. Kurosu, J. Mater. Chem. 6 (1996) 719.

84   M. Ottens, G. Leene, A. Beenackers, N. Cameron, D.C. Sherrington, Ind. Eng. Chem. Res. 39 (2000) 259.

85   J.V. Crivello, J. Org. Chem. 46 (1981) 3056.

86   I.M. Huxham, P. Hainey, L. Tetley, D.C. Sherrington, Trans. R. Microsc. Soc. 1 (1990) 15.

87   K. Jones, B.R. Lothian, A. Martin, G. Taylor, Z. Haq, Eur. Pat. Appl. EP 156541, 1985.

88   K. Jones, B.R. Lothian, A. Martin, G. Taylor, Z. Haq, U.S. 4,612,334, 1986.

89   K. Jones, B.R. Lothian, A. Martin, G. Taylor, Z. Haq, U.S. 4,611,014, 1986.

90   K. Jones, B.R. Lothian, A. Martin, G. Taylor, Z. Haq, U.S. 4,668,709, 1987.

91   B.C. Benicewicz, G.D. Jarvinen, D.J. Kathios, B.S. Jorgensen, J. Radioanal. Nucl. Chem. 235 (1998) 31.

92   S.D. Alexandratos, R. Beauvais, J.R. Duke, B.S. Jorgensen, J. Appl. Polym. Sci. 68 (1998) 1911.

93   R.V. Law, D.C. Sherrington, C.E. Snape, Macromolecules 30 (1997) 2868.

94   A. Mercier, H. Deleuze, O. Mondain-Monval, Actual. Chim. (2000) 56.

95   A. Mercier, H. Deleuze, O. Mondain-Monval, React. Funct. Polym. 46 (2000) 67.

96   A. Mercier, H. Deleuze, O. Mondain-Monval, Macromol. Chem. Phys. 202 (2001) 2672.

97   A. Mercier, S. Kuroki, I. Ando, H. Deleuze, O. Mondain-Monval, J. Polym. Sci., Part B: Polym. Phys. 39 (2001) 956.

98   J.S. Park, E. Ruckenstein, J. Electron. Mater. 21 (1992) 205.

99   E. Ruckenstein, J.H. Chen, J. Appl. Polym. Sci. 43 (1991) 1209.

100 E. Ruckenstein, X. Wang, Biotechnol. Bioeng. 42 (1993) 821.

101 E. Ruckenstein, X. Wang, Biotechnol. Tech. 7 (1993) 117.

102 T.A. Desmarais, PCT Int. Appl. WO 9304113, 1993.

103 T.A. Desmarais, K.J. Stone, PCT Int. Appl. WO 9304115, 1993.

104 T.A. DesMarais, K.J. Stone, U.S. 5,292,777, 1994.

105 M.A. Mitchell, A.S. Tomlin, U.S. 5,900,437, 1999.

106 S. Sotiropoulos, I.J. Brown, G. Akay, E. Lester, Mater. Lett. 35 (1998) 383.

107 I.J. Brown, D. Clift, S. Sotiropoulos, Mater. Res. Bull. 34 (1999) 1055.

108 I.J. Brown, S. Sotiropoulos, J. Appl. Electrochem. 30 (2000) 107.

109 I.J. Brown, S. Sotiropoulos, Electrochim. Acta 46 (2001) 2711.

110 I.J. Brown, S. Sotiropoulos, J. Appl. Electrochem. 31 (2001) 1203.

111 J.M. Williams, D.A. Wrobleski, Langmuir 4 (1988) 656.

112 N.R. Cameron, D.C. Sherrington, L. Albiston, D.P. Gregory, Colloid Polym. Sci. 274 (1996) 592.

113 D.P. Gregory, M. Sharples, I.M. Tucker, Eur. Pat. Appl. EP 299762, 1989

114 J.M. Williams, A.J. Gray, M.H. Wilkerson, Langmuir 6 (1990) 437.

115 A. Barbetta, N.R. Cameron, S.J. Cooper, Chem. Commun. (2000) 221.

116 N.R. Cameron, A. Barbetta, J. Mater. Chem. 10 (2000) 2466.

117 P. Hainey, I.M. Huxham, B. Rowatt, D.C. Sherrington, L. Tetley, Macromolecules 24 (1991) 117.

118 M.A. Hoisington, J.R. Duke, P.G. Apen, Polym. Mater. Sci. Eng. 74 (1996) 240.

119 M.A. Hoisington, J.R. Duke, D.A. Langlois, Mater. Res. Soc. Symp. Proc. 431 (1996) 539.

120 M.A. Hoisington, J.R. Duke, Int. SAMPE Tech. Conf. 28 (1996) 1317.

121 M.A. Hoisington, J.R. Duke, P.G. Apen, Polymer 38 (1997) 3347.

122 J.R. Duke, M.A. Hoisington, D.A. Langlois, B.C. Benicewicz, Polymer 39 (1998) 4369.

123 N.R. Cameron, D.C. Sherrington, Macromolecules 30 (1997) 5860.

124 Z. Haq, Eur. Pat. Appl. EP 110678, 1984.

125 Z. Haq, U.S. 4,473,611, 1984.

F. Švec, T.B. Tennikova and Z. Deyl (Editors)
Monolithic Materials
Journal of Chromatography Library, Vol. 67
© 2003 Elsevier Science B.V. All rights reserved.

Chapter 13

Imprinted Monoliths

Börje SELLERGREN

University of Dortmund, INFU, Otto Hahn Strasse 6, 44221 Dortmund, Germany

CONTENTS

13.1 INTRODUCTION

Recent advances in molecular imprinting have opened up the possibility to custom make robust molecular recognition elements [1-5]. A range of applications has been foreseen for these so-called molecularly imprinted polymers (MIPs). For instance, stable recognition elements capable of strong and selective binding of molecules could be used in areas with urgent needs for methods enabling selective separations, extractions [6,7], and sensing [3] of low molecular weight or macromolecular targets in biological fluids. Alternatively, such recognition elements could be used to separate

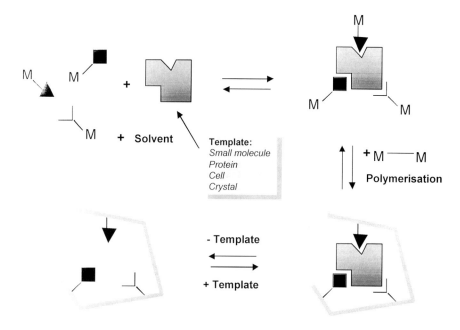

Fig. 13.1. Concept of molecular imprinting using a template and complementary functional monomers to generate recognition sites in network polymers.

undesirable compounds from foods or biological fluids [8,9] for targeted delivery of drugs [10,11], or for demanding separations at preparative level in the fine chemical industry [12]. If these recognition elements can be designed to bind specific proteins, a number of important applications in areas such as biotechnology including down-stream processing, sensors, diagnostics, as well as for enhancing selectivity and sensitivity in the new field of proteomics, can be envisioned [13].

MIPs are highly reticulated network polymers consisting of a common matrix structure, most often poly(methacrylate)-based, and binding sites formed by a template present during the polymer synthesis (Fig. 13.1). In spite of the promising properties of MIPs, few of the above mentioned applications have been explored. This can be attributed partly to difficulties in generating high affinity binding sites while simultaneously controlling the porous properties and morphology of the materials. This is particularly problematic in non-covalent imprinting, where the binding site structure depends more strongly on the polymerization conditions, e.g. solvent, temperature and concentrations [14]. Thus, general methods to prepare MIPs in the shape of beads [15] or to control their porous properties are thus far not available and most work to date has involved the classical monolith approach where the materials are prepared in bulk and obtained as monoliths that need to be crushed and sieved prior to use. In this process typically only 10–20% of the material can be recovered as useful particles, often precluding the imprinting of precious templates. From this aspect, *in situ* preparation techniques would be very attractive [16]. For instance, if the synthesis conditions could be chosen to create macropores in the monoliths, allowing convective mobile phase transport, they could be directly applied in the many chromatographic modes already demonstrated for MIPs. The *in situ* polymerized organic separation media introduced by Svec and Fréchet [17] acted as a catalyst for this work and the first *in situ* prepared MIPs appeared soon thereafter [18-20]. This chapter presents an overview of the work performed to date in this field and will also show some new promising approaches to tailor the porous properties of MIPs.

13.2 APPROACHES TO THE DESIGN OF IMPRINTED BINDING SITES

The three dimensional arrangement of the binding functional groups in MIPs is obtained by linking the functional monomers covalently or non-covalently to the template during polymerization (Fig. 13.2, step (i)). Removal of the template from the formed polymer then generates a structure complementary to the template structure. These sites can be reoccupied by the template or an analogous structure by reformation of the binding interactions present during synthesis or, alternatively, weaker kinetically more favorable interactions (Fig. 13.2, step (ii)). Essentially, three main approaches exist to date to generate high fidelity imprinted sites distinguished by the nature of the linkage during synthesis and during rebinding. The covalent-covalent

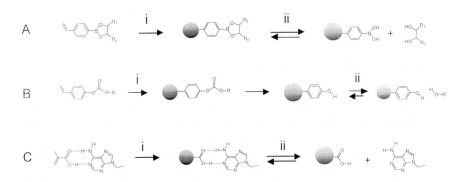

Fig. 13.2. Design of high affinity imprinted binding sites. The approaches differ in the nature of the bonds between the template and the functional monomers during polymerization (i) and between the binding site functional groups and a bound guest molecule (ii): covalent-covalent (A), covalent-noncovalent (B), and noncovalent-noncovalent (C).

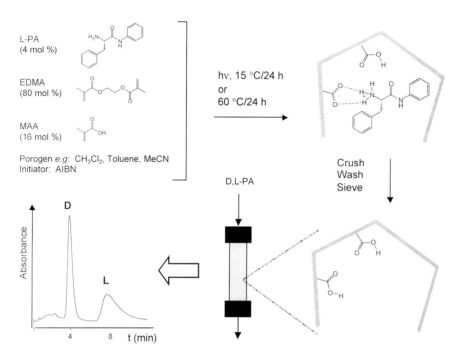

Fig. 13.3. Protocol used for the imprinting of small basic molecules and subsequent assessment of selectivity by liquid chromatography.

approach has the advantage of a known stoichiometry between the functional monomer and the template (Fig. 13.2A) [4,21,22]. Provided that the template can be recovered in high yields, a high density of well-defined sites can be expected. One problem with this approach is the limited number of covalent linkages that satisfy these criteria. Furthermore, considerable synthetic effort may be required to prepare the template and slow kinetics is often observed for rebinding by reformation of the covalent bond. This approach is therefore difficult to combine with chromatographic applications where fast on-off kinetics are required. In this respect, the use of sacrificial spacers has found more widespread use [23-25]. Here the functional monomer is bound to the template through a disposable spacer that is removed after polymerization is completed. This results in a disposition of the functional groups allowing rebinding to occur through hydrogen bonding interactions (Fig. 13.2B). Therefore, this approach can be more amenable to chromatographic applications and furthermore compatible with the conditions needed to create the perfusive pores. However, the most straightforward approach without doubt involves functional monomers that are chosen to associate non-covalently with the template (Fig. 13.2C) [14,26]. Here the template is directly mixed with one or several functional monomers and can be easily extracted from the polymer after polymerization and used repeatedly (Fig. 13.3). Generally, the resulting materials can be directly applied to perform separations with high affinity and selectivity in the liquid chromatographic mode.

For example, a simple commodity monomer such as methacrylic acid (MAA) can be used to create good binding sites for a large variety of template structures containing hydrogen bond- or proton- accepting functional groups (see Fig. 13.3 for the imprinting of L-phenylalanine anilide (L-PA)) [27]. MAA forms complementary hydrogen bonds or hydrogen-bonded ion pairs with the template with individual binding constants ranging from units for weak hydrogen bonds to several hundreds for cyclic hydrogen bonds or hydrogen-bonded ion pairs formed in weakly polar aprotic solvents such as chloroform [28]. Particularly strong binding occurs with amidine or guanidine containing templates which form cyclic hydrogen-bonded ion pairs with the functional monomer (*vide infra*) [19]. In these cases binding constants over 10^6 mol^{-1} have been estimated [29] and the template is here stoichiometrically associated with the monomers during polymerization, leading to a high yield of templated sites. Based on the principle of reciprocity, amidine monomers have also been synthesized for stoichiometric imprinting of acid-containing templates [29,30]. Otherwise, templates containing acids are often well targeted using basic functional monomers, such as vinylpyridine, or amide monomers, such as methacrylamide [31,32]. The latter is also suited for targeting primary or secondary amide functionalities. Finally, several structural motifs can only be bound with sufficient strength using specifically designed functional monomers. A detailed account of this work can be found elsewhere [1]. For templates with multiple functionalities, obtaining the best result may require the use of a multitude of the above functional monomers. Often, the optimum combination is only found after time-consuming trial and error, which has spurred the development of faster synthesis and assessment techniques [33-36] as well as computational approaches [37].

13.3 STRUCTURAL CONTROL OF BINDING SITES AND PORE SYSTEM IN NON-COVALENT IMPRINTING

The degree of template complexation prior to polymerization reflects the amount of binding sites in the resulting MIP and, usually, the goal is to maximize this number. In addition to the relative concentrations and the type of functional monomer and template, this is also affected by the solvent [27], temperature [38], and pressure [39]. In non-covalent imprinting the association is usually based on enthalpically driven formation of hydrogen bonds or ion pairs, and is therefore favored by the use of weakly polar aprotic solvents and low temperatures. However, a number of entropi-

TABLE 13.1

CHROMATOGRAPHIC AND POROUS PROPERTIES OF POLYMERS IMPRINTED USING POROGENS DIFFERING IN HYDROGEN BOND CAPACITY[a] (Reprinted with permission from ref. [27], copyright 1993 Elsevier Sciences B.V.)

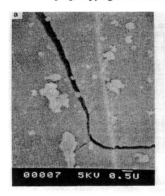 

Dichloromethane   Acetonitrile

Porogen	H-bond capacity[b]	Swelling (ml/ml)	Pore volume (ml/g)	Separation factor
Acetonitrile	P	1.36	0.60	2.6
Dichloromethane	P	2.01	0.007	2.4
Dimethylformamide	M	1.97	0.17	NR
Tetrahydrofuran	M	1.84	0.24	1.5
Acetic acid	S	1.45	0.52	NR

[a]Separation conditions: Mobile phase: MeCN–H$_2$O–HOAc (92.5:2.5:5), injection: 100 nmol D,L-PA. The polymers were prepared using L-PA as template and 83% EDMA as a crosslinking monomer. The scanning electron micrographs were obtained at a magnification of 10 000. NR = not resolved.

[b]P - poor, M - moderate, S - strong.

cally driven associations have also been utilized to generate imprinted sites. Here more polar protic solvents can be used during the polymerization [9,40-42]. However, there is little room to play in choosing the solvent of polymerization for each given system. We observed in our studies of the imprinting of a chiral model template, L-phenylalanine anilide (L-PA), that the hydrogen bond capacity of the porogen was the overriding factor controlling the selectivity of the materials and that the enantioselectivity decreased when more polar protic solvents were used as porogens (Table 13.1) [27]. For instance, porogenic mixture consisting of cyclohexanol and dodecanol,

TABLE 13.2

HPLC EVALUATION OF POLYMERS IMPRINTED USING TERBUTYLAZINE AS TEMPLATE AND DIFFERENT POROGENS a

Porogen	Dichloromethane		Toluene		Toluene–Isooctane 75:25	
Solute	k'_{MIP}	k'_{Blk}	k'_{MIP}	k'_{Blk}	k'_{MIP}	k'_{Blk}
Terbutylazine	3.0	0.4	4.4	0.3	3.3	0.09
Ametryn	0.9	0.5	1.0	0.3	0.6	0.07
Prometryn	0.9	0.5	0.8	0.2	0.6	0.07
Atrazine	2.3	0.5	3.1	0.3	2.2	0.08

[a]The polymers were prepared following a slightly modified procedure of the one shown in Fig. 13.3 using terbutylazine as template and the indicated porogens. k'_{MIP} = retention factor for imprinted polymers, k'_{Blk} = retention factor for blank non-imprinted polymer. Chromatographic conditions: Mobile phase: $CH_3CN–CH_3COOH–H_2O$ (92.5:5.0:2.5); Injection: 10 nmol in 10 μL. Flow rate: 1 ml/min; Wavelength of detection: 260 nm.

commonly used to create the channel-like pores (pore diameter >1000 nm) in organic monoliths, resulted in poor recognition and weak binding [19].

It was also seen that the enantioselectivity of MIPs did not correlate with the dry state morphology of the materials. For example MIPs that were identical except for the porogen used during polymerization, high selectivity was observed for both gel-like non-porous materials and macroporous materials (see Table 13.1). Porogens classified as aprotic and with a low hydrogen bond capacity can therefore be used to control the porous properties of MIPs without seriously compromising the selectivity of the materials. This effect is well demonstrated on triazine imprinted polymers prepared using the binary toluene/isooctane porogen system (Table 13.2). In this case the recognition properties were similar for polymers with very different pore sizes. As will be discussed below, similar systems have been used to create imprinted monolithic chiral stationary phases for CEC [43].

Along with the selectivity of the materials, the affinity and distribution of the binding sites and the kinetic properties of the materials are important measures of the efficiency of the templating process. MIPs commonly contain a distribution of binding sites of various strengths [44,45], which directly reflects the degree of complexation occurring in solution prior to polymerization. However, this comparison assumes that these binding sites are fully accessible which, in turn, requires a high accessible surface area. For monoliths containing convective pores this may be difficult to achieve [46,47].

Chromatographic separations using MIPs often suffer from poor efficiency, resulting in broad and asymmetric peaks (Fig. 13.3). An important contribution to this arises from the slow intraparticle diffusion occurring within materials containing a relatively large volume of micropores [48,49]. This is an additional factor that need to be considered when optimizing MIPs for flow through applications.

In view of the typical conditions used to obtain monolithic separation media [47], it is clear that imprinting is compatible with the formation of convective pores from all respects, except with respect to the porogen.

13.4 APPROACHES TO OBTAIN IMPRINTED MONOLITHS WITH FLOW THROUGH PORES

Following the introduction of the monolithic column format [17], several approaches combining this technology with molecular imprinting have appeared. They differ with respect to how the superpores are generated and their main features and relative benefits will be discussed below.

13.4.1 Solvents as porogens

13.4.1.1 Protic solvents

The first examples of *in situ* prepared imprinted monoliths were reported by Matsui [18] and Sellergren [19]. Similar to the systems developed by Svec and Fréchet [17,51], these relied on the use of mixtures of good and poor solvents to generate the flow-through pores. As discussed in detail by Viklund *et al.* [47], an increase in the relative amount of a poor solvent for the growing polymer chains leads to an increased preference for the polymerization to occur inside monomer-swollen nuclei. This leads to the formation of larger globules that agglomerate and account for the large pore sizes. With MAA as functional monomer, EDMA as crosslinking monomer, and mixtures of cyclohexanol and dodecanol as porogens, templates such as L-phenylalanine anilide, atrazine and diaminonaphtalenes were imprinted *in situ* in stainless steel [18] or glass columns [19]. In the latter case, the polymers could also be prepared by photochemical initiation allowing the polymerization to be performed at low temperatures and in a short period of time. These conditions are beneficial for many non-covalent imprinting systems (*vide supra*). After the preparation, the glass columns could be cut to the desired length and directly applied as chromatographic stationary phases. However, due to the unsuitable porogen they exhibited much poorer recognition prop-

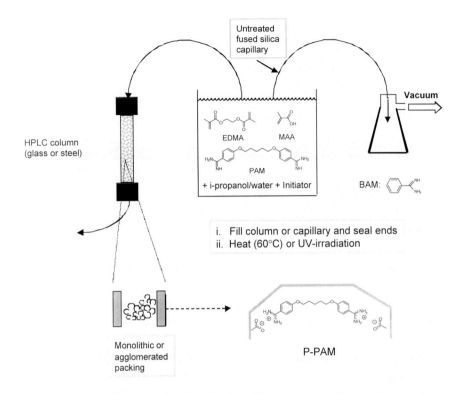

Fig. 13.4. *In situ* preparation of molecularly imprinted polymeric packings for liquid chromatography or capillary electrochromatography (Reprinted with permission from ref. [19], copyright 1994 Elsevier Sciences B. V.). Example of HPLC packings imprinted with pentamidine (PAM): PAM (0.125 mmol) in the free base form was dissolved in 2-propanol (2.8 ml) and EDMA (12 mmol). Addition of MAA (0.5 mmol) caused formation of a precipitate which dissolved after addition of water (1.3 ml). Initiator (AIBN, 12 mg) in 2-propanol (0.5 ml) was added, the solution purged with nitrogen, and heated to 40°C to achieve homogenization. The solution was then transferred under nitrogen to glass tubes (l=150 mm, o.d.=5 mm, i.d.=3 mm), the tubes sealed, and left in an oven at 60°C for 24 h.

erties compared to the conventional crushed monoliths. In contrast, for more strongly associating monomer-template systems, a higher selectivity was observed. This was the case in the imprinting of the bis-benzamidine pentamidine (PAM), that form strong complexes with MAA through hydrogen bonded ion pairs (Fig. 13.4) [19]. In this case, addition of water and increase in the solvent to monomer ratio was necessary to reach a compromise between template solubility and the flow-through properties of the monolith. A low total monomer concentration (20 %) and polar solvents favored the formation of individual large agglomerates of globular micrometer-sized particles (Fig. 13.5), whereas at a total monomer concentration of 40 %, the polymer

Fig. 13.5. Scanning electron micrographs of a polymer imprinted with pentamidine obtained at a magnification of 2000 (a) and 10 000 (b) (Reprinted with permission from ref. [19], copyright 1994 Elsevier Sciences B. V.).

was obtained as a continuous block that could not be dispersed in any solvent. Using a mixture of isopropanol and water as the porogen and a low monomer concentration, dispersible stationary phases for liquid chromatographic applications could thus be prepared. These phases exhibited a very low back pressure of less than 7 MPa even at a flow rate of 5 ml/min. The retention of pentamidine and the reference benzamidine (BAM) (Fig. 13.6) on a pentamidine (P-PAM) and a benzamidine (P-BAM) column were compared using an organic/aqueous mobile phase. While at pH=2 both compounds eluted essentially in the column void volume, pentamidine was 7 times more retained on the pentamidine than on the benzamidine column at pH=5 (Fig. 13.6). Benzamidine on the other hand showed weaker pH dependence and was equally retained on both columns.

The same monomer/porogen system was also used to generate the first *in situ* prepared methacrylate based monolithic packings for CEC [50]. Untreated fused silica capillaries were filled with a mixture of monomers (EDMA and MAA), porogen, and initiator and the polymerization was initiated thermally. This simple procedure resulted in packings exhibiting surprisingly high separation efficiencies. The electrolyte could be mechanically pumped through the capillaries, allowing rapid phase changes and micro-chromatographic possibilities. Due to the weak cation exchange properties of the carboxyl groups containing phase, CEC of basic analytes with organic solvent-buffer mixtures as electrolytes was successful, resulting in plate numbers for D,L-phenylalanine anilide of 150 000 m^{-1} (Fig. 13.7) and for the nucleoside derivative tri-O-acetyladenosin of 53 000 m^{-1}. Analogous to the LC applications, a phase for capillary affinity electrochromatography could be prepared by including pentamidine in the polymerization mixture. The effect of pH was more pronounced in CEC than in LC (Fig. 13.8). At pH=4 the retention times for benzamidine and pentamidine were

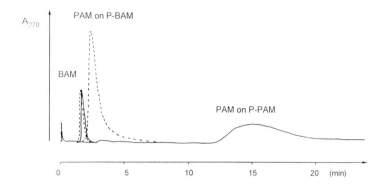

Fig. 13.6. Elution profiles of pentamidine (PAM) and benzamidine (BAM) injected separately (2 nmol) on a PAM imprinted (P-PAM) (straight line) and a BAM imprinted (P-BAM) (dashed line) dispersion polymer prepared in situ in a chromatographic column (Reprinted with permission from ref. [19], copyright 1994 Elsevier Sciences B. V.). Mobile phase: acetonitrile–0.05 mol/l potassium phosphate buffer pH 5, 70:30; flow rate: 0.3 ml/min.

7.2 and more than 50 min, respectively, on a pentamidine-selective capillary, while they were only 6.5 and 7.2 min on a reference capillary imprinted with benzamidine. The pentamidine selective capillary exhibited a significant selectivity for pentamidine that could be controlled by pH of the electrolyte.

Other templates such as nicotine [52], and cinchona alkaloids [53], have also been found to be compatible with the more polar protic porogen system. For example, imprinting of cinchona alkaloids using MAA or 2-(trifluoromethyl)acrylic acid (TFM) as functional monomers and cyclohexanol/dodecanol as porogenic system, resulted in polymers capable of discriminating between cinchonine and its antipode cinchonidine (Fig. 13.9) [53]. Using either of these compounds as a template and MAA or TFM as functional monomers resulted in four imprinted polymer rods. Comparing the separation factors α using acetonitrile as the mobile phase, it was found that the two templates exhibit different preference for the functional monomers. Thus, cinchonidine was best imprinted using MAA (α=2.6), whereas cinchonine was best targeted using TFM (α=5.3).

13.4.1.2 Aprotic solvents

As discussed in the previous paragraph, the use of aprotic less polar porogens may overcome the destabilizing effect on the monomer-template interactions typical of the polar porogen systems. This was demonstrated by Nilsson [43] and Lin [54] with the

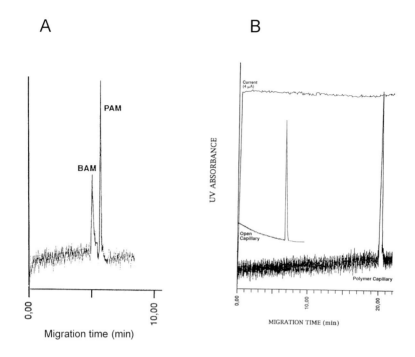

Fig. 13.7. Electropherograms of pentamidine (PAM) and benzamidine (BAM), using a capillary containing in situ prepared packing imprinted with PAM (P-PAM) (A) and D,L-phenylalanine anilide (D,L-PA) on a capillary column containing in situ prepared packing imprinted with L-PA (B) (Reprinted with permission from ref. [50], copyright 1994 Elsevier Sciences B. V.). Conditions: (A) Running buffer: acetonitrile–0.5 mol/l potassium phosphate pH 2, 70:30; Injection: 5 kV, 3 s; Applied voltage: 5 kV; Current, 4.2-4.5 μA; Detection, BAM 254 nm, PAM 280 nm. (B) Column: 25 cm × 100 μm i.d.; Running buffer: acetonitrile–0.5 mol/l potassium phosphate pH 5, 70:30; Injection: 0.33 mg/ml, 6 kV, 3 s; Applied voltage, 6 kV; Current, 3.8 μA; Detection, 254 nm. The appearance of the peak in an identical run using an open, deactivated capillary is shown in the inserted profile.

aim at developing new chiral stationary phases for CEC. Schweitz *et al.* used toluene [43] or isooctane/toluene mixtures [55] to achieve the convective pore system. To obtain the monolith, the template (e.g. (R)-propranolol) initiator, MAA and the crosslinker trimethylolpropane trimethacrylate (TRIM) were dissolved in toluene. After degassing, this mixture was transferred to a fused silica capillary containing a UV transparent protective coating and which inner wall had been pre-treated with [2-(methacryloyloxy) propyl]trimethoxysilane. The ends of the capillary were sealed and the UV-polymerization carried out at –20°C in a freezer. Using toluene as the sole porogen, the polymerization needed to be prematurely terminated after 80 min in order to obtain the desired flow-through pores, whereas in the presence of isooctane, the polymerization was allowed to continue for 16 hours to achieve high conversions.

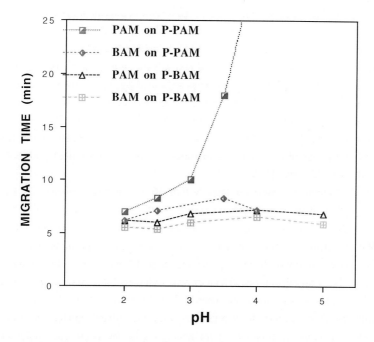

Fig. 13.8. Plot of the migration time of pentamidine (PAM) and benzamidine (BAM) as a function of pH of the electrolyte using capillaries containing packings imprinted with PAM (P-PAM) and BAM (P-BAM). Conditions see Fig. 13.7A.

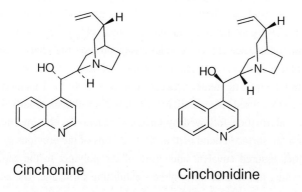

Fig. 13.9. Structures of cinchona alkaloids used as templates for the preparation of imprinted monoliths.

In the former procedure, unreacted monomers were removed by pressurized washing and replaced by the electrolyte to be used in the subsequent CEC separations. A near base line resolution of (R)- and (S)-propranolol could be achieved in less than two minutes and even non-racemic mixtures could be analysed (Fig. 13.10). The packings consisted of micrometer-sized globular particles separated by superpores with diameters of 1–20 μm. Due to the irregular pore structure and the attachment of the polymer to the capillary wall, some of these packings resemble capillary coatings. Brüggemann *et al.* [56] and Tan and Remcho [57] used similar protocols to prepare capillaries for open tubular chromatography and CEC, sometimes resulting in separations with both high selectivity and efficiency [57].

Based on the toluene/isooctane porogen system, [55] Schweitz performed an extensive optimization of the polymers with respect to the performance in CEC. This included use of different crosslinking monomers such as TRIM, EDMA, pentaerythritol triacrylate and pentaerythritol tetraacrylate, and variations in the total monomer concentration, concentration of the (S)-ropivacaine template, and the isooctane content in the porogen. The best separation efficiency was obtained for capillaries prepared using TRIM as crosslinking monomer whereas highest selectivity was seen using EDMA. Assuming from the high back pressure during the solvent replacement step, the TRIM based packing is probably more analogous to the continuous polymer monoliths reported by Svec *et al.* [17,51]. High selectivities were obtained for high ratios of MAA to template, whereas the opposite was true for the resolution. An increase in the flow resistance was observed for lower ratios implying that this factor needs to be optimized when other templates are to be used.

A somewhat different approach to prepare the monoliths was used by Lin *et al.* [54,58]. L-Phenylalanine anilide was imprinted using MAA and 2-vinylpyridine as the functional monomers, EDMA as the crosslinking monomer, chloroform as the porogenic solvent and ammonium-acetate as the additive to provide the charge carriers during subsequent solvent replacement. The molar ratio of functional monomers to crosslinker was 1:5 and the total monomer concentration was *ca* 40 %. This monomer mixture was transferred to a fused silica capillary which walls had been modified with vinyl groups through chlorination followed by Grignard reaction with vinyl magnesium bromide. As with the surface-bound methacrylate groups of the previous system, the vinyl functionalities assured covalent anchoring of the polymer to the capillary wall, which is of particular importance for polymers exhibiting shrinkage. The polymerization reaction took place at 60°C for 24 hours. Details concerning the replacement of the porogen with the electrolyte buffer are unfortunately not revealed. Nevertheless, these capillaries could be used to achieve baseline separation of the enantiomers of the corresponding free amino acid phenylalanine. It should be noted that in bulk format, these polymerization conditions lead to monolithic materials that lack flow

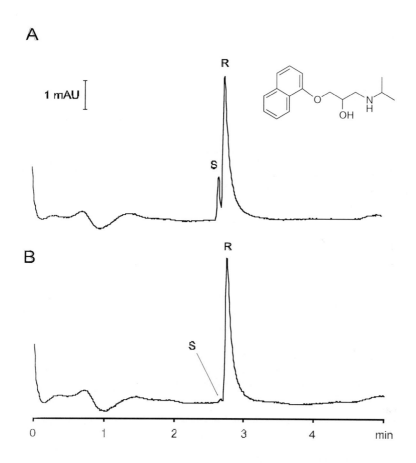

Fig. 13.10. CEC of non-racemic mixtures of propranolol on a monolithic capillary column prepared using (R)-propranolol as template. Separation of a 9:1 (A) and 99:1 (B) mixture of (R)- and (S)-propranolol. (Reprinted with permission from ref. [43], copyright 1997 Elsevier Sciences)

through pores and appear to be gel-like in the dry state (see Table 13.1). The effect of ammonium acetate in this regard is not known.

In spite of the successful demonstrations discussed above, the use of good solvents as porogens suffers from the difficulties in achieving high affinity binding sites and channel like pores at the same time. As demonstrated in many reports, [55,56] small changes in the concentrations or type of the components in the polymerization mixture may affect the pore size distribution leading to excessive flow resistance. Thus, for each new template, it may be necessary to completely re-optimize the conditions. In contrast, approaches where these properties are optimized separately may offer a more general solution to the problem.

13.4.2 Silica as porogen

Beds of porous silica particles may be used as a pore template for polymerizations leading to crosslinked organic polymers [59-62]. After the polymerization is completed, the mold is removed by dissolving it in an aqueous fluoride solution, leaving behind a pore system that resembles the solid part of the mold. Mallouk *et al.* demonstrated this approach using a compressed bed of monodisperse silica nanoparticles with diameters of 15 or 35 nm as templates (Fig. 13.11) [59]. The interstitial volume of the bed was filled with divinylbenzene containing a radical initiator, and the polymerization performed at elevated temperature. The silica was then dissolved by treatment with hydrofluoric acid. This process resulted in a porous organic monolith in which the average pore diameter could be controlled by the size of the silica nanoparticles. A similar approach was also used by Remcho to prepare a new class of monolithic stationary phases for LC and CEC [62]. The columns were packed with the silica particles followed by filling the interstitial void with monomer solution. After polymerization the silica mold was dissolved resulting in a monolithic stationary phase.

Instead of using bare silica particles, the use of porous silica containing template molecules immobilized at the inner surface may lead to imprinting of both pores and molecular binding sites. This hierarchical templating approach allows the structure of the templated binding site to be optimized independently from the pore structure of the monolith. Moreover these sites are confined at the surface of pores for which the size can be tuned depending on the pore size of the original pore templates. Yilmaz *et al.* first demonstrated the feasibility of this approach by using wide pore silica particles containing surface-immobilized theophylline [60]. Simultaneously, our group demonstrated the application of this concept to the preparation of stationary phases for chromatography (Fig. 13.12) [61]. Adenine and triaminopyrimidine derivatives were here used as templates immobilized to silica with 10 nm average pore diameter. Our polymer was the more broadly applicable methacrylate resin based on EDMA/MAA. After dissolving the silica beads, spherical particles with an average diameter close to that of the original template were obtained (Fig. 13.13 A,B). The surface area of 293 m^2/g determined for this material reflected that of the precursor particles (350 m^2/g) and the particles exhibited a narrow pore size distribution with a maximum around 8–9 nm (Fig. 13.14). By post-reacting the polymer-bound carboxylic acid groups with a fluorescent label, the reaction could be clearly visualized (Fig. 13.13 C,D). As seen in Table 13.3 these materials exhibited selectivity in the liquid chromatographic mode for their respective templates.

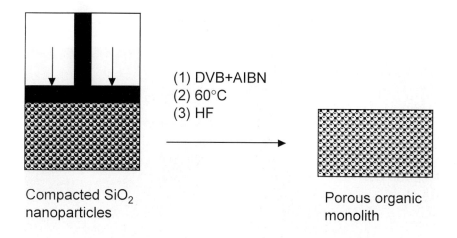

Fig. 13.11. Approach used by Mallouk *et al.* for the preparation of polymeric replicas of the interstitial voids formed between compressed monodisperse silica nanoparticles [59].

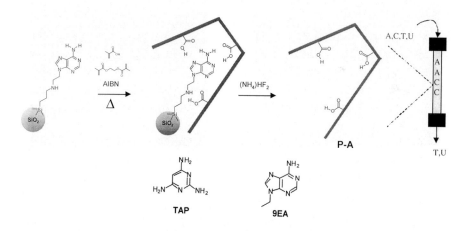

Fig. 13.12. Preparation of affinity stationary phases (P-A) containing surface confined templated sites for adenine (Reprinted with permission from ref. [61], copyright 2002 American Chemical Society).

A   B

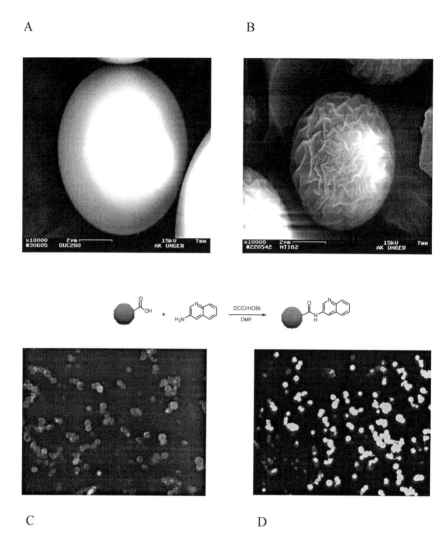

C   D

Fig. 13.13. Scanning electron micrographs of the precursor silica template (A), P-TAP after dissolution of the silica template (B) at a magnification of 10 000, and fluorescence micrographs of P-A before (C), and after (D) dissolution of the silica template. The particles were postlabeled with 3-aminoquinoline as indicated in the equation (Reprinted with permission from ref. [61], copyright 2002 American Chemical Society).

13.4.3 Grafting techniques and particle immobilization

Another means of decoupling the imprinting step from the pore formation process consists in grafting where the MIPs are grown on preformed support materials of known morphology [63-67]. These coatings have been prepared by grafting polymers to a variety of supports. This approach relies on use of surfaces containing polym-

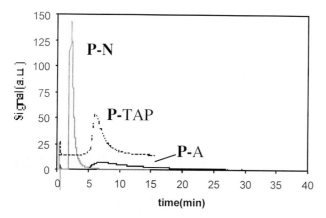

Fig. 13.14. Pore size distribution profiles of the imprinted polymers (P-A, P-TAP, and P-N) and the silica template calculated from the desorption branch of the isotherms obtained from nitrogen adsorption/desorption measurements [61].

TABLE 13.3

RETENTION OF TEMPLATES AND STRUCTURAL ANALOGUES ON COLUMNS PACKED WITH HIARCHICALLY IMPRINTED POLYMERS PREPARED USING ADE-NINE-SILICA AND TRIAMINOPYRIMIDINE-SILICA AS TEMPLATES (Reprinted with permission from ref. [61], copyright 2002 American Chemical Society)

Template	Polymer	Solute	k_P	k_{P-N}	IF $(=k_P/k_{P-N})$
	P-A	9-EA	18	5.3	3.3
		TAP	39	17	2.4
		Thymine	1.1	0.8	1.5
	P-TAP	9-EA	16	5.3	3.1
		TAP	59	17	3.6
		Thymine	1.4	0.8	1.8

Polymers P-A and P-TAP imprinted with adenine (A) or triaminopyrimidine (TAP) were obtained using A or TAP modified silica beads as templates. P-N was synthesized analogous-ly to P-A and P-TAP but using aminopropylsilica beads as template. The selectivity is reflected in the imprinting factor IF which for P-A defined as IF $=k'_{P-A}/k'_{P-N}$. Chromato-graphic conditions: Mobile phase: acetonitrile/acetic acid (99/1); Flow rate: 1.0 mL/min; Injection: 10 nmol in 10L.

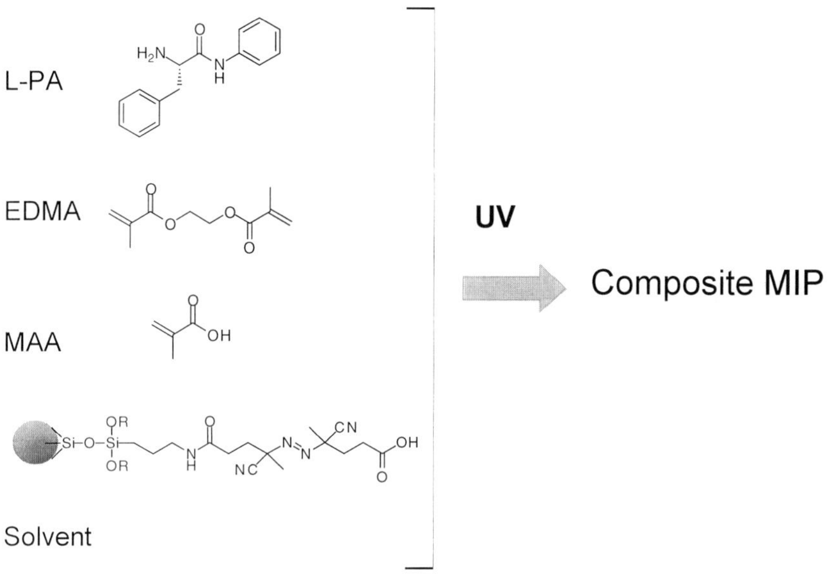

Fig. 13.15. Preparation of supported imprinted polymer films by UV irradiation of initiator-modified porous silica particles suspended in a mixture of template (L-PA), monomers, and solvent. (Reprinted with permission from ref. [71], copyright 2002 American Chemical Society)

erizable double bonds that can add to the growing polymer chains in solution, thus linking them to the surface. Since the initiator is present in solution, propagation occurs in the solution phase and the attachment mainly results from reaction between growing radicals and the surface anchor groups. Therefore the thickness of the result-ing film is determined by the amount of monomers added to the support. The control of homogeneity of these films and their thickness are therefore limited. By using initiator attached to the support however, the initiation occurs at the surface, and the polymer is grafted from this site [68]. Since this involves reactions mainly between monomers and the surface, a higher density of grafted polymer chains can be achieved and the film thickness better controlled [69,70]. We first demonstrated that this tech-nique could be applied to the preparation of imprinted composite beads (Fig. 13.15) [71], using silica particles containing surface-bound free radical initiators as supports. The column efficiency of such MIPs imprinted with L-phenylalanine anilide depended strongly on the amount of grafted polymer. For example, silica having 10 nm average pore diameter, grafted with a 0.8 nm thin film exhibited the highest column efficiency, giving a plate number of about 700 m^{-1} for the imprinted enantiomer and of up to 24 000 m^{-1} for the antipode. This resulted in base line resolutions on a 33 mm long column in less than 5 min (Fig. 13.16). The materials could be prepared in a short time

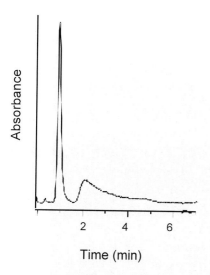

Fig. 13.16. Elution profiles of D,L-phenylalanine anilide (D,L-PA) on a column packed with silica particles modified with a L-PA imprinted film. Conditions: Mobile phase: acetonitrile–0.01 mol/l sodium acetate buffer pH 4.8, 70:30; Injection: 10 μl of a 1 mmol/l solution; Column: 33 × mm i.d.; Support: average pore diameter of: 10 nm, average particle diameter 10 μm; Average film thickness 0.8 nm (Reprinted with permission from ref. [71], copyright 2002 American Chemical Society)

(1-2 hours) and could be applied in HPLC, as well as in capillary electrochromatography (CEC) [72]. By grafting the polymers from the surface of non-porous materials, monolithic composite materials was prepared consisting of nonporous 2.8 μm particles connected to each other via a web-like polymer structure with micrometer-sized interstitial pores (Fig. 13.17) [71]. The grafting step here plays two roles: First, it introduces surface functional properties and, second, it immobilizes the bed of microparticles in columns or capillaries.

13.5 CONCLUSIONS

A number of approaches combining imprinting at the molecular level with the formation of a desired pore structure have led to the development of new separation techniques and sorbents. Materials with convective flow-through pores containing binding sites for small molecules can be prepared *in situ* at the location of their use. Although the single step molding, in which the desired binding sites and pores are formed simultaneously, is attractive, it suffers from the sensitivity of the porous properties on the type and amount of the most important variable, the template. This

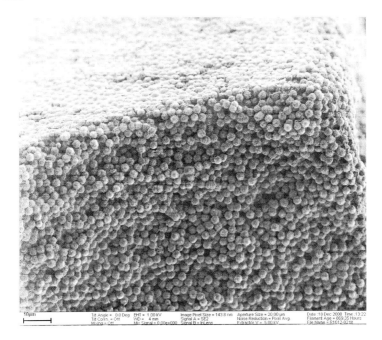

Fig. 13.17. Scanning electron micrograph of non-porous SiO_2 particles with an average particle diameter of 2.8 μm modified with an initiator and used for grafting of imprinted polymer to achieve a carbon content of 16 wt%. Adapted from reference [71].

problem is solved by the use of two-step techniques, such as hierarchical templating or polymer grafting.

13.6 ACKNOWLEDGEMENTS

The author is grateful to Andrew J. Hall for reviewing the manuscript and to Dr. Francesca Lanza for providing Table 13.2.

13.7 REFERENCES

1   B. Sellergren (Ed.), Molecularly imprinted polymers: Man made mimics of antibodies and their applications in analytical chemistry. In: Techniques and instrumentation in analytical chemistry, Vol. 23, Elsevier Science B.V., Amsterdam 2001.

2   B. Sellergren, Angew. Chem. Int. Ed. 39 (2000) 1031-1037.

3   K. Haupt, K. Mosbach, Chem. Rev. 100 (2000) 2495-2504.

4   G. Wulff, Angew. Chem., Int. Ed. Engl. 34 (1995) 1812-32.

5   K.J. Shea, Trends Polym. Sci. 2 (1994) 166-173.

6   L.I. Andersson, J. Chromatogr. B. 745 (2000) 3-13.

7   F. Lanza, B. Sellergren, Chromatographia 53 (2001) 599-611.

8   M.J. Whitcombe, M.E. Rodriguez, E.N. Vulfson, Spec. Publ. - Royal Soc. Chem. 158 (1994) 565-71.

9   B. Sellergren, J. Wieschemeyer, K.-S. Boos, D. Seidel, Chem. Mat. 10 (1998) 4037-4046.

10   P. Bures, Y. Huang, E. Oral, N.A. Peppas, J. Control. Release 72 (2001) 25-33.

11   W.J. Wizeman, P. Kofinas, Biomaterials 22 (2001) 1485-1491.

12   B. Sellergren, J. Chromatogr. A 906 (2001) 227-252.

13   P.K. Dahl, M.G. Kulkarni, R.A. Mashelkar in: B. Sellergren (Ed.), Molecularly imprinted polymers: Man made mimics of antibodies and their applications in analytical chemistry, Vol. 23, Elsevier Science B.V., Amsterdam 2001, p. 271-293.

14   B. Sellergren in: B. Sellergren (Ed.), Molecularly imprinted polymers: Man made mimics of antibodies and their applications in analytical chemistry, Vol. 23, Elsevier Science B.V., Amsterdam 2001, p. 113-184.

15   A.G. Mayes in: B. Sellergren (Ed.), Molecularly imprinted polymers: Man made mimics of antibodies and their applications in analytical chemistry, Vol. 23, Elsevier Science B. V., Amsterdam 2001, p. 305-324.

16   J. Matsui, T. Takeuchi in: B. Sellergren (Ed.), Molecularly imprinted polymers: Man made mimics of antibodies and their applications in analytical chemistry, Vol. 23, Elsevier Science B. V., Amsterdam 2001, p. 325-340.

17   F. Svec, J.M.J. Fréchet, Anal. Chem. 64 (1992) 820.

18   J. Matsui, T. Kato, T. Takeuchi, M. Suzuki, K. Yokoyama, E. Tamiya, I. Karube, Anal. Chem. 65 (1993) 2223-2224.

19   B. Sellergren, J. Chromatogr. A 673 (1994) 133-141.

20   B. Sellergren, Anal. Chem. 66 (1994) 1578-1582.

21   G. Wulff, A. Biffis in: B. Sellergren (Ed.), Molecularly imprinted polymers: Man made mimics of antibodies and their applications in analytical chemistry, Vol. 23, Elsevier Science B. V., Amsterdam 2001, p. 71-111.

22   G. Wulff, Mol. Cryst. Liq. Cryst. Sci. Technol. A 276 (1996) 1-6.

23   M.J. Whitcombe, E.N. Vulfson in: B. Sellergren (Ed.), Molecularly imprinted polymers: Man made mimics of antibodies and their applications in analytical chemistry, Vol. 23 2001, p. 203-212.

24   J.U. Klein, M.J. Whitcombe, F. Mulholland, E.N. Vulfson, Angew. Chem. Int. Ed. 38 (1999) 2057-2060.

25   M.J. Whitcombe, M.E. Rodriguez, P. Villar, E.N. Vulfson, J. Am. Chem. Soc. 117 (1995) 7105-7111.

26   R.J. Ansell, D. Kriz, K. Mosbach, Curr. Opin. Biotechnol. 7 (1996) 89-94.

27   B. Sellergren, K.J. Shea, J. Chromatogr. 635 (1993) 31-49.

28   M. Quaglia, K. Chenon, A.J. Hall, E. De Lorenzi, B. Sellergren, J. Am. Chem. Soc. 123 (2001) 2146-2154.

29   G. Wulff, R. Schönfeld, Adv. Mater. 10 (1998) 957-959.

30   G. Wulff, T. Gross, R. Schönfeld, Angew. Chem. Int. Ed. Engl. 36 (1997) 1962-9164.

31   O. Ramström, L.I. Andersson, K. Mosbach, J. Org. Chem. 58 (1993) 7562-7564.

32   C. Yu, K. Mosbach, J. Org. Chem. 62 (1997) 4057-4064.

33   F. Lanza, B. Sellergren, Anal. Chem. 71 (1999) 2092-2096.

34   T. Takeuchi, D. Fukuma, J. Matsui, Anal. Chem. 71 (1999) 285-290.

35   F. Lanza, A.J. Hall, B. Sellergren, A. Bereczki, G. Horvai, S. Bayoudh, P.A.G. Cormack, D.C. Sherrington, Anal. Chim. Acta 435 (2001) 91-106.

36   T. Takeuchi, A. Seko, J. Matsui, T. Mukawa, Instrum. Sci. Technol. 29 (2001) 1-9.

37   S. Piletsky, personal communication.

38   B. Sellergren, Makromol. Chem. 190 (1989) 2703-2711.

39   B. Sellergren, C. Dauwe, T. Schneider, Macromolecules 30 (1997) 2454-2459.

40   K. Haupt, A. Dzgoev, K. Mosbach, Anal. Chem. 70 (1998) 628-631.

41   H. Asanuma, M. Kakazu, M. Shibata, T. Hishiya, M. Komiyama, Chem. Commun.
      (1997) 1971-1972.

42   S.A. Piletsky, H.S. Andersson, I.A. Nicholls, Macromolecules 32 (1999) 633-636.

43   L. Schweitz, L.I. Andersson, S. Nilsson, Anal. Chem. 69 (1997) 1179-1183.

44   Y. Chen, M. Kele, P. Sajonz, B. Sellergren, G. Guiochon, Anal. Chem. 71 (1999)
      928-938.

45   K.D. Shimizu, R.J. Umpleby, II, Polym. Prepr. 41 (2000) 0032-3934.

46   C. Viklund, E. Pontén, B. Glad, K. Irgum, P. Hörstedt, F. Svec, Chem. Mater. 9 (1997)
      463-471.

47   C. Viklund, F. Svec, J.M.J. Fréchet, Chem. Mater. 8 (1996) 744-750.

48   K. Miyabe, G. Guiochon, Biotechnol. Prog. 16 (2000) 617-629.

49   B. Sellergren, K.J. Shea, J. Chromatogr. A 690 (1995) 29-39.

50   K. Nilsson, J. Lindell, O. Norrlöw, B. Sellergren, J. Chromatogr. A 680 (1994) 57-61.

51   E.C. Peters, M. Petro, F. Svec, J.M.J. Fréchet, Anal. Chem. 70 (1998) 2288-2295.

52   J. Matsui, T. Takeuchi, Anal. Commun. 34 (1997) 199-200.

53   J. Matsui, I. Nicholls, T. Takeuchi, Anal. Chim. Acta 365 (1998) 89.

54   J.-M. Lin, T. Nakagama, K. Uchiyama, T. Hobo, J. Liq. Chromatogr. Relat. Technol.
      20 (1997) 1489-1506.

55   L. Schweitz, L.I. Andersson, S. Nilsson, J. Chromatogr. A 792 (1997) 401-409.

56   O. Brüggemann, R. Freitag, M.J. Whitcombe, E.N. Vulfson, J. Chromatogr. A 781
      (1997) 43-53.

57   Z.J. Tan, V.T. Remcho, Electrophoresis 19 (1998) 2055-2060.

58   J.-M. Lin, T. Nakagama, K. Uchiyama, T. Hobo, J. Pharm. Biomed. Anal. 15 (1997)
      1351-1358.

59   S.A. Johnson, P.J. Ollivier, T.E. Mallouk, Science 283 (1999) 963-965.

60   E. Yilmaz, K. Haupt, K. Mosbach, Angew. Chem., Int. Ed. 39 (2000) 2115-2118.

61   M.M. Titirici, A.H. Hall, B. Sellergren, Chem. Mater. 14 (2001) 21-23.

62   G.S. Chirica, V.T. Remcho, J. Chromatogr. A 924 (2001) 223-232.

63   G. Wulff, D. Oberkobusch, M. Minarik, React. Polym. 3 (1985) 261-275.

64   M. Glad, P. Reinholdsson, K. Mosbach, React. Polym. 25 (1995) 47-54.

65   F.H. Arnold, S. Plunkett, P.K. Dhal, S. Vidyasankar, Polym. Prepr. 36(1) (1995) 97-98.

66   V.P. Joshi, S.K. Karode, M.G. Kulkarni, R.A. Mashelkar, Chem. Engn. Sci. 53 (1998)
      2271-2284.

67   S. A. Piletsky, H. Matuschewski, U. Schedler, A. Wilpert, E.V. Piletska, T.A. Thiele,
      M. Ulbricht, Macromolecules 33 (2000) 3092-3098.

68   E. Carlier, A. Guyot, A. Revillon, React. Polym. 16 (1991/1992) 115-124.

69   O. Prucker, J. Rühe, Macromolecules 31 (1998) 602-613.

70   C. Viklund, F. Svec, J.M.J. Fréchet, K. Irgum, Biotechnol. Prog. 13 (1997) 597-600.

71   C. Sulitzky, B. Rückert, A.J. Hall, F. Lanza, K. Unger, B. Sellergren, Macromolecules
      35 (2002) 79-91.

72   M. Quaglia, E. De Lorenzi, C. Sulitzky, G. Massolini, B. Sellergren, Analyst 126
      (2001) 1495-1498.

F. Švec, T.B. Tennikova and Z. Deyl (Editors)
Monolithic Materials
Journal of Chromatography Library, Vol. 67
© 2003 Elsevier Science B.V. All rights reserved.

Chapter 14

Ordered Inorganic Structures

Peidong YANG

Department of Chemistry, University of California, Berkeley, CA 94720

CONTENTS

14.1 INTRODUCTION

Nature abounds in hierarchical structures that are formed through highly coupled and often concurrent synthesis and assembly process over both molecular and long-range length scales [1]. The existence of these structures such as ablones and diatoms has both biological and evolutionary significance. It has been a long-sought goal to mimic the natural process responsible for these exquisite architectures using different strategies to control the structural organization, which will have implication in preparing ordered inorganic porous monoliths.

One of the principal objectives of materials chemistry is the development of new environmentally benign strategies for the synthesis of materials of desired shapes and structures. In pursuit of this goal, self-assembled organic templates of synthetic or biological origin have been used to organize and pattern inorganic material deposition involving molecular or nanoparticle building blocks. For example, surfactants, block copolymer lyotropic mesophases, latex particles, colloidal crystals, gel fibers, lipid tubules, viral filaments, bacterial membranes, bacterial superstructures, and spider silk fibers have been employed in the synthesis of novel inorganic-organic composites with controlled size, shape, organization, and porosity. Such materials could have important applications in separation and purification processes, catalysis, storage and release systems, and nanoscale magnetic and quantum devices. Several approaches are currently available for the preparation of ordered structures at different length scales. For example, organic molecular templates can be used to form zeolitic structures with ordering lengths less than 3 nm [2]; mesoporous materials with ordering lengths of 3–30 nm can be obtained through inorganic/organic cooperative assembly using surfactants or amphiphilic block copolymers as the structure-directing agents [3-6]; the use of latex spheres affords macroporous materials with ordering lengths of 100 nm– 1 μm [7-19]; and soft lithography can be used to make high-quality patterns and structures with lateral dimensions of about 30 nm to 500 mμm [20-22]. This article reviews recent progress in the synthesis, characterization and potential applications of these ordered porous inorganic materials.

14.2 MICROPOROUS MONOLITHS

Zeolites and the related microporous molecular sieves are widely used in industry as catalysts, absorbents, ion exchanges. They give rise to excessive pressure drops over the reactors if they are charged in fine powder form. To avoid this, the fine zeolite particles are commonly agglomerated into large (~mm) extrudes or granules using inert inorganic binders such as clay minerals, silica, and alumina. However, the use of binders gives rise not only to dilution of the active zeolite species but also to pore blocking, diffusion limitation and the inaccessibility of the active species within the interior of the agglomerated bodies. To overcome such practical problems, zeolite scientists have attempted to develop synthetic methods to prepare rigid, self-supporting zeolite monolith with ramified macropores for easy internal molecular diffusion [23-28].

Silicalite nanoparticles have been extensively used as the primary building unit during these monolith synthesis processes. Silicalite is an archetypal pure silica zeolite that can be routinely synthesized in the form of discrete nanoparticles in the presence of the molecular structure-directing agent, tetrapropylammonium hydroxide (TPAOH). The as-synthesized particles are relatively uniform in size with a mean diameter of 10–50 nm. These TPA-silicalite-1 nanoparticles can be assembled into core-shell structures by layer-by-layer deposition on monodisperse polystyrene (latex) beads [27]. After the zeolite nanoparticles are made negatively charged, multilayered shells can be constructed on the latex beads by alternating the deposition of silicalite

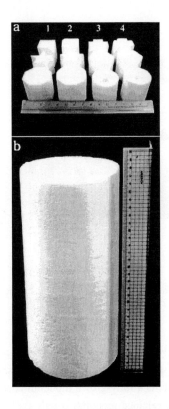

Fig. 14.1. (a). Photographic images of PUF templates in various shapes (colomn 1 and 3) and the corresponding zeolite monoliths (column 2 and 4). (b). Image of a very large zeolite monolith. [Reprinted with permission from Ref. 28, Copyright Wiley-VCH, 2001].

with a polycationic layer of poly(diallyldimethylammonium chloride) (PDADMAC). The preparation of a hierarchical zeolite material with ordered nano-, meso-, and macroporosity can be carried out in a simple stepwise manner. In essence, alternating layers of preformed zeolite (silicalite) nanoparticles and oppositely charged polyionic macromolecules are sequentially adsorbed onto micrometer-sized spherical latex beads to produce nanostructured composite shells of controlled thickness. These pre-fabricated micrometer-sized core/shell building blocks are then assembled into close-packed structures over macroscopic dimensions. Subsequent calcination removes the organic components of the zeolite framework, composite shell, and latex template to produce the hierarchically ordered monolith with controlled wall thickness and pore diameters.

Recently, K. B. Yoon *et al.* developed another interesting microporous monolith synthetic method that uses polyurethane foams (PUFs) as templates [28]. In this process, a clear precursor gel of the silicate-1 was transferred to an autoclave containing the PUF template. The capped autoclave was then heated in an oven for 1-2 days. The as made silicate-1 foam was pale yellow, due to the presence of decomposed fragment of PUF within the struts. The PUF fragments of PUF can be removed by

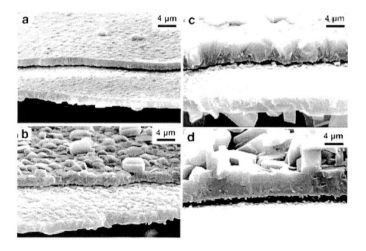

Fig. 14.2. Scanning electron images showing the zeolite shell within the monolith in Fig. 14.1. [Reprinted with permission from Ref. 28, Copyright Wiley-VCH, 2001].

washing with water and acetone or readily through calcinations. The shape and size of the zeolite monolith are nearly identical with those of the PUF templates. They do not shrink even after calcination. The product is a fully-grown monolith consists of pure-silicate-1 (Fig. 14.1). The density of the calcined monolith is only 0.0157 g/mL with surface area of 445 m^2/g. The SEM images (Fig. 14.2) of the cross section of the monolith reveal that the whole body of monolith is made of a single piece of very thin silicate-1 film shaped into a discrete architecture of open macroporous cells and struts. Using this method, production of a very large microporous monolith with a volume of as much as 3.58 L was claimed. Since the PUF manufacturers can provide PUFs with nearly unlimited variation of pore window, size and volume, this template method could be used to synthesize zeolite monolith of desired shape. With some modification, this method can be easily extended to the synthesis of other zeolite monolith (ZSM-5, TS-1) with the similar magnitude of control over size, volume and purity.

14.3 MESOPOROUS MONOLITHS

Over the past few years, a great deal of interest has developed in using hydrophilic-hydrophobic surfactants or block copolymers to organize polymerizable species under conditions where hydrophilic components form continuous regions within a mesostructurally ordered system [3-6]. Noteworthy examples are the M41S family of materials and analogues, in which polymerizable metal oxide species are selectively incorporated into aqueous domains located between low-molecular-weight surfactant aggregates. Mesoscopic order is imparted by typically cooperative self-assembly of the inorganic and surfactant species interacting across their hydrophilic-hydrophobic interfaces. The inorganic species are then cross-linked to form dense, continuous metal oxide networks. The resulting mesostructurally ordered inorganic-organic composite materials have been produced in the form of powder, film, or opaque monolith [29-35].

Considerable effort has been devoted to increase the inorganic mesophase ordering length scale using different organic templates. Biodegradable nonionic alkyl polyoxyethylene surfactants and polyoxyethylene block copolymers are advantageous structure directing agents compared to cationic surfactants because of their low-cost, nontoxicity, and variable hydrophobic/hydrophilic segments size [4,5]. Pinnavaia and co-workers first used nonionic surfactants to synthesize mesoporous silica and alumina in neutral media, and have extensively developed the S^0I^0 (S^0 = neutral organic, I^0 = neutral inorganic precursor species) synthesis route [36]. This approach has resulted in porous multilamellar silica (MSU-V) materials with vesicular particle morphology through the use of diamine bola-amphiphile surfactant such as 1,12-diamino-dodecane. The lamellar material is thermally stable and has a high degree of framework cross-linking, high specific surface area (~ 900 m^2/g) and pore volume (~ 0.5 mL/g). However, in general the S^0I^0 pathway in neutral media results in disordered, worm-like small pore (<6 nm) silica and alumina structures.

Hexagonal and cubic mesoporous silica phases have been synthesized by Attard *et al.* under acidic conditions (pH=1–2) using highly concentrated nonionic surfactant solutions (~ 50 wt %) [37]. More recently, Wiesner *et al.* have used high concentrations of poly(isoprene-b-ethylene oxide) block copolymers (PI-b-PEO) to make lamellar and hexagonal aluminosilicate-polymer mesostructures that are highly ordered on length scales to ~ 40 nm. The synthesis was carried out in an acidic and non-aqueous solution (mixture of CHCl$_3$ and tetrahydrofuran) [38].

Hexagonal and cubic mesoporous silica structures with large (~ 5–30 nm) pores have been synthesized at Santa Barbara using commercially available amphiphilic triblock copolymers (poly(alkylene oxides)), strong acidic conditions and dilute aqueous polymer solutions [4-5]. The hexagonal (plane group p6mm) materials (SBA-15) are highly ordered, silica-copolymer mesophases. Calcination at 500°C gives thermally-stable mesoporous structures with unusually large, uniform pore sizes that can be varied from 4 to over 30 nm, pore volumes to 2.5 mL/g, and pore volume fractions as high as 0.85. SBA-15 silica wall thicknesses can be varied from 3 to 6 nm, which is much thicker than that of MCM-41, and results in a high hydrothermal stability in boiling water. The copolymer can be easily recovered for re-use by solvent extraction with ethanol or removed by heating at 140°C for 3 hours, in both cases yielding a

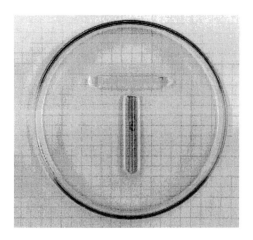

Fig. 14.3. Silica monolith templated with poly(styrene-b-ethylene oxide) copolymer. The background lines are 5 mm apart. [Reprinted with permission from Ref. 30, Copyright Wiley-VCH, 1998].

product that is thermally stable up to 850°C. This polymer templating approach has been extended to the synthesis of many other non-silica mesoporous materials.

Several groups have advanced the processability of such mesoscopically ordered organic/inorganic systems. For example, Ryoo *et al.* have prepared transparent mesoscopically ordered thick films [29], and Marlow *et al.* have reported laser and waveguide properties for fiberlike particles of mesoscopically ordered silica doped with rhodamine dye species [39]. Recently, research groups at Santa Barbara have developed synthesis strategies that use amphiphilic block copolymers as structure-directing agents to produce mesoscopically ordered inorganic-organic composites with large characteristic length scales (up to 30 nm) and improved processability [4-5]. An important benefit of using block copolymers as structure-directing agents for inorganic oxide networks is the attractive processing advantages that these species present. By adjusting the composition, architecture, and molecular weight of the block copolymer species, the phase, ordering length scale (~5–40 nm), and macroscopic morphologies of the self-assembled structures can be controlled. Furthermore, these materials can be processed to obtain high degrees of both mesoscopic organization and macroscopic transparency, for instance in fibers and thin films. Göltner *et al.* have recently demonstrated that transparent inorganic-organic monoliths (Fig. 14.3) with wormlike aggregate mesostructures (Figs. 14.4, 14.5) can be prepared, though the resulting composites cracked upon exposure to dry air [30].

a)

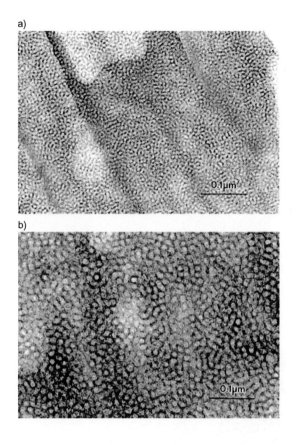

b)

Fig. 14.4. Transmission electron microscopy images of the calcined silica monolith shown in Fig. 14.3. [Reprinted with permission from Ref. 30, Copyright Wiley-VCH, 1998].

Mesoscopically ordered, transparent silica-surfactant monoliths have been prepared using amphiphilic triblock poly(ethylene oxide)-poly(propylene oxide)-poly(ethylene oxide) (PEO-PPO-PEO) copolymer species to organize polymerizing silica networks [31-32]. The polymerization of the silica precursor species under strongly acidic conditions (pH ~1) produces a densely cross-linked silica network that may be mesoscopically organized by the block copolymer species into composites with characteristic ordering length scales of >10 nm. When this is accompanied by slow evaporation of the aqueous solvent, such composite mesostructures can be formed into transparent and crack-free monoliths. An example of a 2.5 cm diameter and 3 mm thick monolith is shown in Fig. 14.6. Distributions and dynamics of the PEO and PPO copolymer blocks within the silica matrix have been investigated in situ

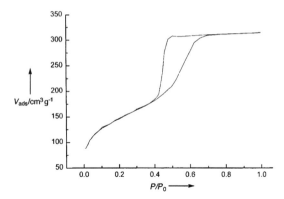

Fig. 14.5. Nitrogen BET adsorption isotherm of the silica monolith shown in Fig. 14.3. P is the equilibrium pressure of the adsorbate, P_0 the saturation pressure of the adsorbate at experimental temperature. [Reprinted with permission from Ref. 30, Copyright Wiley-VCH, 1998].

Fig. 14.6. A mesostructured ordered transparent block copolymer/silica monolith. The monolith is 2.5 cm in diameter and 3 mm thick. Transparency enables clear reading of the text located beneath the monolith. [Reprinted with permission from Ref. 32, Copyright American Chemical Society, 1999].

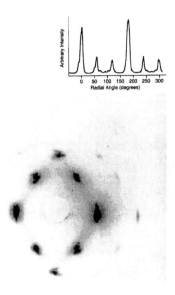

Fig. 14.7. 2-dimensional X-ray diffraction pattern from a hexagonally ordered mesoporous silica monolith after calcinations. The sharp diffraction and higher order peaks indicate that the structure is predominantly a highly ordered single domain. [Reprinted with permission from Ref. 31, Copyright American Chemical Society, 2000].

using ^{29}Si{1H} and ^{13}C{1H} two-dimensional solid-state heteronuclear correlation NMR techniques and 1H NMR relaxation measurements. Mesostructural ordering was determined by X-ray diffraction (Fig. 14.7) and transmission electron microscopy (Fig. 14.8). The degree of microphase separation and the resulting mesostructure of bulk samples were found to depend strongly upon the concentration of block copolymer, with higher concentrations producing higher degrees of order. Two-dimensional X-ray diffraction (XRD) analyses of transparent mesoscopically ordered silica/block copolymer composite monoliths reveal single-crystal-like patterns that correspond to well-ordered hexagonal domains that are greater than $10 \times 1 \times 1$ mm in size. Analyses of the diffraction patterns reveal highly uniform hexagonal domains with narrow distributions of orientational order (Fig. 14.7). Such large single-domain hexagonal and cubic mesostructures were furthermore preserved following removal of the organic species by calcination to produce mesoporous silica monoliths.

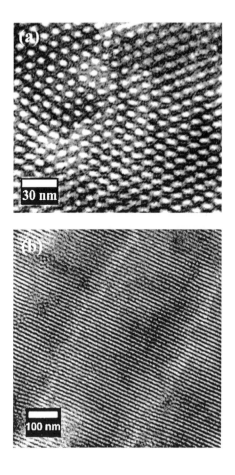

Fig. 14.8. TEM micrographs of the copolymer/silica transparent composite shown in Fig. 14.6 revealing hexagonal array of cylindrical aggregates. [Reprinted with permission from Ref. 32, Copyright American Chemical Society, 1999].

14.4 MACROPOROUS MONOLITHS

A variety of artificial macroporous structures based on colloidal crystals have been synthesized recently using monodispersed microspheres as templates [7-19,40]. In this approach, colloidal crystals are first assembled in order to serve as templates, the voids of which are infiltrated by material that solidifies there. The original colloidal particles are subsequently removed, leaving behind a new material with pores that

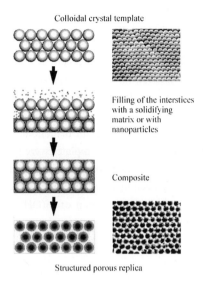

Colloidal crystal template

Filling of the interstices with a solidifying matrix or with nanoparticles

Composite

Structured porous replica

Fig. 14.9. Schematic of the general procedure for replicating the structure of colloidal crystals into porous materials. [Reprinted with permission from Ref. 42, Copyright Elsevier Science, 1999].

preserve the most valuable property of the colloidal crystals — the long-ranged periodic structure.

Latex and silica microspheres are the two major types of particles that are usually used for the colloidal crystal assembly, as they can be obtained both highly monodisperse and relatively cheaply. The simplest method is gravitational sedimentation of the particles (Fig. 14.9). The process can be accelerated and the quality of the materials can be improved via centrifugation. Another way to speed the process up is by filtration, which also allows easy washing and subsequent infusion with different media. Of certain interest is the formation of crystalline sheets of specific thickness, for which the convective assembly method has been used after recent modifications and improvements [18].

The solidified structure in the pores between the microspheres can be formed by polymerization or sol-gel hydrolysis of a liquid precursor. More recently, structures have been grown via electroless or electrochemical deposition or precipitation. A conceptual extension of the method that avoids complex chemistry is to fill the interstices of the colloidal template with smaller colloidal particles, which also leads to a structure with a hierarchical porosity on both nanoscopic, mesoscopic and macroscopic scales. The size of the large pores can be manipulated via the diameter of the templating particles, while the size of the small pores and the overall specific surface area are determined by the size of the small particles. In the final step of the process, the colloidal crystal templates are removed from the composite material either by calcination or by chemical or physical dissolution. The pores left in place of the particles are arranged in three-dimensional ordered arrays that represent a negative replica of the original colloidal crystal. The morphology of the structures is broadly similar irrespective of the type of solid matrix used.

The first demonstration of replicating the ordered structure of colloidal crystals into a stable matrix yielded structured porous silica [7]. The templates were assembled by filtering diluted suspensions of latex microspheres through smooth membranes. The surfaces of the microspheres were functionalized by adsorption of cationic surfactant from solution, which served to initiate the polymerization of $Si(OH)_4$ to solid silica after the crystal was subsequently infused with aqueous silica solution. The latex particles were removed from the composite via calcination. The products of the mineralization are porous low-density silica flakes.

In a separate leading study, Imhof and Pine described how colloidal crystal-like assemblies of densely packed monodisperse non-aqueous emulsion droplets can serve as templates for the formation of porous titania, zirconia, and silica [8]. The pores are less ordered and less uniform than with latex microspheres, but the method has technological potential due to the use of easily obtained templates. In addition, the sol-gel alkoxide hydrolysis method employed for depositing the oxide matrix had later proved useful with latex particle templates.

The versatility of the principle has been demonstrated both in the range of materials made and in the ability to design specific functional and structural characteristics into them [40]. Holland *et al.* assembled latex crystals by both filtration and centrifugation and used them as templates for porous titania, alumina, and zirconia via the sol-gel technique. This group has adapted the method further to form structures of a wide variety of chemical compositions, including oxides of W, Fe, Sb, Zr/Y, aluminophosphates, silicates, carbonates, zeolites, and others [8]. The first study specifically to target the formation of a material with photonic crystal properties, dense porous titania, was that of Wijnhoven and Vos [9]. The templates are assembled via centrifugation, and the titania structure is formed by the sol-gel method. The high refractive index of the titania is a prerequisite for a remarkably wide reflectance peak of the calcined structure, although the samples obtained do not display a full photonic band gap.

Vlasov *et al.* were the first to assemble a secondary structure from nanoparticles and prepare a porous material from semiconductor CdSe quantum dots [12]. The effective refractive index of the material assembled is, however, lower than that of solid semiconductors due to the presence of air in the pores between the nanocrystals.

The problem of creating an optically dense, continuous, semiconductor structure is solved in the electrochemical method reported by Braun and Wiltzius [41]. Here, the template crystals of latex or silica microspheres are assembled via sedimentation onto conductive indium tin oxide (ITO) electrodes, and then CdSe and CdS are grown electrochemically in their interstices via electrodeposition. Structures made by this approach should display improved photonic band gaps and good mechanical stability.

Due to the importance of metals in technological applications, structured porous metals may find application in a variety of areas, particularly electronics and optoelectronics. Yan *et al.* assembled latex crystals by centrifugation and impregnated the template with a nickel salt that was then decomposed to NiO and finally reduced to metallic Ni in a hydrogen atmosphere (Fig. 14.10) [17]. The highly porous nickel obtained can be useful in catalysis and in electrochemical electrodes. Similarly, Jiang *et al.* synthesized a variety of porous metals by electroless deposition (Fig. 14.11) [18]. Silica colloidal crystals were assembled in thin wetting films and functionalized with gold nanocrystals that provide the nucleation sites for electroless deposition. The method has been used to form samples from Ni, Cu, Ag, Au, and Pt, which have remarkable optical properties due to the long-ranged ordered porous structure on the surface. O. Velve *et al.* used a wet method to produce a templated gold structure assembled entirely from a suspension of gold nanoparticles [42]. This is made possible by assembling the latex crystal by filtration through a membrane of pore size small enough to retain both the latex as well as the gold particles subsequently added, while still allowing a reasonably high flux of water. Thus, a mesoscopically porous gold structure is grown in the interstices of the latex crystals.

Zakhidov *et al.* have developed procedures for using crystalline templates to form a family of structured porous carbons with remarkable optical properties [15]. The templates are artificial opals assembled from SiO_2 microspheres, which are embedded into carbon phases by three alternative routes: (i) by infiltrating the crystals with phenolic resin, removing the microspheres, and pyrolyzing the resin structure to glassy carbon; (ii) by chemical vapor deposition (CVD) of graphitic carbon; and (iii) by diamond-seeded CVD from plasma. The carbon 'inverse opals' obtained are highly conductive, show intense opalescence from the ordered arrays of holes and may have a photonic band gap in the infrared region. Recently, Ozin *et al.* developed a CVD infiltration method to produce Si and Ge inversed opal structure [13,14]. Norris *et al.* further fabricated these macroporous silicon on substrate with additional capability of monolithic patterning (Fig. 14.12) [43].

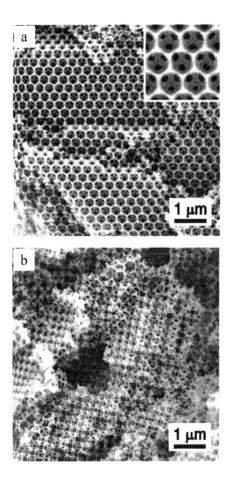

Fig. 14.10. SEM images of (a) mesoporous Ni and (b) Co. The lighter features correspond to the metal walls that surround macropores, and the black circular features correspond to windows between adjacent macropores. The inset in image a) shows an enlarged area of the mesoporous Ni sample to demonstrate the smooth, spherical appearance of the walls in this sample. [Reprinted with permission from Ref. 17, Copyright American Chemical Society, 1999].

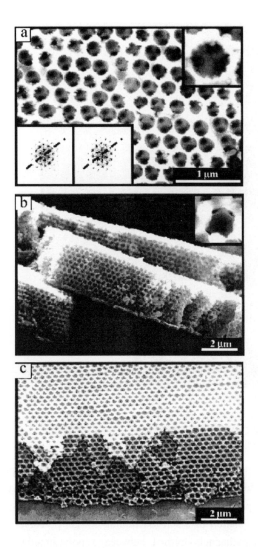

Fig. 14.11. Typical SEM images of three macroporous metal films. (a) Top view of a platinum film with 318 ± 13 nm diameter voids. Bottom left insets show FFTs of two areas (15.75 × 11.75 μm area). Upper right inset shows interconnecting pores and rough surface of the same sample. (b) Cross-sectional view of a copper film with 325 ± 15 nm diameter voids revealing porous morphology throughout the film. The inset shows interconnecting pores and smooth surface of the same sample. (c) A nickel film made from a colloidal template (353 ± 17 nm diameter) with 15 colloidal layers. [Reprinted with permission from Ref. 18, Copyright American Chemical Society, 1999].

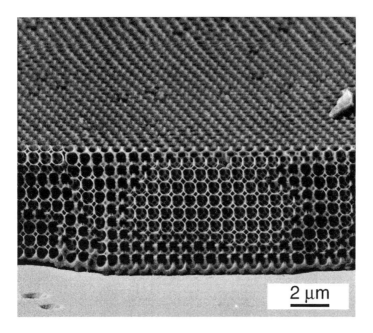

Fig. 14.12. SEM image of ordered macroporous Si monolith on substrate. [Reprinted with permission from Ref. 43, Copyright Macmillan Publishers Ltd, 2001].

14.5 HIERARCHICALLY ORDERED POROUS STRUCTURES

Ordering at multiple length scale is important in order to achieve multiple functionalities from porous materials. The group at Santa Barbara has recently developed a straightforward approach to pattern mesoporous materials on substrates (Fig. 14.13) [44-45]. In this process, the gelation of a self-assembling sol-gel precursor solution was carried out in the microchannels formed between a poly(dimethylsiloxane) (PDMS) mold and a substrate. The area of the patterned surface was typically 1–5 cm^2 with molded feature sizes in the micrometer range. Recently, zeolite nanocrystals were used to fill the microchannels and to form patterned microporous solids [46].

Fig. 14.14 shows several representative scanning electron microscope (SEM) images of the patterned mesoporous materials. The structural ordering observed at the micrometer level is imparted by the microchannel network, while the mesoscopic ordering results from the self-assembly of the sol-gel block copolymer/inorganic solution. This hierarchical ordering process can be further extended to the preparation of other patterned mesoporous metal oxides, such as TiO$_2$ using sol-gel mesophase self-assembly chemistry [5].

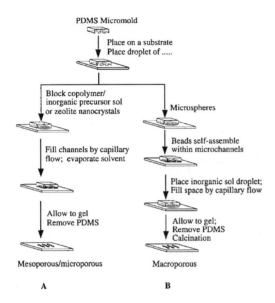

Fig. 14.13. Schematic illustration of synthesis process for making hierarchical ordered porous structure. [Reprinted with permission from Ref. 44, Copyright Wiley-VCH, 2001].

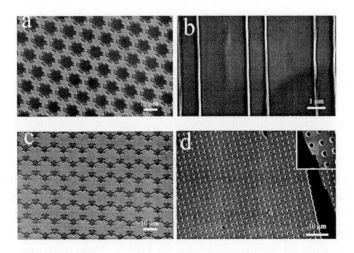

Fig. 14.14. SEM images of the patterned mesoporous silica thin film. [Reprinted with permission from Ref. 44, Copyright American Association for the Advancement of Science, 2000].

References pp. 320-322

When doped with rhodamine 6G, these mesoscopically ordered structures exhibited amplified spontaneous emission [45], with low thresholds of 10–25 kW·cm^{-2}. This threshold is an order of magnitude lower than has been reported for sol-gel glass waveguides [47]. This can be attributed to the ability to incorporate the luminescent species within the high-surface-area ordered mesopores of the waveguides, which results in a homogeneous dye distribution that maintains high dispersion and isolation of the individual dye molecules at relatively high dye concentrations. The gain-narrowed emission (577 nm) is from the end of the mesostructured waveguides and directed parallel to their axes. These highly processible, self-assembling mesostructured host media offer great potential for the one-step fabrication of microlaser arrays and components for integrated optical circuits and microanalytical applications. More recently, researchers at Sandia National Laboratories successfully combined silica-surfactant self-assembly with three rapid printing procedures (pen lithography, ink-jet printing, and dip-coating of patterned self-assembled monolayers) to form functional, hierarchically organized structures in seconds [48]. This high-speed fabrication process may prove useful for directly writing sensor arrays and fluidic or photonic systems.

Ordered macroporous materials with pore sizes in the sub-micrometer range have attracted much interest because of their potential application in fields such as catalysis, separation, and photonics. In order to fabricate these macroporous materials into useful photonic devices such as waveguide channels for microscopic light manipulating purpose, it is important to be able to pattern these photonic crystals into 2-dimensional arrays/circuits on substrate [49]. Current methodologies for patterning photonic crystals involve electron beam lithography/anisotropic etching [50-53], two-photon polymerization, [54] or holographic lithography [55], all of which require major equipments and complex procedures. Patterning macroporous photonic crystals within microchannel networks apparently represents a simple alternative approach.

The fabrication process is similar to that used for patterning mesoporous oxides except colloidal crystallization within microchannel networks is involved in the first place (Fig. 14.13) [44]. Generally, polymer or silica spheres with diameters of 200–2000 nm can be readily assembled and packed into a close-packed lattice within microchannels with depths and widths of 4–10 μm. A drop of inorganic precursor solution was then placed at the same end of the mold and imbibed into the latex-sphere-filled microchannels by capillary actions. Various precursors can be used, including tetraethoxysilane in ethanol, titanium(diisopropoxide) bis(2,4-pentanedianate) in isopropanol, and prepolymer solution for PbTiO$_3$. Upon calcination, patterned macroporous structures with condensed silica, anatase TiO$_2$, and PbTiO$_3$ framework can be obtained.

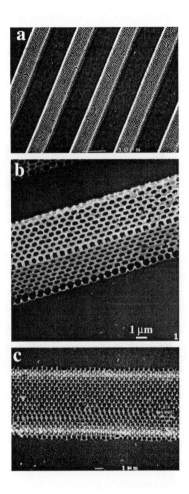

Fig 14.15. SEM image of patterned macroporous silica (a-b) and titania (c). [Reprinted with permission from Ref. 44, Copyright Wiley-VCH, 2001].

The capability of patterning macroporous materials with high refractive index such as TiO_2, $PbTiO_3$, Si and Ge may represent a viable way to fabricating single domain photonic crystals, 3-dimensional photonic crystal devices, or even circuits. Figures 14.15a-c show SEM images for patterned macroporous silica and titania structures. Overall, the orientation and large area single domain crystallinity of the template were replicated with high fidelity. The process can be readily applied to micromolds with depth profiles up to 10 μm. If block copolymer/inorganic precursor sol was used in scheme B of Fig. 14.13, mesoporosity can be introduced into the framework of the macroporous structure. The final materials exhibit hierarchical ordering with discrete characteristic length scales of 10, 100, and 1000 nm. This structural ordering at multiple length scales within one material is reminiscent of those in living systems [1].

14.6 CONCLUSIONS

A number of strategies available for introducing structural organization into inorganic porous materials are reviewed here. The most popular and widely applicable method is template-directed assembly, which affords ordered porous materials at a range of length scales (from microporous to macroporous). The large accessible pore size, high surface areas, and easy functionalization of the inorganic solid wall provide many opportunities to use these highly ordered channels and pores as nano-size reaction vessel for chemical/biological reactions. The successful synthesis of highly ordered large pore (10–50 nm) mesoporous materials using block copolymers opens the avenue to many new materials of relevance including large-molecule catalysis, separation sensors, matrixes for drug delivery, environmental remediation and the fabrication of nanocomposites with useful magnetic, optical and electric properties. Heavy metal environmental remediation is an important application that appears to be on the verge of being realized. The future of these materials in commercial applications rests on the high degree of control over pore properties, processibility, hierarchical structure and function.

14.7 ACKNOELEDGMENT

The author acknowledges the support of National Science Foundation.

14.8 REFERENCES

1   I.A. Aksay, M. Trau, S. Manne, I. Honma, N. Yao, L. Zhou, P. Fenter, P.M. Eisenberger, S.M. Gruner, Science 273 (1996) 892.
2   X. Bu, P. Feng, G.D. Stucky, Science 278 (1997) 2080.
3   C.T. Kresge, M.E. Leonowicz, W.J. Roth, J.C. Vartuli, J.S. Beck, Nature 359 (1992) 710.
4   D. Zhao, J. Feng, Q. Huo, B.F. Chmelka, G.D. Stucky, Science 279 (1998) 548.
5   P. Yang, D. Zhao, D.I. Margolese, B.F. Chmelka, G.D. Stucky, Nature 396 (1998) 152.
6   A. Firouzi, D.J. Schaefer, S.H. Tolbert, G.D. Stucky, B.F. Chmelka, J. Am. Chem. Soc. 119 (1997) 9466.
7   O.D. Velev, T.A. Jede, R.F. Lobo, A.M. Lenhoff, Nature 389 (1997) 447.
8   B.T. Holland, C.F. Blanford, A. Stein, Science 281 (1998) 538.
9   J.E.G.J. Wijnhoven, W.L. Vos, Science 281 (1998) 802.

10   A. Imhof, D.J. Pine, Nature 389 (1997) 948.

11   G. Subramania, K. Constant, R. Biswas, M.M. Sigalas, K.M. Ho, Appl. Phys. Lett. 74 (1999) 3933.

12   Y.V. Vlasov, N. Yao, D.J. Norris, Adv. Mater. 11 (1999) 165.

13   A. Blanco, E. Chomski, S. Grabtchak, M. Lbisate, S. John, S.W. Leonard, C. Lopez, F. Meseguer, H. Miguez, J. Mondia, G.O. Ozin, O.T. Toader, H.M. van Driel, Nature 405 (2000) 437.

14   H. Miguez, F. Meseguer, C. Lopez, M. Holgado, G. Andreasen, A. Mifsud, V. Fornes, Langmuir 16 (2000) 4405.

15   A.A. Zakhidov, R.H. Baughman, Z. Iqbal, C.X. Cui, I. Khayrullin, S.O. Dantas, I. Marti, V.G. Ralchenko, Science 282 (1998) 897.

16   S.H. Park, Y. Xia, Adv. Mater. 10 (1998) 1045.

17   H. Yan, C.F. Blanford, B.T. Holland, M. Parent, W.H. Smyrl, A. Stein, Adv. Mater. 11 (1999) 1003; H. Yan, C.F. Blanford, J.C. Lytle, C.B. Carter, W.H. Smyrl, A. Stein, Chem. Mater. 13 (2001) 4314.

18   P. Jiang, J. Cizeron, J.F. Bertone, V.L. Colvin, J. Am. Chem. Soc. 121 (1999) 7957

19   S.A. Johnson, P.J. Ollivier, T.E. Mallouk, Science, 283 (1999) 963.

20   Y. Xia, G.M. Whitesides, Angew. Chem. Int. Ed. 37 (1998) 550.

21   E. Kim, Y. Xia, G.M. Whitesides, Adv. Mater. 8 (1996) 245.

22   C. Marzolin, S.P. Smith, M. Pretiss, G.M. Whitesides, Adv. Mater. 10 (1998) 571.

23   A.S.T. Chiang, K. Chao, J. Phys. Chem. Solid 62 (2001) 1899.

24   J. Caro, M. Noack, P. Kolsch, R. Schaefer, Microporous and Mesoporous Mater. 38 (2000) 3.

25   G.S. Lee, Y.J. Lee, K. Ha, K.B. Yoon, Adv. Mater. 13 (2001) 1491.

26   M.W. Anderson, S.M. Holmes, N. Hanif, C.S. Cundy, Angew. Chem. Int. Ed. 39 (2000) 2707.

27   K.H. Rhodes, S.A. Davis, F. Caruso, B. Zhang, S. Mann, Chem. Mater. 12 (2000) 2832.

28   Y.J. Lee, J.S. Lee, Y.S. Park, K.B. Yoon, Adv. Mater. 13 (2001) 1259.

29   R. Ryoo, C.H. Ko, S.J. Cho, J.M. Kim, J. Phys. Chem. B 101 (1997) 10610.

30   C.G. Goltner, S. Henke, M.C. Weissenberger, M. Antonietti, Angew. Chem. Int. Ed. 37 (1998) 613.

31   N.A. Melosh, P. Davidson, B.F. Chmelka, J. Am. Chem. Soc. 122 (2000) 823.

32   N.A. Melosh, P. Lipic, F.S. Bates, F. Wudl, G.D. Stucky, G.H. Fredrickson, B.F. Chmelka, Macromolecules 32 (1999) 4332.

33   B. Lebeau, C.E. Fowler, S.R. Hall, S. Mann, J. Mater. Chem. 9 (1999) 2279.

34   O. Dag, A. Verma, G.A. Ozin, C.T. Kresge, J. Mater. Chem. 9 (1999) 1475.

35   S. Pevzner, O. Regev, R. Yerushalmi-Rozen, Curr. Opin. Coll. Inter. Sci. 4 (2000) 420.

36   S.A. Badshaw, E. Prouzet, T.J. Pinnavaia, Science 269 (1995) 1242.

37   G.S. Attard, J.C. Glyde, C.G. Goltner, Nature 378 (1995) 366.

38   M. Templin, A. Franck, A. Du Chesne, H. Leist, Y. Zhang, R. Ulrich, U. Wiesner, Science, 278 (1997) 1795.

39   F. Marlow, M.D. McGehee, D. Zhao, B.F. Chmelka, G.D. Stucky, Adv. Mater. 11 (1999) 632.

40   Y. Xia, B. Gates, Y. Yin, Y. Lu, Adv. Mater. 12 (2000) 693.

41   P.V. Braun, P. Wilzius, Nature, 402 (1999) 603.

42   O.D. Velve, P.M. Tessier, A.M. Lenhoff, E.W. Kaler, Nature, 401 (1999) 548; O. D. Velev, A.M. Lenhoff, Curr. Opin. Coll. Inter. Sci. 5 (2001) 56.

43 Y.V. Vlasov, X. Bo, J.C. Strum, D. Norris, Nature 414 (2001) 289.

44 P. Yang, T. Deng, D. Zhao, P. Feng, D. Pine, B.F. Chmelka, G.M. Whitesides, G. D. Stucky, Science, 282 (1998) 2244; P. Yang, A. H. Rizvi, B. Messer, B.F. Chmelka, G.M. Whitesides, G.D. Stucky, Adv. Mater. 13 (2001) 427.

45 P. Yang, G. Wirnsberger, H.C. Huang, S.R. Cordero, M.D. McGehee, B. Scott, T. Deng, G.M. Whitesides, B.F. Chmelka, S.K. Buratto, G.D. Stucky, Science 287 (2000) 465.

46 L. Huang, Z. Wang, J. Sun, L. Miao, Q. Li, Y. Yan, D. Zhao, J. Am. Chem. Soc. 122 (2000) 3530.

47 H. Yanagi, T. Hishiki, T. Tobitani, A. Otombo, S. Mashiko, Chem. Phys. Lett. 292 (1998) 332.

48 H. Fan, Y. Lu, A. Stump, S.T. Reed, T. Baer, R. Schunk, V. Perez-Luna, G. P. Lopez, C.J. Brinker, Nature 405 (2000) 56.

49 J.D. Joannopoulos, R.D. Meade, J.N. Winn, Photonic Crystals, Princeton University Press, Princeton, NJ, 1995.

50 O. Painter, R.K. Lee, A. Scherer, A. Yariv, J.D. Obrien, P.D. Dapkus, I. Kim, Science 284 (1999) 1819.

51 J.S. Forsei, P.R. Villeneuve, J. Ferrera, E.R. Thoen, G. Steinmeyer, S. Fan, J.D. Joannopoulos, L.C. Kimerling, H.I. Smith, E.P. Ippen, Nature 390 (1997) 143.

52 S.Y. Lin, J.G. Fleming, D.L. Hetherington, B.K. Smith, R. Biswas, K.M. Ho, M.M. Sigalas, W. Zubrzycki, S.R. Kurtz, J. Bur, Nature 394 (1998) 251.

53 S. Noda, K. Tomoda, N. Yamamoto, A. Chutinan, Science 289 (2000) 604.

54 B.H. Cumpston, S.P. Ananthavel, S. Barlow, D.L. Dyer. J.E. Ehrlich, L.L. Erskine, A.A. Heikal, S.M. Kuebler, I.Y. Lee, D. McCord-Maughon, J. Qin, H. Rockel, M. Rumi, X. Wu, S.R. Marder, J.W. Perry, Nature 398 (1999) 51.

55 M. Campbell, D.N. Sharp, M.T. Harrison, R.G. Denning, A.J. Turberfield, Nature 404 (2000) 53.

Applications

F. Švec, T.B. Tennikova and Z. Deyl (Editors)
Monolithic Materials
Journal of Chromatography Library, Vol. 67

Chapter 15

Flow and Mass Transfer

Alírio E. RODRIGUES, Vera G. MATA, Michal ZABKA, and Luís PAIS

Laboratory of Separation and Reaction Engineering, Department of Chemical Engineering, Faculty of Engineering, University of Porto, Porto, Portugal

CONTENTS

15.1 INTRODUCTION

In the last decade, a growing interest in large-pore, permeable, or flow-through particles for engineering applications as catalyst supports, adsorbents, HPLC packings [1-3], ceramic membranes, supports for mammalian cell culture/biomass growth [4]

and building materials can be observed. In particular in the area of separation engineering, perfusion chromatography appeared as a key technique for separation of proteins. Examples of perfusive packings for HPLC separations of proteins are alumina Unisphere (Biotage, USA), Polystyrene PL4000 (Polymer Laboratories, UK), Polystyrene POROS (PerSeptive Biosystems, USA), and Silica gel Daisogel SP 2705 (Daiso Co, Japan).

Continuous efforts are being made to prepare materials with large transport pores to enhance mass transport while simultaneously containing smaller diffusive pores affording adsorption capacity. These materials are in the form of polymeric beads, membranes, and continuous rods or monoliths and have found numerous applications in column and membrane chromatography processes for protein separation. Monoliths differ from membranes in terms of geometry since their longitudinal dimension exceeds the lateral dimension.

15.1.1 Perfusive chromatography

A typical chromatographic packing used in perfusion chromatography has a bi-disperse structure with 700 nm large transport pores, and pores of 50 nm to provide for adsorption capacity. The combined benefits of large pores for convective mass transport and smaller diffusive pores have long been recognized in the literature concerning catalyst manufacture [5,6]. Wheeler discussed the importance of intraparticle convection in a remarkable paper providing order-of-magnitude analysis and the steady-state mass balance equation for intraparticle forced convection, diffusion, and reaction in isothermal catalysts [7]. Unfortunately, he did not solve the model equation, thus missing the point later recognized by Nir and Pismen [8] who discovered the enhancement of catalyst effectiveness by convection in the intermediate range of Thiele modulus, in which diffusion- and reaction rates are of similar magnitude. We quantified the augmented diffusivity by forced convective flow (viscous or Poiseuille flow) due to a total pressure gradient [9,10] and this became the key to understanding perfusion chromatography [11-13] and related membrane chromatographic processes [14,15] suitable for protein separations.

15.1.2 The concept of "augmented diffusivity" by convection

The effort in making flow-through particles aims at reducing in the intraparticle mass-transfer resistance by increasing particle permeability. The other "classic" ways of eliminating or reducing mass transfer inside particles shown in Figure 15.1 are either the use of pellicular packings or a reduction in particle size. Diffusivity augmented by convection inside transport pores is the main concept behind the improved

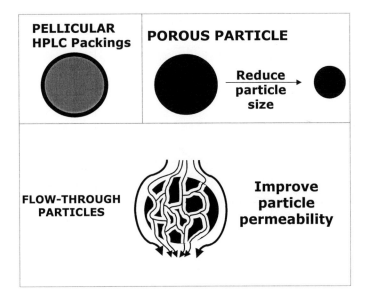

Fig. 15.1. Methods enabling reduction in intraparticle mass-transfer resistance.

performance of flow-through packings [9]. In fact, the augmented diffusivity $\tilde{D}_e$ is related to the effective diffusivity, D_e, by:

$$\tilde{D}_e = D_e \frac{1}{f(\lambda)} \tag{15.1}$$

where the intraparticle Peclet number, λ, is

$$\lambda = \frac{v_0 (R_p/3)}{D_e}, \tag{15.2}$$

defined as the ratio between the time constants for pore diffusion and pore convection where R_p is the particle radius and v_0 is the intraparticle convective velocity inside large pores, given by Darcy's law. The enhancement of diffusivity by intraparticle convection is

$$\frac{1}{f(\lambda)} = \frac{\tilde{D}_e}{D_e} \tag{15.3}$$

with

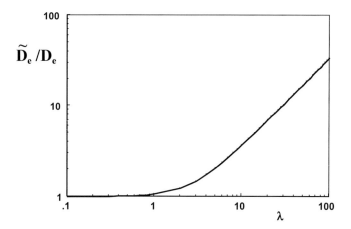

Fig. 15.2. Enhancement factor of diffusivity, $\tilde{D}_e / D_e$, as a function of the intraparticle Peclet number, λ.

$$f(\lambda) = \frac{3}{\lambda} \left[\frac{1}{\tanh \lambda} - \frac{1}{\lambda} \right] \qquad (15.4)$$

Figure 15.2 shows the enhancement factor

$$\frac{1}{f(\lambda)} = \frac{\tilde{D}_e}{D_e} \qquad (15.5)$$

as a function of the intraparticle Peclet number λ. At low superficial velocities, u_0, the intraparticle convective velocity, v_0, is also small and $f(\lambda)=1$; therefore $\tilde{D}_e = D_e$ (diffusion-controlled case). At high superficial velocities, and thus high λ, $f(\lambda)=3/\lambda$ and the augmented diffusivity is then

$$\tilde{D}_e = \frac{v_0 R_p}{9} \ . \qquad (15.6)$$

$\tilde{D}_e$ Then depends on the particle permeability (convection-controlled case).

15.1.3 The first extension of the Van Deemter equation

The classic Van Deemter analysis [16] of column performance is based on the concept of the height equivalent to a theoretical plate (HETP) or, simply, H. It is the peak variance in the axial coordinate, σ_z^2, normalized by the column length, L. For

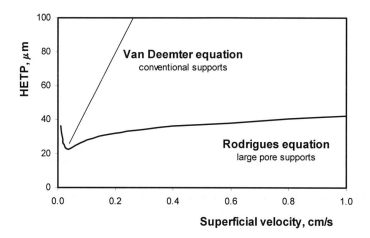

Fig. 15.3. Height equivalent to the theoretical plate, HETP, as a function of the superficial velocity, u_0 calculated using the Van Deemter equation for conventional packings and the Rodrigues equation for large-pore packings.

conventional chromatographic packings the HETP as a function of the superficial velocity is:

$$H = A + \frac{B}{u_0} + C\,u_0 \qquad (15.7)$$

where the A term accounts for the eddy dispersion contribution, the B term for the molecular diffusion contribution, and the C term for mass transfer. For linear isotherms, the term C is given by:

$$C = \frac{2}{15} \frac{\varepsilon_p\,(1 - \varepsilon_b)\,b^2}{\left[\varepsilon_b + (1 - \varepsilon_b)\,\varepsilon_p\,b\right]^2}\,\tau_d \qquad (15.8)$$

with the pore diffusion time constant $\tau_d = \varepsilon_p R_p^2/D_e$ and the retention factor $b = 1 + \{(1 - \varepsilon_p)/\varepsilon_p\}m$, ε_p is the intraparticle porosity, ε_b is the interparticle porosity, and m is the slope of the isotherm.

For large-pore packings, we derived the extended Van Deemter equation [17,18]

$$H = A + \frac{B}{u_0} + C\,f(\lambda)\,u_0 \qquad (15.9)$$

since the "augmented diffusivity" is $\tilde{D}_e = D_e/f(\lambda)$ and therefore the time constant based on the augmented diffusivity is $\tilde{\tau}_d = \tau_d f(\lambda)$.

The plots representing the Van Deemter equation for conventional supports (thin line) and our equation for large-pore supports (full line) are shown in Figure 15.3 for a typical HPLC process. At low velocities both equations afford similar results. However, at high superficial velocities, the last term in our equation becomes constant since the intraparticle convective velocity, v_o, is proportional to the superficial velocity, u_0. The HETP reaches a plateau, which does not depend on the solute diffusivity but only on the particle permeability and pressure gradient (convection-controlled limit).

Use of permeable packings results in two important features: (i), improved column efficiency, since the HETP is reduced compared with conventional packings as the C term in the Van Deemter equation is smaller, and (ii), the speed of separation can be increased without losing column efficiency.

15.1.4 Structured packings and monoliths

One difficulty in perfusion chromatography is associated with the estimation of the convective pore velocity, v_o, used for the calculation of the intra-particle Peclet number λ. According to Komiyama and Inoue [19] the equality of pressure drop across the particle and along the bed

$$\frac{\Delta p}{d_p} = \frac{\Delta P}{L} \tag{15.10}$$

coupled with Darcy's law for the bed pressure drop

$$\frac{\Delta P}{L} = \frac{\eta}{B_b} u_0 \tag{15.11}$$

and across the particle

$$\frac{\Delta p}{d_p} = \frac{\eta}{B_p} v_0 \tag{15.12}$$

where B_b and B_p are the bed and particle permeability, respectively, leads to

$$v_0 = \frac{B_p}{B_b} u_0 \tag{15.13}$$

Although only a small fraction of the flow goes through the packing material, typically less than 1%, this is sufficient for the mass-transfer enhancement.

The development of monolithic columns [20-28] of different types (compressed soft gels, polymeric, and silica monolithic disks as well as columns) is similar to the progress described above in perfusion chromatography. All these approaches aim at an increase in the efficiency of the separation device by means of fast convective mass transfer and small diffusive length. The search for structured packings [29] in separation engineering closely follows the parallel development in catalysts and reaction engineering.

15.2 FLOW IN MONOLITHS

15.2.1 The porous structure of monoliths

The goal of research concerning adsorbent materials for separation processes is to find the correct balance between adsorbent capacity, selectivity, and speed of the separation. A typical silica monolithic column consists of through-channels of 1–6 μm and silica skeletons with mesopores in the range of 10–20 nm. A surface area of 260 m^2/g, a mesopores volume of 1 ml/g, and a packing density of 200 kg/m^3 are the other characteristics. The external porosity, ε, is in the range 0.65–0.75 and the total porosity ε_t is 0.8–0.9. The through-pore size/skeleton size ratio is of the order 3–5. Figure 15.4 shows a scanning electron micrograph of a silica monolith, Chromolith Performance RP-18e 100-4.6 from Merck [30]. Cavazzini *et al.* [31] determined a total porosity of 0.82 for a similar monolithic column using uracil and thiourea, and methanol–water as the mobile phase under unretained conditions. An external porosity of 0.69 was calculated from the retention times of polystyrene standards.

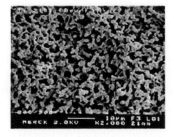 

Fig. 15.4. SEM photographs showing the internal morphology of a silica monolith material.

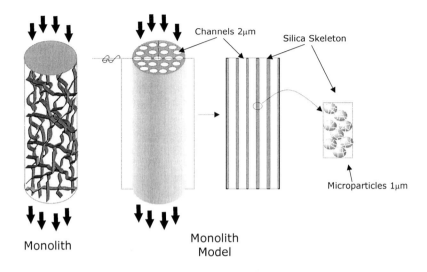

Fig. 15.5. Scheme of the monolith viewed as a bundle of parallel channels.

Figure 15.5 shows the schematics of both a monolith and an idealized model of parallel capillaries separated by the silica skeletons. Attempts have been made to apply network pore models to describe the porous structure of the adsorbent using experimental pore size distribution determined by both mercury porosimetry and image analysis [32,33] and to model the monoliths [34,35].

15.2.2 Permeability

Permeability values reported in the literature for silica monoliths with 2 μm channels are in the range $1-4 \times 10^{-13} m^2$. These values were calculated from the experimental plots of pressure drop, ΔP, versus the superficial velocity, u_0, using Darcy's law:

$$B_0 = u_0\, \eta\, \frac{L}{\Delta P} \qquad\qquad (15.14)$$

The permeability of a monolith with channels of diameter d_{ch} can be estimated [36] from

$$B_0 = \varepsilon \cdot \frac{d_{ch}^2}{32} \qquad\qquad (15.15)$$

TABLE 15.1

PERMEABILITY OF MONOLITHIC AND PERFUSIVE COLUMNS

	Monolith		Packed column	
Through-pore size, μm	Permeability m^2		Particle diameter, μm	Permeability m^2
1	2.18×10^{-14}	5		2.95×10^{-14}
2	8.72×10^{-14}	10		1.18×10^{-14}
3	1.96×10^{-13}	20		4.72×10^{-13}
6	7.78×10^{-13}	40		1.89×10^{-13}

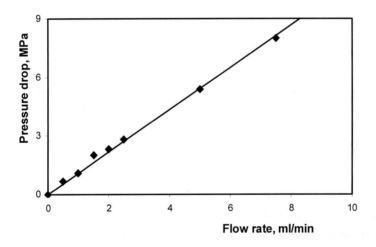

Fig. 15.6. Pressure drop as a function of flow rate for monolithic column. Conditions: monolithic column Chromolith RP 18e 100 × 4.6 mm i.d., mobile phase acetonitrile–water (60:40).

Table 15.1 shows the effect of the channel- or through-pore-size on permeability for an external porosity of $\varepsilon=0.71$. For a Merck monolithic silica column RP-18e 100-4.6 the pressure drop *versus* flow rate is shown in Figure 15.6. From the slope of the straight line, the calculated permeability is 10^{-13} m^2. This is in a good agreement with the value in Table 15.1 for a monolith with a channel size of 2 μm.

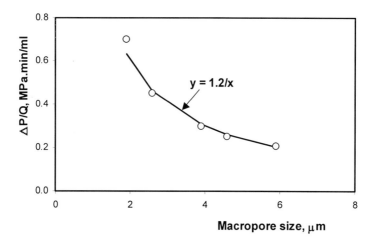

Fig. 15.7. Effect of the monolith macropore size on the slope of ΔP versus Q (experimental points taken from ref. [21]).

The effect of the through-pore (channel) size on the pressure drop, ΔP, of a monolith, as a function of the flow rate, Q, was addressed by Schulte *et al.* [21] for materials having through-pore d_{pore} of different sizes in a range of 2–6 μm. Figure 15.7 was constructed, plotting the slope of ΔP versus Q, *i.e.*, *slope*$_{\Delta P/Q}$, as a function of the through-pore size. The points fit well the equation: *slope*$_{\Delta P/Q}$ = 1.2 / d_{pore}.

Silica-based monoliths with skeleton sizes of 1–2 μm and through-pores of 2–8 μm were prepared by Motokawa *et al.* [37] in fused silica capillaries having an inside diameter of 50–200 μm and derivatized with C_{18} functionalities. The permeability of these capillary columns was in the range 0.08–1.3×10^{-12}m^2 and their external porosity over 0.8. For a silica-based monolith in a fused-silica capillary of diameter 50 μm and a through-pore size of 8 μm, Ishizuka *et al.* [38] found permeabilities in the range of 1–1.3×10^{-12} m^2 which compared favorably with 1.7×10^{-13} m^2 found for monolithic silica prepared in a mold, and a column packed with 5 μm particles with a permeability of 3.9–5.0×10^{-14} m^2.

Monoliths are designed to overcome the pressure-drop limitation found in liquid chromatography using columns packed with very small particles. In fact, a large through-pore size/skeleton size ratio reduces both the diffusion path-length inside microparticles and the flow resistance

$$\phi = \frac{d_{sk}^2}{B_0} \approx \frac{32\, d_{sk}^2}{\varepsilon\, d_c^2} \tag{15.16}$$

The permeability values calculated above for monoliths can be compared with the bed-permeability values, B_b, also shown in Table 15.1, that were obtained for perfusion chromatographic column using Equation 15.14 and calculated from

$$B_b = \frac{\varepsilon^3 d_p^2}{150 (1 - \varepsilon)^2} \qquad (15.17)$$

Equation 15.17 applies to columns with particles of diameter d_p. For packed columns with porosity $\varepsilon = 0.4$, the bed permeability is simplified to

$$B_b \approx \frac{d_p^2}{1000} \qquad (15.18)$$

The permeability of CIM disks was found to be similar to that of columns packed with 45 μm particles [39-41]. It is also interesting to compare permeabilities of the monoliths with those of a column packed with perfusive particles as measured by Pfeiffer *et al.* [42]. For example, a permeability of $7.89 \times 10^{-15} m^2$ was found for 30–50 μm POROS particles with pores in a range of 400–800 nm. This value is 7–15 times higher than that predicted by the Kozeny equation.

15.2.3 Flow and dispersion: the Taylor–Aris model

The visualization of a monolith depicted in Fig. 15.5 as a bundle of parallel channels enables one to use the Taylor–Aris description of flow and dispersion of a solute in a capillary in which a solvent moves in laminar flow fashion [43-47]. Dispersion is an unsteady diffusion process characterized with an effective diffusion or dispersion coefficient, which represents the rate at which the solute spreads axially, and is proportional to the axial convection/radial molecular diffusion ratio.

Figure 15.8 shows the velocity profile, reference coordinates, and band profiles for the solute at various times. The velocity field, $u(r)$, in a channel for laminar flow is parabolic and given by:

$$u(r) = 2 \bar{u} \left[1 - \left(\frac{r}{r_0} \right)^2 \right] \qquad (15.19)$$

where $\bar{u}$ is the average fluid velocity (ratio of the flow rate, Q, and cross-section area of the channel, A), r is the radial coordinate, and r_0 is the radius of the channel or through-pore. The pressure drop in the channel is obtained from Hagen-Poiseuille's law

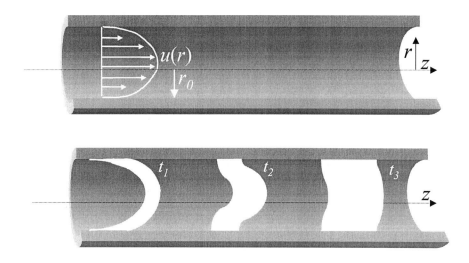

Fig. 15.8. Velocity profile and dispersion of a plug in tube.

$$\frac{\Delta P}{L} = \frac{8\eta}{r_0^2} \bar{u}$$

(15.20)

The plug of a solute introduced at the time zero will, at short times, acquire the parabolic shape imposed by the velocity profile. The axial convection is then the controlling mechanism. However, this leads to radial concentration gradients and, therefore, the radial molecular diffusion contributes to the dispersion process at longer times. The net effect is the compressed mixing zone. At longer times $t \gg r^2/D$, at which the axial convection is important and radial diffusion exceeds axial diffusion, axial concentration gradients can be neglected if $Pe \gg 1$ (Peclet number $Pe = r_0 u/D$). This is called the Taylor dispersion regime and the flow is quasi-steady relative to the axis of the mean fluid velocity.

The mass balance for the solute in a volume element of the tube is

$$\frac{\partial c}{\partial t} + u(r)\frac{\partial c}{\partial z} = D\left[\frac{\partial^2 c}{\partial z^2} + \frac{1}{r}\frac{\partial}{\partial r}\left(r\frac{\partial c}{\partial r}\right)\right]$$

(15.21)

with boundary conditions of, (i), no solute flux at the wall $\partial c/\partial r = 0$ at $r = r_0$ and for, (ii), semi-finite tube $c \to 0$; $z \to \infty$. If a reference frame moving with the mean fluid velocity is chosen as $z' = z - \bar{u}\, t$, the mass balance is

$$\frac{\partial c}{\partial t} + [u(r) - \bar{u}] \frac{\partial c}{\partial z'} = D \left[\frac{\partial^2 c}{\partial z'^2} + \frac{1}{r} \frac{\partial}{\partial r} \left(r \frac{\partial c}{\partial r} \right) \right] \tag{15.22}$$

The asymptotic behavior for the neglected axial diffusion compares to axial convection, *i.e.*,

$$[u(r) - \bar{u}] \frac{\partial c}{\partial z} \gg D \frac{\partial^2 c}{\partial z'^2} \tag{15.23}$$

is reached for long-time distribution of the solute, *i.e.*, $t \gg r_o^2/D$ or $L/r_o \gg Pe$ allowing the quasi-steady state assumption $[\partial c/\partial t]_{z'} \approx 0$, and axial concentration gradient $\partial c/\partial z' \approx \partial \bar{c}/\partial z'$ where $\bar{c}$ is the area-averaged solute concentration. In a typical monolith with a channel size of 2 μm and for liquids with diffusion coefficients of the order of 10^{-5} cm^2/s, the long time behavior is achieved for $t > r_o^2/D \approx 10^{-3}$s. For example, at a flow rate of 10 ml/min through a monolith with internal diameter of 4.6 mm, $Pe \approx 10$ and $L/r_o \gg Pe$ holds.

The concentration profile is related to the average concentration by:

$$c \approx \bar{c} + \frac{r_0^2 \, \bar{u}}{4 \, D} \left[-\frac{1}{3} + \left(\frac{r}{r_0} \right)^2 - \frac{1}{2} \left(\frac{r}{r_0} \right)^4 \right] \frac{\partial c}{\partial z'} \tag{15.24}$$

The conservation equation, in terms of area-averaged axial solute flux, J', representing the sum of the area-averaged axial molecular fluxes

$$J^m = - D \frac{\partial \bar{c}}{\partial z'} \tag{15.25}$$

and area-averaged convective flux relative to the moving observer

$$J^c = \frac{2}{r_0^2} \int_0^{r_0} [u(r) - \bar{u}] \, crdr \tag{15.26}$$

is:

$$\frac{\partial \bar{c}}{\partial t} + \frac{\partial J'}{\partial z'} = 0 \tag{15.27}$$

The area-averaged convective flux J'^c is calculated by using the velocity profile Equation 15.19 and the concentration profile of Equation 15.24 to give

$$J^c = -\frac{r_0^2 \bar{u}^2}{48\,D}\frac{\partial \bar{c}}{\partial z'} \qquad\qquad (15.28)$$

with the Taylor dispersion coefficient given by $D_C = r_0^2 \bar{u}^2/48\,D$.

The one-dimensional macrotransport equation in terms of the fixed reference frame z controlling the area-averaged solute concentration $\bar{c}(z,t)$ is

$$\frac{\partial \bar{c}}{\partial t} + \bar{u}^* \frac{\partial \bar{c}}{\partial z} = D^* \frac{\partial^2 \bar{c}'}{\partial z^2} \qquad\qquad (15.29)$$

where $\bar{u}^* = \bar{u}$, and

$$D^* = D + \frac{r_0^2 \bar{u}^2}{48\,D} \qquad\qquad (15.30)$$

which is the result obtained by Taylor-Aris [46].
The area-averaged concentration profile is:

$$\bar{c}(z,t) = \frac{N/\pi \cdot r_0^2}{\sqrt{4\pi \cdot D^* t}} \; \exp\left[-\frac{(z - \bar{u}\,t)^2}{4\,D^* t} \right] \qquad\qquad (15.31)$$

The validity of Taylor dispersion imposes $D_C \ll r_0^2 \bar{u}^2/48D$ or $Pe \gg 7$. In the region $Pe \ll 7$ and $Pe \gg 7$ where both radial and axial diffusion are important, $D^* = D(1 + Pe^2/48)$.

The limiting case of pure convection holds if the time constants for radial diffusion, r_0^2/D, and axial diffusion, L^2/D, are both higher than the convective time constant $L/\bar{u}$, i.e., $Pe \gg L/r_0$ and $Pe \gg r_0/L$, respectively. Similarly, axial diffusion is controlling for $r_0^2/D \ll L/\bar{u}$ or $L^2/D \ll L/\bar{u}$. These results are condensed in Fig. 15.9, adapted from Aris [46]. It is worth noting that the averaging of the basic Equation 15.21 over the cross-section area leads to:

$$\frac{\partial \bar{c}}{\partial t} + \bar{u}\frac{\partial <c>}{\partial z} = D\frac{\partial^2 \bar{c}}{\partial z^2} \qquad\qquad (15.32)$$

where

$$<c> = \frac{4}{r_0^2}\int_0^{r_0} r\left(1 - \frac{r^2}{r_0^2}\right) c\cdot dr \qquad\qquad (15.33)$$

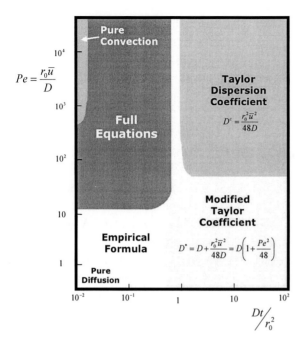

Fig. 15.9. Dispersion of a solute in a tube: Map of controlling mechanisms in terms of Peclet number, *Pe*, versus dimensionless time, $D t / r_0^2$. (Reprinted with permission from ref. [46]. Copyright 1994 Rutherford Aris).

is the cup-mixing mean. An effective dispersion coefficient can thus be obtained in the general case from

$$D^* = D + \bar{u} \frac{\partial}{\partial z} (\bar{c} - <c>) / \frac{\partial^2 \bar{c}}{\partial z^2} \tag{15.34}$$

The Taylor description can be extended to include adsorption at the wall of the channels and longitudinal diffusion at the wall surface, D_s; in that case the zone migrates at a velocity:

$$\bar{u}^* = \frac{\bar{u}}{1 + K} \tag{15.35}$$

and the effective diffusion coefficient will have a form of a Golay equation:

$$D^* = \frac{D + K D_s}{1 + K} + \frac{r_0^2 \bar{u}^2}{48D} \frac{1 + 6K + 11K^2}{(1 + K)^3} \tag{15.36}$$

where K is the adsorption equilibrium parameter for linear isotherms.

The Taylor-Aris method was recently used by McCalley [48] to measure diffusion coefficients in reversed-phase monolithic columns, and flat plots of HETP with $H > 5$ μm *versus* flow rate were found.

15.3 MASS TRANSFER

15.3.1 Mass transfer mechanisms

In linear chromatography the efficiency of a column is usually characterized by the height equivalent to a theoretical plate (HETP). The dependency of HETP on flow rate, presented in Eq. 15.7, was derived by Van Deemter *et al.* [16]. A monolith can be seen as a single large particle and, therefore, the Van Deemter equation can be adapted with pore velocity, v, replacing superficial velocity. Hahn and Jungbauer [49], introduced an intraparticle pore effective diffusivity, D_{pe}, containing a velocity-dependent part, following ideas of Van Kreveld and Van den Hoed [50] developed for gel permeation chromatography with permeable supports, and Afeyan *et al.* [51] for perfusion chromatography

$$D_{pe} = D_p + dv \tag{15.37}$$

where d is the eddy diffusion length. The modified Van Deemter equation is then

$$H = A + \frac{B}{v} + \frac{K}{30\,(1+K)^2}\frac{d_p^2\,v}{D_p + dv} \tag{15.38}$$

It is claimed that the eddy diffusion term A can be minimized in monoliths and that the last term in Eq. 15.38 results from simultaneous association of mass transfer resistances due to diffusion and convection. This follows the ideas presented in Giddings' "coupling theory" for the mobile phase [52]. He pointed out that packed bed dispersion and mass transfer in the mobile phase are coupled, and variances are not additive. Instead, their reciprocals are, and the same holds for their contributions to HETP. Using their model, Hahn and Jungbauer analyzed experiments with retention of BSA in a polymethacrylate-based monolith [49].

Reduction in the particle size of column packings can decrease both the contribution of eddy diffusion and mobile-phase transfer (A-term), and the diffusive path, thus decreasing the C-term. However, in liquid chromatography the extent of size-reduction is limited by pressure-drop constraints. The extension of this reasoning towards monoliths is discussed by Motowaka *et al.* [37]. They consider Giddings' coupling theory and analyze the A-term. They found that the straight throughpores cause peak broadening as a result of slow mobile-phase transfer (large A_m), and inefficient exchange between streamlines, leading to large eddy diffusion (large A_e) in the low flow velocity region. The HETP is then

$$H = \frac{1}{\dfrac{1}{A_e} + \dfrac{D_m}{A_m v}} + \frac{B}{v} + C v \tag{15.39}$$

A similar approach was adopted by Tanaka *et al.* [53] and Ishizuka *et al.* [38]. A dependency of HETP with the slope of the isotherm K was observed for a silica-based monolith containing 7 μm large throughpores, according to the Golay equation:

$$H = \frac{2D_m}{\bar{u}} + \frac{1 + 6K + 11K^2}{(1+K)^2} \frac{d_c^2}{96D_m} \bar{u} + \frac{2K}{3(1+K)^2} \frac{d_s^2}{D_m} \bar{u} \tag{15.40}$$

The importance of the A-term addressed above is in line with the Knox conclusions [54] concerning packed bed chromatography. He claimed that the C-term is not important and the main contribution to peak broadening is effected by the moving zone. Giddings suggested [52] that in a packed column both tortuous flow and transverse diffusion would be combined to reduce dispersion, and postulated that these resistances were acting simultaneously

$$H_A = \frac{1}{\dfrac{1}{A} + \dfrac{1}{C_m u}} . \tag{15.41}$$

The contribution of the flow profile in liquid chromatography was also considered by Sie and Rijnders [55]. The mechanisms contributing to lateral mass transport were molecular diffusion and convective diffusion, $D_r = \lambda_r d_p u_i$. The latter resulted from the repeated division and combination of fluid streams in the channels. They also assumed that molecular diffusion and convective diffusion are statistically independent processes. Thus the profile contribution is

$$H = \frac{2\gamma \cdot R^2 \, d_p \, u_i}{\lambda_r \, d_p \, u_i + \gamma_r \, D} \tag{15.42}$$

In HPLC these two contributions can be of the same order. Similar velocity-dependent terms have also been presented elsewhere [56,57].

A more rigorous approach was used to elucidate the C-term related with intraparticle mass transport and sorption kinetics for large pore packing materials [9]. The total mass flux within pores was considered to be the sum of both diffusive and convective fluxes, the latter being driven by the pressure drop across the particle. This leads to a specific case of the complete Maxwell–Stephan description of mass transfer in porous media concerning diluted solutions in liquid chromatography, and enabled the derivation of Eq. 15.1, which is a velocity-dependent apparent diffusivity combining pore diffusion and convective Poiseuille flow inside pores. This treatment did not require the assumption of additivity of variances or reciprocals of variances. Therefore, the dependency of HETP on flow rate [17,18] could be explained in the whole range of flow rates and not only in the limiting cases of diffusion- and convection-controlled regimes [11]. This approach led to the extended Van Deemter equation 15.9, which was limited to permeable materials with large pores.

We extended the analysis to bidisperse particles containing both transport pores and smaller diffusive pores [13]. The new extended equation for the reduced HETP, $h = H/d_p$ is

$$h = A + \frac{B}{\bar{v}} + \frac{1}{30} \frac{\varepsilon_p \, \tau_p \, b^2 \, (1 - \varepsilon)/\varepsilon}{\left[1 + \varepsilon_p \, b(1 - \varepsilon)/\varepsilon\right]^2} \left[f(\lambda_p) + \frac{b - 1}{b^2 \, T} \right] \tag{15.43}$$

where $\bar{v} = u_i d_p/D_p$ is the reduced velocity and $T = (D_m/r_m^2)/(D_p/R_p^2)$ is the new parameter comparing time constants for diffusion in transport pores and diffusive pores where the subscript m refers to microspheres containing smaller diffusive pores. Another extension was made to include adsorption kinetics and film mass transfer [58].

15.3.2 Characterization of the efficiency

The basic equations addressing the column efficiency in liquid chromatography are [53]:

(i) the relationship between the height equivalent to a theoretical plate, H, and number of plates, N, in the column with length L,

$$H = \frac{L}{N} \tag{15.44}$$

(ii) a Van Deemter type relationship between H and the superficial velocity in the bed, u_0:

$$H = f(u_0) . \tag{15.45}$$

(iii) A flow-resistance parameter expressed as the ratio between the square of particle diameter and bed permeability:

$$\phi = \frac{\Delta P}{L\,\eta\,u_0}\,d_p^2 = \frac{d_p^2}{B_b} \tag{15.46}$$

(iv) Separation impedance, which is the ratio between the square of height equivalent to the theoretical plate, and the bed permeability:

$$E = \frac{\Delta P}{N}\,\frac{L}{u_0}\,\frac{1}{N}\,\frac{1}{\eta} = \frac{H^2}{B_b} \tag{15.47}$$

The separation impedance, E, for columns packed with 5 μm silica particles is 2500–3000 and the number of plates in HPLC is limited to 10,000–25,000 per column because of the pressure-drop limit of 35–40 MPa. For monoliths the above relationships can also be adapted. The flow resistance parameter is $\phi = d_{sk}^2 / B_m$ and the separation impedance is $E = H^2/B_m$ where the permeability of a monolith is given by Eq. 15.15. The separation impedance of a monolith can be as low as 100, much lower than a value of 3,000 observed for columns packed with 5 μm particles. HETP values in a range of 8–12 μm were found for silica-based monoliths at a flow velocity of 1 mm/s.

The need of low flow resistance and high surface area requires materials with large transport pores for convection, and smaller diffusive pores providing adsorption capacity [59]. The efficiency of the monolithic columns is often characterized through the number of theoretical plates, or plate number, and the height equivalent to the theoretical plate [60,61]. HETP can be calculated from experimental pulse responses, $C(t)$, as $HETP = \sigma_t^2 L/t_r^2$, where L is the column length, t_r and σ_t^2 are the mean residence time or retention time, and the variance of the eluted peak, respectively. These quantities are easily evaluated from Eq. 15.48

$$E(t) = \frac{C(t)}{\int\limits_0^\infty C(t)dt} \tag{15.48}$$

from the moments of the residence time distribution for $t_R = \mu_1$ and the peak variance $\sigma_t^2 = \mu_2 - \mu_1^2$, with the moments of order k being

$$\mu_k = \int\limits_0^\infty E(t)t^k dt \tag{15.49}$$

The residence time distribution $E(t)$ is the normalized pulse response.

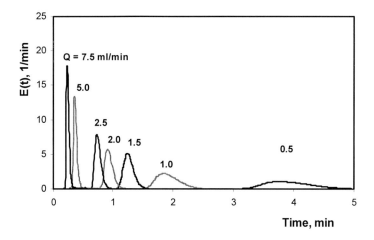

Fig. 15.10. Normalized concentration from pulse response of UV detector for thiourea at different flow rates. Conditions: column Chromolith Performance RP-18e 100 × 4.6 (Merck, Germany), mobile phase acetonitrile–water (60:40), sample concentration 10^{-2} g/l, injected volume 10 µl, temperature 20°C, UV detection at 254 nm.

Pulse responses in a monolithic silica-based column were measured with thiourea at different flow rates, Q. The results are shown in Fig. 15.10. Peak broadening is important only at low flow rate. This is also shown in Fig. 15.11 where the effect of the flow rate on HETP for various compounds such as thiourea, anthracene, toluene and progesteron was analyzed. It was observed that the HETP did not change significantly with the flow rate exceeding 1.5 ml/min.

The plot of the reciprocal retention time as a function of flow rate is shown in Fig. 15.12. It is a straight line with the slope $1/[V\{\varepsilon+(1-\varepsilon)m\}]$, where V is the column volume, m is the slope of the linear isotherm, and ε is the external porosity of the monolith. It allows the calculation of the amount of adsorbed compound related to the volume of silica skeletons, that are treated as being homogeneous.

Mass transfer in a monolithic column can also be characterized using experimental breakthrough curves [62,63]. In a typical breakthrough experiment, a solution of a compound at a concentration of C_{i0}, in a mobile phase, passes through the monolithic column until the solute concentration at the column outlet, $C(t)$, reaches the feed value, C_{i0}. The breakthrough curve enables calculations of important characteristics such as the stoichiometric time, t_{st}, and the total amount of solute retained in the adsorbent bed in equilibrium with the feed concentration, q_{i0}^{*}. One point (C_{i0}, q_{i0}^{*}) of the adsorption equilibrium isotherm can be calculated from each breakthrough experiment. Running various experiments at various feed concentrations enables the construction of the complete adsorption equilibrium isotherm.

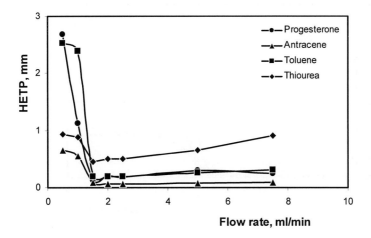

Fig. 15.11. HETP for anthracene, progesterone, toluene, and thiourea as a function of flow rate. Conditions: monolithic column Chromolith RP 18e 100 × 4.6 mm i.d., mobile phase acetonitrile–water (60:40), concentrations of injected compounds 10^{-2} g/l, injected volume 10 μl, ambient temperature, UV detection at 254 nm.

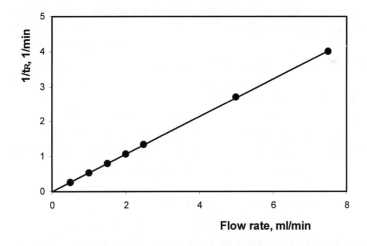

Fig. 15.12. Effect of flow rate on the reciprocal retention time of thiourea.

Figure 15.13 shows the breakthrough curves obtained experimentally (points) for thiourea at inlet concentrations, C_0, of 2.2, 3.9 and 7.9 mg/l. The lines represent the simulated results using a model, which includes flow with axial dispersion and mass transfer in the silica skeleton microparticles. However, as a first approximation, this

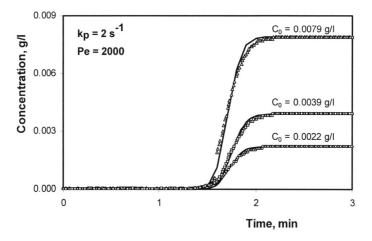

Fig. 15.13 Effect of feed concentration on the breakthrough curves for thiourea using a monolithic column. Conditions: column Chromolith RP 18e 100×4.6 mm i.d., feed concentrations 2.2, 3.9, and 7.9 mg/l. Points experimental, lines calculated using model including axial dispersion flow and LDF mass transfer in skeletons.

mass transfer was assumed to be very fast, approaching an equilibrium situation due to the short diffusive path within the skeletons.

15.3.3 Modeling and simulation of monolithic columns

The model described below assumes plug-dispersed flow in the axial direction, mass transfer resistance in the silica skeleton microparticles, and equilibrium at the interface between fluid/microparticle surface. Model equations are the conservation equation (mass balance) for the fluid phase, the adsorption equilibrium at the fluid/microparticle interface, and the kinetics of mass transfer inside the skeletons.

15.3.3.1 Mass balance for species in the bulk fluid phase

The mass balance in a volume element of the monolithic column is

$$\frac{\partial c_i}{\partial t} = D_L \frac{\partial^2 c_i}{\partial z^2} - u_i \frac{\partial c_i}{\partial z} - \frac{(1-\varepsilon)}{\varepsilon} k_P \left(q_i^* - q_i\right) \tag{15.50}$$

where c_i and q_i are the fluid phase and average adsorbed phase concentrations of species i in the monolithic column, respectively, q_i^* is the adsorbed phase concentration in equilibrium with c_i, z is the axial co-ordinate, t is the time variable, ε is the external porosity of the monolith, u_i is the interstitial fluid velocity, D_L is the axial dispersion coefficient, and k_p is the mass-transfer coefficient in the silica skeletons that are assumed to be homogeneous.

15.3.3.2 Kinetics of mass transfer inside the microparticles

The mass balance in the microparticle according to the simple linear driving force (LDF) model particle is

$$\frac{\partial q_i}{\partial t} = k_P \left(q_i^* - q_i \right) \tag{15.51}$$

15.3.3.3 Adsorption equilibrium isotherm

At the fluid/microparticle interface holds equilibrium, *i.e.*. For linear adsorption it is simply $q_i^* = K\, c_i$

15.3.3.4 Initial and boundary conditions

The initial condition is

$$t = 0: c_i = q_i = 0 \tag{15.52a}$$

while the Danckwerts boundary conditions are

$$z = 0 \quad (t{>}0): \qquad u_i\, c_{i0} = u_i\, c_i - D_L\, \frac{\partial c_i}{\partial z} \tag{15.52b}$$

$$z = 0 \quad (t{>}0): \qquad \frac{\partial c_i}{\partial z} = 0 \tag{15.52c}$$

The model equations in dimensionless form will include the model parameters: the Peclet number and the number of diffusion units $\alpha = k_p\tau$, where τ is the space time, which is the ratio between time constants for convection in throughpores and diffusion in the skeletons. The intraparticle rate constant, k_p, was estimated using Glueckauf analysis of the LDF model to be $k_p = 60 D_s/d_{sk}^2$.

This model was used to simulate the breakthrough curves in monolithic columns shown in Fig. 15.13. Furthermore, it allows the calculation of HETP and their comparison with results obtained in pulse experiments. From this model we obtained

$$H = \frac{2L}{Pe} + \frac{2(1-\varepsilon)\,K}{[\varepsilon + (1-\varepsilon)\,K]^2}\,\frac{1}{k_p}\,u_0 \tag{15.53}$$

This equation results from the definition of $H = \sigma_t^2 L/t_r^2$ and noting that the moments of the residence time distribution can be calculated from the transfer function of the monolith system, $G(s)$ by

$$\mu_k = \left[(-1)^k\,\frac{d^k G(s)}{ds^k} \right]_{s=0} \tag{15.54}$$

where

$$G(s) = e^{\frac{Pe}{2} - \sqrt{\frac{Pe^2}{4} + Pe\tau s\left[\varepsilon + (1-\varepsilon)\frac{k_p K}{s + k_p}\right]}} \tag{15.55}$$

The above model is also the first step in the simulation of processes such as Simulated Moving Bed (SMB) designed for the separation of enantiomers using monolithic columns [64].

15.4 CONCLUSIONS

Monoliths are structured materials which contain large transport pores connected to smaller diffusive pores with larger surface area. A better understanding of flow and mass transfer would involve experimental techniques for monolith characterization, such as pore size distribution and image analysis of sections for 3-D reconstruction, coupled with computational flow dynamics (CFD) studies.

15.5 REFERENCES

1   G. Carta, H. Massaldi, M. Gregory, D. Kirwan, Sep. Technol. 2 (1992) 62.
2   N. Afeyan, S. Fulton, N. Gordon, L. Mazsaroff, L. Varady, F. Regnier, Bio/technology 8 (1990) 203.
3   L. Lloyd, F. Warner, J. Chromatogr. A 512 (1990) 365.
4   C.L. Prince, V. Bringi, M.L. Shuler, Biotech. Progress 7 (1991) 195.
5   A. Nielsen, S. Bergh, B. Troberg, US Patent 3,243,386, March 29 (1966).
6   N. Harbord , U.K. Patent 1,484,864, Sept 8 (1977).
7   A. Wheeler, Adv. Catal. 3 (1951) 250.
8   A. Nir, L. Pismen. Chem. Eng. Sci. 32 (1977) 35.
9   A.E. Rodrigues, B. Ahn, A. Zoulalian, AIChE J. 28 (1982) 541.
10   A.E. Rodrigues, R.Q. Ferreira, AIChE Symp.Series 84 (1988) 80.
11   N. Afeyan, S. Fulton, F. Regnier, J. Chromatogr. A 544 (1991) 267.

12   D. Frey, E. Scheinheim, C. Horvath, Biotech. Prog. 9 (1993) 273.

13   G. Carta, A. Rodrigues, Chem. Eng. Sci. 48 (1993) 3927.

14   T.B. Tennikova, F. Svec, B.G. Belenkii, J. Liquid Chromatogr. 13 (1990) 63.

15   M. Chaara and R. Noble, Sep. Sci. Tech. 24 (1989) 893.

16   J. Van Deemter, F. Zuiderweg, A. Klinkenberg, Chem. Eng. Sci. 5 (1956) 271.

17   A.E. Rodrigues, L. Zuping, J. Loureiro, Chem. Eng. Sci. 46 (1991) 2765.

18   A.E. Rodrigues, LC-GC 6 (1993) 20.

19   H. Komiyama, H. Inoue, J. Chem. Eng. Japan 7 (1974) 281.

20   W.E. Collins, Sep. Purif. Meth. 26 (1997) 215.

21   M. Schulte, D. Lubda, A. Delp, J. Dingenen, J. High Resol. Chrom., 23 (2000) 100.

22   K. Cabrera, A. Lubda, H. Eggenweiler, K. Minakuchi, J. Nakanishi, J. High Resol. Chrom., 23 (2000) 93.

23   N. Tanaka, H. Kobayashi, K. Nakanishi, H. Minakuchi, N. Ishizuka, Anal. Chem., 73 (2001) 421A.

24   F. Svec, J.M.J. Fréchet, Anal. Chem. 64 (1992) 820.

25   D. Josic, A. Strancar, Ind. Eng. Chem. Res. 38 (1999) 333.

26   S. Vogt, R. Freitag, Biotechnol. Prog. 14 (1998) 742.

27   M. Buchmeiser, J. Chromatogr. A 918 (2001) 233.

28   D. Josic, A. Buchacher, A. Jungbauer, J. Chromatogr. B 752 (2001) 191.

29   A. Cybulski, J. Moulijn, Structured Catalysts and Reactors, M. Dekker, New York, 1998.

30   K. Sinz, K. Cabrera, Lab Asia 8 (2001) 1.

31   G. Cavazzini, G. Bardin, K. Kaczmarski, P. Szabelski, M. Al-bokari, G. Guiochon, *J. Chromatogr. A 957 (2002) 111.*

32   V. Mata, J.C. Lopes, M.M. Dias, Ind. Eng. Chem. Res. 40 (2001) 3511.

33   V. Mata, J.C. Lopes, M.M. Dias, Ind. Eng. Chem. Res. 40 (2001) 4836.

34   J. Meyers, A. Liapis, J. Chromatogr. A 852 (1999) 3.

35   A. Liapis, J. Meyers, O. Crosser, J. Chromatogr. A 865 (1999) 13.

36   R. Jackson, Transport in Porous Catalysts, Elsevier, New York, 1977.

37   M. Motokawa, H. Kobayashi, N. Ishizuka, H. Minakuchi, K. Nakanishi, H. Jinnai, K. Hosoya, T. Ikegami, N. Tanaka, J. Chromatogr. A, in press.

38   N. Ishizuka, H. Kobayashi, H. Minakuchi, K. Nakanishi, K. Hirao, K. Hosoya, T. Ikegami, N. Tanaka, J. Chromatogr. A 960 (2002) 85.

39   T. Tennikova, F. Svec, J. Chromatogr. A 646 (1993) 279.

40   L. Berruex, R. Freitag, T. Tennikova, J. Pharm. Biomed. Anal. 24 (2000) 95.

41   N. Ostryanina, G. Vlasov, T. Tennikova, J. Chromatogr. A 949 (2002) 163.

42   J. Pfeiffer, J. Chen, J. Hsu, AIChE J. 4 (1996) 932.

43   G.I. Taylor, Proc. Roy. Soc. A 219 (1953) 186.

44   H. Brenner, D. Edwards, Macrotransport Processes, Butterworth-Heinemann, New York, 1993.

45   R. Probstein, Physicochemical Hydrodynamics. An Introduction, Butterworth, Stoneham, 1989.

46   R. Aris, Mathematical Modeling Techniques, Dover, New York, 1994.

47   H. Brenner, Langmuir 6 (1990) 1715.

48   D. McCalley, J. Chromatogr. A, in press (2002).

49   H. Hahn, A. Jungbauer, Anal. Chem. 72 (2000) 4853.

50   M. van Kreveld, N. van den Hoed, J. Chromatogr. A 149 (1978) 71.

51   N. Afeyan, N. Gordon, I. Mazsaroff, L. Varady, S. Fulton, Y. Yang, F. Regnier, J. Chromatogr. A 519 (1990) 1.

52   J.C. Giddings, Dynamics of Chromatography, Part I Principles and Theory, M. Dekker, New York, 1965.

53   N. Tanaka, H. Kobayashi, N. Ishizuka, H. Minakuchi, K. Nakanishi, K. Hosoya, T. Ikegami, J. Chromatogr. A, in press.

54   J. Knox, J. Chromatogr. A 831 (2000) 3.

55   S.T. Sie, G. Rijnders, Anal. Chim. Acta 38 (1967) 3.

56   Cs. Horvath, H. Lin, J. Chromatogr. A 126 (1976) 401.

57   B. Berdichevsky, U.D. Neue, J. Chromatogr. A 535 (1990) 189.

58   A.E. Rodrigues, A. Ramos, J. Loureiro, M. Diaz, Z. Lu, Chem. Eng. Sci., 47 (1992) 4405.

59   F. Svec, J.M.J. Fréchet, Ind. Eng. Chem. Res., 38 (1999) 34.

60   N. Ishizuka, H. Minakuchi, K. Nakanishi, K. Hirao, N. Tanaka, Coll. Surf. A-Physicochem. Eng. 187 (2001) 273.

61   U.D. Neue, HPLC Columns: Theory, Technology and Practice, Wiley-VCH, New York, 1992.

62   F. Ming, J. Howell, Trans. I. Chem. E., Part C, 71 (1993) 267.

63   S. Ghose, S.M. Cramer, J. Chromatogr. A 928 (2001) 13.

64   M. Schulte, J. Dingenen, J. Chromatogr. A 923 (2001) 17.

F. Švec, T.B. Tennikova and Z. Deyl (Editors)
Monolithic Materials
Journal of Chromatography Library, Vol. 67
2003 Published by Elsevier Science B.V.

Chapter 16

Theoretical Aspects of Separation Using Short Monolithic Beds

Tatiana B. TENNIKOVA[1] and František ŠVEC[2]

[1]*Institute of Macromolecular Compounds, Russian Academy of Sciences, Bolshoy pr. 31, 199 004 St. Petersburg, Russia*
[2]*University of California, Department of Chemistry, Berkeley, CA 94720-1460, USA*

CONTENTS

16.1 INTRODUCTION

In general, all chromatographic methods have one common feature: the dynamic distribution of substances in the flow system. Until recently, chromatography was considered a process that occurs in a column packed with a particulate solid sorbent – the stationary phase. Current HPLC enjoys a rich arsenal of stationary phases, which can effectively separate a wide variety of mixtures consisting of both small molecules and large polymeric compounds. The latter also include the very important natural biopolymers. The sorbents used in HPLC are most often rigid spheres having sizes from single to a few tens of micrometers and span a wide range of porosities from non-porous to those with very large pores that even enable, to a certain extent, convective flow-through.

The adsorptive or interactive modes of HPLC are also those mostly used in biotechnology and related fields. The adsorption/desorption acts that occur between the separated substances and the stationary phase are typical of these separation modes and include *inter allia* Coulombic, hydrophobic, and bioaffinity interactions. These interactions are also included in the respective terms coined for the specific separation modes, such as ion exchange, reversed-phase, hydrophobic, pseudoaffinity, and bioaffinity chromatography.

In addition to the predominantly diffusive transport within their pores, the particulate separation media are unable to fill completely the space within the chromatographic column and at the interparticular voids represent in the theoretical case 28% of the overall column volume. This contributes to peak broadening and decreases column efficiency. Use of media with a higher degree of continuity enables considerable reduction in the void volume [1-6]. Development of separation media that consist of a continuous bed or "monolith", actually a very large single particle with a certain well-defined geometry such as rod, disk, and tube consisting of rigid, highly porous polymer decreases dramatically the undesired void volume inside the separation device. The most important feature characterizing these materials is the flow of *all* of the mobile phase *through* the large pores or channels. As a consequence of this convective flow, the mass transfer within the pores is significantly enhanced and exerts a positive effect on the separation. A detailed theoretical analysis of mass transport effects on the separation of solutes using both conventional packings and monolithic sorbents has been published recently [7,8]. The engineering aspects of the mass transfer are dealt with elsewhere in this book. This chapter concerns mechanism and efficiency of monolithic thin layers used for the separation of a variety of analytes using different interactive modes of chromatography that rely on positive interactions of solute molecules with the functionalized surface of the stationary phase.

16.2 CHROMATOGRAPHIC SEPARATION OF PROTEINS

16.2.1 Specific features of protein molecules subjected to separation processes controlled by mass transfer

It appears useful to recall first some of the important physical parameters that control the behavior of the globular proteins [9]. With respect to chromatography, these specific factors differ significantly from those typical of small and midsize molecules. They affect (i) the kinetics of the adsorption process, (ii) its thermodynamics, and (iii) the character of the interaction between the protein molecules and the surface.

Let us first relate the diffusional mobility or diffusivity of proteins to kinetic of the adsorption. The diffusivity of proteins is 2-3 orders of magnitude lower than that of small molecules. It is worth noting that both solvent and modifier that constitute the eluent are also small molecules participating in the separation process. The low diffusivity translates in the 2-3 times longer residence time of a protein molecule in the mobile phase, which also represents the time between consecutive adsorption/desorption acts.

The thermodynamically based steric factor also differentiates proteins from small molecules. It is understandable that the large molecule of a protein will experience significant steric restrictions in the pores within a porous particle compared to the small molecule counterparts. In addition, it is increasingly difficult for the large molecules to pass through the orifice of the pore. This phenomenon is most important for sorbents with a rigid porous structure.

The interaction factor is also linked to the size of the protein molecule, which can be presented as a large globule. It carries functional groups at its outer surface that participate in the adsorptive interactions with the sorbent. The distribution of these interacting functionalities on the protein surface is not random. They are localized very specifically and may even form larger clusters. This inhomogeneous distribution of strongly interacting sites at the surface of the protein globule defines, at least partially, the stoichiometric parameter of protein/surface interaction (*vide infra*) and varies for each protein. This difference is most important for the separation of proteins using short membrane-like monolithic layers.

In addition to the above-mentioned factors that affect the chromatographic process, some other aspects may also become critical in the specific interaction modes of chromatography. For example, the effect of concentration of the analyte is effective once it reaches values sufficiently high to enable the presence of more than one protein molecule in a given pore. The mutual interactions between individual protein molecules also affect their adsorption. Therefore, it is useful to consider both "synergistic" and concentration effects simultaneously. Finally, it is also necessary to consider the "hysteresis" representing the fact that adsorption and desorption isotherms for many proteins do not coincide. This results from the multiple point interactions of the protein on the sorbent surface. Therefore, desorption step for many proteins requires a higher concentration of the modifier in the mobile phase than that at which the protein has been adsorbed.

16.2.2 Protein separations using membranes and columns: "single-step" vs. "multiple-step" ("multi-plate") adsorption-desorption process

A chromatographic column packed with small porous beads is the most important component of the separation process in conventional HPLC. In the typical scenario, the mobile phase flows through the interstitial voids between the particles where the flow resistance is the smallest. In contrast, the liquid present in the pores of the porous beads does not move and remains stagnant. If a protein or a mixture of proteins is injected into the stream of the flowing mobile phase, they are carried through the voids. However, diffusional transport of these compounds in the pores also occurs because of the concentration gradient between the solution in the voids and the stagnant liquid within the pores. First, the proteins migrate through the layer of stagnant liquid film surrounding each particle to the outer surface of the sorbent bead, then they move into the pores, and finally adsorb on their surface.

The act of desorption, at which the molecules adsorbed on the surface are displaced by competing modifier molecules present in the mobile phase, is critical to achieve a good separation since the differences in desorption rate among various proteins are larger than those characteristic of the adsorption step. Two elution modes are mostly used in interactive types of chromatography: (i) *isocratic elution* where the percentage of modifier in the eluent is kept constant through the entire separation process, and (ii) *gradient elution* in which the concentration of modifier in the mobile phase increases gradually or in steps during the separation. Gradient elution is more common for chromatographic separations of proteins since they vary widely in their adsorption energies and it is often very difficult to find an isocratic mobile phase to achieve complete desorption of all components and their separation.

As a result of a broad range of the desorption rates of proteins caused by significant differences in their adsorption energies, their separation in adsorptive HPLC modes can be achieved in only a single desorption step. It was therefore reasonable to infer that the separation of proteins could be possible even without using long column and repetitive adsorption/desorption events. This theoretical possibility has been materialized in chromatographic technique called *membrane chromatography* that is carried out in specifically designed device that contains a flat layer of the stationary phase through which all the mobile phase must permeate. This device is called *mem-*

brane adsorber. This term originated from the membrane filtration technology and well describes the geometry of the stationary phase - interactive (adsorptive) filter-like membrane [10-12]. Various types of filtration operation such as hollow fibers and cross-flow units have been adapted to separation processes simply by using interactive materials instead of an inert membrane [12,13].

In the late 1980s, we have developed another format of flat beds separation devices – rigid macroporous disks, and coined a term *high-performance membrane chromatography* (HPMC) [2,4,11,14,15]. This has recently been modified to *high-performance monolith chromatography* and finally extended to *high-performance monolithic disk chromatography* (HPMDC) [16,17]. Specifically designed thin monolithic macroporous disks with well-defined homogeneous pore structure lie at heart of this separation technique. Use of these monolithic layers also enabled to derive theoretical model of the dynamic interphase distribution of proteins during the separation.

In addition to the mass transfer enhanced by convection and by the absence of small diffusive pores [7,18-20] typical of all porous particulate packings including even beads for perfusion chromatography that contain very large flow-through pores, the theoretical treatment of the separation using a disk or a thin membrane must take into account the small length of the "separation distance", which *a priori* makes unlikely multiple or repetitive interaction for molecules characterized by slow diffusional mobility. This is an important feature that clearly differentiates HPMDC from column chromatography no matter what type of column (packed, monolithic) is used.

Roper and Lightfoot claimed in their excellent review that "*membrane chromatography is a method based on the selective separation of adsorbed substances as a consequence of the desorption process*" [10]. To get a clear picture of what occurs during interactive chromatography of proteins using the "membrane-like" stationary phase, the separation is assumed to result from a single-step desorption process [11,17]. This does not mean that the adsorption step can be neglected. Since HPMDC is assumed to be a single step process, the initial adsorption plays much larger role in this mode compared to the conventional chromatographic processes. In the latter, the spatial distribution is readily averaged as a result of the multitude of repeated adsorption/desorption steps occurring across the column through which the molecules migrate. The added advantage of chromatography on short monolithic layers is that it provides a unique opportunity to monitor "snapshots" of the distribution of molecules after their initial adsorption that later translates in the peak broadening.

16.2.2.1 Gradient elution and column length in protein separation

The current theory of gradient HPLC [21,22] based on the pioneering work of Snyder [23-25] indicates that the column length often has only a small effect on the

resolution of large molecules such as proteins in all modes of retentive chromatography. Snyder studied in detail the processes typical of both gradient and isocratic
HPLC for molecules of different sizes. The characteristics of separation efficiency
such as the peak capacity *PC*, the resolution factor R_S, and the chromatographic band
broadening expressed as the band width σ in time or volume units or the peak height
PH are functions of gradient parameters such as the gradient time t_G, the size of
packed particles d_p, the volumetric flow rate *F*, and the solute diffusion coefficient
D_m. The most important result of Snyder's studies concerning protein separations is
the understanding of the role played by the mobile phase modifier during the desorption in gradient elution. Snyder introduced the average retention (capacity) factor
corresponding to the retention factors $\overline{k}'$ averaged over the entire column length. This
average retention factor largely depends on the acts of desorption [25].

The classical theory results in the following functions:

$$PC \approx R_S \approx t_G^{1/2} \, F^0 \, L^0 \, d_p^{-1} \qquad\qquad (16.1)$$

$$1/\sigma_t \approx PH \approx t_G^{1/2} \, F^{-1} \, L^0 \, d_p^{-1} \qquad\qquad (16.2)$$

where *L* is the column length. These functions indicate that resolution increases with
t_G. However, it is independent of flow-rate and column length.

Similar observation has also been made in other studies [26-30]. For example,
Moore and Walters demonstrated an impressive separation of a model mixture of three
proteins using columns with lengths ranging from 0.16 to 4.5 cm. Surprisingly, an
almost 30-fold change in the column length had only a small effect on the separation
[28]. Similar effects were also observed for the separation of proteins in reversed
phase mode using capillary columns differing widely in their length [31].

Yamomoto presented very detailed study of protein separations using ion exchange
HPLC in the gradient mode and developed a *quasi-steady state* model based on the
theory of continuous-flow plate [32]. According to his model of gradient elution, the
zone spreading effects in the regions of adsorption lead to an increase in the peak
width, while simultaneously zone compression occurs as a result of the peak sharpening due to the flow. During the early stage of elution, the zone spreading effects
prevail and the width of the protein zone increases. Since contribution of the zone
compression increases with the time, this effect becomes more distinct as the zone
width increases. The compression eventually counterbalances the zone spreading and
a quasi-steady state is achieved. As a result, no further increase in the width of the
protein zone is observed. This occurs at the point at which the equilibrium distribution
constant K_D for the protein equals that for the displacer. The concentration of the
displacer at which these distribution constants are equal is very important since it also

indicates the intrinsic adsorption ability of the protein. In the quasi-steady state, the migration velocity of the protein zone is equal to that of the displacer. As the zones for separated proteins move at equal velocities, the distance between them does not change any longer, and the R_S value remains constant. In other words, at a fixed gradient slope and flow rate, the resolution initially increases with the column length until a certain point. The length required to reach the point beyond which the resolution does not change is shorter for separations in steeper gradients. Therefore, the separation may occur only in a relatively short initial section of the chromatographic column while its rest does not contribute to better resolution.

16.2.2.2 Stoichiometric retention model in protein separation

Ion exchange mode is one of the most popular retentive chromatographic modes typically used for the protein separations. The experimentally determined retention factor k', is often expressed as a function of concentration of the modifier in the mobile phase. Generally, the relationship can be reduced to the following simple equation:

$$k' = e^{f(\phi)} \tag{16.3}$$

where $f(\phi)$ is the polynomic function. However, the function $f(\phi)$ can be in most cases substituted by a single value to yield equation:

$$k' = M e^{-S\phi} \tag{16.4}$$

where M and S are the constants characterizing the protein and ϕ is the concentration of modifier in the eluent. For practical reasons, this equation is mostly used in its logarithmic form:

$$ln\, k' = ln\, M - S\phi \tag{16.5}$$

Regnier carried out a very thorough experimental and theoretical investigation of this relation [22,33-35]. He derived a retention model applying the law of mass action to describe the system that included the competition between protein and modifier, interactions between the protein and the sorbent, as well as interactions between the sorbent and the modifier.

His study resulted in a general expression that relates the retention factor k' to the concentration of the desorbing agent (modifier):

$$k' = N \psi^{-Z} \tag{16.6}$$

where N is the constant, ψ is the molar concentration of the modifier, and Z is the stoichiometric parameter related to the number of molecules of the modifier required to achieve solvation of all of the adsorptive sites located on both protein and stationary phase. This parameter is an important factor since it affects the effectiveness of separation for proteins. The k' value may change in this system from zero at $\psi \to \infty$ to infinity at $\psi \to 0$.

$$ln\ k' = \ln N - Z \ln \psi \tag{16.7}$$

This logarithmic form of equation 16.6 is used successfully in practice for a long time and excellent correlations between the predicted values and the experimental results are often found for both reversed phase and ion exchange chromatography of proteins.

16.3 MAJOR ASPECTS OF PROTEIN SEPARATION USING SHORT MONOLITHIC BEDS

16.3.1 Operative thickness of the adsorption layer

A theoretical model for protein retention and consequently for the separation in HPMDC must also fit equations derived for equilibria discussed above. Since the separation of the proteins according to our concept occurs in a single-step adsorption-desorption act, or more exactly in a single step desorption, the initial distribution of the protein molecules at the adsorptive surface located within the channels of the short monolithic disk is most important. This distribution is controlled by the energy of interaction, which in turn is characterized by the parameter Z. Once desorbed, the protein molecules immediately leave their "landing grounds" and are eluted from the disk by the mobile phase. In this situation the Z parameter is no longer an average as is the case for multistep adsorption-desorption typical of a column.

Correlation of the parameter Z in short column chromatography such as HPMDC to the time and equally to the space (volume) is important. The time correlation controls the efficiency of "a single theoretical plate" [36], while the volume connection indicates the existence of a defined critical length of the separation layer [37, 38] or operative thickness of the adsorption layer (OTAL) X_0 [17]. This term takes into account the non-stationary character of the initial part of the separation process during the gradient elution. The value X_0 is defined as a distance at which the quasi-steady state would be achieved in a column. This value can be calculated from equation [38]:

$$X_0 = \lambda UC_c/(ZB) \tag{16.8}$$

where U is the linear velocity of the mobile phase, C_c is the concentration of the mobile phase modifier at which the solute molecule is desorbed, Z is the effective charge of the solute ion divided by the charge of ion in the mobile phase, B is the steepness of the linear gradient, and λ is the auxiliary parameter with an estimated default value of 0.5.

This equation makes clear that the operative thickness of the adsorption layer depends only on the gradient conditions. The quasi-steady state may be achieved in HPLC at a distance X_0 that is much shorter than the overall length of the conventional column. This applies in particular to separations using steep gradients. This also suggests that ultra-short columns may be a viable alternative for the separation of proteins that have traditionally been separated using standard columns, steep gradients, and short gradient times to afford the desired resolution. This conclusion has led to the idea of an efficient separation of proteins in very short beds no matter what separation medium is used. The dramatically improved mass transfer of large molecules between the mobile and the stationary phase typical of monolithic sorbents, made these materials particularly suitable for this mode of separation.

The major advantage of the parameter X_0 is that it enables comparison of quantitative parameters characterizing separations carried out in devices varying in geometry or at different flow conditions while keeping constant the separation distance [15,39]. In fact, this model corresponds well with the concept of the constant gradient volume that has resulted from our early study of *gradient chromatography* of proteins achieved using thin monolithic layers [5].

16.3.1.1 Description of separation in HPMDC

Equation 16.6 describes the adsorption equilibrium measured experimentally. For a convenience, parameters of that equation can be reduced to a single variable:

$$k' = \tau^{-Z} \tag{16.9}$$

which results from substitution

$$\tau = \psi / \psi_o \tag{16.10}$$

where $\psi_o = N^{1/Z}$ is the value ψ of at $k' = 1$.

According to Regnier's model [34], the separation of proteins by adsorption chromatography is typically achieved using a gradient of the mobile phase, for which k' is a function of ψ (often given as $\psi = \psi(t)$) and depends on the gradient shape. Under these conditions, the adsorption process is unlikely to reach the equilibrium.

The adsorption process can also be described statistically [11,17]. Obviously, the protein resides either in the mobile phase or on the stationary phase. The probability that at the equilibrium the desorbed protein resides in the *mobile phase* P_b is defined using the equilibrium retention factor k':

$$P_b = 1/(1+k') = 1/(1 + \tau^{-Z}) \tag{16.11}$$

This probability is normalized since the sum of both probabilities, i.e. the residence in the *mobile phase* P_b plus the residence in the *stationary phase* P_a ($P_a = k'/(1+k')$) must equal 1. The estimated probability of residence of the protein in the mobile phase in the equilibrium is then obtained by integration of all probabilities of transfer of the protein molecule from the stationary to the mobile phase under non-equilibrium conditions. This is defined as the derivative of P_b over ψ:

$$\rho = (k')'_\psi/(1 + k')^2 = Z\tau^{-Z-1}/(1 + \tau^{-Z})^2 \tag{16.12}$$

The parameter corresponds to the probability density and the value $\rho(\psi)\,d\psi$ to the probability of the transfer of a protein molecule from the stationary to the mobile phase in a certain interval of modifier concentration in the mobile phase. The value $\psi = \psi_o$ relates to the maximum of $\rho(\psi)$.

The concept of probability density enables the use of equation 16.12 to define the moments of the parameter. Thus, the first moment describes the average concentration of modifier at which the protein is desorbed while the second the dispersion of this desorption process.

16.3.2 Peak parameters in desorption chromatography

As a direct consequence of specifics of protein adsorption discussed above, the experimental values of the parameter Z lie within a narrow interval of 10–100. This fact can also be used for the calculations of the moments. By combining equations 16.9, 16.11 and 16.12, the following expression for the first moment is obtained:

$$<\tau> = \int_0^\infty \tau \cdot \rho(\tau)\,d\tau = \frac{\pi/Z}{\sin(\pi/Z)} \tag{16.13}$$

By analogy, the second moment is then:

$$<\tau^2> = \int_0^\infty \tau^2 \cdot \rho(\tau)\,d\tau = \frac{\pi/Z}{\sin(\pi/Z)} \tag{16.14}$$

If $Z \gg 1$, the first moment is

$$< \tau > 16 \ (1 + \ \pi^2 / 6 \ Z^2) \tag{16.15}$$

and the second

$$< \tau^2 > 16 \ (1 + \ 2 \ \pi^2 / 3 \ Z^2) \tag{16.16}$$

The dispersion s is then defined by:

$$\sigma^2 = (< \tau^2 > - < \tau >^2) \ 16 \ \pi^2 / 3 \ Z^2 \tag{16.17}$$

An expression for the parameter Z can be derived from this equation:

$$Z = \pi / \sqrt{3} \ \sigma \tag{16.18}$$

Since the dispersion can be calculated from equation 16.17 and the parameter Z from equation 16.18, it verifies that the elution peak of a given protein monitored after a single step desorption process is mathematically described by the probability density ρ as a function of ψ. A more accurate expression for Z would require including the peak shape for the same protein passed through the specific chromatographic layer in the absence of adsorptive interactions.

16.3.3 Resolution and efficiency in HPMDC

The proposed theoretical model for the separation of proteins in HPMDC mode indicates that the resolving power of this process does not depend on the number of available theoretical plates since only a single adsorption-desorption act is assumed for this separation, but on the respective initial and "momentary" distributions of the protein molecules between both phases corresponding to the distribution of their ener- gies of interactions with surface of the sorbent. Obviously, as the parameter Z in- creases, which means that the plot of log k' as a function of concentration of modifier in the mobile phase is steeper, the distribution of the probability density $\rho(\psi)$ of relocation of the protein to the mobile phase becomes narrower. In addition, the width of the peak depends on the distribution of the parameter Z. It broadens with increasing Z, exceeds significantly the second moment of the probability density , and becomes critical for the peak width actually seen in the chromatogram.

Use of equation 16.18 for the determination of Z from the dispersion requires segregating the contribution of Z alone to the overall distribution from all of the other contributions to the peak shape. This can be achieved by fractionating the eluting zone for a specific protein into small portions using a shallow stepwise gradient consisting of very small increments in the modifier concentration in each step. A series of "components" is obtained that posses basically identical *Z*-values related to a given modifier concentration range [40]. Repeated chromatographic separations of the individual fraction containing compounds with identical *Z*-values can then result in peaks with a width that corresponds to the dispersion of the probability density for the residence of a protein in the mobile phase.

The resolution in HPMDC decreases rapidly for substances with similar values of *Z*. Moreover, small differences between the average values of *Z* can be completely obscured by the dispersion of the Z values for the individual proteins. As a consequence, proteins characterized by similar *log k'* vs. *log ψ* plots, are the most difficult to separate by gradient HPMDC.

Since HPMDC is presented as a single-step desorption process in which, in contrast to the migration of substances through the longer adsorptive layer, the parameters of the separation are not averaged, the topography of the surface within the channels significantly affects the width of the eluted peaks. The presence of pores through which no flow occurs ("connecting drifts") introduces a diffusional motive into the otherwise very simple mechanism of HPMDC. The contribution of diffusion to the distribution of the adsorption energies makes it very broad and decreases the probability of success of the separation. In addition, these pores also exert serious constraints for the desorbed molecules leaving the solid surface. This results in broad asymmetric peaks and low resolution.

The use of very steep gradients of modifier in the mobile phase allows the HPMDC separation of compounds with very similar properties [5]. It should be noted that small peaks occurring in the chromatograms under these conditions might well reflect both the temporary energetic inhomogeneity of the adsorptive interactions resulting from the inhomogeneity of the adsorptive surface and the presence of additional compounds such as impurities in the sample. This dilemma can be resolved by monitoring repeated separation. Since the energetic inhomogeneity has a random character, peaks originating from this phenomenon will be observed in consecutive chromatograms at different elution volumes. In contrast, elution times of impurities are completely reproducible.

16.3.4 Hydrodynamics and mass transfer in flow-through pore channels

16.3.4.1 Residence time and time of free diffusional migration

HPMDC is characterized by the flow of all mobile phase through the pores. The walls of these pores also represent the adsorptive surface. Therefore, understanding the hydrodynamics of this system is important for the control and optimization of the chromatographic process [12,39,41-45]. The linear flow velocity $\overline{U}$ of the mobile phase through the separation unit having a flat geometry can be expressed as:

$$\overline{U} = F\,A^{-1}\varepsilon^{-1} \tag{16.19}$$

where F is the volumetric flow rate, A is the cross section of the disk or membrane, and ε is its porosity.

Obviously, this equation does not reflect the fact that the flow velocity in the individual pores may differ from the "average" velocity U, since it depends on the size of the pores. According to the Hagen-Poiseuille law [46], the average laminar flow velocity through a tube of length L_t is proportional to the square of its diameter r:

$$\overline{U} = \Delta P\,r^2\,/\,8\,\eta\,L_t \tag{16.20}$$

where ΔP is the pressure drop along the liquid path and η is the dynamic viscosity of the liquid.

Although equation 16.20 applies exactly only to the flow through a straight cylindrical tube, it has been demonstrated several times that the flow through a macroporous monolith can also be described using this law regardless the high tortuosity of the flow-through channels [2,47,48].

Rearranging the equation yields:

$$\Delta P\,/\,\overline{U} = 8\,\eta\,L_t\,/\,r^2 \tag{16.21}$$

showing that the pressure drop per unit of the linear flow velocity increases exponentially with the decreasing tube diameter. This translates for the monolithic membranes to a simple claim: the larger the pores, the lower the flow resistance, and the more liquid can flow through these pores at a given pressure. Taking into account a short length of separation layer, the ideal matrix for HPMDC may appear to be such that has large pores with a narrow distribution.

A simple description of the effect of both pore size and its distribution on chromatographic separations can be derived from the ratio of the residence time in the mobile phase within the monolith, t_{res}, and the time t_{film} required to achieve the transfer through the film of stagnant liquid adhering to the pore wall representing the adsorptive surface [41,45]. For a good separation, the value of t_{res} must be larger than that for t_{film}.

The residence time is a simple function of the thickness of the separation layer L and the flow velocity $\overline{U}$:

$$t_{res} \approx L\,/\,\overline{U} \tag{16.22}$$

In contrast, t_{film}, increases exponentially with the pore diameter d_p according to the classic definitions of dynamics in chromatography [49]:

$$t_{film} \approx d_p^2/4D}$$ (16.23)

where D is the diffusion coefficient of the specific protein. Thus, the residence time can be controlled via the flow rate and the thickness of the monolithic support while the transfer time is controlled by its pore size. Obviously, very thin layers in combination with very large pores are not suitable for good separation of proteins because the analyte may pass the layer without coming in contact with the surface. We have studied the effect of pore size of flat monolithic disks on efficiency of the protein separation and found that the best separation can be achieved using monolithic disks with the optimized pore size that is sufficiently small to allow the protein molecule to reach the wall within the given residence time, while also being sufficiently large to facilitate fast flow at the lowest possible back pressure. For example, Fig. 16.1 dem-

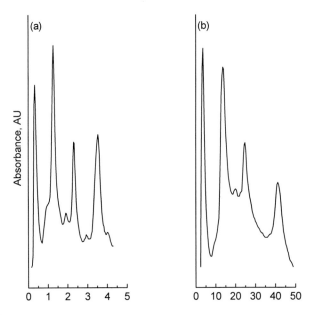

Fig. 16.1. Separation of standard protein mixture by ion exchange HPMDC on discs with a median pore size of 700 (a) and 166 nm (b) (Reprinted with permission from ref. [39], copyright 1998 Elsevier). *Conditions*: disk 25 × 2 mm, proteins: myoglobin, conalbumin, chicken egg albumin, soybean trypsin inhibitor (order of elution), protein concentration 4.5 mg/mL in buffer A, buffer A 0.05 mol/L Tris.HCl, pH 7.0, buffer B 1 mol/L NaCl in buffer A, (a) flow rate 5 mL/min, gradient time 24 min, (b) flow rate 0.5 mL/min, gradient time 240 min.

onstrates the effect of pore size. While very good separation of all four proteins could be achieved using optimized disk with a median pore size of 700 nm, only three peaks were monitored when this separation was attempted using disk with a smaller pore size centered at 166 nm and broad size distribution [39]. Narrowing the pore size distribution also contributed to improvements in the separation. An important characteristic of HPMDC as well as membrane chromatography is an insignificant effect of the flow rate on column efficiency [5,41,42].

16.3.4.2 Enhanced diffusivity

The convective velocity in HPMDC may be assumed equal to the linear flow velocity because all the mobile phase flows through the pores of the slab-like monolithic stationary phase. The permeability B_m of the disk is [18]:

$$B_m = \varepsilon_m^3 \, d_g^2 / 150 \, (1 - \varepsilon_m)^2 = a \, \varepsilon_m^2 \, d_{pore}^2 / 150 \, (1 - \varepsilon_m)^2 \qquad (16.24)$$

where ε_m is the porosity, d_g is the diameter of globule or microsphere clusters forming the macroporous structure that includes the transport channels, d_{pore} is the pore diameter, and a is defined as:

$$a = d_g^2 / d_{pore}^2 \qquad (16.25)$$

Darcy's law relates the convective velocity v_0 to the permeability B_m, fluid viscosity μ, and the pressure drop ΔP across the flat layer of the thickness L:

$$v_0 = - \, (B_m \, \Delta P / 2 \mu L) \qquad (16.26)$$

The values of d_{pore}, d_g, and ε_m, which can be obtained for each monolithic material using standard methods such as scanning electron microscopy and mercury intrusion porosimetry, then allow calculation of B_m [5]. Indeed, using this approach, the calculated convective intramembrane ("intradisk") velocity was found very close to the value of the linear flow velocity thus confirming validity of the assumptions made above.

The convective flow of the mobile phase through the flat bed also increases the diffusivity inside the pores as compared to the free diffusion. The *apparent* or *augmented* effective diffusivity $\overline{D}_e$ is a function of both the effective diffusivity D_e and the Peclet number λ ($\lambda = v_0 \, l \, / \, D_e$) [16]. The high Peclet number typical of disks permits the use of the simplified equation for the calculation of the apparent diffusivity:

$$\overline{D}_e = v_0 l / 3 \tag{16.27}$$

This calculation affords $\overline{D}_e = 6.4 \times 10^{-3}$ cm^2 s^{-1}. In contrast to the free diffusion of proteins characterized by $D_e \approx 10^{-7}$ cm^2 s^{-1}, the diffusivity forced by convection is more than four orders of magnitude higher and therefore the mass transfer is faster. Consequently, protein separation can be faster in HPMC than in conventional column chromatography for which diffusion is the primary mass transfer mechanism. The separation in HPMDC mode occurs mostly in the large channels where the convection plays the major role. Similar observations were also made in the field of perfusion chromatography [19,20,50,51].

16.4 SEPARATION OF DNA

DNA molecules are also biopolymers or, more specifically, biopolyelectrolytes. Similar to proteins, DNAs diffuse slowly and exhibit a significant steric effect. In contrast to proteins for which both positive and negative charges contribute to the overall net charge and which surface most often features patchy distribution of charges and local charge clusters, DNA carries only negative charges evenly distributed across the macromolecule. This makes their chromatographic behavior easier to understand. In contrast to most synthetic polyelectrolytes that share with DNA the homogeneous charge structure, all DNA molecules of a given type are monodisperse and have identical molecular mass. We have already demonstrated earlier in this chapter that the separation of proteins using HPMDC is possible in a long gradient elution mode. Surprisingly, this mode was found not suitable to achieve separation of plasmid DNA since these molecules strongly interacted with the stationary phase and were desorbed only in a mobile phase containing a high concentration of an electrolyte [14].

The current model of HPMDC derived for proteins interprets the separation of large molecules with slow diffusional mobility as a single step desorption process. The resulting peak should closely correspond to the theoretical curve calculated from the probability of the transfer of the molecule desorbed from the stationary in the mobile phase at a certain concentration of the modifier [17]. Sharp peak of DNA separated by gradient chromatography shown in Figure 16.2 may lead to an assumption of a narrow distribution of the adsorption energies (parameter Z) among the molecules of the plasmid preparation. This could be expected given the simple chemistry of DNA.

However, typical supercoiled plasmid preparations contain two "impurities", presumably nicked and open circular DNA that can be detected by agarose gel electrophoresis. It is therefore likely that the adsorption energy of these conformations differ sufficiently from that for the supercoiled plasmid to enable their separation. To assure

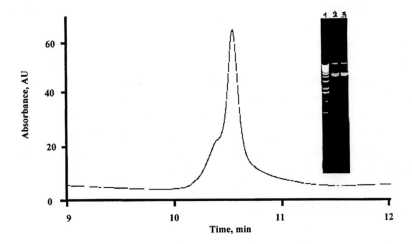

Fig. 16.2. Gradient elution of plasmid DNA in HPMDC mode (Reprinted with permission from ref. [14], copyright 1998 American Chemical Society). *Conditions*: stationary phase, strong anion exchange disk; flow rate, 1 mL/min; buffer A, 20 mmol/L Tris HCl, pH 7.4, buffer B 1 mol/L NaCl in buffer A; gradient profile, buffer A for 5 min, A to B in 5 min, B for 3 min, B to A in 5 min; sample, 10 µL containing 2 µg of plasmid DNA (pCMVβ). Inset: agarose gel electrophoresis. Line 1 DNA ladder; lines 2 and 3 two samples of plasmid pCMVβ.

that none of the possible components remained adsorbed at the disk, the effect of sample concentration on the peak area was determined and an excellent linear correlation found. This linearity can be readily explained by completeness of the desorption process (100 % recovery) and absence of permanent interaction between the nucleic acids and the polymer surface.

Due to the presence of only negative charges, DNA molecules are adsorbed at the surface of an anion-exchange stationary phases more strongly than proteins. In gradient HPMDC of proteins, an adsorption buffer with a low ionic strength is frequently required to achieve complete retention during the loading phase. In contrast, the plasmid DNA interaction with the positively charged stationary phase is much stronger and made efforts to decrease the binding strength feasible. Optimization of the gradient conditions then enabled the separation of all three expected major fractions. However, using larger gradient volumes, additional peaks were emerging in a reproducible manner. Since the supercoiled conformation can exists in several varieties of the dense molecular structure, it is safe to assume that the additional peaks represent these forms. In contrast, the chromatograms became less "smooth" with a number of minor peaks or spikes in separations using shallower gradients. However, the occurrence of these peaks was random and irreproducible. Similar phenomenon was observed previously in HPMDC separations of proteins using very shallow gradients [5] and explained by the inhomogeneity of the adsorptive surface and small temporary fluctuations in energy distribution within the macromolecular analytes at the moment of adsorption [17].

One of the distinct advantages of HPMDC observed already in chromatography of proteins is the independence of the separation efficiency on flow rate. Indeed, this was confirmed again for the separation of DNA since changes in the flow rate ranging from 1 to 4 mL/min did not exert any adverse effect on the separation. With the diffusion effects rather limited, the most prominent outcome of an increase in the flow rate is an increase in the active surface within the matrix. The higher pressure required to achieve the desired flow rate also forces the liquid through smaller pores thus "opening" them for the convective flow. This has recently been demonstrated both theoretically and experimentally [48].

The appearance of three maxima in the chromatogram of the plasmid DNA is consistent with the number of bands in the agarose gel. Provided the three peaks correspond to supercoiled, nicked, and open circular conformations of the same plasmid, its linearization should result in species exhibiting similar mobilities and interaction energies. Indeed, the peak of the linearized plasmid has much less features that that for the original DNA. Interestingly, the elution of this linearized plasmid occurs after a shorter retention time indicating a decrease in the overall interaction energy as a consequence of the linearization. This can be explained by the lower charge concentration within the more flexible linearized molecule resulting in a weaker interactive force. It is worth noting than in contrast to conventional column chromatography based on molecular diffusion, the shear effects cannot be ignore in HPMDC since the adsorption occurs on the walls of pores through which the mobile phase flows. The size of large molecule such as DNA may exceed the thickness of the stagnant layer adhering to the pore surface and reach out into the flow region. Differences in molecular flexibility may then easily affect both the kinetics and the thermodynamics of the interaction.

16.5 HPMDC OF SMALL MOLECULES

A number of differences in chromatographic behavior of proteins and small molecules can be easily discerned. These differences are common for all liquid chromatographic methods including HPMDC and undoubtedly affect the separations [11]. For example, the coefficients of free diffusion for proteins are 3-4 orders of magnitude smaller that those for low molecular mass compounds ensuing significant differences in the diffusional mobility of the separated molecules. In addition, the heterogeneity of surface interaction energies is much lower for the small molecules since their "surface" is smaller, better defined, and does not fluctuate.

We have demonstrated earlier in this chapter that HPMDC of proteins and nucleic acids differs noticeably from their separations using conventional HPLC. We indicated that the separation process in the former is not affected to a significant extent by diffusion through the stagnant pool of the mobile phase located in pores typical of HPLC columns packed with porous particles. Therefore, we only had to consider the diffusive transfer of molecules from the mobile phase flowing through the large pores (flow-through channels) to their walls. This absence of diffusion in and out of the pores enables to achieve the separations of large molecules within very short periods of time. These features might certainly be beneficial also in HPMDC of small molecules.

Combination of equations 16.22 and 16.23 enables to express the number of contacts N_C that a single molecule may experience with the pore surface during a specific period of time:

$$N_c = t_{res}/t_{film} = 4\,DL\,/\,Ud^2 \tag{16.28}$$

For example, assuming $D = 10^{-3}$ cm^2 s^{-1} for a hypothetical low molecular mass compound, disk thickness $L = 3$ mm, average pore size $d = 0.5$ μm, and a flow rate of 8 ml/min, the value of N_c is about 10^5 [15]. As a result of the higher diffusional mobility of small molecules, N_c significantly exceeds that found for globular proteins. Consequently, this high number of interactions indicates that isocratic separation of small molecules in adsorption mode may be achieved using even thin layers of the monolithic of stationary phase.

However, the binding characteristics of various types of small molecules are rather similar and shallow gradients would have to be applied to obtain decent separation. This is particularly important for isocratic separations in which an excessive peak broadening of later eluting components is observed. Upon using the shallow gradient, the peak resolution appears independent of the flow rate and peaks with Gaussian shapes are monitored. Very detailed practical investigation of HPMDC separations of different classes of small molecules mass substances have been published elsewhere [15,52].

16.6 CONCLUSION

This chapter outlines some basic differences between the chromatography carried out in columns and membrane-like devices. Theoretical analysis of the HPMDC processes shown above enables understanding these substantial differences and their impact on the separation of proteins, nucleic acids, and small molecules. This permits the development of separation methods that afford very good and rather fast separations.

References pp. 370-371

We have already noted in the first chapter of this book that high performance chromatography using monolithic disks has often been met with certain level of skepticism. The doubts of potential users were mostly expressed in questions such as "How is it possible to achieve the separation of complex protein mixtures with such a thin layers?" We believe that this chapter helped to scatter the fear from the unknown technique and may help to promote this powerful technology.

16.7 ACKNOWLEDGMENTS

Support of this work by an international grant of Russian Academy of Sciences – German Scientific Society (RAS-GSS, RUS436/555/0) and a grant of the National Institute of General Medical Sciences, National Institutes of Health (GM–48364) are gratefully acknowledged.

16.8 REFERENCES

1   S. Hjertén, J.-L. Liao, R. Zang, J. Chromatogr., 473 (1989) 273.
2   T.B. Tennikova, B.G. Belenkii, F. Svec, J. Liq. Chromatogr. 13 (1990) 63.
3   J.-L- Liao, R. Zang, S. Hjertén, J. Chromatogr. 586 (1991) 21.
4   F. Svec, J.M.J. Fréchet, Anal. Chem. 54 (1992) 820.
5   T.B. Tennikova, F. Svec, J. Chromatogr. 646 (1993) 279.
6   F. Svec, J.M.J. Fréchet, J. Chromatogr. A 702 (1995) 89.
7   J.J. Meyers, A.I. Liapis, J. Chromatogr. A 852 (1999) 3.
8   S. Ghose, S.M. Cramer, J. Chromatogr. A 928 (2001) 13.
9   C. Tanford, The Hydrophobic Effect, 2nd Ed., Wiley, New York, 1980.
10   D.K. Roper, E.N. Lightfoot, J. Chromatogr. A 702 (1995) 3.
11   T.B. Tennikova, R. Freitag, in: H.Y. Aboul-Ehnen (Ed.), Analytical and Preparative Separation Methods of Biomolecules, Dekker, New York, 1999, p.255.
12   E. Klein, J. Membr. Sci. 179 (2000) 1.
13   D.M. Zhou, H.F. Zou, J.Y. Ni, L.Yang, L.Y. Jia, Q. Zhang, Y.K. Zhang, Anal. Chem. 71 (1999) 115.
14   R. Giovannini, R. Freitag, T. Tennikova, Anal. Chem. 70 (1998) 3348.
15   A. Podgornik, M. Barut, J. Jancar, A. Strancar, T. Tennikova, Anal. Chem. 71 (1999) 2986.
16   A. Strancar, M. Barut, A. Podgornik, P. Koselj, Dj. Josic, A. Buchacher, LC-GC Int. 10 (1998) 660.
17   T.B. Tennikova, R. Freitag, J. High Resol. Chromatogr. 23 (2000) 27.
18   A.E. Rodrigues, J.C. Lopez, Z.P. Lu, J.M. Lpureiro, M.M. Diaz, J. Chromatogr. 590 (1992) 93.
19   A.I. Liapis, Y. Xu, O.K. Grosser, A. Tongta, J. Chromatogr. A 702 (1995) 45.
20   G.A. Heeter, A.I. Liapis, J. Chromatogr. A 761 (1997) 35.
21   L.R. Snyder in: K.M. Goodings, F.E. Regnier (Eds.), HPLC of Biological Molecules, Methods and Applications, Dekker, New York, 1990, p. 231.

22   F.E. Regnier, R.M. Chiez in: K.M. Goodings, F.E. Regnier (Eds.), HPLC of Biological Molecules, Methods and Applications, Dekker, New York, 1990, p. 89.

23   L.R. Snyder, M.A. Stadalius, M.A. Quarry, Anal. Chem. 55 (1983) 1412A.

24   M.A. Stadalius, M.A. Quarry, L.R. Snyder, J. Chromatogr. 327 (1985) 93.

25   L.R. Snyder, M.A. Stadalius in: Cs. Horváth (Ed.), High-Performance Liquid Chromatography, Advances and Perspectives, Academic Press, New York, 1988, vol. 4, p. 195.

26   J.D. Pearson, N.T. Lin, F.E. Regnier, Anal. Biochem. 124 (1982) 217.

27   M.T.W. Hearn, F.E. Regnier, C.T. Wehr, Proceedings of the First International Symposium on HPLC of Proteins and Peptides, Academic Press, New York, 1983.

28   R.R.M. Moore, R.R. Walters, J. Chromatogr. 317 (1984) 119.

29   R.S. Blanquet, K.H. Bui, D.W. Armstrong, J. Liq. Chromatogr. 9 (1986) 1933.

30   W.C. Bulton, K.D. Nugent, T.K. Slattery, B.A. Summers, L.R. Snyder, J. Chromatogr. 443 (1988) 363.

31   V.G. Maltzev, D.G. Nasledov, S.A. Trushin, T.B. Tennikova, L.V. Vinogradova, I.N. Volokitina, V.N. Zgonnik, B.G. Belenkii, J. High Resolut. Chromatogr. 13 (1990) 185.

32   S. Yamamoto, K. Nakanishi, R. Matsuno, Ion-Exchange Chromatography of Pro teins, Dekker, New York, 1988, Vol. 4, p. 78.

33   W. Kopaciewicz, M.A.Rounds, F.E. Regnier, J. Chromatogr. 266 (1983) 3.

34   X. Geng, F.E. Regnier, J. Chromatogr. 296 (1984) 15.

35   X. Geng, F.E. Regnier, J. Chromatogr. 332 (1985) 147.

36   J.L. Coffman, D.K. Roper, E.N. Lightfoot, Bioseparations, 4 (1994) 183.

37   B.G. Belenkii, A.M. Podkladenko, O.I. Kurenbin, V.G. Maltsev, D.G. Nasledov, S.A. Trushin, J. Chromatogr. 646 (1993) 1.

38   N.I. Dubinina, O.I. Kurenbin, T.B. Tennikova, J. Chromatogr. A 753 (1996) 217.

39   M.B. Tennikov, N.D. Gazdina, T.B. Tennikova, F. Svec, J. Chromatogr. A 798 (1998) 55.

40   B.G. Belenkii, V.G. Maltsev, BioTechniques 18 (1995) 288.

41   M. Nachman, A.R.M. Azad, P. Bailon, J. Chromatogr. 597 (1992) 155.

42   J.A. Gerstner, R. Hamilton, S.M. Cramer, J. Chromatogr. A 596 (1992) 173.

43   W.F. Weibrenner, M.R. Etzel, J. Chromatogr. 662 (1994) 414.

44   I.A. Adisaputro, Y.-J. Wu, M.R. Etzel, J. Liq. Chromatogr., 19 (1996) 1437.

45   F.T. Sarfert, M.R. Etzel, J. Chromatogr. A 764 (1997) 3.

46   R.B. Bird, W.E. Steward, E.N. Lightfoot, Transport Phenomena; Wiley, New York, 1960.

47   C. Viklund, F. Svec, J.M.J. Fréchet, K. Irgum Chem. Mater. 8 (1996) 744.

48   N.D. Ostryanina, O.V. Il'ina, T.B. Tennikova, J. Chromatogr. B (2002) in press.

49   J.C. Giddings, Unified Separation Science, Wiley, New York, 1991.

50   N.B. Afean, S.P. Fulton, F.E. Regnier, J. Chromatogr. 544 (1991) 267.

51   S.P. Fulton, N.B. Afean, N.F. Gordon, F.E. Regnier, J. Chromatogr. 547 (1992) 452.

52   A. Podgornik, M. Barut, J. Jamcar, A. Strancar, J. Chromatogr. A 848 (1999) 51.

F. Švec, T.B. Tennikova and Z. Deyl (Editors)
Monolithic Materials
Journal of Chromatography Library, Vol. 67
© 2003 Elsevier Science B.V. All rights reserved.

Chapter 17

Monolithic Stationary Phases for the Separation of Small Molecules

Eric C. PETERS and Christer ERICSON

Genomics Institute of the Novartis Research Foundation, 10675 John Jay Hopkins Drive, San Diego, CA 92121 USA

CONTENTS

17.1 IMPETUS BEHIND THE DEVELOPMENT OF MONOLITHIC STATIONARY PHASES

Although the detrimental effects of both slow diffusional mass transfer and flow variations on the chromatographic process were already recognized by Van Deemter *et al.* in 1956 [1], it wasn't until the introduction of HPLC in the early 1970s that significant progress was made in the reduction of these band broadening phenomena. Specifically, the introduction of columns packed with beads 10 µm or smaller in diameter led to an increase in overall column efficiencies due to both smaller eddy diffusion effects as well as shorter diffusional path lengths [2,3]. Over the past 30 years, further improvements in column performance have arisen primarily from the development of spherical stationary phases of ever decreasing particle size and heterogeneity. However, since the pressure required to drive a mobile phase through a packed bed is inversely proportional to the square of the particle diameter, further increases in column performance using this strategy are restricted by the pressure limits of current solvent delivery systems [4]. Other methods for improving column efficiency include open tube chromatography [5], capillary electrochromatography (CEC) [6], and ultrahigh pressure chromatography [7]. However, these methods currently require specialized equipment and/or suffer from operational difficulties, making their wide scale acceptance and implementation problematic.

An alternative approach for addressing this operational conflict of increasing column pressures associated with decreasing particle size packings is the utilization of a monolithic continuous porous solid transversed by relatively large pores as the stationary phase [8,9], or alternatively, continuous beds [10]. The interconnected network of large pores allows all the mobile phase to pass through such stationary phases using significantly lower pressures compared to their packed bed counterparts. The resulting convective flow greatly enhances the rate of mass transfer of analytes in the fluid stream, thus increasing the separation efficiency of the column [11,12]. Although open-pore polyurethane foams detailed in Chapter 1 of this book were investigated as potential stationary phases as early as the 1970s [13,14], it was not until a few decades later that effective monolithic chromatography columns were developed.

The first successful monolithic stationary phases were crosslinked synthetic organic polymers prepared by an *in situ* polymerization process directly within the confines of an empty chromatography column. Since their introduction, numerous different materials have been prepared and tested, all of which fall into two broad categories: fully hydrated polymer beds [15] and rigid macroporous polymers [8]. The enhanced mass transfer properties of these species are clearly evident in their improved separations of both biological [15,16] as well as synthetic macromolecules [17]. These applications are discussed in detail in other chapters. However, these improved efficiencies did not fully extend to the separation of small molecules by HPLC, probably due to the presence of micropores (pores less than 2 nm in diameter) in the polymer matrix that restrict internal diffusion.

The push towards miniaturized capillary columns and separation methodologies such as CEC has provided a second strong impetus for the development of monolithic media. The benefits of miniaturized chromatography are well documented, and include higher sensitivities with both reduced sample and solvent consumption. However, the homogeneous packing of narrow diameter capillaries with small particles and the production of end frits to retain these particles have proven difficult [18,19]. The use of monolithic stationary phases eliminates many of these technical problems. Instead of packing preformed rigid particles, a solution of monomers and additives can easily be introduced into a fused silica capillary [20-22] or the channels of a microchip [23,24]. The monolithic structure is then formed *in situ* using a thermal [25], a chemical [10], or light initiated [26] polymerization process, and can be applied as the stationary phase itself, or alternatively serve as a retaining frit for a more traditional packed column [27]. To date, this technology has mainly been used for CEC separations [28]. These applications are discussed in detail in several other chapters. Although high efficiency separations of small molecules can be achieved in CEC, it would also be desirable to achieve such separations using the physically more robust technique of HPLC. This chapter will focus on the tremendous progress made during the past few years towards this challenge for monolithic stationary phases – the highly efficient HPLC separation of small molecule species.

17.2 MONOLITHIC MACROPOROUS SILICA

17.2.1 Reasons for success

To date, inorganic silica monoliths have demonstrated the most potential as alternative highly efficient stationary phases for the reversed-phase HPLC separation of small molecules. Based primarily on the pioneering work of Minakuchi *et al.* [29], these materials have recently been commercialized under the trade name SilicaRod™ by Merck, KGaA (Germany) [30-32]. The excellent chromatographic performance of these materials is directly related to the high level of control that can be exerted over their porous properties. For example, both the average size and total volume of the macropores (pores larger than 50 nm in diameter) can be varied independently of one another. Whereas the overall pore volume depends primarily on the total volume fraction of non-polymerizable fluids in the initial reaction mixture, the average size of the macropores is a function of the ratio of silica to the additive - poly(ethylene oxide) (PEO), which affects the phase separation conditions of the growing silica chains.

Although active control of the macroporous structure of monolithic materials enables the tailoring of their flow-through properties, it is the fine control that can be exerted over the mesoporous properties of the silica monoliths that is ultimately responsible for their excellent chromatographic performance. Specifically, all micropores can be eliminated, and the average diameter of the mesopores (pores between two and 50 nm in diameter) could be tailored through a partial dissolution and reprecipitation process that occurs during the aging of the initially obtained wet macroporous gel under weakly basic conditions [33].

17.2.2 Applications

As a result of the exquisite control that can be exerted over the porosity of silica-based monolithic columns, these materials enable similarly efficient HPLC separations to be performed at far greater speeds than their packed bed counterparts [30-32]. For example, Fig. 17.1 shows the separations of a test mixture consisting of toluene (as the void volume marker), dimethylphthalate, and dibutylphthalate performed at two different linear flow velocities of 500 and 4800 cm/hr [34]. The first flow rate is typical of packed bed columns with identical dimensions, while the second is nearly ten times higher. Despite the considerable increase in the mobile phase flow rate, the height equivalent to the theoretical plate (HETP) for this separation increased from 29 to only 50 μm. Based on this performance, six of these columns were used to affect a simulated moving bed (SMB) purification of a diastereomeric mixture of a commercial drug substance intermediate. After optimization of the experimental conditions, the use of monolithic stationary phases was found to increase the productivity of a given SMB system by at least a factor of two [34]. Large diameter columns (10 mm or greater) have also been described for use in the preparative scale separation of phthalic acid esters [35].

Similarly, silica monolithic columns have been used to greatly increase the throughput of LC/MS analysis systems [36,37]. For example, Fig. 17.2 shows the LC/MS single ion monitoring chromatograms of six isomeric hydroxylated metabolites (m/z 192) of debrisoquine human liver microsome filtrates following analysis on

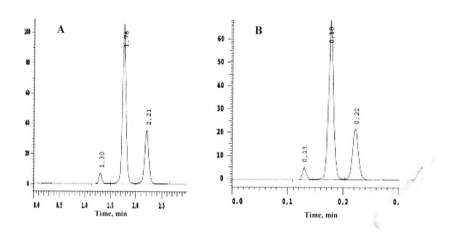

Fig. 17.1. Chromatograms of mixtures of toluene, dimethyl phthalate, and dibutyl phthalate at a flow velocity of 500 (a) and 4800 cm/h (b) using a 100 × 25 mm i.d. monolithic silica column with pore sizes of 2.8 μm (macropores) and 13 nm (mesopores). (Adapted with permission from [34], copyright 2001 Elsevier Science B.V.).

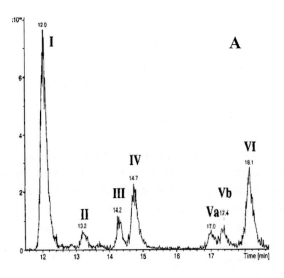

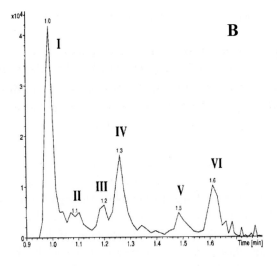

Fig. 17.2. Single ion (m/z 192) monitoring LC/MS of human microsome liver filtrates following analysis on a 2.1 mm × 15 cm Zorbax SB-C-18 5 μm particulate column operated at 0.8 mL/min (A) and a 4.6 mm × 5 cm Chromolith SpeedRod column operated at 4 mL/min (B). The six peaks are hydroxylated metabolite isomers of debrisoquine. (Adapted with permission from [38], copyright 2001 Wiley-VCH).

both a conventional packed as well as a monolithic column. The resolution and selectivity achieved using a 5 cm long monolithic silica column were comparable to those obtained using a 15 cm long analytical column, while the analysis time per sample was reduced from 30 to only 5 min [38].

These columns have also been used for the direct analysis of pharmaceutical compounds in human serum. Approximately 300 20-µL injections of plasma (diluted with an equal volume of water) were performed continuously over a period of two days onto a single column. Despite a total loading of nearly 6 mL of plasma directly onto the column, the retention time of the pharmaceutical compound did not change over all the course of these injections, demonstrating the robustness of the system. Similarly, the analyte response was found to be linear over the range of 5–1000 ng/mL, with a typical correlation efficient being 0.997. Further reduction of the column inner diameter from the originally employed 4.6 mm is assumed to further increase the sensitivity to the sub-ng/mL level. Such a level of sensitivity would be sufficient for the support of most drug development programs [39].

17.3 MONOLITHIC CAPILLARY COLUMNS FOR µHPLC

17.3.1 Silica-based monoliths

The preparation of a continuous silica monolith directly within the confines of a fused-silica capillary could potentially be advantageous in that the resulting long monolith of small diameter could be prepared without the technically difficult cladding step used in the production of larger diameter columns [40-42]. However, the shrinkage associated with the fabrication of monolithic columns in standard formats must practically be addressed by tightly attaching the silica structure to the capillary wall in order to prevent the formation of voids and channels [43]. Although functional reversed-phase and chiral [44] µHPLC and CEC columns can be produced in this manner, their porous structures are significantly different than those of the HPLC-scale columns described previously. Specifically, these capillary silica monoliths possess much greater total porosities and exhibit aggregated structures of globular silica nuclei rather than the network-forming smooth cylindrical structures of the larger diameter silica monoliths [43]. As such, these materials appear structurally more related to synthetic organic polymer monoliths prepared directly within fused-silica capillaries [8,28] rather than their larger scale silica counterparts, and not surprisingly, the differences in column efficiencies of these two types of capillary monoliths for µHPLC and CEC separations are also less pronounced [43,45,46]. For example, both materials exhibited nearly identical column efficiencies for the CEC separation of a homologous series of alkyl benzenes [43,47], and the HPLC efficiencies of silica capillary columns more closely resemble those values reported for organic capillary columns [45] than their larger diameter silica-based counterparts [43].

Alternative approaches towards silica-based monolithic columns for μHPLC have also been investigated, including the sintering of silica particles [48], and the entrapment of conventional particulate silica media in either an organic [49] or inorganic [50,51] matrix. These approaches also produced stationary phases that enabled μHPLC separations, with some of the composite materials being further used in μHPLC/MS studies [52]. However, the effective properties of these columns are often simply composites of the different materials employed, and no dramatic increases in chromatographic efficiencies were observed compared to their *in-situ* sol-gel prepared counterparts.

17.3.2 Synthetic organic polymer-based monoliths

Although they possess no clear advantage in terms of overall separation efficiency, small diameter synthetic polymer monoliths enjoy a considerable advantage in terms of the simplicity of their preparation. For example, monolithic silica columns prepared in 100 cm lengths of fused silica capillary exhibited large voids at both ends, requiring the removal of 10 to 15 cm from each end before usage [42]. Although this phenomenon is not a problem for chromatography columns prepared in capillaries, which can simply be trimmed to size, it would be disastrous for columns prepared within the channels of a microchip. By contrast, organic polymer monoliths can be prepared at precise locations using UV-initiated polymerization processes [53], and have subsequently been used as capillary end frits [27,54], mixers [55], SPE devices [56], and stationary phases in microfluidic devices [23,24]. In addition, a wide range of functionalities can directly be incorporated into a polymer monolith, enabling the one-step preparation of stationary phases for reversed, chiral, and other modes of chromatography [8,57-60]. For example, Fig. 17.3 shows the μHPLC separations of a series of polar aromatic compounds using an optimized acrylamide-based polymer monolith with two different mobile phases [58]. Column efficiencies of up to 150,000 plates/m were reported, and the direct incorporation of monomers of various polarities enabled the facile determination that several different mechanisms of retention were involved in the observed separations. By contrast, silica monoliths often require a post-functionalization step after their formation [42,44]. These significant advantages guarantee that organic polymer monoliths will play an important role in future μHPLC separation methods.

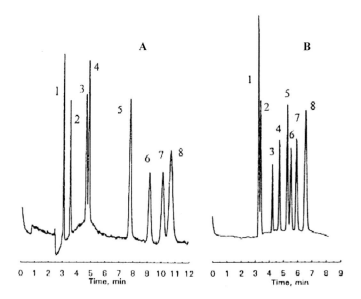

Fig. 17.3. Normal-phase capillary chromatography of polar aromatic compounds. Column: 100 μm i.d. × 125 mm length; 32% total monomer (63.2% *N*-isopropylacrylamide, 10.5% methacrylamide, 26.3% piperazine diacrylamide). Flow rate: 220 nL/min. Mobile phase: hexane–ethanol–methanol (A) or methanol (B). Peaks: pyridine (1), 4-pyridylmethanol (2), 4-methoxyphenol (3), 2-naphthol (4), catechol (5), hydroquinone (6), resorcinol (7), and 2,7-dihydroxynaphthalene (8). (Adapted with permission from [58], copyright 1999 Elsevier Science B.V.)

17.4 SEPARATIONS USING MONOLITHIC SYNTHETIC POLYMERS

In comparison to their analytical-scale (column diameters of 4.6 mm and greater) silica-based counterparts, organic polymer monoliths have generally shown lower efficiencies in the reversed-phase HPLC separation of small molecules [16,61-63]. This reduced performance is primarily due to the presence of micropores in the polymer matrix, which cause slow internal diffusion. In addition to the optimization of current fabrication technologies, this problem may also be addressed by the variety of novel approaches currently being investigated for producing these materials [64-68]. However, certain aspects of the organic polymer monolith fabrication process also make these materials particularly well suited as stationary phases for other modes of chromatography.

17.4.1 Chiral separations

Monoliths possessing a wide variety of surface functionalities can readily be obtained in a single step by the direct copolymerization of the appropriate monomers [69]. This approach has been employed by Sinner and Buchmeiser to prepare a new class of functionalized monolithic stationary phases using a ring-opening metathesis polymerization process that tolerates a wide range of functional monomers. As shown in Fig. 17.4, a monolith produced by incorporating β-cyclodextrin-containing monomer was used to achieve the enantioselective separation of proglumide in less than three minutes [70].

The ability to incorporate a wide range of chemical functionalities during fabrication of the monolithic stationary phase is also critical to the technique of molecular imprinting, which is used for the preparation of tailored polymeric sorbents with

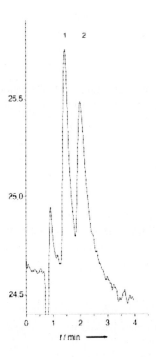

Fig. 17.4. Enantioselective separation of proglumide. Column: 3 mm i.d. × 150 mm length; 40% total monomer (25% norborene, 25% β-cyclodextrin-functionalized norborene). Flow velocity: 360 mm/min. Mobile phase: acetonitrile methanol–trifluoroacetic acid–triethylamine 99.4925:0.25:0.25:0.0075 vol.%. (Reprinted with permission from [70], copyright 2001 Wiley-VCH)

increased affinities for a particular small molecule [72-74]. Although the majority of studies investigating imprinted organic monoliths to date have involved CEC [75], the HPLC separations of diaminonapthalenes [76], nicotine [77], and cinchona alkaloids [78] have also been described. In an alternative application, Steinke *et al.* also reported using the reactive molecule 4,4'-bis(dimethylamino)benzophenone as a template molecule for the preparation of an anisotropic polymer monolith as a potential optical sensor [79].

17.5 GAS CHROMATOGRAPHY

In addition to their more traditional use in liquid-based separations, organic monolithic materials have also been employed as stationary phases for gas chromatography [13,14,80-82]. Since these applications employ a carrier gas for transportation of the analytes, the stationary phases employed must possess permanent macroporous structures that persist even in the dry state. Fig. 17.5 shows a scanning electron micrograph of a monolithic 320 μm i.d. poly(divinylbenzene) column [80]. The removal of the silica capillary surrounding the stationary phase clearly demonstrates the monolithic nature of the polymer matrix, while its general appearance as captured under the operational high vacuum of the SEM demonstrates the permanency of its porous structure. Fig. 17.6 shows the gas chromatographic separation of a mixture of organic solvents utilizing a temperature gradient [80].

In addition to their simplicity of preparation and their easily tailorable functionality, these materials possess the added advantage of being far more tolerant of the presence of water in the injected samples compared to standard liquid stationary phases. Fig. 17.7 shows nearly identical separations of a homologous series of alkyl alcohols injected as a neat mixture or as a 10% aqueous solution [80]. The later conditions would quickly destroy most common liquid-based GC stationary phases.

17.6 CONCLUSIONS

Since their first successful implementation over a decade ago, monolithic materials have proven themselves to be excellent alternatives to their packed bed counterparts for the separation of biological or synthetic macromolecules, as their unique porous structures enable similar separations to be performed at significantly increased flow rates. Other properties of these materials, such as the *in situ* nature of their preparation, make monolithic media uniquely suited to perform a variety of functions in microscale separations in either a capillary or on-chip format. More recently, fundamental studies of both the nature and control of the porous properties of silica monoliths have enabled these materials to finally obtain the original goal behind the devel-

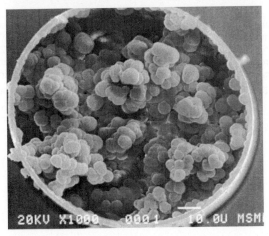

Fig. 17.5. Scanning electron micrograph of a 320 μm diameter monolithic poly(divinylbenzene) capillary column for gas chromatography (Reprinted with permission from [80], copyright 2000 Wiley-VCH).

opment of packed bed columns in the 1970s; the highly efficient separation of small molecules by HPLC. It is expected that additional ongoing research into the nature and properties of these unique monolithic media together with their commercial availability will further fuel their development in the field of chromatography.

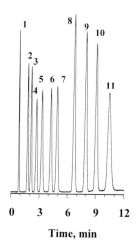

Fig. 17.6. Separation of a mixture of organic solvents using a 50 cm long × 320 μm i.d. monolithic capillary column shown in Fig. 17.5. Conditions: temperature gradient 120–300 °C, 20°C/min; inlet pressure 0.55 MPa; injection split. Peaks: methanol (1), ethanol (2), acetonitrile (3), acetone (4), 1-propanol (5), methyl ethyl ketone (6), 1-butanol (7), toluene (8), ethylbenzene (9), propylbenzene (10), butylbenzene (11). (Reprinted with permission from [80], copyright 2000 Wiley-VCH)

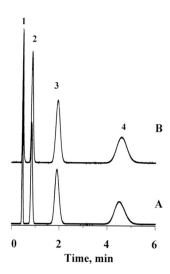

Fig. 17.7. Separation of a straight mixture (A) and 10% aqueous solution (B) of methanol (1), ethanol (2), 1-propanol (3), and 1-butanol (4) using a 50 cm × 320 mm i.d. monolithic column at 180°C. For other condition see Fig. 17.6. (Reprinted with permission from [80], copyright 2000 Wiley-VCH)

17.7 REFERENCES

1   J.J. Van Deemter, F.J. Zuiderweg, A. Klinkenberg, Chem. Eng. Sci. 5 (1956) 271.

2   J.J. Kirkland, J. Chromatogr. Sci. 10 (1972) 593.

3   R.E. Majors, Anal. Chem. 44 (1972) 1722.

4   H. Poppe, J. Chromatogr. A 778 (1997) 3.

5   P.P.H. Tock, C. Boshoven, H. Poppe, J.C. Kraak, J. Chromatogr. A 477 (1989) 95.

6   M.M. Dittmann, G.P. Rozing, J. Chromatogr. A 744 (1996) 63.

7   J.E. MacNair, K.C. Lewis, J.W. Jorgenson, Anal. Chem. 69 (1997) 983.

8   E.C. Peters, F. Svec, J.M.J. Fréchet, Adv. Mater. 11 (1999) 1169.

9   N. Tanaka, H. Kobayashi, K. Nakanishi, H. Minakuchi, N. Ishizuka, Anal. Chem. 73 (2001) 420A.

10   S. Hjertén, Ind. Eng. Chem. Res. 38 (1999) 1205.

11   A.I. Liapis, Modell. Sci. Comput. 1 (1993) 397.

12   E. Rodrigues, Z.P. Lu, J.M. Loureiro, G. Carta, J. Chromatogr. 653 (1993) 93.

13   F.D. Hileman, R.E. Sievers, G.G. Hess, W.D. Ross, Anal. Chem. 45 (1973) 1126.

14   R.D. Ross, R.T. Jefferson, J. Chromatogr. Sci. 8 (1970) 386.

15   S. Hjertén, J.L. Liao, R. Zhang, J. Chromatogr. 473 (1989) 273.

16   Q.C. Wang, F. Svec, J.M.J. Fréchet, Anal. Chem. 65 (1993) 2243.

17   M. Petro, F. Svec, I. Gitsov, J.M.J. Fréchet, Anal. Chem. 68 (1996) 315.

18   B. Boughtflower, T. Underwood, Chromatographia 38 (1995) 329.

19   K. Schmeer, B. Behnke, E. Bayer, Anal. Chem. 67 (1995) 3656.

20   J.L. Liao, Y.M, Li, S. Hjertén, Anal. Biochem. 234 (1996) 27.

21   E.C. Peters, M. Petro, F. Svec, J.M.J. Fréchet, Anal. Chem. 69 (1997) 3646.

22   J.M. Li, J.L. Liao, K. Nakazato, J. Mohammed, L. Terenius, S. Hjertén, Anal. Biochem. 223 (1994) 153.

23   C. Ericson, J. Holm, T. Ericson, S. Hjertén, Anal. Chem. 72 (2000) 81.

24   Y. Fintschenko, W.Y. Choi, S.M. Ngola, T.J. Shepodd, Fresenius J. Anal. Chem. 371 (2001) 174.

25   F. Svec, J.M.J. Fréchet, Macromolecules 28 (1995) 7580.

26   C. Viklund, E. Ponten, B. Glad, K. Irgum, P. Horstedt, F. Svec, Chem. Mater. 9 (1997) 463.

27   J.R. Chen, M.T. Dulay, R.N. Zare, F. Svec, E. Peters, Anal. Chem. 72 (2000) 1224.

28   F. Svec, E.C. Peters, D. Sykora, J.M.J. Fréchet, J. Chromatogr. A 887 (2000) 3.

29   H. Minakuchi, K. Nakanishi, N. Soga, N. Ishizuka, N. Tanaka, Anal. Chem. 68 (1996).

30   K. Cabrera, G. Wieland, D. Lubda, K. Nakanishi, N. Soga, H. Minakuchi, K.K. Unger, Trends Anal. Chem. 17 (1998) 50.

31   B. Bidlingmaier, K.K. Unger, N. von Doehren, J. Chromatogr. A 832 (1999) 11.

32   K. Cabrera, D. Lubda, H.M. Eggenweiler, H. Minakuchi, K. Nakanishi, J. High Resol. Chromatogr. 23 (2000) 93.

33   K. Nakanishi, J. Sol-Gel Sci. Tech. 19 (2000) 65.

34   M. Schulte, J. Dingenen, J. Chromatogr. A 923 (2001) 17.

35   M. Schulte, D. Lubda, A. Delp, J. Dingenen, J. High Resol. Chromatogr. 23 (2000) 100.

36   P. Zollner, A. Leitner, D. Lubda, K. Cabrera, W. Lindner, Chromatographia 52 (2000) 818.

37   J.T. Wu, H. Zeng, Y.Z. Deng, S.E. Unger, Rapid Commun. Mass Spectrom. 15 (2001) 1113.

38   G. Dear, R. Plumb, D. Mallett, Rapid Commun. Mass Spectrom. 15 (2001) 152.

39   R. Plumb, G. Dear, D. Mallett, D. Ayrton, Rapid Commun. Mass Spectrom. 15 (2001) 986.

40   S.M. Fields, Anal. Chem. 68 (1996) 2709.

41   N. Ishizuka, H. Minakuchi, H. Nagayama, N. Soga, K. Hosoya, N. Tanaka, J. High Resol. Chromatogr. 21 (1998) 477.

42   N. Ishizuka, H. Minakuchi, K. Nakanishi, N. Soga, H. Nagayama, K. Hosoya, N. Tanaka, Anal. Chem. 72 (2000) 1275.

43   N. Ishizuka, K. Nakanishi, K. Hirao, N. Tanaka, J. Sol-Gel Sci. Tech. 19 (2000) 371.

44   Z.L. Chen, T. Hobo, Electrophoresis 22 (2001) 3339.

45   I. Gusev, X. Huang, C. Horváth, J. Chromatogr. A 855 (1999) 273.

46   P. Coufal, M. Čihák, J. Suchánková, E. Tesařová, J. Chromatogr. A 946 (2002) 99.

47   E.C. Peters, M. Petro, F. Svec, J.M.J. Fréchet, Anal. Chem. 69 (1998) 2288.

48   R. Asiaie, X. Huang, D. Farnan, C. Horváth, J Chromatogr. A 806 (1998) 251.

49   G.S. Chirica, V.T. Remcho, Anal. Chem. 72 (2000) 3605.

50   Q.L. Tang, N.J. Wu, M.L. Lee, J. Microcolumn Sep. 12 (2000) 6.

51   G.S. Chirica, V.T. Remcho, Electrophoresis 21 (2000) 3093.

52   D.C. Collins, Q.L. Tang, N.J. Wu, M.L. Lee, J. Microcolumn Sep. 12 (2000) 442.

53   C. Yu, F. Svec, J.M.J. Fréchet, Electrophoresis 21 (2000) 120.

54   J.R. Chen, R.N. Zare, E.C. Peters, F. Svec, J.M.J. Fréchet, Anal. Chem. 73 (2001) 1987.

55   T. Rohr, C. Yu, M.H. Davey, F. Svec, J.M.J. Fréchet, Electrophoresis 22 (2001) 3959.

56   C. Yu, M.H. Davey, F. Svec, J.M.J. Fréchet, Anal. Chem. 73 (2001) 5088.

57   C. Ericson, J.L. Liao, K. Nakazato, S. Hjertén, J. Chromatogr. A 767 (1997) 33.

58   A. Maruška, C. Ericson, Á. Végvári, S. Hjertén, J. Chromatogr. A 837 (1999) 25.

59   E.C. Peters, K. Lewandowski, M. Petro, F. Svec, J.M.J. Fréchet, Anal. Commun. 35 (1998) 83.

60   M. Lämmerhofer, E.C. Peters, C. Yu, F. Svec, J.M.J. Fréchet, W. Lindner, Anal. Chem. 72 (2000) 4614.

61   F. Svec, J.M.J. Fréchet, Anal. Chem. 64 (1992) 820.

62   Q.C. Wang, F. Svec, J.M.J. Fréchet, J. Chromatogr. A 669 (1994) 230.

63   J.F. Wang, Z.H. Meng, L.M. Zhou, Q.H. Wang, D.Q. Zhu, Chin. J. Anal. Chem. 27 (1999) 745.

64   M.R. Buchmeiser, Angew. Chem. Int. Ed. 40 (2001) 3795.

65   F.M. Sinner, M.R. Buchmeiser, Macromolecules 33 (2000) 5777.

66   A.D. Martina, J.G. Hilborn, A. Muhlebach, Macromolecules 33 (2000) 2916.

67   A.I. Cooper, A.B. Holmes, Adv. Mater. 11 (1999) 1270.

68   G.S. Chirica, V.T. Remcho, J.Chromatogr. A 924 (2001) 223.

69   J. Mohammad, Y.M. Li, M. El-Ahmed, K. Nakazato, G. Petterson, S. Hjertén, Chirality 5 (1993) 464.

70   F.M. Sinner, M.R. Buchmeiser, Angew. Chem. Int. Ed. 39 (2000) 1433.

71   C.J. Welch, J. Chromatogr. A 666 (1994) 3.

72   G. Wulff, Angew. Chem. Int. Ed. 34 (1995) 1812.

73   K.J. Shea, Trends Polym. Sci. 2 (1994) 166.

74   K. Mosbach, Trends Biochem. Sci 19 (1994) 9.

75   L. Schweitz, L.I. Andersson, S. Nilsson, J. Chromatogr. A 817 (1998) 5.

76   J. Matsui, T. Kato, T. Takeuchi, M. Suzuki, K. Yokoyama, E. Tamiya, I. Karube, Anal. Chem. 65 (1993).

77   J. Matsui, T. Takeuchi, Anal. Commun. 34 (1997) 199.

78   J. Matsui, I.A. Nicholls, T. Takeuchi, Anal. Chim. Acta 365 (1998) 89.

79   J.H.G. Steinke, I.R. Dunkin, D.C. Sherrington, Macromolecules 29 (1996) 407.

80   D. Sýkora, E.C. Peters, F. Svec, J.M.J. Fréchet, Macromol. Mater. Eng. 275 (2000) 42.

81   Q.L. Tang, Y.F. Shen, N.J. Wu, M.L. Lee, J. Microcolumn Sep. 11 (1999) 415.

82   N.J. Wu, Q.L. Tang, J.A. Lippert, M.L. Lee, J. Microcolumn Sep. 13 (2001) 41.

F. Švec, T.B. Tennikova and Z. Deyl (Editors)
Monolithic Materials
Journal of Chromatography Library, Vol. 67
© 2003 Elsevier Science B.V. All rights reserved.

Chapter 18

Separation of Peptides and Proteins

Djuro JOSIĆ

*Octapharma Pharmazeutika ProduktionsgesmbH, Research & Development,
Oberlaaer Strasse 235, A-1100 Vienna, Austria*

CONTENTS

18.1 INTRODUCTION

In the last two decades, the possibility of production of complex biomolecules using new techniques has opened the way to an increase in the quality of human life. One of the most important and expensive steps in this production is the isolation and purification of target molecules [1]. Precipitation, filtration, and chromatographic techniques are most widely used for these purposes. However, only liquid chromatography leads to the level of purity that is acknowledged to be safe for therapeutic use of such substances. Consequently, only the concomitant development of molecular biology and more effective purification techniques — above all, chromatography — will help in the production of complex biomolecules [2,3].

Liquid chromatography of biopolymers on so-called soft supports is typically slow, often causing significant product degradation. Diffusional constraints particularly set an upper limit on the speed of separation because they cause a rapid reduction in resolution with increasing elution velocity in conventional packed columns [1,4]. The evolution in both column and packing designs put strong emphasis on the acceleration of separations, with the following objectives in mind: (i) Miniaturization and micro-preparative analysis; (ii) on-line analysis of production and separation processes necessary for regulation, optimization, and regulatory purposes; (iii) reduction of manufacturing costs; (iv) reduction of losses from degradation of the biopolymers; and (v) scalability of the process.

Apart from the progress made in column chromatography with porous supports, new ways have also been investigated in recent years for furthering improvements in the analytical, micro-preparative and preparative separations of peptides and proteins. The new chromatographic media should allow fast separation combined with both high capacity and low non-specific interactions. A uniform chromatographic behavior of the support is also important since it facilitates scalability.

The first difficulty to be overcome is the rather slow diffusion which impairs the mass-transfer of large, and frequently irregularly shaped, macromolecules such as biopolymers. This problem was first solved by using columns with non-porous supports that cut the time required for chromatographic separations of biopolymers from over 30- to only a few minutes [5-8].

However, columns with non-porous packings have a rather low capacity. This is caused by the small surface of the packing beads and consequently the low ligand density. Also, when supports with small particle diameters of 1 or 2 μm are used, the back-pressure in the column is very high at higher flow rates. For these reasons, short columns packed with non-porous supports are suitable only for fast, analytical separations. Scaling-up and preparative applications are rare exceptions to the rule.

The introduction of packings for perfusion chromatography was another step taken towards the development of media for chromatography of peptides and proteins [9,10]. Because of the size and structure of the particles, mass transfer in perfusion chromatography is faster, the back pressure is low, and separation can be carried out even at high flow rates. Moreover, the media for perfusion chromatography have much higher capacities than the non-porous supports. All of these characteristics allow for fast separation of biopolymers. With neither capacity nor flow rate being a limiting factor, the scaling up of an analytical method to a preparative column is now no longer a problem. The separation properties of a preparative column differ only little from that of an analytical column. This, in turn, allows prompt monitoring of the performance of the preparative separation using the same support as in an analytical scale column.

The same degree of performance has also been observed with other, so-called fast separation media. These media, as with perfusion beads, allow fast mass-transfer and separations, as fast as those carried out on micropellicular stationary phases [11,12]. Here again, neither the capacity nor the flow-rate constitutes a limiting factor, thus allowing simple scale up.

Results similar to those in perfusion chromatography were also obtained using membrane chromatography (MC). The basis for the development of MC in the late 1980s and early 1990s was a fast growth of membrane technologies in the preceding years. The production of membranes with defined, uniform pores has allowed simple separations of components according to their sizes, as in fast desalination and the concentration of biopolymer solutions. These operations could be achieved in a short period of time [13]. If the membrane also contained immobilized ligands, the separation process involved not only filtration but also a specific interaction between the membrane and components of the sample. As a result, the components were first retarded, then subsequently eluted in a gradient of the mobile phase, depending on the strength of the interaction [14-19]. The supports for MC have surface chemistries similar to those typical of existing chromatographic particles such as those used in ion-exchange (IE)-, reversed-phase (RP)-, hydrophobic-interaction (HI)-, and affinity chromatography (AC) [14-18].

Several techniques have already been designed for the construction of membrane units. One strategy is to bundle several thin membranes or hollow fibers, which are made of cellulose or synthetic polymers. Another approach involves compact porous disk-, tube-, or rod-shaped units made of porous silica or polymers [15-17,20,21]. The first successful experiments with membranes in chromatography have been carried out in the affinity- and ion-exchange modes [14,22]. Mandaro *et al.* [22] used membranes composed of vinyl monomers grafted to cellulose, assembled into a filter cartridge (Zeta-Prep). Such membranes with immobilized Protein A or Protein G have been used for isolation of immunoglobulin G (IgG) from different species. Menozzi *et al.* [23] used Zeta-Prep ion-exchange membranes for purification of monoclonal antibodies. Brandt *et al.* [14] used membranes with immobilized gelatin for isolation of fibronectin from human plasma. Already in this early work, the authors have compared conventional chromatography using agarose with the use of immobilized gelatin and the new membrane affinity technology. They demonstrated that membrane-based separation systems largely alleviate the mass-transfer limitations seen in conventional technologies. This statement was later confirmed by Unarska *et al.* [24] in a system with Protein A as the affinity ligand and IgG as the ligate. The mass transfer in the membrane device was much faster than with particulate supports. The experiments performed by this group also documented that with there is practically no diffusional limitation to mass transport. Similar results were obtained shortly afterwards using monolithic disks with affinity ligands [25].

Hjertén *et al.* [26] described in 1989 a monolithic (continuous bed) material produced by strong compression of a soft, swollen poly(acrylic acid-*co*-methylene-*bis*-
acrylamide) gel. Separations in the cation-exchange mode have confirmed that these
monoliths are suitable for the separation of proteins. The same group showed later that
identical technology can also be applied for the preparation of continuous beds for
reversed-phase and chiral-recognition chromatography [27].

At the end of the 1980s, different types of monolithic materials based on poly(glycidyl methacrylate-*co*-ethylene dimethacrylate) emerged [15,28]. The product pos-
sessed the benefits of both membranes (low pressure-drop and ease of handling) and
conventional columns (efficiency and resolution). The authors also coined the name
"high performance membrane chromatography" (HPMC) to distinguish this method
from conventional HPLC. A chromatogram obtained using this method and a DEAE
disk is shown in Fig. 18.1.

Another interesting application of this type of support was demonstrated as early as
1991 by Abou-Rebyeh *et al.* [25]. They carried out the first conversion of a substrate
with the immobilized enzyme – carbonic anhydrase – in flow-through mode. As
shown in Fig. 18.2, rather low flow-rates of up to 1.2 ml/min were first used in these
experiments. However, a higher flow-rate led to an increase in enzymatic activity. The
same filtration unit was also used for installation of an affinity disk. This enabled the
carrying out of kinetic experiments under dynamic conditions.

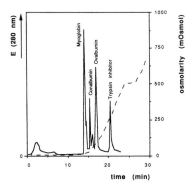

Fig. 18.1. Anion-exchange chromatography of proteins (Reprinted with permission from ref.
[16], copyright 1992 Elsevier Sciences B. V.). The chromatography unit with a DEAE-poly-
-(glycidyl methacrylate-*co*-ethylene dimethacrylate) disk was installed at the bottom of an
Amicon filtration device. The protein mixture in buffer A was applied by use of compressed
air. The container was subsequently filled with buffer A and the proteins eluted in an
exponential salt gradient.

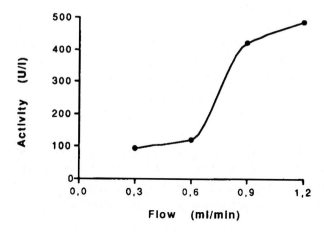

Fig. 18.2. Effect of flow rate on enzymatic activity of carbonic anhydrase immobilized on a monolithic disk, at that time called a "carrier membrane", using 2-chloro-4-nitrophenyl acetate as a substrate. (Reprinted with permission from ref. [25], copyright 1991 Elsevier Sciences B.V.)

The application of monolithic materials for the separations of peptides and proteins was considerably accelerated by an inventive solution to the sample-distribution problem [16], and a newly developed monolithic support based on macroporous polymers was the first product of this kind on the market, distributed under the trade name QuickDisk^{TM} (Knauer Säulentechnik, Berlin, Germany). This monolithic column technology was further developed in two independent directions to a product called CIM^{TM} (Convective Interaction Media, BIA Separations, Ljubljana, Slovenia) in disk- and tube-shape formats [29-31], and to molded monolithic rods [21,32-34], both mainly used for separation of peptides and proteins as well as for enzymatic conversions of low- and high-molecular weight substrates [25,29, 35,36].

Over the last ten years, monoliths have been used in a broad range of separations of peptides and proteins by means of electrochromatography [37], as well as chromatography using microcolumns [38] and capillaries [39]. Since these applications are dealt with in separate chapters of this book, only a few relevant examples will be mentioned here. The ultra-short beds, typically membranes and polymer-based monolithic disks, have been chiefly applied to the separation of biopolymers such as proteins, polynucleotides [36,40-42], and even nanoparticles [43]. In contrast, the silica-based monoliths, which have been developed recently, are mainly used for the separation of small molecules [44-46]. Their use for separation of proteins and peptides is, until now, an exception rather than the rule [47].

18.2 ION-EXCHANGE CHROMATOGRAPHY

Ion-exchange chromatography using monolithic supports is one of the most frequent methods adapted for separation of peptides and, above all, proteins. Monolithic units, such as disks, tubes, and rods with either anion exchange (DEAE or quarternary amine, QA, functionalities) or cation exchange (mostly sulfonic acid groups as ligands) are used for both analytical and preparative purposes [15,16,19,20,29,30,36,40, 41,48-53]. Figure 18.3 shows the very fast separation of standard proteins that has been achieved in less than 15 s in the anion-exchange mode using a monolithic disk [29].

Another example is the semi-preparative separation of the large protein-protein complex containing clotting factor VIII–von Willebrand factor (FVIII–vWF) shown in Fig. 18.4. The complex has a molecular weight of up to 2×10^7. The FVIII molecule is sensitive to degradation and tends to bind unspecifically to various surfaces. Further separation of FVIII–vWF complex was successfully performed with monolithic QA–CIM tubes [54].

Clotting factor IX (FIX), another multi-domain plasma glycoprotein, with an apparent molecular weight of 65×10^3 determined by SDS-PAGE under reducing conditions, was monitored during its purification process by anion-exchange CIM mono-

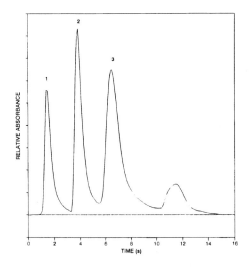

Fig. 18.3. Separation of a mixture of proteins using an anion-exchange QA–CIM disk. Peaks: myoglobin (1), conalbumin (2), and soybean trypsin inhibitor (3). (Reprinted with permission from ref. [30], copyright 1996 American Chemical Society).

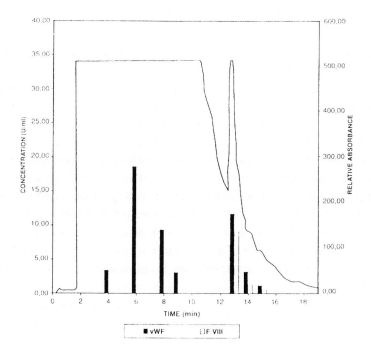

Fig. 18.4. Separation of an in-process sample from production of clotting factor VIII–von Willebrand factor complex (FVIII–vWF) obtained using a QA anion-exchange monolithic tube with a volume of 8 ml. (Reprinted with permission from ref. [54], copyright 1997 Elsevier Sciences B. V.).

lithic disks [30]. Branovic *et al.* [55,56] described the application of recently scaled-up monolithic columns [31] for down-stream processing of various intermediates resulting from the FIX production process. Also, a regeneration procedure using 1 mol/l NaCl and 0.5 mol/l NaOH was proven to be possible. This sanitation feature is very important for sorbents that are intended for production of biologically active peptides and proteins for pharmaceutical use.

The lack of stability at high pH of silica-based supports is one of the main reasons why these media are not frequently used for the isolation of biopolymers from complex mixtures. If sufficiently bonded, they may be resistant against alkaline treatment to some extent. However, they have never reached a stability at high pH that would be even close to that of synthetic polymer monoliths [40].

Three annexins CBP33, CBP35 and CBP65/67 derived from the plasma membranes of rat Morris hepatoma 7777 were separated on a CIM-DEAE disk, applying a linear gradient as well as a step-gradient [57]. Podgornik *et al.* [58] demonstrated the use of CIM ion-exchange monolithic disks for rapid separation of lignin peroxidase

(LiP) isoforms directly from crude rot fungus *Phanochaete chrysosporium* culture filtrate. Four LiP isoenzymes were rapidly monitored in the growth medium during cultivation. This allowed a close to on-line in-process monitoring of LiP isoforms during fermentation [59].

Recombinant human basic fibroblast growth factor (rh-bFGF) was purified from a crude feedstock of a high density *E. coli* cultivation [60]. Figure 18.5 shows the separation of rhbFGF using a strong-cation-exchange UNO-S1 monolithic column and compares it with the separation achieved on the expanded-bed-adsorber packed with Streamline SP beads (Pharmacia Amersham, Sweden). Although the supernatant liquid from the cell homogenate still contained some cell debris and protein aggregates, a complete blockage of the column did not occur, as the precipitates were readily removed by reversal of the flow during cleaning-in-place [CIP] procedures with 0.5 mol/l NaOH. The Rh-bFGF elutes in three consequent peaks, which could be attributed to soluble rh-bFGF aggregates of different sizes, shown in SDS-PAGE in Fig. 18.5. The dynamics of rh-bFGF aggregation and re-aggregation was monitored by fast-gradient-elution chromatography using the same monolithic UNO-S1 column.

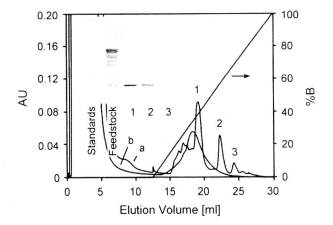

Fig. 18.5. Separation of clarified crude feedstock of *E. coli* cultivation using monolithic cation-exchange columns UNO-S1 (a), and a column packed with cation-exchange beads Streamline SP (b). Insert: SDS-PAGE gels under reducing conditions. Lines: molecular weight standards, clarified rh-b FGF feedstock, and peaks 1, 2, and 3. (Reprinted with permission from ref. [60], copyright 1999 Elsevier Sciences B. V.).

18.3 REVERSED-PHASE AND HYDROPHOBIC-INTERACTION CHROMA-TOGRAPHY

Although the modification of monoliths with weakly hydrophobic groups as ligands was completed rather early [28,61], relatively few applications have been described for the separation of peptides and proteins in this mode. Tennikova *et al.* [28] has demonstrated the separation of a mixture of proteins including myoglobin, ovalbumin, lyzozyme, and chymotrypsinogen on monolithic "membranes", which actually rather been disks of 20 mm diameter and 1 mm thick, bearing C4- or C8-ligands. A separation of 5 mg of proteins was achieved in 30 min.

A recombinant tumor necrosis factor α (TNF-α) from a pre-purified *E. coli* extract was purified by anion-exchange chromatography followed by hydrophobic-interaction chromatography (HIC) on C4 disks [62]. The separation efficiency and recovery of TNF-α was fully comparable to that of a Phenyl-Sepharose column. However, the separation time was considerably shorter. The active form of this protein is a trimer. HIC caused a dissociation of this complex into monomers, with a partial loss of activity as the consequence. The authors of this research assumed that monomers were formed due to the interaction of the trimer with the hydrophobic ligand.

Strancar *et al.* [30] have demonstrated that separation of standard proteins with monolithic disks is possible also in hydrophobic-interaction mode in less than 30 s (Fig. 18.6). Similarly, Xie *et al.* [63] reported hydrophobic-interaction chromatography of proteins on monolithic columns containing polymerized butyl methacrylate. The hydrophobicity of the interacting surface can be controlled readily by the percentage of butyl methacrylate in the polymerization mixture. A mixture of proteins including cytochrome C, ribonuclease, and lyzozyme has also been separated within seconds.

In their early work, Tennikova and Svec [64] investigated the effect of gradient volume, gradient time, gradient profile, and flow-rate on resolution, performing ion-exchange-, hydrophobic-interaction-, and reversed-phase chromatography. According to their calculations, the protein diffusivity enhanced by the convective flow through the monolith is about four orders of magnitude higher than the free diffusion of the protein in the stagnant mobile phase located within the pores of a typical porous bead. Three years later, Strancar *et al.* [30] used this advantage to implement a fast in-process analysis.

Both silica- and polymer-based monoliths can be used for reversed-phase (RP) chromatography. While the use of silica-based monoliths for the separation of proteins and peptides remains less common [47], the use of polymer-based monoliths in RP mode has recently become quite widespread [65-67]. At the beginning, poly(styrene-*co*-divinylbenzene) (PS-DVB) or modified poly(glycidyl methacrylate-*co*-ethylene di-

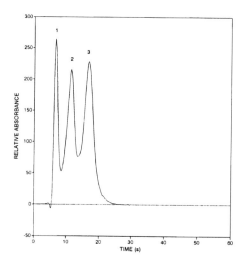

Fig. 18.6. Hydrophobic-interaction chromatography of proteins using monolithic disk containing propyl groups. Peaks: myoglobin (1), conalbumin (2), and soybean trypsin inhibitor (3). (Reprinted with permission from ref. [30], copyright 1996 American Chemical Society).

methacrylate) (GMA-EDMA) monoliths were used for separations in this mode [27,64,68]. Recently, monolithic stationary phases were also prepared by ring-opening metathesis polymerization (ROMP) [69] and applied for the separation of proteins in reversed-phase mode [70].

In an early work, Liao *et al.* [71] used microcolumns filled with a monolithic, C-18 derivatized bed for the separation of proteins in the reversed-phase mode. This group also used monolithic supports in capillary chromatography (CC) and capillary electro-chromatography (CEC) for separating both low- and high molecular weight molecules [72,73]. This field has been reviewed recently by Svec *et al.* [74]. Figure 18.7 shows the separation of a protein mixture using a monolithic capillary in reversed-phase CC and CEC mode.

The production of monoliths and their use in CC and CEC can be regarded as an important step towards miniaturization. The next step is likely to be the production of stationary phases within microfluidic chips [75]. Chromatographic separation in reversed-phase mode will certainly play an important role in this development. This field is treated in detail elsewhere in this book.

In a very simple but impressive application, Moore *et al.* [76] have prepared PS-DVB monoliths in a 150 μm i.d. fused-silica capillary tubing with a pulled 5–10 μm needle tip at one end. Using this monolith, proteins and peptides were chromatographically separated and subsequently analyzed on-line in a mass spectrometer. As

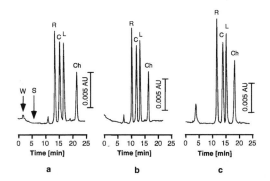

Fig. 18.7. Separation of proteins using gradient electrochromatography (a,b) and reversed-phase micro-HPLC (c) (Reprinted with permission from ref. [72], copyright 1999 American Chemical Society). Peaks: ribonuclease A [R], cytochrome c [C], lysozyme [L], chymotrypsinogen [Ch]. Columns: 8 cm (6 cm effective length) × 50 μm i.d.; mobile phase: a linear gradient from 5 to 80% acetonitrile in 5 mmol/l sodium phosphate, pH 2.0; voltage 5.5 kV (700 V/cm). (a) Moderate-EOF column, (b) high-EOF column, (c) conventional capillary PLC, pressure 5 Mpa.

shown in Fig. 18.8, better separations were achieved with monolithic material than with corresponding microcolumns packed with particles.

Figure 18.9 shows the analytical separation of five proteins on a PS-DVB-based monolith in reversed-phase mode. Separation of the components was achieved within less than 20 s [67]. It is rather clear that chromatography in the RP mode using monoliths has considerable potential. This applies to the fast separations on both micro-analytical and micro-preparative scales as well as to rapid separations of peptides and proteins on a laboratory scale.

18.4 AFFINITY CHROMATOGRAPHY

Although the first published experiments with membrane chromatography were carried out in the affinity mode [14,22], it took some time to design the monolithic units with immobilized affinity ligands [16,30]. The road to the first operational commercially available monoliths with such ligands was even longer [29]. Experimentation with affinity chromatography separations on membranes and monoliths has almost exclusively been carried out using low-molecular-weight ligands [29,77]. Protein A and G were the first high-molecular-weight ligands used for the purification of antibodies. At the beginning, the use of other high-molecular-weight ligands presented some difficulties, especially when membranes were used for their immobilization [78]. These problems were solved by means of improved immobilization chemistry

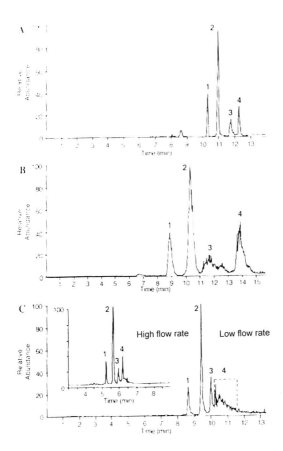

Fig. 18.8. LC/MS separations of protein mixture containing ribonuclease (1), cytochrome C (2), holotransferrin (3), and apomyoglobin (4) using a PS/DVB monolith-filled needle (A), a Vydac C-18-packed needle (B), and a Poros R2 packed needle (C). (Reprinted with permission from ref. [76], copyright 1998 American Chemical Society).

and the introduction of spacers, similar to those used for bead shaped supports [79-81]. Today, it is possible to use to its full extent practically each low- or high-molecular-weight ligands for immobilization- and affinity chromatographic separation in both membrane and monolithic formats [40,82,83].

Klein [84] has recently published a review summarizing the use of affinity membranes over the last ten years. In his opinion, commercialization remains meager in the face of a large number of membrane preparation and process applications described in the literature [40,82]. The situation with the monolithic supports is similar. Despite the recent considerable progress achieved with both monoliths and membranes [31,53,85,86], the decisive breakthrough has not yet been reached.

Fig. 18.9. Reversed-phase separation of proteins using monolithic poly(styrene-*co*-divinylbenzene) column (Reprinted with permission from ref. [67], copyright 1999, Elsevier Sciences B. V.). Peaks: ribonuclease (1), cytochrome C (2), bovine serum albumin (3), carbonic anhydrase (4), and chicken egg albumin (5).

18.4.1 Low-molecular-weight ligands

Low-molecular-weight ligands suitable for affinity chromatography have mainly been attached to poly(glycidyl methacrylate)-based- or agarose monoliths. Monoliths based on silica gel and polyacrylamide (UNO-columns) as well as on other synthetic materials have so far been used much less in the affinity chromatographic mode. An exception was presented in a recently published paper describing a urea-formaldehyde monolith (UF) with immobilized Cibachrom Blue F3 GA ligands. Although the concentration of this immobilized dye and the dynamic capacity of the monolith are low compared to Sepharose-based sorbents, such a unit could be used for separation of standard proteins in the affinity mode [87]. However, further optimization is required before this new packing material can be applied routinely in the laboratory.

As in the typical affinity chromatography using packed bed columns, the most frequently used ligands are dyes, inhibitors, co-enzymes, and other low-molecular-weight substances, which can interact specifically with the components of the sample [25,40,41,82,88]. Increasingly, the use of short peptides and other low-molecular-weight ligands becomes more important. They might even be designed using the methods of combinatorial chemistry, immobilized, and applied for purification of the corresponding target substances [89-91].

Slightly simplified, molecular imprinting represents an approach reciprocal to the method described above. The ligand is not immobilized but, instead, an "artificial receptor" is imprinted in the surface of the support. By combining molecular imprinting with a strategy of combinatorial chemistry, libraries of molecularly imprinted polymers can be prepared and screened. In the early stage, monolithic supports were considered to be well suited for these applications. So far, the use of imprinted supports remains limited to the separation and detection of small molecules. However, it is conceivable that in the near future, once the technical problems have been solved, they will also be available for the separation of larger molecules such as peptides and proteins.

In an early application of affinity chromatography involving low-molecular-weight ligands, monolithic disks with the immobilized inhibitor of carbonic anhydrase, 4-amino methylbenzene sulphonamide as a ligand, were used for isolation of this enzyme from human erythrocytes [25]. This method exploits the fact that the interaction between the enzyme and the inhibitor is pH-dependent. At higher pH, the binding is stable. However, at pH below 6.0, the bond is increasingly unstable, and can dissociate. The protein is not denatured under these conditions, neither through its binding nor through its elution, and therefore retains its enzymatic activity.

In contrast to the widespread use of thin membranes employing immobilized low-molecular-weight affinity ligands for the isolation of a number of proteins [40,41,96-98], monoliths have not been applied often during recent years for such purposes. However, the results of Platonova *et al.* [99] and Amatschek *et al.* [100,101] have demonstrated that peptidic ligands can also be used with monolithic supports. Hahn *et al.* [83] have shown that the choice of spacer and the structure of the matrix, along with immobilization chemistry, are decisive for the performance.

In a recent paper, Gustavsson and Larsson [86] used the continuous superporous agarose with immobilized Cibacron Blue 3GA for the purification of lactate dehydrogenase from a crude bovine heart extract. This extract had a total enzymatic activity of 8000 IU and a total protein content of 600 mg. They achieved purification by a factor of 23 from 13.3 to 308 U/mg in 68 min, and the yield of activity was 73 % (Fig. 18.10).

Other important application which includes low-molecular-weight ligands is immobilized metal affinity chromatography (IMAC), used for the isolation of recombinant proteins containing a polyhistidine sequence at the end of the polypeptide chain [88,96]. A detailed review of these applications has been published by Tennikova and Freitag [36]. Yang *et al.* [102] carried out IMAC with cellulose-based monolithic supports grafted with acrylic polymers. They purified catalase from bovine liver using

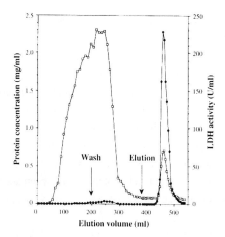

Fig. 18.10. Affinity chromatography purification of lactate dehydrogenase using superporous continuous agarose monolith derivatized with Cibachron Blue 3GA (Reprinted with permission from ref. [86], copyright 1999 Elsevier Sciences B. V.). Column: 65 ml radial flow. Sample: 200 ml of a crude bovine lactate dehydrogenase extract. The binding and washing buffer was 20 mmol/l sodium phosphate, pH 7.0, containing 1 mmol/l EDTH and 2 mmol/l 2-mercaptoethanol; elution was performed with binding buffer containing 1.0 mmol/l NADH.

a disk cartridge similar to that described by Josic *et al.* [16]. The monolithic support enabled enrichment comparable to that achieved with bulk supports but about three times faster.

Continuous rods of poly(glycidyl methacrylate-*co*-ethylene dimethacrylate) with immobilized iminodiacetic acid (IDA) and different metal ions have been used for the IMAC separation of a variety of proteins. The effect of pH on the adsorption capacity of bovine serum albumin on the Cu^{2+}–IDA monolithic column was also investigated [103] and reviewed [41].

Membranes and monoliths with immobilized histidine can also be applied for the purification of antibodies and for the removal of endotoxins from protein solutions [41,104,105]. Strictly speaking, the removal of endotoxins does not fit completely the subject of this chapter, which concerns the separation of peptides and proteins. However, they can be present in protein solutions in extremely low concentrations and are difficult to remove. In the case of therapeutic proteins that are administered intravenously, traces of endotoxins can cause dangerous side effects. A number of papers describe removal of endotoxins from such protein solutions, using chromatography on supports with immobilized histidine. This field has been summarized in recent reviews [41,105].

According to Huang *et al.* [89] and Fassina *et al.* [90,106], synthetic peptides are ideally suited for the purification of high-value therapeutic proteins, since they are

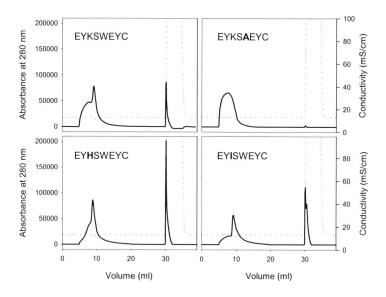

Fig. 18.11. Affinity purification of plasma derived (pd) FVIII with peptide ligands immobilized on a monolithic CIM column. The sequences of the respective peptide ligands are denoted in the upper left corner of each chromatogram. Elution was performed using a step gradient with 2 mol/l guanidine hydrochloride. Protein was detected at 280 nm (full line), conductivity was measured to visualize the step gradient (dotted line). (Reprinted with permission from ref. [100], copyright 2000 Wiley-VCH).

inexpensive, chemically defined, and non-toxic. They do not contain hydrolyzable bonds, are resistant to chemical and biological degradation, and can be sterilized and cleaned *in situ*. Plasma proteins, both recombinant and, in particular, plasma derived, offer numerous options for the use of such combinatorial ligands. Their design has been greatly assisted by increased access to structural data, computer-aided molecular design, and novel combinatorial techniques [89,100,101,106]. Amatschek *et al.* [100] and Pflegerl [101] have immobilized newly designed combinatorial octapeptides with affinity to clotting-factor-VIII [FVIII] on various beads and monoliths. Compared with the bead supports, the monolithic material exhibited much better performance, in terms of both capacity and selectivity. Characteristic separations are shown in Figs. 18.11 and 18.12.

Platonova *et al.* [99] have successfully used synthetic pentadecapeptides, hexadecapeptides, and the nonapeptide hormone bradykinin, all immobilized on monolithic CIM® disks, for the purification of monospecific polyclonal antibodies against corresponding antigens. The model investigations demonstrated that this technology could be used for the fast fractionation of monospecific polyclonal antibodies in serum. This

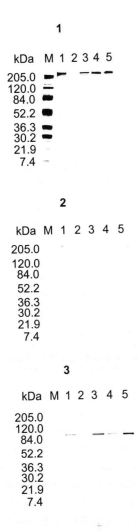

Fig. 18.12. Silver-stained SDS-PAGE (1) and Western blot analysis (2,3) of fractions collected from affinity chromatography of pdFVIII with the peptide ligands shown in Fig. 18.11. Blot (2) was developed with anti-vWF peroxidase conjugate, Blot (3) with Mab 038 directed against the light chain of FVIII and anti-mouse IgG alkaline phosphatase conjugate. Lanes: M - molecular weight marker; 1- starting material; 2 - eluate from the blank column carrying EYKSAEYC; 3 - eluate from the column carrying peptide EYHSWEYC; 4 - eluate from the column carrying peptide EYKSWEYC: 5 - eluate from the column carrying peptide EYISWEYC. (Reprinted with permission from ref. [100], copyright 2000 Wiley-VCH)

result also represents a good basis for recommending this method for medical diagnostics and for the identification of pathogenic immunoglobulins.

The current experience with immobilized combinatorial ligands indicates that their performance is more sensitive to immobilization chemistry, matrix composition, and the type of spacer than that of ligands with high molecular weight. This applies to both beads and monolithic supports. Hahn *et al.* [83] have varied the above-mentioned parameters to determine the effect of immobilization chemistry, matrix composition, and type of spacer, using a model peptide with an affinity to chicken egg lysozyme. With this model ligand, they investigated representative matrices as supports in order to find an optimal ligand–spacer–support combination. Different spacer lengths and types were included in this investigation and the performance of each affinity sorbent was characterized by breakthrough curves. Monolithic CIM disks with immobilized model peptide had the highest capacity per micromole of the ligand. Among the investigated beads, the highest capacity related to volume was observed for "tentacle gel" using a 16-atom-long spacer. However, if the model peptide was immobilized on the tentacle gel without this spacer, the capacity of this medium to lysozyme was considerably reduced. All other beads used in this study had much lower capacity and selectivity than the two supports mentioned above. The results indicated that selectivity and capacity are linked to the optimal exposure of the immobilized ligand. This, in turn, depends on the characteristics of the chosen support.

18.4.2 High-molecular-weight ligands

Affinity chromatography separation using membranes with immobilized Protein A and gelatin were among the first applications of this technology for protein purification [14,22]. Apart from Protein A, Brandt *et al.* [14] also immobilized gelatin on hollow-fiber membranes. The affinity units were used for the isolation of fibronectin from blood plasma. The early experiments with this high-molecular-weight adhesion protein exhibited a surprisingly high speed of separation and high capacity of the membrane device. However, detailed investigations which were carried out shortly afterwards showed that the low capacity of ligands immobilized on cellulose-based membranes can be a problem. The reason for this is probably the low density of reactive functionalities located on the surface of the original support [78].

Membranes with immobilized ligands such as Proteins-A and -G, antibodies, receptors, and other proteins with a variety of biological functions, have been used from the very beginning [36,41,77,82]. For example, Josic *et al.* [16,57,80] have shown that porous poly(glycidyl methacrylate) monolithic disks with an epoxide functionality can be used for the immobilization of high-molecular-weight ligands. However, in the early years, monoliths were not used very often for immobilizations of these ligands. This may result from the unjustified expectation that affinity separations of biopolymers could be achieved within a few seconds [29,30,107]. However, affinity chromatography mode is rarely used for such fast separations. Consequently, only one analyt-

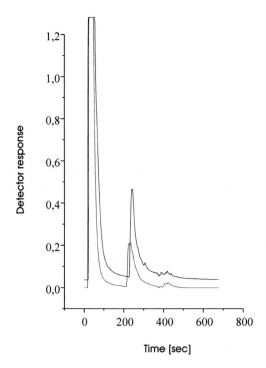

Fig. 18.13. Fast isolation of Protein G from 1:5 (a), and 1:10 (b), diluted *E. coli* cell lysate using a monolithic disk with immobilized IgG. (Reprinted with permission from ref. [79], copyright 1998 Elsevier Sciences B.V.).

ical affinity separation is found among the applications of HPMC in this context. A disk with immobilized heparin was used for fast quality control of preparations of the plasma proteins antithrombin III and clotting factor IX [108]. The option of using monolithic supports for scale-up to semi-preparative and preparative purposes [84] has opened the way for the application of affinity chromatography in these fields. Recently, the use of other monolithic supports for similar applications has also been demonstrated [41,70,86].

Figure 18.13 shows the isolation of Protein G from a cell lysate of *E. coli* by affinity chromatography using a CIM disk with immobilized IgG. Kasper *et al.* [79] have shown that direct immobilization of large proteins and immunoglobulins on poly(glycidyl methacrylate)-based disks is possible without the need for inserting a spacer between the support and the protein. Here again, the considerable power of affinity chromatography was confirmed by the isolation of target substances from highly diluted solutions. The specific advantage inherent to monolithic supports is the significant reduction of time required for separation [109]. Josic *et al.* [80] have shown that monolithic supports, CIM disks, can also be used for immobilization of other ligands such as lectins and various enzymes. For example, a disk with the immobilized lectin concanavalin A (Con A) was used successfully for the purification of highly hydrophobic proteins including proteins from plasma membranes of the liver. Recently, Gustavsson and Larsson have shown that a continuous agarose bed can also be used for immobilization of high-molecular-weight ligands [86].

Early uses of membranes with immobilized Protein-A or -G are summarized in a review by Langlotz and Kroner [110]. Later investigations revealed that some of these supports, specifically the membranes, could not be recommended for such immobilizations and separations because of their low capacity and/or selectivity [78]. Some products such as MemSep cartridges [19] have most likely disappeared from the market for this reason.

In contrast, some other membranes and, in particular monoliths with immobilized Proteins-A and -G, have held their ground and are now the supports of choice for very fast and effective separations and purifications of both monoclonal and polyclonal antibodies [40,99]. Zhou *et al.* [41,111,112] prepared a composite membrane using cellulose fibers grafted with acrylic polymers dispersed in glycidyl methacrylate and polymerized. Protein A was immobilized on this membrane that had then been cut into pieces and packed in a column to enable the isolation of IgG from human serum in less than 30 s [112].

In another interesting study carried out by the same group, a monolithic support with immobilized Protein A was used as a model for immune adsorption therapy by means of hemoperfusion [41,113]. A tangential-flow unit shown in Fig. 18.14 was used in order to facilitate the flow through the support, and IgG and immune complexes were removed from blood and plasma. Experiments *in vitro* and *in vivo* with an animal model confirmed that the Protein A monolith adsorbed mainly IgG and only small amounts of other plasma proteins. According to the authors, the extra-corporeal circulation system proved to be safe and reliable [113].

Obviously, there is still a long way to go from these first successful experiments to the actual application for treatment of autoimmune diseases. First of all, the risk of leakage of a potentially toxic ligand such as Protein A from the support, as well as the possibility that it might reach the blood stream, has to be completely excluded. In addition, the non-specific interactions of the support with components of the blood plasma have not been targeted in these experiments, and cannot be ruled out either [114]. Such interactions can possibly activate the clotting cascade and cause fatal side effects.

The isolation of antibodies by affinity chromatography using Protein-A or -G ligands always carries within a certain risk of denaturation since the elution occurs under rather harsh conditions. Fast separations on monoliths that can be achieved

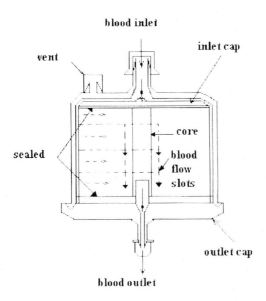

Fig. 18.14. Schematic diagram of tangential-flow chromatographic cartridge with a diameter of 70 mm and a height of 50 mm containing monolith with immobilized Protein A used for therapeutic extra corporeal adsorption of immunoglobulins. (Reprinted with permission from ref. [113], copyright 1999 Elsevier Sciences B. V.)

within several seconds certainly minimize this risk. An alternative approach which has been described above involves the use of ligands from a combinatorial library. Since none of the bacterial proteins used so far is able to bind IgM or IgA specifically, ligands produced by means of combinatorial chemistry appear to offer an ideal solution to this problem [90,105,115].

Multi-dimensional chromatography has been used specifically for the isolation of antibodies. Since the monolithic disks exhibit only a low back-pressure, several units with different ligands can be stacked in a single cartridge. Using this approach, Josic *et al.* [57] have isolated monospecific, polyclonal antibodies against the calcium-binding protein annexin VI. The antibodies that cross-react with similar proteins of the annexin family were removed from the antiserum by passing it through the disk with the corresponding immobilized antigens – low-molecular-weight annexins.

Using immobilized synthetic peptides, Ostryanina *et al.* [116] have isolated the corresponding monospecific polyclonal immunoglobulins from human serum. For this purpose, several stacked disks with different peptide ligands were used, which had been held in a single housing. This allowed the adsorption of various monospecific antibodies in a single step. Fig. 18.15 shows another example of multi-dimensional chromatography using monolithic disks for the isolation of immunoglobulins.

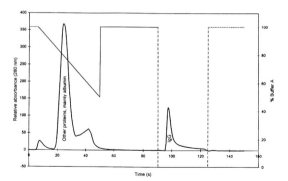

Fig. 18.15. Separation of IgG from other proteins in mouse ascitis fluid using two-dimensional chromatography: 1 ml of mouse ascites fluid was applied onto a monolithic anion-exchange QA CIM disk and Protein A disk in a tandem. The major portion of the proteins binds to the QA disk, while IgG is captured by the Protein A disk. The accompanying proteins are eluted from the QA disk with a salt gradient. In a subsequent step the IgG is recovered from the Protein A disk. (Reprinted with permission from ref. [80], copyright 1998 Elsevier Sciences B.V.)

Monoclonal antibodies from ascites fluid were purified in a device with tandem configuration, consisting of an anion-exchange disk and a disk with immobilized Protein A. Branovic *et al.* [56,117] use a tandem consisting of a Protein G-CIM disk and an anion-exchange disk in order to detect traces of accompanying proteins such as albumin and transferrin in the therapeutic IgG. Whereas the IgG binds to the Protein G-disk in front, albumin and transferrin bind to the following anion-exchange disk. In this way, all traces of the accompanying proteins are concentrated, eluted, and detected in SDS-PAGE [56,117].

The isolated antibodies can then be immobilized and used for the isolation of the corresponding antigen. This applies equally to both monoliths and beaded supports, so that the method used for immobilization of the antibodies, *i.e.*, their exposure on the surface of the support, may be critical. Monoliths bearing epoxide functionalities were often used for the immobilization of protein ligands [57,79,99]. These monoliths can also be modified to include spacers of different compositions and length in order to enhance the exposure of ligand molecules and binding of the antigens [80,83]. Another option is to bind the antibody via the Fc-part to the Protein A immobilized on a monolith, and to attach it covalently by means of a crosslinker [118]. This warrants the desired exposure of the antigen-binding part of the immunoglobulin. However, simple immobilization of antibodies on an epoxy-activated monolith was sufficient for the majority of successful applications [99,79]. In all cases, it was possible to isolate the respective antigens even from very complex biological matrices within a very short period of time.

Schuster *et al.* [119] compared the characteristics of antibodies immobilized on both monoliths and glass beads containing gigapores using immunoaffinity chromatography exploiting the Ca(II)-dependent interaction of anti-Flag antibody and Flag--tagged proteins [120]. They found that both supports were suited for high-throughput immunoaffinity chromatography. The isolation of the antigen was achieved in less than 2 min. For both supports, the efficiency of separation was independent of the flow-rate. However, in contrast to the anti-flag monolithic disk, the dynamic binding capacity for the porous glass beads depended on the flow-rate. The reason for this may be the contribution of the convective mass transport combined with minimal diffusion [40].

18.5 CONCLUSIONS

After emerging in the late 1980s, protein- and peptide chromatography on monoliths and membranes required almost a decade to consolidate. During this period, a number of technical problems was solved and significant progress has been made in terms of polymerization, surface chemistry, and immobilization of the ligands [31,74,83]. The development of this new technology has followed mainly these lines: (i), Miniaturization and subsequent use of the separation units, mainly monoliths, for fast in-process analysis [29,99,121], micro-preparative work [76], and micro-chromatography [74,75] during protein isolation and characterization and, (ii), scaling-up and use of monolithic separation units, typically membranes and large-scale monoliths, for preparative purification of biopolymers from complex biological mixtures [31,86,120].

The miniaturization of monolithic supports and their use for fast analytical and micro-preparative separations has already been brought close to perfection [74-76]. In contrast, the preparative use of such supports on a large scale still suffers from some serious problems. However, considerable progress toward a solution to this problem has also been made recently [31,53,86]. The applications of these systems have confirmed that they are suitable tools for the fast and selective concentration of biopolymers from diluted complex solutions.

18.6 REFERENCES

1   H. Chen, C. Horváth, J. Chromatogr. A 705 (1995) 3.
2   C.R. Lowe, Curr. Opinion Biotechnol. 7 (1996) 1.
3   S.D. Patterson, Curr. Opinion Biotechnol. 11 (2000) 413.
4   M. Nachman, J. Chromatogr. 597 (1992) 167.
5   K.K. Unger, G. Jilge, J.N. Kinkel, M.T.W. Hearn, J. Chromatogr. 359 (1986) 61.

6   D.J. Burke, J.K. Duncan, L.C. Dunn, L. Cummings, C.J. Siebert, G.S. Ott, J. Chromatogr. 353 (1986) 425.

7   K. Kalghatgi, C. Horváth, J. Chromatogr. 398 (1987) 335.

8   T. Hashimoto, J. Chromatogr. 544 (1991) 257.

9   N.B. Afeyan, S.P. Fulton, F.E. Regnier, LC-GC, 9 (1991) 824.

10   S.P. Fulton, N.B. Afeyan, N.F. Gordon, F.E. Regnier, J. Chromatogr. 547 (1991) 452.

11   P. Gagnon, E. Grund, T. Lindbäck, BioPharm April (1995) 21.

12   J. Horvath, E. Boschetti, L. Guerrier, N. Cooke, J. Chromatogr. A 679 (1994) 11.

13   K. Scott, Membrane Separation Technology, STI, Oxford, 1990.

14   S. Brandt, R.A. Goffe, S.B. Kessler, J.L. O'Connor, S.E. Zale, Bio/Technology 6 (1988) 779.

15   T.B. Tennikova, B.G.Belenkii, F. Svec, J. Liq. Chromatogr. 13 (1990) 63.

16   D. Josic, J. Reusch, K. Löster, O. Baum, W. Reutter, J. Chromatogr. 590 (1992) 59.

17   F. Svec, M. Jelinková, E. Votavová, Angew. Makromol. Chem. 188 (1991) 167.

18   B. Champluvier, M.R. Kula, Bioseparation 2 (1992) 343.

19   J.A. Gerstner, R. Hamilton, S.M. Cramer, J. Chromatogr. 596 (1992) 173.

20   B.G. Belenkii, V.G. Malt'sev, BioTechniques 18 (1995) 288.

21   F. Svec, J.M.J. Fréchet, Ind. Eng. Chem. Res. 38 (1999) 34.

22   R. Mandaro, S. Roy, K. Hou, Bio/Technology 5 (1987) 928.

23   F.D. Menozzi, P. Vanderpoorten, C. Defaiffe, A.O.A. Miller, J. Immunol. Methods 99 (1987) 229.

24   M. Unarska, P.A. Davies, M.P. Esnouf, B.J. Bellhouse, J. Chromatogr. 519 (1990) 53.

25   H. Abou-Rebyeh, F. Körber, K. Schubert-Rehberg, J. Reusch, D. Josic, J. Chromatogr. 566 (1991) 341.

26   S. Hjertén, J.-L. Liao, Z. Rong, J. Chromatogr. 473 (1989) 273.

27   S. Hjertén, Y.-M. Li, J. Liao, J. Mohammad, K. Nakazato, G. Pettersson, Nature 356 (1992) 810.

28   T.B. Tennikova, M. Bleha, F. Svec, T.V. Almazova, B.G. Belenkii, J. Chromatogr. 555 (1991) 97.

29   A. Strancar, M. Barut, A. Podgornik, P. Koselj, D. Josic, A. Buchacher, LC-GC 11 (1998) 660.

30   A. Strancar, P. Koselj, H. Schwinn, D. Josic, Anal. Chem. 68 (1996) 3483.

31   A. Podgornik, M. Barut, A. Strancar, D. Josic, T. Koloini, Anal. Chem. 72 (2000) 5693.

32   F. Svec, J.M.J. Fréchet, Anal. Chem. 64 (1992) 820.

33   Q.C. Wang, F. Svec, J.M.J. Fréchet, Anal. Chem. 65 (1993) 2243.

34   F. Svec, J.M.J. Fréchet, J. Chromatogr. A 702 (1995) 89.

35   S. Xie, F. Svec, J.M.J. Fréchet, Biotech. Bioeng. 62 (1999) 30.

36   T.B. Tennikova, R. Freitag, J. High. Res. Chrom. 23 (2000) 3.

37   F. Svec, E.C. Peters, D. Sýkora, J.M.J. Fréchet, J. Chromatogr. A 887 (2000) 3.

38   F.E. Regnier, J. High Resol. Chromatogr. 23 (2000) 19.

39   J. Gusev, X. Huang, C. Horváth, J. Chromatogr. A 855 (1999) 273.

40   D. Josic, A. Buchacher, A. Jungbauer, J. Chromatogr. B Biomed. Appl. 752 (2001) 191.

41   H. Zou, Q. Luo, D. Zou, J. Biochem. Biophys. Methods 49 (2001) 199.

42   R. Giovannini, R. Freitag, T.B. Tennikova, Anal. Chem. 70 (1998) 3348.

43 K. Branovic, D. Forcic, M. Santek, T. Kosutic-Gulija, R. Zgorelec, R. Mazuran, A. Trescec, B. Benko, Poster P023 presented at the 20th Int. Symp. on the Separation and Analysis of Proteins, Peptides and Polynucleotides ISPPP'00, Nov. 2000, Ljubljana, Slovenia.

44 H. Minakuchi, K. Nakanishi, N. Soga, N. Ishuzika, N. Tanaka, Anal. Chem. 68 (1996) 3498.

45 H. Minakuchi, K. Nakanishi, N. Soga, N. Ishizuka, N. Tanaka, J. Chromatogr. A 797 (1998) 121.

46 M. Schulte, D. Lubda, A. Delp, J. Dingenen, J. High Resol. Chrom. 23 (2000) 100.

47 F.C. Leinweber, K. Cabrera, D. Lubda, U. Tallarek, Poster presented at Intern. Symp. HPLC 2001, Maastricht, Holland.

48 G. Iberer, R. Hahn, A. Jungbauer, LC-GC 17 (1999) 998.

49 F. Svec, J.M.J. Fréchet, J. Chromatogr. A 702 (1995) 89.

50 F. Svec, J.M.J. Fréchet, Science 273 (1996) 205.

51 J. Mihelic, T. Koloini, A. Podgornik, A. Strancar, J. High Resol. Chromatogr. 23 (2000) 39.

52 D. Petsch, P. Rantze, F.B. Anspach, J. Molec. Recognit. 11 (1998) 222.

53 M.A. Teeters, T.W. Root, E.N. Lightfoot, J. Chromatogr. A 944 (2002) 129.

54 A. Strancar, M. Barut, A. Podgornik, P. Koselj, H. Schwinn, P. Raspor, D. Josic, J. Chromatogr. A 760 (1997) 117.

55 K. Branovic, A. Buchacher, M. Barut, A. Strancar, D. Josic, J. Chromatogr. A 903 (2000) 21.

56 K. Branovic, PhD Thesis, University of Zagreb, Croatia, 2001.

57 D. Josic, Y.-P. Lim, A. Strancar, W. Reutter, J. Chromatogr. B Biomed. Appl. 662 (1994) 217.

58 H. Podgornik, A. Podgornik, A. Perdih, Anal. Biochem. 272 (1999) 43.

59 H. Podgornik, A. Podgornik, P. Milavec, A. Perdih, J. Biotechnol. 857 (1999) 137.

60 G. Garke, J. Radtschenko, F.B. Anspach, J. Chromatogr. A 857 (1999) 137.

61 F. Svec, T.B. Tennikova, J. Bioact. Compat. Polym. 6 (1991) 393.

62 J. Luksa, V. Menart, S. Milicic, B. Kus, V. Gobec-Porekar, D. Josic, J. Chromatogr. A 661 (1994) 161.

63 S. Xie, F. Svec, J.M.J. Fréchet, J. Chromatogr. A 775 (1997) 65.

64 T.B. Tennikova, F. Svec, J. Chromatogr. 646 (1993) 279.

65 S. Hjertén, K. Nakazato, J. Mohammad, D. Eaker, Chromatographia 37 (1993) 287.

66 M. Petro, F. Svec, J.M.J. Fréchet, J. Chromatogr. A 752 (1996) 59.

67 S. Xie, R.W. Allington, F. Svec, J.M.J. Fréchet, J. Chromatogr. A 865 (1999) 169.

68 F. Svec, J.M.J. Fréchet, Macromol. Symp. 110 (1996) 203.

69 B. Mayr, R. Tessadri, E. Post, M.R. Buchmeister, Anal. Chem. 73 (2001) 4071.

70 M.R. Buchmeister, J. Chromatogr. A 918 (2001) 233.

71 J.L. Liao, Y.M. Li, S. Hjertén, Anal. Biochem. 234 (1996) 195.

72 C. Ericson, S. Hjertén, Anal. Chem. 71 (1999) 1621.

73 C. Ericson, J.-L. Liao, K. Nakazato, S. Hjertén, J. Chromatogr. A 767 (1997) 33.

74 F. Svec, E.C. Peters, D. Sykora, C. Yu, J.M.J. Fréchet, J. High Resol. Chromatogr. 23 (2000) 3.

75 C. Yu, F. Svec, J.M.J. Fréchet, Electrophoresis 21 (2000) 120.

76 R.E. Moore, L. Licklider, D. Schumann, T.D. Lee, Anal. Chem. 70 (1998) 4879.

77   T. Tennikova, R. Freitag in: H.Y. Aboul-Enein (Ed.) Analytical and Preparative Separation Methods of Biomolecules, Marcel Dekker, New York, 1999, pp. 255-300.

78   A. Strancar, Ph.D. Thesis, University of Ljubljana, Slovenia, 1997.

79   C. Kasper, L. Meringova, R. Freitag, T. Tennikova, J. Chromatogr. 798 (1998) 65.

80   D. Josic, H. Schwinn, A. Strancar, A. Podgornik, M. Barut, Y.-P. Lim, M. Vodopivec, J. Chromatogr. A 803 (1998) 61.

81   L.G. Berruex, R. Freitag, T.B. Tennikova, J. Pharm. Biomed. Anal. 24 (2000) 95.

82   D. Josic, A. Buchacher, J. Biochem. Biophys. Methods 49 (2001) 153.

83   R. Hahn, K. Amatschek, R. Schallaun, D. Josic, A. Jungbauer, Int. J. BioChromatogr. 5 (2000) 175.

84   E. Klein, J. Membr. Sci. 179 (2000) 1.

85   R. Hahn, A. Podgornik, M. Merhar, E. Schallaun, A. Jungbauer, Anal. Chem. 73 (2001) 5126.

86   P.-E. Gustavsson, P.-O. Larsson, J. Chromatogr. A 925 (2001) 69.

87   X. Sun, Z. Chai, J. Chromatogr. A 943 (2002) 209.

88   V. Gaberc-Porekar, V. Menart, J. Biochem. Biophys. Methods 49 (2001) 335.

89   P.Y. Huang, G.A. Baumbach, C.A. Dadd, J.A. Bueltner, B.L. Masecar, M. Hentsch, D.J. Hammond, R.G. Carbonell, Bioorg. Med. Chem. 4 (1996) 699.

90   G. Fassina, M. Ruvo, G. Palombo, A. Verdoliva, M. Marino, J. Biochem. Biophys. Methods 49 (2001) 481.

91   K. Sproule, P. Morill, J.C. Pearson, S.J. Burton, K.R. Hejnaes, H. Valore, S. Ludvigsen, C.R. Lowe, J. Chromatogr. B 740 (2000) 17.

92   G. Wulff, Angew. Chem. Int. Ed. Engl. 34 (1995) 1812.

93   T. Takeuchi, D. Fukuma, J. Matsui, Anal. Chem. 71 (1999) 285.

94   T. Takeuchi, J. Matsui, J. High Resol. Chromatogr. 23 (2000)44.

95   T. Takeuchi, J. Haginaka, J. Chromatogr. B 728 (1999) 1.

96   O.W. Reif, V. Nier, U. Bahr, R. Freitag, J. Chromatogr. A 664 (1994) 13.

97   M.Y. Arica, H.N. Testereci, A. Denizli, J. Chromatogr. A 799 (1999) 83.

98   J. Crawford, S. Ramakrishnan, P. Periera, S. Gardner, M. Coleman, R. Beitle, Sep. Sci. Technol. 34 (1999) 2793.

99   G.A. Platonova, G.A. Pankova, J.Y. Il'ina, G.P. Vlasov, T.B. Tennikova, J. Chromatogr. A 852 (1999) 129.

100   K. Amatschek, R. Necina, R. Hahn, E. Schallaun, H. Schwinn, D. Josic, A. Jungbauer, J. High Resolut. Chromatogr. 23 (2000) 47.

101   K. Pflegerl, Ph.D. Thesis, University of Agricultural Sciences, Vienna, 2001.

102   L. Yang, L. Jia, H. Zou, Y. Zhang, Biomed. Chromatogr. 13 (1999) 229.

103   Q. Luo, H. Zou, X. Xiao, Z. Guo, L. Kong, X. Mao, J. Chromatogr. A 926 (2001) 255.

104   C. Legallais, F.B. Anspach, S.M.A. Bueno, K. Haupt, M.A. Vijayalaksmi, J. Chromatogr. B 691 (1997) 33.

105   F.B. Anspach, J. Biochem. Biophys. Meth. 49 (2001) 665.

106   G. Fassina, A. Verdoliva, G. Palombo, M. Ruvo, G. Lassini, J. Mol. Recognit. 11 (1998) 128.

107   D. Josic, A. Strancar, Ind. Eng. Chem. 36 (1999) 333.

108   D. Josic, F. Bal, H. Schwinn, J. Chromatogr. 632 (1993) 1.

109   J. Hagedorn, C. Kasper, R. Freitag, T. Tennikova, J. Biotechnol. 69 (1999) 1.

110   P. Langlotz, K.H. Kroner, J. Chromatogr. 591 (1992) 107.

111   D. Zhou, H. Zhou, J. Ni, H. Wang, L. Yang, Y. Zhang, Chromatographia 50 (1999) 23.

112   D. Zhou, H. Zhou, J. Ni, L. Jia, G. Zhang, Y. Zhang, Anal. Chem. 71 (1999) 115.

113   L. Jia, L. Yang, H. Zou, Y. Zhang, J. Zhao, C. Fan, L. Sha, Biomed. Chromatogr. 13 (1999) 472.

114   D. Josic, P. Schulz, L. Biesert, L. Hoffer, H. Schwinn, M. Kordis-Krapez, A. Strancar, J. Chromatogr. B 694 (1997) 253.

115   C.R. Lowe, A.R. Lowe, G. Gupta, J. Biochem. Biophys. Meth. 49 (2001) 561.

116   N.D. Ostryanina, G.P. Vlasov, T.B. Tennikova, J. Chromatogr. A (2002) in press.

117   K. Branovic, A. Buchacher, D. Josic, M. Barut, A. Strancar, presented at the 21th Int. Symp. on the Separation and Analysis of Proteins, Peptides and Polynucleotides ISPPP'01, Nov. 2001, Orlando, Florida, USA.

118   C. Schneider, R. Newman, D. Sutherland, U. Asser, M. Greaves, J. Biol. Chem. 257 (1982) 10766.

119   M. Schuster, E. Wasserbauer, A. Neubauer, A. Jungbauer, Bioseparation 9 (2000) 259.

120   A. Einhauer, A. Jungbauer, J. Biochem. Biophys. Meth. 49 (2001) 455.

121   M. Petro, F. Svec, J.M.J. Fréchet, Biotech. Bioeng. 49 (1996) 355.

F. Švec, T.B. Tennikova and Z. Deyl (Editors)
Monolithic Materials
Journal of Chromatography Library, Vol. 67
© 2003 Elsevier Science B.V. All rights reserved.

Chapter 19

Nucleic Acid Analysis

Christian G. HUBER and Herbert OBERACHER

Institute of Analytical Chemistry and Radiochemistry, Leopold-Franzens-University, Innrain 52 a, A-6020 Innsbruck, Austria

CONTENTS

19.1 INTRODUCTION

Fractionation, purification, quantitation, and structural analysis of nucleic acids are of utmost importance for numerous applications in the biological and medical sciences. By virtue of its high resolving capability, short cycle times, full automation, and preparative separation capability, liquid chromatography, and especially high-performance liquid chromatography (HPLC), have emerged as very powerful tools for nucleic acid separation and analysis [1,2]. For the efficient chromatographic separation of biological macromolecules having low diffusivities such as nucleic acids the availability of stationary phases with favorable mass transfer properties constitutes a prime prerequisite [3].

Although stationary phases based on microparticles have been successfully utilized as separation media for HPLC for more than three decades [4-8], the relatively large void volume between the packed particles represents a significant factor limiting the separation efficiency of conventional granular packing materials. One means of enhancing mass transfer is the use of monolithic separation media [9,10], in which the chromatographic bed consists of a single piece of a rigid, porous polymer which has no interstitial volume but only internal porosity consisting of micro-, meso-, and macropores [10-13]. Because of the absence of intraparticular volume, all of the mobile phase is forced to flow through the pores of the separation medium [14]. According to theory, mass transport is enhanced by such convection [15-17] and has a positive effect on chromatographic efficiency.

A general problem with all kinds of porous packing materials having diffusive pores is the slow mass transfer of solutes into and out of the stagnant mobile phase present in the micro- and mesopores of the stationary phase, resulting in considerable band broadening particularly with high molecular analytes [18,19]. This drawback can be addressed by the elimination of diffusive pores, which restricts the mass transfer to a thin, retentive layer at the outer surface of the stationary phase, resulting in so-called micropellicular stationary phases [20]. A monolithic column configuration lacking diffusive micro- and mesopores may be adequately described as a micropellicular monolith [21] and has been shown to enable the separation of nucleic acids over a very broad size range with efficiencies significantly better compared to that of columns packed with micropellicular, granular stationary phases [22].

Miniaturized chromatographic separation systems applying capillary columns of 10–500 μm inner diameter are frequently the method of choice for the separation and characterization of nucleic acid mixtures, because the amount of available sample material is usually limited [23]. The concept of monolithic stationary phases is especially suitable for the fabrication of capillary columns, because the chemical immobilization of the monolith at the wall of fused silica capillaries has a positive effect on column stability and eliminates the rather difficult preparation of a frit to retain the stationary phase particles in the capillary tube [23]. Femtomol to attomol amounts of nucleic acids are separable and detectable in such miniaturized systems [24,25]. Moreover, the low flow rates ranging from a few nanoliters to microliters per minute characteristic for capillary HPLC are well suited for direct interfacing the separation process with electrospray ionization mass spectrometry (ESI-MS). Such hyphenation adds another dimension to the analytical process by providing molecular mass and structural data about the separated analytes [26].

Monolithic separation media also offer distinctive advantages for preparative scale separations of nucleic acids [27]. Highly permeable monoliths enable high percolation flow rates at low column backpressure without loss in column efficiency, resulting in fast loading and elution times [28]. Moreover, very stable and uniform chromatographic beds can be manufactured reproducibly even in large-diameter preparative columns [29,30], which is very difficult to achieve with conventional, granular packing materials. Finally, the technology is readily scalable from laboratory devices to multigram loading levels required for manufacturing [31].

The aforementioned advantages render monolithic column technology a real complement to conventional separation columns packed with granular stationary phase materials. The following sections introduce the separation modes, column configurations, and packing materials that have been realized so far with monolithic separation media for the fractionation and analysis of nucleic acids. Examples of application illustrate the potential of monolithic materials in nucleic acid separations and the challenges that have to be met in order to further improve the performance and applicability of this kind of separation media in biochromatography.

19.2 LIQUID CHROMATOGRAPHIC SEPARATION SYSTEMS FOR NUCLEIC ACIDS

Six major chromatographic modes are utilized for the separation of nucleic acid mixtures [32]: size-exclusion chromatography [33], anion-exchange HPLC [5,6], mixed-mode HPLC [34], reversed-phase HPLC, ion-pair reversed-phase HPLC [35], and affinity chromatography [36,37]. It is generally accepted that the non-interactive modes of chromatographic separation do not exhibit the same level of resolution as the interactive modes, which exploit selective interactions between a nucleic acid molecule and the surface of a stationary phase [1]. Hence, the non-interactive modes such as size-exclusion chromatography and slalom chromatography, based solely on differences in size, have seen only limited applications with nucleic acids, while the interactive modes, including anion-exchange HPLC and ion-pair reversed-phase HPLC have been the most successful techniques so far. Affinity chromatography, despite its highly selective interactions, represents a chromatographic mode that is not very frequently applied for nucleic acid separations, mainly because of the need to prepare specifically designed stationary phases for every new separation problem.

19.2.1 Ion-exchange chromatography

19.2.1.1 Separation principle

Since nucleic acid molecules are polyanions in aqueous solution, anion-exchange chromatography is the most obvious interactive mode for their separation. An anion-exchange stationary phase consists of positively charged functional groups immobilized onto a suitable support material (Fig. 19.1a). The electrostatic interactions between the positively charged groups on the surface of the stationary phase and the negatively charged analytes are responsible for adsorption. Desorption takes place

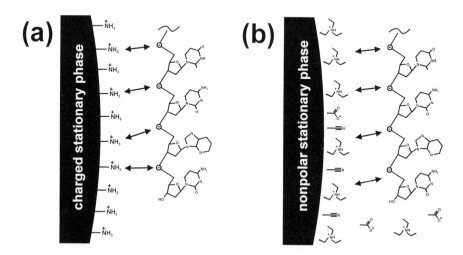

Fig. 19.1. Schematic illustrations of chromatographic retention mechanisms suitable for the separation of nucleic acids. (a) Electrostatic interactions typical of ion-exchange HPLC; (b) electrostatic interactions characteristic of ion-pair reversed-phase HPLC.

when the ionic strength of the mobile phase is increased by applying a gradient of increasing salt concentration. The phosphodiester groups are fully dissociated over a broad pH range (2–14) and the number of negative charges is proportional to the length of DNA. Therefore, the salt concentration required to desorb nucleic acids from an anion-exchange stationary phase increases with the size of the analytes, resulting in a primarily size-dependent separation. However, other mechanisms may play a role in determining the retention, including size-exclusion, hydrogen-bonding, and solvophobic interactions, which, to a certain degree, make retention also sequence-dependent [38,39].

19.2.1.2 Separation of single-stranded oligonucleotides

Fig. 19.2a demonstrates the separation of a ladder of 5′-phosphorylated oligodeoxythymidylic acids [p(dT)$_{12-24}$] by anion-exchange HPLC in a poly(glycidylmethacrylate-*co*-ethylene dimethacrylate) monolith, which has been modified with diethylamine to yield tertiary-amine, weak anion-exchange functionalities [40]. The 13 oligomers present in the mixture were separated almost to baseline with a gradient of 0.35–0.48 mol/l sodium chloride in 16 mmol/l phosphate buffer, pH 7.0, in 90 min. Acetonitrile was added to the mobile phase in order to suppress solvophobic interactions of the oligodeoxynucleotides with the stationary phase, which results in sharper peak shapes and reduced dependency of retention on base composition [38]. In the same study, an increase in the flow-rate from 1 ml/min to 4 ml/min with a concomitant decrease in gradient time from 90 to 22.5 min, essentially keeping the gradient volume constant, enabled the separation of p(dT)$_{12-18}$ with equivalent resolution in one fourth of the time. This result revealed that the separation power in monolithic chromatographic beds is relatively invariant with linear flow velocity.

The influence of the degree of derivatization on chromatographic performance can be deduced from a comparison of the separation of p(dT)$_{12-24}$ in monoliths modified with 1.26 (Fig. 19.2a), 0.67 (Fig. 19.2b), and 0.27 mmol/g 1-N,N-diethylamino-2-hydroxypropyl groups (Fig. 19.2c), respectively. While the two separations in a and b are equivalent, a lower salt concentration is necessary to elute the oligodeoxynucleotides from the column with lower surface charge density. A further reduction in the number of functional groups not only reduced the retentivity of the stationary phase, but also significantly deteriorated the separation efficiency (c). Obviously, this degree of functionalization does not afford sufficient surface coverage of interactive sites required to achieve a good separation.

The separation of oligodeoxynucleotides in the isocratic anion-exchange mode in thin poly(glycidyl methacrylate-*co*-ethylene dimethacrylate) disks modified with diethylaminoethyl groups using an eluent which contained 0.5 mol/l sodium chloride

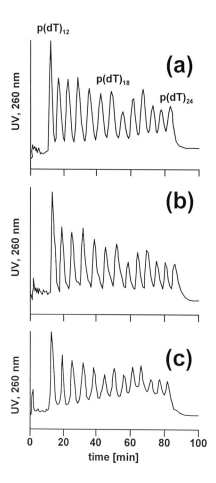

Fig. 19.2. Separation of oligodeoxythymidylic acids on three different 3-diethylamino-2-hydroxypropyl functionalized columns (Reprinted with permission from ref. [40], copyright 1999 Elsevier Sciences B. V.). Monolithic poly(glycidyl methacrylate-*co*-ethylene dimethacrylate) column, 50 × 8 mm i.d.; mobile phase, (A) 16 mmol/l phosphate buffer, pH 7.0, 20% acetonitrile, (B) 16 mmol/l phosphate buffer, pH 7.0, 20% acetonitrile, 1.0 mol/l sodium chloride; linear gradient, (a) 35–48% B in 90 min, (b) 28–41% B in 90 min, (c) 15–25% B in 90 min; flow-rate, 1 ml/min; detection, UV, 260 nm; sample, p(dT)$_{12-24}$

is illustrated in Fig. 19.3 [41]. Different column lengths were obtained through the combination of several monolithic disks, which proves that the stacking of several monolithic layers is a simple and appropriate means to vary the length of the separation bed. The separation improved considerably and retention times of oligodeoxynucleotides increased proportionally with increasing monolith thickness (Fig. 19.3a-c).

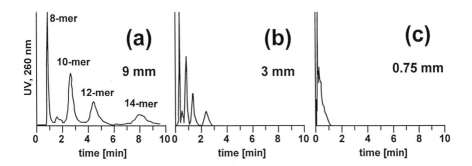

Fig. 19.3. The effect of column length on the separation of oligonucleotides (Reprinted with permission from ref.[41], copyright 1999 Elsevier Sciences B. V.). CIM DEAE disk, (a) 0.75, (b) 3, (c) 9 × 12 mm i.d.; mobile phase, 20 mmol/l Tris HCl buffer, pH 8.5, 0.5 mol/l sodium chloride; flow-rate, 3.0 ml/min; detection, UV, 260 nm; sample, 8-mer, 10-mer, 12-mer, 14-mer, 1 μg of each oligodeoxynucleotide.

This finding is indicative for a multistep adsorption-desorption process effective under the isocratic elution conditions.

Although a satisfactory separation of the four oligomers was not feasible in the 0.75 mm thick disk under the chosen isocratic elution conditions (Fig. 19.3a), they were separable upon a reduction in the ionic strength of the mobile phase from 0.5 to 0.37 mol/l sodium chloride. This rather surprising result demonstrated that multiple adsorption/desorption could be applied for isocratic separation of oligonucleotides in very thin monolithic column beds. The tortuous traveling through the highly interconnected channels of the monolith results in a real path length significantly longer than the physical thickness of the column bed. Moreover, both enhanced mass transfer and absence of micropores facilitate a higher number of rapid adsorption/desorption processes in monoliths compared to packed column beds consisting of microporous materials.

19.2.1.3 Separation of double-stranded DNA

Very few applications report the use of monolithic columns for the anion-exchange HPLC of large nucleic acids such as double-stranded DNA fragments [42,43]. For example, genomic DNA and plasmids were successfully chromatographed on monolithic anion-exchangers [44,45]. One reason for this situation may be the necessity to adapt the pore structure of monolithic materials to the steric requirements of large DNA molecules occurring as rather flexible random-coils in solution in order to achieve highly efficient separations. Future developments can be expected to expand the range of applications for large nucleic acid molecules.

19.2.2 Ion-pair reversed-phase chromatography

19.2.2.1 Separation principle

In ion-pair reversed-phase HPLC, the phase system comprises a hydrophobic stationary phase and a hydroorganic mobile phase modified with an ion-pair reagent which consists of an amphiphilic ion carrying both charge and hydrophobic groups and a small, hydrophilic counterion. Among the various trialkyl- and tetraalkylammonium salts, triethylammonium acetate represents the most commonly applied ion-pair reagent in ion-pair reversed-phase HPLC of nucleic acids. According to the electrostatic retention model [46], the positively charged, hydrophobic triethylammonium ions are adsorbed onto the nonpolar surface of the stationary phase, resulting in the formation of an electric double layer having an excess of positive charges near the surface. Fig. 19.1b presents an idealized view of the electrostatic interactions of a negatively charged nucleic acid molecule with the surface potential generated by triethylammonium ions at the surface of a hydrophobic support material.

In ion-pair reversed-phase HPLC, the magnitude of electrostatic interaction, and thus retention is determined by several factors including hydrophobicity of the column packing, charge, hydrophobicity and concentration of the pairing ion, ionic strength, temperature and dielectric constant of the mobile phase, concentration of organic modifier (see Equation 10 in [46]), and charge and size of the nucleic acid molecule. Elution of the adsorbed nucleic acids is effected by a decrease in the surface potential due to desorption of the amphiphilic ions from the stationary phase with a gradient of increasing organic modifier concentration. Because the number of charges uniformly increases with size, double-stranded DNA molecules are separated according to chain length in ion-pair reversed-phase HPLC [47].

Additional solvophobic interactions between hydrophobic regions of the solutes such as nucleobases, and the hydrophobic surface of the stationary phase are possible in ion-pair reversed-phase HPLC. This kind of interaction contributes considerably to the sequence dependence of retention with single-stranded oligodeoxynucleotides [48,49], RNA, and partially denatured, double-stranded DNA [50,51].

19.2.2.2 Separation of single-stranded oligonucleotides

Fig. 19.4 demonstrates the high resolving power of micropellicular, monolithic capillary columns for the separation of single-stranded oligonucleotides by ion-pair reversed-phase HPLC. The hydrophobic monolithic material is a poly(styrene-*co*-divinylbenzene) copolymer, which has been prepared by radical polymerization using tetrahydrofuran and decanol as porogenic solvents [52]. The sample was generated by

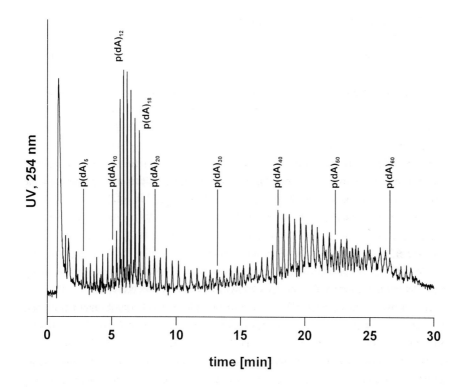

Fig. 19.4. High-resolution capillary ion-pair reversed-phase HPLC separation of a mixture of 120 phosphorylated and dephosphorylated deoxyadenylic acids (Reprinted with permission from ref. [22], copyright 2002 Elsevier Sciences B. V.). Monolithic PS-DVB column, 60 × 0.20 mm i.d.; mobile phase, (A) 100 mmol/l triethylammonium acetate, pH 7.00, (B) 100 mmol/l triethylammonium acetate, pH 7.00, 20% acetonitrile; linear gradient, 5-35% B in 5.0 min, 35–40% B in 5.0 min, 40–45% B in 6.0 min, 45–52% B in 14.0 min; flow-rate, 2.1 l/min; temperature, 50°C; detection, UV, 254 nm; sample, hydrolyzed $p(dA)_{40}$–$p(dA)_{60}$ spiked with 2.5 ng $p(dA)_{12}$–$p(dA)_{18}$.

partial hydrolysis of a commercially available ladder of phosphorylated oligodeoxyadenylic acids (40–60-mers, $p(dA)_{40-60}$), which gave two series of either phosphorylated or dephosphorylated oligonucleotides that may range in size from 1 to 60 nucleotide units. For correct peak assignment the hydrolysate was spiked with $p(dA)_{12-18}$.

The analytes were separated into 98 peaks within 28 min, which meant elution of one peak every 17 s. The peak widths at half height ranged from 1.5–3.2 s for the oligomers up to the 18-mer and did not exceed 11.2 s for oligomers up to the 60-mer. The increase in peak width with increasing size of the oligomers is a consequence of the decreasing gradient steepness that was necessary to ensure adequate resolution of the longer oligonucleotides. From Fig. 19.4 it can be deduced that the whole series both of phosphorylated and dephosphorylated oligodeoxyadenylic acids could be re-solved to baseline with single-nucleotide resolution. However, occasional overlapping of the signals of both series resulted in only partial resolution or coelution of some of the oligonucleotides, which explains why only 98 peaks instead of the maximum of 120 could be observed in the chromatogram. As already found previously, the dephos-phorylated oligomers eluted later than the phosphorylated analogs of the same length [38]. The separation efficiency observed for monoliths clearly surpasses that obtain-able on columns packed with micropellicular C18 functionalized poly(styrene-*co*-divinylbenzene) particles, where baseline resolution of dephosphorylated oligode-oxyadenylic acids was possible up to the 50-mers (see Fig. 7 in ref. [38]).

As shown elsewhere in this book, ring-opening metathesis polymerization has been recently shown to be a very useful polymerization technique to synthesize well-de-fined non-functionalized as well as functionalized monoliths [53]. The hydrophobic monoliths are prepared from norborn-2-ene and 1,4,4a,5,8,8a-hexahydro-1.4.5.8-*exo,endo*-dimethanonaphthalene by copolymerization in the presence of a transition metal catalyst and toluene and 2-propanol as porogens. A representative chroma-togram of the separation of p(dT)$_{12-18}$ by ion-pair reversed-phase HPLC in a norbor-nene-based monolith is shown in Fig. 19.5, in which the seven oligodeoxythymidylic acids were separated with a gradient of 11–16% acetonitrile in 0.1 mol/l triethylam-monium acetate.

The microstructure of the norbornene-based monoliths could be optimized in a reproducible way upon variation of the polymerization conditions, including monomer and porogen stoichiometry and additional phosphine concentration [54]. A study of these effects was carried out with different monoliths, in which the weight ratio of 1:1 for norbornene to hexahydrodimethanonaphthalene was kept constant, while the (nor-bornene + hexahydrodimethanonaphthalene) to 2-propanol ratio was varied from 30:60 to 50:40 [55]. Although the differences in the structures revealed by electron microscopy were insignificant, an increase in the (norbornene + hexahydrodi-methanonaphthalene) to 2-propanol ratio entailed significantly enhanced chroma-tographic retention and resolution.

In a second series of experiments, the influence of amount of crosslinker was elucidated. The best oligonucleotide separation was obtained on a monolith with a 1:1 ratio of norbornene to hexahydrodimethanonaphthalene, whereas higher or lower con-tents of crosslinker lead to a decrease in chromatographic resolution. The addition of triphenylphosphine had a dramatic effect on the microstructure of norbornene-based monoliths. The presence of even small amounts of phosphine (20–80 μg/g) lead to a reduction in the volume fraction of pores and pore volume, an increase in the mean

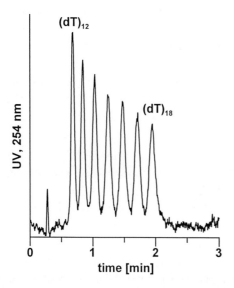

Fig. 19.5. Ion-pair reversed-phase HPLC separation of a oligodeoxynucleotide mixture (Reprinted with permission from ref. [55], copyright 2001 American Chemical Society). Monolithic poly(norbornene-*co*-hexahydrodimethanonaphthalene) column, 100 × 3 mm i.d.; mobile phase, (A) 100 mmol/l triethylammonium acetate, pH 7.00, (B) 100 mmol/l triethylammonium acetate, pH 7.00, 20% acetonitrile; linear gradient, 55–80% B in 10.0 min; flow-rate, 2.0 ml/min; temperature, 20 °C; detection, UV, 264 nm; sample, p(dT)12–p(dT)18 0.1 µg each.

microglobule diameter with a concomitant decrease in resolving power of the synthesized monoliths.

19.2.2.3 Separation of double-stranded DNA

Ion-pair reversed-phase HPLC has been shown to be efficient not only for the rapid separation of single stranded oligodeoxynucleotides, but also for the fractionation of double-stranded DNA fragments up to chain lengths of 2000 base pairs [47]. The applicability of monolithic stationary phases to ion-pair reversed-phase HPLC of double stranded DNA was evaluated by the analysis of DNA restriction fragments ranging in size from 34–622 base pairs [22]. Fig. 19.6 compares the separation of 30 DNA restriction fragments using both a 50×0.2 mm i.d. monolithic capillary column and a conventional 50×4.6 mm i.d. column packed with octadecylated poly(styrene-*co*-divinylbenzene) microparticles. The separation efficiency of both columns is equivalent allowing the separation of DNA fragments that differ less than 5% in size.

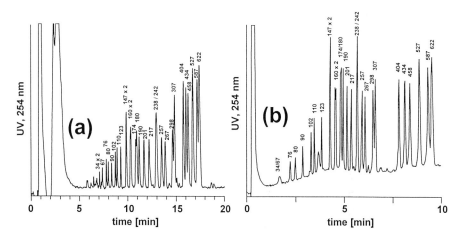

Fig. 19.6. Comparison of high-resolution ion-pair reversed-phase HPLC of double-stranded DNA fragments obtained from *Hae*III and *Msp*I digests of pUC18 and pBR322, respectively, using either a 0.2 mm i.d. monolithic capillary column or a conventional 4.6 mm i.d. column packed with beads (Reprinted with permission from ref. [77], copyright 2001 Elsevier Sciences B. V.). Conditions in (a): monolithic PS-DVB column, 50 × 0.2 mm i.d.; mobile phase, (A) 100 mmol/l triethylammonium acetate, 0.1 mmol/l Na$_4$EDTA, pH 7.0, (B) 100 mmol/l triethylammonium acetate, 0.1 mmol/l Na$_4$EDTA, pH 7.0, 25% acetonitrile; linear gradient, 30–50% B in 4 min, 50–65% B in 13 min; flow rate, 3.0 l/min; temperature, 49.7°C; detection, UV, 254 nm; injection volume, 1 μl. Conditions in (b): 50 × 4.6 mm i.d. column packed with 2 μm PS-DVB-C18 particles (DNASep™, Transgenomic); mobile phase, (A) 100 mmol/l triethylammonium acetate, 0.1 mmol/l Na$_4$EDTA, pH 7.0, (B) 100 mmol/l triethylammonium acetate, 0.1 mmol/l Na$_4$EDTA, pH 7.0, 25% acetonitrile; linear gradient, 38–56% B in 4.0 min, 56–67% B in 6.0 min; flow rate, 0.90 ml/min; temperature, 50°C; injection volume, 8 μl.

A comparison of the gradient profiles applied in Fig. 19.6 reveals that a lower concentration of acetonitrile is required to elute the DNA restriction fragments from the monolithic column. Taking into consideration a gradient delay time of 0.7 min with the conventional HPLC instrument, the DNA fragments elute at acetonitrile concentrations between 11.5 and 16.3% (Fig. 19.6a). On the other hand, with a gradient delay time of 6 min in the capillary HPLC instrument, the fragments elute between 10.0 and 14.7% acetonitrile (Fig. 19.6b). This is due to the higher polarity of the monolithic poly(styrene-*co*-divinylbenzene) stationary phase, which is not alkylated in contrast to the particles made of the same polymer that are derivatized with highly nonpolar octadecyl groups [48].

19.2.3 Column materials and formats

A number of different column formats have been applied to the separation of nucleic acids in monolithic materials. Column inner diameters and column lengths may vary from a few hundred micrometers to several centimeters. Monolithic rods, compressed beds, monolithic disks, and thin membranes described elsewhere in this book have been shown to be suitable as separation media. Anion-exchange stationary phases are usually based on hydrophilic support materials, e. g. methacrylates or methacrylamides, onto which amine-based tertiary or quarternary functional groups are immobilized, whereas stationary phases for ion-pair reversed-phase chromatography are made of hydrophobic polymers such as poly(styrene-*co*-divinylbenzene) or poly(norbornene-*co*-hexahydrodimethanonaphthalene) without further derivatization. Table 19.1 summarizes the monolithic column materials and column configurations that have been used so far for the separation of nucleic acids.

19.2.4 Detection schemes

19.2.4.1 Ultraviolet absorbance and fluorescence detection

Spectrophotometric techniques are among the most widely applied for the detection of nucleic acids. Nucleotides, the monomer units of nucleic acids, are UV active over the range of 180–300 nm, with maxima at 180–200 nm and 250–280 nm. Since most eluents used for chromatographic separation absorb light strongly below 200 nm, nucleic acids are normally detected at wavelengths between 250 and 260 nm. The limits of detection for single- and double stranded nucleic acids with UV detection at 254 nm using 200 μm i.d. capillary columns lie in the low femtomol range [23].

Fluorescence detection generally offers very low limits of detection (30 to 1000 times lower than UV/VIS detection for nucleic acids [56]) but, as nucleic acids are naturally non-fluorescent, tagging with fluorescent dyes by pre- or postcolumn derivatization is indispensable for fluorescence detection. The most common fluorophores for nucleic acids are based on fluorescein, such as 5-carboxyfluorescein (FAM), 2′,7′-dimethoxy-4′,5′-dichloro-6-carboxyfluorescein (JOE), N,N,N′,N′-tetramethyl-6-carboxy rhodamine (TAMRA), 6-carboxy-2′,4′,7′,4,7-hexachlorofluorescein (HEX), or 6-carboxy-X-rhodamine (ROX) with a general excitation maximum at approximately 260 nm and individual maxima at 494, 528, 560, 535, and 580 nm, respectively. The maxima of emission for the five fluorophores conjugated to nucleic acids are at 522, 550, 580, 556, and 605 nm, respectively. Another important fluorescent dye for nucleic acid labeling is N-1-naphthylethylenediamine (NED).

19.2.4.2 Mass spectrometric detection

The molecular mass of a nucleic acid molecule is a very important parameter for its characterization and identification. Since nucleic acids are very polar and nonvola-

TABLE 19.1

MONOLITHIC COLUMN CONFIGURATIONS AND MATERIALS USED FOR THE SEPARATION OF NUCLEIC ACIDS

Support material	Functional group	Trade name/supplier	Dimensions length (thickness) × diameter [mm]	Application	Ref.
Cellulose-membrane	Diethyl-aminoethyl	DEAE-cellulose Mem-Sep/Millipore	0.150 × 19–28 stacks of 40–60 membranes	Fractionation of bacterial lysates	[44]
Cellulose-membrane	Quaternary ammonium	Sartobind Q/ Sartorius	n. a. stacks of 3–60 membranes	Purification of antisense oligodeoxynucleotides	[31]
Poly(methacrylamide-co-piperazine diacrylamide-co-diallyldimethylam-monium chloride)	Quaternary ammonium	UNO Q1 BioRad	35 × 7 68 × 15	AEX HPLC[a] of oligo-deoxynucleotides and Plasmid DNA	[45,71]
Poly(glycidyl methacrylate-co-ethylene dimethacrylate) (GMA-EDMA)	N,N-dimethylamino-2-hydroxypropyl	—	50 × 8	AEX HPLC[a] of oligo-deoxynucleotides	[40]

Polymer	Functional group	Format	Dimensions (mm)	Application	Ref.
Poly(glycidyl methacrylate-co-ethylene dimethacrylate) (GMA-EDMA)	Quaternary ammonium	CIM-QA Disk/ BIA Separations	3 × 12, 3 × 26 (stacking of up to 4 disks is possible)	AEX HPLC[a] of Plasmid DNA	[45]
Poly(glycidyl methacrylate-co-ethylene dimethacrylate) (GMA-EDMA)	Diethyl-aminoethyl	CIM-DEAE Disk/ BIA Separations	3 × 12, 3 × 26 (stacking of up to 4 disks is possible)	AEX HPLC[a] of oligodeoxy-nucleotides	[41]
Poly(styrene-co-divinylbenzene) (PS-DVB)	None	—	30–60 × 0.2	IP-RP-HPLC[b] of oligodeoxy-nucleotides, double-stranded DNA	[22]
Poly(norbornene-co-hexahydro-dimethano-naphthalene) (NBE/DMN-H$_6$)	None	—	30–60 × 0.2, 40–60 × 8	IP-RP-HPLC[b] of oligodeoxy-nucleotides, double-stranded DNA	[55]

[a] Anion-exchange HPLC
[b] Ion-pair reversed-phase HPLC

tile molecules, the classical mass spectrometric ionization methods like electron impact or chemical ionization are not applicable. However, the development of new, soft ionization techniques, such as fast atom bombardment (FAB) [57], electrospray ionization (ESI) [58], and matrix-assisted laser desorption/ionization (MALDI) [59] has made mass spectrometry (MS) useful as a liquid chromatographic detection system for nucleic acids.

While fast atom bombardment ionization is confined to small nucleic acids of up to M_r 15,000 [60], electrospray ionization enables the transfer of nucleic acids larger than 7,000,000 [61] into the gas phase for mass analysis. Moreover, ESI-MS greatly benefits from the multiple charging of nucleic acids resulting in mass-to-charge ratios well within the mass range of most commercial mass spectrometers ($\leq$4,000–6,000 u) even for very large nucleic acid molecules. By virtue of its versatility for ionization of various biopolymers, broad mass range, high mass precision (typically 0.01%), and tolerance for high eluent flow rates, the electrospray ionization interface is the most commonly used to hyphenate liquid chromatography and mass spectrometry. Low femtomol to high attomol amounts of nucleic acids can be readily detected with ESI-MS, and thus, the limit of detection is comparable to that of UV detection [25,62]. Matrix-assisted laser desorption/ionization mass spectrometry (MALDI-MS) is frequently applied to the off-line analysis of collected fractions after chromatographic separation, although prototypes of continuous-flow matrix-assisted laser desorption/ionization interfaces have been reported [63].

In addition to information regarding the molecular mass, tandem mass spectrometry (MS/MS), utilizing collisionally induced dissociation (CID), provides valuable information about the base sequence of oligodeoxynucleotides [64]. For sequence determination by MS/MS, a precursor ion is selected from the series of multiply charged oligodeoxynucleotide ions and subsequently fragmented by energetic collisions with gas atoms. The first fragmentation step under low-energy CID conditions is usually the elimination of a nucleobase, followed by cleavage of the 3'-phosphodiester bond of the nucleotide that suffered the base loss. The resulting series of 5'-terminal and 3'-terminal ions show characteristic mass differences, from which the nucleotide sequence can be deduced manually [65,66] or with the help of computer-based algorithms [67,68].

19.3 EXAMPLES OF APPLICATION

19.3.1 Analysis of mononucleotides

Although the monolithic column configuration turned out to offer substantial advantages mainly for macromolecular analytes such as polypeptides and polynu-

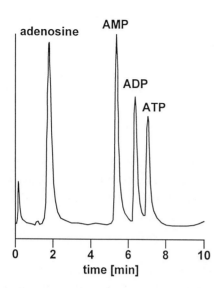

Fig. 19.7. Separation of mononucleotides by ion-exchange chromatography (Reprinted with permission from ref. [42], copyright 1998 Elsevier Sciences B. V.). UNO Q1 column, 35 × 7 mm i.d.; mobile phase, (A) 20 mmol/l Tris buffer, pH 8.2, (B) 20 mmol/l Tris buffer, pH 8.2, 1.0 mol/l sodium chloride; linear gradient, 0% B for 1 min, 0–30% B in 5.0 min, 30% B for 2 min; sample, mixture of adenosine, AMP, ADP, and ATP, 50 ng each.

cleotides, monolithic separation media have also been successfully applied to the separation of small molecules. Fig. 19.7 shows as an example the separation of adenosine, adenosine mono-, di-, and triphosphate by anion-exchange HPLC using a hydrophilic monolith functionalized with quaternary ammonium groups [42]. This separation was utilized to investigate the batch-to-batch reproducibility as well as the lifetime of monolithic columns. Twenty-three different batches of monolithic columns showed essentially identical elution profiles with relative standard deviations in retention time between 0.77 and 1.3% for the four analytes. Moreover, the chromatographic pattern did not change significantly even after 96 injections of the standard mixture proving that column preparation and column stability meet the strict quality criteria for separation columns suitable for routine HPLC.

19.3.2 Quality control and purification of synthetic oligonucleotides

19.3.2.1 Oligodeoxyribonucleotides

Much of the importance of synthetic oligodeoxyribonucleotides in biotechnology, biochemistry, and biomedical research results from their widespread use as hybridization probes, primers for DNA amplification by polymerase chain reaction, DNA se-

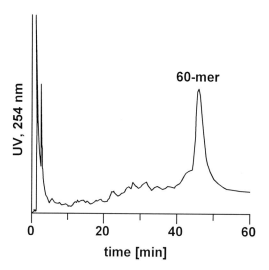

Fig. 19.8. Purification of a synthetic 60mer (Reprinted with permission from ref. [71], copyright 1999 Preston Publications). UNO Q1 column, 35 × 7 mm i.d.; mobile phase, (A) 25 mmol/l triethylammonium acetate, pH 7.4, (B) 25 mmol/l triethylammonium acetate, pH 7.4, 1 mol/l sodium chloride; linear gradient, 40–47% B in 20 min, 47–50% B in 30 min; flow-rate, 1.0 ml/min; detection, UV, 254 nm; sample, 60-mer.

quencing, and primer-extension minisequencing, adaptors for cloning, templates for the construction of deletions, insertions, as well as site specific mutations [69]. Since the introduction of phosphoramidite chemistry, rapid solid-phase synthesis of oligomers up to 100 and more nucleotide units is easily feasible. However, assuming a coupling efficiency of 98-99% per synthesis cycle, the maximum yield of a 60-mer oligodeoxynucleotide will be only 30-55%, and contamination of the target sequence with a number of failure sequences or partially deprotected sequences is generally observed [38,70].

Fig. 19.8 illustrates the anion-exchange purification a crude 60-mer oligodeoxynucleotide in a compressed poly(methacrylamide-*co*-piperazine diacrylamide-*co*-diallyldimethylammonium chloride) 35×7 mm i.d. monolith [71]. With a gradient of 0.40–0.47 mol/l sodium chloride in 25 mmol/l triethylammonium acetate in 20 min, followed by 0.47–0.50 mol/l sodium chloride in 30 min, the main product eluted at 46 min, while a number of small peaks eluting between 3 and 45 min indicated the presence of failure sequences. Up to 14 mg of oligodeoxynucleotide per g of column material could be loaded onto the column and the recovery of pure material was higher than 70%.

The major disadvantage of anion-exchange purification protocols utilizing gradients of non-volatile salt is the necessity to desalt the isolated fractions for recovery of pure product by precipitation or solid-phase extraction, resulting in additional sample losses. However, fractionation and subsequent isolation of a 30-mer oligodeoxynucleotide simply by evaporation of the chromatographic eluent was possible upon replacement of the gradient of sodium chloride with a gradient of 0.2–2 mol/l triethylammonium acetate, which is entirely volatile under vacuum conditions. Not surprisingly, the retention times of oligodeoxynucleotides on the quaternary ammonium monolith in the presence of triethylammonium acetate were strongly affected by the addition of methanol the chromatographic eluent, because electrostatic interaction due to both an anion-exchange as well as an ion-pair reversed-phase mechanism are responsible for retention of the oligodeoxynucleotides.

The identity of the by-products in oligodeoxynucleotide preparations may be readily determined by on-line hyphenation of the separation to mass spectrometry [70,72]. The high resolving power of anion-exchange or ion-pair reversed-phase HPLC usually enables the separation of failure sequences with single-base resolution [22,73]. ESI-MS not only adds another dimension of separation by enabling the distinction of co-eluting analytes in the mass spectra, but also facilitates the positive identification of the components in a mixture on the basis of their molecular masses. Hence, ion-pair reversed-phase HPLC—ESI-MS could be utilized to identify as many as 28 by-products in the raw product of a 32-mer oligodeoxynucleotide (Table 19.2). The target component represented approximately 65% of all compounds in the reaction mixture. Failure sequences ranging from the 10-mer to the 31-mer eluted before, whereas partially protected sequences containing 1–3 isobutyryl-protecting groups eluted after the target product. With analysis times of less than 15 min and its high information content, ion-pair reversed-phase HPLC—ESI-MS analysis is very well suited to monitor the effectiveness of chemical synthesis of nucleic acids, to estimate the synthesis yield on the basis of relative peak areas, and to spot byproducts due to failure of chain elongation or partial deprotection.

19.3.2.2 Phosphorothioate-oligodeoxynucleotides

Antisense oligodeoxynucleotides are oligonucleotides targeted against specific sequences at the genomic or transcriptomic level for therapeutic use [74]. They contain a modified backbone that prevents their rapid degradation by ubiquitous ribonucleases and improves their transport through the cell membranes. Replacement of a non-bridging oxygen atom in the phosphodiester group by sulfur yields oligodeoxynucleotide analogs known as phosphorothioate oligodeoxynucleotides. The therapeutic significance of such modified oligodeoxynucleotides requires a high level of quality control to ensure sample integrity. Moreover, to establish antisense oligonucleotides as a

TABLE 19.2

IDENTIFICATION OF THE COMPONENTS IN THE REACTION MIXTURE FROM THE SOLID-PHASE SYNTHESIS OF A 32-MER BY ION-PAIR REVERSED-PHASE HPLC—ESI-MS[a]

Comp. Nr.	Calculated mass	Identification	Theoretical mass	Comp. Nr.	Calculated mass	Identification	Theoretical mass	Comp. Nr.	Calculated mass	Identification	Theoretical mass
1	3012.2	10-mer	3013	11	6365.4	21-mer	6367	21	9144.4	30-mer	9144
2	3317.0	11-mer	3317	12	6076.8	20-mer	6077	22	9471.2	31-mer	9473
3	3606.3	12-mer	3606	13	7331.2	24-mer	7330	23	9760.4	32-mer	9762
4	3934.8	13-mer	3936	14	7657.6	25-mer	7659	24	9212.4	30-mer + ibu[b]	9215
5	4537.2	15-mer	4538	15	7947.2	26-mer	7948	25	9540.0	31-mer + ibu	9544
6	4247.7	14-mer	4249	16	6670.0	22-mer	6671	26	9830.4	32-mer + ibu	9834
7	4865.4	16-mer	4867	17	7000.4	23-mer	7001	27	9900.4	32-mer + 2 ibu	9905
8	5170.5	17-mer	5172	18	8237.6	27-mer	8237	28	9968.0	32-mer + 3 ibu	9976
9	5475.0	18-mer	5476	19	8525.6	28-mer	8527				
10	5764.2	19-mer	5765	20	8828.4	29-mer	8831				

[a]Experimental conditions: Column, poly(styrene-co-divinylbenzene) monolith, 60 × 0.2 mm i.d.; mobile phase, (A) 25 mM triethylammonium bicarbonate, pH 8.40, (B) 25 mM triethylammonium bicarbonate, pH 8.40, 20% acetonitrile; linear gradient, 10–60% B in 10 min; flow-rate, 2.0 µl/min; temperature, 50°C; sheath liquid, 2.0 µl/min acetonitrile; sheath gas, nitrogen; scan, 1000–3000 u; sample, 65 ng raw product of a 32-mer synthesis

[b]ibu = isobutyryl protecting group

competitive and viable therapeutic option, cost-effective and scalable processes to purify these compounds are mandatory.

While "normal" oligodeoxyribonucleotides are rather easy to separate with single nucleotide resolution, it is more difficult to obtain baseline separation of homologous all-phosphorothioate oligodeoxynucleotides. This is most probably due to the stereogenicity of the phosphorothioate group, causing band broadening as a consequence of partial separation of multiple stereoisomers. Monolithic membrane adsorbers were successfully applied to the analytical separation of all-phosphorothioate oligodeoxynucleotides from their mono-, di-, and tri-phosphodiester analogs [31]. The membrane adsorber used for this application was a thin, quaternary ammonium modified cellulose membrane with 15 cm^2 total surface area. Amounts of a few hundred micrograms of raw products were separated at a flow-rate of 10 ml/min under strongly denaturing conditions with a gradient of 0–2.5 mol/l sodium chloride in 20 mmol/l sodium hydroxide.

This separation system was scaled up by using a large-scale absorber module with 20,000 cm^2 total surface area, representing a 1300 fold increase. 300 mg of a crude phosphorothioate 20-mer were loaded onto the adsorber and eluted in with a gradient of 0–2.5 mol/l sodium chloride in 20 min (Fig. 19.9). Fractions were collected from the column eluate and analyzed by anion-exchange HPLC. The mono-phosphodiester species was adequately separated from the main product peak even on the preparative scale. The pooled fractions as indicated in Fig. 19.9 had a product purity of 91% at a yield of 81%. Even amounts of up to 3 g of raw product could be loaded onto the 20,000 cm^2 absorber module for preparative fractionation. These data indicate that membrane absorbers are highly suitable for the isolation of target products in mixtures rather difficult to separate by other separation methods. The technology is readily applicable to loads ranging from a few micrograms to multigrams without the need for extensive reoptimization of the process at upscaled level.

19.3.3 Analysis of oligoribonucleotides

Three forms of ribonucleic acids are found in cells serving an important function during protein synthesis: messenger RNA, transfer RNA, and ribosomal RNA. RNA molecules are usually single-stranded, but contain loops and/or double-helical regions folding into stable secondary structures. Characteristic for ribonucleic acids is their high affinity to metal ions, particularly bivalent cations, that stabilize proper secondary and tertiary structures. While such complexation of metal ions does not interfere with separations utilizing UV detection, the presence of metal adducts represents one of the major problems in ESI-MS of nucleic acids [75]. Because of cation adduction, the pseudomolecular ions are dispersed among several species of different m/z ratios, resulting in highly complex spectra, decreased sensitivity, and inaccurate mass deter-

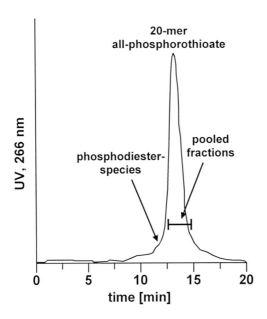

Fig. 19.9. Large-scale purification of an antisense oligodeoxynucleotide on a membrane adsorber (Reprinted with permission from ref. [31], copyright 2000 Elsevier Sciences B. V.). Sartobind Q-20K-60-12 column, 12 cm length; mobile phase, (A) 20 mmol/l sodium hydroxide, (B) 20 mmol/l sodium hydroxide, 2.5 mol/l sodium chloride; linear gradient, 0–100% B in 20 min; flow-rate, 620 ml/min; detection, UV, 266 nm; sample, 300 mg of crude antisense oligonucleotide.

minations. Limbach *et al.* proposed the coaddition of triethylamine and chelating agents to RNA samples after purification by ethanol precipitation to improve the quality of the mass spectra of transfer- and messenger RNAs [75].

Although ion-pair reversed-phase HPLC has been shown to efficiently remove cationic adducts from DNA [52,76], the affinity of bivalent ions, in particular Mg^{2+}, to RNA is still strong enough to endure the chromatographic process without being exchanged by protons. In order to remove those strongly bound cations, ethylenediaminetetraacetic acid (EDTA) is added to the RNA sample before chromatographic purification. Fig. 19.10 shows the analysis of a solution of a 120-nucleotide 5S ribosomal RNA from *E. coli* to which 25 mmol/l EDTA have been added. The Mg-EDTA complex eluted together with the monovalent cations as broad peak in the void volume (Fig. 19.10a). The mass spectrum extracted from the HPLC peak at 3 min showed series of multiply charged ions, which were deconvoluted into molecular masses of 38,815 and 38,852, respectively. These two detected masses are related to transcripts of the two 5S ribosomal RNA genes that occur in *E. coli* and correspond excellently to the theoretical molecular masses of 38,814.4 and 38,852.4 [75].

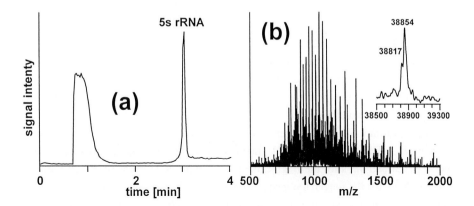

Fig. 19.10. Analysis of 5S ribosomal RNA from *E. coli* by ion-pair reversed-phase HPLC–ESI-MS (Reprinted with permission from ref. [22], copyright 2002 Wiley-VCH). Monolithic PS-DVB column, 60 × 0.2 mm i.d.; mobile phase, (A) 25 mmol/l butyldimethyl-ammonium bicarbonate, pH 8.4, (B) 25 mmol/l butyldimethylammonium bicarbonate, pH 8.4, 80% acetonitrile; linear gradient, 0–100% B in 10 min, flow-rate, 3.0 μl/min; temperature, 70°C; sheath liquid, 3.0 μl/min acetonitrile; scan, 500–2000 u; sample, 75 ng 5S ribosomal RNA in water containing 25 mmol/l ethylenediaminetetraacetic acid.

19.3.4 DNA restriction fragments

Separation and fractionation of DNA restriction fragments is a key element in various molecular biological experiments, including cloning, DNA sequencing, genome fingerprinting, and DNA hybridization. The equimolar mixture of fragments generated by the enzymatic cleavage of DNA with restriction endonucleases may range from a few base pairs to thousands of base pairs, depending on DNA size, DNA sequence and the restriction enzyme used. Anion-exchange and ion-pair reversed-phase HPLC represent fast and reliable alternatives to slab gel electrophoresis for the separation of DNA restriction fragments [39]. While both chromatographic modes have been demonstrated with traditional granular column packings, ion-pair reversed-phase HPLC is the only mode used so far for the separation of large, double-stranded DNA with monolithic columns.

19.3.4.1 High-resolution separation of DNA restriction fragments

The two chromatograms shown in Fig. 19.6 demonstrate that the separation effi-ciency with both micropellicular monolithic and granular stationary phases for ion-pair reversed-phase HPLC of DNA restriction fragments is equivalent [77]. However, under optimized synthesis conditions as well as upon minimization of extracolumn band broadening, the separation efficiency in a monolithic stationary phase was shown to even surpass that of granular stationary phases [22]. Hence, a mixture of the pBR322 DNA-*Hae* III and the pBR322 DNA-*Msp* I digest containing 37 fragments was fractionated at least partially into 33 peaks [22]. The fragments ranged in size from 51 to 622 base pairs and were separated in a poly(styrene-*co*-divinylbenzene) monolith by applying a gradient of 8.75–13.75% acetonitrile in 6.0 min, followed by 13.75–15.75% acetonitrile in 9.0 min in 0.1 mol/l triethylammonium acetate at a flow rate of 2.2 μl/min. The total amount of DNA analyzed was only 9.0 ng, corresponding to approx. 1.7 fmol for each DNA fragment. The peak widths at half height were between 2.3 and 4.3 s for the fragments up to a length of 217 base pairs and between 5.6 and 8.5 s for the longer fragments. Compared to the previously published separations of the same mixture in an analytical [47] or a capillary column [23] packed with poly(styrene-*co*-divinylbenzene)-C18 particles, the separation efficiency is higher, because better or at least equivalent resolution was achieved in a shorter period of time.

19.3.4.2 Micropreparative HPLC of nucleic acids

In order to apply the full range of manipulations necessary to perform in molecular biological experiments, a preparative method for fractionation of nucleic acids is frequently required. Monolithic columns consisting of norbornene and hexahydrodimethanonaphthalene copolymer, synthesized *via* ring-opening metathesis polymerization possess good separation capabilities for oligodeoxynucleotides and proteins [53]. Fig. 19.11 demonstrates that this type of stationary phase is also suitable for the high-resolution separation of double-stranded DNA restriction fragments [78]. The separation of *Hae*III fragments of the pBR322 plasmid was accomplished with a gradient of 4–10% acetonitrile in 5 min, followed by 10–18% acetonitrile in 12 min in 100 mmol/l aqueous triethylammonium acetate at a flow rate of 2 ml/min. The addition of 4% glycerol to the mobile phase significantly improved the resolution of the large 434-587 base pair fragments. Loads of up to 2.5 μg of DNA were separable on the 100×3 mm i.d. norbornene/hexahydrodimethanonaphthalene monoliths without significant loss in chromatographic resolution, which is more than sufficient for most molecular biological experiments involving nucleic acids.

19.3.4.3 Liquid chromatography-mass spectrometry of DNA restriction fragments

Chromatographic retention data can afford valuable information about the size of DNA fragments [47]. However, measurement of the molecular mass certainly represents a very powerful tool for confident identification and characterization of nucleic acids [79,80] as well as for the identification of sequence variations [81]. Neverthe-

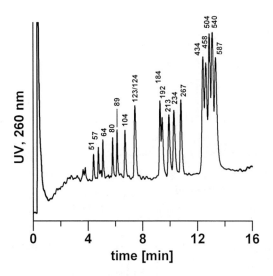

Fig. 19.11. Micropreparative separation of DNA fragments using monolithic columns prepared by ring-opening metathesis polymerization (Reprinted with permission from ref. [78], copyright 2002 Elsevier Science B. V.). Monolithic poly(norbornene-*co*-hexa-hydrodimethanonaphthalene) column, 100 × 3 mm i.d.; mobile phase, (A) 100 mmol/l triethylammonium acetate, pH 7.0, 4% glycerol (B) 100 mmol/l triethylammonium acetate, pH 7.0, 40% acetonitrile, 4% glycerol; linear gradient, 10–25% B in 5.0 min, 25–45% B in 12.0 min; flow-rate, 2 ml/min; temperature, 50 °C; detection, UV, 260 nm; sample, 0.75 µg pBR322 DNA-*Hae*III .

less, the potential to obtain high quality ESI-mass spectra of large, double stranded DNA is essentially determined by both the amount of adducts [82] and the number of different compounds present in the sample introduced into the mass spectrometer [83]. Early attempts to analyze double-stranded DNA fragments longer than 100 base pairs by ion-pair reversed-phase HPLC–ESI-MS were hampered mainly due to the unfavorable detection limits. The addition of acetonitrile as sheath liquid [62] and the use of butyldimethylammonium bicarbonate [25] as ion-pair reagent substantially improved the quality of the measured mass spectra.

The ion-pair reversed-phase HPLC–ESI-MS analysis of 15 fmol of restriction fragments from an *Msp*I digest of the cloning vector pUC19 is illustrated in Fig. 19.12 [84]. High quality mass spectra could be extracted from the chromatogram for characterization of the individual DNA fragments. For example, in the mass spectrum of the 190-mer (Fig. 19.12b), multiply charged ions with charge states from 41- to 86- were monitored and deconvoluted into a molecular mass of 117,425 corresponding very well to the theoretical mass of 117,438.0. Although signal-to-noise ratios decreased as spectrum complexity increased significantly with increasing chain length of the DNA

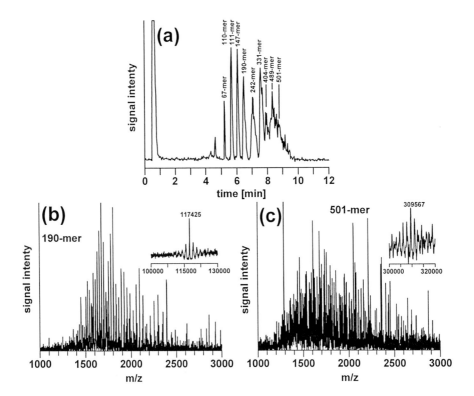

Fig. 19.12. Chromatographic separation and mass analysis of the double-stranded restriction fragments of a pUC19 DNA-*Msp*I digest (Reprinted with permission from ref. [84], copyright 2002 Wiley-VCH). (a) Reconstructed ion chromatogram (b) mass spectrum of double-stranded 190-mer, (c) mass spectrum of double-stranded 501-mer. Monolithic PS-DVB column, 60 × 0.2 mm i.d; mobile phase, (A) 25 mmol/l butyldimethylammonium bicarbonate, pH 8.40, (B) 25 mmol/l butyldimethylammonium bicarbonate, pH 8.40, 40% acetonitrile; linear gradient, 10–40% B in 3 min, 40–50% B in 12 min; flow-rate, 3.0 μl/min; temperature, 25°C; sheath-gas, nitrogen; sheath-liquid, 3.0 μl/min acetonitrile; scan, 1000–3000 u; sample, 15 fmol digest.

molecules, confident mass data were obtained for DNA fragments up to the 501-mer (measured mass, 309,567; theoretical mass, 309,574.2; Fig. 19.12c). This data validates that the on-line hyphenation to liquid chromatography represents an important step towards the applicability of electrospray ionization mass spectrometry for highly accurate mass measurements of low femtomol amounts of nucleic acids over a broad size range.

19.3.5 Polymerase chain reaction products

Today, the polymerase chain reaction (PCR) is one of the most important and widespread tools for nucleic acid studies in biochemistry, molecular biology, and clinical diagnostics [85]. The PCR has also strongly enhanced the usefulness of DNA genotyping techniques, as it allows the targeted *in vitro* amplification of short segments of genomic DNA in order to detect single nucleotide polymorphisms (SNPs) or short tandem repeat polymorphisms (STRs). Following the PCR process, the amplified DNA fragments have to be separated from the other reaction components, including DNA polymerase, oligonucleotide primers, mononucleotides, and buffer components before their identification and quantitation [77,86]. The advantages of HPLC over traditional slab gel or capillary gel electrophoretic methods for the analysis of PCR products include the possibility to analyze the reaction mixtures without any prior sample preparation, the availability of the results in a short period of time, and the automated processing of hundreds of samples for routine analysis.

19.3.5.1 Mutation detection by denaturing ion-pair reversed-phase HPLC

Denaturing high-performance liquid chromatography (DHPLC) has shown great potential for efficient and sensitive detection of single-base substitutions as well as small deletions and insertions in DNA fragments ranging from 100 to 1500 base pairs in size [87,88]. In particular, denaturing HPLC is based on the separation of homo- from heteroduplex species generated by PCR amplification of polymorphic loci containing one or more mismatches [89]. Ideally, four chromatographic peaks corresponding to the two heteroduplexes and the two homoduplexes are observed in ion-pair reversed-phase HPLC at elevated column temperatures in the range of 48–67°C (Fig. 19.13), with the former eluting before the latter because of the perturbation of the double-helical structure in partially denatured DNA fragments resulting in decreased retention [51].

Fig. 19.14 illustrates the DHPLC profiles observed in the chromatograms of a 413 base pair amplicon from the human P-glycoprotein corresponding to a homozygous control (Fig. 14a) and two heterozygous individuals (Fig. 19.14b and c). The peak labels in the chromatogram indicate the type and position of the mismatch from the 5′-end of the forward primer. As can be seen from Fig. 19.14, different mismatches or combinations of mismatches tend to generate different chromatographic profiles, whereas homozygous samples show only one major peak. Consequently, sequence analysis, which is still required to determine the exact location and nature of mismatches, can be limited to a few representative profiles. However, identical profiles

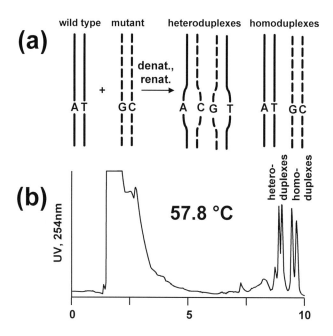

Fig. 19.13. Principle of mutation detection by denaturing HPLC (Reprinted with permission from ref. [77], copyright 2001 Elsevier Science B. V.). (a) Formation of heteroduplices by denaturing and reannealing of a mixture of wild type and mutant; (b) ion-pair reversed-phase HPLC separation of hetero- and homoduplices. Monolithic PS/DVB column, 50 × 0.20 mm i.d.; mobile phase, (A) 100 mM TEAA, 0.1 mM Na₄EDTA, pH 7.0, (B) 100 mM TEAA, 0.1 mM Na₄EDTA, pH 7.0, 25% acetonitrile; linear gradient, 30–50% B in 3.0 min, 50–70 % B in 7.0 min; flow rate, 3.0 μl/min; temperature, 57.8°C; detection, UV, 254 nm; injection volume, 500 nl, sample, 209 base pair amplicon containing a single A→G transition.

for different mutations located within the same melting domain but not at the same nucleotide position have been reported [90].

19.3.5.2 Genotyping by liquid chromatography-mass spectrometry

The high frequency of single nucleotide polymorphisms in the human genome makes them a valuable source of genetic markers for identity testing, mapping of simple and complex traits, genotype-phenotype association studies, and reconstruction of human evolution. Numerous methods have been introduced for determining the allelic state of individual SNPs [91,92]. During the past years, mass spectrometry (MS) [93], specifically matrix-assisted laser desorption-ionization mass spectrometry (MALDI-MS) and electrospray ionization mass spectrometry (ESI-MS), have emerged as powerful analytical tools for the genotyping of SNPs.

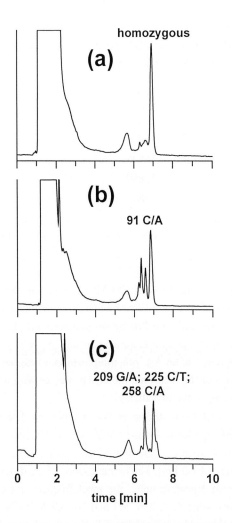

Fig. 19.14. Chromatographic separation profiles obtained for one homozygous control (a) and two heterozygotes (b, c) under partially denaturing conditions (Reprinted with permission from ref. [77], copyright 2001 Elsevier Sciences B. V.). Monolithic PS-DVB column, 60 × 0.2 mm i.d.; mobile phase, (A) 100 mmol/l triethylammonium acetate, 0.1 mmol/l Na$_4$EDTA, pH 7.0, (B) 100 mmol/l triethylammonium acetate, 0.1 mmol/l Na$_4$EDTA, pH 7.0, 25% acetonitrile; linear gradient, 43–50% B in 0.5 min, 50–54% B in 3.5 min; flow rate, 3 μl/min; temperature, 61°C; injection volume, 500 nl.

ESI-MS has been applied to the characterization of SNPs in PCR amplicons after on-line purification by ion-pair reversed-phase HPLC utilizing monolithic poly(styrene-*co*-divinylbenzene) capillary columns [25]. The distinction of the two alleles of an SNP both in homozygous and heterozygous individuals was possible by means of

TABLE 19.3.

ALLELE IDENTIFICATION THROUGH COMPARISON OF THE THEORETICAL POS-
SIBLE WITH THE MEASURED MOLECULAR MASSES

Allele[a]	Theoretical molecular mass	Measured molecular mass
A+dA$_f$	15893.54	15890
A$_r$	15806.42	15806
C+dA$_f$	15869.51	15869
C$_r$	15831.44	15831

[a]The indices $_f$ and $_r$ stand for the reverse and forward strands of the PCR amplicon, + dA is
indicative of non-template addition of a deoxyadenosine commonly observed with Taq
polymerase

completely denaturing HPLC [49] in combination with ESI-MS. For this method,
short DNA double strands (<100 base pairs) were amplified by PCR, then completely
denatured into single strands, chromatographically purified by ion-pair reversed-phase
HPLC, and finally characterized by accurate mass measurements using ESI-MS. Table 19.3 gives an example for the recognition of an A/C polymorphism in a heterozygous individual based on the comparison of the measured molecular masses with the
theoretical molecular masses corresponding to the two alleles.

STRs are characterized by length variation in tandem arrays of repeating 2 to 6
base pair sequences, which are frequently applied to forensic DNA typing. The human
TH01 locus is located in intron 1 of the human tyrosine hydroxylase gene and has a
tetrameric repeating unit of the sequence AATG [94]. The nomenclature of the different alleles is based upon the number of repeating units, the most frequent alleles
observed in white caucasian populations being 6, 7, 8, 9, 9.3 and 10. Allele 9.3 is a
common variant of allele 10 with an 2′-deoxyadenosine deletion in the seventh repeat.
The ion-pair reversed-phase HPLC–ESI-MS analysis of PCR products from a heterozygous TH01 amplicon under denaturing conditions is illustrated in Fig. 19.15,
wherein the alleles 9.3 and 10 were readily identified on the basis of the measured
mass values [95]. The reverse single-strands were chromatographically separated from
the forward with the latter eluting after the reverse due to the higher proportion of
relatively hydrophobic thymidines in the forward strand (Fig. 19.15a). Deconvolution
of the extracted mass spectra yielded molecular masses of 25,782, 26,092, 24,750,
25,056, and 25,366, which correspond well to the masses of the reverse strands, the
forward strands, and the forward strands with an additional deoxyadenosine (Fig.
19.15b and c).

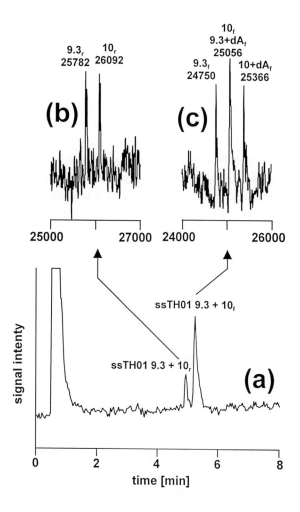

Fig. 19.15. Genotyping of polymorphic short tandem repeat (STR) loci from the human tyrosine hydroxylase gene (humTH01) by ion-pair reversed-phase HPLC–ESI-MS (Reprinted with permission from ref. [95], copyright 2001 American Chemical Society). Monolithic PS-DVB column, 60 × 0.20 mm i.d.; mobile phase, (A) 25 mmol/l BDMAB, pH 8.40, (B) 25 mmol/l BDMAB, pH 8.40, 40% acetonitrile; linear gradient, 15–70% B in 10 min, flow-rate, 3.0 µl/min; temperature, 70°C; scan, 500–2000 u; sheath liquid, 3.0 µl/min acetonitrile; sample, mixture of the TH01 alleles 9.3 and 10, 50 fmol each. The indices $_f$ and $_r$ stand for the reverse and forward strands of the PCR amplicon, + dA is indicative of non-template addition of a deoxyadenosine commonly observed with Taq polymerase.

19.3.6 Analysis and purification of bacterial nucleic acids

Plasmids are frequently used as cloning vectors for recombinant DNA studies or as substrates for digestion by restriction enzymes to generate sizing standards of double-stranded DNA fragments. Thus, the need for plasmid purification reaches from preparing many small samples up to purification of milligram quantities of plasmids. High-performance membrane chromatography, using very thin monolithic disks, has been shown to enable the rapid isolation of microgram amounts of bacterial plasmids. Fig. 19.16 a illustrates the fractionation of the native plasmid pCMVβ on a monolithic, quaternary ammonium poly(glycidyl methacrylate-*co*-ethylene dimethacrylate) disk of 3 mm thickness [45]. The sample was eluted with a gradient of 0.73–1.00 mol/l sodium chloride in 30 mmol/l Tris HCl buffer, pH 7.4, in 4.5 min at a flow rate of 3 ml/min. The UV signal of the eluting plasmid showed several local maxima, most probably corresponding to the different conformations of the plasmid, namely supercoiled, nicked, or open circular. The presence of several conformations of the plasmid was corroborated by the analysis of a linearized plasmid sample, in which the complexity of the signal was significantly reduced (Fig. 19.16b). Optimization of the chromatographic conditions showed that elution of the DNA is possible even under isocratic elution conditions in contrast to protein samples, where usually a solvent gradient is obligatory for elution of the analytes.

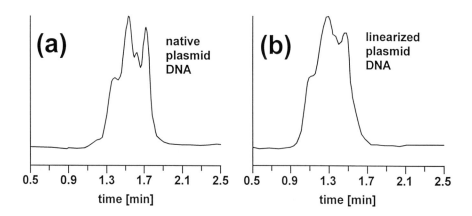

Fig. 19.16. Comparison of the retention behavior of (a) linearized and (b) native plasmid DNA on a modified poly(glycidyl methacrylate-*co*-ethylene dimethacrylate) disk (Reprinted with permission from ref. [45], copyright 1998 American Chemical Society). DEAE CIM disk, 3 × 12 mm i.d.; mobile phase, (A) 20 mmol/l Tris HCl buffer, pH 7.4, (B) 20 mmol/l Tris HCl buffer, pH 7.4, 1.0 mol/l sodium chloride; linear gradient, 73–100% B in 4.5 min; flow-rate, 3.0 ml/min; detection, UV, 260 nm; sample, 5 μg pCMVβ-plasmid DNA.

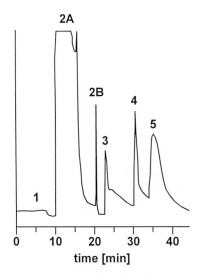

Fig. 19.17. Sequential elution of the components of bacterial lysates adsorbed onto membranes (Reprinted with permission from ref. [44], copyright 1993 Academic Press). DEAE-cellulose membrane, 6 × 19 mm i.d.; stepwise elution, unadsorbed material during injection (1), denatured proteins eluted by phenol-chloroform extraction (2a) and ethanol wash (2b), oligoribonucleotides desorbed by 0.3 mol/l LiCl, 5mM LiOH (3), DNA eluted by 0.5 mmol/l NaCl, 5 mmol/l NaOH (4), DNA-RNA-complexes eluted by 2 mol/l NaCl, 5 mmol/l NAOH (5); flow-rate, 5.0 ml/min; detection, UV, 260 nm; sample, lysate from a 50 ml culture of *Escherichia coli*.

The high-performance membrane chromatographic experiments were compared to separations performed in a monolithic, compressed-bed column (UNO Q1) with a 35 mm separation layer thickness. While under comparable gradient conditions the eluting peak was sharper with the monolithic disk, the peak profile resulting from the monolithic column did not indicate the presence of any subfractions within the plasmid preparation. However, under isocratic elution conditions, the monolithic columns revealed the presence of more than one plasmid species at the cost of more pronounced zone broadening, most probably as a consequence of the longer effective separation length in the monolithic column.

The isolation of plasmid DNA from total cell lysates using membrane adsorbers was attempted with the aid of DEAE-cellulose membranes [44]. The elution profile of a crude cell lysate depicted in Fig. 19.17 shows that the compounds eluted from the column in the order of increasing negative charges. Denatured proteins were eluted utilizing a mixture of phenol and chloroform. A single washing step with about 60 times less phenol (fraction 2A) was sufficient to obtain the same extent of deproteination for DNA adsorbed onto the membranes as compared to DNA in solution. Subsequent washing with ethanol removed some residual proteins (faction 2B). Oligori-bonucleotides were desorbed with 0.3 mol/l lithium chloride, 5 mmol/l lithium hydroxide (fraction 3), before the plasmid DNA was eluted with 0.5 mol/l sodium chloride and 5 mmol/l sodium hydroxide (fraction 4). Finally, DNA-RNA complexes were removed from the membranes with 2 mol/l sodium chloride, 5 mmol/l sodium hydroxide (fraction 5). The amount of DNA recovered from the column was considerably increased upon treatment of the sample with ribonuclease. Several hundred micrograms of plasmid DNA could be purified in cartridges containing stacks of fourty 150 μm thick DEAE cellulose membranes in a cycle time of 40 min. The purified plasmids gave the expected fragments upon treatment with several restriction endonucleases. The DEAE cellulose membranes could withstand more than 1000 cycles of regeneration without any noticeable decrease in their binding capacity.

19.3.7 Multiplexing using monolithic capillary arrays

Although HPLC has seen considerable improvements in terms of separation speed [3,7], sample throughput is limited as HPLC does not lend itself readily to the arraying of separation columns as is the case in modern capillary electrophoretic DNA sequencers. Monolithic capillary columns offer some advantages for the fabrication of capillary arrays, including their easy preparation, robustness, easy mounting and handling, as well as small dimensions. Fig. 19.18 illustrates the scheme of an instrument configuration suitable for the parallel analysis in four separation columns [96]. The low-pressure gradient mixing pump is operated typically at a flow-rate of 100–200 μl/min and a back pressure of 10–20 MPa. A tee and a restriction capillary is used to split the primary flow to create an effective secondary low-flow stream of approx. 10 μl/min representing approx. 2.5 μl/min per column. For the loading of four independent samples onto four different columns, a manifold with a single inlet and four outlets is placed immediately after the splitting tee. The four outlet lines are connected to a custom-made electrically actuated injection valve with four internal 1-μl sample loops. Loading is accomplished by injecting 3 μl of sample to ensure complete filling of the loops. Stainless steel zero dead-volume unions are used to connect four 60×0.2 mm i.d. monolithic capillary columns to the inlet and outlet fused silica lines with an internal diameter of 25 μm to minimize extra-column volume. To enable monitoring of the fluorescent dye-labeled DNA fragments using an argon-ion laser induced fluorescence detector, a detection window is created by burning off the polyimide layer on the outside of the fused silica capillary.

One example of application for the miniaturized column array is illustrated in Fig. 19.19 through the multiplex analysis of polymorphic DNA amplicons by DHPLC [96]. The four chromatograms were obtained simultaneously on four monolithic poly(styrene-*co*-divinylbenzene) columns. The use of four different fluorescent labels

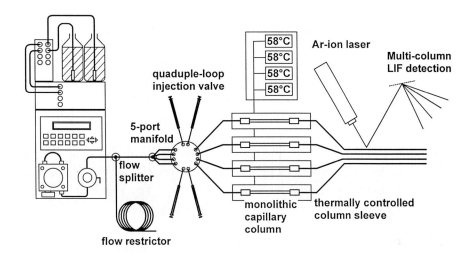

Fig. 19.18. Scheme of an instrument configuration of HPLC arrays for the detection of mutations by heteroduplex analysis under partially denaturing conditions (Reprinted with permission from ref. [96], copyright 2001 Gold Spring Harbor Laboratory Press). Four different samples are injected onto four different columns. The temperatures of the columns can be regulated individually.

(FAM, HEX, NED, and ROX) attached to the amplification primes facilitated a second dimension of multiplexing by the selective measurement of fluorescence emission in a multicolor fluorescence detector. Hence, sixteen different homozygous and heterozygous samples could be analyzed in one single 10 min run on the four-column array. The monolithic column array proved to be very robust, as more than 500 injections were performed without any notable deterioration in column performance. Since a great part of the 100–200 μl/min primary flow from the gradient pump goes to waste, there is plenty of eluent available to run even more columns in parallel, and multiplexing of up to 96 capillary columns should be possible. However, at the present time the simultaneous injection of more than four samples represents a major technical limitation, and progress in injection technology is needed for successful multiplexing of HPLC into large column arrays.

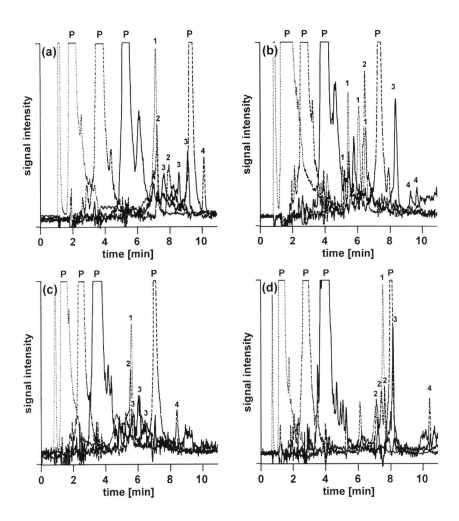

Fig. 19.19. Simultaneous analysis of sixteen different samples homozygous or heterozygous for exon 13, 22, or 26 in MDR1 using four capillary columns (Reprinted with permission from ref. [96], copyright 2001 Gold Spring Harbor Laboratory Press). Four monolithic poly(styrene-*co*-divinylbenzene) columns, 60 × 0.2 mm i.d.; mobile phase, (A) 100 mmol/l triethylammonium acetate, pH 7.0, (B) 100 mmol/l triethylammonium acetate, pH 7.0, 25% acetonitrile; linear gradient, 45–70% B in 10.0 min; flow rate, 2.0–3.0 μl/min; temperature, (a,d) 55°C, (b) 56°C, (c) 57°C; detection, LIF, emission monitored at 525 nm for FAM (···), 555 nm for HEX (·–·), 580 nm for NED (—), and 605 nm for ROX (---); injection volume, 1 μl each; samples, column1, exon 22, FAM, homozygote, exon 13, HEX, 52 T>G and 291 T>C double heterozygote, exon 26, NED, 212 C>T and 346 T>C double heterozygote, exon 26, ROX, homozygote; column 2, exon 22, FAM, 320 G>A heterozygote, exon 13, HEX, homozygote, exon 26, NED, homozygote, exon 26, ROX, 212 C>T and 346 T>C double heterozygote; column 3, exon 22, FAM, homozygote, exon 13, HEX, homozygote, exon 13, NED, 52 T>G and 291 T>C double heterozygote, exon 26, ROX, homozygote; column 4, exon 22, FAM, homozygote, exon 13, HEX, 52 T>G and 291 T>C double heterozygote, exon 13, NED, homozygote, exon 26, ROX, homozygote.

19.4 CONCLUSIONS AND OUTLOOK

Monolithic separation media hold great potential for the chromatographic separation of nucleic acids in amounts ranging from microanalytical to process scale. The major advantage of the monolithic stationary phases rests within their excellent chromatographic separation efficiency for nucleic acids, which significantly exceeds the efficiency of the currently available granular stationary phases. Moreover, the high mechanical and chemical stability of polymeric monoliths greatly contribute to the stability, ruggedness, and longevity of the separation systems. Through the use of miniaturized columns in the capillary format, very small amounts of nucleic acids of biological origin are amenable to separation and on-line mass spectrometric characterization. Future developments in column chemistry and functionalization are expected to broaden the range of applicable chromatographic modes as well as the number of applications both in industrial and academic environments

19.5 ACKNOWLEDGMENTS

This work was supported by grants from the Austrian Science Fund (P-13442-PHY, P-14133-PHY).

19.6 REFERENCES

1   M.T.W. Hearn, HPLC of Proteins, Peptides and Polynucleotides, VCH Publishers, New York, 1991.

2   M. Gilar, D.L. Smisek, A.S. Cohen, in: Z. Deyl, I. Miksik, F. Tagliaro, E. Tesarova (Eds.), Advanced Chromatographic and Electromigration Methods in Biosciences, Elsevier, Amsterdam, 1998, p. 575.

3   H. Chen, C. Horváth, J. Chromatogr. A 705 (1995) 3.

4   C. Horváth, B.A. Preiss, S.R. Lipsky, Anal. Chem. 39 (1967) 1422.

5   E. Westman, S. Eriksson, T. Laas, P.-A. Pernemalm, S.-E. Skold, Anal. Biochem. 166 (1987) 158.

6   Y. Kato, M. Sasaki, T. Hashimoto, T. Murotsu, S. Fukushige, K. Matsubara, J. Chromatogr. 265 (1983) 342.

7   C.G. Huber, P.J. Oefner, G.K. Bonn, Chromatographia 37 (1993) 653.

8   J.J. Kirkland, F.A. Truszkowski, C.H. Dilks Jr., G.S. Engel, J. Chromatogr. A 890 (2000) 3.

9   S. Hjerten, Y.M. Li, J.L. Liao, J. Mohammad, K. Nakazato, G. Pettersson, Nature 356 (1992) 810.

10   F. Svec, J.M.J. Fréchet, Anal. Chem. 64 (1992) 820.

11   L.C. Hansen, R.E. Sievers, J. Chromatogr. 99 (1974) 123.

12   S. Hjertén, J.-L. Liao, R. Zhang, J. Chromatogr. 473 (1989) 273.

13   H. Minakuchi, K. Nakanishi, N. Soga, N. Ishizuka, N. Tanaka, Anal. Chem. 68 (1996) 3498.

14   J.M.J. Fréchet, F. Svec, Science 273 (1996) 205.

15   A.I. Liapis, M.A. McCoy, J. Chromatogr. A 660 (1994) 85.

16   A.E. Rodrigues, Z.P. Lu, J.M. Loureiro, G. Carta, J. Chromatogr. 653 (1993) 189.

17   J.J. Meyers, A.I. Liapis, J. Chromatogr. A 852 (1999) 3.

18   K.K. Unger, in: K.K. Unger (Ed.), Packings and Stationary Phases in Chromatographic Techniques, Marcel Dekker, New York, 1990, p. 75.

19   J.L. Liao, Adv. Chromatogr. 40 (2000) 467.

20   K. Kalghatgi, C. Horváth, in: C. Horváth, J.G. Nikelly (Eds.), Analytical Biotechnology-Capillary Electrophoresis and Chromatography, American Chemical Society, Washington, DC, 1990, p. 163.

21   S.L. Koontz, R.V. Devivar, W.J. Peltier, J.E. Pearson, T.A. Guillory, J.D. Fabricant, Colloid Polym. Sci. 277 (1999) 557.

22   H. Oberacher, C.G. Huber, Trends Anal. Chem, in press (2002).

23   H. Oberacher, A. Krajete, W. Parson, C.G. Huber, J. Chromatogr. A 893 (2000) 23.

24   C.G. Huber, A. Krajete, J. Mass Spectrom. 35 (2000) 870.

25   H. Oberacher, P.J. Oefner, W. Parson, C.G. Huber, Angew. Chem. 40 (2001) 3828.

26   K.B. Tomer, M.A. Moseley, L.J. Deterding, C.E. Parker, Mass Spectrom. Rev. 13 (1994) 431.

27   D. Josic, A. Strancar, Ind. Eng. Chem. Res. 38 (1999) 333.

28   C. Viklund, F. Svec, J.M.J. Fréchet, Chem. Mater. 8 (1996) 744.

29   E.C. Peters, F. Svec, J.M.J. Fréchet, Chem. Mater. 9 (1997) 1898.

30   M. Podgornik, M. Barut, A. Strancar, Anal. Chem. 72 (2000) 5693.

31   R.R. Deshmukh, T.N. Warner, F. Hutchison, M. Murphy, W.E. Leitch II, P. De Leon, G.S. Srivatsa, D.L. Cole, Y.S. Sanghvi, J. Chromatogr. A 890 (2000) 179.

32   C.G. Huber, R.A. Meyers (Ed.), Encyclopedia of Analytical Chemistry, Wiley-VCH, Chichester, 2000.

33   H. Ellergren, T. Laas, J. Chromatogr. 467 (1989) 217.

34   L.W. McLaughlin, Chem. Rev. 89 (1989) 309.

35   J.A. Thompson, R.D. Wells, Nature, 334 (1988) 87.

36   T.A. Goss, M. Bard, H.W. Jarrett, J. Chromatogr. 508 (1990) 279.

37   T.A. Goss, M. Bard, H.W. Jarrett, J. Chromatogr. 588 (1991) 157.

38   C.G. Huber, E. Stimpfl, P.J. Oefner, G.K. Bonn, LC-GC Int. 14 (1996) 114.

39   C.G. Huber, J. Chromatogr. A 806 (1998) 3.

40   D. Sýkora, F. Svec, J.M.J. Fréchet, J. Chromatogr. A 852 (1999) 297.

41   A. Podgornik, M. Barut, J. Jancar, A. Strancar, J. Chromatogr. A 848 (1999) 51.

42   T.L. Tisch, R. Frost, J.-L. Liao, W.-K. Lam, A. Remy, E. Scheinpflug, C. Siebert, H. Song, A. Stapleton, J. Chromatogr. A 816 (1998) 3.

43   A. Strancar, M. Barut, A. Podgornik, P. Koselj, D. Josic, A. Buchacher, LC-GC (1998) 660.

44   N. van Huynh, J.C. Motte, J.F. Pilette, M. Decleire, C. Colson, Anal. Biochem. 211 (1993) 61.

45   R. Giovannini, R. Freitag, T.B. Tennikova, Anal. Chem. 70 (1998) 3348.

46   A. Bartha, J. Stahlberg, J. Chromatogr. A 668 (1994) 255.

47   C.G. Huber, P.J. Oefner, G.K. Bonn, Anal. Chem. 67 (1995) 578.

48   C.G. Huber, P.J. Oefner, G.K. Bonn, Anal. Biochem. 212 (1993) 351.

49 P.J. Oefner, J. Chromatogr. B 739 (2000) 345.

50 P.J. Oefner, P.A. Underhill, Am. J. Hum. Genet. 57 [Suppl.] (1995) A266.

51 C.G. Huber, G.N. Berti, Anal. Chem. 68 (1996) 2959.

52 A. Premstaller, H. Oberacher, C.G. Huber, Anal. Chem. 72 (2000) 4386.

53 F.M. Sinner, M.R. Buchmeiser, Angew. Chem. 39 (2000) 1433.

54 F. Sinner, M.R. Buchmeiser, Macromolecules 33 (2000) 5777.

55 B. Mayr, R. Tessadri, E. Post, M.R. Buchmeiser, Anal. Chem. 73 (2001) 4071.

56 P.J. Oefner, C.G. Huber, F. Umlauft, G.N. Berti, E. Stimpfl, G.K. Bonn, Anal. Biochem. 223 (1994) 39.

57 M. Barber, R.S. Bordoli, R.D. Sedgwick, A.N. Tyler, J. Chem. Soc. Chem. Commun. 1981 (1981) 325.

58 C.M. Whitehouse, R.N. Dreyer, M. Yamashita, J.B. Fenn, Anal. Chem. 57 (1985) 675.

59 M. Karas, F. Hillenkamp, Anal. Chem. 60 (1988) 2299.

60 D.N. Nguyen, G.W. Becker, R.M. Riggin, J. Chromatogr. A 705 (1995) 21.

61 S.D. Fuerstenau, W.H. Benner, Rapid Commun. Mass Spectrom. 9 (1995) 1528.

62 C.G. Huber, A. Krajete, J. Chromatogr. A 870 (2000) 413.

63 K.K. Murray, Mass Spectrom. Rev. 16 (1997) 283.

64 S.A. McLuckey, G.J. Berkel, G.L. Glish, J. Am. Soc. Mass Spectrom. 3 (1992) 60.

65 D.P. Little, D.J. Aaserud, G.A. Valaskovic, F.W. McLafferty, J. Am. Chem. Soc. 118 (1996) 9352.

66 A. Premstaller, C.G. Huber, Rapid Commun. Mass Spectrom. 15 (2001) 1053.

67 J. Ni, S.C. Pomerantz, J. Rozenski, Y. Zhang, J.A. McCloskey, Anal. Chem. 68 (1996) 1989.

68 H. Oberacher, B. Wellenzohn, C.G. Huber, Anal. Chem. 73 (2001) 211.

69 F. Eckstein, Oligonucleotides and Analogues, A Practical Approach, Oxford University Press, Oxford, 1991.

70 C.G. Huber, A. Krajete, Anal. Chem. 71 (1999) 3730.

71 C.H. Lochmüller, Q. Liu, L. Huang, Y. Li, J. Chromatogr. Sci. 37 (1999) 251.

72 A. Apffel, J.A. Chakel, S. Fischer, K. Lichtenwalter, W.S. Hancock, J. Chromatogr. A 777 (1997) 3.

73 Y. Kato, T. Kitamura, A. Mitsui, Y. Yamasaki, T. Hashimoto, T. Murotsu, S. Fukushige, K. Masubara, J. Chromatogr. 447 (1988) 212.

74 E. Uhlmann, A. Peyman, Chem. Rev. 90 (1990) 543.

75 P.A. Limbach, P.F. Crain, J.A. McCloskey, J. Am. Soc. Mass Spectrom. 6 (1995) 27.

76 A. Apffel, J.A. Chakel, S. Fischer, K. Lichtenwalter, W.S. Hancock, Anal. Chem. 69 (1997) 1320.

77 C.G. Huber, A. Premstaller, W. Xiao, H. Oberacher, G.K. Bonn, P.J. Oefner, J. Biochem. Biophys. Meth. 47 (2001) 5.

78 S. Lubbad, B. Mayr, C.G. Huber, M.R. Buchmeiser, J. Chromatogr. A (2002) submitted.

79 N. Portier, A. Van Dorsselaer, Y. Cordier, O. Roch, R. Bischoff, Nucl. Acids Res. 22 (1994) 3895.

80 D.C. Muddiman, A.P. Null, J.C. Hannis, Rapid Commun. Mass Spectrom. 13 (1999) 1201.

81 M.T. Krahmer, Y.A. Johnson, J.J. Walters, K.F. Fox, A. Fox, M. Nagpal, Anal. Chem. 71 (1999) 2893.

82 M.J. Greig, R.H. Griffey, Rapid Commun. Mass Spectrom. 9 (1995) 97.

83   C.G. Huber, M.R. Buchmeiser, Anal. Chem. 70 (1998) 5288.

84   C.G. Huber, H. Oberacher, Mass Spectrom. Rev. (2002) in press.

85   K.B. Mullis, F. Ferre, R.A. Gibbs, The Polymerase Chain Reaction, Springer Verlag, Heidelberg, 1994.

86   P.J. Oefner, C. G. Huber, E. Puchhammer-Stöckl, F. Umlauft, G.K. Bonn, C. Kunz, BioTechniques 16 (1994) 898.

87   A.C. Jones, J. Austin, N. Hansen, B. Hoogendoorn, P.J. Oefner, J.P. Cheadle, M.C. O'Donovan, Clin.Chem. 45 (1999) 1133.

88   T. Wagner, D. Stoppa-Lyonnet, E. Fleischmann, D. Muhr, S. Pagès, T. Sandberg, V. Caux, R. Moeslinger, G. Langbauer, A. Borg, P.J. Oefner, Genomics 62 (1999) 369.

89   W. Xiao, P.J. Oefner, Hum. Mutat. 17 (2001) 439.

90   M.C. O'Donovan, P.J. Oefner, C.S. Roberts, J. Austin, B. Hoogendoorn, C. Guy, G. Speight, M. Upadhyaya, S.S. Sommer, P. McGuffin, Genomics 52 (1998) 44.

91   V.N. Kristensen, D. Kelefiotis, T. Kristensen, A.L. Borresen-Dale, Biotechniques 30 (2001) 318.

92   I. G. Gut, Hum. Mutat. 17 (2001) 475.

93   E. Nordhoff, F. Kirpekar, P. Roepstorff, Mass Spectrom. Rev. 15 (1996) 76.

94   A. Edwards, H. A. Hammond, L. Jin, C. T. Caskey, R. Chakraborty, Genomics 12 (1992) 241.

95   H. Oberacher, W. Parson, R. Mühlmann, C. G. Huber, Anal. Chem. 73 (2001) 5109.

96   A. Premstaller, W. Xiao, H. Oberacher, M. O'Keefe, D. Stern, T. Willis, C. G. Huber, P. J. Oefner, Genome Res. 11 (2001) 1944.

F. Švec, T.B. Tennikova and Z. Deyl (Editors)
Monolithic Materials
Journal of Chromatography Library, Vol. 67

Chapter 20

Synthetic Polymers

David SÝKORA[1], František ŠVEC[2]

[1]*Department of Analytical Chemistry, Institute of Chemical Technology, Technická 5, 166 28 Prague, Czech Republic*
[2]*University of California, Department of Chemistry, Berkeley, CA 94720, U.S.A.*

CONTENTS

20.1 INTRODUCTION

Synthetic polymers present many unique separation challenges because they consist of macromolecules featuring a distribution of structurally different chains that can vary in both chain length and end groups. Separation of copolymers is even more difficult since in addition to the previous two variables they also may differ in ratios of individual repeat units and their sequence distribution. Therefore, a variety of methods have been developed for the characterization of molecular parameters of both natural and synthetic polymers.

20.2 SIZE EXCLUSION CHROMATOGRAPHY

Synthetic polymers are currently primarily characterized by size exclusion chromatography (SEC). Within the last four decades, SEC has almost completely replaced the tedious fractionation methods that were used for the determination of the molecular weight distribution of synthetic macromolecules [1,2]. Size exclusion chromatography is based on differences in the accessibility of pores of the separation medium to molecules of varying hydrodynamic volume. This method provides accurate, reliable, and reproducible data on the molecular weight distribution of soluble polymers within a reasonably short period of time. In general, the separation medium for SEC must be inert, because all interactions between the solutes and the stationary phase have to be avoided. Modern SEC employs small, rigid, polymeric or silica-based beads of controlled pore size to separate molecules differing in hydrodynamic volume, and to calculate average molecular weights and molecular weight distribution information for desired polymers.

Typical SEC separation run times are often in the range of several tens of minutes since the flow rate is low (typically 1–2 mL/min) to achieve the highest column efficiency and a series of rather large columns (typically 2–4 columns, each 300 × 8 mm i.d.) with large void volumes are often used to achieve the desired resolution. Despite only recently reported success [3], substantial acceleration of true SEC separations remains difficult to achieve.

20.3 ALTERNATIVE METHODS FOR SEPARATION OF SYNTHETIC POLYMERS

Several other methods that complement SEC, such as matrix-assisted laser desorption/ionization – mass spectrometry (MALDI-MS), electrophoresis, field-flow fractionation (FFF), and hydrodynamic liquid chromatography have also been used for characterization of molecular parameters of polymers [4-6]. In addition to molecular weight characteristics some of these methods may also provide information on the chemical composition of the polymers.

In consideration of the application of monolithic stationary phases for the separation of synthetic polymers, the most important alternative to SEC is interactive high-performance liquid chromatography (HPLC). HPLC, which was introduced in the late 1960s, has sparkled an enormous interest in chromatographic separation process. In particular, HPLC in the reversed-phase mode is used most frequently for the separation of small hydrophobic molecules and molecules of medium polarity. Although only marginal compared to SEC, interactive HPLC can also be used for the selective separation and characterization of synthetic polymers. These separations rely on differences in enthalpy resulting from the interaction of a polymer with the stationary phase [7-9] or solubility [7,8,10-12].

The literature contains several separation mechanisms for the chromatographic separation of synthetic polymers in interactive modes. Stadalius *et al.* [7] have indicated that only two of them including the conventional retention resulting from adsorption at the surface of the stationary phase and separations using precipitation/redissolution processes in a gradient of the mobile phase appear to be really relevant [7-11]. The latter mode also called precipitation/redissolution HPLC [8,10-12] involves mobile phase typically consisting of two solvents that differ significantly in their solvating strength with respect to the polymer. Use of columns packed with small pore media that cannot be permeated by the polymer molecules facilitates the separation.

20.3.1 Adsorption HPLC

As a rule, the retention factor in a fixed solvent increases exponentially with increasing molecular weight of polymers and the retention time for large molecules might be too long for practical use. Therefore, the use of isocratic elution with a single solvent is limited to the separations of oligomers [10,13,14], although the isocratic separation of high molecular weight polymers has also been attempted [15-17]. In contrast, elution using a mobile phase with increasing solvency [11,18-21] or a gradient of temperature [22-25] enables separations of both oligomers and polymers in a much shorter period of time. In addition to the determination of molecular weight, interactive HPLC can also separate polymers according to their chemical composition [26]. In contrast to SEC, interactive HPLC commonly offers much higher resolution and a larger number of variables such as stationary phase properties, mobile phase composition, flow rate, gradient profile, as well as temperature can be used to control the chromatographic process and fine-tune the separation.

Because of the specific requirements including the need to optimize separation conditions for each individual polymer, re-equilibration between consecutive runs, and method validation for each chromatographic system, HPLC currently only complements the most popular SEC.

20.3.2 Precipitation/redissolution HPLC

The concept of using polymer precipitation followed by subsequent redissolution was first used for the fractionation of polymers in the early 1950s [27]. However, this approach was exploited for the high-performance liquid chromatography of polymers only recently [12,26,28]. In this technique, the polymer solution is injected into a stream of the mobile phase, in which the polymer is not soluble. Therefore, the macromolecules precipitate and form a separate gel phase that adsorbs at the surface of the packing and does not move along the column. The solvency of the mobile phase is then increased gradually until it reaches a point at which some of the macro-molecules start to dissolve again and travel with the stream. If the column is packed with a support that contains pores smaller than the size of the polymer molecules, the mobile phase penetrates the pores, while the dissolved molecules move with the stream only through the interstitial voids. As a result, the polymer solution moves forward faster than the solvent gradient, reaches point at which the solvency of the mobile phase is no longer good enough to hold the polymer in solution, and the polymer precipitates again. This newly formed precipitated gel phase will then redis-solve only when the solvent strength is enhanced. A multitude of such precipitation/re-dissolution steps is repeated over and over again until the macromolecule finally leaves the column. The solubility of each polymer molecule in the mobile phase depends on both molecular weight and composition. As a result, separation of macro-molecules differing in these properties is achieved. Under ideal conditions, the column packing should serve only as a support for the precipitated phase and the separation should not depend on its chemistry. However, this is often not the case, and, in reality, interactions often play a significant role, even in the precipitation/redissolution separation processes [28,29].

20.4 COLUMN TECHNOLOGIES FOR THE SEPARATION OF POLYMERS

In SEC the stationary phase has usually a format of rigid porous polymeric or silica-based beads. The separation results from the distribution of the polymer molecules between the moving mobile phase and the stagnant portion of the mobile phase retained within the porous structure of the stationary phase. Stationary phases used for interactive HPLC may also have the same shape of porous beads. However, a slow mass transfer within the pores is a serious problem in the chromatographic separations of macromolecules such as synthetic polymers using packed columns [26,30]. Reducing the particle size of porous beads can offset the negative effect of diffusion [31]. However, their use may create difficulties with their packing in columns and very high back pressure they would generate. Obviously, the restricted diffusion in pores that is characteristic of porous media is completely eliminated in non-porous separation media [32-34]. However, for the characterization of synthetic polymers they have seldom been used [29,35]. Small nonporous particles (1.5–4 µm) are preferred because they have sufficient surface area for the separation of *detectable* amounts of components in the polymer sample. Yet, only relatively short columns (3–5 cm) can be used to keep the back pressures within limits tolerated by typical instrumentation.

Another way to solve the slow diffusion of macromolecules is to accelerate mass transfer by convection. Regnier designed a technique called perfusion chromatography, in which a small fraction of the mobile phase is driven through the very large pores (600–800 nm) of chromatographic particles, thus increasing the diffusivity of large molecules [36,37]. However, only the recent development of monolithic columns enabled the complete exploitation of the potential of rapid determination of molecular parameters of synthetic polymers using precipitation/ redissolution separations.

20.4.1 Monolithic Columns

In contrast to the standard beaded separation media with typical porous properties, the monolith provides almost no separation in the size exclusion mode. SEC experiments confirm the lack of separation of polystyrene standards in the range of molecular weights between 10^4 and 10^6 and clearly document the absence of medium-sized pores (Fig. 20.1). However, this is a substantial advantage if the monolithic column is to be used for separations that rely on the interaction of the solutes with the stationary and mobile phases rather than on their sizes (hydrodynamic volumes), because the size exclusion effect does not interfere with the separation. The most often used poly(styrene-*co*-divinylbenzene) monoliths are sufficiently hydrophobic for use in reversed-phase chromatography without any further modification [38].

20.5 RAPID SEPARATIONS

In order to achieve a good separation, classical chromatographic columns are typically packed with micrometer-sized particles. It is well known that resolution increases with a decrease in the particle size of the sorbent. However, the reduction of particle diameter leads simultaneously to a rapid build up of a back pressure in the

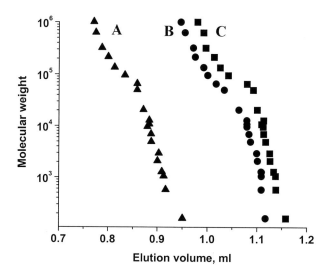

Fig. 20.1. Size exclusion calibration curves for monolithic poly(styrene-*co*-divinylbenzene) (A), poly(glycidyl methacrylate-*co*-ethylene dimethacrylate) (B), and poly(2,3-dihydroxy-propyl methacrylate-ethylene dimethacrylate) (C) columns (Reprinted from ref. [44]. Copyright 1996 Elsevier). Separation conditions: Solvent tetrahydrofuran, flow rate of 0.25 mL/min, polystyrene standards, UV detection 254 nm.

column. Consequently, a trade-off is required to achieve a reasonable resolution under acceptable flow rate and analysis time. Hence, column packings of particle size 3–10 μm are the most common at the present time in analytical HPLC.

One of the most significant advantages of the specifically designed monolithic columns is their capability to facilitate fast separations using very high flow rates of the mobile phase while still affording moderate back pressure. For example, Fig. 20.2 shows fast separations of three polystyrene standards using poly(styrene-*co*-divinyl-benzene) monolithic column at a flow rate of 20 mL/min and two different very steep gradients of the mobile phase. The back pressure of the 50 × 8 mm i.d. column using pure tetrahydrofuran (THF) was only 2.6 MPa despite such a high flow rate. The separation was excellent at a gradient time of 30 s, and three well resolved peaks are obtained within 15 seconds (Fig. 20.2a). An application using an even faster gradient of 12 s still leads to a good separation, which is achieved within a mere 4 seconds (Fig. 20.2b).

Figure 20.3 shows the chromatograms obtained using a column packed with po-rous monodispersed 7 μm poly(styrene-*co*-divinylbenzene) beads. The separation con-ditions were the same as those used for the separation using monolithic column. In contrast, a back pressure of 9.6 MPa at 20 mL/min was significantly higher than that observed for the monolithic column of the same dimensions. The separations using

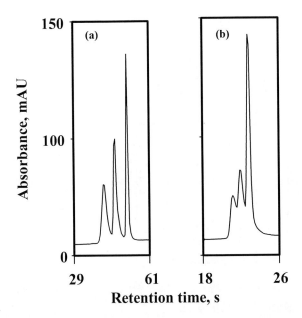

Fig. 20.2. Rapid separations of polystyrene standards on monolithic column (Reprinted from ref. [35]. Copyright 1996 Elsevier). Conditions: column: 50 × 8 mm i.d, poly(styrene-*co*-divinylbenzene) monolith; mobile phase, linear gradient from 0 to 100% tetrahydrofuran in methanol within 30 (a), and 12 s (b); flow rate, 20 mL/min; analytes, mol. weight 9 200 (1), 34 000 (2), and 980 000 (3), 3 mg/mL of each standard in tetrahydrofuran; injection volume 20 µL; UV detection, 254 nm.

30 s gradients are comparable with columns of both types (Figs. 20.2a, 20.3a). However, the polystyrene standards are co-eluted from the packed column in a single peak if a 12 s gradient is used (Fig. 20.3b). Obviously, an increase in the gradient steepness while keeping the flow rate unchanged, leads to rapid deterioration of the separation. Such result is not completely unexpected and can be easily explained by the slow mass transfer within the stagnant mobile phase located in the pores of the packing. In contrast, the total convection of the entire mobile phase through the large pores of the monolithic stationary phase facilitates very fast separations with only moderate decrease in resolution.

20.5.1 Optimization of Speed and Solvent Consumption

In general, an increase in the resolution of an SEC system can be achieved using a better column packing and/or a longer column, which represent a "hardware approach", rather than by changing the chromatographic conditions such as the flow rate. In contrast to SEC, which is always performed in an isocratic mode using an

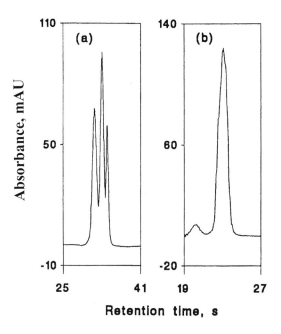

Fig. 20.3. Rapid separation of polystyrene standards on conventional particle packed column (Reprinted from ref. [35]. Copyright 1996 Elsevier). Conditions: column, 50 × 8 mm i.d., porous poly(styrene-*co*-divinylbenzene) beads; for other conditions see Fig. 20.2.

optimal flow rate that cannot be changed without an adverse effect on the separation efficiency, several variables controlling the resolution are available for the gradient elution technique. For example, the speed of the separation may be increased considerably using high flow rates and steep gradients. However, high flow rates are not possible for columns packed with typical small microparticles as they often exhibit prohibitively high flow resistance. In contrast, molded rod columns allow the application of very high flow rates at a reasonable back pressure, thus making very fast chromatographic runs possible and the method optimization more flexible.

According to Equation 20.1, the average retention factor in the gradient elution, $\bar{k}$, depends on the gradient time, t_G, the volume flow rate, F, the change in composition of the mobile phase $\Delta\phi$, the column dead volume V_m, and the constant, S, that is calculated for each solute from the retention data in isocratic systems and characterizes the strength of the interaction between the solute and the stationary phase [30]:

$$\bar{k} = (t_G \cdot F) / (\Delta\phi \cdot V_m \cdot S) \tag{20.1}$$

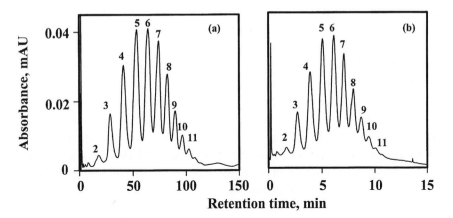

Fig. 20.4. Effect of flow rate and gradient time on the separation of styrene oligomers using monolithic poly(styrene-*co*-divinylbenzene) column (Reprinted from ref. [40]. Copyright 1996 American Chemical Society). Conditions: column, 50 mm × 8 mm i.d.; (a) mobile phase, linear gradient from 30 to 60% tetrahydrofuran in water within 200 min; flow rate, 1 mL/min; (b) mobile phase, linear gradient from 30 to 60% tetrahydrofuran in water within 20 min; flow rate, 10 mL/min; analyte, 15 mg/mL in tetrahydrofuran; injection volume 20 μL; UV detection, 254 nm; peak numbers correspond to the number of styrene units in the oligomer.

For a constant range of mobile phase composition and for a specific column and solute, the denominator in Equation 20.1 remains constant and the average retention factor, $\bar{k}$, depends only on the gradient time and the flow rate. Because the product of these variables, $t_G \cdot F$, is the gradient volume, V_G, equal peak capacities that are independent of flow rate and gradient steepness should be achieved within the same gradient volume. Figure 20.4 shows separations of styrene oligomers obtained on 50 × 8 mm i.d. monolithic column with gradient times of 200 and 20 min and flow rates of 1 and 10 mL/min, respectively. Because the gradient volume of 200 mL is the same for both separations, no significant differences can be seen between these two chromatograms. This result supports the validity of Equation 20.1 and clearly shows that a remarkable acceleration of the separation process is feasible using the monolithic stationary phase.

The speed of the analysis itself is only one of a number of important parameters concerning optimization of the gradient separations. Several other aspects are just as important. For example, much of the present efforts are focused on the reduction of solvent consumption in fast gradient techniques because of both environmental concerns and waste solvent disposal costs. The original separations of synthetic polymers were carried out using columns with, for today's standards, a rather wide internal diameter of 8 mm. Under these conditions, the requirement to achieve sufficiently large gradient volume to maintain the desired resolution requires high volumetric flow rates typically in the range of 10–20 mL/min. Consequently, the daily consumption of the mobile phase can easily reach up to 6–12 L just for a single 8 mm i.d. column. In addition, these high flow rates are also very close to the upper limit of the typical analytical HPLC pumps, which is 10 mL/min. Thus, the application of semipreparative HPLC pumps would be needed.

These problems provided a strong motivation to reduce the size of the monolithic columns while retaining their excellent separation characteristics. This has also led to the recent transfer of the concept of the rapid separations of synthetic polymers to smaller monolithic column format [39]. Almost identical chromatographic results were obtained using newly developed commercial 4.6 mm i.d. monolithic rod columns (ISCO Inc., Lincoln, NE) with a threefold decreased cross section compared to the previously used 8 mm i.d. columns (*vide infra*). This modification enables reduction in the solvent consumption by more than three times while maintaining the linear flow velocity at the same level as in the 8 mm i.d. monoliths. Although further reduction of the column size is feasible, the separation of detectable amounts of polymer molecules in the interactive mode requires sufficiently large surface area and column overloading can become a serious limitation.

20.6 RETENTION MECHANISM

20.6.1 Oligomers

The separation of styrene oligomers by HPLC on reversed-phase octadecyl silica columns in a gradient of the mobile phase has been studied as early as 1983 [10]. These experiments demonstrated that the separation followed the normal course for reversed-phase chromatography of small molecules and that retention depended on both the composition of the mobile phase and the number of repeat units in the oligomer. Larger oligomer molecules, having longer hydrophobic chains, exhibited longer retention times or required a higher percentage of good solvent in the mobile phase used for their elution. This means that shorter oligomers eluted prior to longer ones, quite unlike size exclusion chromatography for which large molecules elute first.

Figure 20.5 shows the separations of a commercial sample of polystyrene oligomers with a number average molecular weight of 630 in a short monolithic column using the gradient HPLC mode and compares it with the separation in a series of four high-performance SEC columns. Despite a very narrow polydispersity of 1.07, both chromatograms exhibit a number of peaks that can be assigned to the individual styrene oligomers. Addition of a small amount of the dimer (2,4-diphenyloctane) to

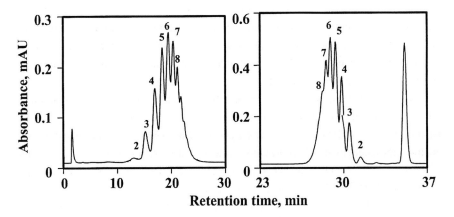

Fig. 20.5. Separation of styrene oligomers by reversed-phase (a) and size-exclusion chromatography (b) (Reprinted from ref. [40]. Copyright 1996 American Chemical Society). Conditions: (a) monolithic poly(styrene-*co*-divinylbenzene) column, 50 mm × 8 mm i.d.; mobile phase, linear gradient from 30 to 60% tetrahydrofuran in water within 20 min; flow rate 1 mL/min; injection volume 20 μL; UV detection, 254 nm; (b) series of four 300 × 7.5 mm i.d. PL Gel columns (100 Å, 500 Å, 10^5 Å and Mixed C); mobile phase tetrahydrofuran; flow rate, 1 mL/min; injection volume 100 μL; toluene added as a flow marker; UV detection, 254 nm; temperature 25°C; peak numbers correspond to the number of styrene units in the oligomer.

this sample led to an increase in the height of peak number 2, thus confirming clearly its identity. As expected, because of the opposite order of elution, the chromatograms in Fig. 20.5 are almost mirror images. The resolution achieved with the monolithic column is good and the chromatogram in Fig. 20.5a even indicates the presence of an undecamer. Unlike SEC separation in Fig. 20.5b, the resolution of gradient HPLC shown in Fig. 20.5a can be further improved by simply increasing gradient volume (Fig. 20.4).

A simple calculation [40] provides the composition of the mobile phase in which the oligomers are eluted. For alkylbenzenes this composition is a linear function of molecular weight, which obeys equation 20.2 for a process driven by hydrophobic interaction [10]:

$$C_{THF} = a + b \cdot M \tag{20.2}$$

In this equation, C_{THF} is the percentage of tetrahydrofuran in the mobile phase, a and b are constants, and M is the molecular weight.

In contrast, the data obtained for styrene oligomers is a better fit for equation 20.3 [40]:

$$C_{THF} = a' + b' \cdot log\ M \qquad\qquad (20.3)$$

This indicates that for the separation of styrene oligomers, selectivity decreases as the molecular weight of the oligomers increases, and therefore, the rules typical of reversed-phase chromatography of small molecules no longer completely apply.

Earlier literature often reports the large effect of sample load on peak shapes and retention for the gradient elution of macromolecules [8,11,12,41,42]. In contrast to packed columns, monolithic columns tolerate much higher sample injections and the separation of 0.05–6.25 mg of styrene oligomers using a 50 × 8 mm i.d. poly(styrene-*co*-divinylbenzene) monolith could easily be achieved [40].

20.6.2 Polymers

Reversed-phase separation of polystyrenes differs from that of small molecules (alkylbenzenes) and oligomers [7,8,10]. In particular, the dependence of the logarithm of the retention factor, k, on temperature is much steeper, and it is not a linear function of molecular weight. However, the general elution pattern remains unchanged and, once again, the species with lower molecular weights elute prior to those with higher molecular weights.

Tetrahydrofuran was found to be a suitable good solvent for the gradient elution of polystyrene standards. Various poor solvents such as water, methanol, and acetonitrile were evaluated as precipitants [40]. Provided the contents of the relevant solvents were properly adjusted, similar separations were obtained for all three precipitants. The amount of tetrahydrofuran in the different mobile phases used to elute the peaks was then plotted against $M^{-1/2}$ to afford straight lines that were fitted to the equation [40]:

$$C_{THF} = a'' + b'' \cdot M^{-1/2} \qquad\qquad (20.4)$$

where a'' and b'' are constants and M is again the molecular weight of the polymer. This finding confirms that precipitation of the polymer in the mobile phase is the driving force for the separation. Water was found to be the most powerful precipitant but the selectivity of system tetrahyrofuran–water was poor. Acetonitrile had the highest selectivity but it was the poorest precipitant and up to about 80% of acetonitrile had to be added to the mobile phase in order to observe retention of polystyrene with a molecular weight of 9,200. An extrapolation revealed that styrene oligomers with a molecular weight of less than about 2,600 could be completely soluble in pure acetonitrile. This effect disqualifies acetonitrile from use for the chromatographic analysis of unknowns or broad molecular weight samples because the low molecular weight part of the sample may not precipitate. Methanol proved to be the precipitant of choice as it provides sufficient selectivity and still precipitates even the oligomers.

20.6.3 Mixed mechanism

The results obtained from the separations of alkylbenzenes, styrene oligomers and polymers discussed above revealed different retention behavior for each of these three families of compounds [40]. The percentage of the good solvent needed for the elution is a linear function of molecular weight for low molecular weight compounds, of logarithm of molecular weight for oligomers and of square root of molecular weight for polymers. The appropriate S-shaped curve that depicts the percentage of tetrahydrofuran required for the elution of solutes that have similar chemical composition but that differ in their molecular weights can be clearly divided into three sections marked out in Fig. 20.6. This indicates that the separation mechanism is different for each class of compounds and changes from a typical reversed-phase mechanism to one involving solubility differences, i.e. precipitation/redissolution. Thus, for the separation mechanism of polystyrenes on monolithic columns it can be concluded that indeed two different retention mechanisms, adsorption and precipitation/redissolution, control the separation of high-molecular weight solutes simultaneously.

From the extensive study of Quarry *et al.* [8] follows that the molecular weight of the polymer, its solubility, and the injected amount are the most important variables in determining which retention mechanism is operative in the specific separation. For example, the composition of the mobile phase required for elution remains constant within a broad range of the amount of sample injected for the adsorption mode, while the percentage of good solvent necessary for elution increases in the case of the precipitation/redissolution mechanism. Measurements of isocratic retention also provide insight into the retention mechanism. Thus, in adsorption mechanism the retention time depends on the composition of the mobile phase and polymer analytes can be always eluted as a distinct near-Gaussian peak. In contrast, it is impossible to elute peaks at retention times exceeding the retention time of the standard under non-retained conditions in the precipitation/redissolution mode. Experiments with poly(styrene-*co*-divinylbenzene) monolithic columns clearly indicated the strong effect of precipitation in the separation process of polystyrenes even for standard with a relatively low molecular weight of 9,680 in the tetrahydrofuran–methanol system [35]. Although the adsorption mechanism of retention cannot be excluded completely, the experimental data suggest that the precipitation/redissolution effect is quite strong and contribute considerably to the retention of synthetic polymers in the monolithic column. This is expected as the monolithic columns have a limited surface area in the large pores available for the interactions [43] and, in contrast to some of the silica based separation media, these columns also contain very small pores that are not

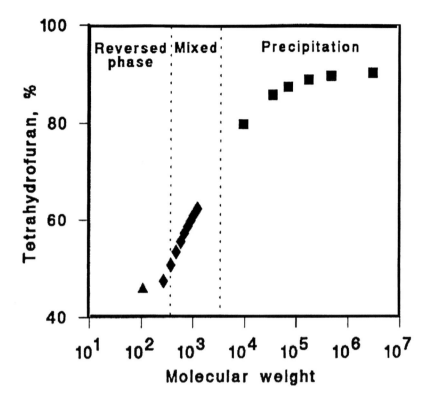

Fig. 20.6. Composition of the mobile phase for elution of ethylbenzene, styrene oligomers, and polystyrenes from the monolithic poly(styrene-*co*-divinylbenzene) column as a function of their molecular weight (Reprinted from ref. [40]. Copyright 1996 American Chemical Society). Conditions: column, 50 mm × 8 mm i.d.; mobile phase, linear gradient from 30 to 60% tetrahydrofuran in water within 20 min; flow rate, 1 mL/min; analytes, injection volume 20 μL; UV detection, 254 nm.

permeable by polymer solutes. Therefore, these solutes migrate during the gradient elution through the column faster than the mobile phase zone is capable of dissolving them. As a result, the polymer molecules precipitate as they move from the zone of stronger solvent into the poorer solvent and dissolve only when the solvent strength is again sufficient to dissolve them.

20.7 SEPARATIONS USING NON-LINEAR GRADIENTS

20.7.1 Polystyrenes

The precipitation/redissolution technique using monolithic columns has a great potential for high throughput characterization of synthetic polymers. However, precipitation/redissolution HPLC is a relative method and, therefore must be calibrated for determination of actual molecular weights. Typically, the calibration is performed by determining retention times of polymer standards with narrow molecular weight distribution. Logarithm of the nominal molecular weight is then plotted against the retention time or volume to obtain the desired calibration curve. The calibration curve should span as broad a range of molecular weights as possible to ensure that the molecular parameters of the future "unknown" samples will fall within this range. Since no "universal" calibration currently exists for the gradient HPLC technique, standards of the same chemical composition as the analyzed samples must be used for the calibration. This requirement restricts the precipitation/redissolution analysis to polymers for which standards with well-defined molecular weights are available.

Use of the linear gradient of THF in methanol discussed in previous sections affords for polystyrene standards within the molecular weight range 3,000 to 980,000 a non-linear calibration curve with an exponential shape (Fig. 20.7) [39]. This shape is not favorable since the resolution for polymer molecules of higher molecular weight is lower than that for molecules with lower molecular weight. This "non-linearity" also affects the peak shape of samples of higher molecular weight. The peak shape is not Gaussian and appears to exhibit fronting.

In order to avoid this undesired effect, a non-linear gradient was developed with a profile that compensates for the decreasing resolution in the high molecular weight region and leads to a linear calibration curve [39]. The gradient shape can be inferred from cloud point curves that are obtained by titration of a THF solution of polystyrene standards with methanol. These curves were found well-suited guide to the design of the gradient shape. The final optimization has led to a gradient composed of five linear ramps that are steeper at the beginning and shallower at the end. An excellent separation of a mixture of 9 polystyrene standards, shown in Fig. 20.8a, was achieved using this "tailored" non-linear gradient within 6.5 minutes at a flow rate of only 1 mL/min. In contrast, Fig. 20.8b shows the SEC separation of the same mixture of PS standards using a tandem of 2 columns at the same flow rate that required 18 min. This figure also depicts calibration curves for both methods. The HPLC calibration is linear over the entire range of measured molecular weights.

The accuracy of the HPLC method featuring non-linear gradient elution for molecular weight determinations was compared to that of HPLC with linear gradient and

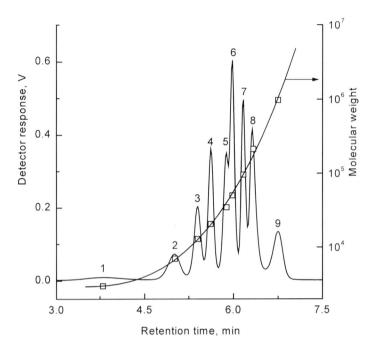

Fig. 20.7. Separation of a mixture of nine polystyrene standards on a monolithic 50 × 4.6 mm i.d. column using linear gradient and the corresponding calibration curve obtained from the elution data (Reprinted from ref. [39]. Copyright 2000 Wiley). Separation conditions: linear gradient 0–60% THF in methanol in 5 min; flow rate 1 mL/min.; sample volume 10 μL; overall sample concentration 18 mg/mL (2 mg/mL of each standard) in THF; ELSD detection. Molecular weights of polystyrene standards: 3,000 (1), 7,000 (2), 12,900 (3), 20,650 (4), 34,800 (5), 50,400 (6), 96,000 (7), 214,500 (8), and 980,000 (9).

SEC using the binary mixtures of two narrow molecular weight polystyrene standards. These binary mixtures were used to validate the absence of mutual effects of polymer molecules of different molecular weights on both the precipitation and redissolution processes. Table 20.1 summarizes the experimental results. Using the linear gradient approach, both M_w and M_n calculated for standards with lower molecular weight were slightly higher than those calculated from SEC. Simultaneously, the molecular weight values of high molecular weight standards were generally much smaller because of the peak tail compression effect resulting from the non-linearity of the calibration curve. In contrast, the agreement of M_w and M_n obtained using the non-linear gradient HPLC elution and SEC was much better, their chromatographic traces were very close and matched well the nominal molecular weights provided by manufacturer. This demonstrates that the optimized precipitation/redissolution chromatography can afford data that are in a good agreement with those obtained using SEC.

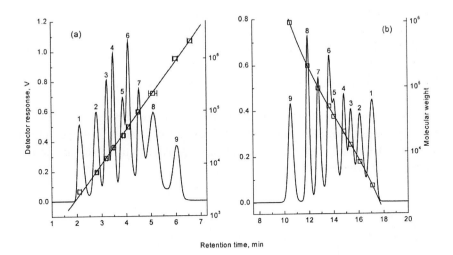

Fig. 20.8. Separation of a mixture of nine polystyrene standards using (a) non-linear gradient in a monolithic 50 × 4.6 mm i.d. poly(styrene-*co*-divinylbenzene) column or (b) SEC in two 300 × 7.5 mm i.d. PL gel columns together with the corresponding calibration curves obtained from the elution data (Reprinted from ref. [39]. Copyright 2000 Wiley). Separation conditions: flow rate 1 mL/min; overall sample concentration 18 mg/mL (2 mg/mL of each standard) in THF; ELSD detection; (a) HPLC eluent: non-linear gradient consisting of 0–35% THF in methanol in 0.5 min, 35–50% in 1.5 min, 50–55% in 1min, 55–59% in 1 min, and 59–60% in 1 min; (b) SEC eluent: THF. Molecular weights of polystyrene standards: 3,000 (1), 7,000 (2), 12,900 (3), 20,650 (4), 34,800 (5), 50,400 (6), 96,000 (7), 214,500 (8) and 980,000 (9).

The use of non-linear gradients also helps to accelerate the separation. For example, Fig. 20.9 shows the rapid and efficient separation of polystyrene standards with molecular weight ranging from 3,000 to 980,000 using the poly(styrene-*co*-divinylbenzene) monolithic column at only a moderate volumetric flow rate of 2 mL/min.

Similar to the styrene oligomers, overloading the monolithic columns with polystyrenes is of much less concern compared to conventional columns containing particulate packings. Figure 20.10 shows the separations of mixture of five polystyrene standards on a 50 × 4.6 mm i.d. monolithic column at overall concentrations ranging from 2.5 to 50 mg/mL. The retention times for all of the peaks remain almost constant over the entire load range and demonstrate the broad dynamic range of the monolithic column. This result also suggests that even small monolithic columns may be useful for the preparative separation of polymers. Higher column loading can also be achieved by injecting larger volumes of samples. However, comparative experiments document that this approach leads to a larger shift in retention times for high molec-

TABLE 20.1

MOLECULAR PARAMETERS OF INDIVIDUAL POLYSTYRENE STANDARDS WITH NARROW MOLECULAR WEIGHT DISTRIBUTION AS CALCULATED FOR EACH PEAK AFTER THE SEPARATIONS OF THEIR BINARY MIXTURES DETERMINED BY HPLC AND SEC

Chromatographic conditions: HPLC – 50 × 4.6 mm i.d. monolithic poly(styrene-*co*-divinylbenzene) column (ICSO, Inc.; Lincoln, NE), evaporative light scattering detector (Polymer Laboratories), conditions of both linear and five-step gradient were described in Figs. 20.7 and 20.8a, respectively. SEC – two 300 × 7.5 mm i.d. 10^3Å and 10^5Å PL gel columns (Polymer Laboratories), THF, flow rate 1 mL/min. Injections 20 μL of polymer solution, concentration 1–2 mg/mL

Binary polymer mixture[a]	$M_w \times 10^{-3}$			$M_n \times 10^{-3}$			M_w/M_n		
	HPLC		SEC	HPLC		SEC	HPLC		SEC
	Linear	Stepwise		Linear	Stepwise		Linear	Stepwise	
18.1k	18.9	17.7	17.4	18.5	17.6	17.2	1.01	1.01	1.01
150k	139.8	153.4	148.5	137.8	152.1	146.1	1.01	1.01	1.02
30k	32.6	28.9	30.1	32.3	28.7	29.9	1.01	1.01	1.01
200k	168.9	211.4	211.6	166.5	209.4	209.9	1.01	1.01	1.01

[a]Values provided by manufacturer

ular weight polymers. In addition, larger injection volumes introduce larger volumes of the good solvent into the system thereby triggering undesired pre-elution [11,26,35].

20.7.1.1 Effect of surface chemistry

Using an optimized non-linear five-step gradient of THF in methanol affords a very good separation of a diverse set of polystyrene standards with molecular weights ranging from 3,000 to 980,000, as shown in Fig. 20.11A. In contrast to the styrenic monolith, a monolithic poly(glycidyl methacrylate-*co*-ethylene dimethacrylate) column only afforded very poor results using identical elution conditions (Fig. 20.11B). This result suggests that the more polar surface chemistry of the methacrylate-based column with its numerous epoxide groups might not be well suited for the separation of the highly non-polar polystyrene samples. However, as shown in Fig. 20.12 the peaks of polystyrene standards injected and eluted individually using the same separa-

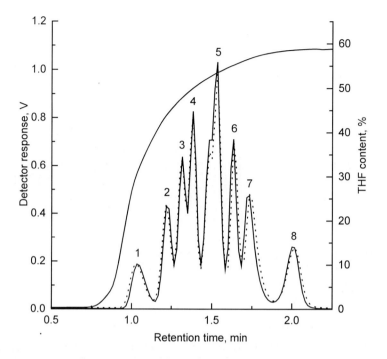

Fig. 20.9. Rapid separation of a mixture of eight polystyrene standards using a monolithic 50 × 4.6 mm i.d. poly(styrene-*co*-divinylbenzene) column and the corresponding gradient profile monitored by the UV detector (Reprinted from ref. [39]. Copyright 2000 Wiley). Separation conditions: flow rate 2 mL/min; 1.25 min non-linear gradient of THF in methanol consisting of 0–35% THF in methanol in 0.12 min, 35–50% in 0.38 min, 50–55% in 0.25 min, 55–59% in 0.25 min, and 59–60% in 0.25 min; overall sample concentration 16 mg/mL (2 mg/mL of each standard) in THF; ELSD detection. Molecular weights of polystyrene standards: 3,000 (1), 7,000 (2), 12,900 (3), 20,650 (4), 50,400 (5), 96,000 (6), 214,500 (7), and 980,000 (8). Solid and dotted lines represent the chromatograms of the same polymer mixture recorded about 400 injections apart.

tion conditions are narrow and have a symmetrical shape. Also, this retention is a linear function of its molecular weight and the retention pattern is very similar to that observed using polystyrene-based monolith. The poor separation only occurs when a mixture of standards is injected [44].

Similarly, a monolithic poly(2,3-dihydroxypropyl methacrylate-ethylene dimethacrylate) column, prepared by hydrolysis of the glycidyl methacrylate-based column afforded a slightly better separation of the mixed polystyrene standards although with a lower resolution than observed for the polystyrene-based monolith (Fig. 20.11C). Given the higher polarity of this diol column, these results clearly suggest that the difference in separation observed for the different monolithic columns is unlikely to be solely due to differences in polarity of surface chemistry.

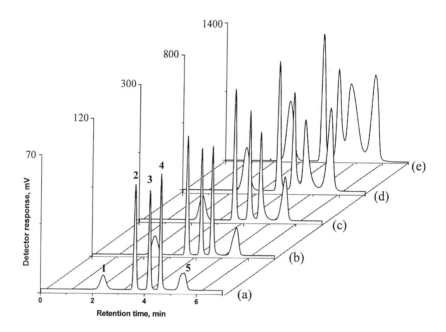

Fig. 20.10. Effect of sample concentration on retention times of polystyrene standards (Reprinted from ref. [39]. Copyright 2000 Wiley). Column: 50 × 4.6 mm i.d. poly(styrene-*co*-divinylbenzene) monolith; 5 min non-linear gradient consisting of 0–35% THF in methanol in 0.5 min, 35–50% in 1.5 min, 50–55% in 1 min, 55–59% in 1 min, and 59–60%in 1 min; flow rate 1 mL/min.; sample volume 10 µL; overall concentration: 2.5 (a), 5.0 (b), 10.0 (c), 25.0 (d) and 50 mg/mL (e) in THF; ELSD detection. Molecular weights of polystyrene standards: 4,950 (1), 30,300 (2), 66,000 (3), 135,000 (4), and 629,000 (5).

Although some results such as those shown in Fig. 20.11 may indicate an effect of surface chemistry, it is more likely that the significant differences in porous properties demonstrated in Fig. 20.1 are responsible for the failure to achieve good separations with methacrylate-based monoliths in situations where the corresponding poly(styrene-*co*-divinylbenzene) monolithic columns perform well. Porous structure with both smaller pores and larger pore volume, characteristic of monolithic methacrylate columns, do not create a suitable "scaffold" for the desired precipitation/redissolution process. In contrast, the very large pores that are almost exclusively present within the polystyrene monolith enable the rapid exchange of solvents near the precipitated polymer layer and the fast release of the re-dissolved species in the flow of the solvent allowing them to form sharp zones thereby leading to good separations. In contrast, the release of dissolved polymers from the smaller, less accessible, pores in methacry-late columns is slower. However, since the solvent exchange is augmented by convec-

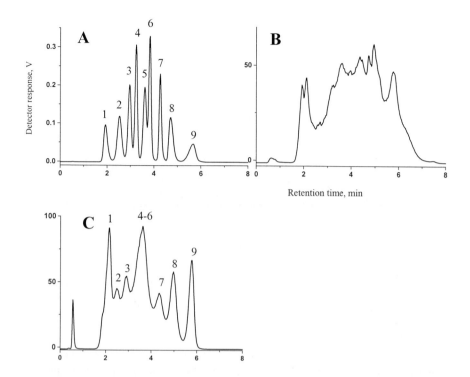

Fig. 20.11. Separation of a mixture of nine polystyrene standards using monolithic poly(styrene-*co*-divinylbenzene) (A), poly(glycidyl methacrylate-*co*-ethylene dimethacrylate) (B), and poly(2,3-dihydroxypropyl methacrylate-ethylene dimethacrylate) (C) columns. Separation conditions: Five steps non-linear 5 min long gradient consisting of 0–35% THF in methanol in 0.5 min, 35–50% in 1.5 min, 50–55% in 1min, 55–59% in 1 min, and 59–60% in 1 min at a flow rate of 1 mL/min, sample volume 10 µL, overall sample concentration 18 mg/mL (2 mg/mL of each standard) in THF, ELSD detection. Molecular weights of polystyrene standards: 3,000 (1), 7,000 (2), 12,900 (3), 20,650 (4), 34,800 (5), 50,400 (6), 96,000 (7), 214,500 (8), and 980,000 (9).

tion, it remains rapid. The combination of fast change in solvency with slow liberation of the dissolved macromolecules from the smaller pores results in massive co-elution and poor separation. This hypothesis is supported by retentions and peak shape data for standards injected separately on methacrylate-based columns as shown in Fig. 20.12. These results lead to a conclusion that monolithic columns must be tailored for each type of separations and optimized for various families of analytes.

20.7.2 Poly(methyl methacrylates)

Methacrylate polymers are more polar than polystyrene and their solubilities are therefore also different. In order to characterize poly(methacrylates) it is again neces-

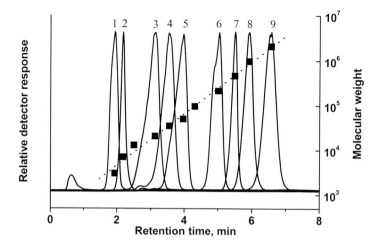

Fig. 20.12. Separation of nine polystyrene standards injected individually using monolithic poly(glycidyl methacrylate-*co*-ethylene dimethacrylate) column. For separation conditions and peak assignment see Fig. 20.11.

sary to perform all of the steps that include cloud point titration, optimization of the gradient shape, and calibration with standards. These experiments indicated that a simple two-step gradient of THF in methanol was sufficient to "linearize" the calibration curve for six poly(methyl methacrylate) standards with molecular weights in the range of 5,750–910,000 [39]. Rapid separation of this mixture accomplished within 1.2 min at a flow rate of only 1 mL/min is demonstrated in Fig. 20.13.

20.7.3 Poly(vinyl acetates)

In contrast to polystyrene and poly(methyl methacrylate), methanol is a solvent for poly(vinyl acetate). Therefore, a series of solvent-precipitant pairs were tested [39]. Molecular weight dependent elution was achieved with THF–water, dichloromethane-n-hexane, and THF–*n*-hexane. The last appeared in these tests to afford the best results. Optimization of the gradient elution was then carried out with this pair. Figure 20.14 shows the HPLC traces for "secondary" standards with a broader molecular weight distribution. Unlike the separations of polystyrene and poly(methyl methacrylate), an acceleration of the poly(vinyl acetate) elution could not be achieved using means such as changes in both flow rate and gradient steepness without a concomitant rapid deterioration of the selectivity.

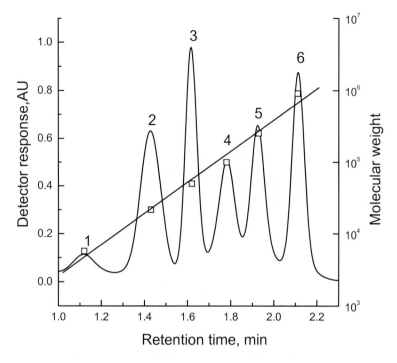

Fig. 20.13. Rapid separation of a mixture of six poly(methyl methacrylate) standards using a monolithic 50 × 4.6 mm i.d. poly(styrene-*co*-divinylbenzene) column and the corresponding calibration curve obtained from the elution data (Reprinted from ref. [39]. Copyright 2000 Wiley). Separation conditions: 2 min gradient of THF in methanol consisting of 0–35% THF in methanol in 1 min and 35–50% in 1 min; flow rate 1 mL/min.; sample volume 10 µL; overall sample concentration 30 mg/mL in THF; UV detection at 220 nm. Molecular weights of polymer standards: 5,750 (1), 21,650 (2), 50,000 (3), 100,000 (4), 254,000 (5), and 910,000 (6).

20.7.4 Polybutadienes

Since THF–methanol did not perform well in the separation of polybutadienes in HPLC mode, dichloromethane was used as the solvent in combination with methanol (precipitant) to achieve the molecular weight-dependent separation [39]. The optimized five step gradient profile once again affords a linear calibration curve for all five polybutadiene standards that were available. The separation of these polybutadienes is shown in Fig. 20.15. The run time was about 6 min at a flow rate of 1 mL/min. However, doubling the flow rate to 2 mL/min and halving the gradient time enabled the separation to be achieved within 3 min. A further acceleration was possible by running the gradient within only 1.25 min. Although the resolution of peaks 2 and 3 is

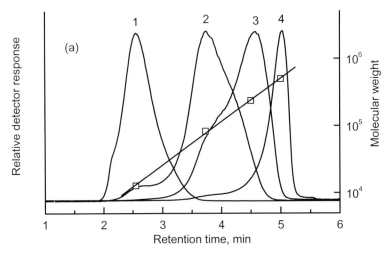

Fig. 20.14. HPLC elution traces of individually injected secondary poly(vinyl acetate) standards using a monolithic 50 × 4.6 mm i.d. poly(styrene-*co*-divinylbenzene) column and the corresponding HPLC calibration curve obtained from the elution data (Reprinted from ref. [39]. Copyright 2000 Wiley). Separation conditions: 5 min gradient consisting of 0–46% THF in hexane in 1 min, 46–60% in 1.2 min, 60–67% in 1.8 min, and 67–100% in 1 min; flow rate 1 mL/min.; sample volume 10 μL; sample concentration: 5 mg/mL; ELSD detection. Molecular weights: 12,800 (1), 83,000 (2), 129,000 (3), and 500,000 (4).

slightly lower in this rapid run, the complete separation of the mixture of polybutadiene standards is achieved within 1 min.

20.8 DETERMINATION OF MOLECULAR PARAMETERS OF "REAL LIFE" POLYMERS

A number of different polymers can been subjected to rapid separation using monolithic columns. Typical examples include polystyrene, poly(methacrylates), poly(vinyl acetate), and poly(butadiene). Moreover, as has already been described, a rigorously designed gradient profile affords a linear calibration curve for well characterized standards, which enables the determination of M_w, M_n, and polydispersity and results in good agreement with the data obtained by SEC and/or supplied by the polymer manufacturers (Table 20.1).

Although the results obtained for narrow molecular weight polystyrene standards had been encouraging, the real suitability test of the method was the separation of common polymers with a broader molecular weight distribution. For example, a variety of polystyrenes from different sources were analyzed using both precipitation/redissolution HPLC and SEC and their molecular characteristics calculated [39]. Table 20.2 compares results obtained by both chromatographic modes. Despite the differ-

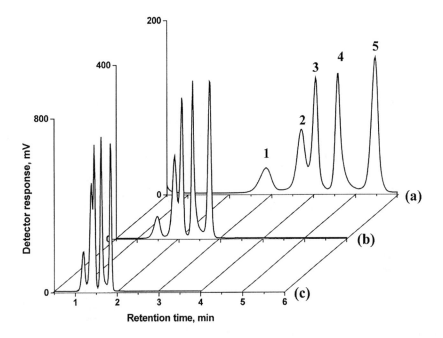

Fig. 20.15. Effect of gradient time and flow rate on the separation of a mixture of five polybutadiene standards using a 50 × 4.6 mm i.d. monolithic poly(styrene-*co*-divinyl-benzene) column (Reprinted from ref. [39]. Copyright 2000 Wiley). Separation conditions: non-linear gradients of dichloromethane in methanol (a) overall gradient time 5.5 min consisting of 0–75% in 3 min, 75–79% in 0.5 min, 79–81% in 0.5 min, 81–84% in 1 min, and 84–100% in 0.5 min; (b) overall gradient time 2.75 min consisting of 0–75% in 1.5 min, 75–79% in 0.25 min, 79–81% in 0.25 min, 81–84% in 0.5 min, and 84–100% in 0.25 min, (c) overall gradient time 1.38 min consisting of 0–75% in 0.75 min, 75–79% in 0.13 min, 79–81% in 0.12 min, 81–84% in 0.25 min, and 84–100% in 0.13 min); flow rate 1 (a) and 2 mL/min (b,c); sample volume 10 μL; overall sample concentration: 25 mg/mL (each standard 5 mg/mL) in THF; ELSD detection. Molecular weights of polymer standards 520 (1), 2,820 (2), 6,000 (3), 24,800 (4), and 215,800 (5).

ences in separation mechanisms, an excellent match was obtained. Lower polydisper-sity values calculated from HPLC data may result from peak compression in the HPLC mode and/or band broadening in the SEC mode. Moreover, it is important to emphasize that the determination using interactive HPLC in gradient mode is com-pleted in about 6 min with the possibility of a further acceleration, while the complete SEC trace is eluted only after 20 minutes.

Similar determinations of molecular parameters for two isotactic poly(methyl methacrylates) using the optimized gradient is summarized in Table 20.3 [39]. Al-though the values of molecular weight and polydispersity determined by both HPLC

TABLE 20.2

MOLECULAR PARAMETERS OF VARIOUS POLYSTYRENES AS DETERMINED BY HPLC AND SEC

Chromatographic conditions: HPLC – 50 × 4.6 mm i.d. monolithic poly(styrene-*co*-divinyl-benzene) column (ICSO, Inc.; Lincoln, NE), evaporative light scattering detector (Polymer Laboratories), conditions of five-step gradient were described in Fig. 20.8a. SEC – two 300 × 7.5 mm i.d. 10^3Å and 10^5Å PL gel columns (Polymer Laboratories), THF, flow rate 1 mL/min. Injections 20 μL of polymer solution, concentration 1–2 mg/mL

Origin	$M_w \times 10^{-3}$		$M_n \times 10^{-3}$		M_w/M_n	
	HPLC	SEC	HPLC	SEC	HPLC	SEC
Rhodia	73.4	72.5	67.2	67.7	1.09	1.08
Sp^2	211.3	208.1	159.8	147.3	1.32	1.41
Aldrich 1	266.1	285.1	202.3	202.6	1.32	1.41
Bayer 1	290.2	297.5	213.4	206.5	1.36	1.44
Kodak	293.1	311.8	227.9	221.2	1.29	1.41
Aldrich 2	294.0	330.5	204.5	211.4	1.44	1.56
Bayer 2	302.3	319.9	225.7	223.2	1.34	1.43

Table 20.3

MOLECULAR PARAMETERS OF ISOTACTIC POLY(METHYL METHACRYLATES) DETERMINED BY HPLC AND SEC

Chromatographic conditions: HPLC – 50 × 4.6 mm i.d. monolithic poly(styrene-*co*-divinyl-benzene) column (ICSO, Inc.; Lincoln, NE), UV detection, two step gradient of THF in methanol (0–35% in 1 min, 35–50% in 1 min), gradient time 2 min, flow rate 1 mL/min; SEC – two 300 × 7.5 mm i.d. 10^3Å and 10^5Å PL gel columns (Polymer Laboratories), THF, flow rate 1 mL/min. Injections 20 μL of polymer solution, concentration 1-2 mg/mL

Origin	$M_w \times 10^{-3}$		$M_n \times 10^{-3}$		M_w/M_n	
	HPLC	SEC	HPLC	SEC	HPLC	SEC
PMMA 1	31.2	33.0	29.1	29.4	1.07	1.12
PMMA 2	128.5	125.7	111.0	95.7	1.16	1.31

and SEC do not match exactly, they are again very close. However, the HPLC data are obtained in about one tenth of the time required to record the complete SEC trace.

20.9 HIGH-THROUGHPUT DETERMINATION OF MOLECULAR PARAMETERS

Perhaps the best demonstration of performance of the monolithic columns is the rapid screening of a polystyrene library for differences in molecular weights. This library has been prepared at Symyx Technologies using a living free radical polymerization technique (LFRP). The molecular weight of polymers prepared using LFRP is affected by a number of factors including polymerization conditions and the composition of the polymerization mixture. In addition to monomer, this mixture typically consists of an LFRP control agent and possibly other additives. A series of polymerizations varying the polymerization recipe was carried out to produce a library of polymers. Screening of this library then enabled selection of the most active control system. Using a heated 96-well glass lined aluminum reactor agitated with an orbital shaker, a large number of polystyrenes was created in a short period of time. High-throughput characterization technique was required to determine molecular weight of all of these polymers at the same rate at which they were prepared [3,45].

Precipitation/redissolution HPLC was a good candidate to achieve this goal. Using a robotic device, the homogeneous polymer solution formed in each well was diluted with THF and injected directly in the chromatographic system [45,46]. Alternatively, the polymer was first purified using an automated routine. This was achieved by precipitation triggered by addition of methanol to the mixture, separation of the solid polymer by decantation, and dissolution in THF. The chromatographic procedure included an automatic filling of a loop from which the sample was injected in a 50 × 8 mm i.d. monolithic poly(styrene-*co*-divinylbenzene) column. Since the mobile phase at this stage was pure methanol, the injected polystyrene precipitated and was dissolved only after a 45 s long gradient of 0–30% THF was applied. The system was equilibrated after each analysis by changing the mobile phase to pure methanol in 6 s that was then pumped through the column for the next 30 s. Using this approach, injections occurred each 110 s. This frequency enables analysis of more than 30 samples per hour and the complete screening of all polymers created in the 96-well polymerization reactor is achieved in about 3 hours. The bar diagram shown in Fig. 20.16 demonstrates the results of this rapid screening [46].

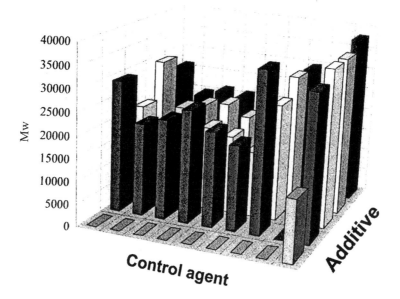

Fig. 20.16. Molecular weights of polymers prepared by ATRP upon simultaneous variation of both copper(I) salt and organic halide determined using high throughput precipitation-redissolution HPLC (Reprinted from ref. [46]. Copyright 2002 Kluwer Publisher). Separation conditions: 50 × 8 mm i.d. monolithic poly(styrene-*co*-divinylbenzene) column; mobile phase gradient of 0–30% tetrahydrofuran in methanol in 45 s, followed by 30–0% gradient in 6 s and washing with methanol for 30 s; injection frequency 110 s. Organic halides (from left to right): phenethyl chloride, phenethyl bromide, ethyl bromoacetate, ethyl 2-bromopropionate, ethyl 2-bromoisobutyrate, diethyl 2-bromo-2-methylmalonate, benzenesulfonyl chloride, 4-nitrobenzenesulfonyl chloride, and 4-methoxybenzenesulfonyl chloride.

20.10 STABILITY OF MONOLITHIC COLUMNS

The monolithic columns are surprisingly rugged. The solid and the dotted lines in Fig. 20.9 represent the chromatograms of the same polymer mixture recorded more than two months and about 400 injections apart [39]. Even after such a long time, the very small difference that can be observed between these two runs lies within the experimental error of the measurements. Moreover, it should be considered that in each gradient run one component of the mobile phase is a good swelling agent for the monolithic material of the column while the other is a precipitant. Although the high level of crosslinking does not allow extensive swelling of the monolith, even small volumetric changes of the matrix constitute a periodic stress for the column. However, this repeated stress had no effect on long-term column performance. Great column stability was also demonstrated by the use of several different solvents such as THF, dichloromethane, methanol, hexane, and water with repeated changes in gradient composition without any adverse effects on the separations. An occasional low flow rate flush with THF was the only "maintenance" carried out on the column. Similarly, frequent changes of the flow rate, which is necessary mainly in method optimization, have no detrimental effect on long-term column performance and no change in back pressure was observed.

20.11 CONCLUSIONS

The simplicity of the preparation, unique flow properties, and enhanced mass transport ability of monolithic columns makes them attractive as an alternative to particulate column packings for the separation of macromolecules. The rapid and efficient determination of molecular weight parameters of synthetic polymers in a gradient elution HPLC mode enabled by use of small monolithic poly(styrene-*co*-divinylbenzene) columns constitute a viable, less expensive, and much faster alternative to the SEC separations using a costly sets of sophisticated columns. Although the use of monolithic columns has so far been reported for the separations of only a few types of polymers, the short molded monoliths are likely to be useful for the high-throughput separations of many polymers using optimized gradients of suitable pairs of solvents and non-solvents. The sample loading affects the separation much less than in the case of packed beds. This is a promising feature for future high-throughput separations of polymers on a preparative scale.

20.12 ACKNOWLEDGMENTS

Support of this work by research grant of the Ministry of Education of the Czech Republic CB MSM 223400008 and by grant of the National Institute of General Medical Sciences, National Institutes of Health (GM–48364) is gratefully acknowledged. Thanks are also due to ISCO Inc. for the gift of monolithic columns and to Dr Miroslav Petro, Symyx Technologies, for valuable discussion and a number of useful information concerning the high-throughput separations of polymers.

20.13 REFERENCES

1   M. Cantow. Polymer Fractionation. Academic Press, New York, 1967.
2   F. Francuskiewicz. Polymer Fractionation. Springer, Berlin, 1994.
3   M. Petro, A.L. Safir, R.B. Nielsen, Polym. Prepr. 40 (1999) 702.
4   B.J. Radota. Analytical Isotachophoresis. VCH:Weinheim, Germany, 1988.

5   J. Janča. Field-Flow Fractionation. Wiley, New York, 1984.

6   G. Stegeman, J.C. Kraak, H. Poppe, R. Tijssen, J. Chromatogr. A 657 (1993) 283.

7   M.A. Stadalius, M.A. Quarry, T.H. Mourey, L.R. Snyder, J. Chromatogr. 358 (1986) 17.

8   M.A. Quarry, M.A. Stadalius, T.H. Mourey, L.R. Snyder, J. Chromatogr. 358 (1986) 1.

9   K. Kaczmarski, W. Prus, T. Kowalska, J. Chromatogr. A 869 (2000) 57.

10   J.P. Larmann, J.J. DeStefano, A.P. Goldberg, R.W. Stout, L.R. Snyder, M.A. Stadalius, J. Chromatogr. 255 (1983) 163.

11   R.A. Shalliker, P.E. Kavanagh, I.M. Russel, J. Chromatogr. 543 (1991) 157.

12   G. Glöckner, Pure Appl.Chem. 55 (1983) 1553.

13   P. Jandera, M. Holčapek, L. Kolářová, J. Chromatogr. A 869 (2000) 65.

14   H. Pasch, B. Trathnigg. HPLC of Polymers. Springer-Verlag, Berlin, 1998.

15   C.H. Lochmuller, M.B. McGranaghan, Anal. Chem. 61 (1989) 2449.

16   D.M. Northrop, D.E. Martire, R.P.W. Scott, Anal. Chem. 64 (1992) 16.

17   C.H. Lochmuller, C. Jiang, M. Elomaa, J. Chromatogr. Sci. 33 (1995) 561.

18   D.W. Armstrong, K.H. Bui, Anal. Chem. 54 (1982) 706.

19   C.H. Lochmuller, C. Jiang, Q. Liu, V. Antonucci, M. Elomaa, Crit. Rev. Anal. Chem. 26 (1996) 29.

20   K. Rissler, J. Chromatogr. A 786 (1997) 85.

21   B. Klumperman, H.J.A. Philipsen, LC-GC 17 (1999) 118.

22   C.H. Lochmuller, M.A. Moebus, Q. Liu, C. Jiang, J. Chromatogr. Sci. 34 (1996) 69.

23   W. Lee, H.C. Lee, T.Y. Chang, S.B. Kim, Macromolecules 31 (1998) 344.

24   H.C. Lee, T.H. Chang, S. Harville, J.W. Mays, Macromolecules 31 (1998) 690.

25   W. Lee, H.C. Lee, T. Park, T. Chang, J.Y. Chang, Polymer 40 (1999) 7227.

26   G. Glöckner. Gradient HPLC of Copolymers and Chromatographic Cross-Fractionation. Springer, Berlin, 1991.

27   V. Desreux, M.C. Spiegels, Bull. Soc. Chim. Belg. 59 (1950) 476.

28   G. Glöckner, J.H.M.van den Berg, J. Chromatogr. 352 (1986) 511.

29   J. Bullock, J. Chromatogr. 694 (1995) 415.

30   L.R. Snyder, K.A. Stadalius, M.A. Quarry, Anal. Chem. 55 (1983) 1412A.

31   L.R. Snyder, M.A. Stadalius, in Cs. Horváth (Ed.), High-Performance Liquid Chromatography, Advances and Perspectives, Vol. 4, Academic Press, New York, 1986.

32   L.F. Colwell, R.A. Hartwick, J. Liq. Chromatogr. 10 (1987) 2721.

33   D.P. Lee, J. Chromatogr. 443 (1988) 143.

34   C.G. Huber, P.J. Oefner, E. Preuss, G.K. Bonn, Nucleic Acid Res. 21 (1993) 1061.

35   M. Petro, F. Svec, J.M.J. Fréchet, J. Chromatogr. A 752 (1996) 59.

36   F.E. Regnier, Nature 350 (1991) 634.

37   N.B. Afeyan, N.F. Gordon, I. Maszaroff, L. Varady, S.P. Yang, F.E. Regnier, J. Chromatogr. 519 (1990) 1.

38   Q.C. Wang, F. Svec, J.M.J. Fréchet, J. Chromatogr. A 669 (1994) 230.

39   M. Jančo, D. Sýkora, F. Svec, J.M.J. Fréchet, J. Schweer, R. Holm, J. Polym. Sci. A, Polym. Chem. 38 (2000) 2767.

40   M. Petro, F. Svec, J.M.J. Fréchet, Anal. Chem. 68 (1996) 315.

41   H. Engelhardt, M. Czok, R. Schultz, E. Schweinheim, J. Chromatogr. 458 (1988) 79.

42   G. Glockner, H. Engelhardt, D. Wolf, R. Schultz, Chromatographia 42 (1996) 185.

43   C. Viklund, F. Svec, J.M.J. Fréchet, K. Irgum, Chem. Mater. 8 (1996) 744.

44   M. Janco, S. Xie, D.S. Peterson, R.W. Allington, F. Svec, J.M.J. Fréchet, J. Chromatogr. A submitted.

45   M. Petro, A.L. Safir, R.B. Nielsen, G.C. Dales, E.D. Carlson, T.S. Lee, US Pat. 6,260,407 (July 17, 2001).

46   D. Sýkora, F. Svec, J.M.J. Fréchet, M. Petro, A.L. Safir, in: R.A. Potarailo, E.J. Amis (Eds.), High Throughput Analysis: A Tool for Combinatorial Materials Science, Kluwer Academic/Plenum Publishers, New York, in press.

F. Švec, T.B. Tennikova and Z. Deyl (Editors)
Monolithic Materials
Journal of Chromatography Library, Vol. 67

Chapter 21

Capillary Electrochromatography

Michael LÄMMERHOFER and Wolfgang LINDNER

*Christian Doppler Laboratory for Molecular Recognition Materials, Institute of
Analytical Chemistry, University of Vienna, Währingerstrasse 38, A-1090 Vienna,
Austria*

CONTENTS

21.1 INTRODUCTION

The quest for analyzing ever more complex mixtures, in ever shorter times, in fields such as the life sciences, pharmaceutical- and bio-analysis, combinatorial chemistry, environmental analysis, and many others, has placed new demands on modern separation science, requiring significant advances in terms of selectivity, efficiency, throughput, and structural confirmation of the eluted compounds. Capillary electrochromatography (CEC) [1] holds great promise to enable at least several of the above requirements. In CEC, an electric field is applied across a capillary column with a typical inner diameter of less than 100 µm that contains the stationary phase. The resulting electroosmotic flow (EOF) drives the mobile phase and analytes through the separation bed. Electroosmotic flow results from the presence of an electrical double layer at the solid–liquid interface of the ionized surface of the stationary phase that is in contact with the electrolyte solution. Upon application of an electric field, the counterions of the diffuse layer adjacent to the solid surface begin to migrate to the corresponding electrode and generate bulk liquid flow. The linear velocity of the EOF u_{eo} is given by the relationship

$$u_{eo} = \mu_{eo}\,E = -\,\frac{\varepsilon_0\,\varepsilon\,\zeta}{\eta}\,E \qquad\qquad (21.1)$$

where μ_{eo} is the electroosmotic mobility, ε_0 is the permittivity of vacuum, ε_r is the relative permittivity of the medium (dielectric constant), ζ is the zeta-potential at the surface of shear, that depends on the surface charge density and the double layer thickness which, in turn, is indirectly proportional to the square root of the ionic strength of the surrounding liquid, η is the viscosity of the medium, and E is the electric field strength. EOF is an advantageous driving force for the flow because of its uniform plug-like velocity profile. Therefore, CEC affords significantly higher column efficiencies than corresponding methods in which the flow is driven by pressure, such as in HPLC. The absence of a pressure gradient enables the use of longer and more densely packed columns than would be permeable for hydrodynamic flow.

Neutral solutes move through the CEC column solely with EOF driven flow and, as in HPLC, they are separated as a result of their differential distribution between both stationary and mobile phases, that is expressed as the chromatographic retention factor, k. In contrast, electrophoretic migration occurs additionally for ionized solutes and is characterized by a velocity u_{ep}, which, depending on their charge, may be in either direction relative to the EOF, resulting in a co- or counter-directional separation process. In this case, differences in electrophoretic mobility of specific solutes also contribute to the overall selectivity of the separation. In some electrochromatographic modes, a pressure is superimposed over the EOF, leading to pressurized flow with an additional velocity, u_{press}, that may accelerate the separation but often with distortion of the favorable flow profile [1]. Consequently, the observed net migration velocity u_{obs} in CEC is

$$u_{obs} = \frac{(u_{eo} + u_{ep} + u_{press})}{(1 + k)} \tag{21.2}$$

Pretorius *et al.* [2] were the first to demonstrate CEC using particle-packed columns, in 1974. Painstaking problems with packed columns, mainly related to laborious and tedious column packing and fabrication of retaining frits that stabilized the bed, contributed to the slow start of CEC and its current lower popularity. However, it appears that the monolithic technologies could be the necessary fuel for an acceptance of CEC as an equal partner in the broad family of current well established chromatographic methods. The novel CEC column technologies may also help to overcome many of the problems of this separation technique. It is even likely that CEC and monoliths inspired each other in the past few years. This chapter illustrates the great potential of monolithic columns in CEC, demonstrating applications in various chromatographic modes for the separations of low-molecular weight compounds, pharmaceuticals including chiral compounds, biomolecules such as peptides and proteins, and synthetic polymers.

21.2 SPECIFICS OF MONOLITHIC COLUMNS IN CEC

21.2.1 Fritless design

Essentially all monolithic technologies described in previous chapters of this book, such as those based on soft gels, rigid polymers, inorganic materials, and monoliths prepared from particles, have also been adapted for CEC or were developed specifically for this method. Capillary columns filled with all these monoliths have one important feature in common: they do not require retaining frits to keep the bed in a specific location. Therefore monolithic columns are sometimes termed, "fritless". The monolith itself is anchored to the wall of the fused-silica capillary, thus creating a single column body.

This binding to the wall is accomplished for organic polymer monoliths by their preparation within the confines of a fused-silica capillary treated on the inner surface

with 3-trimethoxysilylpropyl methacrylate (Bind-Silane) [3-5]. In contrast, no pre-
treatment is required while producing silica-based monolithic columns. This also applies to approaches in which chromatographic particles are entrapped in a silica matrix. Immobilization then occurs by co-condensation of the fused silica surface into the
sol-gel network [6,7]. Similarly, sintering of packed particulate beds forms consolidated packings in which particles are fused to each other and to the silica wall [8].

The immobilization of the monoliths within the columns eliminates most of the
problems typical of packed CEC columns that are associated with sintered retaining
frits. These include limited stability or insufficient permeability, susceptibility to bubble formation (outgassing), and additional band spreading. Formation of voids at the
interface monolithic columns/capillary wall, that was observed in packed beds and
significantly reduced their efficiency, is absent in well-prepared monolithic CEC columns. Similarly, outgassing occurs to a lesser extent, if at all, and monolithic columns
can be used without external pressurization of the buffer reservoirs. Thereby, their
operational flexibility is greatly improved.

The ability to be coupled to a mass spectrometer is an important criterion enabling
a separation technique to be broadly accepted and used routinely. In general, CEC can
favorably meet this request since its features such as column size, flow-rate range, and
mobile phase compositions are well compatible with electrospray ionization mass
spectrometry (ESI-MS). With no need for pressurization of the electrolyte vessels,
coupling of monolithic CEC columns to ESI-MS is facilitated. In addition, the risk of
accidental carry-over of particles into the MS detector, that occurs for packed columns
with broken end frits, is completely avoided.

Interestingly, monoliths have also been used as in alternative technology to form
frits [9,10]. Photopolymerization allows one to place a short plug of a porous polymer
such as poly(glycidyl methacrylate-*co*-trimethylolpropane trimethacrylate) at any desired position along the column. The porosity, and thus the permeability and stability
of the frit, can be readily adjusted by choosing suitable porogens.

21.2.2 EOF in monolithic columns

In CEC, the stationary phase plays a dual role: in addition to providing the chromatographic selectivity, it also generates EOF or, in other words, it is a part of the
"pumping" system. Equation 21.1 indicates that the EOF velocity is directly proportional to the zeta potential of the stationary phase, which in turn depends on the
surface charge density. A strong EOF is a highly desired property. Therefore, the
monolith should carry a sufficiently high number of ionized groups on its surface.

One of the strengths of the organic polymer monolith technology is the ease of
control of the surface chemistry. This allows one to adjust the EOF of a column, with
respect to both its magnitude and direction, in a broad range through simple copolym-

erization of a specific type and percentage of an ionizable comonomer. Weakly acidic (meth)acrylic acid and the strongly acidic vinylsulfonic acid (VSA) or 2-acrylamido--2-methylpropanesulfonic acid (AMPS) are suitable EOF-generating monomers for columns, affording cathodic flow (EOF directed from anode to cathode). Both the latter monomers are preferred, since they afford high EOF for separations even at low pH values, owing to their permanent charge irrespective of pH. On the other hand, copolymerization of a positively ionizable monomer such as [2-(acryloy-loxy)ethyl]trimethylammonium methyl sulfate (AETMA) leads to monolithic columns generating anodic EOF directed from cathode to anode. The EOF direction, adjusted for the given separation problem and the electrophoretic mobility of the ionized analytes, is an important variable while one is designing a CEC separation method. Therefore, it is of the utmost importance to control the surface charge of the stationary phases.

Several studies have documented that the EOF velocity and speed of analysis can be enhanced by increasing the percentage of the ionizable monomer. For example, Fujimoto observed for polyacrylamide gel columns with a low crosslinking of 5% that the migration velocity increased linearly over the range of 1–13% AMPS of the total monomers to reach a maximum of 1.6 cm/min [11]. In another study, with 60% crosslinked monolithic polyacrylamide columns, the EOF leveled off after a rapid increase between 0 and 15% VSA in the monomer mixture reaching a maximal velocity of 6 cm/min [12]. A similar trend was also reported by Peters *et al.* for rigid macroporous 40% crosslinked poly(butyl methacrylate-*co*-ethylene dimethacrylate--*co*-AMPS) monoliths [13]. AMPS at a level as low as 0.3% afforded strong EOF, which could be boosted considerably at a higher percentage, achieving the saturation level at 1.8% AMPS (Fig. 21.1a). Although direct comparison is difficult due to different eluent conditions, the about 10-times higher electroosmotic mobilities observed for a polymethacrylate column compared to swollen gel monolith suggest that rigid monolithic columns promote strong flow, even at low concentrations of ionizable monomer. This may be attributed to different polymer morphologies of these stationary phases and also indicates an effect of another variable on EOF, the pore structure (*vide supra*).

As expected, completely uncharged or neutral monolithic columns afford poor, but measurable EOF [4,14]. For example, electroosmotic mobilities of only 0.106×10^{-8} and 0.034×10^{-8} m^2 V^{-1} s^{-1} at pH 8 and 3, respectively, have been observed for neutral poly(lauryl methacrylate-*co*-ethylene dimethacrylate) monoliths [14]. Obviously, these columns are useful only for separations of ionized solutes featuring a significant electrophoretic migration, such as peptides [4,14]. However, a higher EOF in these neutral columns can be achieved by addition of surfactants such as sodium dodecylsulfate (SDS) or cetyltrimethylammonium bromide (CTAB) to the electrolytes

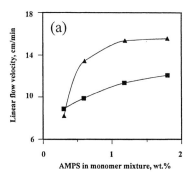
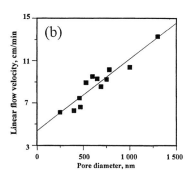

Fig. 21.1. Effect of percentage of AMPS (a) and mode pore diameter (b) on flow velocity in rigid polymethacrylate-based monolithic columns (Reprinted with permission from ref. [13], Copyright 1998, American Chemical Society). Conditions: EOF marker, thiourea; capillary, 38.5 cm × 100 μm I.D., 30 cm active length; poly(butyl methacrylate-*co*-ethylene dimethacrylate-*co*-AMPS) monolithic columns, with differing mode pore size (constant porogen composition) (▲) and constant mode pore size (adjusted by porogen composition to 0.7 μm in dry state) (■); mobile phase, acetonitrile–5 mmol/L phosphate buffer pH 7 (80:20); applied voltage, 25 kV.

at concentrations below the critical micelle concentration (CMC), which create dynamically modified ion-exchangers with reasonable ζ-potential and EOF [15]. The electroosmotic mobility of a poly(butyl methacrylate-*co*-ethylene dimethacrylate) monolith thus could be increased from 0.1×10^{-8} up to 1.5- and 1.1×10^{-8} m^2 V^{-1} s^{-1} after addition of 2 mmol/L SDS and CTAB, respectively. However, surfactants are not a welcome component of the effluent as they may interfere with MS detection.

Independent studies with monolithic systems found that the pore size of polymethacrylate monoliths also significantly affects the flow [13,16,17]. The linear EOF velocity was directly proportional to the size of macropores that could be controlled by the choice of porogens (Fig. 21.1b). This can be explained readily using Rathore-Horvath's theory [1]. Differences in the tortuosity of the monoliths varying in pore size, and corresponding variations in conductivity and effective field strength in the packed structure, translate into changes in flow velocities [18]. Not surprisingly, this phenomenon is not restricted to polymethacrylate monoliths. A similar trend was also observed for polyacrylamide-based continuous beds, for which increasing concentration of ammonium sulfate in the reaction mixture led to wider flow channels [19-21] and promoted stronger electroosmotic flow [21].

The EOF in inorganic silica monolith columns depends widely on the presence of silanol groups. In octadecylated materials, a significant portion of silanols originally available on surface of the silica matrix is derivatized and no longer contributes to the

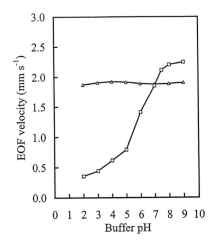

Fig. 21.2. Effect of mobile phase pH on electroosmotic flow for sol-gel bonded Spherisorb ODS1 (3 μm, 80 Å), (□) and Spherisorb ODS/SCX (3 μm, 80 Å) (△), monolithic columns (Reprinted with permission from ref. [24], copyright 2000 Wiley-VCH). Columns, 34 cm × 75 μm I.D., 25 cm effective length; mobile phase, acetonitrile–water–50 mmol/L phosphate buffer (70:25:5); applied voltage, 30 kV; thiourea as EOF marker.

electroosmotic flow [22]. As a consequence, the EOF velocity of these monolithic materials is low [23]. Moreover, due to the weakly acidic nature of the silanols, substantial EOF is only observed at pH values above 5–6. An effective EOF modulation can also be achieved for monoliths prepared by the sol-gel technique. For example, by incorporation of a quaternary ammonium precursor in the sol-gel solution, monolithic columns with anodic EOF were prepared [7].

The EOF in monoliths prepared from particles (particle-fixed monoliths), such as sol-gel bonded silica columns, is dictated by the surface chemistry of the particles rather than of the matrix. For example, a strong decrease in EOF at low pH was observed for monolithic columns prepared from the sol-gel bonded Spherisorb ODS1. In contrast, the corresponding mixed-mode Spherisorb ODS/SCX containing both C_{18} and sulfopropyl groups afforded a desirable EOF over a much broader pH-range of 2–9 (Fig. 21.2). About 30% higher EOF velocities than those of corresponding packed columns were found for sol-gel monoliths with immobilized particles. This effect was attributed to the sol-gel matrix itself that can dissociate at high pH to liberate negatively charged silanols that contribute to the overall EOF.

21.3 SEPARATION MODES

21.3.1 Reversed-phase CEC

21.3.1.1 Polyacrylamide-based continuous beds and rigid monoliths

The pioneering work of Hjertén's [25] and Fujimoto's [26] groups with continuous monolithic beds, based on crosslinked polyacrylamides modified with hydrophobic alkyl ligands for reversed-phase CEC, aimed at adjustment of the hydrophobic selectivity of the polymers, which was evaluated with neutral non-polar test mixtures (Table 21.1). Finding the optimal polymerization conditions to obtain matrixes with hydrophobic interactive ligands was difficult, owing to the lack of solubility of hydrophobic monomers in aqueous reaction mixtures. Although "plain" 5.1% crosslinked poly(acrylamide-*co*-N,N'-methylene bisacrylamide) gel containing AMPS for EOF generation afforded some selectivity and the elution order for neutral ketones coincided with their hydrophobicity, excessively long run times were observed. The retention was attributed to "aromatic adsorption" [11,27]. Incorporation of C_3 ligands through copolymerization of moderately hydrophobic N-isopropylacrylamide (NIPAAm), which is soluble in water, enabled separations of steroids and polyaromatic hydrocarbons (PAHs) [26]. However, this monomer did not solve the problem of hydrophobic selectivity satisfactorily. Hjertén later re-approached this problem with better success by incorporation of more lipophilic monomers in poly(methacrylamide--*co*-piperazine diacrylamide)-based monoliths. However, their preparation necessitated multi-step procedures [19,25,28]. For example, small amounts of water-insoluble stearyl- or butyl methacrylate monomers were emulsified in the reaction mixture using a surfactant [25]. The inadequate selectivity of these monoliths for the separation of PAHs was offset by elution using a step-wise gradient, which sharpened the bands and improved resolution or, alternatively, by the addition of SDS at concentrations below CMC (Fig. 21.3a and b). Hjertén argued that the C_{12} alkyl chains of SDS were inserted between the ligands of the polymer resulting in a denser hydrophobic layer that improved both retention and selectivity. The run times remained almost unchanged, because the higher retention was outweighed by stronger EOF resulting from the contribution of dynamically coated SDS to the overall surface charge. Despite some improvements, hydrophobic selectivities typical of octadecylated silica have never been achieved. Further improvement included the use of polyacrylamide-based monolithic continuous beds prepared by a three-step procedure comprising the creation of a rigid polymer matrix in the first step, followed by embedding of dextran sulfate to generate EOF and monomers with reactive hydroxyl or epoxide groups into

TABLE 21.1

APPLICATIONS OF POLYACRYLAMIDE GELS AND CONTINUOUS BEDS FOR CEC SEPARATIONS IN REVERSED-PHASE MODE [a]

Analytes	Reaction mixture [b]		Mobile phase (applied voltage)	Maximum efficiency (theor. pl/m)	Ref.
	Monomers	Solvents/Porogen/Additive			
Poly(acrylamide-co-N,N'-methylenebisacrylamide) gels					
Ketones	13.7% T (AAm, BIS, AMPS), 5.1% C, 10% S	100 mmol/L Tris–150 mmol/L boric acid (pH 8.1)	100 mmol/L Tris–150 mmol/L boric acid (pH 8.1); 20 kV		[11]
Ketones	4% T (AAm, BIS, AMPS), 10% C, 10% S	100 mmol/L Tris–150 mmol/L boric acid (pH 8.1)	100 mmol/L Tris–150 mmol/L boric acid (pH 8.1); 15 kV		[26]
Ketones, steroids	4% T (AAm, NIPAAm, BIS, AMPS), 10% C, 10% S			160,000	
PAHs	6.9% T (AAm, NIPAAm, BIS, AMPS), 5.8% C, 5.5% S	100 mmol/L Tris–150 mmol/L boric acid (pH 8.1)	35% (v/v) ACN in 100 mmol/L Tris–150 mmol/L boric acid (pH 8.1); 20 kV		[26]
Poly(acrylamide-co-N,N'-piperazine bisacrylamide) continuous monolithic beds					
Aromatic hydrocarbons	37% T (MAAm, PDA, VSA, stearyl methacrylate or BMA), 32% C, 13.5% S, 34% C18	(NH₄)₂SO₄ in Tris buffer (15 mmol/L, pH 8.5) with Triton-X	50% (or 70%) (v/v) ACN in 4 mmol/L phosphate buffer		[25]

Continued on the next page

TABLE 21.1 (continued)

Analytes	Reaction mixture[b]		Mobile phase (applied voltage)	Maximum efficiency (theor. pl/m)	Ref.
	Monomers	Solvents/Porogen/Additive			
Aromatic hydrocarbons	Step 1: rigid matrix, 18% T (HEMA, PDA), 44% C Step 2: immobilization of dextran sulfate, 33% T (AGE, HEMA, PDA), 30% C + 3% dextran sulfate (w/v) step 3: derivatization with C18	$(NH_4)_2SO_4$ (75 mg/mL) in potassium phosphate buffer (80 mmol/L, pH 7) $(NH_4)_2SO_4$ (48 mg/mL) in potassium phosphate buffer (80 mmol/L, pH 7)	62% (v/v) ACN in 5 mmol/L borate buffer (pH 8.7); 26 kV	120,000	[19]
Proteins (ribonuclease A, lysozyme, α-chymo-trypsinogen, cytochrome C)	Step 1: PDA, MAAm, stearyl methacrylate Step 2: polymer of Step 1 + DMDAAC, PDA	DMF–50 mmol/L phosphate buffer (pH 7) 50 mmol/L phosphate buffer (pH 7)	Linear gradient elution; 5–80% ACN in 5 mmol/L sodium phosphate, adjusted to pH 2.0		[28]
Tricyclic antidepressants (nortriptyline, amitriptyline, N-methyl-amitriptyline)	35% T (PDA, MAAm, NIPAAm, VSA), 50% C, 25% NIPAAm, 0.25% VSA	$(NH_4)_2SO_4$ in sodium phosphate buffer (50 mmol/L, pH 7)	10% phosphate buffer (pH 2.75), 20% water, 70% ACN (0.012 mol/L ionic strength); 20 kV	200,000	[29]
Polyaromatic hydrocarbons and phenols; nonsteroidal antiinflammatory drugs including profens and diclofenac, mefenamic acid	15 or 29% T (PDA, VSA, DMAA, MAAm or hexylacryl-amide), 52% C, 1.5 or 3% VSA	$(NH_4)_2SO_4$ in 50 mM phosphate buffer (pH 7) or phosphate buffer-FA (1:1, v/v)	ACN–MeOH (3:1) without supporting electrolytes	110,000	[20]

Analyte	Composition	Mobile phase	Conditions	Plates	Ref.
Alkyl benzoates	21.4% T (PDA, MAAm, NIPAAm), 41.7% C + charged β-CD-derivatives	(NH$_4$)$_2$SO$_4$ in 50 mM phosphate buffer (pH 7)	50% aqueous MeOH buffered with TEMED (0.05%)–AcOH (0.1%)	140,000	[21]

Poly(acrylamide-co-N,N″-methylenebisacrylamide)-poly(ethylene glycol) monoliths

Analyte	Composition	Mobile phase	Conditions	Plates	Ref.
Aniline and ketones (aceto-, propio-, butyro-, 2',5'-dihydroxyaceto-, and 2',5'-dihydroxy--propiophenone)	5% T (AAm, BIS, AA, BA), 60% C, 3% (w/v) PEG, 10% AA, 29% BA	50% (v/v) FA in 100 mM Tris–150 mM boric acid (pH 8.2)	20% (v/v) ACN in tris (10 mM)–boric acid (15 mM) (pH 8.2); 22.5 kV	400,000	[30]
Aromatic hydrocarbons	5% T (AAm, BIS, AA, BA), 60% C, 3% (w/v) PEG, 10% AA, 58% BA	80% (v/v) FA in 100 mM Tris–150 mM boric acid (pH 8.2)	50% (v/v) ACN in tris (10 mM)–boric acid (15 mM) (pH 8.2)		[30]
Tyr-containing peptides including Leu-enkephalirs	5% T (AAm, BIS, AA, LA), 60% C, 3% (w/v) PEG, 10% AA, 29% LA	95% (v/v) FA in 100 mM Tris–150 mM boric acid (pH 8.2)	47% (v/v) ACN in tris (10 mM)–boric acid (15 mM) (pH 8.2); 22.5 kV		[30]
Oligosaccharide mixtures (Glc$_1$-Glc$_6$ and Glc$_4$-Glc$_{10}$)	5% T (AAm, BIS, VSA, BA), 60% C, 3% (w/v) PEG, 10% VSA, 14% BA	50% (v/v) FA in 100 mM Tris–150 mM boric acid (pH 8.2)	20% (v/v) ACN in tris (10 mM)–boric acid (15 mM) (pH 8.2); 22.5 kV	230,000	[30]
Bile acids and glycine- and taurine conjugates	5% T (AAm, BIS, VSA, LA), 60% C, 3% (w/v) PEG, 10% VSA, 15% LA	95% (v/v) NMF in 100 mM Tris–150 mM boric acid (pH 8.2)	ACN–water–240 mM ammonium formate (pH 3) 55:40:5, v/v/v; E: 550 V/cm	300,000	[31]
Ketosteroids and conjugates (sulfates and glucuronides)	5% T (AAm, BIS, VSA, LA), 60% C, 3% (w/v) PEG, 10% VSA, 15% LA	95% (v/v) NMF in 100 mM Tris–150 mM boric acid (pH 8.2)	ACN–water–240 mM ammonium formate (pH 3) 55:40:5, v/v/v; E: 600 V/cm	200,000	[32]

Continued on the next page

TABLE 21.1 (continued)

Analytes	Reaction mixture[b] Monomers	Solvents/Porogen/Additive	Mobile phase (applied voltage)	Maximum efficiency (theor. pl/m)	Ref.
Homologous alkylphenones	5% T (AAm, BIS, VSA, LA), 60% C, 3% (w/v) PEG, 10% VSA, 15% LA	95% (v/v) NMF in 100 mM Tris–150 mM boric acid (pH 8.2)	Linear gradient elution; A: 10 mM Tris–15 mM boric acid (pH 8.7)–MeOH (70:30, v/v); B: same as A but (30:70)		[33]
Lignans (from seeds of *Schisandra chinensis*)	5% T (AAm, BIS, VSA, LA), 60% C, 3% (w/v) PEG, 10% VSA, 15% LA	95% (v/v) NMF in 200 mM Tris–300 mM boric acid (pH 8.2)	30% ACN in 10 mM Tris–15 mM boric (pH 8.2); 22.5 kV	300,000	[34]
Alkyl phenones, urea herbicides, carbamate insecticides	4%T (AAm, BIS, VSA, LA), 62.5%C, 3% (w/v) PEG, 15% VSA, 10% LA	NMF–100 mM Tris–150 mM boric acid buffer (pH 8.2)	25% ACN in 7.5 mM Tris–11.25 mM boric acid (pH 8.2); 30 kV		[12]

[a] For abbreviations see list at end of Chapter

[b] Polymerization started with APS–TEMED initiator system

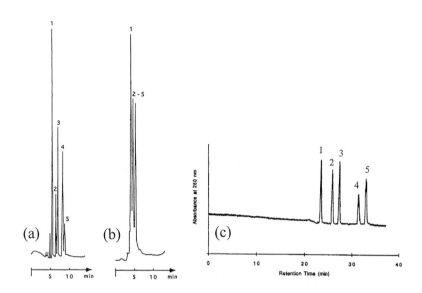

Fig. 21.3. CEC separation of polycyclic aromatic hydrocarbons using a polyacrylamide monolith with C₁₈-ligand and negative surface charge in the presence (a) and absence (b) of SDS in the mobile phase (Reprinted with permission from ref. [25], copyright 1996 American Chemical Society) and the effect of preparation technique (c) that affords higher C₁₈-ligand density in a three step procedure (Reprinted with permission from ref. [19], copyright 1997 Elsevier Science B.V.). Conditions: column, 14.0 cm (10.0 cm effective length) × 100 μm I.D.; eluent, 60% (v/v) acetonitrile in 4 mmol/L sodium phosphate (pH 7.4), 1 mmol/L SDS (a) and with no SDS (b); applied voltage, 3.0 kV. (c) column, 55 cm (52.5 cm effective length) × 25 μm I.D.; eluent, 62% (v/v) acetonitrile in 5 mmol/L borate buffer (8.7); applied voltage, 26 kV. Peaks: (1) naphthalene, (2) 2-methylnaphthalene, (3) fluorene, (4) phenanthrene, (5) anthracene.

the growing gel, and the subsequent derivatization of the monolith with C₁₈ ligands. Indeed, these monoliths exhibited suitable selectivity (Fig. 21.3c) [19].

Palm and Novotny achieved a breakthrough in the preparation of polyacrylamide-based monoliths for reversed phase CEC [30]. They performed the polymerization in mixtures of water with an organic solvent, typically formamide or N-methylfor-mamide, in the presence of poly(ethylene glycol). The hydrophobicity of the resulting matrixes could be adjusted by the type and concentration of the hydrophobic monomer such as butyl-, hexyl-, and lauryl acrylate.

These monoliths deserve particular attention since they enabled the achievement reproducibly of very high efficiencies of up to 600,000 plates/m. Plate height vs. flow velocity plots (Van Deemter plots), shown in Fig. 21.4, are very flat with minima at

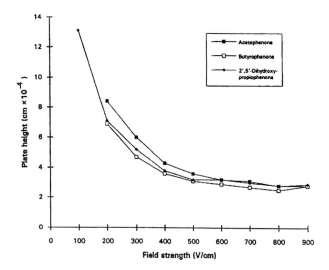

Fig. 21.4. Effect of field strength on plate height for acetophenone (■), butyrophenone (□), and 2′,5′-dihydroxypropiophenone (♦) (Reprinted with permission from ref. [30], copyright 1997 American Chemical Society). Conditions: Monolith composition, 5.0% T, 60% C, 10% acrylic acid, 29% C4 ligand (butyl acrylate), 3% (w/v) poly(ethylene glycol); capillary, 25 cm (20.5 cm effective length) × 100 μm I.D.; mobile phase, 20% (v/v) acetonitrile in 10 mmol/L Tris–15 mmol/L boric acid buffer (pH 8.2); temperature, 22°C.

about 3 μm, indicating an absence of mass transfer resistance and very low flow-dispersion. This enables operation at both high field strengths and linear flow velocities without loss in efficiency. The exceptional column performance may be attributed to the positive effect of PEG in combination with N-methylformamide on the homogeneity and morphology of the monolith. Monoliths prepared in the absence of PEG exhibited much lower plate counts. Palm and Novotny [30] also demonstrated that CEC might compete with HPLC in terms of reproducibility of separations and column fabrication (Table 21.2). The simple column fabrication procedure, excellent column-to-column reproducibility, and the stability lasting for several months, indicated that monoliths might soon out-perform the packed CEC columns.

The great potential of these monoliths for CEC has been demonstrated in a variety of separations including ketones, PAHs, tyrosine-containing peptides, oligosaccharides [30], urea herbicides and carbamate insecticides [12]. For example, Palm and Novotny separated oligomaltoses containing up to seven glucose units (Glc_{1-7}) (Fig. 21.5) and an oligosaccharide ladder Glc_4-Glc_{10} tagged with 2-aminobenzamide, using LIF detection and a highly efficient, moderately hydrophobic, C4 polyacrylamide monolithic column [30]. Good separations were obtained only at pH values less

TABLE 21.2

REPRODUCIBILITY OF MIGRATION TIMES, RETENTION FACTORS, AND EFFI-
CIENCIES ON POLYACRYLAMIDE-PEG MONOLITH COLUMNS [30] [a]

Solute	t_r		k		Column efficiency	
	min	% RSD		% RSD	plates/m	% RSD
Run-to-run (n = 6)						
Aniline	3.38	1.0	0.08	2.8	331,900	4.2
Acetophenone	3.49	0.9	0.12	2.7	353,300	4.5
Propiophenone	3.64	0.8	0.17	2.4	363,500	5.3
Butyrophenone	3.83	0.7	0.23	3.0	377,000	5.4
2′,5′-Acetophenone	4.20	0.7	0.34	1.9	365,700	5.1
2′,5′-Propiophenone	4.43	0.6	0.42	2.5	361,300	1.2
Day-to-day (9 days, n = 9)						
Aniline	3.42	2.2	0.08	4.2	332,000	4.6
Acetophenone	3.55	2.0	0.12	5.0	356,500	4.3
Propiophenone	3.70	1.9	0.17	5.8	370,600	2.7
Butyrophenone	3.92	1.9	0.24	6.7	376,100	7.9
2′,5′-Acetophenone	4.30	2.2	0.36	6.8	373,400	4.2
2′,5′-Propiophenone	4.56	2.5	0.44	8.2	372,800	4.0
Column-to-column (n = 3) (including batch-to-batch)						
Aniline	3.30	1.8	0.08	0.0	322,700	2.1
Acetophenone	3.40	1.8	0.12	4.0	356,900	4.2
Propiophenone	3.54	2.1	0.16	5.1	364,600	0.8
Butyrophenone	3.72	2.2	0.22	3.7	380,300	3.5
2′,5′-Acetophenone	4.09	2.1	0.34	2.4	361,300	2.0
2′,5′-Propiophenone	4.31	2.2	0.42	3.0	362,800	2.2

[a] Conditions: Monolith composition, 5.0% T, 60% C, 10% acrylic acid, 29% butyl acrylate,
3% poly(ethylene glycol); capillary, 25 cm (20.5 cm effective length) × 100 μm i.d.; mobile
phase, 20% acetonitrile in 10 mmol/L Tris–15 mmol/L boric acid buffer (pH 8.2); temperature, 22°C; applied voltage, 22.5 kV.

References pp. 556-559

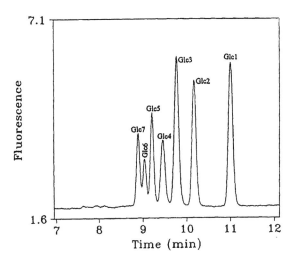

Fig. 21.5. Isocratic electrochromatography of malto-oligosaccharides (glucose=Glc$_1$, maltohexaose=Glc$_6$) labeled with 2-aminobenzamide, enabling laser-induced fluorescence detection in a capillary filled with a macroporous polyacrylamide monolith derivatized with 25% C4 ligands and containing 10% vinylsulfonic acid (Reprinted with permission from ref. [30], copyright 1997 American Chemical Society). Conditions: capillary, 32 cm × 100 μm I.D., 25 cm effective length; eluent, mobile phase, 20% (v/v) acetonitrile in 10 mmol/L Tris–15 mmol/L boric acid buffer (pH 8.2); temperature, 22°C; field strength, 900 V/cm; sample concentration, 5–10 μmol/L.

than 4.0, at which the amine groups of the tags were fully protonated and strongly retained, supposedly owing to electrostatic interaction with the negatively charged polymer matrix. Off-gel detection, as close to the end of the monolith as possible, afforded five-times higher sensitivity than the in-bed detection.

These polyacrylamide-PEG based monolithic capillary columns were also evaluated in challenging separations of "real life" samples and complex biological mixtures. For example, isocratic and gradient CEC were used for steroid profiling in typical biological matrices such as human urine [32]. The potential of CEC–LIF and CEC–ESI/MS using monolithic columns was demonstrated on separations of the ketosteroids, androstan-17-one derivatives that lack strong chromophores. Seven neutral ketosteroids derivatized with dansylhydrazine could be separated within 16 min using a monolithic column with an efficiency of 200,000 plates/m. The corresponding sulfate and glucuronide derivatives required longer run-times, owing to their counter-directional electrophoretic migration. The limit of detection for the labeled steroids was in the attomolar level. In contrast, the sensitivity of CEC–ESI/MS was only at the femtomolar level, about three orders of magnitude higher.

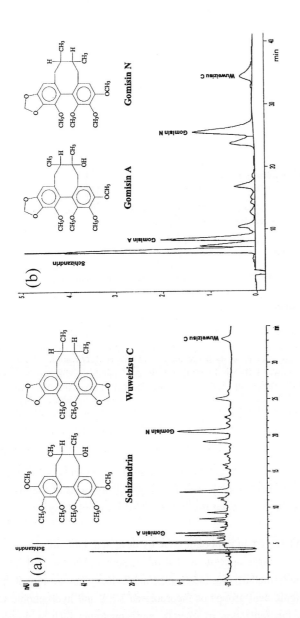

Fig. 21.6. Separations of *Schisandra extract* using CEC with a polyacrylamide-based monolithic column, (a), and a standard HPLC column, (b). (Reprinted with permission from ref. [34], copyright 2001 Elsevier Science B.V.). Conditions: (a) Macroporous polyacrylamide monolith, 33 cm (24.5 cm effective length) × 75 μm I.D.; mobile phase, 30% (v/v) acetonitrile in 10 mmol/L Tris–15 mmol/L boric acid (pH 8.2); applied voltage, 22.5 kV; temperature, 25°C; UV detection, 200 nm. (b) Column, Separon SGX C18, 5 μm, 150 × 3 mm I.D.; mobile phase, methanol–water (75:25, v/v); flow rate, 0.3 mL/min; UV detection, 254 nm.

A monolithic polyacrylamide column was also used for the CEC analysis of complex lignans from extracts of *Schisandra chinensis* seeds [34]. The CEC method was compared to a standard reversed-phase HPLC assay. Representative CEC and HPLC chromatograms are shown in Fig. 21.6. It is evident that the CEC method with the monolithic column is superior in terms of efficiency, and therefore also peak capacity, which is critical for the separation of a complex mixture. In contrast, no significant

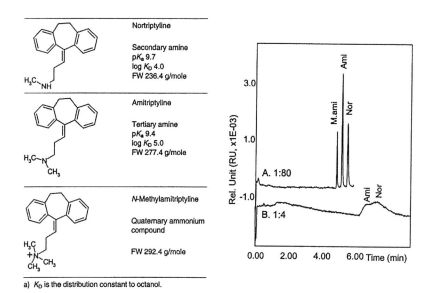

a) K_D is the distribution constant to octanol.

Fig. 21.7. Effect of percentage of vinylsulfonic acid (VSA) on column efficiency in the CEC separation of the tricyclic antidepressant drugs amitriptyline (Ami), nortriptyline (Nor), and N-methylamitriptyline (M.ami) using polyacrylamide-based monolith. (Reprinted with permission from ref. [29], copyright 2001 Wiley-VCH). (a) VSA-NIPAAm ratio in polymerization mixture = 1:80, (b) VSA-NIPAAm ratio = 1:4. Mobile phase, phosphate buffer (pH 2.75)–water–acetonitrile (10:20:70, v/v/v) (0.012 M ionic strength); applied voltage, 20 kV; column, L_{eff} = 40 cm, L_{tot} = 57.5 cm, 50 μm i.d.

differences in selectivities were noticed. The major lignan constituents, schizandrin, gomisin A and N, and wuweizisu, representing 0.2–1.0% of dry mass of the extract, could be determined quantitatively in the range of 0.025–1.0 mg/ml using UV detection.

Westerlund's group addressed an important issue by designing monoliths for CEC separation of analytes with charge opposite to that of the sorbent surface [29]. Their goal was to optimize a poly(methacrylamide-*co*-piperazine diacrylamide) monolithic bed containing both sulfonic acid groups of incorporated VSA and hydrophobic C_3 ligands of NIPAAm for the separation of tricyclic antidepressants such as nortriptyline, amitriptyline, and N-methylamitriptyline. These basic pharmaceuticals are difficult to separate on silica-based stationary phases. The undesired interactions with silanol groups lead to severe peak tailing. A similar problem may also arise with polymer monoliths that contain negatively charged surface functionalities. In order to avoid electrostatic interactions between oppositely charged analytes and sorbent, which lead to asymmetrical peak shapes, as shown in Fig. 21.7b, the amount of ionizable monomer must be kept to the minimum required to achieve sufficiently strong EOF. Indeed, a monolith with wide pores and a low surface concentration of ionized groups already provided efficient flow. The chemistry of the polyacrylamide monolith was varied systematically by changing the ratio of sulfonic acid and C3 groups. The best separations shown in Fig. 21.7a were obtained with a monolith containing NIPAAm and VSA in a ratio of 80:1, which corresponds to only 0.25% VSA in the total monomer mixture. The limit of detection was in a range of 50 pg/mL. The other option, elimination of undesired Coulombic interactions by reversal of the surface charge, was regarded as less suitable, because the resulting counter-directional separation would be slow.

21.3.1.2 Polyacrylate and polymethacrylate-based monoliths

The fabrication procedure for the rigid poly(meth)acrylate monoliths comprising a single molding step within the capillary is very simple. Systematic studies using both thermally [13,35-37]- and UV initiated polymerizations [5,16,38] have led to optimal polymer compositions and reaction conditions that afforded monolithic columns with the desired properties for CEC separations.

The hydrophobic character of the methacrylate-type monomers enables the fabrication of monoliths from monomers with C4 ligands, *i.e.*, butyl methacrylate [13,16,35-37] or butyl acrylate [5,38] and their direct applications for reversed-phase CEC that are summarized in Table 21.3. Hydrophobic selectivity values $\alpha(CH_2)$, calculated from the slopes of plots of log k *vs.* the carbon number of alkylbenzene substituents, were derived for the poly(butyl methacrylate-*co*-ethylene dimethacrylate) monoliths containing 0.3% AMPS. Their comparison with corresponding selectivity factors for a column packed with standard octadecylated silica indicated that the monoliths exhibited similar selectivity (Table 21.4) [36]. Apparently, the lower hydrophobicity of the C4 chains, compared to the highly hydrophobic C18 strands typical of ODS phases, was offset by the high concentration of the incorporated C4 ligands forming a dense hydrophobic surface layer and by the contribution of the hydrocarbon polymer backbone itself. Further increase in hydrophobicity can be achieved by using methacrylates with longer ester alkyl chain length and has been demonstrated by Ngola *et al.* [5].

Detailed studies of polymethacrylate monoliths, concerning the effect of polymer morphology and pore structure on their performance in CEC, have also been carried out [13,16,35,37]. Pore-size distributions obtained by mercury intrusion porosimetry in the dry state have been related to the column efficiency. Generally, monoliths with large pores exceeding 1 μm afforded lower efficiencies. For example, Peters *et al.* [13] found that thermally polymerized poly(BMA-*co*-EDMA-*co*-AMPS) monolith

TABLE 21.3

REVERSED-PHASE CEC APPLICATIONS OF POLY(METH)ACRYLATE-TYPE AND STYRENIC MONOLITHS [a]

Analytes	Reaction mixture		Mobile phase (applied voltage)	Maximum efficiency (theor. pl/m)	Ref.
	Monomers	Porogens (Solvents)	Monomer-to-porogen ratio		
Poly(meth)acrylate-type monoliths					
Neutral benzene derivatives	40% EDMA, 60% BMA + AMPS (0.3–5%) (w/w)	10% water, 90% 2-propanol and 1,4-butanediol (various ratios) (w/w)	40:60 (w/w)	ACN–5 mM phosphate buffer (pH 7) (80:20, v/v)	210,000	[13, 16, 35, 36]
Tyr-containing peptides including enkephalins	40% EDMA, 60% BMA + AMPS (8%) (w/w)	10% water, 90% 2-propanol and 1,4-butanediol (various ratios) (w/w)	40:60 (w/w)	10% of 10 mmol/L aqueous 1-octane-sulfonate and 90% of mixture of ACN–5 mM phosphate buffer (pH 7) (80:20, v/v)	43,000	[16]
Benzene derivatives	40% EDMA, 60% BMA + AMPS (0.3%) (w/w)	10% water, 90% 2-propanol and 1,4-butanediol (various ratios) (w/w)	47–67 (vol%) porogen	ACN–5 mM phosphate buffer (pH 7) (80:20, v/v)	149,000	[37]

Analyte	Stationary phase	Porogen	Ratio	Mobile phase	Efficiency	Ref.
Benzene derivatives and PAHs	30% BDDA, 70% BA + AMPS or AETMA (0.5%)	EtOH (20%), ACN (60%), 5 mM phosphate (pH 6.8) (20%)	33:67	ACN (70-80%)–5 mM Tris (pH 8)	200,000	[5]
PTH-amino acids	30% BDDA, 70% (BA + LA + AMPS)	EtOH (20%), ACN (60%), 5 mM phosphate (pH 6.8) (20%)	33:67	ACN–25 mM phosphate (pH 7.3) (40:60, v/v)	268,000	[38]
NDA-amino acids				ACN–20 mM phosphate (pH 7.2) (15:85, v/v)	371,000	[38]
3 Bioactive peptides as NDA derivatives				ACN–12.5 mM phosphate (pH 7) (35:65, v/v)	11,000	[38]
(Phenanthrene-like) impurities of heroin, *e.g.*, acetylthebaol				ACN–5 mM Tris (pH 9) (50:50, v/v)	63,000	[39]
Bioactive peptides (pI 8.5-8.9) (various enkephalins, α-casein fragment, β-lipoprotein fragment, t:ymopentin, splenopentin, and kyotorphin)	30% BDDA, 68% BA + AETMA (1%) + 1% cellulose	EtOH (20%), ACN (60%), 5 mM phosphate (pH 6.8) (20%)	33:67	ACN–12.5 mM phosphate (pH 2.8) (15:85, v/v)	330,000	[38]
di- and tripeptides	45% EDMA and 55% LMA (v/v)	45% 1-propanol and 55 % 1,4-butanediol (w/w)	50:50 (v/v)	40 mM phosphate buffer (pH 2.1)	150,000	[14]

Continued on the next page

TABLE 21.3 (continued)

Analytes	Reaction mixture			Mobile phase (applied voltage)	Maximum efficiency (theor. pl/m)	Ref.
	Monomers	Porogens (Solvents)	Monomer-to-porogen ratio			
Test mixture consisting of aromatic bases, acids, neutral compounds and dipeptides	50% BMA and 50% EDMA (w/w); dynamically coated with SDS or CTAB	50% 1-propanol and 50 % 1,4-butanediol (w/w)	50:50 (v/v)	35% ACN in 5 mM phosphate buffer (pH 2.5) containing 2.5 mM SDS or CTAB	140,000	[15]
Peptides (angiotensin I, angiotensin II, [Sar1,Ala8]angiotensin II, and bradykinin) proteins (ribonuclease A, insulin, α-lactalbumin, myoglobin)	25% GMA, 25% MMA and 50% EDMA; derivatized with N-ethylbutylamine	16.7% n-propanol and 83.3% FA (v/v)	40:60	40% ACN in 50 mM aqueous phosphate buffer (pH 2.5); −25 kV	150,000	[40]
				30% ACN in 60 mM aqueous phosphate buffer (pH 2.5); −25 kV	140,000	
Styrenic monoliths						
Alkylbenzenes, phenols, chlorobenzenes, aromatic amines, phenylenediamine isomers	25% S, 50% DVB, 25% methacrylic acid (v/v)	100% toluene	20:80 (v/v)	70 or 80% ACN in 4 mM Tris buffer (pH 8.8); 20 kV	140,000	[41]

Analyte	Composition	Porogen	Ratio	Mobile phase	Ref.
Insulin and peptides (angiotensin I, angiotensin II, [Sar^1,Ala^8]angiotensin II)	50% VBC, 50% DVB (v/v); derivatized with N,N-dimethyloctylamine	n-propanol and FA	40:60 (v/v)	50 mM NaCl in 5 mM phosphate buffer with 25% ACN, pH 3.0	[4]
Gly-Gly-Phe, Phe-Gly-Gly	67% S, 33% DVB (v/v)	EtOH, water, MeOH	40:60 (v/v)	10 mM phosphate buffer, pH 6.8	[4]
Proteins (ribonuclease A, insulin, α-lactalbumin, myoglobin)	50% VBC and 50% EDMA (v/v); derivatized with N,N-dimethylbutylamine	67% n-propanol and 33% FA (v/v)	40:60 (v/v)	30% ACN in 60 mM phosphate buffer, pH 2.5; -30 kV	[42]
Insulin and peptides (angiotensin I, angiotensin II, [Sar^1,Ala^8]angiotensin II)				30% ACN in 60 mM phosphate buffer, pH 2.5	[42]
Proteins (insulin, α-lactalbumin, myoglobin)				40% ACN in 70 mM phosphate buffer, pH 2.5	[42]
Tryptic digest of cytochrome C				40% ACN in 50 mM phosphate buffer, pH 2.5	[42]

a AIBN as initiator, polymerization started by free radicals obtained by thermal decomposition of AIBN or UV irradiation

TABLE 21.4

COMPARISON OF HYDROPHOBIC SELECTIVITIES $\alpha(CH_2)$ FOR POLY(BUTYL METHACRYLATE)-TYPE MONOLITHIC COLUMN AND OCTADECYLATED SILICA PACKED COLUMN [36]

Mobile phase	αCH_2 for different acetonitrile-buffer mixtures [a]				
	90:10	80:20	70:30	60:40	50:50
Monolithic column [b]	1.14	1.21	1.28	1.36	1.46
Packed column [c]	1.21	1.32	1.34	1.48	1.56

[a] Mobile phase, acetonitrile–5 mmol/L phosphate buffer (pH 7)

[b] Column, 100 μm I.D. × 30 cm active length; monolith, 0.3 wt% AMPS, 59.7 wt% BMA, 40 wt% EDMA, pore size 750 nm.

[c] Column, 100 μm I.D. × 25 cm active length; stationary phase, 3 μm Hypersil C18

with a pore diameter of about 700 nm afforded a maximum plate number for alkylbenzenes of around 170,000/m, while Yu *et al.* [16] found an optimum at ca. 300 nm for UV polymerized monoliths of identical composition to obtain an efficiency of 210,000 pl/m. It is of utmost importance to optimize the pore structure, for example by varying the composition of porogens, to fully exploit the efficiency potential in CEC. This was clearly demonstrated by Peters *et al.* [13]. Uncontrolled changes of pore size as a result of variations in the monomer composition that could be used to adjust the hydrophobicity or EOF may easily lead to the loss of resolution as shown in Fig. 21.8. Re-adjustment of the optimal pore size recovered the maximum efficiency (Fig. 21.8c).

Separations of alkylbenzenes were also demonstrated using polymethacrylate monoliths with C_4 ligands and negative surface charge (AMPS) or positive charge ([2-(acryloyloxy)ethyl]trimethylammonium methyl sulfate, AETMA). Monoliths with a neutral surface containing no ionizable comonomer were used for the separations of various small molecules (Table 21.3). For example, a poly(BMA-*co*-EDMA) monolithic column simultaneously separated mixtures of neutral, basic, and acidic aromatic analytes in a single run, using 2.5 mmol/L SDS or CTAB in the eluent, leading to dynamic ion-exchange systems [15].

All twenty native amino acids, in the form of phenylthiohydantoin (PTH) derivatives, were separated in a single 75 min run on a negatively charged poly(butyl acrylate-*co*-1,3-butanediol diacrylate) with 0.5% AMPS and a low percentage of C_{12} ligands (Fig. 21.9) [38]. Since most PTH derivatives are neutral, separation was

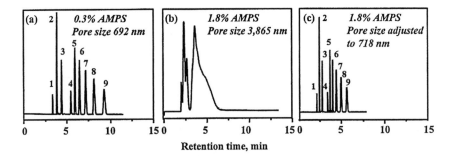

Fig. 21.8. Effects of percentage of 2-acrylamido-2-methyl-1-propanesulfonic acid (AMPS) and pore size on the electrochromatographic properties of monolithic capillaries (Reprinted with permission from ref. [18], copyright 2000 Elsevier Science B.V.). Conditions: capillary column, 30 cm effective length × 100 μm i.d.; stationary phases with 0.3 (a), 1.8 (b), and 1.8 wt.% AMPS (c); pore size of 692 nm (a), 3,865 nm (b), and 718 nm (c) in dry state; mobile phase, acetonitrile–5 mmol/L phosphate buffer, pH 7 (80:20, v/ v); UV detection, 215 nm; applied voltage, 25 kV; Peaks: thiourea (1), benzyl alcohol (2), benzaldehyde (3), benzene (4), toluene (5), ethylbenzene (6), propylbenzene (7), butylbenzene (8), and amylbenzene (9).

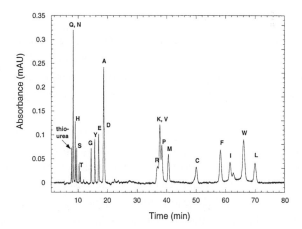

Fig. 21.9. Isocratic separation of PTH-amino acids by CEC using a negatively charged poly(butyl acrylate-*co*-1,3-butanediol diacrylate-*co*-lauryl acrylate) monolith with 0.5% AMPS (Reprinted with permission from ref. [38], copyright 2001 Elsevier Science B.V.). Conditions: mobile phase, acetonitrile–25 mmol/L phosphate (pH 7.3) (40:60, v/v); UV detection, 214 nm; field strength, 175 V/cm; capillary, 28.5 cm total length, 17.5 cm effective length, 100 μm I.D.

largely based on chromatographic interaction, as could be deduced from the elution order that followed the increasing hydrophobicities of the PTH-derivatives. However, the electrophoretic mobility contributed to the observed retention of negatively (Asp,

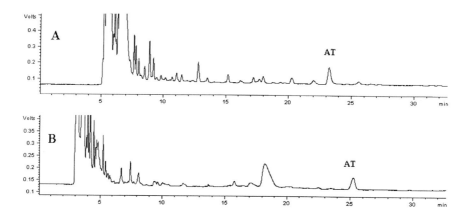

Fig. 21.10. CEC of acidic-neutral extract of refined Southeast Asian heroin (Reprinted with permission from ref. [39], copyright 2001 Elsevier Science B.V.). Conditions: (A) mobile phase, 5.0 mmol/L Tris (pH 9.0)–acetonitrile (70:30, v/v); stationary phase, 1.5 μm non-porous ODS II column; temperature, ambient; applied voltage, 15 kV. (B) mobile phase, 5.0 mmol/L Tris (pH 9.0)–acetonitrile (50:50, v/v); stationary phase, porous butyl acrylate-based monolithic column with C12 and sulfonic acid ligands; temperature, ambient; applied voltage, 20 kV; LIF detection.

Glu)- and positively charged (Arg)-derivatives. In addition, the Arg-derivative was also retained owing to electrostatic interactions with the stationary phase. Similarly, a series of fifteen naphthalene-2,3-dicarboxaldehyde (NDA) labeled amino acids in 10^{-8} mol/L solution were separated with efficiencies between 65,000–371,000 plates/m in a still long run time of 40 min [38].

The same monolithic column was used in forensic analysis for CEC profiling of phenanthrene-like impurities in heroin with LIF detection [39]. The resolving power of this monolithic column was similar to that of a column packed with 1.5 μm non-porous ODS that reportedly afforded performances among the best ever achieved in CEC (Fig. 21.10). Compared to the ODS beads, the monolithic phase required a higher percentage of organic modifier to achieve the elution of the analytes in similar times. Acetylthebaol, which exhibits a native fluorescence excited at a wavelength of 258 nm, was identified as the most abundant impurity and eluted with an efficiency of 63,000 plates/m. To further improve the peak capacity of the monolithic CEC column, multi-step gradient elution was suggested that resolved 2.5–3 times more peaks than the standard HPLC or MEKC techniques used earlier for these analyses. The LOD for CEC of acetylthebaol was an impressive 66 pg/mL, 30–15 times lower than that for other methods.

A number of studies dealt with the separation of peptides and proteins using poly(meth)acrylate-type monoliths. The CEC separation of these solutes is a challenging task. First, the physicochemical nature of both peptides and proteins, and their retention behavior, may vary over a wide range often requiring a gradient elution, which is difficult to implement in CEC. The stationary phase should therefore allow separation upon isocratic elution. Secondly, peptides and proteins are usually ionized under the separation conditions, which in CEC superimposes their electrophoretic migration over the chromatographic process. Finally, and most importantly, permanent charges on the surface of the stationary phase, required for driving the mobile phase by electroosmosis, may interact with ionized solutes and lead to a severe band dispersion (Fig. 21.7). Therefore, a very careful design of the separation system is required, that includes the hydrophobicity and surface charge of the stationary phase as well as the composition and pH of the mobile phase.

For example, Shediac *et al.* have demonstrated the separation of eight bioactive peptides, including various enkephalins, α-casein fragment, β-lipoprotein fragment, thymopentin, splenopentin, and kyotorphin, with pI values between 8.5 and 8.9 [38]. In order to avoid electrostatic interactions between oppositely charged sorbent and the analytes, separations were performed at pH 2.8, which is below the pI of the peptides, using a monolithic stationary phase carrying quaternary ammonium groups. Gusev *et al.* [4] and Ericson *et al.* [28] earlier suggested similar positively charged monolithic columns for the peptide and protein separations. The positively charged peptides were repelled from the equally charged surface sites thus circumventing attractive Coulombic interactions. Unfortunately, the electrophoretic migration opposite to the direction of EOF reduced the speed of the separations, yet might positively affect the selectivity. It appeared that under these conditions the retention times were mostly determined by the electrophoretic migration. Only three poorly resolved peaks were monitored using CE in a coated empty capillary, demonstrating the positive effect of the stationary phase on selectivity. A further improvement in resolution was achieved by incorporating low amounts of cellulose ester derivative into the monolith, and a separation of seven out of the eight peptides was achieved.

Other concepts enabling CEC separation of peptides on polymethacrylate-based monolithic columns include the addition of ion-pairing agents such as 1-octanesulfonate to preclude ion-exchange interactions, exemplified in the separations of Tyr-containing peptides using a negatively charged monolith at pH 7 [16], and the avoidance of surface-bound charges using neutral monolithic columns at low pH, at which the peptides are positively charged and driven through the column solely by means of electrophoretic migration [14].

Problems encountered with peptides apply to an even higher extent to the separations of proteins. Zhang *et al.* designed a specific monolithic phase for isocratic separations of proteins and peptides at low pH using water-rich hydro-organic eluents [40]. A poly(glycidyl methacrylate-*co*-ethylene dimethacrylate) monolithic column

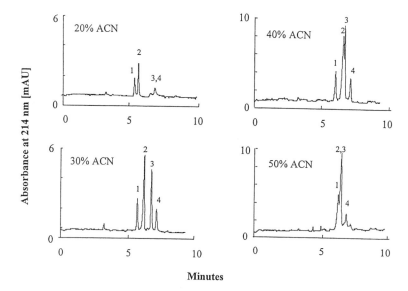

Fig. 21.11. Effect of acetonitrile concentration in the eluent on the separation of proteins (Reprinted with permission from ref. [40], copyright 2000 Elsevier Science B.V.). Column, poly(glycidyl methacrylate-*co*-ethylene dimethacrylate) monolith functionalized with N-ethylbutylamine, 39 cm (29 cm effective length) × 50 μm I.D.; mobile phase, acetonitrile–60 mmol/L sodium phosphate buffer (pH 2.5); applied voltage, –25 kV; UV detection, 214 nm; Peaks, ribonuclease A (1), insulin (2), β-lactalbumin (3), myoglobin (4).

was functionalized with N-ethylbutylamine to obtain a high-capacity weak anion-ex-changer, that was positively charged and afforded strong EOF at low pH [43,44]. Since electrostatic protein–sorbent interactions are repulsive at this pH, the separations were thought to be controlled by solvophobic interactions with the moderately lipo-philic butyl ligands, together with the contribution of electrophoretic mobility. N-Ethylbutylamine was selected after preliminary experiments with a series of secondary alkylamines. Amines with longer alkyl chains, such as N-methyldodecylamine and N-methyloctadecylamine, allegedly afforded both poor separations and recovery. A mixture of ribonuclease A, insulin, α-lactalbumin, and myoglobin was resolved using a mobile phase containing 30% acetonitrile in phosphate buffer pH 2.5 (Fig. 21.11). The selectivity and resolution were very sensitive to small changes in the concentra-tion of organic modifier in the eluent, while the migration did not change consider-ably. It is likely that faster elution resulting from the weaker solvophobic interactions at higher contents of modifier is counterbalanced by enforced electrophoretic counter-directional migration, clearly suggesting a dual separation mechanism. Similarly, the separation of a peptide mixture consisting of angiotensin I, angiotensin II,

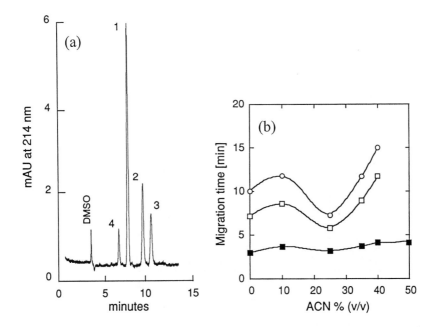

Fig. 21.12. Separation of peptides using poly(vinylbenzyl chloride-*co*-divinylbenzene) monolith modified with N,N-dimethyloctylamine (Reprinted with permission from ref. [4], copyright 2000 Elsevier Science B.V.). (a) Electrochromatogram of a mixture consisting of angiotensin II (1), angiotensin I (2), [Sar[1],Ala[8]]-angiotensin II (3), and insulin (4). (b) Effect of acetonitrile on migration times of DMSO (■), angiotensin II (□), angiotensin I (O). Conditions: column, 31 cm (21 cm effective length) × 75 μm I.D.; mobile phase, 50 mmol/L NaCl in 5 mmol/L phosphate buffer (pH 3.0) with 25% acetonitrile (a), and varying percentage of organic modifier (b); reversed polarity.

[Sar[1],Ala[8]]angiotensin II, and bradykinin was accomplished on the same weak anion exchange monolithic column [40].

21.3.1.3 Styrenic polymer monoliths

So far, only a limited number of studies using styrene-based monolithic columns in CEC has been published (Table 21.3) [4,41,42]. Following the requirements discussed above for electrochromatographic separation of peptides, a quaternary anion-exchange column prepared from a poly(chloromethylstyrene-*co*-divinylbenzene) monolith by an on-column reaction with N,N-dimethyloctylamine was successfully used to separate peptides (Fig. 21.12a) [4]. Although the yield of on-column quaternization has not been determined, a strong anodic EOF of up to 4 mm/s confirms a reasonable surface charge. Very flat van Deemter plots enabled the elution of peptides with plate heights

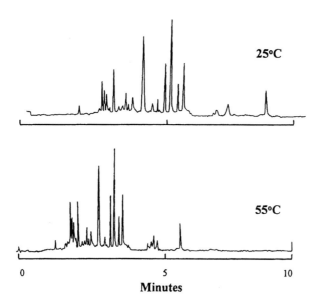

25°C

55°C

0   5   10

Minutes

Fig. 21.13. Capillary electrochromatograms illustrating the separation of tryptic digest of cytochrome C using isocratic elution at 25°C and 55°C (Reprinted with permission from ref. [42], copyright 2001 Elsevier Science B.V.). Conditions: Column, poly(vinyl-benzylchloride-*co*-ethylene dimethacrylate) monolith modified with N,N-dimethyl-butylamine; 40 cm (effective length 30 cm) × 75 μm I.D.; mobile phase, 40% acetonitrile in 50 mmol/L phosphate buffer, pH 2.5; applied voltage, –30 kV; UV detection, 214 nm.

of less than 10 μm within the entire flow range. Once again, the complex interplay of EOF, hydrophobic interaction, Coulombic repulsion, and electrophoretic migration in this counter-directional CEC separation resulted in an atypical correlation of migration times and acetonitrile content in the mobile phase (Fig. 21.12b), which does not follow the typical reversed phase elution pattern.

Temperature is a very effective, though often neglected, variable that enables tuning of separations in CEC. Besides the common effects on thermodynamic and kinetic parameters of the chromatographic process, an increase in temperature also accelerates the separations through improved transport properties, such as enhanced diffusivity and lower viscosity, that promotes EOF. For example, the analysis of a tryptic cytochrome C digest via isocratic CEC using a mixed styrenic-methacrylic poly(chloromethylstyrene-*co*-ethylene dimethacrylate) monolith quaternized with N,N-dimethylbutylamine enabled reduction of the run-time from about 10 min at 25°C to 6 min at 55°C without a loss in resolution (Fig. 21.13) [42]. This might be ascribed to the strong reduction in the viscosity of the mobile phase that led to an almost linear increase in EOF velocities from 1.95 at 25°C to 2.90 mm/s at 55°C.

21.3.1.4 Monolithic silica columns

Monolithic silica columns are typically prepared by a sol-gel process from alkoxysilane precursors. For the reversed-phase separations, the resulting silicious monolithic structures are subsequently octadecylated by continuously pumping a toluene solution of octadecyldimethyl-N,N-diethylaminosilane through the capillary column at 60°C for several hours. Since these capillary columns are still under development, and their properties need to be optimized, the applications have mostly included only model mixtures consisting of neutral aromatic compounds (Table 21.5). The attachment of the monolithic structure to the inner wall of the fused silica tube prevented its shrinkage during the sol-gel process. Therefore, the morphology of skeletons was different, the through-pores were larger (up to 8 μm) and the total porosity higher than with similar monoliths applied in HPLC and prepared in wider tubes, which allowed shrinkage [23]. The large through-pores affect chromatographic performance and were thought to be responsible for the poor efficiency of early CEC capillary columns used under pressure-driven flow (Fig. 21.14). The slope of the van Deemter plot was steep and a larger A-term contribution to band spreading was also observed. However, this term appeared less important in CEC. The efficiency of the equivalent column was 4–5 times better in CEC than in the pressure driven mode. For example, an increase in efficiency from 12,000 for HPLC to 58,000 plates corresponding to a plate height of about 7 μm for the CEC mode was observed for the separation of alkylbenzenes using a 25 cm octadecylated monolithic silica column (Fig. 21.14). However, the maximum flow rate that could be achieved with this column, using a commercial CEC instrument with a maximum upper voltage limit of 30 kV, was slow and prohibited fast separations that were possible in the HPLC mode. This resulted from the high purity of the silica and the efficient modification of silanol groups. This was also confirmed by high hydrophobic selectivity constants $\alpha(CH_2)$, reaching up to 1.51. Several attempts were made to increase the EOF and thus the speed of separations [45]. For example, the EOF increased after the reduction of the extent of surface coverage with octadecyl groups, that had left more residual silanols available. However, this was achieved at the expense of hydrophobic selectivity. Reaction with octadecyltrichlorosilane led to a two-fold higher EOF. This together with the pressure-assisted flow which was enabled by the high permeability of these monolithic columns even at a very low pressure of 0.14 MPa — that can be easily achieved in most CE instruments — also accelerated the analyses.

TABLE 21.5

REVERSED-PHASE CEC APPLICATIONS OF SILICA MONOLITHS

Analytes	Reaction mixture	(Post-)treatment	Mobile phase	Maximum efficiency (theor. pl/m)	Ref.
Alkylbenzenes; polyaromatic hydrocarbons	TMOS, PEO (M_w 10,000) in AcOH	Ammonia wash, heat treatment (330°C), octadecylation	ACN–Tris.HCl (50 mmol/L, pH 8) (90:10, v/v)	232,000	[6, 23, 46]
Alkyl phthalates	TMOS, urea, PEO (M_w 10,000) in AcOH	Ammonia wash, heat treatment (330°C), octadecylation	ACN–Tris.HCl (5 mmol/L, pH 8) (80:20, v/v)	4.4-7.4 μm (CEC), 5.5-13 μm (pressure-assisted CEC), 12-15 μm (LC)	[45]
PAHs; benzene derivatives; aldehydes and ketones	TMOS, N-octadecyldimethyl-[3-(trimethoxysilyl)propyl]-ammonium chloride, phenyldimethylsilane, TFA	Thermal treatment up to 150°C	70–80% (v/v) ACN in Tris.HCl (5 mmol/L, pH 2.34); –15 or –25 kV	175,000	[7]
Alkylbenzenes; PAHs; alkyl-phenones	3-(methacryloyloxy)propyl-trimethoxysilane, HCl; toluene (porogen), photoinitiator	Irradiation at 365 nm for 5 min	50–60% (v/v) ACN in ammonium acetate (5 or 10 or 12.5 mmol/L); 15 or 10 kV		[47, 48]

Analytes	Stationary phase	Mobile phase / conditions	Ref.
Peptides (bradykinin, angiotensin II, Gly-Gly-Gly, Val-Tyr-Val, methionine enkephaline)		40% (v/v) ACN in 10 mmol/L phosphoric acid; 12kV	[48]
Alkyl phenones	3-(methacryloyloxy)propyl trimethoxysilane, HCl; toluene (porogen), photoinitiator (parent monolith) Irradiation at 365 nm for 5 min, silanization with 3,3,3-trifluoropropyltrichlorosilane, pentafluorophenyldimethylchlorosilane, pentafluoro phenyltriethoxysilane, *n*-octyldimethylchlorosilane, tridecafluoro-octyldimethylchloro- silane	Ammonium acetate (50 mmol/L, pH 6.5)–water–ACN (10:50:40, v/v/v); 769 V/cm	[49]
Nucleosides (inosine, uridine, guanosine, cytidine)	Parent monolith silanized with aminopropyltriethoxysilane	Phosphate (50 mmol/L, pH 8)–water–ACN (5:35:60, v/v/v); –577 V/cm	[49]
Peptides (angiotension I and II, bradykinin, Gly-Gly-Gly, Val-Tyr-Val, Met-enkephalin)	Parent monolith silanized with pentafluorophenyltriethoxysilane	Ammonium acetate (50 mmol/L, pH 4.3)–water–ACN (10:40:50, v/v/v); 385 V/cm	[49]
Taxol derivatives (baccatin III, taxol, acetylbaccatin)	Parent monolith silanized with 3,3,3-trifluoropropyltrichlorosilane	Ammonium acetate (50 mmol/L, pH 6.5)–water–ACN (10:40:50, v/v/v); 385 V/cm	[49]

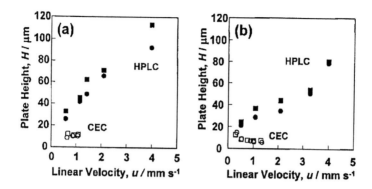

Fig. 21.14. Van Deemter plots obtained for a C18 monolithic silica column using a capillary in HPLC (closed symbols) and CEC (open symbols) with hexylbenzene (●,○) and benzo[a]pyrene (■,□) (Reprinted with permission from ref. [23], copyright 2000 American Chemical Society). Conditions: mobile phase, HPLC: acetonitrile–water; CEC: 80/20 (a), 90/10 (b) acetonitrile–Tris.HCl buffer, 50 mmol/L pH 8.

Two distinct shortcomings of the silica-based monolithic columns — the necessity of an additional octadecylation step and the poor EOF — were circumvented by using the co-precursor N-octadecyldimethyl-[3-(trimethoxysilyl)propyl]ammonium chloride in the sol-gel process (Fig. 21.15a) [7]. This precursor provides the alkoxysilane groups that undergo hydrolysis and subsequent condensation affording the 3-dimensional network and, in addition, both interactive octadecyl moieties and quaternary ammonium functionalities supporting EOF. The counteracting residual silanol groups were eliminated by their capping using the deactivation reagent phenyldimethylsilane, which was added to the sol-gel solution, and was expected to react with the residual silanols during the thermal treatment (Fig. 21.15b). The EOF velocity of this monolith was higher than that of octadecylated negatively charged silica monoliths. Theoretical plate numbers of the order of 170,000 plates/m could be achieved for neutral aldehydes and ketones (Fig. 21.15c).

Dulay *et al.* [47] prepared another type of hybrid organic–inorganic columns consisting of porous photopolymerized sol-gel monolith. A sol consisting of 3-methacryloyloxypropyl trimethoxysilane and hydrochloric acid was mixed with the porogen toluene, and then photopolymerized by irradiation at 365 nm for 5 min. This preparation involved no heat treatment so that the organic part of the polymer remained intact and provided the desired hydrophobic selectivity for reversed phase separations of neutral analytes such as PAHs, alkylbenzenes, and alkyl phenyl ketones with efficiencies of up to 100,000 plates/m [47]. Bonded photopolymerized sol-gel monolithic columns, prepared by subsequent silanization of the parent phase with octyl-, fluori-

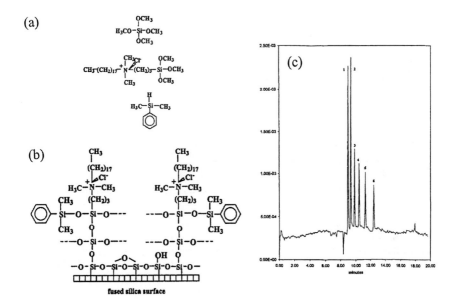

Fig. 21.15. Sol-gel reagents (a), sol-gel monolith with *in situ* incorporated octadecyl and quaternary ammonium groups (b), and representative separation of a mixture of aldehydes and ketones (Reprinted with permission from ref. [7], copyright 2000 American Chemical Society). Column, 50 cm (46.1 cm effective length) × 50 μm I.D.; mobile phase, acetonitrile–Tris.HCl (5 mmol/l, pH 2.34) (70:30, v/v); applied voltage, –25 kV; Peaks: benzaldehyde (1), *o*-tolualdehyde (2), butyrophenone (3), valerophenone (4), hexaphenone (5), heptaphenone (6).

nated octyl-, or pentafluorophenylalkyl silane substantially improved the resolution and enabled acceptable separations of nucleosides, peptides, and taxol derivatives [49].

Lack of detection sensitivity in CEC could be alleviated by chromatographic and, for ionized solutes, electrophoretic on-line sample preconcentration. This was demonstrated with diluted solutions of neutral alkyl phenones and peptides test mixtures [48]. In contrast to typical injections of sample plugs 0.1 mm long, 2-cm-long plugs were injected using pressure or voltage, thus increasing the limit of detection (LOD) by a factor of 100 for neutral analytes under isocratic elution conditions using equal organic content of modifier in both the injection solution and eluent, and up to 1000 using a gradient elution in which the sample was dissolved in a solution with less organic modifier and thus lower elution strength [48]. Electrophoretic pre-concentration similar to sample stacking can be used for ionized compounds. For example, a 1000-fold increase in LOD was observed for a peptide mixture [48]. The electrochromatograms shown in Fig. 21.16a and b depict the separation of a peptide test mixture

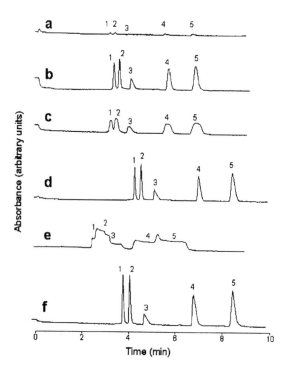

Fig. 21.16. CEC separation of bradykinin (1), angiotensin II (2), Gly-Gly-Gly (3), Val-Tyr-Val (4), and methionine enkephalin (5) using a photopolymerized sol-gel monolith (Reprinted with permission from ref. [48], copyright 2001 American Chemical Society). Conditions: Sample plug lengths, 0.1 mm (a) and 1.8 cm (b–f); mobile phase, 10 mmol/L phosphoric acid in 40% acetonitrile; sample dissolved in the mobile phase (a,b), 10 mmol/L phosphoric acid in 20% acetonitrile (c), 10 mmol/L phosphoric acid in 70% acetonitrile (d), 50 mmol/L phosphoric acid in 40% acetonitrile (e), 0.05 mmol/L phosphoric acid in 40% acetonitrile (f); peptide concentrations, 16.7 µg/mL each; applied voltage, 12 kV; detection, 214 nm; temperature, 20°C.

obtained after injection of 0.1- and 1.8-mm-long sample plugs, respectively. Both peak focusing and broadening were observed for the peptides dissolved in a solvent with a composition differing from that of the mobile phase. Focusing (stacking) occurred as a result of a reduction in electrophoretic velocity at the concentration boundary that separates the region of low conductivity sample matrix and the high conductivity separation solution. Accordingly, sample stacking could be achieved either by lowering the conductivity of the sample matrix using a higher percentage of acetonitrile (Fig. 21.16d) or by reducing the ionic strength of the buffer (Fig. 21.16f). Broadening or destacking were observed when the electrophoretic velocity at the concentration boundary increased (Fig. 21.16c and e).

21.3.1.5 Particle-loaded monolithic columns and consolidated packed beds

A variety of studies reported the use of monolithic columns that have been pre-pared by loosely embedding chromatographic particles in a sol-gel derived matrix simply by suspending particles in the sol-gel solution before filling the capillary [50-52], by packing particulate stationary phases into a capillary followed by fixation of the packed bed, either through sintering of the entire bed [8,53], or entrapment of the packed bed in a polymer network such as a silicate matrix [54], sol-gel matrix [24,55-61], or macroporous organic polymer [62] (Table 21.6). Detailed descriptions of these technologies can be found elsewhere in this book.

Sol-gel bonded 180 μm i.d. columns packed with wide-pore Nucleosil ODS (7 μm, 4000 Å) were used for the CEC separation of retinyl esters, including all-*trans*-retinyl-acetate, -palmitate, -heptadecanoate, -stearate, -oleate, and -linoleate, which are stor-age forms of vitamin A in the liver [58]. The limited solubility of these solutes even in methanol and acetonitrile required the use of a non-aqueous mobile phase consisting of 2.5 mmol/L lithium acetate as an electrolyte in N,N-dimethylformamide–acetoni-trile–ethanol (20:70:10). Reported run-to-run and day-to-day variations in retention times and peak areas of all-*trans*-retinyl palmitate were less than 0.3, with an RSD less than 2. Excellent column stability exceeding 1 month allowed an extensive study of liver extracts from arctic seals. The selectivity of the method could be significantly improved by use of sol-gel bonded C_{30}-modified Nucleosil 5 μm 4000 Å that, in contrast to the above C_{18} column, also afforded shape selectivity for cis/trans isomers and even enabled the resolution of 13-*cis*- from all-*trans*-retinyl palmitate [59]. The higher hydrophobicity and stronger retentivity were compensated for by temperature programming to accelerate the analysis (Fig. 21.17). The small capillary format en-ables fast temperature changes, which makes this approach a viable alternative to solvent gradients that require more sophisticated instrumentation. Figure 21.17 shows that high resolution was maintained for the early eluted components, as in the isother-mal method, while the peak-shape of later eluted components such as all-*trans*-retinyl palmitate (II) and stearate (VI) was significantly improved by the temperature gradi-ent.

21.3.1.6 Gradient-elution reversed-phase CEC with monolithic columns

Many real-life sample mixtures are fairly complex and contain constituents that differ widely in their hydrophobicities and therefore also in retentions. This applies in particular to biomolecules like proteins and peptides, for which the isocratic mode may fail to elute all components, or may require long run-times. Then, gradient elution is routinely utilized in standard and μ-HPLC to optimize the resolution of early eluted

TABLE 21.6

REVERSED-PHASE CEC APPLICATIONS OF PARTICLE-LOADED- OR CONSOLIDATED PACKED-BED MONOLITHS

Analytes	Matrix	Particles	Mobile phase (applied voltage)	Maximum efficiency (theor. pl/m)	Ref.
	Sintered bed monoliths				
PAHs	Sintered bed	Zorbax-ODS (6 μm, 80 Å)	ACN–sodium borate (10 mmol/L, pH 8) (80:20, v/v); 30 kV	160,000	[8]
PTH amino acids			ACN–sodium phosphate (5 mmol/L, pH 7.5) (70:30, v/v); 30 kV		
Parabenes and PAHs	Sintered bed	Hypersil ODS, 3 μm	ACN–Tris.HCl (25 mmol/L, pH 8) (80:20, v/v); 25 kV		[53]
Triazine herbicides		Spherisorb ODS I, 3 μm	ACN–Tris.HCl (25 mmol/L, pH 8) (56:44, v/v); 30 kV		
PAHs		Spherisorb ODS I, 3 μm	ACN–Tris.HCl (50 mmol/L, pH 8.5) (80:20, v/v); 30 kV		
	Particle-loaded monoliths				
PAHs	TEOS, EtOH, HCl	3 μm or 5 μm ODS particles + 4% silica	ACN–phosphate (50 mmol/L, pH 6.5) (80:20, v/v); 10 kV	100,000	[50-52]

Polymer-bonded packed-bed monoliths

Analytes	Precursors	Stationary phase	Mobile phase	Efficiency	Ref.
Neutral benzene derivatives and PAHs	TMOS, ETMOS, MeOH, TFA, water, FA	Vydac ODS (5 μm, 90 Å)	ACN–Tris (5 mmol/L, pH 8) (60:40, v/v); 15 kV	130,000	[55]
Neutral aromatic compounds, aromatic amines		Nucleosil ODS (7 μm, 4000 Å)	60% ACN–2.5 mmol/L Tris (pH 8)	220,000	[56]
Neutral benzene derivatives		Spherisorb ODS/SCX (3 μm, 80 Å)	80% ACN containing 1.5 mmol/L phosphate buffer (pH 3)	75,000	[57]
Corticosteroids (ouabain, 4-pregnene-6β,11β,21-triol-3,20-dione, strophantidin)		Spherisorb ODS/SCX (3 μm, 80 Å)	ACN–water–50 mmol/L phosphate buffer (pH 3) (65:30:5); 25 kV		[24]
Neutral aromatic compounds, aromatic amines		Spherisorb ODS (3 μm, 80 Å)	ACN–water–50 mmol/L Tris buffer (pH 8) (80:10:10); 20 kV		
PAHs in coal tar		Spherisorb ODS (3 μm, 80 Å)	ACN–50 mmol/L Tris buffer (pH 8) (90:10); 20 kV		
Alkaloids (mitraphylline, pteropodine, isopteropodine)		Nucleosil ODS (7 μm, 1400 Å)	ACN–water–50 mmol/L phosphate buffer (pH 8) (40:50:10); 25 kV		

Continued on the next page

TABLE 21.6 (continued)

Analytes	Matrix	Particles	Mobile phase (applied voltage)	Maximum efficiency (theor. pl/m)	Ref.
Retinyl esters (liver extracts from arctic seal)		Nucleosil ODS (7 µm, 4000 Å)	2.5 mmol/L lithium acetate in DMF–ACN–MeOH (20:70:10, v/v/v)	75,000	[58]
Retinyl esters (liver extracts from arctic seal)		Nucleosil C30 (5 µm, 4000 Å)	2.5 mmol/L lithium acetate in DMF–ACN–MeOH (20:70:10, v/v/v)		[59, 60]
PAHs and parabens	TEOS, *tert*-butyl- or *n*-octyltriethoxysilane, EtOH, acetic acid	Nucleosil ODS (5 µm)	ACN–Tris (25 mmol/L, pH 8.0) (80:20, v/v)	h_{red} = 1.1–1.4	[61]
PAHs	Silicate-network (formed from Kasil 2130 solution purged through packed bed followed by heat treatment)	Nucleosil ODS (5 µm)	ACN–0.1 mol/L acetate buffer (pH 3) (80:20, v/v); 30 kV	h_{red} = 1.2	[54]
PAHs	BMA–EDMA (water–1,4-butanediol–1-propanol as porogen)	Nucleosil ODS (5 µm)	ACN–Tris (25 mmol/L, pH 8) (80:20, v/v)	h_{red} = 1.1–1.7	[62]
Nonsteroidal antiinflammatory drugs (acetylsalicylic acid, naproxen, flurbiprofen, ibuprofen) and caffeine			ACN–Tris (10 mmol/L, pH 3) (70:30, v/v)		

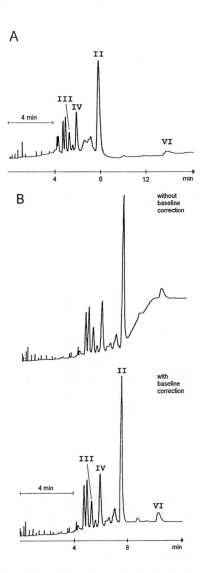

Fig. 21.17. CEC seal-liver extract profile obtained using (a) isothermal separation at 30°C and (b) temperature programming without (middle) and with baseline correction (bottom) (Reprinted with permission from ref. [60], copyright 2001 Wiley-VCH). Temperature program: 30°C for 4 min, 17°C/min until 70°C, 70°C for 5 min. Column, effective length 35 cm, total length 55 cm × 180 μm I.D.; mobile phase, 2.5 mmol/L lithium acetate in N,N-dimethylformamide–acetonitrile–methanol (20:70:10). Peaks: all-*trans*-retinyl palmitate $RC_{16:0}$ (II), linoleoate $RC_{18:2}$ (III), oleate $RC_{18:1}$ (IV), and stearate $RC_{18:0}$ (VI). Unlabeled peaks are degradation products and isomeric species.

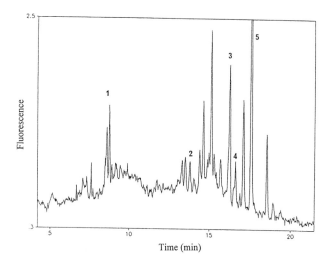

Fig. 21.18. Gradient electrochromatogram of derivatized urinary neutral steroids extracted from pregnancy urine (Reprinted with permission from ref. [32], copyright 2000 Elsevier Science B.V.). Gradient, acetonitrile–water–0.24 mol/L ammonium formate (pH 3) (35:60:5–65:30:5); current, 13–8 µA; monolith, polyacrylamide-based with C_{12} and sulfonate ligands; capillary, 35 cm × 100 µm ID, 25 cm column length; LIF detection. Peaks: labeling reagent (1), 11β-hydroxyandrosterone (2), dehydroisoandrosterone (3), estrone (4), spiked androsterone (5).

components, sharpen the bands, and accelerate the analysis by faster elution of late eluting compounds.

Since suitable commercial instrumentation for CEC enabling elution in a continuous gradient of the mobile phase is currently not available to most CEC practitioners, stepwise gradients, which can easily be programmed even with commercial CE/CEC instruments, have been used more often [25,39].

Linear gradients [28,32,33] require specifically designed devices that deliver the solvent mixtures to the inlet electrolyte vessel. For example, a home-made gradient CEC system was used for analysis of ketosteroids including androsterone and its structural analogs, derivatized with danyslhydrazine enabling laser-induced fluorescence (LIF) detection [32]. The highly efficient method affording up to 200,000 plates/m was also used for detection of free neutral ketosteroids extracted from pregnancy urine (Fig. 21.18).

Ericson and Hjertén showed that the gradient elution of ionized components such as proteins exhibiting significant electrophoretic migration in the CEC mode, was rather complex [28]. They carried out separations of a mixture consisting of ribonuclease A, cytochrome c, lysozyme, and α-chymotrypsinogen using C_{18}-modified polyacrylamide-based monolithic columns. These columns contained ionizable monomer, diallyldimethylammonium chloride, at two different concentrations to achieve either moderate or high EOF. An acidic mobile phase with pH 2.0, which was lower than the pI values of the proteins, was used to avoid electrostatic solute–sorbent interactions, thus making the reversed-phase-type chromatographic retention and selectivity operative simultaneously. In addition, the electrophoretic migration of the proteins was always opposite to the direction of the bulk liquid flow in the system.

In the moderate EOF column, the net migration velocity, and therefore the protein elution, were largely controlled by their electrophoretic migration towards the cathode located at the column outlet side, while the minor EOF velocity vector had the opposite direction, towards the injection (inlet) end, and reduced overall migration velocity (counter-flow gradient) (Fig. 21.19a). The sample was introduced at the anode end and the gradient, quite unusually, from the outlet side. Owing to the different acetonitrile percentages along the column, neither the electrophoretic or electroosmotic velocities was constant, and both increased near the outlet at which the eluent conductivity is lower. A representative electrochromatogram of a separation under these conditions is shown in Fig. 21.19c.

In contrast, the EOF velocity outweighed countercurrent electrophoretic migration in the high-EOF column so that net migration direction coincided with the EOF direction (normal-flow gradient). Figure 21.19b shows that following the common practice in chromatography, the sample and gradient were introduced at the same side of the capillary column. Lower background conductivity at the column inlet end translated into faster electroosmotic and electrophoretic migration velocities at the inlet end, presumably leading to a focusing. The separation achieved using this approach (Fig. 21.19d) compared well with that of counter-flow gradient.

Both separation modes discussed above afforded high resolution. Their comparison with the results of corresponding pressure-driven reversed-phase HPLC (Fig. 21.19e) clearly indicated that the separations were controlled by the reversed-phase mechanism, while the differences in electrophoretic mobility had only a minor effect on selectivity. The same elution orders in both CEC and HPLC runs (compare Fig. 21.19c–e) and lack of selectivity in the corresponding CE separations confirmed that the separations had a chromatographic character.

21.3.2 Normal-phase CEC

Separation of highly hydrophilic compounds in the reversed-phase CEC mode may fail easily, if the solvophobic interactions with the hydrophobic surface or ligands are not sufficiently strong. Then, CEC in the normal-phase mode (Table 21.7), that ex-

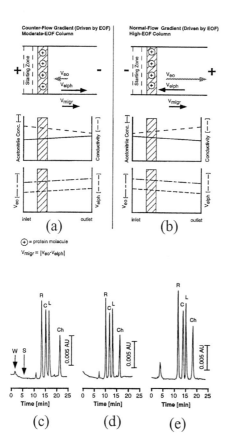

Fig. 21.19. Diagram illustrating the counter-flow, (a), and normal-flow, (b), gradient elution of proteins using positively charged polyacrylamide-based monolithic columns with low, (a), and high anodic EOF, (b), and corresponding electrochromatograms for counter-flow (c), and normal-flow separation (d), as well as pressure-driven separation (e). (Reprinted with permission from ref. [28], copyright 1999 American Chemical Society). Linear gradient from 5 to 80% acetonitrile in 5 mmol/L sodium phosphate, pH 2.0. On-tube detection at 280 nm. Columns: 8.0 cm (6 cm effective length) × 50 μm I.D. Applied voltage, ±5.5 kV (c,d). Pressure, 5 MPa (e). Peaks: ribonuclease A (R), cytochrome c (C), lysozyme (L), α-chymotrypsinogen (Ch).

ploits the hydrophilic nature of the stationary phases and hydrophilic analyte–sorbent interactions such as hydrogen bonding and dipole-dipole interactions, represents a suitable alternative that may offer better selectivity.

Novotny's group [31] designed a normal-phase polyacrylamide-based monolithic column containing trimethylammonium quaternary salt and 3-aminopropoxy ligands, affording anodic EOF for the separation of acidic bile acids and their corresponding glyco- and taurocholates. Simultaneous analysis of these compounds using a typical hydrophobic phase with negatively charged surface at pH 3 was unsuccessful, since the strongly acidic taurocholates exhibited fast countercurrent electrophoretic mobility. In contrast, the hydrophilic amino phase, with reversed polarity, afforded the co-directional electrophoretic mobility and facilitated the separation in a single run — and thus allowed determination of the "total profile". A hydro-organic mobile phase mode with 60% acetonitrile was used for elution. Retention of free bile acids, as well as its glycine and taurine conjugates, increased with the increasing number of hydroxyl groups in the cholan core, in accordance with a normal-phase chromatography mode (Fig. 21.20a). It is worth noting that extremely high efficiencies, of over 600,000 plates/m have been observed, the highest so far published for CEC using monolithic stationary phases. This separation method also allowed straightforward coupling to an electrospray ion-trap mass spectrometer and acquisition of ESI-MS spectra in the negative-ion mode, and enabled analysis of bovine bile (Fig. 21.20b). Detection limits in the femtomole level, and the possibility of identification of the analytes through tandem MS, clearly demonstrated the merits of this method (Fig. 21.20c–e), which clearly turned out to be more powerful than a corresponding reversed-phase system.

As demonstrated above, the polar solute–sorbent interactions are preserved even in strongly polar water-based mobile phases such as acetonitrile–water. This separation mode is often referred to as hydrophilic interaction chromatography (HILIC). In contrast, the separations in typical normal-phase mode involve hydrocarbon-based mobile phases (hexane, heptane) with polar modifiers such as alcohols. Maruška's group separated alkyl benzoates in the order of decreasing lipophilicity using a poly-acrylamide-based monolithic bed with incorporated 6-amino-β-cyclodextrin and hexane–methanol–propanol containing TEMED and acetic acid as electrolytes [21]. Although feasible, these hexane-based mobile phases appear to be less suitable for CEC owing to their limited ability to dissolve electrolytes in a sufficient concentration, leading to low conductivity and system stability for electrokinetic separations. Therefore, the use of polar-organic acetonitrile–alcohol mobile phases which dissolve electrolytes well, is preferred. This was exemplified by separations of mixtures of phenols and xanthines (caffeine, theobromin, theophylline) on polymethacrylate-based monoliths with pendant hydroxyethyl ligands [66]. A similar separation in the polar organic-phase mode using plain acetonitrile–methanol eluent without supporting electrolytes was reported by Hoegger and Freitag [20] for the separation of phenols on

TABLE 21.7

CEC APPLICATIONS OF ORGANIC POLYMER MONOLITHS IN VARIOUS CHROMATOGRAPHY MODES

| Analytes | Reaction mixture | | Monomer-to-porogen ratio | Mobile phase (applied voltage) | Maximum efficiency (theor. pl/m) | Ref. |
	Monomers	Porogens (Solvents)				
Normal-phase CEC						
Bile acids and glycine and taurine conjugates	5% T (AAm, BIS, charge-able co-monomers), 60% C, 3% (w/v) PEG, 30% AETMA, 30% APVE	50% (v/v) FA in 100 mmol/L Tris–150 mmol/L boric acid (pH 8.2)		ACN–water–240 mmol/L ammonium formate (pH 3) 60:35:5, v/v/v; E: 400 V/cm (neg. polarity)	610,000	[31]
Phenols; xanthines (caffeine, theobromine, theophylline)	DMAEMA (20%), HEMA (60%), EDMA (20%) (w/w); (on-column quaternized)	Cyclohexanol (33%), dodecanol (67%) (w/w)	40:60 (w/w)	0.4 mol/L acetic acid and 4 mmol/L triethylamine in ACN–MeOH (60:40, v/v): –25 kV; 50°C	80,000	[66]

Sample	Monolith composition	Porogen / buffer	Mobile phase	Efficiency	Ref.
Alkyl benzoates	Polyrotaxane-based polyacrylamide continuous beds; 21.4% T (PDA, MAAm, NIPAAm), 41.7% C + 6-amino β-CD-derivative	$(NH_4)_2SO_4$ in 50 mmol/L phosphate buffer (pH 7)	Hexane–MeOH–propanol (42.5/50/7.5) + 0.05% TEMED + 0.05 AcOH; –5 kV		[21]
Phenols	15 or 29% T (PDA, VSA, DMAA, MAAm or hexyl-acrylamide), 52% C, 1.5 or 3% VSA	$(NH_4)_2SO_4$ in 50 mmol/L phosphate buffer (pH 7) or phosphate buffer–FA (1:1, v/v)	ACN–MeOH (3:1) without supporting electrolytes	110,000	[20]
Ion-exchange CEC					
Substituted benzoic acids; profens (ibuprofen, naproxen, ketoprofen, suprofen)	DMAEMA (20%), HEMA (60%), EDMA (20%) (w/w); (on-column quaternized)	Cyclohexanol (33%), dodecanol (67%) (w/w) 40:60 (w/w)	0.4 mol/L acetic acid and 4 mmol/L triethylamine in ACN–MeOH (60:40, v/v); –25 kV; 50°C	230,000	[66]
Size-exclusion CEC					
Polystyrene standards	40 wt.% EDMA, 60 wt% BMA + AMPS (0.3%) (w/w)	10% water, 90% 2-propanol and 1,4-butanediol (w/w) 40:60 (w/w)	2% (v/v) water in THF		[36]

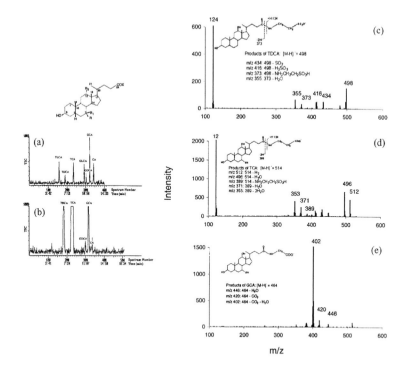

Fig. 21.20. Normal-phase CEC/negative-ion ESI-MS analysis. (Reprinted with permission from ref. [31], copyright 2000 American Chemical Society). Experimental conditions: capillary, 100 μm I.D., column length 30 cm; polymer, normal-phase (amino phase) polyacrylamide-based matrix containing quaternary ammonium groups and 3-aminopropyl ether moieties; mobile phase, acetonitrile–water–0.24 mol/L ammonium formate buffer (pH 3) (60:35:5); field strength, 400 V/cm; injection, 6 kV for 5 s. (a) A mixture of standards containing free bile acids together with glycine and taurine conjugates (taurolithocholic acid, TLCA: 0.2 mg/ml, taurodeoxycholic acid, TDCA: 0.1 mg/ml, taurocholic acid, TCA: 0.05 mg/ml, glycolithocholic acid, GLCA: 0.5 mg/ml, glycodeoxycholic acid, GDCA: 0.2 mg/ml, glycocholic acid, GCA: 0.2 mg/ml, and cholic acid, CA: 0.1 mg/ml). (b) Bile acids in bovine bile. (c,d,e) CEC/MS/MS of constituents of bovine bile in (b) identified as TDCA (c), TCA (d), and GCA (e).

a polyacrylamide-based monolithic column prepared from moderately hydrophobic N,N-dimethylacrylamide and vinylsulfonic acid.

21.3.3 Ion-exchange CEC

The stationary phases dedicated for CEC carry, by default, fixed charges on their surface to generate the desired strong EOF. Therefore, they possess intrinsic ion-exchange capabilities. However, CEC is seldom operated in the ion-exchange mode since the peak-shape is poor, owing to unbalanced ionic solute–sorbent interactions resulting from low ionic strength of the eluents. Therefore, only a few studies have addressed CEC separations in the ion-exchange selectivity mode using monolithic columns (Table 21.7) [17,66-68]. In this case, the ionizable ligands needed to be present in a higher surface density than was necessary to achieve the desired EOF. Typically, 5–20% of the ionizable monomer had to be included in the polymerization mixture to obtain a monolith with 0.2–1.0 mmol/g of ion-exchange functionalities. This high content is required since these functionalities fulfilled a dual function: they generated EOF to create flow, and simultaneously interacted with the analytes.

For example, the tertiary amine functionalities of a directly polymerized poly(2-di-methylaminoethyl methacrylate-*co*-2-hydroxyethyl methacrylate-*co*-ethylene di-methacrylate) monolith were quaternized *in situ* to afford a capillary column with high selectivity for the separation of substituted benzoic acids. The elution order followed the increase in the acidity of the analytes, as expected for an ion-exchange mechanism. Two factors appeared to be key for the efficient separation: (i) The use of the non-aqueous mobile phase, which resulted in significant shifts in pK_a of both analytes and fixed charges of the ion exchanger, as well as in change in the pH value. This, despite the lower polarity of non-aqueous solvents and thereby stronger electrostatic interactions in this mobile phase, favorably reduced both the dissociation and effective strength of the ionic interaction. (ii) High concentrations of counter-ions (acetic acid) could be present in the mobile phase without being the source of high current and Joule heating. As a result, a mixture of profens including ibuprofen, naproxen, keto-profen, and suprofen could be separated in the ion-exchange mode, affording a column efficiency of up to 230,000 plates/m and good peak symmetry. In a related study, enantiomers of N-derivatized amino acids could be resolved in the non-aqueous ion-exchange mode using a chiral analog of this monolithic column [17,67,68] (*vide supra*).

21.3.4 Size-exclusion CEC

Size-exclusion CEC is a non-interactive chromatographic mode which use differences in the hydrodynamic volume for the separation of macromolecules — typically neutral synthetic polymers such as polystyrenes, polyacrylates, polycarbonates, and polyurethanes. The selectivity is provided only by the pore-size distribution of the stationary phase, while any interactions of the analyte molecules with the stationary phase must be eliminated. Poor solubility of these synthetic polymers in hydro-organic mobile phases typical of CEC has led to almost exclusive use of tetrahydrofuran (THF) as eluent. Since pure THF afforded negligible EOF, 2% water had to be added to THF to achieve the desired EOF for the CEC separation of polystyrene standards using a hydrophobic monolithic column (Table 21.7) [36]. Polystyrenes with a molecular mass as high as 980,000 were still soluble in this mixture and the elution pattern, with the largest molecules eluting first, clearly confirmed the size-exclusion chromatography mode. A benefit of this, as-yet seldom used, approach is the high efficiency of CEC.

21.3.5 Enantioselective CEC

In recent years, the separations of enantiomers in CE and MEKC, in which a chiral selector or surfactant is simply added to the background electrolyte, became popular and increasingly replaced stereoselective HPLC assays, owing to their high efficiency, technical simplicity, flexibility in method development, and low cost [70]. However, these methods have a narrow dynamic range and, in particular, the presence of the soluble selector in the effluent interferes in UV and MS detection. In contrast, enantioselective CEC with monolithic chiral stationary phases (CSP) involves chiral selectors immobilized on the support or incorporated into the porous polymer network, and eliminates the presence of the selector in the effluent. The CEC mode also expands the dynamic range, thereby facilitating the analysis under overload conditions, which is usually required to reach the detection limits for minor enantiomeric components — often amounting to less than 0.1%.

Early developments of stereoselective CE with gel-filled columns or completely homogeneous separation media with chiral selectors such as cyclodextrins [71-73] or bovine serum albumin [74,75], physically or chemically incorporated into slightly cross-linked neutral gel matrixes, paved the way for enantioselective CEC with chiral monolithic separation media (Table 21.8) [76].

21.3.5.1 Monolithic polyacrylamide beds with chiral moieties

The originally used neutral β-cyclodextrin-modified polyacrylamide gels [71-73] afforded only a low EOF, thus leading to long run-times, bubble formation, short column lifetime, and poor reproducibility [72]. This has been overcome, at least in part, by inclusion of ionizable monomers in the monolith. For example, a polymeric selector with β-cyclodextrin (β-CD) moieties was simply admixed to the polymerization mixture and physically entrapped in the negatively charged poly(acrylamide-*co*-2-acrylamido-2-methylpropanesulfonic acid-*co*-methylenebisacrylamide) network [77,78]. This column then enabled separation of cationic drugs such as terbutaline, propranolol, and chlorpheniramine, as well as neutral chiral compounds such as benzoin and 1,2-diphenylethanol. In contrast, acidic analytes such as DNS-amino acids, which could not be eluted owing to counter-directional flow, required anodic EOF created in a positively charged monolithic gel containing N-(2-acrylamidoethyl) triethylammonium iodide (AMTEA) (Table 21.8). Since the pore size of these monoliths was kept small to minimize "bleeding" of the selector from the column, elution times for several analytes were excessively long. The undesired bleeding was later

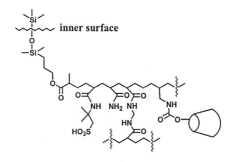

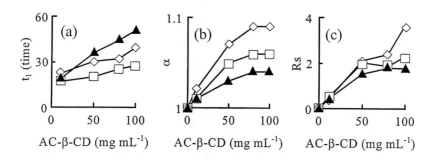

Fig. 21.21. Effect of concentration of chiral monomer (allylcarbamoylated β-cyclodextrin, AC-β-CD) in the reaction mixture on (a) retention times of first eluted enantiomers of terbutaline (◇), metaproterenol (□), and propranolol (▲), (b) enantioselectivity, and (c) resolution (Reprinted with permission from ref. [80], copyright 2000 Elsevier Science B.V.). Conditions: capillary, 75 μm I.D., 35 cm effective length; polymer, 5% T (AA, BIS, AMPS), 10% C, 5.6% S; mobile phase, 0.2 mol/L Tris–0.3 mol/L boric acid (pH 9); electric field strength, 214 (terbutaline) and 257 V/cm (metaproterenol, propranolol).

eliminated by covalent attachment of the cyclodextrin selector via copolymerization (Fig. 21.21) [79-81].

The cyclodextrin-containing monoliths enabled the separations in mobile phases that did not contain any organic modifier that would interfere with the inclusion of aromatic groups in the hydrophobic β-CD cavity and decrease the enantioselectivity. A hydrophilic polyacrylamide backbone is paramount for these separations, because it minimizes non-stereoselective hydrophobic interactions between hydrophobic portions of the solute and polymer chains. In contrast to secondary and tertiary aromatic amines, including β-sympathomimetics and β-blockers, the separation of primary aromatic amines was unsuccessful using the typical aqueous buffers. However, their separation could be achieved after the addition of an achiral 18-crown-6, presumably owing to the formation of a sandwich complex by simultaneous inclusion of the

TABLE 21.8

ENANTIOSELECTIVE CEC WITH MONOLITHIC CHIRAL STATIONARY PHASES

Analytes	Chiral selector or monomer (template)	(Co)monomers	Solvent (Porogens)	Mobile phase (applied voltage)	Maximum efficiency (theor. pl/m)	Ref.
Polyacrylamide gels and monoliths						
benzoin, terbutaline, chlorpheniramine, 1,2-diphenylethanol, propranolol	Poly-β-CD (50 mg/mL) and CM-β-CD (50 mg/mL) (physically incorporated)	8% T (AAm, BIS, AMPS), 5% C, 5.5% S	0.1 mol/L Tris–0.15 mol/L boric acid (pH 8.1), 5% (v/v) Tween 20	0.2 mol/L Tris–0.3 mol/L boric acid (pH 9), 5% (v/v) Tween 20; 20 kV (12 μA)	127,000	[77, 78]
DNS-amino acids, phenylmercapturic acid, warfarin	Poly-β-CD (80–120 mg/mL) (physically incorporated)	8.6% T (AAm, BIS, AMTEA), 4.7% C, 5.6% S	0.1 mol/L Tris–0.15 mol/L boric acid (pH 8.1)	0.2 mol/L Tris–0.3 mol/L boric acid (pH 8.1)	240,000	[78]
Propranolol, terbutaline, benzoin	AC-β-CD (50 mg/mL)	5% T (AAm, BIS, AMPS), 10% C, 5.6% S	0.1 mol/L Tris–0.15 mol/L boric acid (pH 8.1)	0.2 mol/L Tris–0.3 mol/L boric acid (pH 9)	84,000	[79]
Terbutaline, metaproterenol, isoproterenol, propranolol, pindolol, chlorpheniramine, TrpOMe, TrpOEt, α-methyltryptamine, clenbuterol; 1-(1-naphthyl)ethanol, methyl mandelate	AC-β-CD (50–100 mg/mL)	5 or 7% T (AAm, BIS, AMPS), 5 or 10% C, 5.5, 5.6 or 6% S	0.1 mol/L Tris–0.15 mol/L boric acid (pH 8.1)	0.2 mol/L Tris–0.3 mol/L boric acid (pH 7 or 9)	144,000	[80]

Analytes	Chiral selector	Matrix	Buffer	Mobile phase / additive	Efficiency	Ref.
1-Aminoindan, 1,2,3,4-tetrahydro-1-naphthylamine, 1-(1-naphthyl)ethylamine, primaquine	AC-β-CD (50 mg/mL)	5 % T (AAm, BIS, AMPS), 5% C, 5.5% S	0.1 mol/L Tris–0.15 mol/L boric acid (pH 8.1)	0.2 mol/L Tris–0.3 mol/L boric acid (pH 7) + 5–10 mmol/L 18-crown-6	150,000	[80]
DNS-amino acids, phenylmercapturic acid, warfarin, 2-phenoxypropionic acid, FMOC-Val, benzoin, 1-(1-naphthyl)ethanol	AC-β-CD (10–50 mg/mL)	5.4% T (AAm, BIS, AMTEA), 9.3% C, 5.6% S	0.1 mol/L Tris–0.15 mol/L boric acid (pH 8.1)	0.2 mol/L Tris–0.3 mol/L boric acid (pH 8.1)	150,000	[81]
Hexobarbital, mephobarbital, warfarin, tropicamide, ibuprofen, propranolol, mephenytoin, hydrobenzoin	Allyl-β-CD (37.5–150 µmol/mL)	5% T (AAm, BIS, AMPS), 3% C, 17% S	0.1 mol/L Tris–0.15 mol/L boric acid (pH 8.2)	0.1 mol/L Tris–0.15 mol/L boric acid (pH 8.2)	570,000	[82]
Ibuprofen, warfarin, mephobarbital	Allyl-β-CD (50 µmol/mL)	5% T (AAm, BIS, DMDAAC), 3 % C, 17% S	0.1 mol/L Tris–0.15 mol/L boric acid (pH 8.2)	0.1 mol/L Tris–0.15 mol/L boric acid (pH 8.2)	65,000	[82]
Ibuprofen	Polyrotaxane-based polyacrylamide continuous bed formed from sulfated β-cyclodextrin	21.4% T (MAAm, PDA, NIPAAm), 41.7% C	$(NH_4)_2SO_4$ in 50 mmol/L phosphate buffer (pH 7)	50% aqueous MeOH buffered with AcOH–TEMED (0.05% each)		[21]
1-Aminoindan, 1-(1-naphthyl)ethylamine, 1-phenylethylamine, alanine-2-naphthylamide, α-methyltryptamine, 1,2-diphenylethylamine, tryptophanol, 2-amino-1,2-diphenylethanol, TrpOMe	Mono or tetraallyl 18-crown-6-tetra-carboxylate (20–40 mmol/L)	5 % T (AAm, BIS, AMPS), 5% C, 5.5% S	0.1 mol/L Tris–0.15 mol/L borate (pH 8.1)	0.2 mol/L triethanol-amine–0.3 mol/L boric acid (pH 6 or 7) with 0–50% ACN (v/v)	135,000	[83]

Continued on the next page

TABLE 21.8 (continued)

Analytes	Chiral selector or monomer (template)	(Co)monomers	Solvent (Porogens)	Mobile phase (applied voltage)	Maximum efficiency (theor. pl/m)	Ref.
Amino acids (Asn, DOPA, α-methyl-DOPA, α-methyl-Phe, Tyr, Phe, Thr, Trp, Ser	N-(2-hydroxy-3-allyl-oxypropyl)-L-4-Hyp (10–20%) (25 μL)	PDA (22 mg), MAAm (18 mg), VSA (30%) (3 μL)	10 mg $(NH_4)_2SO_4$ in 175 μL 50 mmol/L phosphate buffer (pH 7–8)	50 mmol/L NaH_2PO_4/0.1 mmol/L Cu(II) pH 4.6; 7 kV; 12 bar on inlet		[84]
Hydroxy acids such as mandelic acid (MDA), 3-hydroxy-MDA, 4-hydroxy-MDA, 4-bromo-MDA, 4-methoxy-MDA, 3,4-dihydroxy-MDA, 3-hydroxy-4-methoxy-MDA, 4-hydroxy-3-methoxy-MDA, atrolactic acid, 3-(4-hydroxyphenyl)lactic acid, 3-phenyllactic acid	N-(2-hydroxy-3-allyl-oxypropyl)-L-4-Hyp (10–20%) (25 μL)	PDA (22 mg), MAAm (18 mg)	10 mg $(NH_4)_2SO_4$ in 175 μL 50 mmol/L phosphate buffer (pH 8)	20 mmol/L NaH_2PO_4/0.1 mmol/L Cu(II) pH 4.5; −7 to −15 kV		[85]
Thalidomide, warfarin, coumachlor, felodipine	Vancomycin immobilized on epoxy-polymer	HMAA (0.16 (g/mL), PDA (0.08), AGE (0.08), VSA (30%) (40 μL)	$(NH_4)_2SO_4$ in 50 mmol/L phosphate buffer (pH 7)	ACN–triethylammonium acetate (pH 5) (15/85, v/v)	120,000	[86]

Brush-type polymethacrylate monoliths

Analyte	Selector/Template	Monomers	Porogen	Mobile phase	Plates	Ref.
DNB-Leu-diallylamide	2-hydroxyethyl methacrylate (N-L-Val-3,5-dimethyl-anilide) (19.7 wt%)	GMA (40%), EDMA (40%), AMPS (0.3%)	70% (w/w) of reaction mixture; 1-propanol (75%), 1,4-butanediol (15%), water (10%)	ACN–5 mmol/L phosphate buffer (pH 7) (80:20, v/v)	61,000	[87]
N-derivatized amino acids (DNB, DNZ, DNP, FMOC, Z, Bz, Ac), mecoprop, fenoprop	[O-2-(methacryloyl-oxy)ethylcarbamoyl]-10,11-dihydroquinidine (20 wt%)	HEMA (60-70%), EDMA (10-20%)	60% (w/w) of reaction mixture; cyclohexanol, dodecanol (various ratios)	0.4 mol/L AcOH + 4 mmol/L NEt$_3$ in ACN-MeOH (80/20, v/v)	240,000	[17, 67, 68]

MIP-type polymethacrylate monoliths

Analyte	Selector/Template	Monomers	Porogen	Mobile phase	Plates	Ref.
Propranolol	(R)-propranolol (0.03 mol/L)	MAA (0.24 mol/L), TRIM (0.24 mol/L)	Toluene	ACN–4 mol/L ammonium acetate (pH 3) (80/20, v/v)		[88]
Metoprolol	(S)-metoprolol (0.03 mol/L)	MAA (0.24 mol/L), TRIM (0.24 mol/L)	Toluene	ACN–2 mol/L ammonium acetate (pH 3) (80/20, v/v)		[88]
Ropivacaine, mepivacaine, bupivacaine	(S)-ropivacaine (0.02–0.24 mol/L)	MAA (0.24–0.48 mol/L), TRIM (0.24–0.48 mol/L)	Isooctane (1–25%) in toluene	ACN–25 mmol/L phosphate or citrate buffer (pH 2–6.5) (80/20, v/v)		[89]
β-Blockers (propranolol, pindolol, prenalterol, atenolol)	various	various		various		[90-92]

Continued on the next page

TABLE 21.8 (continued)

Analytes	Chiral selector or monomer (template)	(Co)monomers	Solvent (Porogens)	Mobile phase (applied voltage)	Maximum efficiency (theor. pl/m)	Ref.
Amino acids (Phe, Tyr, Phg)	(S)-Phe-anilide (0.16 mol/L)	MAA (0.35 mol/L), 2-vinylpyridine (0.35 mol/L), EDMA (3.3 mol/L)	Ammonium acetate (0.1 mol/L)– chloroform	ACN–acetic acid– water (80:10:10, v/v/v); T, 60°C; 400 V/cm		[93, 94]
Silica monoliths						
DNS-amino acids	L-Phe-amide bonded to epoxy-silica monolith	TMOS, PEO (Mr 10,000) in AcOH; ammonia wash, heat treatment (up to 300°C), epoxy-functionalization		ACN–50 mmol/L NH$_4$Ac with 0.5 mmol/L Cu(Ac)$_2$ pH 5.5 (70:30, v/v); –300 V/cm	90,000	[95]
DNS-amino acids and hydroxy carboxylic acids like *m*- and *p*-hydroxy-mandelic acid (MDA), 3-hydroxy-4-methoxy-MDA, 4-hydroxy-3-methoxy-MDA, 3-(4-hydroxyphenyl)lactic acid, 3-phenyllactic acid, and indole-3-lactic acid	L-Pro-amide bonded to epoxy-silica monolith			ACN–50 mmol/L NH$_4$Ac with 0.5 mmol/L Cu(Ac)$_2$ pH 6.5 (70:30, v/v); –13.6 kV	17,000	[96]

Particle-fixed monoliths

Analytes	Stationary phase	Mobile phase	N	Ref.
Barbiturates, benzoin, α-methyl-α-phenylsuccinimide, MTH-proline, mecoprop methyl, fenoxaprop methyl, carprofen, ibuprofen	Chira-Dex immobilized on sintered silica-packed bed (Na$_2$CO$_3$ flush, 380°C)	MeOH–20 mmol/L MES, pH 6 (30:70, v/v)	68,000	[97]
Phe, Tyr, Phg, Phe-anilide	L-Phe, L-Phe-anilide directed MIP; MIP entrapped in polyacrylamide matrix	0.05 mol/L Tris-citric acid (pH 2.5) ; ACN–acetic acid–water (90:5:5, v/v/v); T, 60°C, 350 V/cm		[98, 99]
DNS-Phe, DNS-Leu	DNS-L-Leu			[100]
DNS-Phe	L-DNS-Phe directed MIP; Entrapped in potassium silicate	ACN–acetate (pH 3.0) (89/20, v/v); 30 kV		[54]
NBD-amino acids	(S)-DNB-1-naphthyl-glycine-silica (5 μm) and (S)-N-(3,5-dinitrophenyl- aminocarbonyl) valine-silica (5 μm); Embedded in sol-gel matrix (TEOS, EtOH, HCl)	ACN–5 mmol/L phosphate buffer (pH 2.5) (70/30, v/v)		[101]

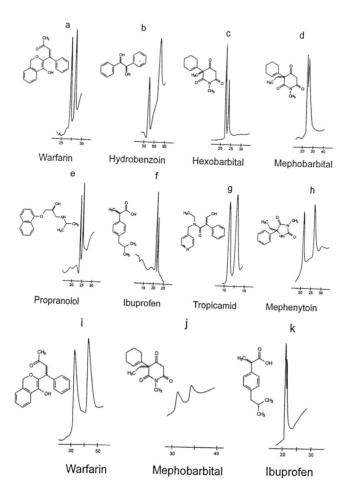

Fig. 21.22. Separations of enantiomers of drugs using negatively- and positively charged homogeneous poly(acrylamide-*co*-2-hydroxy-3-allyloxypropyl-β-cyclodextrin) monoliths (Reprinted with permission from ref. [82], copyright 2000 Wiley-VCH). Conditions: capillary, 25 μm I.D., total length 12–16.5 cm, effective length 8.7–13.5 cm; allyl-β-CD, 50–75 mmol/l; mobile phase, 0.1 mol/L Tris–0.15 mol/L boric acid (pH 8.2); applied voltage, ±2–4 kV.

aromatic moiety of the analyte in the CD-cavity and the amino group in the crown ether [80]. These monolithic columns were found to be stable over several months and exhibited good run-to-run reproducibility.

Hjertén's group also developed both negatively and positively charged homogeneous polyacrylamide monolithic gels with copolymerized 2-hydroxy-3-allyloxypropyl-β-cyclodextrin (allyl-β-CD). The latter, with a multiplicity of allyl functionalities, also served as a crosslinker [82]. Therefore, an increase in the allyl-β-CD percentage in the polymerization mixture afforded a monolith that was more rigid, had narrower pores, and exhibited a lower EOF. Fig. 21.22 shows the separation of a variety of neutral, basic, and acidic compounds including warfarin, barbiturates, hydrobenzoin, propranolol, tropicamide, and ibuprofen (Table 21.8). The major advantage of these homogeneous monoliths was their enhanced performance resulting from the almost completely absent eddy diffusion and lower resistance to mass-transfer effected by shorter diffusion distances between the interaction sites. The larger homogeneity compared to the rigid macroporous monoliths resulted in very good column efficiency for several solutes. While the HETP did not depend on the flow velocity for the first eluted enantiomer of tropicamide, this, as expected, was not true, for the stronger retained enantiomer, for which the mass transfer effect was significant.

A number of other selectors operating upon different chiral recognition principles has also been covalently bonded to polyacrylamide monoliths [21,83,84,86]. For example, a negatively charged chiral crown ether containing monolithic stationary phases incorporating the mono- or tetra-allyl ester of (+)-18-crown-6-tetracarboxylic acid as a selector, separated a variety of chiral primary amines such as L-alanine-2--naphthylamide containing only 0.1% D-enantiomer (Fig. 21.23) [83]. Obviously, detection of this minute amount of enantiomeric impurity requires separation under overload conditions, using a system affording high resolution ($R_S > 2. 0$). This condition is easily met in the separation of alanine-2-naphthylamide shown in Fig. 21.23, for which $R_S = 7$.

Polyacrylamide monoliths with copolymerized N-(2-hydroxy-3-allyloxypropyl)-L--4-hydroxyproline as chelating selector were used for enantioselective CEC of underivatized amino acids [84] and hydroxy carboxylic acids [85] in the ligand-exchange mode, using copper-ion-containing eluents. Vinylsulfonic acid was included in the polymerization mixture to promote strong a EOF in the co-directional mode for the separation of amino acids. Since this sulfonic acid chemistry would afford a counter-directional separation of hydroxy acids, VSA was eliminated from the reaction mixture used for the preparation of monolith suitable for their separation. The analyte transport of the negatively charged hydroxycarboxylic acids was then accomplished by virtue of electrophoretic mobility alone. A selector–solute binding model shown in Fig. 21.24 was suggested for this separation, which is based on the reversible formation of mixed ternary chelate complexes involving selector, copper ion, and both enantiomers. These complexes are diastereomeric and therefore may differ in their stability, which thus enables the separation. Low back-pressures typical of the monolithic columns allowed the use of pressure-assisted CEC in a conventional CE system with an external pressurization of up to 1.2 MPa, which enabled much faster separations and resolutions similar to those in the authentic CEC mode (Fig. 21.24a–c). The

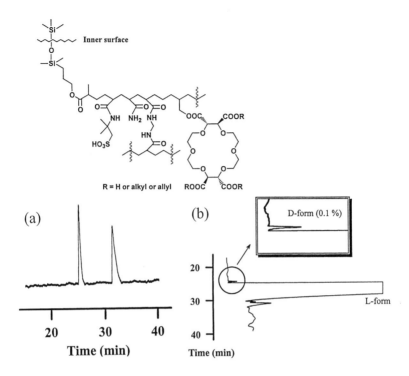

Fig. 21.23. Separation of racemate, (a), and determination of D-enantiomer (enantiomeric impurity) in L-alanine-2-naphthylamide, (b), using a crown ether-bonded negatively charged monolithic polyacrylamide gel-filled capillary. (Reprinted with permission from ref. [83], copyright 2001 Elsevier Science B.V.). Conditions: capillary, 80 cm × 75 μm I.D., 35 cm effective length; polymer, 5.0% T, 5.0% C, 5.5% S, selector, 20 mmol/L (+)-tetra-allyl 18-crown-6-tetracarboxylate; mobile phase, 200 mmol/L triethanolamine-300 mmol/L boric acid buffer (pH 6.0)–acetonitrile (80:20); electric field strength, 239 V/cm; UV detection 254 nm.

CEC separation using a monolith was faster, and the system more stable, than CEC column packed with silica-based ligand-exchange beads.

21.3.5.2 Enantioselective brush-type polymethacrylate-based monoliths

The practically unlimited number of functional monomers that can be copolymerized to afford organic polymer monoliths enables fine adjustments in selectivity, which is particularly important for enantioselective CEC. While certainly the most important structural element of any chiral phase, the selector structure itself is not the only determinant of the chemical character of the surface, the chromatographic properties and, most importantly, the enantioselectivity. A major effect is also exerted by the comonomers used for the preparation of the monolith, that are often regarded as

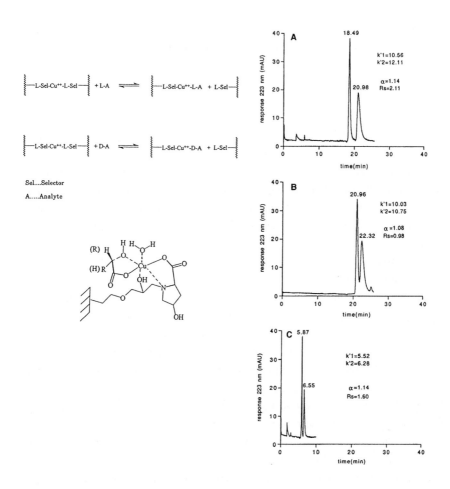

Fig. 21.24. Equilibria in chiral ligand-exchange, selector–solute binding model for hydroxy acids, and separation of phenylalanine enantiomers by CEC: (a), pressure-driven nano-HPLC, (b), and pressure-assisted CEC, (c). (Reprinted with permission from refs. [84] and [85], copyright 2000 and 2001 Wiley-VCH). Conditions: mobile phase, 50 mmol/L sodium dihydrogenphosphate–0.1 mmol/L Cu(II), pH 4.6; capillary, 34.5 cm × 75 μm I.D., 26 cm effective length; monolithic column, poly(methacrylamide-*co*-piperazine diacrylamide-*co*--N-(2-hydroxy-3-allyloxypropyl)-L-4-hydroxyproline-*co*-VSA); 30 kV (a), 1.2 MPa at the inlet (b), and 30 kV plus 1.2 MPa at the inlet (c).

less important constituents. Pendant groups of the co-monomer may help to hide the unfavorable nature of the polymer backbone or, in contrast, may reduce the efficiency and enantioselectivity by creating the undesired non-specific interactive sites. Obviously, these non-specific interactions must be minimized through both suitable design of the surface chemistry of the stationary phase, and the eluent composition, to achieve highly efficient and selective enantioseparations.

References pp. 556-559

The importance of adjustment of the chemical nature of the surface was demonstrated in studies by Peters *et al.* [87] and Lämmerhofer *et al.* [67] using polymethacrylate-based monoliths with pendant low-molecular-weight selectors. For example, although a 'brush-type' chiral polymethacrylate-based monolith obtained by copolymerization of N-[(2-methacryloyloxyethyl)oxycarbonyl]-(S)-valine-3,5-di-methylanilide, EDMA, AMPS, and butyl methacrylate exhibited enantioselectivity for the separation of N-(3,5-dinitrobenzoyl)leucine diallylamide enantiomers under reversed phase conditions, the efficiency of only 160–600 plates/m was very poor. In order to minimize detrimental non-specific interactions, seemingly responsible for the low plate count, the polarity of the monolith surface was increased. First, a monolith with more polar epoxypropyl functionalities, prepared by substitution of butyl methacrylate with glycidyl methacrylate, led to higher efficiencies. However, practically useful levels of efficiency of up to 60,000 plates/m were achieved only when the epoxide groups were hydrolyzed on-column with dilute sulfuric acid to afford even more hydrophilic diols.

In another example, the ionizable chiral monomer O-[2-(methacryloyl-oxyethyl)carbamoyl]-10,11-dihydroquinidine was copolymerized with ethylene di-methacrylate crosslinker and an additional monomer, using both thermal and photoinitiation (Fig. 21.25) [17,67,68]. Advantageously, the chiral monomer eliminated the need for the additional ionizable monomer that had to be used with neutral selectors. The quinidine moieties were positively charged in acidic mobile phases and afforded both anodic EOF for co-directional separation of acidic analytes and enantioselectivity via an anion-exchange mechanism. A more hydrophilic surface, prepared by using 2-hydroxyethyl methacrylate as a comonomer, was superior to that containing gly-cidyl methacrylate, in terms both of enantioselectivity and column efficiency. Monoliths comprising 10 and 20 wt.% ethylene dimethacrylate exhibited better mass transfer characteristics and less flow dispersion, thus affording higher column efficiencies than those with a typical 40% crosslinking, despite the equal, 'dry state' pore size of about 1 μm. Since the less crosslinked materials swelled more, and the overall space available within the column was fixed, the pores were partly filled with the swollen polymer chains, resulting in more homogeneous beds. This translates into lower A- and C- terms contributions to the Van Deemter plots. Hjertén's group also reported similar findings [82]. The trade-off for this improvement in the chromatographic properties was the concomitant decrease in the flow velocity.

Efficiencies reaching levels typically reserved for capillary electrophoresis could be obtained with these quinidine-functionalized monolithic columns in CEC separations of acidic chiral compounds, both in aqueous reversed-phase and non-aqueous polar organic-phase modes. For example, column efficiencies of 242,000 and 194,000 plates/m were obtained for the separation of N-2,4-dinitrophenyl valine enantiomers

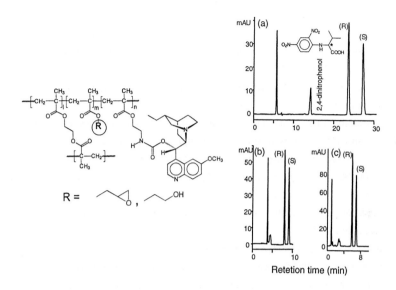

Fig. 21.25. Enantioseparation of DNP-(R,S)-Val using 15 cm (a)-, 8.5 cm (b)-, and 8.5 cm (short-end injection) (c),-long quinidine-functionalized polymethacrylate monolithic columns (Reprinted with permission from ref. [17], copyright 2000 American Chemical Society, and from ref. [68], copyright 2000 Wiley-VCH). Conditions: capillary, 33.5 cm × 100 μm I.D., effective length 25 cm (a,b) and 8.5 cm (c); mobile phase, 0.4 mol/L acetic acid and 4 mmol/L triethylamine in acetonitrile–methanol (80:20, v/v); temperature, 50°C; applied voltage, -25 kV (a,b) and +25 kV (c); injection, –15 kV for 5 s from inlet end (a,b), and +15 kV for 5 s from outlet end.

using a 15 cm-long optimized chiral monolith (Fig. 21.25a). Similarly, other chiral acids such as N-benzoyl-leucine and the α-aryloxycarboxylic acid herbicide, Fenoprop, were also separated efficiently. Due to both high enantioselectivity and resolution, monolithic segments only 8.5 cm long proved to be sufficient for baseline separations of a wide variety of chiral acids, with R$_S$ values still exceeding 2.0. This reduction in length also enabled substantial acceleration of the CEC separations (Figs. 21.25b and c).

21.3.5.3 Molecular imprinted polymer monoliths

Early studies on enantioselective CEC with molecular imprinted polymer (MIP) monoliths [102] failed to separate the target racemic analytes owing to improper choice of porogens, but clearly showed that it might not be easy to prepare rigid monolithic beds with both high enantioselectivities and suitable polymer morphology to afford useful resolutions in this separation mode. Later, Nilsson's group [88] and Lin *et al.* [93] succeeded with the *in situ* preparation of porous molecularly imprinted chiral monoliths using photo- and thermally-initiated polymerizations. In both approaches, they used methacrylic acid as the functional monomer that associated strongly with the basic template molecules via ion-pairing, and also effected EOF in the resulting monolithic columns. Suitable porogens were the aprotic solvents toluene and isooctane, that did not interfere with the non-covalent binding between complementary functionalities. The macroporous monolithic MIP columns were well suited for the enantioseparation of the racemic template in the CEC mode. For example, chiral β-adrenergic antagonists and local anesthetics were separated with efficiencies between 35,000–70,000 and 5,000–20,000 plates/m for the first- and second-eluted enantiomer, respectively (Fig. 21.26) [88,89,91]. These efficiencies are quite remarkable compared to the generally poor values typical of MIP columns in HPLC applications.

A number of other studies on enantioselective CEC with monolithic molecularly imprinted polymer columns has been published (Table 21.8) [90-94] and compiled in excellent reviews [103,104]. Although electrically driven flow significantly enhanced the column efficiencies compared to HPLC separations with similar stationary phases, the deleterious effect of both polydispersity of the binding sites created by the imprinting process, and the extensive crosslinking of the monoliths, on peak performance persists. Therefore, brush-type chiral stationary phases prepared from porous polymers are currently superior for analytical applications.

21.3.5.4 Chiral modified silica monoliths and particle-fixed monolithic beds

Chen and Hobo [95,96] adopted Tanaka's concept for the preparation of monolithic silica-based columns [6,23], which they derivatized for enantioselective ligand exchange CEC. The silica monolith was modified with a spacer group by on-column silanization with 3-(glycidyloxypropyl)trimethoxysilane, followed by reaction of the epoxy groups with L-phenylalanine amide or L-proline amide chelating selectors. The column was then loaded with Cu(II) ions. DNS amino acids were separated with a resolution of 1.5–5.4 on the L-phenylalanine amide-modified ligand-exchange column [95] while the L-proline amide-modified silica monolith also separated enantiomers of hydroxycarboxylic acids, such as mandelic- and 3-aryllactic acids [96]. A non-optimized pore structure was the likely reason for the lower than expected efficiency of 90,000 plates/m.

Other approaches include as the consolidation of a packed silica bed by sintering and subsequent coating with selectors such as Chirasil-Dex, a dimethylpolysiloxane

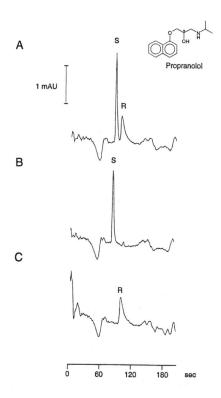

Fig. 21.26. CEC separation of propranolol enantiomers using a capillary column containing macroporous monolith imprinted with (R)-propranolol. (Reprinted with permission from ref. [88], copyright 1997 American Chemical Society). Conditions: capillary, 35 cm × 75 μm I.D., 26.5 cm effective length; mobile phase, acetonitrile–4 mol/L acetate buffer, pH 3.0 (80:20); applied voltage, 30 kV (857 V/cm); temperature, 60°C; UV detection, 214 nm.

derivative with permethylated β-cyclodextrin, covalently attached through a C8 spacer [97], or entrapping irregular MIP particles in a porous polyacrylamide [98-100] or silicate matrix [54], and the loosely embedded 5 μm (S)-N-(3,5-dinitrobenzoyl)-1-naphthylglycine and (S)-N-(3,5-dinitrophenylaminocarbonyl)valine silica particles in sol-gel matrix [101], though all to some extent successful, appear to have only limited future potential.

21.4 SUMMARY AND OUTLOOK

In a number of studies published to date, both the potential and advantages of monolithic columns for CEC have been clearly documented. They include simple column preparation, which in particular applies for the organic polymer monoliths, the flexibility in tailoring surface chemistry and thus selectivity for specific separation problems, the departure from retaining frits which are typical of packed bed column technology, and efficiencies of up to 600,000 plates/m, which are typical of CE but much higher than those commonly achieved in HPLC. In addition, monolithic CEC columns also have higher sample loading capacity than open-tubular CEC columns and capillary electrophoresis, and their capability of interfacing with mass spectrometers are striking advantages and characteristics required of a modern powerful separation technique. Run-to-run repeatability and column-to-column reproducibility are satisfactory, and encourage the application of such monolithic columns on a routine basis.

Unfortunately, CEC is not yet widely accepted, mainly due to the problems that have been reported with packed columns and frits included in these columns, but partly also owing to the current lack of suitable commercially available CEC columns. This can change in the future, and a broad choice of monolithic columns with various chemistries might be available. The practical applicability for solving real-life analytical problems has convincingly and successfully been demonstrated, although more tests with such samples remain to be done. This may not be completely straightforward. Upon injection of real samples, the surface of the stationary phase can become dynamically modified with ionized components of the matrix which may vary the retention behavior and also the EOF. Regular regeneration of the surface by flushing with strongly acidic or strongly basic buffer solutions may help to solve this problem. Monolithic columns made of specific organic polymers easily resist these conditions and have great potential in this respect. Alternatively, more thorough sample cleanup may help to avoid excessive surface contamination with ionized constituents from complex matrixes. Eluents with higher buffer concentrations would be preferred, in view of the uncontrolled adsorption of ionic matrix components, although they would slow EOF.

Moreover, many of the monolith technologies appear to be easily transferable from the capillary format to analytical microfluidic chip devices [5,38,105-107], which would be difficult to achieve using other technologies such as packed particles or pressure-driven chromatography.

21.5 ACKNOWLEDGEMENTS

Financial support from the Austrian Christian Doppler Research Society and the Austrian Science Fund (project no. P13965-CHE) are gratefully acknowledged.

21.6 ABBREVIATIONS

% T	Total acrylamide concentration %T = 100 (a + b + c) / V (a,b,c are the masses of monomer, crosslinker, and co-monomers, and V is the volume of the reaction solvent in mL, mostly Tris–boric acid buffer) [27]
% C	Percentage (w/w) of crosslinker related to total monomers [27]
% S	Percentage (w/w) of chargeable co-monomer (mostly AMPS) related to total monomers [27]
AA	Acrylic acid
AAm	Acrylamide
Ac	Acetyl
AC-β-CD	Allylcarbamoylated β-cyclodextrin
ACN	Acetonitrile
AETMA	[2-(acryloyloxy)ethyl]trimethylammonium methyl sulfate
AGE	Allyl glycidyl ether
allyl-β-CD	2-hydroxy-3-allyloxy-propyl-β-cyclodextrin
AMPS	2-acrylamido-2-methylpropane sulfonic acid
AMTEA	N-(2-acrylamidoethyl)triethylammonium iodide
APS	Ammonium persulfate
APVE	3-Amino-1-propanol vinyl ether
BA	Butyl acrylate
BDDA	1,3-butanediol diacrylate
BIS	N,N′-methylenebisdiacrylamide
BMA	Butyl methacrylate
BSA	Bovine serum albumin
Bz	Benzoyl
β-CD	β-Cyclodextrin
CEC	Capillary electrochromatography
CGE	Capillary gel electrophoresis
CSP	Chiral stationary phase
CTAB	Cetyl trimethylammonium bromide
DMAA	Dimethylacrylamide
DMAEMA	2-Dimethylaminoethyl methacrylate
DMDAAC	Dimethyldiallylammonium chloride
DMF	N,N-Dimethylformamide
DNB	3,5-Dinitrobenzoyl
DNP	2,4-Dinitrophenyl
DNS	(5-dimethylaminonaphthalenesulfonyl)
DNZ	3,5-Dinitrobenzyloxycarbonyl
DVB	Divinylbenzene
EDMA	Ethylene dimethacrylate
EOF	Electroosmotic flow
ESI-MS	Electrospray ionization-mass spectrometry
ETMOS	Ethyltrimethoxysilane
EtOH	Ethanol
FA	Formamide
FMOC	9-Fluorenylmethoxycarbonyl
GMA	Glycidyl methacrylate
HEMA	2-Hydroxyethyl methacrylate
HETP	Height equivalent of a theoretical plate
HMAA	N-(Hydroxymethyl)acrylamide
Hyp	Hydroxyproline
LA	Lauryl acrylate
LE-CEC	Ligand exchange CEC
LIF	Laser induced fluorescence detection
LMA	Lauryl methacrylate

LOD	Limit of detection
MAA	Methacrylic acid
MAAm	Methacrylamide
MEKC	Micellar electrokinetic chromatography
MeOH	Methanol
MES	2-(1-Morpholino)ethanesulfonic acid
MIP	Molecularly imprinted polymer
MMA	Methyl methacrylate
NBD-F	4-Fluoro-7-nitro-2,1,3-benzoxadiazole
NDA	Naphthalene-2,3-dicarboxaldehyde
NIPAAm	N-Isopropylacrylamide
NMF	N-methylformamide
ODS	Octadecylsilica
PDA	Piperazine diacrylamide
PEG	Poly(ethylene glycol)
PEO	Poly(ethylene oxide)
PTH	Phenylthiohydantoin
S	Styrene
SDS	Sodium dodecyl sulfate
SO	Chiral selector
TEMED	N,N,N',N'-tetramethylethylenediamine
TEOS	Tetraethoxysilane (tetraethyl orthosilicate)
TFA	Trifluoroacetic acid
TMOS	Tetramethoxy silane
TRIM	Trimethylolpropane trimethacrylate
VBC	Vinylbenzyl chloride
VSA	Vinylsulfonic acid

21.7 REFERENCES

1   Z. Deyl, F. Svec (Eds.), Capillary Electrochromatography. Elsevier, Amsterdam, 2001.

2   V. Pretorius, B. Hopkins, J.D. Schieke, J. Chromatogr. 99 (1974) 23.

3   S. Hjertén, J. Chromatogr. 347 (1985) 191.

4   I. Gusev, X. Huang, C. Horváth, J. Chromatogr. A 855 (1999) 273.

5   S.M. Ngola, Y. Fintschenko, W.-Y. Choi, T.J. Shepodd, Anal. Chem. 73 (2001) 849.

6   N. Tanaka, H. Nagayama, H. Kobayashi, T. Ikegami, K. Hosoya, N. Ishizuka, H. Minakuchi, K. Nakanishi, K. Cabrera, D. Lubda, J. High Resol. Chromatogr. 23 (2000) 111.

7   J.D. Hayes, A. Malik, Anal. Chem. 72 (2000) 4090.

8   R. Asiaie, X. Huang, D. Farnan, C. Horváth, J. Chromatogr. A, 806 (1998) 251.

9   J.-R. Chen, M.T. Dulay, R.N. Zare, F. Svec, E. Peters, Anal. Chem. 72 (2000) 1224.

10   J.-R. Chen, R.N. Zare, E.C. Peters, F. Svec, J.M.J. Fréchet, Anal. Chem. 73 (2001) 1987.

11   C. Fujimoto, Anal. Chem. 67 (1995) 2050.

12   M. Zhang, Z. El Rassi, Electrophoresis 22 (2001) 2593.

13   E.C. Peters, M. Petro, F. Svec, J.M.J. Fréchet, Anal. Chem. 70 (1998) 2288.

14   R. Wu, H. Zou, M. Ye, Z. Lei, J. Ni, Anal. Chem. 73 (2001) 4918.

15   R. Wu, H. Zou, M. Ye, Z. Lei, J. Ni, Electrophoresis 22 (2001) 544.

16   C. Yu, F. Svec, J.M.J. Fréchet, Electrophoresis 21 (2000) 120.

17   M. Lämmerhofer, F. Svec, J.M.J. Fréchet, W. Lindner, Anal. Chem. 72 (2000) 4623.

18 F. Svec, E.C. Peters, D. Sýkora, J.M.J. Fréchet, J. Chromatogr. A 887 (2000) 3.
19 C. Ericson, J.-L. Liao, K. Nakazato, S. Hjertén, J. Chromatogr. A 767 (1997) 33.
20 D. Hoegger, R. Freitag, J. Chromatogr. A 914 (2001) 211.
21 O. Kornyšova, E. Machtejevas, V. Kudirkaite, U. Pyell, A. Maruška, J. Biochem. Biophys. Meth. 50 (2002) 217.
22 K.K. Unger, Porous Silica. Its Properties and Use as Support in Column Liquid Chromatography. Elsevier, Amsterdam, 1979.
23 N. Ishizuka, H. Minakuchi, K. Nakanishi, N. Soga, H. Nagayama, K. Hosoya, N. Tanaka, Anal. Chem. 72 (2000) 1275.
24 Q. Tang, M.L. Lee, J. High Resolut. Chromatogr. 23 (2000) 73.
25 J.-L. Liao, N. Chen, C. Ericson, S. Hjertén, Anal. Chem. 68 (1996) 3468.
26 C. Fujimoto, Y. Fujise, E. Matsuzawa, Anal. Chem. 68 (1996) 2753.
27 C. Fujimoto, J. Kino, H. Sawada, J. Chromatogr. A 716 (1995) 107.
28 C. Ericson, S. Hjertén, Anal. Chem. 71 (1999) 1621.
29 A.M. Enlund, C. Ericson, S. Hjertén, D. Westerlund, Electrophoresis 22 (2001) 511.
30 A. Palm, M.V. Novotny, Anal. Chem. 69 (1997) 4499.
31 A.H. Que, T. Konse, A.G. Baker, M.V. Novotny, Anal. Chem. 72 (2000) 2703.
32 A.H. Que, A. Palm, A.G. Baker, M.V. Novotny, J. Chromatogr. A 887 (2000) 379.
33 A.H. Que, V. Kahle, M.V. Novotny, J. Microcol. Sep. 12 (2000) 1.
34 L. Kvasničková, Z. Glatz, H. Štěrbová, V. Kahle, J. Slanina, P. Musil, J. Chromatogr. A 916 (2001) 265.
35 E.C. Peters, M. Petro, F. Svec, J.M.J. Fréchet, Anal. Chem. 69 (1997) 3646.
36 E.C. Peters, M. Petro, F. Svec, J.M.J. Fréchet, Anal. Chem. 70 (1998) 2296.
37 T. Jiang, J. Jiskra, H.A. Claessens, C.A. Cramers, J. Chromatogr. A 923 (2001) 215.
38 R. Shediac, S.M. Ngola, D.J. Throckmorton, D.S. Anex, T.J. Shepodd, A.K. Singh, J. Chromatogr. A 925 (2001) 251.
39 I.S. Lurie, D.S. Anex, Y. Fintschenko, W.Y. Choi, J. Chromatogr. A 924 (2001) 421.
40 S. Zhang, X. Huang, J. Zhang, C. Horváth, J. Chromatogr. A 887 (2000) 465.
41 B. Xiong, L. Zhang, Y. Zhang, H. Zou, J. Wang, J. High Resolut. Chromatogr. 23 (2000) 67.
42 S. Zhang, J. Zhang, C. Horváth, J. Chromatogr. A 914 (2001) 189.
43 F. Svec, J.M.J. Fréchet, J. Chromatogr. A 702 (1995) 89.
44 D. Sýkora, F. Svec, J.M.J. Fréchet, J. Chromatogr. A 852 (1999) 297.
45 H. Kobayashi, C. Smith, K. Hosoya, T. Ikegami, N. Tanaka, Anal. Sci. 18 (2002) 89.
46 N. Ishizuka, K. Nakanishi, K. Hirao, N. Tanaka, J. Sol-Gel Sci. Technol. 19 (2000) 371.
47 M.T. Dulay, J.P. Quirino, B.D. Bennett, M. Kato, R.N. Zare, Anal. Chem. 73 (2001) 3921.
48 J.P. Quirino, M.T. Dulay, R.N. Zare, Anal. Chem. 73 (2001) 5557.
49 M.T. Dulay, J. Quirino, B.D. Bennett, R.N. Zare, J. Sep. Sci. 25 (2002) 3.
50 M.T. Dulay, R.P. Kulkarni, R.N. Zare, Anal. Chem., 70 (1998) 5103.
51 C.K. Ratnayake, C.S. Oh, M.P. Henry, J. High Resolut. Chromatogr. 23 (2000) 81.
52 C.K. Ratnayake, C.S. Oh, M.P. Henry, J. Chromatogr. A 887 (2000) 277.
53 T. Adam, K.K. Unger, M.M. Dittmann, G.P. Rozing, J. Chromatogr. A 887 (2000) 327.
54 G.S. Chirica, V.T. Remcho, Electrophoresis 20 (1999) 50.
55 Q. Tang, B. Xin, M.L. Lee, J. Chromatogr. A 837 (1999) 35.

56   Q. Tang, N. Wu, M.L. Lee, J. Microcol. Sep. 11 (1999) 550.

57   Q. Tang, M.L. Lee, J. Chromatogr. A 887 (2000) 265.

58   L. Roed, E. Lundanes, T. Greibrokk, J. Chromatogr. A 890 (2000) 347.

59   L. Roed, E. Lundanes, T. Greibrokk, J. Microcol. Sep. 12 (2000) 561.

60   L. Roed, E. Lundanes, T. Greibrokk, J. Sep. Sci. 24 (2001) 435.

61   G.S. Chirica, V.T. Remcho, Electrophoresis 21 (2000) 3093.

62   G.S. Chirica, V.T. Remcho, Anal. Chem. 72 (2000) 3605.

63   Q. Tang, M.L. Lee, Trends Anal. Chem. 19 (2000) 648.

64   R. Stol, H. Poppe, W.T. Kok, Anal. Chem. 73 (2001) 3332.

65   P.T. Vallano, V.T. Remcho, Anal. Chem. 72 (2000) 4255.

66   M. Lämmerhofer, F. Svec, J.M.J. Fréchet, W. Lindner, J. Chromatogr. A 925 (2001) 265.

67   M. Lämmerhofer, E.C. Peters, C. Yu, F. Svec, J.M.J. Fréchet, W. Lindner, Anal. Chem. 72 (2000) 4614.

68   M. Lämmerhofer, F. Svec, J.M.J. Fréchet, W. Lindner, J. Microcol. Sep. 12 (2000) 597.

69   P.B. Wright, A.S. Lister, J.G. Dorsey, Anal. Chem. 69 (1997) 3251.

70   M. Lämmerhofer, F. Svec, J.M.J. Fréchet, W. Lindner, Trends Anal. Chem., 19 (2000) 676.

71   A. Guttman, A. Paulus, S. Cohen, N. Grinberg, B. L. Karger, J. Chromatogr. 448 (1988) 41.

72   I.D. Cruzado, G. Vigh, J. Chromatogr. 608 (1992) 421.

73   J.M. Lin, T. Nakagama, H. Okazawa, X.-Z. Wu, T. Hobo, Fresenius J. Anal. Chem. 354 (1996) 451.

74   S. Birnbaum, S. Nilsson, Anal. Chem. 64 (1992) 2872.

75   S. Hjertén, Á. Végvári, T. Srichaiyo, H.-X. Zhang, C. Ericson, D. Eaker, J. Capil. Electrophor. 5 (1998) 13.

76   C. Fujimoto, Anal. Sci. 18 (2002) 19.

77   T. Koide, K. Ueno, Anal. Sci. 14 (1998) 1021.

78   T. Koide, K. Ueno, Anal. Sci. 16 (2000) 1065.

79   T. Koide, K. Ueno, Anal. Sci. 15 (1999) 791.

80   T. Koide, K. Ueno, J. Chromatogr. A, 893 (2000) 177.

81   T. Koide, K. Ueno, J. High Resolut. Chromatogr. 23 (2000) 59.

82   Á. Végvári, A. Földesi, C. Hetenyi, O. Kocnegarova, M.G. Schmid, V. Kudirkaite, S. Hjertén, Electrophoresis 21 (2000) 3116.

83   T. Koide, K. Ueno, J. Chromatogr. A 909 (2001) 305.

84   M.G. Schmid, N. Grobuschek, C. Tuscher, G. Gübitz, Á. Végvári, E. Machtejevas, A. Maruška, S. Hjertén, Electrophoresis 21 (2000) 3141.

85   M.G. Schmid, N. Grobuschek, O. Lecnik, G. Gübitz, Á. Végvári, S. Hjertén, Electrophoresis 22 (2001) 2616.

86   O. Kornyšova, P.K. Owens, A. Maruška, Electrophoresis 22 (2001) 3335.

87   E.C. Peters, K. Lewandowski, M. Petro, F. Svec, J.M.J. Fréchet, Anal. Commun. 35 (1998) 83.

88   L. Schweitz, L.I. Andersson, S. Nilsson, Anal. Chem. 69 (1997) 1179.

89   L. Schweitz, L.I. Andersson and S. Nilsson, J. Chromatogr. A 792 (1997) 401.

90   L. Schweitz, L.I. Andersson, S. Nilsson, Chromatographia 49 (1999) S93.

91   L. Schweitz, L.I. Andersson, S. Nilsson, Anal. Chim. Acta 435 (2001) 43.

92 L. Schweitz, I. Andersson Lars, S. Nilsson, Analyst 127 (2002) 22.

93 J.-M. Lin, T. Nakagama, K. Uchiyama, T. Hobo, J. Pharm. Biomed. Anal. 15 (1997) 1351.

94 J.-M. Lin, T. Nakagama, X.-Z. Wu, K. Uchiyama, T. Hobo, Fresenius J. Anal. Chem. 357 (1997) 130.

95 Z. Chen, T. Hobo, Anal. Chem. 73 (2001) 3348.

96 Z. Chen, T. Hobo, Electrophoresis 22 (2001) 3339.

97 D. Wistuba, V. Schurig, Electrophoresis 21 (2000) 3152.

98 J.M. Lin, T. Nakagama, K. Uchiyama, T. Hobo, Chromatographia 43 (1996) 585.

99 J.-M. Lin, T. Nakagama, K. Uchiyama, T. Hobo, J. Liq. Chromatogr. Rel. Technol. 20 (1997) 1489.

100 J.M. Lin, T. Nakagama, K. Uchiyama, T. Hobo, Biomed. Chromatogr. 11 (1997) 298.

101 M. Kato, M.T. Dulay, B. Bennett, J.-R. Chen, R.N. Zare, Electrophoresis 21 (2000) 3145.

102 K. Nilsson, J. Lindell, O. Norrlöw, B. Sellergren, J. Chromatogr. A 680 (1994) 57.

103 L. Schweitz, L.I. Andersson, S. Nilsson, J. Chromatogr. A 817 (1998) 5.

104 L. Schweitz, P. Spégel, S. Nilsson, Electrophoresis 22 (2001) 4053.

105 C. Ericson, J. Holm, T. Ericson, S. Hjertén, Anal. Chem. 72 (2000) 81.

106 Y. Fintschenko, W.-Y. Choi, S.M. Ngola, T.J. Shepodd, Fresenius J. Anal. Chem. 371 (2001) 174.

107 D.J. Throckmorton, T.J. Shepodd, A.K. Singh, Anal. Chem. 74 (2002) 784.

F. Švec, T.B. Tennikova and Z. Deyl (Editors)
Monolithic Materials
Journal of Chromatography Library, Vol. 67
© 2003 Elsevier Science B.V. All rights reserved.

Chapter 22

Large Scale Separations

Alois JUNGBAUER and Rainer HAHN

Institute for Applied Microbiology, University of Agricultural Sciences, Vienna, Austria

CONTENTS

22.1 INTRODUCTION

Monoliths are continuous stationary phases cast as homogenous columns in a single piece of various dimensions [1]. Typical examples are polymethacrylate polymers produced by thermally or UV irradiation initiated radical polymerization [2-7], monoliths with templated pores [8,9], compressed polyacrylamide gels [10,11], silica prepared as a single block using a sol-gel process [12-14], silica xerogels [15], emulsion derived polyHIPE® foams [16], monoliths prepared by metathesis polymerization [17], continuous urea-formaldehyde resins [18], monoliths cast from polysaccharides [19,20], rolled woven fabrics [21,22], and adsorptive membranes of various types [23-34]. Used as a single sheet, membranes can be considered monoliths with an extreme geometry, which dimension in the axial direction is very short [1]. Since membranes are prepared as very thin layers, the scale up is not possible while preserving their monolithic structure. However, they can be stacked to provide an additional volume and consequently also more capacity.

Monoliths of various types became very popular for the separation of biopolymers such as proteins, peptides, oligonucleotides, and DNA. Scale up to large industrial scale has not been reported yet, since the monolithic format is rather new and commercialization commenced only a few years ago. However, monoliths have already been used for preparative separations of biomolecules that are summarized in Table 22.1. The most popular are poly(glycidyl methacrylate) based monoliths (CIM® disks and CIM tubes®) from BIA Separations in Ljubljana, Slovenia. The silica-rods (PrepROD®) from Merk KgA (Darmstadt Germany) are currently in a pilot phase and have not reached the market yet. It is worth noting that the major challenge in scale up of chromatography with monoliths is the production of the material in large dimensions and its sealing at the column walls. As earlier reported by Lightfoot and coworkers [24-26], the stacked membranes largely failed since liquids bypassed the stacks along the wall. Modern design of stacked membrane modules with improved headers avoids this problem [34].

Furthermore substantial differences exist between monoliths used for catalysis and separation. For catalysis, especially for the gas-solid reactions, monoliths with channels throughout the whole block (honeycomb type monoliths) are preferred [35-40]. For separation purposes, a high degree of interconnection between the flow channels is desirable. Here we describe the scale up of monoliths for the separation of biomolecules. Monoliths have also been frequently used as packings in capillary electrochromatography. Currently this method is used almost exclusively for analytical separations [41]. The dissipation of the Joule's heat is a significant problem and prevents electrokinetic methods to be applied for large-scale separations, although the efficiency of these methods is extremely high.

TABLE 22.1

MONOLITHS USED FOR THE SEPARATION OF BIOMOLECULES.

Brand name	Manufacturer	Material	Pore size	Dimensions
CIM tube	BIA Separations	Polymethacrylate	1500 nm	45 × 15 mm i.d. 80 × 35 mm i.d.
UNO	Bio-Rad	Polyacrylamide	1000 nm	35 × 7 mm i.d. 53 × 12 mm i.d. 68 × 15 mm i.d.
PrepROD	Merck KgA	Silica	Macropores 2 μm Mesopores 14 nm	100 × 25 mm i.d.
Seprasorb	Sepragen	Cellulose	Not known	10 ml
CB-silica	Conchrom	Silica [a]	Mesopores 30 nm Micropores 5 nm	10 × 28 mm i.d.

[a] This material included with graphite, activated carbon, and ferrit is also available

Progress in development of modern chromatographic material and modern equipment turned scale-up to a less critical issue than it was in the past. Usually non-chromatographic parameters such as inconsistency or limited stability of starting material created problems, while working with labile biomolecules such as proteins. The scale up of purification is often linked with the scale up of fermentation. Typically different reactor types are used for both small and large scale. This may lead to a different composition of the impurities, product concentration, and even change in the molecular structure. Maiorella *et al.* [42] demonstrated that bioreactor types and culture conditions affect the isoelectric pattern of monoclonal antibodies. This may have a large impact on purification. Different isoelectric patterns may lead to different solubility and binding to both ion-exchangers and sorbents partly operating through ion-interaction. Similar effects were also reported by Schenerman *et al.* [43]. Biochemical and functional testing of a humanized monoclonal antibody directed against Respiratory Syncytial Virus (Synagis®) has been performed to evaluate cell line stability, support process validation, and demonstrate "comparability" during the process development. Using a variety of analytical methods, products manufactured at different production sites and in bioreactors with a volume of 20–10,000 l were shown to be biochemically and functionally equivalent. The biochemical testing for microheterogeneity found in Synagis® included evaluation of changes in post-translational modifications such as deamidation, truncation, and carbohydrate structure. Studies were also performed to support assessment of cell line stability and cell culture process-validation. Cell culture conditions were intentionally varied to determine their impact on the microheterogeneity of the product. In these studies, Synagis® was produced from cells cultured beyond the population doublings achieved at the maximum manufacturing scale, under conditions of low glucose, and using harvest times outside of the historical manufacturing operating range. Results showed a different pattern of glycosylation during the early stages of bioreactor culture. However, no other changes in microheterogeneity were apparent for the other culture conditions studied.

Here we also want to emphasize the effect of residence time of the biomolecules in the separation unit and the process time during scale up. When the residence time is changed during scale up, then the recovered product may have different properties depending on the scale. Usually the residence time increases with the scale. This also implies that a protein has longer contact-time for adsorption and molecular reorientation on the surface of a chromatography sorbent. It can also diffuse much deeper into the particle or the porous layer of a monolith. As a consequence the protein may bind tighter and is not released under the same conditions as worked out for the laboratory scale process. The rule of thumb is to keep the residence time constant if possible and not to change the temperature, since temperature also affects binding strength [44,45].

22.1.1 Packing of large-scale columns

Packing of large-scale columns is a delicate process for both silica-based and synthetic materials, which tend to form large particle agglomerates. In contrast this is much easier while using chromatographic media based on synthetic or natural polymers. The beads can be packed into the column as dry powder or slurry is poured or pumped into the empty column. These procedures are called dry and slurry packing, respectively [46]. The most frequently used dry packing method is also called dynamic axial compression. The dry powder is filled into the column and then a piston (adapter) is slowly brought upwards while the column is vibrated. This can be simply done by tapping with a hammer. The whole procedure has to be properly standardized [47]. During the dry packing procedure air is included into the packed bed. This air has to be completely removed by the liquid phase. Slurry packing is much easier than dry packing. Modern chromatographic materials can be pumped readily as slurry. The bottom of the chromatography column is covered with liquid phase. The slurry is pumped or poured (when smaller columns are used) into the column. The column is then closed and the liquid phase pumped through the column using a flow rate higher than that used later for the separation. The slurry is rapidly compacted to the final packed bed height and then the adapter of the column is lowered to touch the surface of the packed bed. Finally the pump is turned off.

The attractiveness of monoliths is that packing of a column is no longer required. The remaining challenge is the production of large-scale monoliths. In this case the user's task is reduced to mere check of system performance. This is accomplished in a way analogous to conventional packed bed chromatography. Preparative monoliths are usually delivered together with the housing (cartridge). After connecting the cartridge to the chromatographic workstation, the performance of the system must be tested. This can be done by injection of a small pulse of salt or a small organic molecule such as an amino acid. The selection of the solute depends on the nature of the chromatographic material. The number of theoretical plates N is then determined. It is also recommended to determine the asymmetry factor. The peak is symmetrical if back mixing is prevented and the bed does not promote flow irregularities. This is achieved if (i) injection technique and connection to the chromatographic workstation is good, (ii) dead volume in the solute flow path is very low, (iii) the solute concentration is low in order not to overload the detector, (iv) the residence time of the solute is sufficiently long to achieve dynamic equilibrium, and (v) the solute adsorption isotherm is linear.

Asymmetry can be expressed as the tailing factor or tailing coefficient

$$Tailing \ factor = \frac{w_{0.5}}{2\,A} = \frac{B+A}{2\,A} \tag{22.1}$$

where $w_{0.5}$ is the peak width at the half height, A is the width from the start to the peak maximum, and B is the width from the peak maximum to the end at the respective height [48]. At the top of the peak, values of B/A are relatively small and close to unity. It is worth noting that values of asymmetry ratio are one of the system suitability factors (Fig. 22.1).

For higher accuracy, a more sophisticated analysis is recommended. This includes for instance approximation of the peak data using the exponentially modified Gaussian function [49,50] and/or moment analysis [51]. In that case, electronic peak data acquisition is required. Recently, a hybrid of exponential and Gaussian function has also been developed to approximate asymmetric peak profiles [52]. This exponential-Gaussian hybrid function (EGH) is mathematically simple, numerically stable, and its parameters are readily determined by using graphical measurements and applying simple equations. Furthermore, the statistical moments of the EGH function can be accurately approximated within an error margin of $\pm 0.15\%$ at any level of asymmetry using formulae that are easily programmed into a computer. These features make EHG very easy to implement.

The asymmetry factor together with N is a very sensitive measure of the quality of packing. The number of plates is defined as a ratio of square of the retention time t and peak variance σ^2

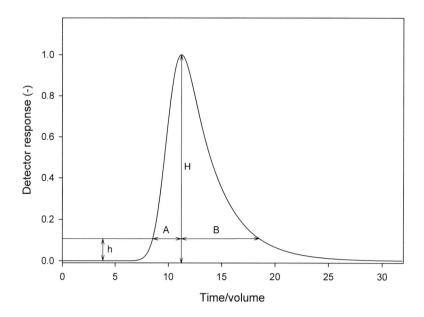

Fig. 22.1. Asymmetry of a chromatographic peak.

$$N = \frac{t^2}{\sigma^2} \tag{22.2}$$

for negligible dead volume to dead time (t_0) ratio. In preparative chromatography this is often not the case and an effective plate number N_{eff} must be defined (Fig. 22.2)

$$N_{eff} = \left(\frac{t - t_0}{\sigma} \right)^2 \tag{22.3}$$

Measurement of the elution volume is recommended for scale up. Different scales of separations should be compared on the basis of total column volumes. Furthermore, it is assumed that the pulse used for the determination of the plate count is infinitely small. This assumption does not often apply for preparative chromatography. In order to correct for the finite width of the pulse t_{pulse} the variance of the pulse is best approximated by

$$\sigma_{pulse}^2 = \frac{t_{pulse}^2}{12} \tag{22.4}$$

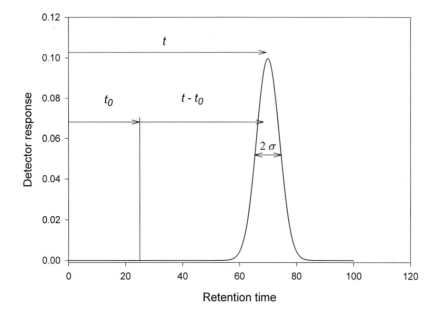

Fig. 22.2. Definition of theoretical (N) and effective plate number (N_{eff}).

The peak width is determined either at the base w, or at the point of inflection w^*, or at the half height w^{**}, and the plate number is then calculated as

$$N = 16 \cdot \left(\frac{t}{w}\right)^2 = 4 \cdot \left(\frac{t}{w^*}\right)^2 = 5.434 \cdot \left(\frac{t}{w^{**}}\right)^2 \tag{22.5}$$

Whenever the pulse and the extra column volume are not negligible the results must be corrected (Fig. 22.3).

Peers *et al.* [53] have demonstrated for conventional packed beds that the analysis of the breakthrough curves is more sensitive regarding to indicate irregularities of packed beds. Information such as N and the asymmetry can also be extracted from breakthrough curves. Assuming the chromatography column is an ergotic space [54], the first derivative of the breakthrough curve is equivalent to a peak produced by the injection of a pulse of tracers. The peak function $C(t)$ is

$$C(t) - \frac{dF(t)}{dt} \tag{22.6}$$

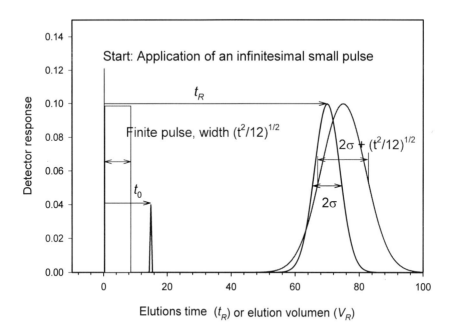

Fig. 22.3. Determination of the theoretical plate number for the width of the pulse having a finite dimension that cannot be neglected.

Rosenberg *et al.* [55] and Lettner *et al.* [56] showed that the peak width could be directly extracted from the breakthrough curve (Fig. 22.4). N determined at the point of inflection is

$$N = 2\pi \cdot \left[\frac{t_{inf}}{w'}\right]^2 \tag{22.7}$$

or

$$N = 4 \cdot \left(\frac{t_{0.5}^2}{\left(t_{0.85} - t_{0.15}\right)^2}\right) \tag{22.8}$$

Equation (22.8) can be used for calculation of the theoretical plate number from regeneration, equilibration, or washing step, when the change of the mobile phase composition is continuously recorded. Additional pulse experiments become then unnecessary, while sensitivity is improved.

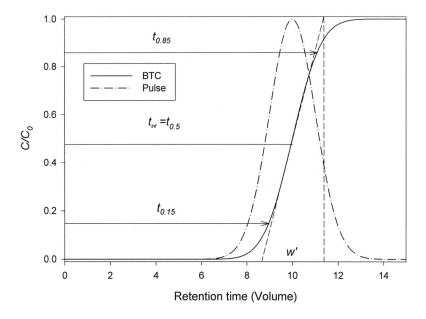

Fig. 22.4. Determination of the theoretical plate number by breakthrough curve experiments.

The packed bed may also undergo changes in its "structure" and at a certain time the column can even collapse. It is not easy to predict the collapsing point since this is a chaotic process [57]. Currently, information is not available on the life cycles of preparative monolithic columns. However, up to several hundred cycles were reported for conventional columns [58,59].

Several authors have reported that the quality of packing measured in terms of HETP decreases with increasing diameter [60,61]. They have accounted wall effects for the higher separation efficiency of columns with the smaller bore, since the wall supports the chromatographic beads. The larger the particle the less pronounced this effect appears. Only particles with large diameters and/or polymer-based materials are used at very large scales. In both cases the packing procedure is less critical. To which extent wall effects play a role in the preparative separations using monolithic columns, cannot be currently rated due to the lack of information.

22.2 PRODUCTION OF LARGE SCALE MONOLITHS

Compared to conventional chromatography, monoliths are rather novel materials for chromatographic separations. Pioneering work has been done by Hjertén

[10,11,62] and Svec at the end of the 1980s [63-67]. Currently, four companies are manufacturing monoliths for separation purposes. Monoliths for large-scale separations can be produced in three different geometric formats: disks, tubes, and rods (columns). The disks can be easily stacked in order to increase the binding capacity. Disks with a sheet-like geometry are often called membranes, although their morphology is different [1].

22.2.1 Tubes

The production of large-scale monoliths is not trivial. In the case of polymethacrylate monoliths the heat generation is the most critical factor. Since the pore size is controlled by the polymerization temperature [9,68-70], an efficient dissipation of heat during the polymerization process is the key to homogenous large-scale monoliths. Detailed description of their preparation is presented elsewhere in this book.

22.2.2 Rods

Continuous 300 × 8 mm rods of porous poly (glycidyl methacrylate-*co*-ethylene dimethacrylate) have been prepared by free radical polymerization [71]. The epoxy groups of these monoliths were modified by a reaction with diethylamine that affords ionizable functional groups required for ion-exchange chromatography. This material was then tested for separation of proteins. A dynamic capacity of 15 mg chicken egg albumin per one ml of the bed volume could be obtained at a flow velocity of 200 cm/min. The linear gradient elution mode enabled an excellent selectivity. However, a further scale up of monolithic columns is complicated. Therefore a modified method for the preparation of large polymethacrylate monoliths has been developed by Peters *et al.* [68]. In order to achieve the heat dissipation, they have used a slow polymerization procedure involving a gradual addition of the polymerization mixture into the reaction vessel. The gradual addition of the polymerization mixture resulted in very homogeneous monoliths. The pore volume in different sections of the monolith was examined and a high degree of homogeneity observed (Table 22.2). A similar approach was also tested for styrenic systems.

Silica-based chromatography columns with continuous gel skeletons and through-pores together with open mesopores produced by the sol-gel method have been developed for both analytical and preparative separation purposes [12-14]. Columns for the latter, which are currently in testing phase, will be marketed by Merck KgA (Darmstadt, Germany) under the trade name PrepROD®. The general features of silica rod columns are shown in Table 22.3. Due to the high content of mesopores, a typical van-Deemter curve is observed. The macropores together with the mesopores affect efficiency [72]. The smaller the macropore size, the lower the HETP. Any

TABLE 22.2

POROUS PROPERTIES OF MACROPOROUS POLY(GLYCIDYL METHACRYLATE-
CO-ETHYLENE DIMETHACRYLATE) POLYMER MONOLITHS PRODUCED BY THE
GRADUAL ADDITION OF POLYMERIZATION MIXTURE [68]. CONDITIONS: PO-
LYMERIZATION MIXTURE: GLYCIDYL METHACRYLATE 24%, ETHYLENE
DIMETHACRYLATE 16%, CYCLOHEXANOL 60%, BENZOYL PEROXIDE 1 wt %
WITH RESPECT TO MONOMERS, 12 h FEED AT 20 ml/h, NOMINAL TEMPERATURE
55°C

Sample taken at		$V_p{}^a$	Pore volume, % b				$D_{p.\,mode}{}^c$
position	portion	mL/g	<100	−500	−1000	>1000	μm
bottom	inner	1.40	15.37	7.08	11.39	66.15	1.77
	outer	1.51	13.78	7.48	10.68	68.05	1.66
middle	inner	1.34	17.51	6.70	10.70	65.09	1.76
	outer	1.48	15.37	6.59	10.03	68.00	1.76
top	inner	1.43	18.94	6.41	8.63	66.02	2.04
	outer	1.53	17.42	7.12	8.59	66.87	1.92

[a]Total pore volume

[b]Percentage of pore volume in the pores less than 100; 100–500; 500–1000, and over 1000
nm in size

[c]Pore diameter at the highest peak in the pore size distribution profile

difference cannot be observed for rods with a pore size in the range of 1–3 μm (Fig.
22.5). At high velocity, the band spreading is mass transport controlled. Such data
currently are not available for large molecules. The advantage of there monoliths is
their low pressure drop (Fig. 22.6) [12-14]. Use of these monolithic columns in
simulated moving bed (SMB) chromatography has also been suggested. SMB is very
sensitive to the packing quality of the interconnected columns. With silica monoliths,
an identical pressure drop for all columns can be easier achieved. As an example, the
separation of tocopherols from vegetable oil on preparative scale has been demonstrated [73]. The separation parameters for χ and δ tocopherol by SMB were calculated from this experiment. Monolithic columns in SMB mode have also been used for
separation of diastereomers [72]. Unfortunately, the nature of the compounds has not
been disclosed. Since both the selectivities of the diastereomers on the packed and on
the monolithic column and the saturation capacity for a given column volume are
comparable, the productivity of a SMB system only depends on the linear velocity of

TABLE 22.3

GENERAL FEATURES OF THE SILICA BASED COLUMNS SILICAROD® AND
PREPROD™ [73,113,114].

	SilicaROD™	PrepROD™
Dimension	100 × 4.6 mm i.d.	100 × 25 mm i.d.
	50 × 4.6 mm i.d.	
Silica type	High purity	Not available
Macropore size (μm)	2	3.98
Mesopore size (nm)	13	14
Mesopore volume (ml/g)	n.a.	0.94
Surface area (m²/g)	300	260
Surface modification	C-18 endcapped	n.a.
Surface coverage (%)	17	n.a.
Total porosity (%)	81	n.a.

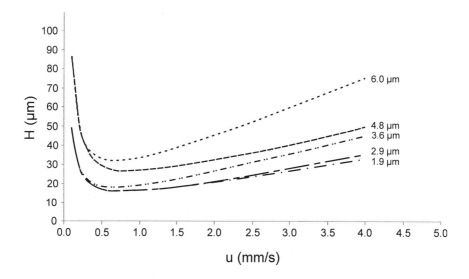

Fig. 22.5. Dependence of the HETP on the mesopore structure of silica type monoliths, the
PrepRODs™ (Reprinted with permission from ref. [72], copyright 2001 Elsevier Sciences B.
V.).

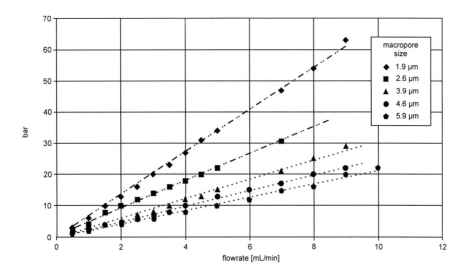

Fig. 22.6. The influence of the macropore size on the pressure drop of silica type monoliths, the PrepRODs™ (Reprinted with permission from ref. [72], copyright 2001 Elsevier Sciences B. V.).

the mobile phase. While SMB using monolithic phases could be operated at a linear velocity of 2000 cm.h^{-1}, the maximum velocity for a system including columns packed with particulate sorbents was only in the range of 700–1000 cm.h^{-1}. Therefore the monolithic stationary phase doubled the productivity of the SMB system.

22.3 SCALE UP MODELS

The goal of the scale up is to achieve identical separation conditions at each scale. The question arises what does it mean identical? First of all, the recovered product should be identical. Since the size of the column has to be changed and consequently a number of other parameters, it is not trivial to achieve identical conditions at each scale. The relative band spreading of each component should remain constant. Band spreading in a chromatography column is caused by (i) non-uniformity of flow in the packed bed, (ii) non-linearity of adsorption isotherm, (iii) mass transfer (diffusion), (iv) kinetics of adsorption, and (v) extra column contributions. The non-uniformity of the flow itself is determined by the column inlet design (affected by scale up), packing quality (not affected by scaling in case of monoliths), column diameter, viscosity difference between the feed and the eluent (viscous fingering), and the flow velocity.

The design of the column inlet and outlet is extremely important in radial flow chromatography. The mass transfer should not be a problem for monoliths without mesopores such as CIM tube®. The contribution of diffusion can be neglected for the band spreading dominated by the axial dispersion. The diffusional path is very small and the mass transport of the solute to the internal surface is achieved by convection. The contribution of the adsorption kinetics may also be of importance. The adsorption kinetics for ion exchange chromatography is usually very fast and can be neglected. In contrast the adsorption kinetic can be slow in other modes such as affinity chromatography and the high flow regime through a monolith cannot be exploited.

The extra column contribution to band spreading is extremely important for the scale up. Equation 22.9 indicates that the total band spreading is related to the extra column band spreading σ_{ex} and to the square of the retention volume V_R divided by the plate number N.

$$\sigma_{total}^2 = \frac{1}{N} V_R^2 + \sigma_{ex}^2 = \frac{HETP}{L} V_R^2 + \sigma_{ex}^2 \tag{22.9}$$

where L is the column length and *HETP* is the height equivalent to the theoretical plate. Different pipes, connectors, and injection devices are required for the different scales. The dead space responsible for the exponential washout kinetic of solutes may also vary. The extra column band spreading has to be kept small. This is particularly important in the isocratic mode. However, if the feed is adsorbed on the stationary phase, especially when a very steep adsorption isotherm applies, the compounds are concentrated. A sharpening effect can be observed during step elution [74]. Obviously, the extra column band spreading in front of the column is less critical than that occurring after the bands leaves the column [75].

A number of variables have to be considered in optimization of a chromatographic separation. The column size and geometric shape including the height (L), diameter ($2r$), aspect ratio, and the application of the feed in radial or axial direction are the most obvious. Composition of the mobile phase and the modifier, as well as the rate of change of composition of mobile phase (gradient shape) are also important. The other variables are then the flow rate u during loading, washing, elution, and regeneration, as well as the residence time L/u, feed composition, volume, concentration, feed flow rate, physical condition of the packing such as packing quality, packing density, and age of the packing. With monoliths the packing quality should not be a problem. However, fouling is a serious drawback of the monoliths. If the top layer of a conventional bed is fouled, it can be easily scrapped off and replaced by a layer of a new packing. In case of monoliths, the whole bed is lost. Therefore it is advisable to use a stack of monolithic units or use a precolumn.

The stationary phase is characterized by the ligand type, ligand density, type of support, porosity, tortuosity, and connectivity of the pores. The effect of connectivity on performance of monolithic columns is not yet clear. Although it is not completely correct, chromatographic separations are often treated as isothermal processes. It is worth noting that the heat of adsorption can be a substantial for separations in which hydrophobic interaction is involved. Tsai *et al.* [76] reported heat of adsorption of up to 20 kJ/mol for protein adsorption on a hydrophobic surface.

The response values for the optimization and scale up are productivity P, yield Y, dynamic capacity DBC, purity Q_R, concentration factor CF, and process time t_C. Some of the response factors are interrelated.

The productivity is defined as

$$P = \frac{Q_R\, C_0\, V_F}{V_t\, t_C} \cdot t_{life} \tag{22.10}$$

where Q_R is the purity ratio, C_0 is the sample concentration, V_F is the feed volume, V_t is the column volume, t_C is the cycle time (process time) and t_{life} is the column life time. In preparative chromatography the binding capacity controls the productivity since both high feed concentration and volume are preferred. High productivity can be achieved with monoliths because they can be operated at extremely high velocities without loosing their dynamic binding capacity (Fig. 22.7B).

It is obvious that a high yield is desirable. The yield Y is defined as

$$Y = \frac{C_E\, V_E}{C_F\, V_F} \tag{22.11}$$

where C_E is the concentration of the product in the eluate, V_E is the volume of the eluate, C_F is the feed concentration, and V_F is the feed volume. Discussing all the effects affecting the yield would exceed the scope of this chapter. The most important task is to find operation conditions at which the biological activity of the separated molecules is preserved. Transfer losses on large scale may also be responsible for a lower yield of the product [77].

The dynamic binding capacity DBC that can be determined using breakthrough experiments is defined as

$$DBC = V_{BTC} \cdot C_F \tag{22.12}$$

where V_{BTC} is the breakthrough volume and C_F is the feed concentration (Fig. 22.7A). In monolithic columns with no mesopores, the breakthrough curve is independent of the flow rate within a broad range of flow rates [78,79]. This allows an easier optimization of chromatography and gives more space for the scale up compared to conventional chromatography with packed columns. In certain cases, the static binding capacity $q*$ defined in equation (22.13) is also of interest.

$$q* = \frac{1}{V_t}\left[C_F (V_e - V_0) - \int_{V_0}^{V_e} C(V)\, dV \right]$$

(22.13)

In this equation V_t is the column volume, V_e the elution volume, V_0 the void volume, and C_F the feed concentration. The purity Q_F is very important and determines the choice of the stationary phase.

$$Q_F = \frac{C_E}{C_{total}}$$

(22.14)

Here C_E is the concentration of the product in the eluate and C_{total} is the total protein concentration in the feed. It is desirable to elute the product at a high concentration. The concentration factor CF is defined as

$$CF = \frac{C_E}{C_F}$$

(22.15)

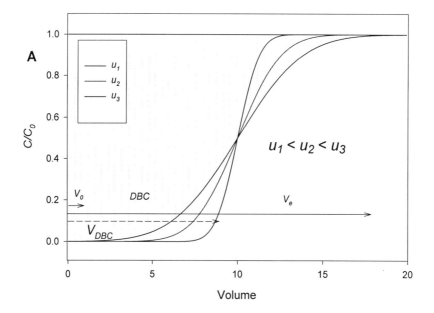

Fig. 22.7. – continued on the next page

where C_F is the product concentration in the feed. The process time t_c includes all processes required to purify the product and to regenerate the column. The process time is calculated as

$$t_C = \frac{V_{equil} + V_{feed} + V_{wash} + V_{elut} + V_{reg}}{F}$$

(22.16)

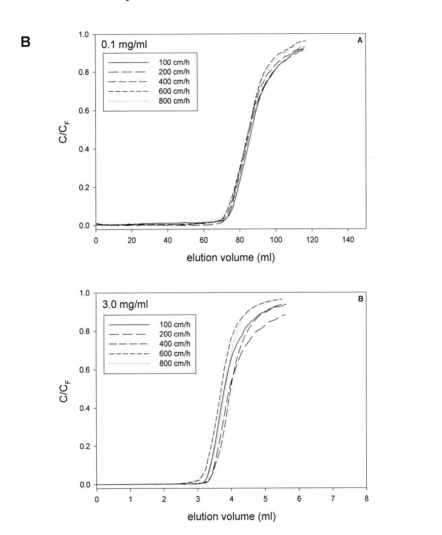

Fig. 22.7 (continued from previous page). Definition of the dynamic and static binding capacity. (A) The hatched area is the static binding capacity including the amount of feed in the void fraction. (B) For monoliths, the shape of the breakthrough curve is invariant with the velocity and protein concentration. Breakthrough curves of lysozyme on CIM-S0₃ disks™. Data obtained from ref. [79]

where V_{equil} is the volume required for the equilibration, V_{feed} is the feed volume, V_{wash} is the volume required for washing, V_{elut} is the volume required for elution, and V_{reg} is the volume for regeneration. V_{elut} is not identical with V_E, since the volume in which the product is recovered is smaller than the volume of buffer required for elution. The response values have to be prioritized for each case. For example, it might be possible scarify some product in order to obtain higher purity. Under other circumstances, purity is less important while productivity and dynamic binding capacity are an important issue [80].

22.3.1 Constant height

Keeping the column height during the scale-up constant also keeps both process and residence times unchanged. Under such conditions, the separation efficiency, yield, peak width, and buffer consumption normalized with respect to the total column volume should remain constant at all scales. The best way to proceed is to carry out the conventional optimization routine at a small scale. Then the height of the column is increased to the required value. Pragmatic issues such as availability of column hardware and required throughput also determine this height.

When the column height is changed, both dynamic capacity and column efficiency have to be determined again, since their values may change as well. Once these values are known, the further scale-up is very simple. The feed volume of the large column $V_{F,2}$ is calculated as

$$V_{F,2} = V_{F,1} \cdot \frac{A_2}{A_1} \tag{22.17}$$

where $V_{F,1}$ is the optimized feed volume of the smaller column and A_2 and A_1 are the superficial areas for the large and small column. All other volumes such as the peak volume, elution volumes, and the overall buffer consumption will change according to the same relationship. With increasing diameter the separation efficiency may slightly decrease. However, increased column height can compensate for this, although on the account of changes in residence time.

22.3.2 Constant residence time

The scale up at constant residence time allows the operation using different variables such as linear velocity, column height, and diameter. The residence time of the solute is the only variable that is kept constant during this scale up procedure. The simplified van Deemter equation [81]

$$HETP = a + \frac{b}{u} + c \cdot u \qquad (22.18)$$

where a, b and c are the empirical parameters. *HETP* can be expressed as

$$HETP = \frac{L}{N} \qquad (22.19)$$

and the residence time $\bar{t}$ is

$$\bar{t} = \frac{L}{u} \qquad (22.20)$$

The molecular diffusion (second term in equation 22.18) can be neglected for large molecules. Insertion of equations 22.19 and 22.20 in 22.18 yields

$$N = \frac{1}{\dfrac{a}{L} + \dfrac{c}{\bar{t}}} \qquad (22.21)$$

This equation shows, that despite variations in column length and velocity, the number of plates can be kept constant. Consequently, an unchanged resolution is obtained. Feed and all other variables have to be scaled with respect to the total column volume V_t.

22.3.3 Constant Biot number

The Biot number Bi expresses the ratio of film mass transfer rate to the time required for a molecule to diffuse from the surface to the center of the chromatographic particle. The particle diameter must be replaced for monoliths by the diffusional length [78]. The Biot number is defined as

$$Bi = \frac{k_f \cdot d_p}{D_{eff}} \qquad (22.22)$$

where k_f is the film mass transfer coefficient, d_p is the particle diameter, and D_{eff} is the effective diffusion coefficient. The k_f value of bovine serum albumin in silica monoliths was calculated to be about 2×10^{-5} ms^{-1} and the effective pore diffusivity about 2×10^{-12} m^2 s^{-1} [82]. Assuming a diffusional distance shorter than 1 μm, Bi is less than 10. Such situation can be expected for monoliths. In contrast, Bi numbers in a range of 170–2600 have been calculated for conventional packed beds [83].

Gu *et al.* [84] reported the relationship between Bi and velocity for radial flow chromatography. The bed volume V_b is

$$V_b = \pi L \left(r_1^2 - r_0^2 \right) \tag{22.23}$$

Furthermore they have also defined a dimensionless constant V_0'

$$V_0' = \frac{\pi L \cdot r_0^2}{V_b} = \frac{r_0^2}{r_1^2 - r_0^2} \tag{22.24}$$

and a dimensionless volumetric coordinate V'.

$$V' = \frac{\pi L \cdot \left(r^2 - r_0^2 \right)}{V_b} = \frac{r^2 - r_0^2}{r_1^2 - r_0^2} \tag{22.25}$$

The variation in *Bi* values observes the following relationship for radial flow chromatography

$$Bi \propto k_f \propto u^{1/3} \propto \left(\frac{1}{r} \right)^{1/3} \propto \left(V' - V_0' \right)^{1/6} \tag{22.26}$$

If the $Bi_V=1$ values are known, *Bi* values anywhere else in the bed can be calculated using equation 22.27.

$$Bi_V = \left(\frac{1 + V_0'}{V + V_0'} \right) \cdot \frac{1}{6} \cdot Bi_{V=1} \tag{22.27}$$

This above set of equations can be used if the transport mechanism substantially changes with scale. However, the *Bi* number is not affected in monoliths without mesopores.

22.3.4 Constant L/d_p^2 ratio

Chromatography can also be scaled-up while keeping constant ratio between L and d_p^2 [85]. In preparative chromatography it might be desirable to increase the particle diameter in order to reduce the back pressure, which determines the equipment costs. The empirical relationship between the number of plates N, the velocity u and the particle diameter d_p is according to Ladisch *et al.* [86] given by

$$L_1 = \left(\frac{N_1}{N_2}\right)\left(\frac{u_1}{u_2}\right)\left(\frac{d_{p1}}{d_{p2}}\right)^2 L^2 \tag{22.28}$$

The subscripts 1 and 2 denote the columns of different sizes. According to this equation, the column length is proportional to the square of the particle diameter to maintain the same separation efficiency. Therefore, the efficiency changes with the column length. This also applies for monoliths. However, the particle diameter must be replaced by the diffusional length, which changes with the column length. Therefore, it is not recommended to use this parameter for scale-up. The same applies to constant L/r ratios. The residence time changes in both cases and the separation conditions cannot be predicted [87].

22.3.5 Radial chromatography

Instead of running a column in the axial direction, radial flow chromatography has been developed to enhance speed of the separation [84,88,89]. In this mode the linear velocity u changes with the column radius as shown in equation 22.29:

$$u = \frac{F}{2 \pi r L} \tag{22.29}$$

where F is the volumetric flow rate, r is the radius, and L is the column height. The average flow $\bar{u}$ is described as

$$\bar{u} = \int_{r_1}^{r_2} u(r)\, dr \tag{22.30}$$

Inserting equation 22.29 into 22.30 and solving the integral for r_1 to r_2 yields the average velocity $\bar{u}$

$$\bar{u} = \frac{F}{2 \pi L} \cdot \frac{\ln r_2 / r_1}{r_2 - r_1} \tag{22.31}$$

Filter company Cuno manufactured specifically designed columns. A layer of chromatographic separation medium was placed on a grid and wound on a central tube. The eluent streams have to perfuse through this cartridge in axial direction and collected through the central tube. Although the performance of these cartridges was excellent, they did not survive on the market since the housings designed for filtration was used for chromatography. The large dead space contributed to a significant extra-column band spreading and completely lost of resolution observed for some applica-

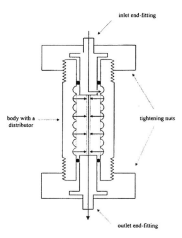

inlet end-fitting

body with a
distributor

tightening nuts

outlet end-fitting

Fig. 22.8. Schematic drawing of the housing for a large volume polymethacrylate monolith (CIM-tube™). The housing comprises inlet and outlet end fittings, two tightening nuts, and a body with distributor. Two O-rings (black dots) ensure proper sealing. Solid arrows indicate the direction of the flow (Reprinted with permission from ref. [69], copyright 2000 American Chemical Society).

tions using the cartridge [88]. Another radial chromatography system has been designed by Sepragen Inc. This column allows loading and operation in radial direction. The bed is stabilized in the specifically designed hardware optimal for soft gels [89].

Monoliths have also been scaled-up for radial chromatography. The material is synthesized in the shape of concentric rings and then assembled in a telescope-like way [69]. The schematic drawing of the chromatography column is shown in Fig. 22.8. This design is suited only for centripedal (inward) flow. This system can be easily scaled, only when the capacity/efficiency does not change significantly with the velocity, which is the case for CIM-tubes as shown earlier [78,79]. Assuming that the critical velocity must not be exceeded to avoid losses in efficiency, the optimal bed usage can be simply calculated by rearranging equation 22.29. The critical radius r_{crit} at which the critical flow velocity is reached is:

$$r_{crit} = \frac{F}{u_{crit} \cdot 2\pi \cdot L} \tag{22.32}$$

The optimal bed usage is then calculated for the given critical velocity:

$$bed\ usage = V_t - V_{crit} = 1 - \frac{V_t}{V_{crit}} \tag{22.33}$$

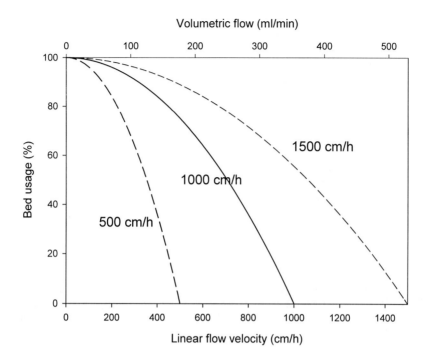

Fig. 22.9. Relationship between optimal bed usage and flow-rate in radial flow chromatography. Data were calculated for a 45 mm × 15 mm i.d. CIM-tube™ assuming a flow rate of 35.4 ml/min. This value corresponds to a linear velocity of 100 cm/h at the outer surface.

Inserting equation 22.32 in 22.33 and rearranging yields:

$$bed\ usage = 1 - \left(\frac{u_1}{u_{crit}}\right)^2 \tag{22.34}$$

Allowing for an internal critical velocity of 1500 cm/h, an 8 ml tube with a diameter of 1.5 cm and length 4.5 cm can be operated at a flow rate equivalent to 40 column volumes per minute (Fig. 22.9). This simple calculation demonstrates the speed of operation in preparative scale with monoliths used in the radial mode.

Another type of radial flow column consisted of a Plexiglas cylinder (61 mm I.D., 80 mm O.D. and 100 mm length), Plexiglas bottom and top plates, and a central flow distributor made of stainless steel (8 mm diameter and 150 mm long) [20]. The flow distributor distributed the flow in and out of the column and also enabled adjustment of the top plate. A Nylon net positioned between the bed and the cylinder helped to center the bed in the column and improved the flow distribution. The column was constructed from transparent material to allow inspection of the bed during operation

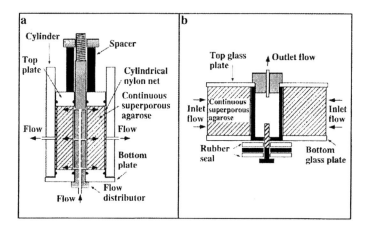

Fig. 22.10. Radial flow columns used with continuous superporous agarose beads. (A) Type 1 column was constructed to work in both flow directions (centrifugal and centripedal flow). (B) Type 2 column with simplified design used with a centripedal flow. The figures are not drawn to scale. (Reprinted with permission from ref. [20]. Copyright 2001 Elsevier Sciences B. V.).

(Fig. 22.10). This system contains superporous agarose monolith was used for separations and catalysis as described earlier in this book.

22.3.6 Compact disk HPLC

The Compact-Disc-HPLC (CD-HPLC) also called annular disk chromatography is a genuine continuously working chromatographic separation system based on the True-Moving-Bed (TMB) concept (Fig. 22.11). The stationary phase is enclosed in a circular cartridge (Compact-Disc) and turns relative to the point at which the sample is continuously injected. Due to the rotating cartridge, which is specially designed for the CD-chromatograph, the "cross-current-flow" is achieved. Main part of the cartridge is the stationary phase made of silica monolith (CB-SILICA). The cartridge consists of an upper plate and a lower plate both connected together with a cylinder. The continuous supply of the solvent is achieved using 59 inlet capillaries. They are arranged around the outer radius of the upper part of the stationary phase. First, the solvent moves from the outer edge of the plate towards the center and then from the center down through the cylinder into the lower plate of the stationary phase. From the center it is moving towards the outer part where it is drained into 60 outlet capillaries, which are arranged around the lower disc circumference. The sample passes through the cartridge in the same way as the solvent. Due to the relative

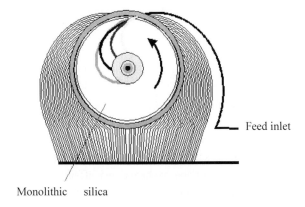

Feed inlet

Monolithic   silica

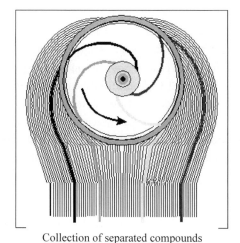

Collection of separated compounds

Fig. 22.11. Schematic of the supply unit for the mobile phase (A) and collection unit for the separated fractions (B) of a compact disk HPLC

movement of the cartridge with respect to the inlet and outlet capillaries, the track of the separated single components looks helical. The separated components leave the cartridge at its outer circumference via different outlet channels. Providing the settings for the pumps and the rotation speed of the cartridge are not changed, each separated component continuously leaves the system at the same position. Due to the rotation of the stationary phase, the discontinuous, "time related" separation typical of column chromatography is transferred into a continuous, "location related" separation. Both target solution feed and collection of separated compounds are collected continuous.

As a result, multi-component mixtures can be separated continuously thus enabling to achieve a high throughput.

22.4 MEMBRANE ADSORBERS

22.4.1 Stacking of membranes and monoliths

As already mentioned, adsorptive membranes can also be considered monoliths [1]. In order to increase the capacity, the surface of these membranes can be grafted. Such membranes have been primarily used for ion exchange and affinity chromatography [23-34,90-94]. The removal of minor impurities is the key application of such membranes with product collected in the flow through mode. Endotoxin and DNA-removal from pharmaceutical protein solutions are typical examples [95].

Casting large sheets of homogenous membranes is difficult. Therefore, more popular is to stack the thin sheets of membranes, wind them around a central drainage tube, or cast membranes as hollow fibers [90]. The typical formats of adsorptive membranes used for bioseparations are shown in Fig. 22.12. The advantage of the stacking approach is the gain in capacity while keeping the compact format, the drawback can be the loss of resolution when the hydrodynamics of the system is not properly designed. Environmental scanning electron microscopy and nuclear magnetic resonance microscopy have been used to characterize the flow pattern in stacked mem-

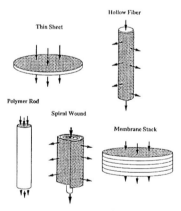

Fig. 22.12. Comparison of the configuration that have been proposed for membrane-based adsorption. Arrows illustrate the direction (S) of bulk flow. Pattern indicates membrane cross-sectional area which lies roughly perpendicular to bulk flow. (Reprinted with permission from ref. [25], copyright 1995 Elsevier Sciences B. V.).

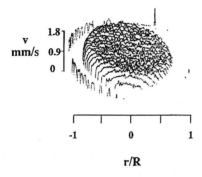

$$\begin{array}{c} v \\ mm/s \end{array} \quad \begin{array}{c} 1.8 \\ 0.9 \\ 0 \end{array}$$

-1   0   1

r/R

Fig. 22.13. Velocity distribution in a MemSep 1000. Shown is a velocity distribution over a 0.5 mm slice transverse to the axis and 10 % of the distance form the stack inlet toward the outlet. The column I.D. is 25 mm and the water flow-rate is 15 mm/min. (Reprinted with permission from ref. [24], copyright 1995 Elsevier Sciences B. V.).

brane units in order to avoid the ambiguities associated with efluent analysis [24,25]. Both techniques afford high-resolution.

Once efficient analysis is achieved, it is not possible to assign the effects of the extra column volume, the header, and the column itself to the distortion of the flow profile. Therefore, Roper *et al.* [26] have proposed to measure the peak profile after the flow has been reversed. This procedure enabled to separate the effect of the header on band spreading from the other contributions [26,34].

Magnetic resonance analysis allows non-destructive and non-invasive visualization of static and operating units using the contrast sensitivity. For example, the magnetic resonance images for a stacked membrane cartridge Memsep 1000 (Millipore, Bedford, MA, USA) are shown in Fig. 22.13. The dimensions of this unit were 5 × 17 mm i.d. constituting a nominal bed volume of 1.4 ml. Fig. 22.13 demonstrates that liquid flow is excluded from the periphery of the bed. Extra-column band broadening is also more significant in the adsorptive membrane systems compared to the chromatographic columns owing to the large dead space. Experimental evidence suggests that extra-column band broadening changes little with flow velocity. Its effects can be significant relative to other contributions to band spreading. The reported plate heights of membrane systems vary to a large extent (Table 22.4). One explanation is the design of the chromatography skid. Our measurements with adsorptive membranes revealed that a single sheet affords efficiency expressed as HETP close to theoretical. With increasing number of membranes in the stack the efficiency decreased (Fig. 22.14). The extra column band spreading and the flow profile in the stack were likely to be responsible for the loss of efficiency. A typical feature of a membrane bed is its large cross sectional area relative to its length. These short, wide beds allow high

TABLE 22.4

EXPERIMENTAL PLATE HEIGHTS (HETP) DETERMINED FOR ADSORPTIVE
MEMBRANES[a]

HETP, μm	Velocity range, cm/min	Estimation method	Extra-column mixing included	Ref.
3.3–0.59	0.52–3.8	reversed-flow	no	[26]
3–7	0.07–0.22	unreported	yes	[115]
25	0.1–2	unreported	n.a.	[116]
50–110	1.5–4	EMG	yes	[24]
80–160	1–45	unreported	n.a.	[24]
400	0.04–1	unreported	yes	[24]
250–800	0.035–6.5	moments	no	[24]
7	0.01–4	moments	yes	[117]
40–50	0.01–4	moments	yes	[117]
<15	1–58	reversed-flow	no	[34]

[a] Modified from ref. [25]

velocities and large volumetric capacities upon only modest pressure drops. These features lead to increased throughputs and short residence times, thus reducing the danger of protein degradation and denaturation. A large diameter to length ratio, however, presents a challenge to achieve uniform flow distribution across the whole surface of the membrane. This is a significant problem that may reduce the membrane efficiency to the level typical of packed beds. Optimized design of flow distributors reported recently largely eliminates this problem [96]. This rational design involving CFD resulted in distributors that afforded uniform flow distribution in both numerical simulation and laboratory prototypes (Fig. 22.15).

Adequate flow distribution is also necessary to maintain the column efficiency in scaled-up devices. Frequently, the scale-up is done by optimizing the separation using a small column followed by an increase in the column diameter to meet the desired throughput. For columns with shape avoiding variations in the radial dimension, the flow is evenly distributed across the column cross-section and identical behavior for the scaled-up separation can be expected. However, for devices in which the diameter to length ratio increases, uniform flow distribution is difficult to maintain. Header design and performance are more significant in large system design.

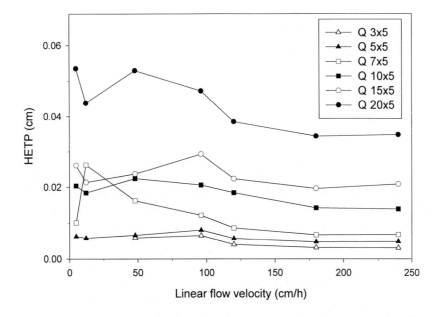

Fig. 22.14. Comparison of the efficiency measured as HETP versus the velocity for stacks of 3, 5, 7, 10, 15, and 20 Sartobind anion exchanger membranes functionalized with quaternary ammonium groups. Prototype cartridges were kindly supplied by Sartorius GmbH (Göttingen, Germany).

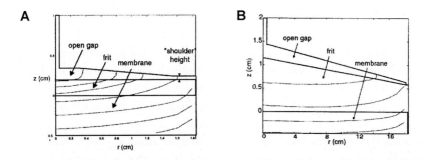

Fig. 22.15. Simulation of the macroscopic flow through a 10 ml (A) and 1 l Mustang membrane absorber (B). Each line represents the position of the centre of mass of the solute band at constant residence time.

Another adsorptive membrane, which has been recently commercialized, is the Mustang membrane produced by Pall Corporation (Ann Arbor, MI, USA). The Mustang cartridges contains polyethersulfone (PES) membranes chemically modified to produce a sulfonic acid or a quaternary ammonium surface functionalities. The nominal pore size of the PES membranes is 0.8 μm, and the thickness of each membrane layer 70–90 μm. The membrane layers are housed in a stainless steel ring with porous frits on each side. The 10 ml Mustang modules are rated to resist pressures of up to 0.17 MPa, while 1 l Mustang modules resist 0.8 MPa. The housing was designed to achieve plug flow. The 10 ml unit consists of a tapered open gap region above a flat porous frit, while the 1 l unit includes this tapered open gap region above a tapered porous frit.

In order to avoid the shortcomings of poor flow distribution in stacked membranes, spiral wound cartridges have also been developed. Microfiltration membranes have been grafted with long spacers in order to increase the capacity of the membrane. However, this was achieved at the expense of reduction in the water flux [31]. These membranes are marketed under the trade name Sartobind® by Sartorius (Göttingen, Germany). The membranes are wound around a central tube. In order to increase the capacity, the cartridges can be connected in series or in parallel. The advantage of these membranes is their high capacity. The Sartobind® large scale membrane adsorber module consists of the membrane, reeled to form a hollow cylinder, with screens of stainless steel on the inner and outer sides embedded in plastic caps on both ends. The length of the standard module is 50 cm and they are available in configurations with 15, 30, and 60 membrane layers corresponding to a nominal membrane surface area of 2, 4, and 8 m^2, respectively. They all have equal outer diameter of 96 mm and fit into the same housings. Three different cores are available to accommodate the different module geometries. The inner and outer channel volumes are equal and add up to 600 ml of total channel volume. For geometric reasons, doubling the membrane area does not exactly correspond to two times higher number of layers. The thickness of the membranes also varies to a certain extent. So, these different configurations are better characterized by the membrane volumes. Similarly, the flux through the different configurations does not correspond directly to the number of layers. The absolute values of the flux are highly dependent on the medium and the type of the membrane. For example, the magnitude of flux with 10 mmol/l potassium phosphate buffer, pH 7.0 is referred to as "normal flux". An overview of these values is given in Table 22.5. Sartobind® modules are also available in lengths of 25, 12.5, 6.2, and 3.1 cm, the smallest unit representing 1/16 of the standard module.

22.4.2 The transsectional flow

Long before monoliths have been considered as chromatographic material, the flow path of solutes in a monolith has been theoretically compared to that in columns packed with pellicular material. The solute can take a variety of routes of different

TABLE 22.5

CHARACTERISTICS OF 50 cm SARTOBIND STANDARD MODULES ACCORDING
TO DEMMER AND NUSSBAUMER [30]

Code	Nominal membrane area, m^2	Nominal number of layers	Membrane volume, l	Nominal flux l/min.kPa
MA-20K-15-50	2	15	0.55	15
MA-40K-30-50	4	30	1.1	7.0
MA-80K-60-50	8	60	2.0	2.5

lengths when it travels through the packed bed. The residence time is then given as the average path length. The differences of the various flow paths through the column are reflected in the residence time distribution, which is equal to the peak of an injected tracer pulse. It can be assumed that the residence time distribution will increase, when the solute is forced to stay in a single individual flow channel. In contrast, when the solutes have the possibility to change flow channels, the residence time distribution will decrease. This contribution to peak broadening due to this process was termed by Giddings "transchannel" contribution [97], a term derived from the fact that the inter-change occurs merely across the channels.

We performed pulse experiments with a stack of 3 × 12 mm i.d. CIM-disk. The peak width was compared to pulse response observed for rods of the same height as that of the stacked disks. Interestingly, lower band spreading was observed for the stacked disk (Fig. 22.16). This is likely due to the interchange of the flow path in each section. This transsectional flow appears to lead to a lower band spreading.

22.5 APPLICATIONS

22.5.1 Removal of contaminants by negative chromatography

Membrane absorbers have been successfully used for the removal of host cell proteins and DNA from IgG prepurified by Protein A affinity chromatography [98]. This process was performed in the radial flow-through mode. The feed solution had a high conductivity of up to 40 mS/cm. Due to the high charge density, the capacity for DNA did not vary with pH. Therefore conditions could be found at which the anti-body passed the adsorptive membrane while the DNA was bound at a capacity of 9 g per liter of the sorbent. Hypothetical extrapolation and comparison between a typical

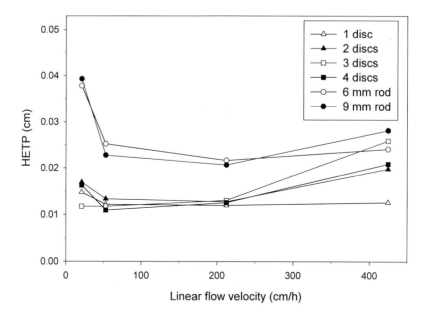

Fig. 22.16. Comparison of band spreading with stacked monolithic disks (CIM-disk) and a rod. The prototype rod was kindly supplied by BIA Separations (Ljubljana, Slovenia).

particulate medium (Q-Sepharose FF) and the Sartorius adsorptive membrane suggests the possibility of the removal of host cell proteins to a limit lower than 2 ng/ml. The flow rate for the adsorptive membrane was about 10 fold higher and 300 times more solution could be processed with monolithic sorbent compared to the same volume of a conventional packed bed. (Table 22.6). The drawback of the adsorptive membrane was removal of viruses to a lower degree.

Anspach and Petch [99] used dextran coated nylon membranes with immobilized Polymixin B, desoxycholate, poly-L-lysine, or polyethyleneimine for removal of endotoxins from protein solutions such as blood serum and plasma. Endotoxin levels could be reduced below the threshold required for parentherial administration after a residence time in the device of only 6 s. Although the protein binding capacity of the adsorptive membranes is only 6–15 mg/ml membrane volume, almost 100 % protein could be recovered. Pitiot *et al.* [100] described a system designed for extracorporeal removal of IgG using poly(ethylene glycol-vinyl alcohol) hollow fiber membranes grafted with histidin.

TABLE 22.6

REMOVAL OF HOST CELL PROTEIN USING Q-SEPHAROSE FF COLUMN AND THE SARTORIUS Q15 MEMBRANE [98]

	Host cell protein ng/ml antibody	Antibody loaded g antibody/l sorbent	Flow-rate cm/h
Load	10.6	–	–
Q-Sepharose FF	< 2	50	76
Sartorius Q15	< 2	15 000	620

22.5.2 Purification of oligonucleotides and DNA

Oligonucleotides have been separated using polymethacrylate based monoliths functionalized with ionizable ligands or poly(styrene-*co*-divinylbenzene) based monoliths modified with alkyl ligands [101-103]. Although Podgornik *et al.* [103] worked with an ultra-short column only 0.75 mm long, they were able to resolve single oligonucleotides under isocratic elution conditions. This small column enabled reduction of the separation time to less than 80 s.

Similarly, resolution of oligonucleotides on a 50 × 8 mm i.d. poly(glycidyl methacrylate-*co*-ethylene dimethacrylate) monolith with diethylamine functionalities was extremely good and 12 to 24-mer oligonucleotides could be baseline separated [102].

Deshmukh *et al.* [104] reported a large scale purification of antisense oligonucleotides using a 550 ml radial flow membrane unit. 3 g of crude material were processed at a flow rate of 2.1 l/min. The 20-mer oligonucleotides could be separated from deletion sequences and other byproducts using a linear 0–2.5 mol/l gradient of NaCl. They also demonstrated purification of 3-9 kg/day using a 21 m^2 unit. Demmer and Nussbaumer described this pilot scale unit in more detail [30]. They have hypothetically compared conventional chromatographic media with monoliths assuming equal binding capacity. While the process time for the purification of 3 g of crude oligonucleotides was 115 min for the conventional medium, it was only 17 min for the adsorptive membrane. However, the drawback of the membrane absorber was an earlier breakthrough of the nucleotides. Therefore the feed had to be partly recycled (Table 22.7).

Plasmid DNA separation has been demonstrated by Giovannini *et al.* [105]. Both gradient and isocratic elution were applied to separate a 7.2 kb predominately super-

TABLE 22.7

HYPOTHETICAL COMPARISON OF PARTICULATE CHROMATOGRAPHY MEDIA
AND MEMBRANE ABSORBERS WITH EQUAL DYNAMIC BINDING CAPACITY
[104]

	Conventional column	MA module
Load, g	3	3
Column	100 ml (25 × 100 mm)	$2 \, m^2$
Flow rate, ml/min	20 (0.2 CV/min)	700 (1.3 CV/min)
Sample volume, ml	300	300
Loading time, min	15	0.6 (1.8 with 3x feed recirculation)
Elution 20 CVs, min	100	15

coiled plasmid using both polymethacrylate and a polyacrylamide based monoliths. Under optimized conditions, all these approaches have led to the separation of super-coiled, linear, and circular forms. The binding capacity of monolithic media for plasmids was very high. Due to the high porosity and the large diameter of the channels, the mass transfer of these large molecules was not restricted.

22.5.3 Purification of antibodies

The adsorptive membranes have also been applied for the purification of antibodies in the positive mode [98]. The breakthrough capacity increased with the number of layers. This can be explained by better flow-distribution of the devices used on the process scale. One membrane layer bound 7 g antibodies while 60 layers absorbed 40 g. The shape of the breakthrough curve did not depend on the flow velocity. A pure axial dispersion mechanism was assumed to dominate the band spreading.

22.5.4 Purification of enzymes, growth factors, and other proteins

Garke *et al.* [106] described the isolation of recombinant human basic fibroblast growth factor from *E. coli* using an UNO-Q column. A fast gradient elution was possible and the performance could be maintained up to a linear flow velocity of 935 cm/h. Dirty feedstock could be directly processed using cellulose-type monoliths [19]. Micrococcus lysates, cheese and milk were applied without pre-processing. Purification of lactate dehydrogenase from bovine heart has been accomplished to demonstrate the functionality of continuous superporous agarose monolith in an affinity mode [20]. The heart extract was pre-purified by ammonium sulfate precipitation. The feed was applied onto the monolith with immobilized Cibachrom Blue. Elution was then achieved using NADH, the co-factor of lactate dehydrogenase.

Immunoaffinity chromatography exploiting the Ca^{2+} dependent interaction of the anti-Flag antibody and Flag-tagged proteins [107] was investigated by Schuster *et al.* [108]. The antibody was immobilized on both gigaporous glass beads (Prosep) and monolithic Convective Interactive Media (CIM) columns that had a ligand density of 2 mg/g or 10 mg/ml. The performance of the columns was assessed by applying clarified yeast culture supernatant containing overexpressed Flag-human serum albumin. Dynamic binding capacity and purity were determined at various flow rates ranging from 100 to 800 cm/h and 95% purity could be obtained. Compared to the Prosep column, anti Flag-CIM monoliths exhibited a higher non-specific adsorption and requiring a longer wash cycle to obtain the same purity. However, the anti Flag-CIM afforded a flow independent performance. In contrast, a decrease in dynamic binding capacity with flow was observed for anti-Flag-Prosep columns. Both columns were suited for purification of milligram quantities of protein from the yeast culture supernatant within a few minutes.

Displacement chromatography of model proteins has also been explored [109,110]. Resolution was considerably better with the monolithic column compared to conventional packed beds. Ghose and Cramer [111] carried out a dimensionless group analysis to determine the relative importance of mass transfer and kinetic resistance of the polyacrylamide-based stationary phase. They found that the kinetic of adsorption was the dominating resistance.

22.6 CONCLUSION AND FUTURE OUTLOOK

Monoliths are a chromatography medium of choice when high productivity is required. The excellent mass transfer properties allow separation at high flow velocities. In addition, a lower pressure drop is often observed compared to conventional chromatography material due to the high porosity of the monoliths. The use of these novel stationary phases for preparative separation of biomolecules is in its infancy. Thus, real-life large-scale applications are not yet available. The manufacturing of large scaled up monoliths is currently under progress using three different concepts: (i) conventional enlarging of the geometry of a column, (ii) radial flow chromatography, and (iii) stacking of membranes. Stacked membranes with an improved header design for flow distribution and monolithic tubes operated in the radial mode enable use of very high flow rates and consequently high productivity. It appears likely that once large monoliths with a volume of up to 100 l and more will be available, they will be rapidly accepted for large-scale industrial bioseparations. Apparently, the demand for large separation units is currently more obvious in the field of protein purification. In contrast, smaller units with a volume of up to a maximum of 10 l might be sufficient for oligonucleotide and plasmid purification.

22.7 References

1   G. Iberer, R. Hahn, A. Jungbauer, LC-GC 11 (1999) 998.

2   D. Josic, A. Strancar, Ind. Eng. Chem. Res. 38 (1999) 333.

3   F. Svec, J.M. Fréchet, Ind. Eng. Chem. Res. 38 (1999) 34.

4   D. Josic, A. Buchacher, A. Jungbauer, J. Chromatogr. B 752 (2001) 191.

5   D. Josic, A. Buchacher, J. Biochem. Biophys. Meth. 49 (2001) 153.

6   H. Zou, Q. Luo, D. Zhou, J. Biochem. Biophys. Meth. 49 (2001) 199.

7   M. Grasselli, E. Smolko, P. Hargittai, A. Safrany, Nuclear. Instr. Meth. Phys. Res. Sec. 185 (2001) 254.

8   G.S. Chirica, V.T. Remcho, J. Chromatogr. A 924 (2001) 223.

9   R. Hahn, A. Podgornik, M. Merhar, E. Schallaun, A. Jungbauer, Anal. Chem. 73 (2001) 5126.

10   S. Hjertén, J. Liao, R. Zhang, J. Chromatogr. 473 (1989) 273.

11   S. Hjertén, M. Li, J. Mohammed, K. Nakazato, G. Pettersson, Nature 356 (1992) 810.

12   H. Minakuchi, K. Nakanishi, N. Soga, N. Ishizuka, N. Tanaka, Anal. Chem. 68 (1996) 3498.

13   H. Minakuchi, K. Nakanishi, N. Soga, N. Ishizuka, N. Tanaka, J. Chromatogr. A 762 (1997) 135.

14   N. Ishizuka, H. Minakuchi, K. Nakanishi, N. Soga, N. Tanaka, J. Chromatogr. A 797 (1998) 133.

15   S.M. Fields, Anal. Chem. 68 (1996) 2709.

16   A. Mercier, H. Deleuze, O. Mondain-Monval, React. Funct. Polymers 46 (2000) 67.

17   B. Mayr, R. Tessadri, E. Post, M.R. Buchmeiser, Anal. Chem. 73 (2001) 4071.

18   X. Sun, Z. Chai, J. Chromatogr. A 943 (2002) 209.

19   R. Noel, A. Sanderson, L. Spark in: J.F. Kennedy, G.O. Philips, P.A. Williams (Eds.), Cellulosics: materials for selective separations and other technologies, E. Horwood, New York, 1993, p. 17.

20   P.-E. Gustavsson, P.-O. Larsson, J. Chromatogr. A 925 (2001) 69.

21   Y. Yang, A. Velayudhan, C.M. Ladish, M.R. Ladish, J. Chromatogr. A 598 (1992) 169.

22   K. Hamaker, S.-L. Rau, R. Hendrickson, J. Liu, C.M. Ladisch, M.R. Ladisch, Ind. Eng. Chem. Res. 38 (1999) 865.

23   K.-G. Briefs, M.-R. Kula, Chem. Eng. Sci. 47 (1991) 141.

24   E.N. Lightfoot, A.M. Athalye, J.L. Coffman, D.K. Roper, T.W. Root, J. Chromatogr. A 707 (1995) 45.

25   D.K. Roper, E.N. Lightfoot, J. Chromatogr. A 702 (1995) 3.

26   D.K. Roper, E.N. Lightfoot, J. Chromatogr. A 702 (1995) 69.

27   K.H. Gebauer, J. Thommes, M.R. Kula, Chem. Eng. Sci. 52 (1996) 405.

28   S.A. Camperi, M. Grasselli, A.A. Navarro del Canizo, E.E. Smolko, O. Cascone, J. Liquid Chromatogr. 21 (1998) 1283.

29   S.A. Camperi, A.A.N. Del Canizo, F.J. Wolman, E.E. Smolko, O. Cascone, M. Grasselli, Biotech. Prog. 15 (1999) 500.

30   W. Demmer, D. Nussbaumer, J. Chromatogr. A 852 (1999) 73.

31   M. Grasselli, A.A. Navarro del Canizo, S.A. Camperi, F.J. Wolman, E.E. Smolko, O. Cascone, Rad. Phys. Chem. 55 (1999) 203.

32   X. Zeng, E. Ruckenstein, Biotech. Prog. 15 (1999) 1003.

33   F.J. Wolman, M. Grasselli, E.E. Smolko, O. Cascone, Biotech. Lett. 22 (2000) 1407.

34   M.A. Teeters, T.W. Root, E.N. Lightfoot, J. Chromatogr. A 944 (2002) 129.

35   G. Bercic, Catalysis Today 69 (2001) 147.

36   A.K. Heibel, T.W.J. Scheenen, J.J. Heiszwolf, H. Van As, F. Kapteijn, J.A. Moulijn, Chem. Eng. Sci. 56 (2001) 5935.

37   J.J. Heiszwolf, M.T. Kreutzer, M.G. van den Eijnden, F. Kapteijn, J.A. Moulijn, Catalysis Today 69 (2001) 51.

38   M.T. Kreutzer, P. Du, J.J. Heiszwolf, F. Kapteijn, J.A. Moulijn, Chem. Eng. Sci. 56 (2001) 6015.

39   M. Vaarkamp, W. Dijkstra, B.H. Reesink, Catalysis Today 69 (2001) 131.

40   P. Woehl, R.L. Cerro, Catalysis Today 69 (2001) 171.

41   J.R. Chen, R.N. Zare, E.C. Peters, F. Svec, J.M. Fréchet, Anal Chem 73 (2001) 1987.

42   B.L. Maiorella, J. Winkelhake, J. Young, B. Moyer, R. Bauer, M. Hora, J. Andya, J. Thomson, T. Patel, R. Parekh, Biotechnology (NY) 11 (1993) 387.

43   M.A. Schenerman, J.N. Hope, C. Kletke, J.K. Singh, R. Kimura, E.I. Tsao, G. Folena-Wasserman, Biologicals: J. Internat. Assoc. Biol. Standardization 27 (1999) 203.

44   G.M. Finette, Q.M. Mao, M.T. Hearn, J. Chromatogr. A 763 (1997) 71.

45   D.J. Roush, D.S. Gill, R.C. Willson, J. Chromatogr. A 653 (1993) 207.

46   J.J. Stickel, A. Fotopoulos, Biotech. Prog. 17 744.

47   J. Klawiter, M. Kaminski, J.S. Kowalcyk, J. Chromatogr. 243 (1982) 207.

48   A. Williams, K. Taylor, K. Dambuleff, O. Persson, R.M. Kennedy, J. Chromatogr. A 944 (2002) 69.

49   J.P. Foley, J.G. Dorsey, J. Chromatogr. Sci. 22 (1984) 40.

50   M.S. Jeansonne, J.P. Foley, J. Chromatogr. Sci. 29 (1991) 258.

51   D.W. Morton, C.L. Young, J. Chromatogr. Sci. 33 (1995) 514.

52   K. Lan, J.W. Jorgenson, J. Chromatogr. A 915 (2001) 1.

53   D. Peers, R. Arnold, E. Ford, J. Cacia, J. Frenz in: G. Guiochon (Ed.), PREP 2000, Washingtion D.C., 2000.

54   P. Naor, R. Shinnar, I&EC Fund. 2 (1963) 278.

55   D.U. Rosenberg, AIChE J. 2 (1956) 55.

56   H.P. Lettner, O. Kaltenbrunner, A. Jungbauer, J.Chromatogr. Sci. 33 (1995) 451.

57   R. Freitag, D. Frey, C. Horváth, J. Chromatogr. A 686 (1994) 165.

58   I. Drevin, B. Johansson, J. Chromatogr. 547 (1991) 21.

59   R.M. O'Leary, D. Feuerhelm, D. Peers, Y. Xu, G.S. Blank, BioPharm 14 (2001) 10.

60   J.L. Rocca, J.W. Higgins, R.G. Brownlee, J. Chromatogr. Sci. 23 (1985) 106.

61   A.M. Wilhelm, J.P. Riba, G. Muratet, A. Peyrouset, R. Prechner, J. Chromatogr. 363 (1986) 113.

62   S. Hjertén, J.L. Liao, J. Chromatogr. 457 (1988) 165.

63   T.B. Tennikova, D. Horák, F. Svec, J. Kolář, J. Čoupek, S.A. Trushin, V.G. Maltzev, B.G. Belenkii, J. Chromatogr. 435 (1988) 357.

64   T.B. Tennikova, M. Bleha, F. Svec, T.V. Almazova, B.G. Belenkii, J. Chromatogr. 555 (1991) 97.

65   T. Tennikova, F. Svec, B.G. Belenkii, J. Liq. Chromatogr. 13 (1990) 63.

66   F. Svec, J.M. Fréchet, Anal. Chem. 64 (1992) 820.

67   Q.C. Wang, F. Svec, J.M. Fréchet, Anal. Chem. 65 (1993) 2243.

68   E.C. Peters, F. Svec, J.M. Fréchet, Chem. Mater. 9 (1997) 1898.

69   A. Podgornik, M. Barut, A. Strancar, D. Josic, T. Koloini, Anal Chem 72 (2000) 5693.

70   I. Mihelic, M. Krajnc, T. Koloini, A. Podgornik, Ind. Eng. Chem. Res. 40 (2001) 3495.

71   F. Svec, J.M. Fréchet, J. Chromatogr. A 702 (1995) 89.

72   M. Schulte, J. Dingenen, J. Chromatogr. A 923 (2001) 17.

73   M. Schulte, D. Ludba, A. Delp, J. Dingenen, J. High Resol. Chromatogr. 23 (2000) 100.

74   A. Jungbauer, S. Hackl, S. Yamamoto, J. Chromatogr. 658 (1994) 399.

75   O. Kaltenbrunner, A. Jungbauer, S. Yamamoto, J. Chromatogr. A 760 (1997) 41.

76   Y.-S. Tsai, F.-Y. Lin, W.-Y. Chen, C.-C. Lin, Colloids Surfaces A 197 (2002) 111.

77   A. Jungbauer, W. Schönhofer, D. Pettauer, G. Gruber, F. Unterluggauer, K. Uhl, E. Wenisch, F. Steindl, Biosciences 8 (1989) 21.

78   R. Hahn, A. Jungbauer, Anal. Chem. 72 (2000) 4853.

79   R. Hahn, M. Panzer, E. Hansen, J. Mollerup, A. Jungbauer, Sep. Sci. Tech. 37 (2002) 1545.

80   A. Jungbauer, O. Kaltenbrunner, in M. Flickinger, S. Drew (Eds.), Encyclopedia of Bioprocess Technology: Fermentation, Biocatalysis & Bioseparation, Wiley, New York, 1999, p. 585.

81   E. Katz, K.L. Ogan, R.P. Scott, J. Chromatogr. A 1983 (1983) 51.

82   A.I. Liapis, J.J. Meyers, O.K. Crosser, J. Chromatogr. A 865 (1999) 13.

83   J.R. Conder, B.O. Hayek, Biochem. Engng. J. 6 (2000) 225.

84   T. Gu, G.-J. Tsai, G.T. Tsao, Chem. Eng. Sci. 46 (1991) 1279.

85   J.H. Knox, J. Chromatogr. Sci. 15 (19977) 353.

86   M.R. Ladisch, R.L. Hendrickson, E. Firouztale, J. Chromatogr. 540 (1991) 85.

87   A. Jungbauer, J. Chromatogr. 639 (1993) 3.

88   A. Jungbauer, F. Unterluggauer, K. Uhl, A. Buchacher, F. Steindl, D. Pettauer, Wenisch, Biotechnol. Bioeng. 32 (1988) 3.

89   T. Gu, in: M.C. Flickinger, S.W. Drew (Eds.), Encyclopedia of Bioprocess Technology: Fermentation, Biocatalysis & Bioseparation, Wiley, New York, 1999, p. 627.

90   S. Brandt, R.A. Goffe, J.L. O'Connor, S.E. Zale, Biotechnol. 6 (1988) 799.

91   H. Iwata, K. Saito, S. Furusaki, T. Sugo, J. Okamoto, Biotechnol Prog 7 (1991) 412.

92   D. Lutkemeyer, M. Bretschneider, H. Buntemeyer, J. Lehmann, J Chromatogr. 639 (1993) 57.

93   O.W. Reif, R. Freitag, J Chromatogr. A 654 (1993) 29.

94   O.W. Reif, R. Freitag, Bioseparation 4 (1994) 369.

95   F.B. Anspach, J. Biochem. Biophys. Meth. 49 (2001) 665.

96   Q.S. Yuan, A. Rosenfeld, T.W. Root, D.J. Klingenberg, E.N. Lightfoot, J. Chromatogr. A 831 (1999) 149.

97   J.C. Giddings, Dynamics of Chromatography, Dekker, New York, 1965.

98   H.L. Knudsen, R.L. Fahrner, Y. Xu, L.A. Norling, G.S. Blank, J. Chromatogr. A 907 (2001) 145.

99   F.B. Anspach, D. Petsch, Proc. Biochem. 35 (2000) 1005.

100 O. Pitiot, C. Legallais, L. Darnige, M.A. Vijayalakshmi, J. Membr. Sci. 166 (2000) 221.

101 A. Podgornik, M. Barut, J. Jancar, A. Strancar, J. Chromatogr. A 848 (1999) 51.

102 D. Sýkora, F. Svec, J.M. Fréchet, J. Chromatogr. A 852 (1999) 297.

103 M. Merhar, A. Podgornik, M. Barut, S. Jaksa, M. Zigon, A. Strancar, J. Liquid Chromatogr. Rel. Techn. 24 (2001) 2429.

104 R.R. Deshmukh, T.N. Warner, F. Hutchison, M. Murphy, W.E. Leitch, P. De_Leon, G.S. Srivatsa, D.L. Cole, Y.S. Sanghvi, J. Chromatogr. A 890 (2000) 179.

105 R. Giovannini, R. Freitag, T.B. Tennikova, Anal. Chem. 70 (1998) 3348.

106 G. Garke, I. Radtschenko, F.B. Anspach, J. Chromatogr. A 857 (1999) 137.

108 A. Einhauer, A. Jungbauer, J. Biochem. Biophys. Meth. 30. (2001) 455.

109 M. Schuster, E. Wasserbauer, A. Neubauer, A. Jungbauer, Bioseparation 9 (2000) 259.

110 S. Vogt, R. Freitag, Biotechol. Prog. 14 (1998) 742.

111 R. Freitag, S. Vogt, J. Biotechnol. 78 (2000) 69.

112 S. Ghose, S.M. Cramer, J. Chromatogr. A 928 (2001) 13.

113 K. Cabreara, G. Wieland, D. Lubda, K. Nakanishi, N. Soga, H. Minakuchi, K.K. Unger, Trends Anal. Chem. 17 (1998) 50.

114 K. Cabrera, D. Lubda, H. Eggenweiler, H. Minakuchi, K. Nakanishi, J. High Resol. Chromatogr. 23 (2000) 93.

115 P. Langlotz, K.H. Kroner, J. Chromatogr. 591 (1992) 107.

116 J.A. Gerstner, R. Hamilton, S.J. Cramer, J. Chromatogr. 596 (1992) 173.

117 A. Jungbauer, R. Hahn, unpublished results.

F. Švec, T.B. Tennikova and Z. Deyl (Editors)
Monolithic Materials
Journal of Chromatography Library, Vol. 67
2003 Published by Elsevier Science B.V.

Chapter 23

Immunoaffinity Assays

Galina A. PLATONOVA and Tatiana B. TENNIKOVA

*Institute of Macromolecular Compounds, Russian Academy of Sciences,
Bolshoy pr. 31, 199 004 St. Petersburg, Russia*

CONTENTS

23.1 INTRODUCTION

It is well known that the biological interactions occurring *in vivo* are based on the formation of specific complexes or *complementary biological pairs*. Textbook examples of these pairs are enzyme–substrate, enzyme–inhibitor, antigen–antibody, hormone–receptor, etc. Among these complements, the antigen-antibody complex is most often used in the development of different biotechnological processes [1,2]. Additionally, a wide range of diagnostic procedures is based on these highly specific interactions [3].

Many immunoaffinity assays (IAA) can be defined as analytical methods that rely on the detection and measurement of the antigen–antibody interaction. These methods can be classified in two major groups: (i), dynamic and, (ii), non-dynamic. The most popular non-dynamic technique based on immunointeraction is the *enzyme-linked immunosorbent assay* (ELISA) [4]. In contrast, *immunoaffinity chromatography* (IAC) is a dynamic approach in which the stationary phase includes an antibody or an antigen and the separated substance is distributed between both stationary and mobile phases. Campbell reported the first preparative IAC in 1951 using antigens immobilized on *p*-aminobenzyl cellulose for the purification of antibodies [5]. At present, this chromatographic method is widely applied in the isolation of a variety of biological materials [6-9], as well as for bioanalytical procedures that, for example, enable the determination of the accumulation of antibodies in stressed human or animal organisms (immunomonitoring) [10]. In addition, IAC appears to be suitable for both practical and theoretical immunology studies. Extensive immunological investigations have led to a better understanding of the mechanism of specific binding between antibodies and their complementary antigens [1,11-13].

23.2 BASIC CONCEPTS OF IMMUNOAFFINITY PAIRING

Antibodies are glycoproteins produced by living organisms in response to the intrusion of foreign agents (antigens). The molecule of the most common immunoglobulin G (IgG) consists of four polypeptides (two identical light chains and two identical heavy chains) held together by disulfide bonds. Each IgG contains two equivalent antigen-binding sites consisting of the variable regions of a single light chain combined with the structural part of a single heavy chain, called Fab units, and a single large fragment with heavy chains called an Fc unit. The amino acid sequence located within the binding sites is highly variable, and specific for each antibody. This difference allows the organism to produce antibodies with a variety of different binding affinities and specificities against the foreign agents [14]. A number of combined weak electrostatic-, hydrogen-bond-, hydrophobic-, and van der Waals interactions leads to a strong and specific binding.

Immunoaffinity interactions between an antibody and its corresponding antigen lead to the formation of a highly specific equimolar pair. Typically, either the antibody or the antigen is tagged with a detectable component that can be used to identify or quantify the analyte in a solution. In most cases, the substance of interest is either the antibody or the antigen molecule itself, and the immune complex that forms as the result of their interaction creates the foundation for analytical/preparative application.

23.3 IMMUNOAFFINITY SYSTEMS

Two essential elements are very important in immunoaffinity system: reagents and techniques. Reagents include antibodies and antigens, in addition to other compounds used to visualize the primary binding reaction. The technique then refers to a device, or system in which the reagents are placed, to perform the assay in either a non-dynamic or dynamic mode [15].

As examples of non-dynamic approaches, the following methods can be mentioned. First, one of the early semi-quantitative immunoanalytical techniques was based on immunoprecipitation that led to an insoluble antigen–antibody complex resulting from the reaction of both species at appropriate concentrations [16]. Latex diagnostics or latex agglutination tests belong to this group of immunoaffinity assays [17]. Radioimmunoassays (RIA) based on isotopic labeling of the antigen or antibody significantly advanced the field of quantitative analysis, contributing substantially to an increase in sensitivity [18].

Later, enzyme immunoassay (EIA) has been developed as the alternative to RIA. In this method, detection of the substance of interest (antigen or antibody) is achieved using an enzyme — which eliminates the problems of safety and short isotope lifetime typical of RIA [19]. A heterogeneous mode of EIA – ELISA – involves the separation of the unbound antibody or antigen residing in a liquid phase from the antigen–antibody complex bound to the solid phase (adsorbent) [20].

Much research effort has also been directed towards developing biosensors or immunosensing techniques. An immunosensor can be defined as a specifically constructed device that explores the immune binding response [21-23].

The dynamic immunoaffinity mode such as flow injection analysis (FIA) is based on concepts of chromatography, and represents a technique that allows automatic analysis of a large number of samples and/or continuous monitoring, and affords solution to a variety of analytical problems [24,25]. Nilsson *et al.* proposed a flow-injection ELISA [26]. A number of different labels have been suggested for use in these assays, including enzymes [27,28] and fluorescent tags [29,30].

An other interesting and prospective method – the flow-injection type of ELISA – can also be achieved in a way similar to *dot-blot* assays [31]. The final colored product is precipitated on the surface of a solid-phase membrane.

23.4 STATIONARY PHASES

The stationary phase for IAA needs to meet the following criteria [32]: it must, be (i), resistant to high flow rates; (ii), perform practically independently of the flow rate; (iii), have low mass-transfer resistance; (iv), be compatible with samples typical in biotechnology processes and cleaning procedures (filtered through 0.2 μm filters, high protein loads); (v), have long-term stablility, and; (vi), be able to immobilize biologically active ligands such as antibodies and antigens.

Taking into account all these critical requirements, a great number of chromatographic stationary phases for IAA have been developed over the last decade [33]. An extensive range of particulate supports is currently available and includes natural polymers (agarose, dextran, cellulose), perfluorinated hydrocarbon polymers, crosslinked synthetic polymers (polyacrylamide, polymers of 2-hydroxyethyl- and other methacrylates, polystyrene), inorganic porous materials (silica, glass, titania), as well as composites such as polysaccharide-coated silica and polyacrylamide/agarose. All these materials are popular in the field of both affinity chromatography and immunoaffinity assays [34].

In contrast to natural supports, their synthetic counterparts easily resist attack by microorganisms and, as a result of their rigid structure, allow much higher flow rates to be used, thus reducing the time required for the separation. Conventional porous or partially porous particles packed in columns, membranes, and monoliths can be used as the solid phases to perform the immunoaffinity chromatography.

23.4.1 Conventional packings

Both gel-type and rigid porous particles are used as stationary phases in conventional immunoaffinity columns [35,36]. It is obvious that applications of these discrete porous particles, even those having gigaporous morphology that enables partial flow-through (perfusion) [37-39], may suffer from slow diffusional mass-transfer of the molecules into the space within the pores. This slow process increases the time required for the separation process, and also involves bio-complementary pairing between immobilized and dissolved complements. As a result, the throughput (productivity) of such an affinity chromatography may be low. An increase in the flow rate is likely to lead to a dramatic deterioration in the separation in these systems.

23.4.2 Membrane adsorbers

In contrast to "classical" immunoaffinity chromatography in packed columns, affinity filtration employs microfiltration membranes (membrane adsorbers) with grafted ligands [33,40-46]. These membranes may be configured in thin sheets, hollow fibers, radial flow layers, and other geometries. The main difference between membrane adsorbers and the conventional columns is in the hydrodynamic flow through the porous membrane. This convective flow substantially increases the mass--transfer rate of the compounds dissolved in a mobile phase to the adsorptive sites of the porous support. In contrast to particulate adsorbents for perfusion chromatography, membranes are totally permeable for liquids. This dramatically improves the technological parameters of bioseparations. These membranes can be considered part of the family of monoliths.

23.4.3 Macroporous ultra-short monolithic beds

Macroporous poly(glycidyl methacrylate-*co*-ethylene dimethacrylate) (GMA–EDMA) monolithic disks were introduced in the late 1980's, first as novel stationary phases for chromatography of biopolymers [47,48]. These disks are promising stationary phases for a variety of dynamic processes based on interphase mass distribution [49,50].

As with the membrane absorbers, these monolithic separation media afford an enhanced accessibility of immobilized ligands located on the open surface of the flow-through channels to their specific counterparts dissolved in the mobile phase. The advantageous hydrodynamic properties of monolithic adsorbents also enable the observation of the real-time kinetics of formation of bioaffinity pairs under dynamic conditions. In other words, the mass transfer enhanced by convection allows one to consider the biospecific reaction as a process controlled only by time [50]. High performance monolithic disk chromatography (HPMDC) combines the advantages of both membrane technology (simple scale-up, low pressure-drop across a monolith, high bed stability) and column chromatography (high selectivity and efficiency of separation, high loading and capacity).

23.5 IMMOBILIZATION OF LIGANDS

Support materials for IAA have to be chemically and mechanically stable, easily activated to allow coupling, and hydrophilic to minimize non-specific interactions. The binding of an antigen or antibody to a solid support affords a specific immunoadsorbent with the following desirable properties [51]. (i), It should be capable of adsorbing the complementary antibody from a mixture of components; (ii), liberation of the adsorbed antibody from the specific adsorbent should be quantitative and carried out under conditions that are innocuous for the specific activity; (iii), it should possess a high capacity for the adsorption of the specific antibody and; (iv), it should retain its biological activity even after repeated use and storage.

Many different methods of ligand-immobilization have been developed for IAA. One common approach involves direct covalent attachment of antibodies or antigens to a solid phase. Protein molecules contain a number of different functionalities suitable for such a binding, including the guanidynyl group of arginine, the γ- and β-carboxylic acid groups of glumatic and aspartic acids, the sulfhydryl of cysteine, the imidazolyl of histidine, the amine of lysine, the thioether of methionine, the indolyl of tryptophan, and the phenolic group of tyrosine. In addition, many proteins also contain prosthetic groups, of which carbohydrates are by far the most important for chemical attachment of proteins since they are commonly not involved in biological activity and are reactive, especially in their oxidized form. Extensive descriptions of the various methods used for protein immobilization can be found in the literature [52].

The immobilization methods most often used include activation of a matrix using CNBr [53], the formation of N-hydroxysuccinimide esters (NHS) [54,55], imidazolyl-carbamate groups (CDI) [56], and aldehyde groups obtained after periodate oxidation of sugar moieties, or the use of glutaraldehyde [57], hydrazide groups [58], and others such as epoxide and amine groups [59].

In specific cases, some supports such as poly(styrene-divinylbenzene) have no active sites or functional groups available for the direct immobilization of ligands. These can be introduced by coating them with a polymer that enables reactions and also shields the hydrophobic polystyrene surface [60]. In contrast to the sorbents that require activation, GMA-EDMA polymers contain epoxy groups. Depending on the pH, these groups can react with amine groups of the biological ligands in a single reaction step under very gentle, "biocompatible" conditions.

Protein molecules are typically bound to the support in a random fashion via a single — but more likely through several — covalent bonds. As a result, the active site might face toward the support surface, and its accessibility can be significantly reduced. This results in a reduction in biological activity and, consequently, in a lower binding capacity. To avoid this, routes for oriented immobilization of IgGs via; (i), a carbohydrate moiety [58,61,62]; (ii), sulfhydryl groups [63]; and, (iii), Protein A or Protein G, have been developed [51,64,65].

Many investigators advocate the use of a spacer between the matrix and the ligand. The major advantage is that this provides ligand accessibility to the binding site of a target molecule. As a result, the ligand is separated from the matrix by a distance derived from the length of the spacer. In addition, the spacer molecule provides an increase in protein flexibility compared with a direct coupling to the matrix. Finally, this fact might improve the biological activity [66].

23.6 MACROPOROUS MONOLITHS IN IMMUNOAFFINITY ASSAYS

Macroporous ultra-short monoliths (disks and membranes) with immobilized immunoaffinity ligands have been used for the separation of monospecific antibodies, as described elsewhere [33,45,46,50]. At present, commercial GMA–EDMA monolithic CIM® disks (BIA Separations d. o. o., Ljubljana, Slovenia) are suitable for use in IAA (Table 23.1). It is worth noting that the most impressive applications of CIM® disks were demonstrated just in affinity separation based on specific recognition of natural biological complements [31,32,67-77]. Only a few fundamental studies of the effect of conditions on chromatographic immunoaffinity separations have been published [43,78,79]. In contrast, the most detailed studies of experimental parameters have been carried out with monolithic CIM® disks.

TABLE 23.1

APPLICATION OF MACROPOROUS MONOLITHIC CIM®- DISKS IN IMMUNOAF-FINITY ASSAYS

Target molecule	Immunoaffinity ligand	Ref.
Polyclonal IgGs from rat liver	Annexin	[67]
Polyclonal IgGs from rabbit antiserum	Protein A	[67]
Recombinant Protein G from *E. coli* cell lysate	Human IgG	[68,74]
Protein G from cell lysate of recombinant *E. coli*	Human IgG	[32]
Antibradykinin polyclonal IgGs from rabbit blood serum	Bradykinin	[69,71]
Polyclonal bovine IgGs	Protein A	[70]
Recombinant human IgGs	Protein G	[70]
Recombinant human IgGs	Protein L	[70]
Flag-HSA	Anti-Flag monoclonal IgG	[72,81]
Human IgGs	Protein A	[73]
Human blood coagulation factor VIII	Peptides from a combinatorial library	[31]
Polyclonal bovine IgGs	BSA	[75]
Human IgGs	Recombinant Protein G	[76,77]

23.6.1 Effect of experimental conditions on immunoaffinity binding

Various factors that affect binding in the affinity mode of HPMDC have been investigated using frontal analysis for several biocomplementary pairs [75]. The affinity interactions were evaluated from linearized adsorption isotherms and expressed in terms of both dynamic dissociation constants of the complexes, K_{diss}, and theoretical adsorption capacities, Q_{max}. Bovine serum albumin (BSA) and human immunoglobulin G (hIgG) were used as the immobilized ligands in these experiments and the role of surface-coverage density of immobilized ligands, flow rate, temperature, and the spacer on dynamic immunoaffinity binding was explored.

23.6.1.1 Surface density of immobilized ligands

Using modified experimental conditions recommended recently [68], series of CIM®-BSA immunoaffinity supports with varying ligand density were prepared. For example, a surface coverage of 13 nmol of immobilized BSA per square meter of the pore surface can be obtained after 8 h reaction with GMA–EDMA disks. This amount doubles after extending the reaction time to 16 h (Table 23.2). Using simple mathematics and the cross-section of protein molecules immobilized on the accessible surface, it is possible to estimate the total surface area covered by the immobilized ligands. In this way, the covered surface, most likely the surface of the flow-through channels, was found to equal 1.0 m^2/ml. This value roughly corresponds to 60–90% of the total surface area determined from mercury intrusion porosimetry. We consid-

TABLE 23.2

Q_{MAX} AND K_{DISS} FOR THE IMMOBILIZED BSA – antiBSA ANTIBODY PAIR MEASURED AT DIFFERENT LIGAND DENSITIES

Reaction time, h	BSA ligand density nmol/m^2	Q_{max} nmol/m^2	K_{diss} µmol/L
8	13	2.4	3.00
16	25	5.6	3.80

Conditions: Solutions of polyclonal antibodies with a concentration of 0.05–1 mg/ml were loaded on CIM®-BSA disks at a flow rate of 2 mL/min to obtain the breakthrough curves; the intermediate washing procedure was performed using 2 mol/L NaCl and the desorption step by using 0.01 mol/L HCl (pH 2.0); desorbed fractions of proteins were analyzed using the Lowry test; all experiments were carried out in triplicate, with RSD 5–10% [75].

ered these data as indirect confirmation of the maximum immobilization capacity upon the conditions used for the covalent coupling. The second conclusion to be drawn is that all flow-through pores or convective channels of the monolithic layer are almost completely covered by affinity ligands that are later involved in the separation process. The two-fold increase in the ligand concentration does not affect K_{diss} of the affinity complex whereas the adsorption capacity increases. The Q_{max} vs. ligand density data of Table 23.2 also indicate that only 20–25% of immobilized ligands are accessible for the soluble counterpart to form the specific pair. This result is in good agreement with those published previously [68,69].

23.6.1.2 Effect of flow rate on binding capacity

The effect of geometry of flow-through channels on chromatographic separations using thin membrane-shaped adsorbent layer has been demonstrated several times [42,43,45,78-80]. The residence time of a solute within the porous space of the sorbent, t_{res}, and the time required for a solute molecule to transverse the distance from its location within the stream to the pore wall by free diffusion through the stagnant layer of a liquid adhering to the surface, t_{film}, control the process [43,79]. The residence time depends on the flow rate of the mobile phase and the thickness of the disk, whereas t_{film} depends on the size of the flow-through pores. Obviously, t_{res} must exceed t_{film} to allow the molecule of a protein to diffuse from the stream to the wall of a channel and to bind to the adsorbing functionality located on the surface of a pore.

HPMDC experiments in the affinity mode revealing the effect of flow rate on binding parameters have been carried out using defined (fixed) volumes of loaded molecules [75]. Specifically, a pool of polyclonal antibodies against BSA was chosen as a soluble part of the pair, and BSA immobilized on a CIM® epoxy disk as the specific ligand. The binding has been studied at flow rates of 0.5–10 ml/min for both adsorption and desorption steps.

Table 23.3 shows the amounts of anti-BSA antibodies eluted at decreasing t_{res} values obtained as a result of an increase in the flow rate. Only a two-fold reduction in the specific adsorption capacity was observed upon a 20-fold reduction in the residence time from 24.4 to 1.2 s. It is worth noting that even the shortest $t_{res} = 1.2$ s was much longer than the calculated value of t_{film}, that equals 2.5×10^{-2} s.

23.6.1.3 Frontal analysis at different flow rates

Frontal analysis has been used to evaluate both the maximum binding capacity and equilibrium dissociation constant of affinity complexes, which are important thermodynamic characteristics of the affinity interaction at different flow rates [75]. The

TABLE 23.3

EFFECT OF FLOW RATE ON THE AMOUNT OF ELUTED ANTIBODIES AT A FIXED
VOLUME OF LOADED SAMPLE

Flow rate ml/min	Residence time, t_{res}, s	Eluted antibodies mg/ml sorbent
0.5	24.4	0.59
2.0	6.1	0.41
5.0	2.5	0.27
10.0	1.2	0.24

Conditions: PBS solution (10 ml) containing 5 mg of polyclonal antibodies was loaded on CIM®-BSA disk at various flow rates [75]. For other conditions see Table 23.2.

loading of the sorbent continued until the breakthrough had been observed and the absorbances of the protein solution at both the outlet and inlet were equal. At this point, all accessible affinity sites were assumed to be occupied with target molecules. Table 23.4 shows the use of affinity CIM®-*h*IgG disks for direct isolation of recombinant Protein G (*r*Prot G) from *E. coli* cellular lysate. At unfavorable t_{res}/t_{film} ratios typical of high flow rates, a considerable part of the target molecules may flow through the adsorptive layer without being retained. To achieve high loading, an increased volume of the loaded sample has to be used to compensate for the lower number of effective interactions. This suggests that extremely high flow rates are not desirable in immunoaffinity HPMDC. Flow rates in a range of 2–5 ml/min appear most attractive to achieve the time benefit without significant loss of valuable biological material, yet allowing completion of the separation process within only a few minutes.

Schuster *et al.* [72] compared CIM® disks and gigaporous glass beads, both with immobilized antibodies. Exploiting the Ca^{2+}-dependent interaction of the anti-Flag antibody and Flag-tagged proteins they found that both supports are suited for the application [81]. The isolation of the antigen was achieved in less than 2 min. However, the dynamic binding capacity of porous glass beads largely depended on the flow-rate, while this was not observed for the anti-flag monolithic disk. The reason for this may be the convective mass transport that is faster than diffusional transfer [82].

TABLE 23.4

EFFECT OF FLOW RATE ON RECOVERY OF RECOMBINANT PROTEIN G FROM *E. COLI* CELLULAR LYSATES

Flow rate ml/min	Loading volume, ml	C^a mg/ml	Q^b mg/ml disk
0.5	7	0.32	0.68
		0.17	0.41
2.0	15	0.32	0.76
		0.17	0.41
5.0	35	0.32	0.78
		0.17	0.38
10.0	55	0.32	0.74
		0.17	0.38

^{a} Concentration of *r*Prot G in lysate

^{b} Desorbed amount of *r*Prot G.

Conditions: The initial concentration of *r*Prot G was estimated using ELISA. Cellular lysates were diluted (1:3), filtered and pumped through a CIM$^®$-*h*IgG disk to afford the breakthrough curve [75]. For other conditions see Table 23.2.

23.6.1.4 Temperature

The effect of temperature in the range of 0–40°C on the immunoaffinity binding was studied for the pair *r*Prot G–*h*IgG [75]. Experiments like these are rather rare in the field of affinity adsorption [83]. Table 23.5 shows that K_{diss}, characterizing the thermodynamic stability of affinity complexes, depends less on temperature whereas the adsorption capacity, Q_{max}, passes through a maximum located at 15–20°C. This suggests that the mechanism of strong affinity protein–protein pairing is more complex than that typical of weak pseudoaffinity interactions. The low adsorption capacity at 40°C cannot be explained by denaturation because no irreversible changes of the affinity adsorbent were observed even after long-term use.

23.6.1.5 Spacer

Proteins can be immobilized directly on poly(glycidyl methacrylate-*co*-ethylene dimethacrylate) disks using the nucleophilic addition of amino groups of the ligand to

TABLE 23.5

EFFECT OF TEMPERATURE ON AFFINITY PAIRING

Temperature °C	*r*Prot G (immob.) – *h*IgG (solub.)	
	Q_{max}, mg/ml disk	K_{diss}, µmol/L
0	1.29	0.25
15	1.85	0.33
20	2.00	0.37
40	0.91	0.24

Conditions: Breakthrough curves were obtained by pumping 0.05–0.5 mg/ml *h*IgG solutions through CIM®-*r*Prot G disks at a flow rate of 2 ml/min [75]. K_{diss} and Q_{max} values are averages calculated from linearized forms of the Langmuir equation. For other conditions see Table 23.2.

the epoxide functionalities of the matrix [32,68-71,74,75]. Korol'kov *et al.* [71] immobilized specifically prepared synthetic peptides with biological activity imitating that of the immunoglobulin binding sites of various proteins and used them as selective ligands instead of native proteins. They noted that the immobilization of peptide ligands could be carried out even without a spacer which is usually recommended in conventional affinity chromatography [84]. It is likely that the –COO–CH$_2$–CH(OH)––CH$_2$– segment that is formed by the opening of the epoxide ring of the glycidyl functionality plays the role of a spacer. The length of this bridge (about 0.9 nm) appears sufficient to facilitate the optimal interaction of the peptide ligand with an antibody.

The current literature suggests a number of linkers that can be inserted between the solid surface of the stationary phase and the immobilized ligand [51,85,86]. We have studied the effect of the spacer on immunoaffinity binding in more detail [76]. Comparing the isolation of antibodies against BSA and bradykinin using CIM® disks with protein and peptide ligands immobilized without any spacer, or utilizing a specifically synthesized nonapeptide or tripeptide as a linker, we were able to calculate and compare the affinity constants for all the interactions involved. Although these dissociation constants calculated from the adsorption isotherms are apparent since their magnitude is affected by the adsorption of different antibodies (monospecific + cross reactive), this quantitative approach is suitable for the comparative study. Table 23.6 shows that the dissociation constants of affinity complexes reveal no effect, negative or positive, of either of the spacers. Table 23.7 demonstrates that the experimental

TABLE 23.6

EFFECT OF PEPTIDE SPACERS ON AFFINITY CHARACTERISTICS (K_{diss} AND Q_{max}) OF DISK-IMMOBILIZED LIGAND–ANTIBODY PAIRS

Ligand	Spacer	Immobilized ligand mg/mL disk	μmol/mL disk	Q_{max} mg/mL disk	K_{diss} μmol/L
Bovine serum albumin	none	5.9	0.09	4.2	2.4
	GGG	2.9	0.04	1.4	3.8
	GVVKNNFVP	1.5	0.02	0.6	3.0
Bradykinin	none	5.0	4.7	1.4	2.6
	GGG	1.5	1.4	0.4	1.5
	GVVKNNFVP	1.2	1.1	0.2	1.2

Conditions: flow rate 2 mL/min; detection 278 nm; concentration range of 0.05–1 mg/mL [76].

TABLE 23.7

EFFECT OF FLOW RATE IN IMMUNOAFFINITY HPMDC ON K_{diss} VALUES MEASURED FOR BSA–*anti*-BSA ABS COMPLEX [75]

Flow rate, mL/min	0.5	2	6	10
K_{diss}, μmol/L	3.0	2.4	3.1	3.1

For conditions see Table 23.6

K_{diss} values, that characterize the thermodynamic stability of affinity complexes, are not even affected by the flow rate used during the elution step. Also, adsorption of both antibodies and model protein (BSA) does not occur on the disks modified with the nonapeptide or tripeptide spacers with no affinity ligands attached.

23.6.2 Multifunctional fractionation

Studies concerning multifunctional chromatography using conventional HPLC such as the application of several columns with different interacting functionalities combined in series are rare [87]. This is most likely due to the complexity of implementation that would require an automated system including switching among various pumps and columns.

One of the significant advantages of HPMDC is the possibility of simultaneously using several disks with different functionalities placed in a single cartridge. A term "conjoint liquid chromatography" (CLC) has recently been coined for this operational mode [49]. This approach enables separation and purification processes that are difficult to achieve using standard methods. We have developed a method that reaches beyond the procedures published earlier since it enables fractionation of a pool of polyclonal antibodies in *a single step* using conjoint immunoaffinity HPMDC [76]. Using this method, we separated a pool of polyclonal antibodies obtained by immunization of rabbits with the covalent conjugate of bradykinin–succinylated bovine serum albumin (BK–BSA-S) used as a model for complex natural mixtures [69]. Such pools typically contain both antibodies monospecific against each part of the conjugate used for immunization, and some so-called "cross-reactive" antibodies that have epitopes for complementary binding to all parts of the complex antigen [88,89].

To achieve a well controlled fractionation of this polyclonal pool of antibodies, each structural part of the antigen, *i.e.*, bradykinin, bovine serum albumin, succinylized bovine serum albumin (BSA-S), and the complete conjugate (BSA-S-BK) were immobilized on individual disks (Fig. 23.1). These affinity disks were installed in a single housing and adsorption experiments were carried out. The quantitative results are shown in Table 23.8. The simplicity of the commercially produced cartridge allows easy rearrangement of the sequence of the disks within the stack, as well as re-insertion of individual disks separately for subsequent desorption.

Table 23.9 shows the fractionation of the serum using individual disks with immobilized on them the different components of the complex immunogen, as well as the results obtained using the procedure that includes the stack of disks. The adsorption capacity of the individual disks appears to be rather different from that of the stacked set. The reason for this is the adsorption of cross-reactive antibodies with affinities close to the specific antigen, that together with monospecific antibodies contribute to the overall binding. Changing the sequence of the disks with different immobilized antigens can be used to detect these cross-reactive antibodies. For this, the disk installed at the top of the stack adsorbs both its "own" and cross-reactive immunoglobulins. Changing the order of disks in the cartridge then makes it possible to separate quantitatively all types of antibodies from the whole serum fraction. This method is highly reproducible and enables quantification of antibodies in the blood serum. Figure 23.2 compares two affinity chromatograms obtained at different flow rates in order to demonstrate the time-scale of the process. A single disk with bound BSA was used for the isolation of anti-BSA- and cross-reactive immunoglobulins fractions from blood serum.

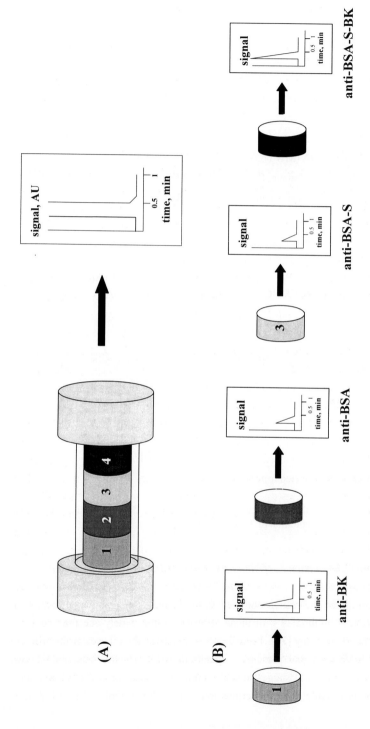

Fig. 23.1. Schematic of multifunctional HPMDC immunoaffinity fractionation (reprinted with permission from ref. [76], copyright 2002 Elsevier Sciences B. V.). (A) Single-step specific adsorption; disk 1 contains immobilized bradykinin (BK); disk 2, bovine serum albumin (BSA); disk 3, succinylated BSA (BSA-S); and disk 4, covalent BSA-S-BK conjugate. (B) Desorption of monospecific antibodies from each single disk following disassembly of the stack. Chromatographic conditions: Mobile phase: adsorption; 0.01 mol/L phosphate buffer, containing 0.15 mol/L NaCl, pH 7.0; desorption 0.01 mol/L HCl (pH 2.0).

TABLE 23.8

IMMOBILIZATION OF AFFINITY LIGANDS [76]

Ligand	Immobilized amount	
	mg/mL disk	μmol/mL disk
Bradykinin, BK (MW 1,060)	1.5	1.40
Bovine serum albumin, BSA (MW 66,000)	3.0	0.05
Succinylated bovine serum albumin, BSA-S (MW 72,000)	3.0	0.04
Bradykinin–bovine serum albumin conjugate, BK–S–BSA (MW 92,000)	7.4	0.08

TABLE 23.9

SERUM FRACTIONATION USING DISKS WITH IMMOBILIZED ANTIGENS

Antibody	Q_{Abs}, μg/mg serum [a]	
	Monospecific	Monospecific + cross-reactive
Anti-BK	21	61
Anti-BSA	13	53
Anti-BSA-S	14	55
Anti-BSA-S-BK	58	100

[a] Quantity of antibodies recovered from serum.

Conditions: Flow rate 2 mL/min; detection 278 nm; loading 500 μL serum precipitated fraction with a protein concentration of 2 mg/mL [76].

23.6.2.1 Comparison of immunoaffinity assays and ELISA

We also compared the results obtained from the multifunctional fractionation of pools of polyclonal antibodies with those afforded by the widely accepted enzyme-linked immunosorbent assay [76]. Table 23.10 summarizes the recoveries for different monospecific antibodies and presents a quantitative relationship between the two methods. These data clearly suggest that the multifunctional approach to immunoaffinity HPMDC is suitable for the fractionation of polyclonal pools. The ELISA tests

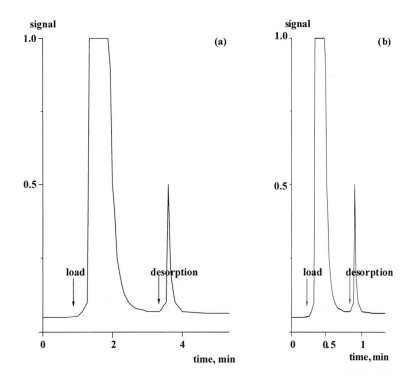

Fig. 23.2. Isolation of monospecific anti-BSA antibodies using immunoaffinity HPMDC at different flow rates (Reprinted with permission from ref. [76], copyright 2002 Elsevier Sciences B. V.). Conditions: stationary phase; CIM® BSA-disk with BSA immobilized via an intermediate nonapeptide GVVKNNFVP spacer: mobile phase; buffer A, 0.01 mol/L sodium phosphate buffer, pH 7.0, containing 0.15 mol/L NaCl; buffer B, 0.01 mol/L HCl, pH 2.0: loading; 500 µL serum solution with protein concentration of 2 mg/mL: detection 278 nm: flow rate 1 mL/min (a), and 6 mL/min (b).

also demonstrate that the antibodies isolated by HPMDC keep their biological activity while their purity is more than ten times better. It may not be completely correct to directly compare these two methods since ELISA is more sensitive and allows a large number of parallel assays to be run simultaneously. However, in contrast to the indirect ELISA method, the HPMDC separations enable the direct determination of antibodies or antigens even in complex biological mixtures such as crude blood serum. In addition, the high speed of the HPMDC permits many separations to be performed in series within a short period of time.

TABLE 23.10

COMPARISON OF FRACTIONATIONS OF MONOSPECIFIC ANTIBODIES FROM SERUM, ACHIEVED BY IMMUNOAFFINITY HPMDC AND ENZYME-LINKED IMMUNOSORBENT ASSAY

Antibody	IA HPMDC	ELISA titer	
	µg/mg serum Ig-fraction	serum	eluate
Anti-BK	21	1,300 – 2,600	—
Anti-BSA	13	1,100 – 2,200	46,400 – 92,800
Anti-BSA-S	14	1,050 – 2,100	—
Anti-BSA-S-BK	58	4,400 – 8,800	—

Conditions: flow rate 2 mL/min; detection 278 nm; loading 500 µL serum precipitated fraction with a protein concentration of 2 mg/mL. ELISA titers were determined according to the literature procedure [4]. Ten subsequent HPMDC runs and ten corresponding ELISA tests were carried out with standard deviations of ± 5% and ± 15%, respectively [76].

23.7 CONCLUSIONS

Dynamic modes of immunoaffinity assays using macroporous ultra-short mono-liths as stationary phase are a very promising approach. Their application appears to provide an attractive tool for fast and highly selective analyses of antibodies and antigens in diluted, complex solutions, as well as reliable means to gain extended knowledge concerning the behavior of biological molecules.

23.8 REFERENCES

1   P.G. Schultz, R.A. Lerner, S.J. Benkovic, Chem. Eng. News 68 (1990) 26.
2   R.D. Davies, E.A. Padlan, S. Sheriff, Ann. Rev. Biochem. 59 (1990) 439.
3   D.S. Hage, Clin. Chem. 45 (1999) 593.
4   T.T. Ngo, H.M. Lenhoff (Eds); Enzyme-mediated Immunoassay, Plenum Press, New York, 1985.
5   D.H. Campbell, E. Luescher, L.S. Lerman, Proc. Nat. Acad. Sci. USA 37 (1951) 575.
6   M. Wilchek, T. Miron, J. Kohn, Meth. Enzymol. 104 (1984) 3.
7   K.E. Howell, J. Gruenberg, A. Ito, G.E. Palade, Prog. Clin. Biol. Res. 270 (1988) 77.
8   H. Ehle, A. Horn, Bioseparations 3 (1990) 47.
9   M.M. Rhemrev-Boom, M. Yates, M. Rudolph, M. Raedts, J. Pharm. Biomed. Anal. 24 (2001) 825.

10   R.L. Fahrner, G.S. Blank, Biotechnol. Appl. Biochem. 29 (1999) 109.

11   D.A. Dmitriev, Yu.S. Massino, M.B. Smirnova, O.L. Segal, E.V. Pavlova, G.I. Kolyaskina, A.P. Osipov, A.M Egorov, A.D. Dmitriev, Bioorg. Khim. (Russ.) 27 (2001) 265.

12   P.G. Schultz, Science 240 (1988) 426.

13   R.A. Lerner, S.J. Benkovic, P.G. Schultz, Science 252 (1991) 659.

14   D.P. Stites, J.D. Stobo, H.H. Fudenberg, J.V. Wells, Basic and Clinical Immunology, 5th Ed., Lange Medical Publications, Los Altos, CA, 1984.

15   R. Puchades, A. Maquieira, J. Atienza, A. Montoya, Crit. Rev. Anal. Chem. 23 (1992) 301.

16   R.F. Ritchie, C.A. Alper, J.A.Graves, Arthritis Rheum. 12 (1969) 693.

17   H. Kawaguchi, Prog. Polym. Sci. 25 (2000) 1171.

18   M. Itoh, G. Kominami, J. Immunoassay Immunochem. 22 (2001) 213.

19   E. Engvall, P. Perlmann, Immunochemistry 8 (1971) 871.

20   K. Mikai, in: C.P. Price, D.J. Newman (Eds.), Principles and Practices of Immunoassay, M. Stockton Press, Basingstoke, Great Britain, 1991, p. 246.

21   G.A. Robbinson, Biosen. Bioelectr. 6 (1991) 183.

22   V. Regnault, J. Arvieux, L. Vallar, T. Lecompte, J. Immunol. Meth. 211 (1998) 191.

23   K. Nord, O. Nord, N. Uhlen, B. Kelley, C. Ljungqvist, P.A. Nygren, Eur. J. Biochem. 268 (2001) 4269.

24   J. Ruzicka, E.H. Hansen, in: J.D. Winedfordner (Ed.), Flow Injection Analysis, 2nd Edition, Wiley, New York, 1988.

25   M. Reineke, T. Scheper, J. Biotechnol. 59 (1997) 145.

26   M. Nilsson, G. Mattiasson, B. Mattiasson, J. Biotechnol. 31 (1993) 381.

27   M.A. Johns, L.K. Rosengarten, M. Jackson, F.E. Regnier, J. Chromatogr. A 743 (1996) 195.

28   M.F. Katmeh, A. J.M. Godfrey, D. Stevenson, G.W. Aherne, Analyst 122 (1997) 481.

29   C.H. Pollema, J. Ruzicka, G.D. Christian, A. Lernmark, Anal. Chem. 64 (1992) 1356.

30   C.H. Pollema, J. Ruzicka, Anal. Chem. 66 (1994) 1825.

31   K. Amatschek R. Necina, R. Hahn, E. Schallaun, H. Schwinn, D. Josic, A. Jungbauer, J. High Resol. Chromatogr. 23 (2000) 47.

32   J. Hagedorn, C. Kasper, R. Freitag, T. Tennikova, J. Biotechnol. 69 (1999) 1.

33   D.K. Roper, E.N. Lightfoot, J. Chromatogr. A 702 (1995) 3.

34   R. Arshady, J. Chromatogr. 586 (1991) 181.

35   Y.D. Clonis, Bio/Technology 5 (1987) 1290.

36   S. Ohlson, L. Hansson, M. Glad, K. Mosbach, P.O. Larsson, Trends Biotechnol. 7 (1989) 179.

37   N.B. Afeyan, N.F. Gordon, I. Mazaroff, L. Varady, Y.B. Yang, S.P. Fulton, F.E. Regnier, J. Chromatogr. 519 (1990) 1.

38   N.F. Gordon, D.H. Whitney, T.R. Londo, T.K. Nadler, Meth. Mol. Biol. 147 (2000) 175.

39   Z. Yan, J. Huang, J. Chromatogr. B 738 (2000) 149.

40   E. Klein, Affinity Membranes, Wiley, New York, 1991.

41   B. Champluvier, M.-R. Kula, J. Chromatogr. 539 (1991) 315.

42   P. Langlotz, K.H. Kroner, J. Chromatogr. 591 (1992) 107.

43   M. Nachman, J. Chromatogr. 597 (1992) 167.

44   C. Kasper, O.-W. Reif, R. Freitag, Bioseparations 6 (1997) 373.

45   E. Klein, J. Membr. Sci. 179 (2000) 1.

46   H. Zou, Q. Luo, D. Zhou, J. Biochem. Biophys. Meth. 49 (2001) 199.

47   T.B. Tennikova, B.G. Belenkii, F. Svec, J. Liq. Chromatogr. 13 (1990) 63.

48   T.B. Tennikova, B.G. Belenkii, F. Svec, M. Bleha, US Pat. 4,889,632 (1989); 4,923,610 (1990); 4,952,349 (1990).

49   A. Štrancar, M. Barut, A. Podgornik, P. Koselj, D. Josic, A. Buchacher, LC-GC Int. 11 (1998) 660.

50   T.B. Tennikova, R. Freitag, J. High Resol. Chromatogr. 23 (2000) 27.

51   J. Turková, in: H.Y. Aboul-Enein (Ed.), Analytical and Preparative Separation Methods of Biomacromolecules, Marcel Dekker, Inc, New York - Basel, 1999, p. 109.

52   R.F. Taylor, Protein Immobilization: Fundamentals and Application, Marcel Dekker Inc., New York, 1991.

53   R. Axen J. Porath, S. Ernback, Nature 214 (1967) 1302.

54   P. Cuatrecasas, I. Parikh, Biochemistry 11 (1972) 2291.

55   A.P. van Sommeren, P.A. Machielsen, W.J. Schielen, H.P. Bloemers, T.C. Gribnau, J. Virol. Meth. 63 (1997) 37.

56   G.S. Bethell, J.S. Ayers, W.S. Hancock, M.T.W. Hearn, J. Biol. Chem. 254 (1979) 2572.

57   C.J. Sanderson, D.V. Wilson, Immunology 20 (1971) 1061.

58   K. Hoffman, D.J. O'Shannessy, J. Immunol. Meth. 112 (1988) 113.

59   L. Sundberg, J. Porath, J. Chromatogr. 90 (1974) 87.

60   A.R. Neurath, N. Strick, J. Virol. Meth. 3 (1981) 155.

61   J. Turkova, J. Chromatogr. B 722 (1999) 11.

62   W. Clarke, J.D. Beckwith, A. Jackson, B. Reynolds, E.M. Karle, D.S. Hage, J. Chromatogr. A 888 (2000) 13.

63   B. Lu, J. Xie, C. Lu, C. Wu, Y. Wei, Anal. Chem. 67 (1995) 83.

64   L. Bjorck, G. Kronvall, J. Immunol. 133 (1984) 969.

65   M. Nisnewitch, M.A. Firer, J. Biochem. Biophys. Meth. 49 (2001) 467.

66   A. Podgornik, T.B. Tennikova, in: T. Scheper, R. Freitag (Eds.), Advances in Biochemical Engineering & Biotechnology, Springer Ferlag, 2002, p. 165.

67   D. Josic, Y-P. Lim, A. Strancar, W. Reutter, J. Chromatogr. 662 (1994) 217.

68   C. Kasper, L. Meringova, R. Freitag, T. Tennikova, J. Chromatogr. 798 (1998) 65.

69   G.A. Platonova, G.A. Pankova, I. Ye. Il'ina, G.P. Vlasov, T.B. Tennikova, J. Chromatogr. A, 852 (1999) 129.

70   L.G. Berruex, R. Freitag, T.B. Tennikova, J. Pharm. Biomed. Anal. 24 (2000) 95.

71   V.I. Korol'kov, G.A. Platonova, V.V. Azanova, T.B. Tennikova, G.P. Vlasov, Lett. Pept. Sci. 7 (2000) 53.

72   M. Schuster, E. Wasserbauer, A. Neubauer, A. Jungbauer, Bioseparations 9 (2001) 259.

73   M. Barut, A. Podgornik, H. Podgornik, A. Štrancar, D. Josic, J.D. MacFarlane, Amer. Biotechnol. Lab. 17 (1999) 11.

74   L.F. Meringova, G.F. Leont'eva, T.V. Gupalova, T.B. Tennikova, A.À. Totolian, Biotechnologia (Russ.) 4 (2000) 45.

75   N.D. Ostryanina, O.V. Il'ina, T.B. Tennikova, J. Chromatogr. B 770 (2002) 35.

76   N.D. Ostryanina, G.P. Vlasov, T.B. Tennikova, J. Chromatogr. A 949 (2002) 163.

77   T.V. Gupalova, O.V. Lojkina, V.G. Palagnuk, A.A. Totolian, T.B. Tennikova, J. Chromatogr. A 949 (2002) 185.

78   J.A. Gerstner, R. Hamilton, S.M. Cramer, J. Chromatogr. A 596 (1992) 173.

79   F.T. Sarfert, M.R. Etzel, J. Chromatogr. 764 (1997) 3.

80   K.-G. Brief, M.-R. Kula, Chem. Eng. Sci. 47 (1992) 141.

81   A. Einhauer, A. Jungbauer, J. Biochem. Biophys. Meth. 49 (2001) 455.

82   D. Josic, A. Buchacher, A. Jungbauer, J. Chromatogr. B 752 (2001) 191.

83   G.M.S. Finette, Q.-M. Mao, M.T.W. Hearn, J. Chromatogr. 763 (1997) 71A.

84   M. Petro, F. Svec, J.M.J. Fréchet, Biotechnol. Bioeng. 49 (1996) 355.

85   G.T. Hermanson, A.K. Mallia, P.K. Smith, Immobilized Affinity Ligand Techniques, Academic Press, New York, 1992.

86   N. Labrou, Y.D. Clonis, J. Biotechnol., 36 (1994) 95.

87   D.S. Hage, R.R. Walters, J. Chromatogr. 386 (1987) 37.

88   J.P. Briand, S. Muller, M.H.V. van Regenmortel, J. Immunol. Meth. 78 (1985) 59.

89   H.J. Geerligs, W.J. Wejler, G.W. Welling, S. Welling-Wester, J. Immunol. Meth. 124 (1989) 95.

F. Švec, T.B. Tennikova and Z. Deyl (Editors)
Monolithic Materials
Journal of Chromatography Library, Vol. 67
© 2003 Elsevier Science B.V. All rights reserved.

Chapter 24

Survey of Chromatographic and Electromigration Separations

Ivan MIKŠÍK and Zdeněk DEYL

Institute of Physiology, Academy of Sciences of the Czech Republic, Vídeňská 1083, 142 20 Prague 4, Czech Republic

CONTENTS

24.1 INTRODUCTION

As indicated in a number of places throughout this book, monolithic sorbents in their wide diversity have brought new dimensions to both chromatographic and (perhaps even more) to electromigration separations. While electromigration separations are used preferably on the analytical scale, various chromatographic operation modes (typically membrane separations) can be exploited also for preparative purposes, in particular if bioaffinity separation mechanisms are used. As with any developing technique the first applications are always directed towards complex artificial test samples, while the practical aspects and applications to naturally occurring mixtures will surely come at a later stage. Nevertheless, the separations of test mixtures show clearly the potential of this methodology and pave the way for its practical use in the future. As even detailed information about working procedures is unlikely to save the reader having to consult the original papers, we have decided to present an overview in a (rather extensive) tabular form (Table 24.1), and then to give some general summarizing comments.

24.2 HYDROCARBONS

Hydrocarbons, in particular the aromatic ones, are frequently used as test mixtures for monolithic packings: naphthalene, chlorobenzene and a number of polycyclic hydrocarbons were successfully separated by CEC on reversed-phases monolithic sorbents (C8, C18) [3-5], frequently sol-gel bonded, using acetonitrile and different proportions of aqueous phase (typically Tris or MOPS buffers, the proportion of which in the eluent usually does not exceed 10 %) as the mobile phase. For purely chromatographic separations (without the involvement of the electrodriven flow) C3 or C18 monoliths with acetontrile–Tris buffer, acetonitrile–water or methanol–water mobile phases were used for the separation of alkylbenzenes (an extensive study on this subject can be found in ref. [6]). Tetraethoxysilane–*n*-octyltriethoxysilane hybrid gels also appear applicable for CEC separations of this category of compounds [10]. Further, aromatic hydrocarbons like toluene and its homologues can also be separated on methacrylate-based monoliths, typically methyl methacrylate or butyl methacrylate–ethylene dimethacrylate copolymers [16,17,20,21]. These separations are generally electrokinetic in rather long capillaries (up to 1 m), and exploited phosphate- or borate buffers of pH 7.0–9.2 containing specific proportions of acetonitrile as background electrolyte. In order to ensure adequate electroosmotic flow, 2-acrylamido-2-methyl-1-propanesulfonic acid was incorporated in the monolithic packing.

Reversed-phase-based separations can be effected not only with hydrophobized silica based monoliths but also with 2-hydroxyethyl methacrylate–piperazine diacrylamide copolymers possessing C18 ligands, as reported in ref. [8].

Styrene–divinylbenzene copolymers are also applicable for HPLC and CEC separations with acetonitrile–water mobile phases [2,15].

TABLE 24.1

APPLICATIONS (compositions of mobile phases are described using volume proportions, *i.e.*, *v/v*; CEC – capillary electrochromatography; HPLC – high performance liquid chromatography; HPMC – high performance monolithic disk chromatography

Compound	Separation mode and detection	Mobile phase/background electrolyte	Column dimensions	Note	Ref.
Hydrocarbons					
Naphthalene 2-methyl-naphthalene, fluorene, phenanthrene anthracene	CEC. Methacrylate-piperazine diacrylamide copolymer with C18 ligands	60% Acetonitrile in 4 mmol/L sodium phosphate, pH 7.4, 1.0 mmol/L SDS	Capillary 14/10 cm (effective length) × 100 μm i.d., 3.0 kV		[1]
Chlorobenzenes (2,4-dinitro-chlorobenzene, chlorobenzene, 1,4-dichlorobenzene, 1,2,4-trichlorobenzene)	CEC. Poly(styrene-*co*-divinyl-benzene-*co*-methacrylic acid). UV detection at 200 nm	Acetonitrile–water (80:20, v/v) containing 4 mmol/L Tris, pH 8.82	Capillary 27/20 cm (effective length), 75 μm i.d., 30 kV		[2]
Alkylbenzenes (benzene, toluene ethylbenzene, propyl-benzene, butylbenzene)	CEC. Poly(styrene-*co*-divinyl-benzene-*co*-methacrylic acid). UV detection at 200 nm	Acetonitrile–water (80:20, v/v) containing 4 mmol/L Tris, pH 8.82	Capillary 27/20 cm (effective length), 75 μm i.d., 20 kV		[2]
Polycyclic aromatic hydrocar-bons in coal tar	CEC. Sol-gel bonded ODS (3 μm, 80 Å), UV detection at 254 nm	Acetonitrile–50 mmol/L Tris buffer, pH 8.0 (90:10)	Capillary 39/30 cm (effective length), 75 μm i.d., 20 kV		[3]
Polynuclear hydrocarbons (phenanthrene, naphthalene, pyrene)	CEC. Particle loaded (3 μm octadecyl silica) monolithic sol-gel column	Acetonitrile–25 mmol MOPS (morpholino ethanesulfonic acid) (95:5) pH 6.2	Capillary 30/10 cm (effective length), 75 μm i.d., 5–30 kV		[4]

Continued on the next page

TABLE 24.1 (continued)

Compound	Separation mode and detection	Mobile phase/background electrolyte	Column dimensions	Note	Ref.
Polynuclear hydrocarbons (naphthalene, phenanthrene, pyrene)	CEC. Particle loaded (3 μm octadecyl silica) monolithic sol-gel column	Acetonitrile–25 mmol MOPS (morpholino ethanesulfonic acid) (95:5) pH 6.2	Capillary 30/10 cm (effective length), 75 μm i.d., pressure driven (10–100 psi)		[4]
Butylbenzene, *o*-terphenyl, amylbenzene, triphenylene	HPLC. Purospher RP 18e, 5 μm and silica rod column RP 18e, UV detection at 254 nm	Methanol–water (80:20, v/v)	Column 125 × 4 mm for Purospher RP 18e or 100 × 4.6 mm for the silica rod column RP 18e	Comparison of conventional packed and monolithic column	[5]
Alkylbenzenes ($C_6H_5C_nH_{2n+1}$, n= 0–6)	HPLC. a) Inertsil ODS-3 (5 μm particles), packed b) SR-PEEK-(A) column (C-18 monolith) c) SR-FS-(A) column (C-18 monolith)	Methanol–water (80:20, v/v)	Columns a) 150 × 4.6 mm b) 100 × 4.6 mm c) 335/250 mm (effective length) × 100 μm i.d.	Flow rates a) 0.8 or 4.9 mm/s b) 1.2 or 4.6 mm/s c) 1.0 or 5.1 mm/s	[6]
Alkylbenzenes ($C_6H_5C_nH_{2n+1}$, n= 0–6)	HPLC and CEC. Comparison of two types of SR-FS columns, C-18 monoliths	A) Acetonitrile–water (90:10, v/v) B) Acetonitrile–50 mmol Tris-HCl, pH 8 (90:10, v/v)	A) 33.5/25 cm (effective length) 100 μm i.d. B) 33.5/25 cm (effective length)	A) 1.9–2.9 kg/cm^2 overpressure depending on the column type used B) 750–900 v/cm depending on the column type used	[6]

Analytes	Conditions	Mobile phase	Capillary	Notes	Ref.
Alkylbenzenes (toluene, ethylbenzene, propylbenzene, butyl benzene)	CEC. 40% ethyl acrylate, 50% methacrylic acid, 10% lauryl methacrylate, 4% polymer concentration	Acetonitrile–water (40:60, v/v), pH 9.1	Capillary 50/42 cm (effective length), 50 μm i.d.	Replaceable media	[7]
Polyaromatic hydrocarbons	CEC. 40% ethyl acrylate, 50% methacrylic acid, 10% lauryl methacrylate, 3.99–4.01 polymer concentration	Acetonitrile–10 mmol/L sodium borate buffer or 10 mmol/L phosphate (40:60, v/v), pH 9.2	Capillary 30/25 cm (effective length), 50 μm i.d.	Replaceable media	[7]
Aromatic hydrocarbons (naphthalene, 2-methylnaphthalene, fluorene, phenanthrene, anthracene)	CEC. 2-hydroxyethyl methacrylate-piperazine diacrylamide copolymer with C18 ligands and immobilized dextran sulfate	Acetonitrile–5 mmol/L borate (62:38, v/v), pH 8.7	Capillary 55/52.5 cm (effective length), 25 μm i.d.		[8]
Polyaromatic hydrocarbons (2-methylnaphthalene, fluorene, phenanthrene, anthracene)	CEC. 2-hydroxyethyl methacrylate-piperazine diacrylamide copolymer with C18 ligands	50% Acetonitrile in 4 mmol/L sodium phosphate pH 7.4 (60%); Acetonitrile–buffer (60:40) or gradient from 60:40 to 70:30 acetonitrile–buffer	16/20 cm (effective length), 75 μm i.d.		[8]
Benzene derivatives (benzyl alcohol, benzaldehyde, benzene, toluene, ethylbenzene, propylbenzene, acrylbenzene)	CEC. 0.3 % 2-acrylamido-2-methyl-1-propanesulfonic acid, pore size 405 nm	Acetonitrile–4 mmol/L phosphate buffer, pH 7.0 (80:20)	25/25 cm (active length), 100 μm i.d.		[9]
Aromatic hydrocarbons (naphthalene, phenanthrene, pyrene)	CEC. Tetraethoxysilane-*n*-octyltriethoxysilane hybrid sol-gels	Acetonitrile–water (1:1, v/v)	47/40 cm (effective length)		[10]

Continued on the next page

TABLE 24.1 (continued)

Compound	Separation mode and detection	Mobile phase/background electrolyte	Column dimensions	Note	Ref.
Aromatic hydrocarbons, benzyl alcohol, benzaldehyde, benzophenone, biphenyl, benzene, toluene, ethylbenzene, butylbenzene	CEC. Particle-loaded 3 μm ODS/SCX sol-gel column	Acetonitrile–1.5 mmol/L phosphate buffer, pH 3.0	34/25 cm (effective length), 75 μm i.d.		[11]
Polynuclear aromatic hydrocarbons (naphthalene, phenanthrene, pyrene)	CEC. Particle loaded (3 μm, C18) monolithic sol-gel column	Acetonitrile–25 mmol/L MOPS (morpholinoethanesulfonic acid) (95:5, v/v), pH 6.2	30/10 cm (effective length), 75 μm i.d.		[12]
Polycyclic aromatic hydrocarbons (naphthalene, biphenyl, fluorene, phenanthrene, fluoranthene, *m*-terphenyl)	CEC. Sintered Zorbax ODS, 6 μm, monolithic packing, UV detection at 214 nm	Acetonitrile–10 mmol/L sodium borate, pH 8.0 (80:20 or 60:40)	33/23 (effective length), 75 μm i.d., 30 kV	The sorbent was re-octadecylated	[13]
Polyaromatic hydrocarbons (benzene, naphthalene, fluorene, phenanthrene, anthracene, fluoranthene, pyrene, benzo[a]pyrene)	CEC. Hybrid sol-gel ODS, UV detection (wavelength not specified)	Acetonitrile–5 mmol/L Tris-HCl, pH 2.34 (80:20)	50/46.1 cm effective length; 50 μm i.d., 12 kV		[14]
Alkylbenzenes	HPLC. Macroporous poly(styrene-*co*-divinylbenzene) rod	Acetonitrile–water (70:30); gradients of acetonitrile	50 × 8 mm i.d.		[15]

Analytes	Method	Mobile phase	Conditions	Ref.
Benzene derivatives (benzaldehyde, benzene, toluene, ethylbenzene, propylbenzene, butylbenzene, amylbenzene)	CEC. Hybrid sol-gel ODS, UV detection (wavelength not specified)	Acetonitrile–5 mmol/L Tris-HCl pH 2.34 (75:5)	50/46.1 cm effective length; 50 μm i.d., 15 kV	[14]
Benzylalcohol, benzaldehyde, benzene, toluene, ethyl-, propyl-, butyl-, amylbenzene	CEC. Ethylene dimethacrylate-butyl methacrylate containing 2-acrylamido-2-methyl-1-propanesulfonic acid. Polymerization by UV light, UV detection at 215 nm	Acetonitrile–5 mmol/L phosphate buffer, pH 7.0	50 cm (35 cm active length), 25 kV	[16]
Toluene, propylbenzene, amylbenzene, octylbenzene	CEC. Butyl methacrylate (60%), ethylene dimethacrylate (39.7%), 2-acrylamido-2-methyl-1-propanesulfonic acid, UV detection at 205 nm	20% sodium phosphate buffer, pH 7.0, 80% acetonitrile	1 m × 25 μm i.d. capillary column, 10 kV	[17]
Alcohols and oxo compounds				
Benzyl alcohol, benzaldehyde dimethyl phthalate, benzophenone, biphenyl	CEC. Sol-gel bonded ODS (3 μm, 80 Å), UV detection at 254 nm	Acetonitrile–water–50 mmol/L Tris buffer, pH 8.0 (80:10:10, v/v/v)	Capillary 39/30 cm (effective length), 20 kV	[3]
Acetophenone, valerophenone	CEC. Fritless packing, Sol-gel C18 silica. UV detection at 254 nm	Acetonitrile–50 mmol/L HEPES, pH 6.6	Capillary 45.8/25.0 cm (effective length), 75 μm i.d., 218 or 415 V/cm	[18]

Continued on the next page

TABLE 24.1 (continued)

Compound	Separation mode and detection	Mobile phase/background electrolyte	Column dimensions	Note	Ref.
Acetophenone, propiophenone, butyrophenone, 2',5'-dihydroxyacetophenone, 2',5'-dihydroxypropiophenone	CEC. Macroporous polyacrylamide matrix derivatized with C4 ligand	10 mmol/L Tris + 15 mmol/L boric acid (pH 8.2)–acetonitrile (80:20, v/v)	Capillary 25/20.5 cm (effective length), 100 μm i.d., 22.5 kV		[19]
Alkylphenones	CEC. 40% Ethyl acrylate, 50% methacrylic acid, 10% lauryl methacrylate, 4% polymer concentration	Acetonitrile–water (60:40), pH 9.1	50/42 cm (effective length), 50 μm i.d.	Replaceable gels	[7]
Benzyl alcohol, benzaldehyde, benzophenone, biphenyl	CEC. Sol-gel bonded 3 μm ODS/SCX, 80 Å pores	Acetonitrile–1.5 mmol/L phosphate buffer (80:20), pH 3.0	34/25 cm (effective length), 75 μm i.d.		[11]
Benzaldehyde, o-tolualdehyde, butyrophenone, valerophenone, hexaphenone, heptaphenone	CEC. Sol-gel ODS UV detection (wavelength not specified)	Acetonitrile–5 mmol/L Tris HCl buffer, pH 2.34 (70:30)	50/46.1 cm effective length, 50 μm i.d., 25 kV		[14]
Benzyl alcohol, benzyl ethanol, aniline, pyridine, benzyl propanol, benzaldehyde, benzonitrile, quinoline, peptides, phenylacetic acid, o-toluic acid, p-cresol	CEC. Butyl methacrylate-ethylene dimethacrylate copolymer prepared with a porogenic solvent	35% Acetonitrile in 5 mmol/L phosphate buffer, pH 7.5, containing 2.5 mmol/L cetyltrimethylammonium bromide	27/20 cm (effective length), 100 μm i.d., 20 kV		[20]

Analyte	Method	Mobile phase	Column	Notes	Ref.
Phthalates (with toluene as void volume marker)	HPLC. Monolithic sorbent preparation, see ref. [21]. Sol-gel polymers of alkoxysilanes, UV detection at 254 nm	Methylcyclohexane or *n*-heptane–ethylacetate	100 × 25 mm. Different columns (varying in diameter) used	Both preparative and analytical separations	[22]
Pyridine, pyridylmethanol, 4-methoxyphenol, 2-naphthol, catechol, hydroquinone, resorcinol, 2,7-dihydroxynaphthalene	HPLC. 63.2% N-*iso*-propylacrylamide, 10.5% methacrylamide, 26.3% piperazine diacrylamide, UV detection at 220 nm	Hexane–ethanol–methanol (A), methanol (B); gradient elution	125 mm × 100 μm i.d.		[23]
1-(1-Naphthalene)ethanol	CEC. Negatively charged polyacrylamide, carbamoyl β-cyclodextrin bonded, UV detection at 240 nm	200 mmol/L Tris–300 mmol/L boric acid buffer, pH 9.0	70/35 cm effective length, 75 μm i.d., 267 V/cm	Chiral separation	[24]
Phenols					
Methylated phenols and other derivatives (quinol, phenol, 2,4-dimethylphenol, 2,3,5-trimethylphenol)	CEC. Poly(styrene-*co*-divinylbenzene-*co*-methacrylic acid), UV detection at 20 nm	Acetonitrile–water (70:30) containing 4 mmol/L Tris, pH 8.82	Capillary 23/17 cm (effective length), 20 kV		[2]
Chloro- and nitro-phenols (2,4,6-trichlorophenol, 2,4-dinitrophenol, 4-chlorophenol, 2-chlorophenol, 2-methyl-4,5-dinitrophenol)	HPLC. Silica rod column RP 18e. Detection UV at 254 mm	Gradient of acetonitrile (A) and 0.02 mol/L phosphate buffer (B), pH 3.0. Gradient timing 0 min – 65% A, 35% B, 1.8 min – 54% A, 46% B, 2.5 min – 20% A, 80% B	Column 100 × 4.6 mm. Flow rate 0–2.5 min 3 mL/min; 2.5 min and later 5 mL/min		[5]

Continued on the next page

TABLE 24.1 (continued)

Compound	Separation mode and detection	Mobile phase/background electrolyte	Column dimensions	Note	Ref.
Carbohydrates					
Maltooligosaccharides, glucose(Glc1)–malto-hexaose(Glc6)	CEC. Macroporous polyacrylamide matrix derivatized with C4 ligand (15%), laser induced fluorescence detection	10 mmol/L Tris + 15 mmol/L boric acid (pH 8.2)–acetonitrile (80:20, v/v)	Capillary 32/25 cm (effective length) × 100 μm i.d., 900 V/cm	Oligosaccharides derivatized with 2-aminobenzamide	[19]
Carboxylic acids and esters					
Phthalic acid esters (diethyl-, dibutylphthalate), toluene	HPLC (preparative). Silica gel rods (Prep ROD) 3.98 μm, 140 Å, UV detection at 254 nm	n-Heptane–ethyl acetate (90:10, v/v)	Column 100 × 25 mm, 390 mL/min. Flow rate 5 to 390 mL/min	Preparative procedure	[25]
Alkyl benzoates	CEC. 40% Ethyl acrylate, 50% methacrylic acid, 10% lauryl methacrylate 0–3.75 wt % polymer concentration	Acetonitrile–water 40:60, pH 11.3	50/42 cm (effective length), 50 μm i.d.	Replaceable medium	[7]
Steroids					
Corticosteroids (ouabain, 4-pregnene-6b,11b-21-triol--3,20-dione, strophanthidin)	CEC. Sol-gel bonded ODS/SCX (3 μm, 80 Å), UV detection at 240 nm	Acetonitrile–water–50 mmol/L phosphate buffer, pH 3.0 (60:30:5, v/v/v), 25 kV	Capillary 34/25 cm (effective length), 75 μm i.d.		[3]

Compounds	Method	Mobile phase	Capillary	Remarks	Ref.
Steroids (neutral or conjugated) and their dansylhydrazine derivatives	CEC. Polyacrylamide gel with C12 ligands. Laser induced fluorescence detection or coupling with ion-trap mass spectrometry	Acetonitrile–water–240 mmol/L ammonium formate buffer pH 3 (55:40:5), also gradient elution	35/25 cm (effective length)		[26]
Hydrocortisone, prednisolone, hydrocortisone-21-acetate, testosterone	CEC. Poly(2-acrylamido-2--methyl-1-propanesulfonic acid-*co*-N-*iso*propylacryl-amide) hydrogel	100 mmol/L Tris–150 mmol/L boric acid, pH 8.1	65/50 cm (effective length), 75 μm i.d.		[27]
Amines					
Phenylene diamines (*o*-, *m*- and *p*-isomers)	CEC. Poly(styrene-*co*-divinyl-benzene-*co*-methacrylic acid), UV detection at 200 nm	Acetonitrile–water (70:30), containing 4 mmol/L Tris, pH 8.82	Capillary 23/17 cm (effective length), 20 kV		[2]
Anilines (*p*-nitroaniline, methyl-aniline, naphthylamine, diphenylamine)	CEC. Poly(styrene-*co*-divinyl-benzene-*co*-methacrylic acid), UV detection at 200 nm	Acetonitrile–water (70:30), containing 4 mmol/L Tris, pH 8.82	Capillary 23/17 cm (effective length), 20 kV		[2]
Anilines (aniline, N-methyl-aniline, N,N-dimethylaniline, N-propylaniline	CEC. Sol-gel bonded ODS (3 μm, 80 Å), UV detection at 254 nm	Acetonitrile–water–50 mmol/L Tris buffer, pH 8.0 (80:10:10, v/v/v)	Capillary 39/30 cm (effective length), 20 kV		[3]
1-(1-Naphthylethylamine), alanine 2-naphthylamide, tryptophanol α-methyl-tryptamine, 1,2-diphenyl-ethylamine	CEC. Negatively charged polyacrylamide, 18-crown-6-bonded, UV detection at 254 nm	300 mmol/L Boric acid–acetonitrile (50:50)	80/35 cm effective length, 75 μm i.d., 125 V/cm	Chiral separation	[28]

Continued on the next page

TABLE 24.1 (continued)

Compound	Separation mode and detection	Mobile phase/background electrolyte	Column dimensions	Note	Ref.
1-Phenylethylamine, alanine-2-naphthylamide, methionine-2-naphthylamide, 1-amino-indan, tryptophol, 2-amino-1,2-diphenylethanol, tryptophan methyl ester 1,2,3,4-tetrahydro-1-naphthylamine	CEC. Negatively charged polyacrylamide, 18-crown-6-bonded, UV detection at 254 nm	200 mmol/L Triethanolamine–300 mmol/L boric acid buffer pH 7.0; 20 mmol/L 18-crown-6	70/35 cm effective length, 75 μm i.d., 200 V/cm	Chiral separation	[28]
1-(1-Naphthyl)ethylamine	CEC. Negatively charged polyacrylamide, carbamoyl β-cyclodextrin bonded, UV detection at 240 nm	200 mmol/L Tris–300 mmol/L boric acid buffer, pH 7; alternatively containing 10 mmol/L 18-crown 6	62/35 cm effective length, 75 μm i.d., 242 V/cm	Chiral separation	[24]
Proteins, peptides and amino acids					
Amino acids, dansylated	CEC. β-CD bonded positively charged polyacrylamide gel, UV detection at 245 or 250 nm	200 mmol/L Tris, 300 mmol/L boric buffer pH 8.1	55-70/35 cm (effective length) × 75 μm i.d.; -269 to -286 V/cm	Sample concentration 2 mmol/L	[29]
PTH amino acids	CEC. Sintered Zorbax ODS, 6 μm, monolithic packing, UV detection at 214 nm	Acetonitrile–5 mmol/L sodium phosphate (30:70) pH 7.5	33/23 (effective length), 75 μm i.d., 15 or 20 kV	The monolith was re-octadecylated	[13]

Analyte	Method	Conditions	Mobile phase	Application	Ref.
N-(3,5-Dinitrobenzoyl) leucine dialkylamide enantiomers	CEC. Chiral monoliths containing (N-L-valine-3,5-dimethylanilide) carbamate as chiral selector	30 cm (effective length) × 100 μm i.d, 25 kV	80:20 (v/v) Mixture of acetonitrile and 5 mmol/L phosphate buffer pH 7.0	Chiral separation	[30]
Amino acids (4-fluoro-7-nitro-2,1,3-benzoxadiazole (NBD-F) derivatized)	CEC. Tetraethylorthosilicate with added silica particles (sol-gel packing). Fluorescence detection	30/15 cm effective length, 75 μm i.d., 0.83 kV/cm	5 mmol/L Phosphate buffer pH 2.5–acetonitrile (30:70)	Separation of enantiomers on D- and L-amino acid-modified sorbent	[31]
DL-Phenylalanine	CEC. Copolymers of methacrylamide, piperazine, diacrylamide, vinylsulfonic acid and N-(2-hydroxy-3-allyloxypropyl)-L-4-hydroxy-proline	26 cm effective length, 75 μm i.d., 30 kV	50 mmol/L Sodium dihydrogen phosphate–100 mmol/L Cu(II), pH 4.6	Chiral separation. Pressure-assisted separation, 1.2 MPa	[32]
DNB-(R,S)-Leucine	CEC. Monolith prepared from quinidine-functionalized chiral monomer, ethylene dimethacrylate and glycidyl methacrylate or 2-hydroxyethyl methacrylate UV detection 250 nm	33.5/25 cm effective length, 100 μm i.d., 25 kV	400 mmol/L Acetic acid and 4 mmol/L triethylamine in acetonitrile–methanol (80:20)	Chiral separation	[33]
DNZ-(R,S)-Leucine, DNP-Valine	CEC. Quinidine-functionalized chiral monoliths (different compositions)	33.5/25 cm effective length, 100 μm i.d., 25 kV	0.4 mol/L Acetic acid and 4 mmol/L triethylamine in acetonitrile–methanol (80:20)	Chiral separation	[34]

Continued on the next page

TABLE 24.1 (continued)

Compound	Separation mode and detection	Mobile phase/background electrolyte	Column dimensions	Note	Ref.
PTH peptides (BSA digest)	EC. Chip separation, collocated monolithic structure, Sylgard chip, poly(dimethylsiloxane) polymer, Fluorescence detection	1 mmol/L Carbonate pH 8.7	76 nl chip volume, 5 μm wide 10 × 10 μm monolith dimensions, 1000 V per chip		[35]
Gly-tyr, val-tyr-val, methionine enkephalin, leucine enkephalin	CEC. Ethylene dimethacrylate–butyl methacrylate copolymers containing 2-acrylamide-2--methyl-1-propanesulfonic acid. Polymerization by UV light, UV detection at 215 nm	Acetonitrile–5 mmol/L phosphate buffer, pH 7.0	50 cm (35 cm active length), 25 kV		[16]
Peptides (gly-tyr, val-tyr, methionine and leucine enkephalins)	CEC. Methylacrylate ester--based column prepared with a ternary porogen system (water–1-propanol–1,4-butanediol)	80% of a 1:9 mixture of 10 mmol/L sodium 1-octane-sulfonate and 5 mmol/L phosphate buffer pH 7.0 and 20% acetonitrile	Capillary 28 cm effective column length × 100 μm i.d.	Pore size 492 nm	[16]
Peptides (synthetic)	CEC. Macroporous polyacrylamide monolith derivatized with C12 ligands (29%); UV detection at 270 nm	47% Acetonitrile in 10 mmol/L Tris–15 mmol/L boric acid, pH 8.2	Capillary 32/25 cm (effective length) × 100 μm i.d., 22.5 kV (900 V/cm)	Sample concentration 4–10 mg/L	[19]

Analyte	Method	Mobile phase	Dimensions	Ref.
Peptides (angiotensin, insulin)	CEC. Porous styrenic monolith with dimethyloctylammonium functionalities, UV detection at 214 nm	25% Acetonitrile in 5 mmol/L phosphate buffer, pH 3.0 containing 50 mmol/L NaCl	Capillary 31/21 cm (effective length) × 75 µm i.d., 5 kV	[36]
Peptides (tryptic digest of ovalbumin)	EC. Microfabricated C18 colocated monolith support structures (COMOSS), 32 channels, 45 mm C18	10 mmol/L Phosphate buffer pH 9.0–acetonitrile, 3:1, v/v	Distribution of channels and their size not specified (most likely ~ 10 × 10 µm), 770 V/cm aligned in chip separation	[37]
Peptide mapping, ovalbumin tryptic digest	EC. C18 COMOSS microfabricated column	Acetonitrile–10 mmol/L potassium phosphate, pH 9.0 (1:3)	1.5 µm wide, 10 µm deep rectangular channel 4.5 cm long, 770 V/cm	[38]
Proteins (a mixture of standards consisting of soybean trypsin inhibitor, conalbumin and myoglobin)	High performance monolithic disk chromatography (HPMC), anion exchangers CIM disk	Gradient from 0–50% B in 6 s, then 50% B for 3 s. A: 20 mmol/L Tris-HCl, pH 7.4, B: 1 mol/L NaCl in A. Flow rate 10 mL/min	3 × 10 mm disc	[39]
Standard proteins	HPMC. Poly(glycidylmethacrylate-co-ethylene dimethacrylate)	Gradient elution 0–100% B in 30 s A: 10 mmol/L Tris-HCl, pH 7.4, B: 1 mol/L NaCl in buffer A. Flow rate 5 mL/min	3 × 25 mm disc	[39]

Continued on the next page

TABLE 24.1 (continued)

Compound	Separation mode and detection	Mobile phase/background electrolyte	Column dimensions	Note	Ref.
Standard proteins (cytochrome c, myoglobin, ovalbumin)	HPLC. Macroporous poly(styrene-*co*-divinyl-benzene)	Linear gradient 20–60% acetonitrile–water, flow rate 5, 10, 15, 25 mL/min	50 × 8 mm i.d.		[40]
Peptides, bradykinin and [D-Phe^7]-bradykinin	HPLC. Macroporous poly(styrene-*co*-divinyl-benzene) rod	Acetonitrile–water (16:84), 1 mL/min	50 × 8 mm i.d.		[15]
Standard proteins (myoglobin, conalbumin, ovalbumin, trypsin inhibitor)	High performance monolithic disk chromatography, UV detection at 280 nm	Gradient elution A: 10 mmol/L Tris-HCl, pH 7.6, B: 1 mol/L NaCl in buffer A flow rate 1 or 3 mL/min	CIM DEAE disk 2 × 25 mm	Different flow rates compared	[41]
Basic proteins (α-chymo-trypsinogen, ribonuclease, cytochrome C, lysozyme)	CEC. Polychloromethylstyrene-based PLOT column	20% Acetonitrile in 20 mmol/L phosphate buffer, pH 2.5	Capillary 47/40 cm (effective length) × 20 μm i.d., Inner polymer layer 2 μm, voltage, 30 kV		[42]
Ribonuclease A, cytochrome C, lysozyme, chymotrypsinogen	CEC. Polyacrylamide-based monolithic column with C18 ligands, UV detection at 280 nm, 5 MPa overpressure	Linear gradient from 5 to 80% acetonitrile in 5 mmol/L sodium phosphate, pH 2.0	Capillary 8/6 cm (effective length) × 50 μm i.d., 5.5 kV	Protein concentration 0.6 mg/mL	[43]

Sample	Method	Column	Conditions	Notes	Ref.
Annexins from Morris hepatoma 7777 plasma membranes	High performance monolithic disk chromatography, UV detection at 280 nm	Quick Disk DEAE 3 × 25 mm	Step-gradient A: 10 mmol/L Tris-HCl, pH 7.8, B: 1 mol/L NaCl in buffer A (both buffers contained 1 mmol/L EDTA). Flow rate 3 mL/min		[44]
Factor VIII	HPLC and HPMC. Epoxy activated Sepharose-immobilized peptides (Pharmacia) and CIM columns (BIA, Slovenia), Pro Sys workstation, FPLC system, absorbance at 280 nm	HR 5 × 0.5 mm column; elution at linear velocity 60 cm/h (for Pharmacia columns) and 160 cm/h (for CIM columns)	10 mmol/L HEPES pH 7.4 containing 5 mmol/L $CaCl_2$, 100 mmol/L NaCl and 0.1% Tween 20 for conditioning the column. Elution with 2 mol/L guanidine hydrochloride containing 5 mmol/L $CaCl_2$ and 0.1% Tween	Affinity chromatography on sorbents bearing peptides capable of specific interaction with F VIII	[45]
Myoglobin, conalbumin, soybean trypsin inhibitor	HPMC. Glycidyl methacrylate–ethylene dimethacrylate disks and tubes, UV absorbance at 280 nm		Gradient 0–100 B over 300 mL eluted volume. A: 20 mmol/L Tris-HCl, pH 7.4, B: 1 mol/L NaCl in buffer A		[46]
Myoglobin, conalbumin, soybean trypsin inhibitor	HPMC. CIM disks and tubes (BIA Separations)		A: 20 mmol/L Tris-HCl, pH 7.4; B: 1 mol/L NaCl in buffer A. Linear gradients over 39–180 s		[47]

Continued on the next page

References pp. 656-658

TABLE 24.1 (continued)

Compound	Separation mode and detection	Mobile phase/background electrolyte	Column dimensions	Note	Ref.
Glycoproteins (Factor IX)	HPMC. CIM disks and tubes	Disks: gradient 0–100% B Tubes: 55–100% B A: 20 mmol/L Tris-HCl buffer (pH 7.4) B: 20 mmol/L Tris-HCl buffer with 1 mol/L NaCl (pH 7.4)	0.36 mL monolithic matrix in the disk format or 8 mL monolith in the tube format	Separation from contaminating serum proteins. Short separation times.	[48]
Myoglobin, conalbumin, soybean trypsin inhibitor	HPMC. QA and DEAE CIM disk, UV detection at 280 nm	20 mmol/L Tris-HCl buffer (pH 7.4)	10 mm diameter, 3 mm thickness	Separation of Factor IX from contaminating serum proteins. Short separations (less than 15 s). The role of NaCl added to the run buffer was investigated.	[49], [39]
Nucleic acids, oligonucleotides and their constituents					
Oligo 8–14	DEAE CIM disk, UV detection at 260 nm	Gradient elution A: 20 mmol/L Tris-HCl, pH 8.5; B: 1 mol/L NaCl in A, flow rate 5 mL/min	3 × 10 mm disk	Comparison of gradient- and isocratic elution	[50]

Oligo 8–14	DEAE CIM disk (glycidyl methacrylate-ethylene dimethacrylate monoliths), UV detection at 260 nm	Different Tris-HCl buffers containing NaCl, pH 8.5. Tris-HCl 20 mmol/L, NaCl 0.37–0.48 mmol/L. Flow rates ranging from 3 to 9 mL/min	3×10 mm disk	Examination of the effect of monolithic disc thickness on separation	[51]
Double-stranded DNA	CGE in microchip. Photopolymerized polyacrylamide gel 8% (7%) acrylamide-bisacrylamide (REPRO gel formation), 4040-1 labelled fragments (intercalated dyes); fluorescence detection	Tris–borate–EDTA, BioRad buffer	Rectangular channel 400×40 μm, 20 V/cm	Sample stacking by programmed voltage	[52]
Oligodeoxyadenylic acids, oligothymicylic acids	HPLC. Poly(glycidyl methacrylate-*co*-ethylene dimethacrylate), UV detection at 260 nm	Mobile phase for oligodeoxyadenylic acids: gradient from 23 to 33% buffer B (buffer A – 20% acetonitrile and 80% 20 mmol/L phosphate buffer, pH 7.0, B – 1 mol/L sodium chloride in A). For oligothymidylic acids: gradient from 35 to 48% (or from 24 to 48%, or from 15 to 25%) B in A (buffer A – 20% acetonitrile and 80% 20 mmol/L phosphate buffer pH 7.0; buffer B – 1 mol/L sodium chloride in A)	50×8 mm i.d.	Diethylamine functionalities	[53]

Continued on the next page

TABLE 24.1 (continued)

Compound	Separation mode and detection	Mobile phase/background electrolyte	Column dimensions	Note	Ref.
Phosphorylated and dephosphorylated deoxyadenylic acids [hydrolyzed p(dA)$_{40}$–p(dA)$_{60}$ spiked with p(dA)$_{12}$–p(dA)$_{18}$]	HPLC. Poly(styrene–divinylbenzene) column, UV detection at 254 nm	Linear gradient 5–35% B in 5 min; B in 5 min; 40–45% B in 6 min; 45–52% B in 14.0 min; A: 100 mmol/L triethylammonium acetate pH 7.0; B: 100 mmol/L triethylammonium acetate, pH 7.0, 20% acetonitrile	60 × 0.20 mm i.d.	Temperature 50°C; ion-pair reversed-phase separation	[54]
Oligohydroxynucleotides [p(dT)$_{12}$–p(dT)$_{18}$]	HPLC. Poly(norbornene-co-hexahydrodimethano-naphthalene monolith, UV detection at 264 nm	Linear gradient 55–80% B in 10 min; A: 100 mmol/L triethylammonium acetate, pH 7.0; B: 100 mmol/L triethylammonium acetate, pH 7.0, 20% acetonitrile	100 × 3 mm i.d.	Temperature 20°C; ion-pair reversed-phase separation	[55]
5S Ribosomal RNA from E. coli	HPLC. Poly(styrene–divinylbenzene) column, ESI-MS detection; sheath liquid acetonitrile	Linear gradient 0–100% B in 10 min; A: 25 mmol/L butyldimethylammonium bicarbonate pH 8.4, B: 25 mmol/L butyldimethylam-monium bicarbonate pH 8.4, 80% acetonitrile	60 × 0.2 mm i.d.	Ion-pair reversed-phase separation	[54]

Sample	Method	Conditions	Dimensions	Notes	Ref.
DNA fragments	HPLC. Poly(norbornene-co-hexahydrodimethano-naphthalene) monolith, UV detection at 260 nm	Linear gradient 10–25% B in 5 min, 25–45% B in 12 min; A: 100 mmol/L triethylammonium acetate, pH 7.0, 4% glycerol, B: 100 mmol/L triethylammonium acetate, pH 7.0, 40% acetonitrile, 4% glycerol	100 × 3 mm i.d.	Micropreparative separation	[56]
DNA fragments	HPLC. Poly(styrene–divinylbenzene) column, UV detection at 254 nm	Linear gradient 43–50% B in 0.5 min, 50–54% B in 3.5 min; A: 100 mmol/L triethylammonium acetate, 0.1 mmol/L Na$_4$EDTA, pH 7.0, B: 100 mmol/L triethylammonium acetate, 0.1 mmol/L Na$_4$EDTA, pH 7.0, 25% acetonitrile	60 × 0.2 mm i.d.	Temperature 61°C, partially denaturing conditions	[57]
Short tandem repeats in hum TH01 (tyrosine hydroxylase gene)	HPLC. Poly(styrene–divinylbenzene) column, ESI-MS detection, sheath liquid acetonitrile	Linear gradient 15–70% B in 10 min; A: 25 mmol/L BDMAB, pH 8.4, B: 25 mmol/L BDMAB, pH 8.4, 40% acetonitrile	60 × 0.2 mm i.d.		[58]

Continued on the next page

TABLE 24.1 (continued)

Compound	Separation mode and detection	Mobile phase/background electrolyte	Column dimensions	Note	Ref.
pUC19-*MspI* digest	HPLC. Poly(styrene–divinylbenzene) monolith, MS detection, sheath liquid acetonitrile	Linear gradient 10–40% B in 3 min, 40–50% B in 12 min; A: 25 mmol/L butyldimethylammonium bicarbonate pH 8.4, B: 25 mmol/L butyldimethylammonium bicarbonate pH 8.4, 40% acetonitrile	60 × 0.2 mm i.d.		[59]
Mutation detection	HPLC. Poly(styrene–divinylbenzene) monolith	Linear gradient 30–50% B in 3 min, 50–70% B in 7 min; A: 100 mmol/L triethylammonium acetate, 0.1 mmol/L Na$_4$EDTA, pH 7.0, B: 100 mmol/L triethylammonium acetate, 0.1 mmol/L Na$_4$EDTA, pH 7.0, 25% acetonitrile	50 × 0.2 mm i.d.	Ion-pair reversed phase separation	[57]

Sample	Method	Conditions	Column dimensions	Notes	Ref.
Hae and *MspI* digests of pUC18 and pBR 322	HPLC. Poly(styrene–divinylbenzene) monolith, UV detection at 254 nm	Linear gradient 30–50% B in 13 min; A: 100 mmol/L triethylammonium acetate, 0.1 mmol/L Na₄EDTA, pH 7.0, B: 100 mmol/L triethylammonium acetate, 0.1 mmol/L Na₄EDTA, pH 7.0, 25% acetonitrile	50 × 0.2 mm i.d.	Temperature 49.7°C, comparison with packed beads	[57]
Mononucleotides (adenosine, AMP, ADP, ATP)	HPLC. UNO Q1 column, detection at 254 nm	Linear gradient 0% B for 1 min, 0–30% B in 5 min, 30% B for 2 min; A: 20 mmol/L Tris buffer, pH 8.2, B: 20 mmol/L Tris buffer, pH 8.2, 1.0 mol/L sodium chloride	35 × 7 mm i.d.		[60]
Plasmid DNA	HPMC. Poly(glycidyl methacrylate-*co*-ethylene dimethacrylate) disk	Linear gradient 70–100% B in 4.5 min; A: 20 mmol/L Tris-HCl buffer, pH 7.4, B: 20 mmol/L Tris-HCl buffer, pH 7.4, 1 mol/L sodium chloride	3 × 12 mm i.d.		[61]

Continued on the next page

TABLE 24.1 (continued)

Compound	Separation mode and detection	Mobile phase/background electrolyte	Column dimensions	Note	Ref.
Oligonucleotides	Membrane chromatography. Stacked DEAE cellulose membranes, UV detection at 260 nm	Stepped gradient to elute (step-wise), a) proteins (phenol, chloroform + ethanol), b) oligoribonucleotides (0.3 mol/L LiCl, 5 mmol/L LiOH), c) DNA (0.5 mmol/L NaCl, 5 mmol/L NaOH), d) DNA–RNA complexes (2 mol/L NaCl, 5 mmol/L NaOH)	6 × 19 mm i.d.		[62]
Synthetic oligomers	HPLC. UNO Q1 column, UV detection at 254 nm	Linear gradient 40–47% B in 20 min, 47–50% B in 30 min; A: 25 mmol/L triethylammonium acetate, pH 7.4, B: 25 mmol/L triethylammonium acetate, pH 7.4, 1 mol/L sodium chloride	35 × 7 mm i.d.		[63]

Antisense nucleotides	Membrane chromatography. Sartobind Q-20K-60-12 column, UV detection at 266 nm	Linear gradient 0–100% B in 20 min; A: 20 mmol/L sodium hydroxide, B: 20 mmol/L sodium hydroxide, 2.5 mol/L sodium chloride	Column of 12 cm length	Preparative separation	[64]
Alkaloids					
Initraphylline, pteropodine, isopteropocine	CEC. Sol-gel bonded ODS (7 μm, 1400 Å). UV detection at 254 nm	Acetonitrile–water–50 mmol/L phosphate buffer pH 8.0	Capillary 39/30 cm (effective length) × 75 μm i.d., 25 kV		[3]
Sulphur-containing compounds (except amino acids)					
Phenylmercapturic acid	CEC. β-CD bonded positively charged polyacrylamide gel, UV detection at 250 nm	200 mmol/L Tris, 300 mmol/L borate pH 8.1	55/35 cm (effective length) × 7.5 μm i.d.	286 V/cm	[29]
Synthetic polymers					
Polystyrene standards ($7.10^3 – 980.10^3$ rel. mol. mass)	CEC. 59.7 wt. % butyl methacrylate, 0.3% wt. 2-acrylamido-2-methyl-1-propanesulfonic acid, UV detection at 215 nm, 0.2 MPa overpressure	Tetrahydrofuran containing 2 vol. % of water	30 cm (effective length), 100 μm i.d., 25 kV	Size exclusion chromatography, pore size 750 nm	[17]

Continued on the next page

TABLE 24.1 (continued)

Compound	Separation mode and detection	Mobile phase/background electrolyte	Column dimensions	Note	Ref.
Drugs					
β-Blocking drugs (atenolol, pindolol, metoprolol, celiprolol, bisoprolol)	HPLC. Silica rod RP 18e column, UV detection at 254 nm	Acetonitrile–0.1% trifluoroacetic acid in water (20:80, v/v)	Column 50 × 4.6 mm, flow rate 1–9 mL/min		[5]
β-Blocking drugs (bisoprolol hemifurate, pindolol, nadolol, pafenolol, metoprolol, celiprolol, carazolol, bisoprolol, alprenolol, propranolol)	HPLC. Silica rod RP 18e column, UV detection at 254 nm	Methanol–0.02 mol/L phosphate buffer, pH 3.0 (30:70, v/v)	Column 100 × 4.6 mm. Flow gradient 0–2 min 2 mL/min; 2–3 min 2–5 mL/min; 3–6 min 5 mL/min		[5]
Cyclosporin A	HPLC. Lichrospher Si 60 and Prep Rod, UV detection at 220 nm	n-Heptane–methyl ethyl ketone (45:55, v/v)	Column 200 × 25 mm i.d., flow rate 20 mL/min or 52 × 25 mm, flow rate 20 mL/min.		[25]
Tocopherols (and δ tocopherols from vegetable oil)	HPLC. Prep Rod	n-Heptane–ethyl acetate (98:2, v/v)	Column 100 × 25 mm i.d., flow rate 40 or 160 mL/min		[25]
Terbutaline, propranolol, benzoin	CEC. β-Cyclodextrin bonded charged polyacrylamide gel	200 mmol/L Tris–300 mmol/L boric acid buffer, pH 9.0	70/35 cm (effective length), 75 μm i.d.	Chiral separation	[65]

Analyte	Conditions	Buffer	Dimensions	Notes	Ref.
Metaproterenol, clenbuterol	CEC. Negatively charged polyacrylamide gel, carbamoylated β-cyclodextrin bonded, UV detection at 240 nm	200 mmol/L Tris–300 mmol/L boric acid buffer, pH 9.0	75/70 cm effective length, 75 μm i.d., 270 V/cm	Chiral separation	[24]
Terbutaline	CEC. Negatively charged polyacrylamide gel, carbamoylated β-cyclodextrin bonded, UV detection at 240 nm	200 mmol/L Tris–300 mmol/L boric acid buffer, pH 7.0	70/35 cm effective length, 75 μm i.d., 214 V/cm	Chiral separation	[24]
Hexobarbital, mephobarbital, warfarin, tropicamid, ibuprofen, propranolol, mephenyto:n, hydrobenzoin	CEC. Acrylamide gels with 2-acrylamido-2-methyl-propanesulfonic acid and alkyl β-cyclodextrin (various proportions), UV detection at 200 nm	Tris-borate pH 8.2, sodium phosphate pH 7.5 and 8.2, 5 mmol/L	14.4–16.5/13.5–8.7 cm effective length, 50–75 μm i.d., 2–4 kV	Chiral separation. Positively charged gels and negatively charged gels compared.	[66]
Mephobarbital	CEC. Chirasil-Dex monolith, UV detection at 230 nm	20 mmol/L MES, pH 6.0–methanol 7:3	40/20 cm effective length, 100 μm i.d., 25 kV	Up to 1.2 MPa overpressure	[67]
Basic drugs: N-methylamitriptyline, amitriptyline, nortriptyline	CEC. Piperazine diacrylamide, methacrylamide, N-isopropyl-acrylamide, vinyl sulfonic acid, ammonium sulfate, 115 mmol/L phosphate buffer pH 2.75 in various proportions, UV detection	115 mmol/L phosphate buffer pH 2.75–water–acetonitrile, 10:20:70 (Tween 20 added to some runs as selector)	57.5/40 cm efficient length, 50 μm i.d., 20 kV		[68]

Continued on the next page

TABLE 24.1 (continued)

Compound	Separation mode and detection	Mobile phase/background electrolyte	Column dimensions	Note	Ref.
Drug manufacturing intermediates, toluene, dimethyl- and dibutyl phthalate	HPLC. Sol-gel from alkoxysilanes, UV detection at 254 nm	Methylcyclohexane	250 × 4 mm i.d.	Preparative separation of diastereomers, SMB procedure	[22]
Debrisoquine hydroxylated metabolites in human liver filtrate and potential drugs	HPLC. Chromolith SpeedRod column, LC/MS monitoring		4.6 × 5 mm i.d.		[69], [70]
Proglumide	HPLC. 25% norbornene, 25% β-cyclodextrin functionalized norbornene, UV detection	Acetonitrile–methanol–trifluoroacetic acid–triethylamine (99.4925:0.25:0.25:0.0075 % v/v)	150 × 3 mm i.d.		[71]
Fluorescent dyes					
Rhodamine, fluorescein	EC. Collocated monolithic structure, Sylgard chip, polydimethylsiloxane polymer, fluorescence detection	1 mmol/L Tris, pH 8.0	76 nl chip volume, 5 μm wide, 10 × 10 μm monolith dimension, 2000 V per chip		[35]

Varia and mixtures of compounds of different chemical natures

Benzoin	CEC. β-CD bonded positively charged polyacrylamide gel, UV detection at 245 nm	200 mmol/L Tris, 300 mmol/L borate pH 8.1	55/35 cm (effective length), 7.5 μm i.d., 269 V/cm	[29]
Parabens (methyl-, ethyl- and propyl paraben)	CEC. Fritless packing C18 silica gel. UV detection at 254 nm	Acetonitrile–50 mmol/L HEPES (40:60), pH 6.6	45.8/25 cm (effective length), 75 μm i.d., 284 V/cm	[18]
Lignans (schizandrin, gomisin, deoxyschizandrin), stimulants, anticancer drugs	CEC. Macroporous polyacrylamide gel containing vinylsulfonic acid and lauryl acrylate, UV detection at 220 nm	20–35% Acetonitrile in 10 mmol/L Tris–15 mmol/L borate, pH 8.2	33/24.5 cm effective length, 75 μm i.d., 22.5 kV	Positive polarity	[72]
Uracil, phenol, benzene, toluene, ethylbenzene	HPLC. Butyl methacrylate–ethylene dimethacrylate monolith, UV detection at 214 nm	Acetonitrile–water, 65:35	19 cm × 350 μm i.d.	[73]

24.3 ALCOHOLS, PHENOLS, OXO COMPOUNDS, CARBOHYDRATES AND ORGANIC ACIDS

This section may appear diversified with regard to the chemical nature of the separated analytes. However, the reason is twofold: first, in the respective practical applications these categories of compounds are frequently separated side-by-side and, second, there are — so far — not many papers describing the use of monolithic packings for the individual chemical entities listed in the heading.

The main type of monolithic sorbents used for alcohol separations are sol-gel bonded ODS or ODS/SCX phases [3,11]. These have been used successfully for the separation of benzyl alcohol, dimethyl phthalate, benzophenone and benzaldehyde. Concomitant separation of hydrocarbons such as biphenyl is possible. The separations are generally effected electrokinetically in capillaries of standard length and i.d. at a voltage of about 20 kV. A number of separations using sol-gel bonded phases can be found in refs. [11] and [14].

The separation of alkyl phenones can be achieved on a number of other monolithic phases, typically using methacrylic acid–methacrylate sorbent containing 10% lauryl methacrylate as EOF-creating component. Such electrokinetic separations run in Tris–borate buffer–acetonitrile or even water–acetonitrile mobile phases in standard dimension capillaries, typically 50–100 µm i.d., 20–50 cm long. The recommended pH is mostly within the alkaline range (8.2–9.1) [3,11]. However, for acetophenone and valerophenone the acetonitrile–HEPES buffer, pH 6.6 was used.

Separation of phthalic acid derivatives has been obtained with monolithic sol-gel polymers of alkoxy silanes, on both the preparative and analytical scale, using 100 × 25 mm columns and methylcyclohexane–ethyl acetate as mobile phases [22].

Separation of phenols in the CEC mode can be achieved with poly(styrene-*co*-divinylbenzene-*co*-methacrylic acid) monoliths in acetonitrile–Tris buffer [2]. Reversed-phase chromatography on C18 rod columns was reported for the separation of chlorophenols. Silica gel rods enable a wide range of applications. They can also be used in HPLC for the separation of carboxylic acids and their esters [7,25]. Alternatively, similar separations can be achieved in the electrokinetic mode using acrylic acid polymers with 10% lauryl methacrylate.

Separation of carbohydrates using monoliths is seldom reported. CEC separation of malto-oligosaccharides (Glc1-Glc6) using a macroporous polyacrylamide matrix derivatized with C4 ligand was described [19].

24.4 STEROIDS

Owing to the nearly general applicability of sol-gel bonded ODS/SCX monolithic packings, it is not surprising that steroids were also separated using this type of stationary phase [3]. Alternatives are macroporous acrylamide polymers [26] and poly(AMPS-*co*-IPAA) hydrogels [27]. Acetonitrile–water–phosphate (or ammonium phosphate) buffers are used as mobile phases at acid pH. Hydrogels can be operated also at alkaline pH (typically at 8.1). The separated analytes comprise both free and conjugated steroids (selected androstanes) and their dansyl derivatives.

24.5 AMINES

Poly(styrene-*co*-divinylbenzene-*co*-methacrylic acid)- and sol-gel bonded ODS columns have so far been used for the separation of amines such as positional isomers of phenylenediamines [2] and aniline-related compounds [3]. Separations were achieved in the CEC mode at alkaline pH (8.0–8.8) using the acetonitrile–aqueous buffer system [2]. The buffer concentration in the mobile phase can vary. The use of negatively charged polyacrylamide monoliths represents another option [28].

24.6 AMINO ACIDS, PEPTIDES AND PROTEINS

Reports on the separation of amino acids using monolithic columns concern practically all typical amino acid derivatives. For example, dansylated amino acids were separated on a positively charged polyacrylamide monolith containing β-cyclodextrin [29]. Phenylthiohydantoin amino acids are best separated on sintered Zorbax ODS monolithic columna. A tetraethyl orthosilicate column with embedded silica particles was used for NBD-F derivatives [31]. Buffer–acetonitrile mixtures were used as mobile phase. For example, the separation of NBD-F derivatives was achieved in 5 mmol/L phosphate, pH 2.5–acetonitrile (30:70). This approach also offers the possibility of separating amino acid enantiomers. Separation of amino acid enantiomers can be effected even without derivatization, using a methacrylamide–piperazine diacrylamide copolymer containing vinylsulfonic acid, that drives the electroosmotic flow, and N-(2-hydroxy-3-allyloxypropyl)-L-4-hydroxyproline as chiral selector in the ligand exchange CEC mode. A 50 mmol/L dihydrogen phosphate buffer containing Cu(II) at pH 4.6 was recommended as the mobile phase for pressure assisted CEC separation [32]. For separations of DNB derivatives, monolithic polymers prepared from quinidine were reported [33]; 0.4 mol/L acetic acid containing 4 mmol/L triethylamine in acetonitrile and methanol in a ratio of 80:20 served as the mobile phase. Alternatively, 0.4 mol/L acetic acid and 4 mmol/L ethanolamine in acetonitrile with methanol (80:20) can also be used for enantiomeric separations of DNZ-leucine and DNP-valine.

Simple peptides such as gly-tyr, val-tyr, methionine and leucine enkephalins were separated by CEC on a methacrylate ester-based monolithic column [16]. Aceto-nitrile–buffer at pH 7.0 served as the mobile phase. For peptide mapping and more complex peptide mixtures, C18 COMOSS microfabricated devices were reported [38].

C18 does not represent the only reversed-phase used. Macroporous polyacrylamide monoliths with C12 ligands or porous polystyrene monoliths with dimethyloctyl func-tionalities can be used as well. Buffer–acetonitrile mixtures optimized according to the nature of the separated peptides, mainly by varying their pH, are generally used.

Monolithic microfluidic chips were reported for the separations of PTH peptides prepared from bovine serum albumin using 1 mmol/L carbonate buffer, pH 8.7, and electrokinetic operation [35]. A different microchip system enables separation of simple underivatized tripeptides containing aromatic residues and leucine- and methionine enkephalins in 5 mmol/L phosphate buffer pH 7.0 and acetonitrile. Detec-tion was achieved either via fluorescence or UV absorbance at 215 nm [16].

Polychloromethylstyrene-based monoliths were used for the separation of basic proteins such as α-chymotrypsinogen, ribonuclease, cytochrome c, and lysozyme [42]. As with the standard capillary electrophoresis of proteins, careful optimization of conditions is necessary for separation in the CEC mode. This regards the choice of the eluent, which almost invariably represents a mixture of acetonitrile and an aque-ous buffer. Sometimes gradient elutions were also applied [43]. Admittedly, this represents a considerable complication in standard CEC systems. Step gradients can also be applied. This is particularly of advantage in operations with preparative disks, such as in the isolation of annexins from hepatoma 777 plasma membranes [44].

Affinity HPLC separations with monolithic beds bearing peptides capable of specific adsorption of compounds such as Factor VIII have also been reported [45].

Methacrylate-based DEAE CIM disks and tubes were applied for fast-, and even preparative, separations of proteins, annexins, $\alpha 1$ antitrypsin, Factor IX, myoglobin, conalbumin, and soybean trypsin inhibitor [39,41,48,49].

24.7 NUCLEIC ACIDS AND THEIR CONSTITUENTS

The separation of nucleic acids, their fragments, and constructs, represents a cate-gory of separations using monolithic columns that is widely explored today. The applications range from the analysis of oligonucleotides [50,51] up to intact nucleic acid [52]. Disk operations appear attractive for isolating fragments from a complex mixture as a pre-separation step. The separation mode is purely chromatographic, and employs gradient elution [50,51].

Methacrylate copolymers appear suitable for the separation of oligonucleotides and DNA using Tris–borate buffers [51,53,61]. Successful separations can also be obtained with Tris-HCl buffers. Polymethacrylate 50 × 8 mm monoliths were applied for the separations of oligodeoxyadenylic and oligothymidylic acids in the HPLC mode [53]. In order to achieve good separations of the individual species, gradient elution was applied. The eluent was 20% acetonitrile–80% phosphate buffer (20 mmol/L) superimposed with a gradient of NaCl.

Poly(norbornene-*co*-hexahydrodimethanonaphthalene) monolithic columns were reported also for HPLC separations of oligodeoxynucleotides. In this case, separation is effected using an acetonitrile gradient in triethylammonium acetate buffer, pH 7.0, containing in some cases 4% glycerol [55,56].

PS-DVB columns were successfully used for the separations of poly(adenylic) acids, both phosphorylated and dephosphorylated DNA digests or ribosomal RNA [54,57–59]. UNO Q1 columns appear suitable for HPLC separation of mono-nucleotides and synthetic oligomers using a gradient of NaCl in Tris buffer, pH 8.2 or 7.4 [60,63].

Some rather exotic applications can be mentioned at the end of this section. A Sartobind Q-20K-60-12 unit operated with a sodium chloride gradient in 20 mmol/L sodium hydroxide was reported for the preparative separation of antisense nucleotides [64]. For preparative separation of proteins, oligonucleotides, DNA, and DNA-RNA complexes, DEAE membranes operated in a step gradient may represent the method of choice [62].

24.8 ALKALOIDS

In our literature search we came across only a single paper dealing with capillary electrochromatographic separation of pteropodine, isopteropodine and metraphylline [3]. A sol-gel ODS-particle-loaded CEC column and acetonitrile–water system were used to achieve the separation.

24.9 FLUORESCENT DYES

To our best knowledge there is only a single report regarding this category of compounds. Rhodamine and fluorescein were separated using a microfluidic chip with collocated monolithic structures [35].

References pp. 656-658

24.10 DRUGS

The reports on drug separation with monolithic packings reflect the diversity in the chemical structures of these compounds. The elution conditions in HPLC separations can be isocratic for drug intermediates, or a wide variety of gradients can be used for established drugs and their metabolites [68-71]. The chemical nature of the monoliths used for these separations reaches from those including acrylic acid, vinylsulfonic acid, and alkoxysilanes, to norbornene polymers along with commercially available Chromolith Speed Rod columns [5,69,70]. A less typical approach is the normal-phase separation of cyclosporin A using heptane–methyl ethyl ketone as eluent [25].

The CEC separation mode is clearly preferred for chiral separations of drugs [24,45,66]. Negatively charged polyacrylamide gels with bonded cyclodextrin also appear very promising.

24.11 CONLUSIONS

Obviously, this survey cannot embrace all of the chromatographic separations achieved using monolithic separation media published during the previous years. In fact, this was not even the aim of this Chapter. The extensive Table, which is its major part, is supposed to illustrate the multiplicity of targets, a wide variety of monolithic structures, and a plethora of conditions that have been demonstrated to work in this area. The data shown in the Table may also help the potential user to accelerate the method-development for a specific separation that is facilitated by monolithic columns.

24.12 REFERENCES

1   J.L. Liao, N. Chen, C. Ericson, S. Hjertén, Anal. Chem. 68 (1996) 3468.
2   B. Xiong, L. Zhong, Y. Zhang, H. Zou, J. Wang, J. High. Resolut. Chromatogr. 23 (2000) 67.
3   Q. Tang, M.L. Lee, J. High Resolut. Chromatogr. 23 (2000) 73.
4   C.K. Ratnayake, C.S. Oh, M.P. Henry, J. High Resolut. Chromatogr. 23 (2000) 81.
5   K. Cabrera, D. Lubda, H.-M. Eggenweiler, H. Minakuchi, K. Nakanishi, J. High Resolut. Chromatogr. 23 (2000) 93.
6   N. Tanaka, H. Nagayama, H. Kobayashi, T. Ikegami, K. Hosoya, N. Ishizuka, H. Minakuchi, K. Nakanishi, K. Cabrera, D. Lubda, J. High Resolut. Chromatogr. 23 (2000) 111.
7   M.R. Schure, R.E. Murphy, W.L. Klotz, W. Law, Anal. Chem. 70 (1998) 4985.
8   C. Ericson, L. Liao, K. Nakazato, S. Hjertén, J. Chromatogr. A 767 (1997) 33.
9   E.C. Peters, M. Petro, F. Svec, J.M. Fréchet, Anal. Chem. 70 (1998) 2296.
10   S. Constantin, R. Freitag, J. Chromatogr. A 887 (2000) 253.

11   Q. Tang, M.L. Lee, J. Chromatogr. A 887 (2000) 265.

12   C.K. Ratnayake, C.S. Oh, M.P. Henry, J. Chromatogr. A 887 (2000) 277.

13   R. Asiaie, X. Huang, D. Farnan, C. Horváth, J. Chromatogr. A 806 (1998) 251.

14   J.D. Hayes, A. Malik, Anal. Chem. 72 (2000) 4090.

15   Q.C. Wang, F. Svec, J.M.J. Fréchet, J. Chromatogr. A 669 (1994) 230.

16   C. Yu, F. Svec, J.M.J. Fréchet, Electrophoresis 21 (2000) 120.

17   E.C. Peters, M. Petro, F. Svec, J.M.J. Fréchet, Anal. Chem. 70 (1998) 2288.

18   C. Fujimoto, J. High Resolut. Chromatogr. 23 (2000) 89.

19   A. Palm, M.V. Novotny, Anal. Chem. 69 (1997) 4499.

20   R. Wu, H. Zou, M. Ye, Z. Lei, J. Ni, Electrophoresis 22 (2001) 544.

21   D. Josic, A. Štrancar, Ind. Eng. Chem. Res. 38 (1999) 333.

22   M. Schulte, J. Dingenen, J. Chromatogr. A 923 (2001) 17.

23   A. Maruska, C. Ericson, A. Vegvari, S. Hjertén, J. Chromatogr. A 837 (1999) 25.

24   T. Koide, K. Ueno, J. Chromatogr. A 893 (2000) 177.

25   M. Schulte, D. Lubda, A. Delp, J. Dingenese, J. High Resolut. Chromatogr. 23 (2000) 100.

26   A.H. Que, A. Palm, A.G. Baker, M.V. Novotny, J. Chromatogr. A 887 (2000) 379.

27   C. Fujimoto, Y. Fujise, E. Matsuzawa, Anal. Chem. 68 (1996) 2753.

28   T. Koide, K. Ueno, J. Chromatogr. A 909 (2000) 305.

29   T. Koide, K. Ueno, J. High Resol. Chromatogr. 23 (2000) 59.

30   E.C. Peters, K. Lewandowski, M. Petro, F. Svec, J.M.J. Fréchet, Anal. Commun. 35 (1998) 83.

31   M. Kato, M.T. Dulay, B. Bennett, J.-R. Chen, R.N. Zare, Electrophoresis 21 (2000) 3145.

32   M.G. Schmid, N. Grobuschek, C. Tuscher, G. Güblitz, A. Vegvari, E. Machtejevas, A. Matuska, S. Hjertén, Electrophoresis 21 (2000) 3141.

33   M. Lämmerhofer, E.C. Peters, C. Yu, F. Svec, M. Fréchet, W. Lindner, Anal. Chem. 72 (2000) 4614.

34   M. Lämmerhofer, F. Svec, M. Fréchet, W. Lindner, Anal. Chem. 72 (2000) 4623.

35   B.E. Slentz, N.A. Penner, E. Lugowska, F. Regnier, Electrophoresis 22 (2001) 3736.

36   I. Gusev, X. Huang, C. Horváth, J. Chromatogr. A 855 (1999) 273.

37   F. Regnier, J. High Resolut. Chromatogr. 23 (2000) 19.

38   B. He, J. Ji, F.E. Regnier, J. Chromatogr. A 853 (1999) 257.

39   A. Štrancar, P. Koselj, H. Schwinn, D. Josic, Anal. Chem. 68 (1996) 3483.

40   Q.C. Wang, F. Svec, J.M.J. Fréchet, Anal. Chem. 65 (1993) 2243.

41   D. Josic, J. Reusch, K. Lester, O. Baum, W. Reutter, J. Chromatogr. A 590 (1992) 59.

42   X. Huang, J. Zhang, C. Horváth, J. Chromatogr. A 858 (1999) 91.

43   C. Ericson, S. Hjertén, Anal. Chem. 71 (1999) 1621.

44   D. Josic, Y.-P. Lim, A. Štrancar, W. Reutter, J. Chromatogr. A 662 (1994) 217.

45   K. Amatschek, R. Necina, R. Hahn, E. Schallaun, H. Schwinn, D. Josic, A. Jungbauer, J. High Resolut. Chromatogr. 23 (2000) 47.

46   D. Josic, H. Schwinn, A. Štrancar, A. Podgornik, M. Barut, Y.-P. Lim, M. Vodopivec, J. Chromatogr. A 803 (1998) 61.

47   A. Podgornik, M. Barut, I. Mihelic, A. Štrancar, in F. Švec, Z. Deyl, T. Tennikova (Editors), Monoliths, Elsevier, Amsterdam, 2002.

48 K. Branovic, A. Buchacher, M. Barut, A. Štrancar, D. Josic, Chromatogr. A, 903 (2000) 21.

49 A. Štrancar, M. Barut, A. Podgornik, P. Koselj, D. Josic, A. Buchacher, LC-GC Int. 11 (1998) 660.

50 A. Podgornik, M. Barut, J. Jančar, A. Štrancar, T. Tennikova, Anal. Chem. 71 (1999) 2986.

51 A. Podgornik, M. Barut, J. Jančar, A. Štrancar, J. Chromatogr. A 848 (1999) 51.

52 N.S. Brahmasandra, V.M. Ugaz, D.T. Burke, C.H. Mastrangelo, M.A. Burns, Electrophoresis 22 (2001) 300.

53 D. Sýkora, F. Svec, J.M.J. Fréchet, J. Chromatogr. A 852 (1999) 297.

54 H. Oberacher, C.G. Huber, Trends Anal. Chem, in press.

55 B. Mayr, R. Tessadri, E. Post, M.S. Buchmeiser, Anal. Chem. 71 (2001) 4071.

56 S. Lubbad, B. Mayr, C.G. Huber, M.R. Buchmeiser, J. Chromatogr. A 959 (2002) 121.

57 C.G. Huber, A. Premstaler, W. Xiao, H. Oberacher, G.K. Bonn, J. Biochem. Biophys. Meth. 47 (2001) 5.

58 H. Oberacher, W. Parson, R. Mühlmann, C.G. Huber, Anal. Chem. 73 (2001) 5109.

59 C.G. Huber, H. Oberacher, Mass Spectrom. Rev. (2002) in press.

60 T.L. Tisch, R. Frost, J.L. Liao, W.-K. Lam, A. Remy, E. Scheinpflug, C. Siebert, H. Song, A. Stapleton, J. Chromatogr. A 816 (1998) 3.

61 R. Giovanni, R. Freitag, T.B. Tennikova, Anal. Chem. 70 (1998) 3348.

62 N. van Huynh, J.C. Motte, J.F. Pilette, M. Decleire, C. Colson, Anal. Biochem. 211 (1993) 61.

63 C.H. Lochmüller, Q. Liu, L. Huang, Y. Li, J. Chromatogr. Sci. 37 (1999) 251.

64 R.R. Desmuth, T.N. Warner, F. Hutchinson, M. Murphy, W.E. Leitch II, P. De Leon, G.S. Srivatsa, D.L. Cole, Y.S. Sanghoi, J. Chromatogr. A 890 (2000) 179.

65 T. Koide, K. Ueno, Anal. Sci. 15 (1999) 791.

66 A. Vegvari, A. Földesi, C. Hetenyi, O. Kocnegarova, M.G. Schmid, V. Kudirkaite, S. Hjertén, Electrophoresis 21 (2000) 3116.

67 D. Wistuba, V. Schurig, Electrophoresis 21 (2000) 3152.

68 A.M. Enlund, C. Ericson, S. Hjertén, D. Westerlund, Electrophoresis 22 (2001) 511.

69 G. Dear, R. Plumb, D. Mallett, Rapid Commun. Mass Spectrom. 15 (2001) 152.

70 R. Plumb, G. Dear, D. Mallett, D. Ayrton, Rapid Commun. Mass Spectrom. 15 (2001) 986.

71 F.M. Sinner, M.R. Buchmeiser, Angew. Chem. Int. Ed. 39 (2000) 1433.

72 L. Kvasničková, Z. Glatz, H. Štěrbová, V. Kahle, J. Slanina, P. Musil, J. Chromatogr. A 916 (2001) 271.

73 P. Coufal, M. Čihák, J. Suchánková, E. Tesařová, Z. Bosáková, K. Štulík, J. Chromatogr. A 946 (2002) 99.

F. Švec, T.B. Tennikova and Z. Deyl (Editors)
Monolithic Materials
Journal of Chromatography Library, Vol. 67

Chapter 25

Miniature and Microchip Technologies

Yolanda FINTSCHENKO[1], Brian J. KIRBY[1], Ernest F. HASSELBRINK[2],
Anup K. SINGH[1], Timothy J. SHEPODD[1]

[1]*Sandia National Laboratories, Livermore, CA 94551*
[2]*University of Michigan, Ann Arbor, MI 48109-2121*

CONTENTS

25.1 INTRODUCTION

The effectiveness of a chemical analysis method is generally measured by five figures of merit: sensitivity, limit of detection, accuracy, precision, and throughput or speed of analysis. While the name suggests that portability or size is the driving force, lab-on-a-chip, or micro total analysis systems (μTAS), science and technology is driven by the belief that miniaturization of (primarily) separation techniques will lead to an improvement in these figures of merit. Widmer and Manz first outlined how different separation and detection methods scale down [1]. This led to the proposal of miniaturizing and integrating sample introduction, separation, and detection, elevating a separation technique to a quasi-continuous generic chemical sensor [1]. Manz and Harrison first reported capillary zone electrophoresis (CZE) in open channels etched in a glass chip in 1992 [2]. However, it was the demonstration of the on-chip baseline separation of two fluorescent dyes within 150 ms by Jacobson *et al.* that best illustrates the impact of lab-on-a-chip technology on chemical analysis [3].

The chip format provides a platform for integrated functions because of the ability to use photolithography to define small volume channel networks that integrate all of the different functions required for the μTAS. For example, in the case of CZE, the integration of the injection channel with the separation channel provides a platform for a rapid valveless injection method. This requires the electronic control of the voltages applied to each channel rather than the physical movement of the separation channel from sample to waste as in a typical capillary-based CZE system [4]. An illustration of a simple injection cross for electrokinetically driven separations is shown in Fig. 25.1a. A diagram of the equivalent system in capillaries is shown in Fig. 25.1b. It is seen that a capillary-based system similar to those etched in chips requires a 4-way interconnect, which presents the problem of dead volume and serves as an additional point of stress, often resulting in capillary breakage.

There are a number of challenges to using a chip-based system. Interconnects to the chip can pose the greatest problem. This has largely been solved by using pipette tips and more complicated fittings that allow for low and high-pressure interfaces [5-7]. Evaporation is minimized in these small volume reservoirs by covering the reservoirs, while problems like pH changes and bubbles can be addressed by using salt-bridges between the solution and channel [8]. Detection issues are similar to those in a fused silica capillary, with both the reduced path length and the detection volume resulting in a reduced limit of detection if using absorbance detection [1]. As a result, most systems employ laser-induced fluorescence detection [3,9-20]. Alternatives, including eletrochemical and chemiluminescence detection, have also been reported [21-29].

In order to realize the full potential of integrated microfluidic systems a number of components at dimensions from 10 – 100 microns have to be developed. For example, flow control has been valveless, an advantage in some cases, but largely a result of the absence of reliable valves for microfluidic systems. CZE and other low-pressure techniques have been pursued in the absence of a small high-pressure pump suitable for either integration with or connection to a chip-based system. High-pressure liquid

A B

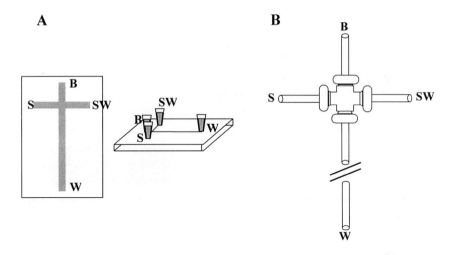

Fig. 25.1. Injector integrated with separation channel. (A) Injector microfabricated in a chip. Chip dimensions: 8.5 × 2.0 cm. Typical channel dimensions are 50 μm wide and 25 μm deep. The separation channel is 7.5 cm long, and the injection channel is 1 cm long. B = Running Buffer, W = Waste, S = Sample, SW = Sample Waste. Sample was introduced by applying voltage across S (100 % total V applied) and SW (0%) with B at 40% and BW at 60 %, and injected onto the separation column by switching voltage to B = 100 % and W = 0 % with S and SW at 60%. A switch box or computer is typically used to control voltages. (B) Injector formed from capillaries. The equivalent integrated injector and capillary column using capillaries and a commercially available cross fitting. Typical capillary sizes are 365 μm o.d. and 50 to 100 μm i.d.

chromatography (HPLC), an industry standard, has been largely untapped in chip separation systems due to the lack of valves and pumps [7].

Porous polymer monoliths (PPM) have great potential as components in planar analytical devices. The greatest advantage of the PPM lies in their tunability. Chemistry, charge, polarity, reactivity, pore size, surface area, hydrophobicity, as well as mechanical strength, can all be varied. By altering these combinations, the function of the PPM changes. Out of these changes devices and device components evolve. Fig. 25.2 illustrates the breadth of the PPM multi-space by associating function with pore diameter. PPM derived components include separation media, mixers, valves, salt bridges, and pumps. The PPM provides practical advantages when used in chips because of their ease of introduction, polymerization and photopatternability. The surface properties of the PPM, such as charge and hydrophobicity, add nuances to their function. In the following sections the monolith as on-chip separation media, valves, pumps, mixers, and pre-concentrators will be described and evaluated.

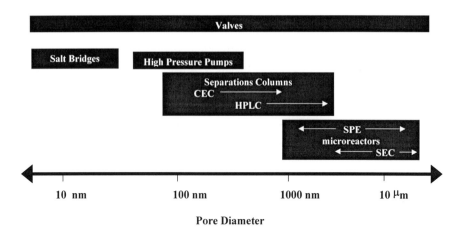

Fig. 25.2. The porous polymer monolith (PPM) multi-space. The PPM devices are shown as a function of pore diameter.

25.2 SEPARATIONS

Electrophoretic techniques are limited to systems where a difference in charge to size ratio exists . For many compounds of interest, for example explosives, polycyclic aromatic hydrocarbons, and drugs, a difference in hydrophobicity or hydrophilicity is the best or only physical handle for separations. Other charged targets, such as proteins, peptides, and amino acids are sufficiently similar in charge to size ratio that chromatography must be used to improve separation selectivity and resolution. Open column methods can be used effectively to separate these compounds, but there is a sample capacity limitation [31]. Consequently, separations that make use of a stationary phase, via a packed or coated channel, have been investigated for on-chip separations. The advantages of using porous polymer monoliths as a stationary phase have been described for capillary methods [32]. The challenges of the planar format of on-chip separations make full use of the unique advantages of porous polymer monoliths in a way that capillary systems do not.

There are several examples of separations in open channels on a chip. Rapid open channel separations of explosives by micellar electrokinetic chromatography (MEKC) on-a-chip have recently been reported [18]. While MEKC has the advantage that the conditions, including the stationary phase, can be rapidly adjusted, it also has two inherent limitations: a dependence on solubility of compounds and a limited elution window [33]. In MEKC there is an optimum elution time outside which maximum resolution cannot be achieved because strongly retained compounds are pushed to the end of the chromatogram and elute after or close to the micellar zone. Organic solvents added to the running buffer, typically used to extend the elution range of MEKC, can disrupt the micellar pseudostationary phase [34]. Gradient elution MEKC, demonstrated on chip by Kutter *et al.* shows promising results with coumarin dyes; however, relatively low organic (30%) buffers were used, limiting the sample range to less hydrophobic compounds [35]. Coated open channels in chips have given good results both with gradient and isocratic conditions [36,37]. Reduction of the channel depth to dimensions on the order of 3 to 5 μm afforded higher efficiencies characterized by low plate heights, but channels of these dimensions proved difficult to make in a controlled and reproducible fashion [35,37]. Additionally, the issue of capacity remains for any open channel technique, particularly as dimensions decrease.

Capillary electrochromatography (CEC) in capillaries packed with octadecylsilane coated silica beads has proven to give comparable efficiencies to high performance liquid chromatography and open column chromatography. By using electroosmotic flow to move the mobile phase, high-pressure drops are avoided, and packings with much smaller particle sizes can be used. Increased surface area of these packed beds provides increased column capacity over open channels. The physical stability of the stationary phase to organic solvents gives CEC an advantage over MEKC for the analysis of neutral molecules.

Packing channels etched in chips with beads is difficult because frits must be added to the structure, pressure attachments must be made from a pump or syringe to the chip, and packing is difficult to compress or remove once the chip is packed [7]. Recent work by Oleschuk *et al.* indicates that packing columns electrokinetically does not yield beds longer than 5 mm [38]. Microfabrication of the stationary phase for CEC has been described by He and Regnier [39] but this technology necessitates very specialized fabrication methods in fused silica, requires post fabrication derivatization of the substrate, and gives a relatively low surface area compared to bead packing or porous materials [39].

Monolithic acrylamide gel supports for a variety of on-chip CEC separations have been reported [40,41]. These supports used additives to impart the desired chromatographic interaction, and use redox chemistry for the initiation of polymerization. Ericson *et al.* described microchip separations using polyacrylamide gels that were modified to support hydrophobic interactions for the separation of neutral compounds such as uracil reporting efficiencies of up to 350 000 plates/m [40]. Koide and Ueno reported an acrylamide gel with covalently attached β-cyclodextrins (β-CD) derivatives for enantiomeric CEC separations of dansyl-DL-amino acids with efficiencies of up to 150 000 plates/m [41]. These gels show promise for separation; however, their gel-like structure limits their structural stability.

Porous polymer monoliths (PPM) for capillary electrochromatography such as those described by Peters *et al.* offer the combined advantage of ease of introduction, patternability, and tunability [42,43]. The channels were filled with a solution of monomer using capillary action, so no pressure attachments are necessary. The PPM form as the growing polymer phase-separates from the solvent to create a continuous monolith. Once polymerized, no frits are required which eliminates surfaces for bubble aggregation, and increases overall ruggedness of the bed [31]. Pretreatment of the chip with silanes ensures the monolith is covalently bonded to the walls. The monolith mean pore size and distribution can be changed by adjusting the monomer/solvent mixture composition [42,43]. The zeta potential, charge, and charge density can be adjusted by changing the identity and concentration of the ionizable monomer [44].

Two methods are commonly used to initiate the phase separation and polymerization. Thermal polymerization, first described for CEC by Peters *et al.* [43], is the most often found in the literature [42,43]. Complications were reported when moving to the chip, where evaporation of the porogenic solvent from the reservoirs mounted on the channel was observed during the polymerization process [45].

Photoinitiation is more desirable for monoliths in chips because it avoids polymerization at elevated temperatures and any change in phase behavior caused by the change in temperature. More importantly, photolithography can be used to define the regions where the monolith forms. This has recently been reported by Fintschenko *et al.* and Singh *et al.* [45.46]. Acrylate monomers were used successfully with UV-initiated polymerization that took only 20 minutes to complete [44]. In summary, a monomer solution in a porogenic solvent was introduced into the channels of a chip using capillary action after overnight pretreatment of the walls with 3-(trimethoxysilyl)-propyl acrylate. The chip was then exposed to 365 nm light from an ultraviolet (UV) lamp after the injection arms were masked with black tape [45]. The photopatterned monolith is shown in Fig. 25.3.

25.2.1 Chromatographic performance

Sandia National Laboratories has a long history of synthesizing microporous materials [47,48]. The synthesis and chromatographic characterization of a photoinitiated acrylate monolith used for chip electrochromatography (ChEC) were first reported by Ngola *et al.* in 2001 [44]. A SEM of a typical monolith cast in an isotropically etched chip is shown in Fig. 25.4 [46]. Three polycyclic aromatic hydrocarbons (PAH) were separated in 20 min under isocratic conditions (80% acetonitrile/ 5 mol/l Tris buffer, pH 8.5) with on-column, off-packing laser-induced fluorescence detection at 257 nm [44]. Shown in Fig. 25.5 is an isocratic ChEC separation of 13 out of 15 PAH in 15 minutes under similar conditions [45]. Van Deemter plots for early, middle, and late eluting compounds yielded a minimum plate height of 5 μm (representing an effi-

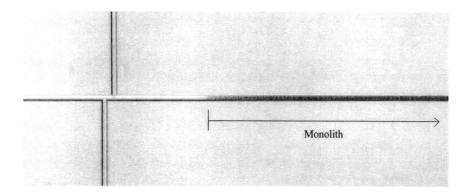

Fig. 25.3. Injection cross with separation channel fabricated in fused silica. The monomer mixture was introduced into all channels in the chip but the polymerization was photoinitiated through a mask of black tape leaving PPM only in the separation channel.

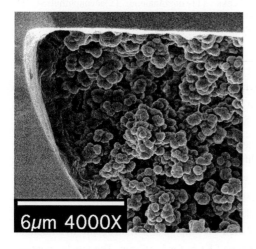

Fig. 25.4. A partial cross section of monolith cast in a channel isotropically etched in glass. This view is shown to highlight the PPM-edge interface. Channel dimensions: width 80 μm and depth 20 μm.

ciency of up to 200,000 plates/m) [45]. These results were comparable to those observed by Yu *et al.* who reported a maximum column efficiency of 210,000 plates/m for an UV-initiated porous bed in a capillary [49].

Chips with channels containing polymer monoliths have also been used for the separation of biological molecules such as amino acids and peptides [46,50]. Peptide separations are important not only in the field of proteomics, but also in drug discov-

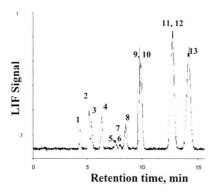

Fig. 25.5. Electrochromatographic separation of 13 PAH on-chip [45]. Channel length to detection point 7 cm, max. voltage 2 kV, running buffer 80:20 (v/v) acetonitrile–20 mmol/L Tris pH 8.5, injection 2kV, 5 min. Peaks: (1) naphthalene, (2)acenaphthalene, (3) acenapthene, (4) fluorene, (5) phenanthrene, (6) anthracene, (7) fluoranthene, (8) pyrene, (9) benz[a]anthracene, (10) chrysene, (11) benz[b]fluoranthene, (12) benzo[k]fluoranthene, (13) benzo[a]pyrene.

ery and detection of potential biological warfare agents. The analysis of proteins from a cell frequently involves the step of proteolytic digestion of protein into peptides followed by their analysis using LC/MS or CE/MS. A number of reports have appeared on separation of peptides by CZE or MEKC in microchips [19,51,52]. While these open-channel electrokinetic techniques are quite powerful, they do not offer the versatility, reproducibility, and robustness of chromatographic techniques.

Lauryl acrylate porous polymer monoliths containing sulfonic acid groups on the surface were prepared in the channels of a chip by polymerization initiated by an UV-lamp. Bioactive peptides were labeled off-chip with a fluorogenic dye naphthalene dicarboxaldehyde (NDA) in presence of KCN for 5–10 min. The labeled peptides were diluted to concentrations of 10 to 100 nmol/l in running buffer prior to injection. The running buffer consisted of acetonitrile–25 mmol/l borate (30:70 v/v) with 10 mmol/l octane sulfonate. Table 25.1 lists the bioactive peptides, their isoeletric points and molecular weights. Fig. 25.6 shows the separation of peptides in the microchip by isocratic reversed-phase electrochromatography [50]. The fastest separation (Fig. 25.6 inset) was achieved at an applied voltage of 5kV (766 V/cm) where all five peptides were baseline separated in about 50 s. Separation of the same peptides in capillary electrochromatography using the same polymer monolith took over 9 min. The 10-fold improvement in speed, without loss of resolution, underlines the advantages of microchip-based analysis. Similar improvements in analysis time have also been re-

TABLE 25.1

BIOACTIVE PEPTIDES USED FOR CHIP SEPARATION SHOWN IN FIG. 25.6

	Peptide	Sequence	Molecular Weight	pI (theoretical)
1	Papain inhibitor	GGYR	451.5	8.75
2	Proctolin	RYLPT	648.8	8.75
3	Casein fragment 90-95	RYLGYL	784	8.59
4	Ile-Angiotensin III	RVYIHPI	897	8.75
5	Angiotensin III	RVYIHPF	931	8.75

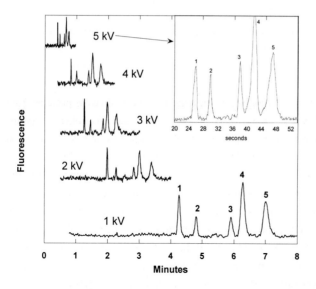

Fig. 25.6. Electrochromatography of peptides in a microchip containing porous polymer monoliths [50]. The peptides are listed in Table 25.1. The isocratic separation was carried out using acetonitrile: 25 mmol/L borate (30:70 v/v) containing 10 mmol/L octane sulfonate. Peptides were labeled with NDA and detected by LIF using 413 nm line of Kr-ion laser.

ported by a number of other researchers [53,54]. The increase in speed of analysis is primarily due to ability to use higher field strengths, which in turn is enabled by better heat dissipation in glass microchips compared to fused-silica capillaries [9,55]. The high separation power of chip-chromatography using polymer monolith is demonstrated by clear resolution of peptides 4 and 5, even at high flow-rates, that have similar pI and only differ in molecular weight by 34 g/mol.

The chip-based separations exhibited approximately 2-fold better efficiency than capillary-based separations under the same conditions [56]. The higher efficiency probably results from the ability to make extremely small injections with a fine control over injection size in a microchip. As electric field strength is increased, electroosmotic flow increases as well, leading to a decrease in separation time from 7.5 min at 1 kV to 50 s at 5 kV. However, a maximum in separation efficiency is achieved at intermediate (or lowest) field strength, and a significant loss in efficiency is experienced with high field-strengths. Peptides 1-3 have maximum efficiency at 2 kV while peptides 4 and 5 achieve their maxima at 1 kV, the lowest voltage used. The highest separation efficiency was 6×10^5 plates/m obtained for peptide 3 at 2 kV. For separation in miniaturized devices, speed is of utmost importance and hence the time-based efficiency (N/t) of separation as a function of field strength was evaluated as well. For peptides 1 and 5, the N/t increase with field strength attaining the maximum of 300 plates/s for peptide 1 at 4 kV, while for peptides 2-4, maximum in N/t is achieved at 2 kV (220 plates/s for peptide 2).

For the separation of complex peptide mixtures such as proteolytic digests and eventually for separation of proteins, gradient chromatography appears to be necessary. A solvent gradient in a microchip can be generated either electrosmotically [57] or by low flow-rate pumps [58,59]. Microscale capillary-based electrokinetic pumps have recently been developed using porous polymer monoliths and may provide a method for gradient separation in microchips.

Rohr *et al.* has reported a PPM mixer, providing a necessary component for any gradient scheme [60]. The mixer is photodefinable, and provides the fine structure necessary for mixing, without requiring any complicated structures in the glass substrate itself. Rohr's mixer takes advantage of the bimodal property of the PPM, and also incorporates a different photoinitiator, 2,2-dimethoxy-2-phenylacetophenone, for very rapid *in situ* photoinitiated polymerization [60]. The best results were obtained using a very heterogeneous PPM [60].

In conclusion, a PPM for on-chip isocratic separations has been demonstrated and characterized for neutral molecules and biomolecules. These on-chip chromatographic separations are comparable to, if not better than; separations achieved in HPLC and capillary-based CEC systems. They are fast amounting to less than 1 min for peptides and less than 15 min for PAH, provided good resolution, and offered high efficiency (up to 600 000 plates/m for peptides). However, further work to create the components necessary for gradient separations is required. Promising results for an on-chip PPM mixer and a PPM low flow rate pump have already been reported. A detailed treatment of the PPM valves and pumps follows.

25.3 VALVES

Microfluidics holds great promise for making portable, inexpensive, highly integrated chemical synthesis and analysis systems [61]. Unfortunately, complex functionality is required to take advantage of this potential, and the fundamental components such as valves required for this functionality are not available for many of the platforms most useful for microscale chemical analysis. Hence, a means of fabricating fundamental flow-control devices is a central problem.

The diversity of microfluidic systems leads to a wide variety of requirements in a microvalve. Microchip-based HPLC systems require high-pressure seals with low leak rates as well as insensitivity to solvent gradients. Electrophoresis systems have lesser pressure sealing requirements, but may require current hold-off and direct voltage control may be difficult. Continuing interest in autonomous microfluidic systems motivates flow control systems based on valve actuation in response to fluid properties. In all cases, goals for microvalve design include inexpensive fabrication (for use in disposable devices) and compatibility with existing substrate materials, such as plastic, glass, and silicon.

Extraordinary valve performance in terms of high-pressure hold-off and leak rates, current hold-off and leak rates, solvent resistivity, repeatability and robustness are necessary for truly useful microvalve systems. Many demonstration systems have been presented which are functionally complex but do not quantitatively evaluate valve performance that includes leak rates and solvent compatibility. Work in the coming years will require that valve performance be more thoroughly and rigorously tested. Despite extensive microelectromechanical systems (MEMS) capability in silicon, glass, and metals [61], the recent most promising flow control devices have been produced using polymeric materials [62-65]. Polymers show advantages over MEMS techniques in cost, repeatability, and performance. A number of valve architectures have been designed around "smart" polymers, which respond to external stimuli such as pH [66-68], temperature [69-77] or light [78,79].

In microchips, pH-sensitive polymers are the class of responsive polymers that has been used most extensively to control flow [64,80]. Hydrogels have been polymerized in-situ and patterned using UV-lithography. Since these hydrogels can be chosen to shrink or swell in response to acids or bases, they can be used to make flow decisions based on the acidity of the solution. The primary engineering challenge is the temporal response of the system, which is typically slow. The advantages of polymer monoliths in this system are clear. First, microfluidic channels are easily filled with solvent/monomer mixtures, and polymer monoliths are easily patterned using UV-lithography. The cost in time and infrastructure associated with these techniques is quite minor. Furthermore, these valves have the compelling property of allowing autonomous control of flow properties [80]. While most analytical systems already control pH straightforwardly through the use of buffers, the ability to autonomously control flow properties is essential for microfluidic systems designed for portability.

While pH-sensitive polymers have perhaps been the most-applied class of smart polymers, temperature-sensitive polymers have the advantage of being more straight-forwardly actuated. Much of the initial work has been motivated by the goal of thermally controlled isocratic HPLC separations, in which a temperature change rather than a solvent gradient results in the resolution of different species. The most common temperature-sensitive polymer in use is poly(N-isopropylacrylamide) (PNIPAM). PNIPAM undergoes a helix-random coil phase transition near 32°C [69-71], which results in drastic changes in volume [72], chemical permeability [73], and water solubility [74]. This has been used for pore control [72,74] and separation selectivity [74,75] in HPLC systems, and for control of drug delivery in membrane systems [73,76,77]. In capillaries, PNIPAM has been used for on/off valving by grafting PNIPAM onto the pores of a phase-separated porous polymer monolith [65]. The resulting temperature-sensitive pores may be opened by heating above the transi-tion point called lower critical solution temperature (LCST), and closed by cooling below the LCST. From the data presented in Ref. [65], it may be inferred that the pressure drop through these pores changes by a factor of at least 25, and probably more. The limits of temperature-controlled valving include the following: (a) the slow time response of thermal systems; (b) adverse effects on analysis systems (particularly electrokinetic systems) caused by temperature changes, particularly conductivity changes; (c) decreasing resolution in protein analysis systems caused by temperature-sensitive hydrophobic interactions; and (d) engineering requirements for temperature actuation, sensing, and control. Work is still required to address solvent effects on temperature-sensitive polymers, especially PNIPAM.

Response to light [78,79] is perhaps the most appealing polymer response, due to the ease of incorporating light-actuation into microfluidic channels in a non-intrusive manner. To our knowledge, work on light-responsive polymers has not been devel-oped for use in microfluidic chips.

25.3.1 Mechanical Actuation

In contrast to techniques that involve "smart" polymers, some techniques rely on the polymer insensitivity to most external stimuli such as temperature and solvent and use mechanical actuation to effect valve opening and closing. Soft polymeric sub-strates such as poly(dimethylsiloxane) (PDMS) have been used as microfluidic sub-strates for low-pressure, aqueous systems since they can be used to straightforwardly fabricate multilayer structures. By patterning multiple layers of intersecting channels, these devices have used an external pressure source to open and close microfluidic channels [63]. The simplicity of soft lithography fabrication techniques allows for the large-scale integration of pressure-driven flow-control devices on a single substrate. Since PDMS is soft and sensitive to many solvents, its application has been limited to low-pressure, aqueous systems. Leak rate performance in these valves has not been extensively presented, and its utility in chromatography has yet to be demonstrated, but the potential applications are numerous and clear.

Work at Sandia National Laboratories [62] has focused on developing a new family of microfluidic control devices by polymerizing freely mobile monoliths on-chip. Stops defined in the substrate prevent the part from moving when the monomer/solvent solution is flushed, and also provide sealing surfaces for valves and other flow-control devices. This architecture shows promise for microvalve applications since it can be quickly actuated, seals to high pressures, and is capable of holding off flow of both fluid and electrical current. Under certain actuation conditions, the devices show essentially no leakage, owing to the favorable sealing characteristics of soft polymer material against the glass substrate. The valves withstand pressures exceeding 20 MPa on-chip, and can be actuated in milliseconds.

25.3.2 Polymer formulations

Formulations typically include an approximately 1:1 ratio of bulk monomer chosen primarily for attractive surface energy properties and cross-linker selected primarily for its mechanical properties. Fluorinated bulk monomers are used to minimize the surface energy. Porogenic solvents are chosen to control pore size and swelling/shrinkage performance. Monomer/solvent ratios ranging from 60/40 to 80/20 are used to generate closed-shell structures that allow little or no flow through the pores. A typical formulation used in the examples presented here is as follows: 80% monomer mixture containing trifluoroethylacrylate and 1,3 butanediol diacrylate with 20% solvent, typically a mixture of 2-methoxyethanol, 1,3 dioxolane, and 5 mmol/l TRIS buffer. Azobisisobutyronitrile is used as a photoinitiator.

25.3.3 Laser microfabrication

Phase-separated polymer monoliths are photopolymerized inside wet-etched silica microfluidic channels. First, a monomer-solvent-initiator mixture is introduced into the channel. Polymerization and crosslinking are initiated within a selected region of the channel using about 45 s UV exposure from a 355 nm Nd:YAG laser at 4 mJ cm^{-2}/pulse at a frequency of 10 Hz through a chrome lithographic mask. Resolution is limited by diffraction and the competition between radical diffusion and initiation/recombination reactions. Typical devices range from 50 to 500 μm in size.

25.3.4 Applications and performance

Unlike most monolith applications, in which the monolith need adhere to the wall, valve applications require that the monolith be freely mobile. Low-surface-energy fluorine groups are used so that the monolith slips along the channel walls and creates a small piston that can be mobilized by applying pressure. This method produces a polymer part that conforms to the cross-sectional shape of the microchannel. The use of the microchannel as the "mold" for the part automatically confers tight sealing characteristics on the piston, making it particularly attractive for microfluidic control applications.

Polymer formulation may be varied in many different ways, leading to mobile monoliths with different structural and mechanical properties. The experience is that modest variations in formulation, especially in the amount of certain surface modifiers, can have dramatic effects on monolith porosity and other properties. Not surprisingly, the degree of polymerization and crosslinking directly affects the mechanical strength and porous nature of the polymer.

Critical parameters determining the behavior of these parts in microfluidic control applications include porosity, pore size, pore size distribution, elastic modulus, and friction coefficient with the wall. Using simple piston-type devices such as those described in the previous section, some of these properties are easily measured by applying pressure and observing the piston under a microscope. For example, by polymerizing a simple piston upstream of a contraction in the channel, and applying high pressure, the compression of the piston gives an estimate of its compressive strength. Compressive tests conducted in this way indicate that, depending on the degree of polymerization, elements have a modulus of elasticity similar to a common rubber stopper $(0.01–0.1 \text{ GN/m}^2)$. This result is strongly dependent on the degree of polymerization, which is easily changed by changing the UV exposure.

A critical parameter governing the utility of these devices is their coefficient of static friction. Tests were performed to determine this coefficient by slowly increasing the pressure applied across simple pistons, polymerized in commercially available fused silica capillary tubes, until the piston began to move. The axial force on the piston is the product of the applied pressure and cross-sectional area of the piston A_X. The contact surface area between the capillary and piston, A_s, is determined by the diameter and length of the piston. The normal force is proportional to A_s, and writing the coefficient for static friction as c_{sf}, we may assume $P_s A_X \sim A_s c_{sf}$, where P_s is the pressure required to move the piston. We also expect that c_{sf} is only a function of polymer formulation and the degree of polymerization. Since A_X is proportional to the square of the piston diameter, d_X, and A_s is proportional to d_X, we then expect $P_s d_X$ to be independent of d_X for a specific formulation and exposure. These expectations

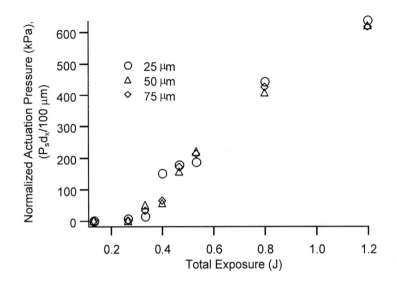

Fig. 25.7. Normalized actuation pressure for polymer pistons as a function of initiation energy. Piston length is 1 mm in all cases [62].

were investigated by polymerizing 600 μm length pistons using a range of UV exposures, and in three different diameters of capillary (25, 50, and 75 μm). Fig. 25.7 shows data for the typical polymer formulation described above, plotted as $P_s d_x/100$ μm versus exposure. The normalized actuation pressures collapse to a single curve, verifying the scaling equation $P_s A_x \sim A_s c_{sf}$. The dependence of c_{sf} on exposure dose can be inferred from the shape of this curve. This test does not, though, address the physical nature of the friction. It has been noticed that friction coefficients are higher in etched channels than in silica capillaries, suggesting that nanometer-scale roughness in the etched channels may be important.

Fig. 25.8 shows a check-valve (one-way valve) configuration, which only allows unidirectional flow due to the location of the piston relative to a bypass channel. Fig. 25.8a shows a three-dimensional schematic of both the substrate (clear) and the piston (gray). The cylindrical channels were constructed by wet etching mirror images of the channel geometry in two glass wafers, and then bonding them together using electrostatic bonding followed by thermal annealing. The two radii of the channels were created by etching the chip twice using a thick positive photoresist and exposing with a mask that only includes the large-radius features before the second etch. Fig. 25.8b and c show the check valve in operation. In Fig. 25.8b the piston allows flow past the piston via a bypass channel, while the piston is seated against the stop at right. In Fig. 25.8c the piston is seated against the stop at left, preventing any flow. Initial pressure

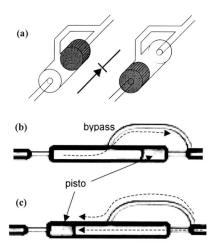

Fig. 25.8. Passive check valve on-chip [62]. (a) Schematic of valve architecture. The concentric-cylinder (with bypass) geometry is created in the substrate (shown clear) by wet-etching mirror images of half-cylinders in two substrates, then aligning and bonding. The free piston (gray) is polymerized *in situ*. Both glass wafer surfaces are HF-etched (in mirror-image) to produce interconnecting channels of concentric cylinders when bonded. (b) Flow from left-to-right bypasses the piston seated against the right stop. (c) Flow in opposite direction is prevented when the piston seats against the left stop, checking the flow. Cylinder diameter is 100 μm with 25 μm weirs. Flow direction is indicated by dashed lines. The typical working fluid is 90% acetonitrile/10% water.

and leakage tests have been performed in silica capillaries and glass chips in this configuration. Maximum pressures exceed 20 MPa bar on-chip, and actuation times are less than 1 video frame (33 ms). Actuation pressure can be less than 7 kPa, depending strongly on UV exposure during polymerization. The leakage rates of these devices vary between 0–50 pL/s over a wide range of pressures (0–20 MPa). As the pressure increases, the sealing behavior improves (leakage decreases), as the polymer deforms into imperfections in the sealing surface.

Fig. 25.9 shows a decision-making valve (two-inputs, one-output) created by polymerizing a piston in a region at the intersection of three channels. The piston is mobilized by any pressure difference between the two inputs, with the result being that the input with the higher pressure will flow through the valve to the output. This configuration has the advantage of zero dead volume and has similar sealing characteristics to the check valve.

More complex functions can be performed with a sequence of dependent elements as shown in Fig. 25.10, where a piston/cylinder combined with two check-valves can be used to make a 10 nL pipette. The fluorescence images show a 10 nL aliquot of

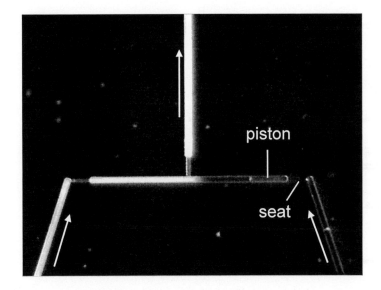

Fig. 25.9. Decision making valve. Input at lower left includes dye solution while the input at lower right contains no dye. The input with higher pressure exits at the top.

sample being introduced into a running (pressure-driven) LC column. Volumes ranging from picoliters to microliters can be defined by chip geometry.

25.4 ELECTROKINETIC PUMPS

Electrokinetic pumps (EKPs) are a relatively new micropump technology that uses electroosmotic flow (EOF) to generate pressure [58,59,81,82]. In the most generic sense, an EKP is simply a microchannel with electrodes situated across it; however, by packing the microchannel with porous media thereby making many small "channels", an EKP can generate significant pressures – up to 34 MPa (10,000 psi) [58]. These pumps provide a practical method to transport fluids on a microscale without moving parts, and so they have many possible on-chip applications, including on-chip HPLC. Making EKP media with porous polymer monoliths (PPMs) is highly scalable, as shown in Fig. 25.11. Both on-chip and macroscale devices are easily produced in the same general manner, and they have a $>10^4$ ratio in cross-sectional area. PPMs may be the best way to pack the channels, if their performance in this application can match that of packed silica beads.

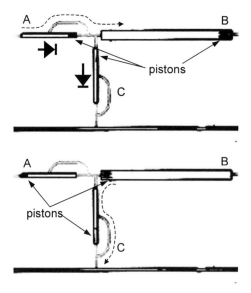

Fig. 25.10. A 10 nL pipette for aliquoting sample into a channel [62]. Check valves are placed at inlet A and outlet C, and a simple piston/cylinder reservoir at inlet B. Dashed lines indicate fluid flow direction. Top: Suction is applied at inlet B; in response, the piston in cylinder B moves to the right, drawing sample from inlet A through the open check valve, until the reservoir at inlet B is filled. The check valve at outlet C remains closed. Bottom: Pressure is applied at inlet B; the check valve at inlet A closes, the valve at outlet C opens, and the fluid in cylinder B is injected into the channel.

Fig. 25.11. Electrokinetic pumps in chip format (left) and in macroscale disk formats (right). The porous polymer monolith in the chip is visible as a white stripe; the polyimide wells glued to the top of the chip mate to standard HPLC fittings. The large disk format has produced flow rates in the range of 1 mL/min.

25.4.1 Polymer formulation

Porous polymer monolith formulations for EK pumping that have been tested to date have included a charged species in addition to the porogen, typically acetonitrile (60%) and ethanol (20%) and a 5 mol/L aqueous phosphate buffer at pH 6.8 (20%), together with methacrylate monomers such as butyl methacrylate, ethylene dimethacrylate, and tetrahydrofurfuryl methacrylate and the initiator, 2,2'-azobis-isobutyronitrile. Ionizable monomers used were [2-(methacryloyloxy)ethyl] trimethyl-ammonium methyl sulfate (MOE), 2-acrylamido-2-methyl-1-propanesulfonic acid (AMPS), and 2-(N,N-dimethylaminoethyl methacrylate (DMA). MOE and DMA may carry a positive charge, while AMPS carries a negative charge. Thus, the resulting pumps are correspondingly called "positive" or "negative". The effect of these ionizable monomers is dramatic, and they are needed in fairly low concentrations (0.5–3%) to yield relatively high surface potentials. Excess of these monomers leads to undesirably large pores. Porosimetry results show that peak pore diameters range from 49 nm in 0.5% MOE to 124 nm in 3% DMA and 3% AMPS. In general, peak pore diameters increase with increasing the content of ionizable species; all 3% pumps produced to date have larger pores than their 0.5% counterparts. Also, AMPS monoliths tend to have larger peak pores than DMA, which in turn are slightly larger than MOE columns.

Because EKPs generate high pressures, the pre-treatment of chip/capillary walls is an important step to achieve strong binding of the monolith. A pretreatment solution consisting of deionized water (50%), glacial acetic acid (30%), and binding agents either *N*-[3-(trimethoxysilyl)propyl]-*N'*-(4-vinylbenzyl)ethylenendiamine hydrochloride for positive pumps, or 3-(trimethoxysilyl)propyl acrylate for negative pumps appears to be adequate for this purpose. The solution is flushed through the glass substrate and allowed to react for 16 h before flushing with acetonitrile and air.

Polymerization of pumps is achieved with thermal or UV initiation. In the case of thermal initiation, the entire wafer is placed in an oven and kept at 65°C for approximately 18–24 h. Photopolymerization of on-chip EKPs is achieved using a 257 nm Ar+ laser (frequency-doubled), 355 nm (tripled) Nd:YAG laser, or a broadband UV lamp. As usual for porous monoliths, the PPM is purged by flushing the pump overnight with a solvent containing a high percentage of organic solvent, using both pressure-driven flow from an HPLC pump and by electroosmosis. Pumps are typically evaluated using 5 mM acetic acid buffer (pH 4.5) with 1-propanol (20%) as the running buffer. This mixture is not optimal for maximum pressure – pure acetonitrile generally results in the best performance.

25.4.2 Performance: theory and applications

What physical properties of the PPM affect pump performance? EKPs operate on a principle that is simple for the case of a round capillary, and only slightly more complicated for porous media. The speed of EOF through a capillary depends only on the permittivity (ε) and dynamic viscosity (η) of the fluid, the zeta potential (ζ) of the wall, and the local electric field (E):

$$q_{eof} = \frac{\varepsilon}{\eta} \frac{\zeta}{} \frac{\Delta \varphi}{L} \tag{25.1}$$

Here q is the volume flow rate per unit area of the capillary (q has the units of velocity such as mm/s), and $\Delta\varphi/L$ is the electric field, where φ is the applied potential (in volts). The velocity profile is approximately flat across the capillary for this component of the flow rate. It is clear from this equation that the only physical property which can be largely influenced by the pump surface chemistry is ζ, and in fact the EOF flow rate is proportional to ζ. For glass in contact with aqueous 1 mM electrolyte at neutral pH, a typical value of ζ is 60 mV, ε is 78, and $\eta \sim 10^{-3}$ Pa·s. So at an electrical field of 250 V/cm, q_{eof} is about 1.0 mm/s.

If there is also a pressure differential across the capillary, one must also account for the pressure-driven flow, which depends strongly on the diameter of the capillary (d), the applied pressure differential per unit length, $\Delta P/L$, and the viscosity:

$$q_{pressure} = \frac{1}{32} \frac{d^2}{\eta} \frac{\Delta P}{L} \tag{25.2}$$

In this expression we have assumed fully-developed flow through round capillary. These flow rates may be linearly added when potential and pressure differences are both applied across the capillary. For example, if an EOF-driven flow through a capillary is dead-ended into a sealed vial, the pressure builds up until the net flow rate is zero. By solving $q_{eof} = - q_{pressure}$, the pressure that is developed is

$$\Delta P_{max} = \left(32 \, \varepsilon \, \zeta / d^2\right) \Delta \varphi \tag{25.3}$$

The maximum *pressure* depends very strongly on the *diameter* of the capillary. Using the values for water/glass shown above, a 5 μm-diameter capillary, which is the smallest commercially available off-the-shelf, gives a maximum pressure of only 68 Pa/V. However, the capillary may be packed with silica beads or filled with a porous polymer monolith that supports EOF such that the pores are as small as 200 nm. In this case, the maximum pressure becomes of the order of 7000-70,000 Pa/V, owing to the proportionality of the pressure with d^{-2}.

Generalizing equations (25.1) and (25.2) to a porous medium, and superposing the two contributions to total flow rate by adding (25.1) and (25.2), gives the following relationship [58]:

$$q = \left(\frac{\varepsilon \, \zeta'}{\eta} \frac{\Delta\varphi}{L} + M \frac{\Lambda^2}{\eta} \frac{\Delta P}{L} \right) \frac{1}{F} = \left(v' \frac{\Delta\varphi}{L} + k' \frac{\Delta P}{L} \right) \frac{1}{F} \tag{25.4}$$

Here we have introduced a formation factor, F, which accounts for the obstruction of current flow due to the packing. The formation factor is the ratio of the current drawn through the open capillary, divided by the current drawn through the packed capillary, filled with the same buffer. It is so defined in the absence of surface conduction, hence it is always greater than unity. The value of v'/F then directly determines the maximum flow rate per unit area through the PPM due to an applied voltage differential, and the value of k'/F determines the flow rate through the PPM due to an applied pressure differential.

The equation 25.4 above shows explicitly that the factor v' is an "effective" electroosmotic mobility that should be proportional to ε/η (fluid properties), but also to ζ', which increases with the surface charge density of the surface [83] but decreases with increasing overlap of the electric double layers on the surface of the pore. The factor k' is an "effective" Darcy permeability that is also proportional to $1/\eta$, but it is strongly dependent on the pore diameter. The desirable properties of a pump are that it have large flow rate and high maximum pressures. These directly translate to the desirable characteristics of a PPM for EK pumping, namely that it should have *small* pores for high pressure generation and a large ζ (*e.g.*, surface charge density) for maximum flow rate. Pore size in the range of 100–200 nm is ideal since the overlap of double layers is less effective.

To characterize these properties of the PPM, the first test is to directly measure v'/F by measuring the maximum flow rate per unit area per unit electric field, when no back pressure opposes the flow. EK pumping flow rates can be as high as 700 nL/s at 1kV/cm·mm^2. The second test is to measure the maximum pressure, per 1 V, with

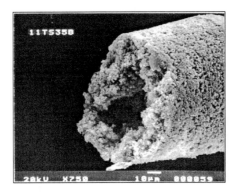

Fig. 25.12. Scanning electron micrograph of a typical porous monolith used for electrokinetic pumping.

no net flowrate. This means that the pump is dead-ended into a pressure gauge. The best results achieved to date have been up to 7.5 kPa/V.

A variety of other tests such as the direct measurement of the Darcy constant, streaming potential measurements [83], BET surface area analysis, porosimetry, and SEM images, as shown in Fig. 25.12 can be used to determine these and other characteristics of the medium as well. Unpublished efforts to correlate these fundamental properties to pump performance, however, were somewhat frustrating due to the wide pore distribution typically encountered with PPM and the non-negligible effects of overlapping double-layers in mesopores.

Although silica beads have given the best performance for high-pressure EKPs to date amounting to 60 kPa/V [58] as compared to 20 kPa/V for PPM, porous polymer monoliths offer some intriguing advantages in terms of manufacturability and adaptability to microchip integration via *in situ* photopolymerization. Unlike silica beads, they do not require micromachining of frits to hold them in place nor do they require special fritting techniques. Large-scale integration of complex EKP-driven fluid handling does not seem feasible with packed silica beads – the additional channels and fritting required to pack a chip with one hundred EKPs would be an interesting engineering challenge. On the other hand, scale-up of UV polymerization to thousands of devices per wafer should be straightforward.

25.5 MEMBRANES

Polymeric membranes have seen significant development for macroscale application in recent years, especially in applications for ion sensing and potentiometry.

Materials such as PVC and commercially available ionophores [84] such as Nafion® [#161] membranes are in common use. Polymeric layers are also routinely used for electrode surface modification [85]. Because polymers have shown such versatility in macroscale systems due to the wide variety of properties and specificity that they exhibit, it seems logical to adapt them for use in microscale systems.

Work is emerging in the area of integrating membranes within more complex microsystems for bioanalysis. A recent review by Wang et al. [86] focuses on several technologies that use a "sandwich" architecture in which conventional polymeric membranes are sandwiched between wafers containing microfabricated devices and/or channel substrates. Devices have been demonstrated for a wide variety of bioanalytical applications, including sample desalting using molecular-weight cutoff membranes [87], sensitive detection applying affinity microdialysis and ultrafiltration [88,89], and membrane chromatography. On-chip porous *glass* membranes in "sandwich" architectures have also been demonstrated for DNA preconcentration, using spin-on silicates [90] to create a porous glass layer.

An alternative architecture, based on *in situ* photopolymerization of membranes within microfluidic channels, has also been fruitful. Bratov *et al.* have developed a photocurable polyurethane composition that is suitable for microscale ion-selective electrodes. They have demonstrated the device for sensing of ammonium and potassium [91,92]. Furthermore, Yu et al. recently demonstrated the copolymerization of several acrylates in the presence of hexane/methanol porogens, to create a porous polymer, which is suitable for solid-phase extraction [93]. This class of photopolymerized porous methacrylate polymers appears to be finding a wide range of applicability for bioanalysis due to the wide range of tailorability of its properties.

A variety of ion-selective electrodes (ISE) suitable for adaptation to microchip format were reviewed by Antonisse and Reinhoudt [95]. These include (i) the conventional electrode, in which a wire is immersed in an internal reference solution, that is then separated from the sample solution, (ii) the covered wire electrode, in which the internal reference solution between the membrane and wire is eliminated, and (iii) the ISFET [96,97], in which case the ion-selective polymer is deposited over the gate region between source and drain on a modified field-effect transistor. Each of these has advantages and limitations. The ISFET, for example, has inherent drift that must be taken into account. Covered-wire electrodes have mechanical durability problems due to imperfect attachment of the membrane to the conductor, and drift associated with leaching of membrane plasticizer. Significant effort has been put forth to reduce the plasticizer content recently, with some success. Microscale versions of conventional electrodes have not been vigorously pursued as of yet; however, the mechanical stability of photopatterned acrylate-based porous polymers makes them likely candidates for development in this area.

References pp. 682-685

25.6 CONCLUSIONS

Chip-based devices hold great promise because of the potential for integration. The ability to use photolithography to define connected structures and a single substrate, such as the glass wafer used to hold these structures, has tremendous appeal. The dimensions that can be manufactured in a variety of materials are on the order of a few microns, dimensions that are appealing to analytical chemists because they translate to the improvement in separation speed and efficiency. However, there are a great many challenges to integrating the components required for a successful μTAS.

The porous polymer monolith has an astonishing versatility in structure. Tuning its chemistry has been reported to change relatively minor features, such as charge, to major structural and functional shifts, such as high-pressure pumps and valves. This is exciting, because the PPM, like the chip, is amenable to photolithography. It is introduced as a monomer, usually by capillary action, eliminating the hurdles of chip connections for fabrication. However, there are still areas of significant work to be done. The PPM used for separation media, valves, and pumps are still in the early stages of research and development. The properties of the PPM match the requirements of the planar chip structure to such a degree it is possible that every component required for complex μTAS incorporating SPE, chromatography, mixers, pumps, and valves could be fashioned solely based on PPM chemistry.

25.7 ACKNOWLEDGEMENT

This work was financially supported by Laboratory Directed Research and Development program at Sandia National Laboratories. Sandia is a multiprogram laboratory operated by Sandia Corporation, a Lockheed Martin Company, for the United States Department of Energy under contract DE-AC04-94AL85000. The authors gratefully acknowledge the assistance of Dan Throckmorton, Werr Yee Choi and George Sartor of Sandia National Laboratories.

25.8 REFERENCES

1   A. Manz, N. Graber, H.M. Widmer, Sens. Actuat. B 1 (1990) 244 .

2   A. Manz, D.J. Harrison, E.M.J. Verpoorte, J.C. Fettinger, A. Paulus, H. Ludi, H.M. Widmer, J. Chromatogr. 593 (1992) 253.

3   S.C. Jacobson, R. Hergenroder, L.B. Koutny, J.M. Ramsey, Anal. Chem. 66 (1994) 1114.

4   S.C. Jacobson, R. Hergenroder, L.B. Koutny, R.J. Warmack, J.M. Ramsey, Anal. Chem. 66 (1994) 1107.

5   D.J. Harrison, A. Manz, Z.H. Fan, H. Ludi, H.M. Widmer, Anal. Chem. 64 (1992) 1926.

6   A. Manz, Y. Miyahara, J. Miura, Y. Watanabe, H. Miyagi, K. Sato, Sens. Actuat. B 1 (1990) 249.

7   G. Ocvirk, E. Verpoorte, A. Manz, M. Grasserbauer, H.M. Widmer, Anal. Meth. Instrum. 2 (1993) 74.

8   S.R. Wallenborg, C.G. Bailey, P.H. Paul, in: A. van den Berg, W. Olthuis, P. Bergveld, (Eds.) Micro Total Analysis Systems 2000, Proceedings of the μTAS Symposium, Enschede, The Netherlands, 14-18 May 2000, Kluwer, Dordrecht, 2000, p. 355.

9   K. Seiler, D.J. Harrison, A. Manz, Anal. Chem. 65 (1993) 1481.

10 A. Manz, E. Verpoorte, C.S. Effenhauser, N. Burggraf, D.E. Raymond, H.M. Widmer, Fresenius J. Anal. Chem. 348 (1994) 567.

11 S.C. Jacobson, J.M. Ramsey, Electrophoresis 16 (1995) 481.

12 N. Chiem, D.J. Harrison, Anal. Chem. 69 (1997) 373.

13 A.G. Hadd, D.E. Raymond, J.W. Halliwell, S.C. Jacobson, J.M. Ramsey, Anal. Chem. 69 (1997) 3407.

14 C.S. Effenhauser, G.J.M. Bruin, A. Paulus, M. Ehrat, Anal. Chem. 69 (1997) 3451.

15 C.L. Colyer, T. Tang, N. Chiem, D.J. Harrison, Electrophoresis 18 (1997) 1733.

16 H.J. Crabtree, M.U. Kopp, A. Manz, Anal. Chem. 71 (1999) 2130.

17 G.F. Jiang, S. Attiya, G. Ocvirk, W.E. Lee, D. J. Harrison, Biosens. Bioelec. 14 (2000) 861.

18 S.R. Wallenborg, C.G. Bailey, Anal. Chem. 72 (2000) 1872.

19 N. Gottschlich, C.T. Culbertson, T.E. McKnight, S.C. Jacobson, J.M. Ramsey, J. Chromatogr. B 745 (2000) 243.

20 K.F. Schrum, J.M. Lancaster, S.E. Johnston, S.D. Gilman, Anal. Chem. 72 (2000) 4317.

21 A. Hilmi, J.H.T. Luong, Envir. Sci. Tech. 34 (2000) 3046.

22 R. Tantra, A. Manz, Anal. Chem. 72 (2000) 2875.

23 J. Wang, M.P. Chatrathi, B.M. Tian, Anal. Chim. Acta 416 (2000) 9.

24 J. Wang, B.M. Tian, E. Sahlin, Anal. Chem. 71 (1999) 5436.

25 J.S. Rossier, M.A. Roberts, R. Ferrigno, H.H. Girault, Anal. Chem. 71 (1999) 4294.

26 J. Wang, M. Jiang, B. Mukherjee, Anal. Chem. 71 (1999) 4095.

27 J. Wang, B.M. Tian, E. Sahlin, Anal. Chem. 71 (1999) 3901.

28 M. Hashimoto, K. Tsukagoshi, R. Nakajima, K. Kondo, A. Arai, J. Chromatogr. A 867 (2000) 271.

29 G.C. Fiaccabrino, N.F. deRooij, M. Koudelka-Hep, Anal. Chim. Acta 359 (1998) 263.

30 R. Weinberger, Practical Capillary Electrophoresis, Academic Press, Inc., New York 1993.

31 A.L. Crego, A. Gonzalez, M.L. Marina, Crit. Rev. Anal. Chem. 26 (1996) 261.

32 F. Svec, E.C. Peters, D. Sýkora, G. Yu, J.M.J. Fréchet, J. High Resolut. Chromatogr. 23 (2000) 3.

33 M.G. Khaledi, J. Chromatogr. A 780 (1997) 3.

34 A.T. Balchunas, M.J. Sepaniak, Anal. Chem. 59 (1987) 1466 .

35 J.P. Kutter, S.C. Jacobson, J.M. Ramsey, Anal. Chem. 69 (1997) 5165.

36 J.P. Kutter, S.C. Jacobson, N. Matsubara, J.M. Ramsey, Anal. Chem. 70 (1998) 3291.

37 S.C. Jacobson, R. Hergenroder, L.B. Koutny, J.M. Ramsey, Anal. Chem. 66 (1994) 2369.

38   R.D. Oleschuk, L.L. Shultz-Lockyear, Y. Ning, D.J. Harrison, Anal. Chem. 72 (2000) 585.

39   B. He, F. Regnier, J. Pharm. Biomed. Anal. 17 (1998) 925.

40   C. Ericson, J. Holm, T. Ericson, S. Hjertén, Anal. Chem. 72 (2000) 81.

41   T. Koide, K. Ueno, J. High Resolut. Chromatogr. 23 (2000) 59.

42   E.C. Peters, M. Petro, F. Svec, J.M.J. Fréchet, Anal. Chem. 70 (1998) 2288.

43   E.C. Peters, M. Petro, F. Svec, J.M.J. Fréchet, Anal. Chem. 69 (1997) 3646.

44   S.M. Ngola, Y. Fintschenko, W.-Y. Choi, T.J. Shepodd, Anal. Chem. 73 (2001) 849.

45   Y. Fintschenko, W.-Y. Choi, S.M. Ngola, T.J. Shepodd, Fresenius J. Anal. Chem. 371 (2001) 174.

46   A.K. Singh, D.J. Throckmorton, T.J. Shepodd, in: J.M. Ramsey, A. van den Berg (Eds.) Micro Total Analysis Systems, Proceedings of the μTAS 2001 Sympsium, Monterey, CA, USA, 21-25 October 2001, Kluwer, Dordrecht, 2001, p. 649.

47   J.N. Aubert, R.L. Clough, Polymer 26 (1985) 2047.

48   J.H. Aubert, R.L. Claugh, US. Patent 4,673,695.

49   C. Yu, F. Svec, J.M.J. Fréchet, Electrophoresis 21 (2000) 120.

50   D.J. Throckmorton, T.J. Shepodd, A.K. Singh, Anal. Chem. in press.

51   J.H. Chan, A.T. Timperman, D. Qin, R. Aebersold, Anal. Chem. 71 (1999) 4437.

52   J.J. Li, J.F. Kelly, I. Chemushevich, D.J. Harrison, P. Thibault, Anal. Chem. 72 (2000) 599.

53   D. Schmalzing, L. Koutny, A. Adourian, P. Belgrader, P. Matsudaira, D. Ehrlich, Proc. Natl. Acad. Sci. USA 94 (1997) 10273.

54   S. Yao, D. Anex, W. Caldwell, D. Arnold, K. Smith, P. Schultz, Proc. Natl. Acad. Sci. USA 96 (1999) 5372.

55   Z. Fan, D. Harrison, Anal. Chem. 66 (1994) 177.

56   R. Shediac, S.M. Ngola, D.J. Throckmorton, D.S. Anex, T.J. Shepodd, A.K. Singh, J. Chromatogr. A 925 (2001) 251.

57   C. Yan, R. Dadoo, R.N. Zare, D.J. Rakestraw, D.S. Anex, Anal. Chem. 68 (1996) 2726.

58   P.H. Paul, D.W. Arnold, D.W. Neyer, K.B. Smith, in: A. van den Berg, W. Olthuis and P. Bergveld, (Eds.) Micro Total Analysis Systems 2000, Proceedings of the μTAS Symposium, Enschede, The Netherlands, 14-18 May 2000, Kluwer, Dordrecht, 2000 p. 583.

59   P.H. Paul, D.W. Arnold, D.J. Rakestraw, in: J. Harrison, A. van den Berg (Eds.) Micro Total Analysis Systems 1998, Proceedings of the μTAS Symposium, Banff, Canada, Kluwer, Dordrecht, 1998, p. 49.

60   T. Rohr, C. Yu, M.H. Davey, F. Svec, J.M.J. Fréchet, Electrophoresis 22 (2001) 3959.

61   S. Shoji, M. Esashi, J. Micromech. Microeng. 4 (1994) 157.

62   E.F. Hasselbrink, T.J. Shepodd, J.E. Rehm, Science, submitted.

63   M.A. Unger, H.P. Chou, T. Thorsen, A. Scherer, S.R. Quake, Science 288 (2000) 113.

64   D.J. Beebe, J.S. Moore, J.M. Bauer, Q. Yu, R.H. Liu, C. Devadoss, B.H. Jo, Nature 404 (2000) 588.

65   E.C. Peters, F. Svec, J.M.J. Fréchet, Adv. Mat. 9 (1997) 630.

66   Y. Okahata, K. Ozaki, T. Seki, J. Chem. S., Chem. Com. (1984) 519.

67   T. Seki, Y. Okahata, Macromolecules 17 (1984) 1880.

68   Y. Okahata, T. Seki, Chem. Let. (1984) 1251.

69   M. Heskins, J.E. Guillet, J. Macromol. Sci., Chem. A2 (1968) 1441.

70   S. Fujishige, K. Kubota, I. Ando, J. Phys. Chem. 93 (1989) 3311.

71   Y.H. Bae, T. Okano, S.W. Kim, J. Polymer Sci. B 28 (1990) 923.

72   M. Gewehr, K. Nakamura, N. Ise, H. Kitano, Macromol.Chem. Phys. 193 (1992) 249.

73   Y.M. Lee, S.Y. Ihm, J.K. Shim, J.H. Kim, C.S. Cho, Y.K. Sung, Polymer 36 (1995) 81.

74   K. Hosoya, K. Kimata, T. Araki, N. Tanaka, J.M.J. Fréchet, Anal. Chem. 67 (1995) 1907.

75   H. Kanazawa, K. Yamamoto, Y. Matsushima, N. Takai, A. Kikuchi, Y. Sakurai, T. Okano, Anal. Chem. 68 (1996) 100.

76   H. Feil, Y.H. Bae, J. Feijen, S.W. Kim, J. Membr. Sci. 64 (1991) 283.

77   T. Ogata, T. Nonaka, S. Kurihara, J. Memb. Sci. 103 (1995) 159.

78   H. Finkelman, E. Nishikawa, G.G. Pereira, M. Warner, Phys. Rev. Lett. 87 (2001) 5501.

79   A. Suzuki, T. Tanaka, Nature 346 (1990) 345.

80   D.T. Eddington, R.H. Liu, J.S. Moore, D.J. Beebe, in: J.M. Ramsey, A. van den Berg (Eds.) Micro Total Analysis Systems, Proceedings of the µTAS 2001 Sympsium, Monterey, CA, USA, 21-25 October 2001; Kluwer, Dordrecht, 2001, p. 486.

81   W.-E. Gan, L. Yang, Y.-Z. He, R.-H. Zeng, M.L. Cervera, M. de la Guardia, Talanta 51(2000) 667.

82   S. Zeng, C.-H. Chen, J.C. Mikkelsen, J.G. Santiago, Sens. Actuat. B 79 (2001) 107.

83   H. Lyklema, Fundamentals of Interface and Colloid Science; Academic Press, London, San Diego, 1991.

84   K.A. Mauritz, C.J. Hora, A.J. Hopfinger, in: A. Eisenberg (Ed.), ACS Advances in Chemistry, American Chemical Society, Washington, D.C., Vol. 187, 1980, p. 124.

85   A.J. Bard, Integrated Chemical Systems, Wiley, New York, 1994.

86   P.-C. Wang, D.L. DeVoe, C.S. Lee, Electrophoresis 22 (2001) 3587.

87   N. Xu, Y. Lin, S.A. Hofstadler, D. Matson, C.J. Call, R.D. Smith, Anal. Chem. 70 (1998) 3553.

88   Y. Jiang, P.-C. Wang, L.E. Locascio, C.S. Lee, Anal. Chem. 73 (2001) 2048.

89   J. Gao, J.D. Xu, L.E. Locascio, C.S. Lee, Anal. Chem. 73 (2001) 2648.

90   J. Khandurina, S.C. Jacobson, L.C. Waters, R.S. Foote, J.M. Ramsey, Anal. Chem. 71 (1999) 1815.

91   A. Bratov, N. Abramova, J. Munoz, Anal. Chem. 67 (1995) 3589.

92   A. Bratov, N. Abramova, J. Munoz, J. Electrochem. Soc. 144 (1997) 617.

93   C. Yu, M.H. Davey, F. Svec, J.M.J. Fréchet, Anal. Chem. 73 (2001) 5088.

94   F. Svec, J.M.J. Fréchet, Science 273 (1996) 205.

95   M.M. G. Antonisse, D.N. Reinhoudt, Electroanal. 11 (1999) 1035.

96   P. Bergveld, IEEE Trans. Biomed. Eng. 19 (1972) 342.

97   P. Bergveld, A. Sibald, Analytical and Biomedical Applications of Ion-Selective Field-Effect Transistors, Elsevier, London, 1988.

F. Švec, T.B. Tennikova and Z. Deyl (Editors)
Monolithic Materials
Journal of Chromatography Library, Vol. 67

Chapter 26

Solid-Phase Extraction

Shaofeng XIE[1], Tao JIANG[1], František ŠVEC[2], Robert W. ALLINGTON[1]

[1] *ISCO Inc., 4700 Superior Street, Lincoln, NE 608504-1328, USA*
[2] *Department of Chemistry, University of California, Berkeley, CA 94720-1460, USA*

CONTENTS

26.1 INTRODUCTION

Today, solid-phase extraction (SPE) is the most popular sample preparation method for the extraction and preconcentration of analytes, as well as cleanup of matrix interferences and undesired compounds from analytical samples [1-3]. SPE not only affords better separation, lower detection limit, and both higher accuracy and precision, but also extends lifetime of the separation columns. Compared with the traditional extraction techniques such as liquid-liquid extraction, SPE is much more attractive since it is faster, better reproducible, consumes less organic solvents, and its automation is easier. During the past few years, characterized by the introduction of new sorbent phases and formats, SPE technique has been significantly improved and widely applied in the determination of pesticides, drugs, and additives in environmental, biological, food, and medical samples [4-16].

Current SPE sorbents are mainly based on chemically bonded silica, crosslinked polymers, and graphitized carbon with silica materials are clearly dominating the field [17]. Due to their inherent limitations, such as the presence of polar silanol groups, relatively low surface area, and the narrow pH-stability range, polymer based supports are becoming a viable alternative. They often enable high analyte recovery, in particular for compounds with aromatic rings due to their specific π-π interactions with the support. The polymer matrix can be chemically modified to afford a variety of functional moieties and extend the application to the analytes with a wide range of polarity. Thus, more polar sorbents and ion exchangers have been used that exhibit very good performance in a number of application fields [3,18]. In the last decade, the SPE materials with highly selective affinity have become very attractive and novel sorbents for immunoaffinity adsorption as well as molecularly imprinted polymers for SPE have also been developed [1-2].

Currently, most sorption materials for SPE are in the bead shape. Typically, these beads are packed in a cartridge or column [19]. However, the intrinsic problem of all particulate media is their inability to completely fill the available space. In addition, the channeling between particles reduces the extraction efficiency and short columns cannot be applied at high flow rates. This has led to the development of new SPE formats such as disks with embedded small sorbent particles [20-24] and thin membranes coated on a fiber, stirrer, or the inner wall of a capillary tube. The last format represents a new SPE technique, called solid-phase microextraction (SPME) [13-16,25]. By using the particle-loaded disks, a good extraction recovery has been achieved at relatively high flow rates because of the absence of channeling and the fast mass transfer that benefited from the use of small particles. SPME has gained widespread acceptance in many areas, specifically in the analysis of volatile and semi-volatile compounds, due to its speed and convenient operation. Typically, SPME is on-line coupled with GC or LC systems. [25-26].

In the early 1990s, the rigid porous monoliths based on organic polymers were introduced into the field of HPLC as alternative separation media [26-28]. They were prepared by in-situ polymerization of monomers in the presence of porogenic solvent using methods shown earlier in this book. These monoliths consisted of interlinked microglobules and highly interconnected pores, had high permeability, and enabled convectional flow through the channels thus considerably enhancing the mass transfer. In addition, these monoliths could be readily made into any shape. Monolithic supports thus circumvent shortcomings typical of particulate media and appear almost ideal for SPE in any of the typical formats. This application has emerged only recently. Based on the adsorption mechanism, the monolith used for SPE can be classified into three categories: hydrophobic, ion exchange, and immunoaffinity materials.

26.2 HYDROPHOBIC SORBENTS

The first porous monolithic polymer used for SPE was poly(styrene-*co*-divinylben-zene) (PS-DVB) [29]. PS-DVB monoliths were prepared from a high-grade divinyl-benzene with a purity of up to 80% in the presence of suitable porogenic solvents. These monoliths have specific surface areas of up to 400 m²/g and a reasonable permeability for liquids. Their high capacity has been demonstrated on adsorption of phenols. For example, the sorption capacity for 2-nitrophenol was 23 mg/g at a flow velocity of 10 cm/min, which is typically used for SPE media in thin disk format. Moreover, the excellent mass transfer properties of this monolithic adsorbent resulted in an acceptable capacity of 2.6 mg/g even at the remarkably high flow velocity of 300 cm/min (150 bed volumes/min). This high-speed adsorption of many substituted phenols together with a high average recovery of 80% compares favorably to devices containing PS-DVB beads. Therefore, the monoliths are key candidates for SPE at high flow rates. In addition to the study of phenol extraction, Huck *et al.* investigated the adsorption of organochlorine and organophosphorous pesticides using a disinte-grated PS-DVB monolith [3]. The results are summarized in Table 261 and indicate that PS-DVB sorbent affords better recoveries than the commercially available C18 silica, which is normally used in standard procedures. The average recovery using the ground monolith was about 77% compared to 69% found for silica-based sorbents.

PS-DVB copolymers are well suited for the extraction of nonpolar compounds due to their hydrophobicity. However, more polar compounds are less retained and may even rapidly break through during the sorption step. This then leads to a decrease in recovery and errors in their quantitation [30-31]. In order to improve both wetting and adsorption of polar compounds on polymer-based sorbents, the surface of the polymer beads was provided with a variety of polar functionalities such as acetyl, hy-droxymethyl, benzoyl, and carboxybenzoyl [3,18,32-36]. These chemically modified and more hydrophilic sorbents have a higher capacity for polar compounds than the typical solid-phase materials. Similar idea has also been implemented with monoliths using polymerization mixtures including polar monomer – 2-hydroxyethyl methacry-late (HEMA) to tune both polarity and the wettability of the PS-DVB [29]. The positive effect of the polar comonomer has been clearly confirmed by the higher recoveries of substituted phenols on PS-DVB-HEMA monolith compared to that com-prising only PS-DVB as shown in Table 26.2. The new monolithic DVB-HEMA sorbent could be readily used for the difficult SPE of polar organic compounds from very dilute aqueous solution.

Poly(butyl methacrylate-*co*-ethylene dimethacrylate) (BMA-EDMA) monoliths were used as SPE sorbents on-chip [37]. They were prepared by UV initiated polym-erization within the channels of a microfluidic chip and enabled a 1600-fold increase

TABLE 26.1

COMPARISON OF RECOVERIES FOR PESTICIDES USING POLY(STYRENE-DIVINYLBENZENE) MONOLITH AND REVERSED-PHASE SILICA-BASED BEADS [3]

Pesticide	Symbol	Recovery (%)	
		PS-DVB monolith	Speed ODS [a]
α-Hexachlorocyclohexane	α-HCH	100	90
β-Hexachlorocyclohexane	β-HCH	99	95
Lindane	Lin	99	93
Quintozine	Quin	75	74
δ-Hexachlorocyclohexane	δ-HCH	98	92
Heptachloroepoxide	HCE	82	67
o,p-Dichlorophenyldichloroethane	*o,p*-DE	53	51
α-Endosulfane	AEN	86	72
p,p-Dichlorodiphenyldichloroethane	*p,p*-DDE	50	46
o,p-Dichlorodiphenyldichloroethane	*o,p*-DDE	72	56
p,p-Dichlorodiphenyldichloroethane	*p,p*-DDD	67	62
o,p-Dichlorodiphenyltrichloroethane	*o,p*-DDT	57	51
p,p-Dichlorodiphenyltrichloroethane	*p,p*-DDT	57	49
Average recovery (%)		77	69

[a] ODS - octadecyl silica

in the concentrations of highly diluted aqueous solutions of coumarin 519 (10 nmol/L), and over 1000-fold concentration enhancement for C-519 labeled Phe-Gly-Phe-Gly peptide, and the green fluorescent protein (GFP). These results clearly demonstrated the potential of monolithic extractor for micro-total analysis system (μ-TAS).

Quirino *et al.* recently reported porous monoliths prepared by photopolymerization of (methacryloyloxypropyl)-trimethoxysilane gel (PSG) [38-39]. These monoliths were used for on-line preconcentration of different compounds such as polycyclic aromatic hydrocarbons (PAHs), alkyl phenyl ketones (APK), and peptides. The monolith acted as both solid-phase extractor and stationary phase for the separation in

TABLE 26.2

RECOVERY OF PHENOLS FROM POROUS POLY(DIVINYLBENZENE) (DVB) AND POLY(2-HYDROXYLETHYL METHACRYLATE-CO-DIVINYLBENZENE) (HEMA-DVB) MONOLITHS [29]

Compound	Recovery %	
	DVB	HEMA-DVB
Phenol	58	92
4-Nitrophenol	77	90
2-Chlorophenol	82	97
2-Nitrophenol	88	96
2,4-Dinitrophenol	76	91
2,4-Dimethylphenol	5	95
4-Chloro-3-methylphenol	88	99
2,4-Dichlorophenol	79	97
4,6-Dinitro-2-methylphenol	80	94
2,4,6-Trichlorophenol	82	96
Pentachlorophenol	91	97
Average	80	95

capillary electrochromatographic mode. The preconcentration was achieved due to the flow rate-independent kinetics of analyte-PSG interaction, resulting from the rapid mass-transfer within the porous monolith. Therefore, injection of large volumes of sample solutions was possible that considerably increased the detection limit. The extent of the preconcentration strongly depended on the retention factor k' of the analyte. A higher accumulation can be achieved for more retained compounds. An increase in concentration of about 100, 30 and 20 for PAHs, APKs, and peptides, respectively, was observed [38]. Figure 26.1 shows an example of the preconcentration and separation of five peptides using a monolithic PSG column modified with pentafluorophenylpropyl groups. This Figure demonstrates that the peak heights increased after longer sample plug injection. This approach was also successfully used to test a real urine sample shown in Fig. 26.2 demonstrating the usefulness of this approach in analysis of biofluids. Furthermore, this study also revealed that an even higher enrichment by a factor of over 1000 can be achieved by using solvent gradient,

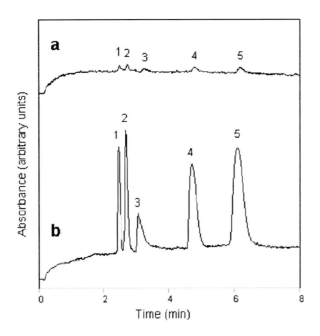

Fig. 26.1. Electrochromatographic separations of peptides (Reprinted from ref. [38]. Copyright 2001 American Chemical Society). Peptide concentrations in solution 16.7 μg/mL each, injection plug 0.1 (a) and 12 mm (b) long, mobile phase 50 mmol/L phosphoric acid–water–acetonitrile (1:5:4), applied voltage 15 kV, detection 214 nm, temperature 20°C. Peaks: bradykinin (1), angiotensin II (2), tripeptide I (3), tripeptide II (4), and methionine enkephalin (5).

i.e. dissolving the sample in a solvent containing a high percentage of non-eluting solvent such as water followed by an increase in the content of stronger solvent or by field enhanced sample injection [39]. For example, Figure 26.3 shows result of injection of a 91.2 cm long sample plug that represents 3.56 times the total column length. No significant loss of the resolution has been monitored while the peak heights of decanophenone and pyrene were increased by a factor of 1,118 and 1,104, respectively.

26.3 ION-EXCHANGE SORBENTS

Charged or ionizable analytes such as some biomolecules can be extracted using ion exchangers. Again, these sorbents are most often particles. Recently, Yu *et al.* reported SPE in ion-exchange mode using monoliths [37]. A strong anion exchange

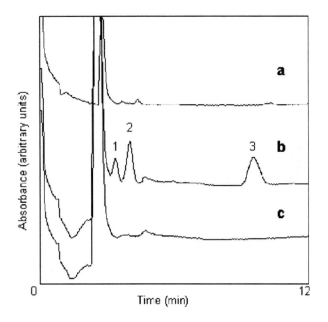

Fig. 26.2. Electrochromatographic separations of steroids in urine (Reprinted from ref. [38]. Copyright 2001 American Chemical Society). Spiked urine contained 0.1 mmol/L hydrocortisone (1), 0.3 mmol/L progesterone (2), and 0.2 mmol/L cortisone (3). Four parts of spiked or unspiked urine were mixed with 6 parts of acetonitrile and centrifuged to remove the proteins. Each supernatant was mixed with an equal volume of 50 mmol/L ammonium acetate–water–acetonitrile (1:7:2) solution before injection; mobile phase 50 mmol/L ammonium acetate/water/acetonitrile (1/5/4), applied voltage 17 kV, detection 254 nm. Injection plug of spiked urine 0.1 (a) and 21.4 mm (b) long; plug of unspiked urine 21.4 mm (c).

monolith was prepared by photoinitiated copolymerization of 2-hydroxyethyl methacrylate, [2-(methacryloyloxy)ethyl]trimethylammonium chloride, and ethylene dimethacrylate within channels of a microfluidic device. SPE ability of this device was then demonstrated on sorption and release of Coumarin 519 that has led to an increase in concentration by a factor of 190. The complete elution of the probe was achieved using pulses of a good solvent and a lower flow rate.

26.4 IMMUNOAFFINITY SORBENTS

The co extraction of analytes and matrix interferences on nonselective sorbents can present a significant problem for analytes at the trace levels in presence of interferences at a higher concentration. Therefore, clean-up procedures are required prior

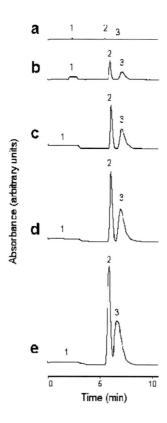

Fig. 26.3. Electrochromatographic separations of thiourea (1), decanophenone (2), and pyrene (3) (Reprinted from ref. [39]. Copyright 2001 American Chemical Society). Injection plug 0.023 (a), 7.6 (b), 22.8 (c), 45.6 (d), and 91.2 cm (e) long, mobile phase 5 mmol/L ammonium acetate in 60% acetonitrile in (a) and 5 mmol/L ammonium acetate in 40% acetonitrile in (b-e), applied voltage 22 kV; detection 254 nm.

to the chromatographic analysis [40]. Typically, the sample treatment includes several steps. Consequently, the risk of sample loss or contamination increases and the reliability of the results is reduced. Therefore, it is very important to obtain extracts free of the matrix interferences in a minimum number of steps. This can be achieved by extraction using immunoaffinity sorbents (immunosorbents). The absorption is based on molecular recognition and mostly employs antigen-antibody interactions. These sorbents exhibit very high affinity and selectivity to a single analyte or a class of structurally related analytes [41-43]. The extraction, concentration, and cleanup of the analytes from complex liquid samples can be accomplished in only a single step thus avoiding the problems mentioned above. This approach has been widely used for the pretreatment of complex biological, food, and environmental samples [41-49]. Once again, vast majority of immunosorbents are prepared by immobilization of affinity ligands on silica, Sepharose, and hydrophilic organic polymer beads [44,50].

Owing to the ease of their preparation as well as their beneficial mass transfer and high permeability [51], the poly(glycidyl methacrylate)-based monoliths with immobilized protein A, protein G, and immunoglobulin G (IgG), have been successfully used as immunosorbents for protein enrichment and purification [52-54].

Recently, thin monolithic disks with immobilized antibodies were also used as an ultrafast immunoextractor [55]. For example, a 95% extraction of fluorescein was achieved in less than 100 ms using a monolith with immobilized anti-fluorescein IgG. This device can be coupled with HPLC or CE for the analysis of free drug fractions in blood fractions and for the fast monitoring of pesticides in environmental samples [56].

Detailed description of immunoadsorption using functionalized monolithic disks is described in the specific chapter of this book.

26.5 CONCLUSION

During the last few years, monoliths have been successfully used as sorbents for SPE in hydrophobic, ion exchange, and immunoadsorption modes. The ability of concentrating highly dilute compounds makes these materials suitable for the pretreatment of different environmental, food and biological samples. The unique properties of monoliths including high permeability for liquids, operation at high flow rates without lost of efficiency, low cost, and ease of preparation together with simple means available to control shape, porosity, and selectivity, meet the requirements of modern SPE materials specifically designed for the rapid, sensitive, accurate, and miniaturized devices operating in an automatic mode. The monolithic SPE devices are likely to be widely commercialized in the near future and their application is expected to grow rapidly.

26.6 REFERENCES

1   M.C. Hennion, J. Chromatogr. A 856 (1999) 3.
2   I. Liska, J. Chroamtogr. A 885 (2000) 3.
3   C.W. Huck, G.K. Bonn, J. Chromatogr. A 885 (2000) 51.
4   H. Sabik, R. Jeannot, B. Rondeau, J. Chromatogr. A, 885 (2000) 217.
5   M.C. Bruzzoniti, C. Sarzanini, E. Mentasti, J. Chromatogr. A, 902 (2000) 289.
6   D. Martinez, M.J. Cugat, F. Borrull, M. Calull, J. Chromatogr. A 902 (2000) 65.
7   E. Hogendoorn, P. van Zoonen, J. Chromatogr. A 892 (2000) 435.
8   J.L. Tadeo, C. Sanchez-Brunete, R.A. Perez, M.D. Fernandez, J. Chromatogr. A 882 (2000) 175.

9   J.M. Soriano, B. Jimenez, G. Font, J.C. Molto, Crit. Rev. Anal. Chem. 31 (2001) 19.

10   J.L. Luque-Garcia, M.D. Luque de Castro, J. Chromatogr. A 935 (2001) 3.

11   H. Tian, A.F.R. Huhmer, J.P. Landers, Anal. Biochem. 283 (2000) 175.

12   Q. Song, L. Putcha, J. Chromatogr. B 763 (2001) 9.

13   S. Ulrich, J. Chromatogr. A 902 (2000) 167.

14   B. Zygmunt, A. Jastrzebska, J. Namiesnik, Crit. Rev. Anal. Chem. 31 (2001) 1.

15   J. Beltran, F.J. Lopez, F. Hernandez, J. Chromatogr. A, 885 (2000) 389;

16   N.H. Snow, J. Chromatogr. A, 885 (2000) 445. [each ref individually]

17   J.S. Fritz, M. Macka, J. Chromatogr. A 902 (2000) 137.

18   M.E. Leon-Gonzalez, L.V. Perez-Arribas, J. Chromatogr. A 902 (2000) 3.

19   R.E. Majors, D.E. Raynie, LC-GC, 15 (1997) 1106.

20   I. Urbe, J. Ruana, J. Chromatogr. A 778 (1998) 337.

21   D. Barcelo, G. Durand, V. Bouvot, M. Neilen, Environ. Sci. Technol. 27 (1993) 271.

22   E. Viana, M.J. Redond, G. Font, J.C. Molto, J. Chromatgr. A 733 (1996) 267.

23   V. Pichon, M. Charpak, M.C. Hennion, J. Chromatogr. A 795 (1998) 83.

24   T.R. Dombrowski, G.S. Wilson, E.M. Thurman, Anal. Chem. 70 (1998) 1969.

25   H. Lord, J. Pawliszyn, J. Chromatogr. A 885 (2000) 153.

26   T.B.Tennikova, M.Bleha, F.Svec, T.V.Almazova, B.G.Belenkii, J. Chromatogr. 555 (1991) 97.

27   F. Svec, J.M.J. Fréchet, Anal. Chem. 54 (1992) 820.

28   F. Svec, J.M.J. Fréchet, Science 273 (1996) 205.

29   S. Xie, F. Svec, J.M.J. Fréchet, Chem. Mater. 10 (1998) 4072.

30   P. Mussmam, K. Levsen, W. Radeck, Fresenius J. Anal. Chem. 348 (1994) 654.

31   J. Schulein, D. Martens, P. Spitzauer, A. Kettrup, Fresenius J. Anal. Chem. 352 (1995) 565.

32   J.J. Sun, J.S. Fritz, J. Chromatogr. 590 (1992) 197.

33   J.S. Fritz, P.J. Dumont, L.W. Schmidt, J. Chromatogr. A 691 (1995) 133.

34   N. Masque, M. Galia, R.M. Marce, F. Borrull, J. High Resolut. Chromatogr. 22 (1999) 547.

35   A.D. Alder, F.R. Longo, J.D. Finarelli, J. Org. Chem. 32 (1967) 476.

36   P.J. Dumont, J.S. Fritz, J. Chromatogr. A 691 (1995) 123.

37   C. Yu, M.H. Davey, F. Svec, J.M.J. Fréchet, Anal. Chem. 73 (2001) 5088.

38   J.P. Quirino, M.T. Dulay, B.D. Bennett, R.N. Zare, Anal. Chem. 73 (2001) 5539.

39   J.P. Quirino, M.T. Dulay, B.D. Bennett, R.N. Zare, Anal. Chem. 73 (2001) 5557.

40   V. Pichon, C. Cau Dit Coumes, L. Chen, M.C. Hennion, Int. J. Environ. Anal. Chem. 65 (1996) 11.

41   D.H. Thomas, M. Beck-Westermeyer, D.S. Hage, Anal. Chem. 66 (1994) 3823.

42   V. Pichon, L. Chen, M.C. Hennion, R. Daniel, A. Martel, F. Le Goffic, J. Abian, D. Barcelo, Anal. Chem. 67 (1995) 2451.

43   S. Ouyang, Y. Xu, Y.H. Chen, Anal. Chem. 70 (1998) 931;.

44   M. Cichna, D. Knopp, R. Niessner, Anal. Chim. Acta 339 (1997) 241. [each ref individually]

45   A. Martin-Esteban, P. Fernandez, C. Camara, Fresenius J. Anal. Chem. 357 (1997) 927.

46   V. Pichon, M. Bouzige, C. Miege, M.C. Hennion, Trends Anal. Chem. 18 (1999) 219.

47   R.K. Bentsen-Farmen, I.V. Botnen; H. Note, J. Jacob, S. Ovrebo, Int. Arch. Occup. Environ. Health 72 (1999) 161.

48 A. Weston, E.D. Bowman, P. Carr, N. Rothman, P.T. Strickland, Carcinogenesis 14 (1993) 1053.

49 M. Bouzige, V. Pichon, Analusis 26 (1998) M112.

50 N. Delaunay, V. Pichon, M.C. Hennion, J. Chromatogr. B 745 (2000) 15.

51 D. Josic, A. Buchacher, A. Jungbauer, J. Chromatogr. B, 752 (2001) 191.

52 G.A. Platonova, G.A. Pankova, I.Ye. Il'ina, G.P. Vlasov, T.B. Tennikova, J. Chromatogr. A 852 (1999) 129.

53 C. Kasper, L. Meringova, R. Freitag, T. Tennikova, J. Chromatogr. A 798 (1998) 65.

54 D. Josic, Y.P. Lim, A. Štrancar, W. Reutter, J. Chromatogr. B 662 (1994) 217.

55 T. Jiang, D.S. Hage, Anal. Chem., submitted.

56 W. Clarke, A.R. Chowdhuri, D.S. Hage, Anal. Chem. 73 (2001) 2157.

F. Švec, T.B. Tennikova and Z. Deyl (Editors)
Monolithic Materials
Journal of Chromatography Library, Vol. 67

Chapter 27

Catalysts and Enzyme Reactors

Alois JUNGBAUER and Rainer HAHN

Institute for Applied Microbiology, University of Agricultural Sciences, Vienna, Austria

CONTENTS

27.1 REACTION KINETICS IN PARTICLES

Monoliths are frequently used as catalysts or supports for catalysts in chemical reactions and combustion. These catalysts differ significantly from monoliths used for separation purposes and enzyme reactors. The former monoliths are designed for reaction engineering and comprise of capillaries in which the reactions occurs. The advantages of monoliths as catalysts for the heterogeneous reactions will be evident from the following text.

27.1.1 Classification of reactions

Chemical reactions can be classified in many ways. We use classification introduced by Levenspiel [1]. Probably the most useful scheme in chemical reaction engineering is the breakdown according to the number and types of phases involved with the major dividing line being between the *homogeneous* and *heterogeneous* systems. A reaction is homogeneous if it takes place in only a single phase. In contrast, a heterogeneous reaction requires the presence of at least two phases to proceed. It is immaterial whether the reaction takes place in one, two, or more phases, at an interface, whether the reactants and products are distributed among the phases or are all contained within a single phase. All what really matters is that at least two phases are necessary for the reaction to proceed.

27.1.2 Reaction in a porous catalyst

Porous particles are commonly used to provide a high surface area in order to increase the reaction rate per volume unit of the reactor. Disadvantages of packed bed reactors are tendency to fouling and a high pressure drop. Thus, monoliths have been developed to provide high surface area at a lower pressure drop. Assuming a first order reaction, $A \to product$, occurring in a single cylindrical pore at its walls and a product diffusing out of the pore, the reaction rate r is

$$-r = \frac{1}{S} \cdot \frac{d\,N_A}{dt} = k\,C_A \tag{27.1}$$

where S is the surface area, N_A is the amount of compound A (mol), t is the time, k is the reaction rate constant, and C_A is the concentration of A. At steady state, the material balance for A is: *output – input + disappearance by reaction* = 0.

For an infinitesimally small zone in the pore with the pore radius R and the diffusion coefficient D the steady state is described as

$$\frac{d^2\,C_A}{dx^2} - \frac{2k}{D\,R} \cdot C_A = 0 \tag{27.2}$$

The solution of equation 27.2 is

$$C_A = M_1\,e^{mx} + M_2\,e^{-mx} \tag{27.3}$$

where M_1 and M_2 are the constants and m is defined as

$$m = \sqrt{\frac{2k}{D \cdot R}} \tag{27.4}$$

This term can be grouped with L the length of the pore to yield the Thiele modulus ϕ

$$\phi = L \sqrt{\frac{k}{D \cdot R}} \tag{27.5}$$

The concentration of reactant within the pore is

$$\frac{C_A}{C_{AS}} = \frac{cosh\left(\phi - \sqrt{\frac{k}{D}} \cdot x\right)}{cosh\ \phi} \tag{27.6}$$

The concentration profile within a single pore is shown in Fig. 27.1. To measure the effect of the resistance to pore diffusion on reaction rate, the effectiveness factor η is defined:

$$\eta = \frac{actual\ mean\ reaction\ rate\ within\ pore}{reaction\ rate\ if\ not\ solved\ by\ pore\ diffusion} = \frac{r_A\ with\ diffusion}{r_A\ without\ diffusional\ resistance}$$

For the first order reaction, η is calculated as

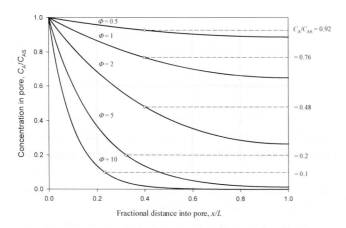

Fig. 27.1. Concentration profiles in a single pore assuming a first order reaction of A → product. C_A is the bulk concentration and C_{AS} the concentration of A at the surface.

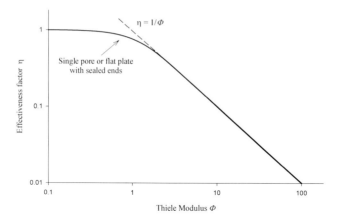

Fig. 27.2. Effectiveness factors related to Thiele modulus calculated for a single pore.

$$\eta_{first\ order} = \frac{C_A}{C_{AS}} = \frac{tanh\ \phi}{\phi} \tag{27.7}$$

For small ϕ the value of $\eta \approx 1$ and the concentration of reactant does not decrease significantly within the pore (Fig. 27.2). This situation is operative for short pores, slow reactions, or rapid diffusion. All three factors lower the resistance to diffusion. Short pores and rapid diffusion are typical features of monoliths. For large ϕ the efficiency is approximately $1/\phi$. The concentration of the reactant decreases rapidly to zero when moving into the pore, hence the diffusion strongly affects the reaction rate. This regime is called strong pore resistance. Monoliths offer certain advantages as supports and catalysts. The diffusional path is short and mass transport is enhanced by convective flow within the pores.

27.1.3 Reactors for heterogeneous reactions

Heterogeneous reactions can be categorized in two-phase and three-phase reactions. The wall of the monolith itself often acts as the catalyst and the reactions occur between the liquid and gas phase while passing the monolith. Two-phase (liquid/gas or solid/gas) or three-phase reactions (gas/liquid/solid) can be carried out in stirred tank reactors (slurry reactor), packed beds, fluidized beds [2,3], trickle beds, bubble columns, monoliths, or polyliths [4]. Monoliths are preferred to achieve rapid contact between both catalyst surface and gaseous reactants assuming that an appropriate material is available [5].

27.2 MONOLITHIC CATALYSTS

27.2.1 Shape and geometries

Monoliths are widely used for automotive emission control, diesel particle filters, stationary emission control, woodstove combustors, molten metal filters, natural gas storage, indoor air purification, ozone abatement, catalytic incineration, industrial heat recovery, ultrafiltration, catalyst supports in chemical processes, water filtration and water purification, as well as many other areas [6]. In contrast, applications of mono-liths in enzyme reactors in biotechnology are still in the exploratory phase.

Various types of monolithic catalysts shown in Fig. 27.3 have been developed in the past. They can be considered as a bundle of capillaries and have a honeycomb-like shape. Therefore, they are also called honeycomb catalysts. These monoliths are noticeably different from monoliths used for separation and enzymatic catalysis since they are produced by extrusion of support material featuring low thermal expansion, often synthetic cordierite ($2MgO \cdot 2Al_2O_3 \cdot 5SiO_2$), from a paste containing catalyst particles such as zeolites or from a precursor of the final product, typically synthetic polymers for carbonaceous monoliths. Alternatively, catalysts, supports, or their pre-cursors can be coated onto a monolithic support structure using a method called 'washcoating'. The corners in the polygonal channels are generally coated more which results in rounded shapes shown in Fig. 27.4 [7]. Zeolites can be coated by growing them directly on the support during the synthesis [8]. Even the extruded monoliths can be partially converted into zeolites by various procedures [5]. All major catalyst support materials, i.e. ceramic and polymeric, have been extruded as mono-lithic structures [9,10], with ceramic materials used mostly. Slurry catalysts often

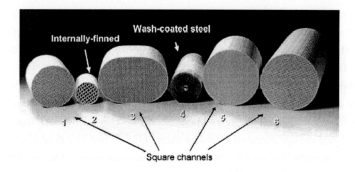

Fig. 27.3. Monolithic structures of various shapes (Reprinted from ref. [5]. Copyright 2001 Elsevier). Square channel cordierite structures (1, 3, 5, 6), internally finned channels (2), and washcoated steel monolith (4)

Model: uniform film   Reality

Fig. 27.4. Schematic representation for measured slower mass transfer compared to model predictions (Reprinted from ref. [7]. Copyright 2001 Elsevier). Thicker film in monolith channel corners due to surface tension (A) or to an uneven surface (B).

consist of Ni-coated carbon [11,12]. Carbon has some distinct advantages over other (ceramic) support materials, the most attractive being its stability in acidic and alkaline media [13]. Moreover, the carbon surface is relatively inert thus preventing the undesired side reactions catalyzed by the support surface. However, it is difficult to produce a structured carbon catalyst. Thus, a double layer consisting of carbon and Ni were coated [11]. Cordierite monolithic supports were dip-coated using partially polymerized furfuryl alcohol mixtures. After solidification of the polymer, the coating was carbonized to obtain a carbon-coated monolithic support. The carbonized polymer is a microporous material. When coated onto the cordierite monolithic support, macropores result from coating of the cordierite pores and carbonization. The polymer tends to shrink during carbonization. Since the overall shrinkage is prohibited by the monolithic support, the coating cracks. Oxidizing prior to carbonization in combination with the use of pore formers yields mesopores. The mesoporous structure affords higher surface area but also contributes to diffusional resistance. While the large channels are readily reached by convection, the transport into mesopores is solely a diffusional process.

Metallic support structures are only used for automotive applications [14]. The choice of a certain catalyst type strongly depends on the balance between maximization of the catalyst inventory and catalyst effectiveness. For slow reactions, a high catalyst loading is desired and the pure catalyst-type monolith is favored, while for fast reactions or for processes characterized by slow diffusion, a thin coating with a maximum geometric area is preferred.

Why are such structures so popular? The catalyst consists of a single piece and no abrasion due to the movement of particles upon vibration occurs. The large open frontal area (OFA) and straight channels afford an extremely low pressure drop essential for end-of-pipe applications such as exhaust pipes and stack gases. The straight channels prevent the accumulation of dust in demanding applications such as in power stations burning powdered coal.

Cell configurations and properties of monoliths have been described using geometric and hydraulic parameters [6]. These properties can be defined in terms of cell spacing L, the distance measured from the center of one cell wall in a square channel to the next wall, and wall thickness t. The cell density N is defined as the number of cells per unit of cross-sectional area and is expressed in units of cells per square inch (cpsi) or per square centimeter:

$$N = \frac{1}{L^2} \tag{27.8}$$

The open frontal area defined in Equation 27.9 is a function of wall thickness, cell spacing and cell density:

$$OFA = N(L - t)^2 \tag{27.9}$$

For a channel, the hydraulic diameter D_h defined by Equation 27.10 decreases as the cell density increases:

$$D_h = L - t \tag{27.10}$$

The hydraulic diameter is different for uncoated and washcoated monoliths since washcoating of catalysts or ceramic materials changes the wall thickness (Fig. 27.4).

During the last decade, the use of monolithic catalysts has also been extended to multiphase processes operated in the co-current mode and recently also in the counter-current mode [15]. A new type of monolith, the so-called finned monolith (IFM), was especially designed for counter-current operation [16]. This structure, which is based on a conventional ceramic monolith having longitudinal fins incorporated in the wall of the channels, combines a high geometric surface area with a high catalyst loading. By shaping the catalyst, it is possible to create different paths for the gas and the liquid thus reducing momentum transfer between the two phases and shifting the flooding limits to higher flow rates. The packing is operated counter-currently in the wavy film flow regime, which means that liquid flows down as a wavy falling film while gas moves up through a central core under a low pressure gradient. Examples of a few channel geometries are shown in Fig. 27.5 and Table 27.1.

Balance between geometric surface area and pressure drop must be also taken into consideration while designing monolithic catalysts. The pressure drop across the monolith depends linearly on its length and flow velocity:

$$\Delta P = \frac{2 f l \, \rho \, u^2}{G_c \, D_h} \tag{27.11}$$

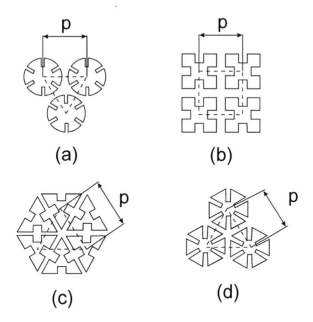

Fig. 27.5. Examples of different cross sectional channel geometries (Reprinted from ref. [55]. Copyright 1998 American Chemical Society) circular (a), square (b), triangular (c) and hexagonal (d).

TABLE 27.1

CHARACTERISTIC PROPERTIES OF SOME INTERNALLY FINNED MONOLITHS [15]

Basic channel cross section	Circular	Triangular	Square	Hexagonal
Arrangement[a]	T	T	S	T
Wall thickness	0.1p[b]	0.2p	0.2p	0.2p
Number of fins	6	3	4	6
Fin height	0.2p	0.1p	0.2p	0.2p
Fin thickness	0.1p	0.2p	0.2p	0.2p
Solid fraction	0.57	0.62	0.60	0.68
Surface-to-volume ratio	10.0/p	8.5/p	8.1/p	8.8/p

[a]T = triangular; S = square

[b]For definition of the distance p see Fig. 27.5

where f is the friction factor, D_h is the hydraulic diameter, G_c is the gravitational constant, l is the length of the monolith, u is the velocity in the channel, and ρ is the gas density.

The fundamental equations allow designing monoliths with geometric parameters such as cell density or wall thickness meeting the constraints of external processing requirements such as space velocity, flow rates, pressure drop, etc. [6].

The pressure drop $-\Delta P/l$ can be expressed for laminar incompressible flow using the Hagen-Poisseuille equation:

$$-\frac{\Delta P}{l} = 2\,B\,\frac{\mu_G\,u_{GS}}{D_h^2} \qquad (27.12)$$

where B is the constant which depends on the velocity profile and therefore on the cross-sectional geometry of the channel. For channels of a finned monolith, B equals 15.4 [15]. This value is slightly higher than 14.23, which was found for squared, non-finned channels [17]. The pressure drop of one meter internally finned monolith is about 70 Pa for single phase gas flow (air) at a velocity of 1 m/s. Computational fluid dynamics (CFD) has been used to calculate the flow in ceramic monoliths [18]. This approach can be extended to counter-current flow in structured packings. Flow within the cross-sectional area varies significantly with length and gas flow in the case of two-phase flow.

27.2.2 General properties

Monoliths can be used both for co- and counter-current operations in gas–liquid reactions. They can combine the advantages of the slurry and trickle bed reactor and eliminate their disadvantages [19]. According to Kreutzer *et al.* [20] these features can be summarized in the following way:

- The catalyst can be coated as a thin layer on the channel walls and described as a "frozen slurry reactor".
- Larger channel geometries (e.g. in the internally finned monolith channels) enable counter-current operation of gases and liquids.
- The inventory of catalysts can be increased by using thicker coatings or using a monolith extruded from the catalyst support such as all-silica or consisting of carbon.
- The high cell density of the monoliths affords a high geometric surface area. Using a packed bed, unrealistically small particles would be required to achieve similar surface areas. This would result in a high pressure drop.
- The monolithic reactor has a negligible pressure drop [15]. For laminar incompressible flow, the pressure drop can be described by the Hagen-Poisseuille equation.

TABLE 27.2

ACTIVITIES OF BOTH A MONOLITHIC AND TRICKLE BED REACTOR IN THE
MASS TRANSFER LIMITED HYDROGENATION OF α-METHYLSTYRENE USING
AN EGG-SHELL NICKEL CATALYST AT 373 K AND 1 MPa H_2 [5]

Activity per	Units	Monolith (400 cpsi)[a]	Trickle bed	Ratio
Reactor volume	$mol/m^3_{reactor}$·s	21.0	4.6	4.6
Nickel amount	mol/g_{nickel}·s	3.4×10^{-3}	8.4×10^{-5}	40

[a]cpsi = cells per square inch

- Scale-up is straightforward and stacking possible.
- Monolithic reactors are intrinsically safer. The monolith channels have no radial communication in terms of mass transport and the runaway by local hot spots in a trickle bed reactor cannot occur. Moreover, when the feed of liquid or gas is stopped, the channels are quickly emptied. They are by a factor of 3–4 more efficient than trickle bed operations.

The hydrogenation of α-methylstyrene is an example of better active phase utilization and exemplifies the good mass transfer properties. Comparison of a trickle bed and a monolithic reactor presented in Table 27.2 shows that the activity of the monolithic reactor was 4.6 times higher calculated on the basis of reactor volume or 40 times higher related to the catalyst weight [5].

27.3 MASS AND HEAT TRANSFER IN MONOLITHIC CATALYSTS

27.3.1 Flow in capillaries

If both gas and liquid flow downwards in channels with a small diameter, a number of different flow regimes shown in Fig. 27.6 can occur. Among these, the important are: annular flow in which the liquid flows as a film along the wall and gas is transported through the central core, slug flow featured by waves of the liquid film occasionally blocking the channel, Taylor flow or bubble train flow intermittently filling the channel with gas bubbles and liquid plugs, and dispersed bubble flow in which the diameter of the bubbles d_b is substantially smaller than that of the tube d_c. Because of the capillary forces, slugging of the channel due to the formation of liquid bridges occurs already at a low liquid loading. Since the channels in the monolith are small ($0.5 < d_c < 5$mm), dispersed bubble flow ($d_b \ll d_c$) is not likely to occur. As a

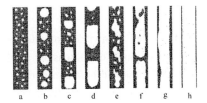

Fig. 27.6. Different types of two-phase flow in a channel (Reprinted from ref. [56]. Copyright 2001 Elsevier) (a) dispersed bubble flow, (b) bubble flow, (c) elongated bubble flow, (d) Taylor flow, (e) churn flow, (f) slugging flow, (g) annular flow, and (h) mist flow.

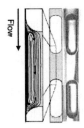

Fig. 27.7. Taylor flow in a capillary (Reprinted from ref. [7]. Copyright 2001 Elsevier). The left CFD picture shows the patterns of liquid circulation.

result, Taylor flow can be achieved in monoliths over a wide range of operating conditions. In this type of flow, the gas bubbles have a diameter nearly equal to that of the channel leaving only a thin liquid film between the bubbles and the wall. Since the gas bubbles almost block the channels, the liquid flow between two consecutive liquid slugs may be neglected. Due to the stationary wall, a circulating flow is induced in a moving liquid slug (Fig. 27.7). This circulation flow continuously refreshes the gas–liquid interface of the gas bubble, leading to a high gas–liquid mass transfer rate. For a given liquid hold-up, Taylor flow with short liquid slugs has a higher mass transfer rate than flow in which larger liquid slugs occur. This results from the increased interfacial area and the more intense mixing typical of short liquid slugs. Due to the small channel diameter of monoliths, the flow in the channel remains laminar. The hydrodynamics of counter-current liquid-gas flow in monolithic catalysts have been studied by ^{1}H-NMR microimaging and the flow pattern in the two phase-systems was found highly dependent on the gas flow [21].

27.3.2 Important heat and mass transfer numbers

Heat and mass transfer have been studied in detail for a variety of situations. The most relevant ones are briefly described below; additional definitions are available in common textbooks [1,22]. Since coefficients for mass and heat transport are sometimes difficult to determine precisely, it is convenient to correlate these coefficients using dimensionless numbers. For example, the Sherwood number *Sh*

$$Sh = \frac{k_f L}{D} \tag{27.13}$$

where k_f is the mass transport coefficient, L is the characteristic length, and D is the coefficient of molecular diffusion, expresses the ratio of mass transport occurring by means of convection to that achieved by molecular diffusion. Another physical meaning of this number is the ratio of a concentration gradient at a surface to an overall concentration gradient. The third interpretation is that this number represents the ratio of a characteristic length to the theoretical film thickness.

In analogy to the Sherwood number, the Nusselt number *Nu* is the number corresponding to heat transport. The physical meaning of the Nusselt number can be interpreted in a way similar to the Sherwood number as the ratio of convective heat transport to molecular heat transport, or as the ratio a temperature gradient at a surface to an overall temperature gradient.

$$Nu = \frac{h L}{k} \tag{27.14}$$

with h is the heat transport coefficient and k the thermal conductivity. Other important dimensionless groups used in transport correlation are summarized in Table 27.3. The Peclet number *Pe* is defined by grouping together *Re·Sc* for mass transfer and *Re·Pr* for heat transfer.

$$Pe_M = Re \cdot Sc = \frac{u L}{D} \tag{27.15}$$

$$Pe_H = Pr \cdot Sc = \frac{u L}{\alpha} \tag{27.16}$$

where the thermal diffusivity $\alpha = k/\rho C_p$.

The physical meaning of the Peclet number is the ratio of convective to diffusive transport. The Graetz number *Gz* is the ratio of thermal capacity to convective heat transfer and is defined as

TABLE 27.3

IMPORTANT DIMENSIONLESS TRANSPORT NUMBERS AND THEIR PHYSICAL MEANINGS

Number	Formula	Physical meaning	Equation
Reynolds	$Re = \dfrac{u\,L}{v}$	Convective momentum transport/diffusive momentum transport	1
Schmidt	$Sc = \dfrac{v}{D}$	Momentum diffusivity/mass diffusivity	2
Prandtl	$Pr = \dfrac{C_p\,\mu}{k}$	Momentum diffusivity/thermal diffusivity	3

$$Gz = \frac{1}{D_h} \cdot Re \cdot Pr \tag{27.17}$$

where l/D_h is the aspect ratio of the channel. Each flow geometry requires different correlation to obtain the transport coefficient. Many correlations have the for

$$Sh = a\,Re^b\,Sc^c \tag{27.18}$$

and

$$Nu = a\,Re^b\,Pr^b \tag{27.20}$$

where a, b, c are the correction factors obtained from non-linear curve fitting. Another common correlation has the form

$$Sh = a + Re\,Sc^b \tag{27.21}$$

27.3.3 Modeling heat and mass transfer

Depending on the Reynolds number, flow in the channels may be in the (i) laminar region for $Re < 2\,100$, (ii) turbulent region with $Re > 10\,000$, or (iii) transition region for which $2100 < Re < 10\,000$. The flow region also indicates the difficulty of modeling and design. For example, the transition region is not well understood and is usually avoided. For the laminar region in a circular channel, Hawthorn [23] has developed correlations for Sh and Nu

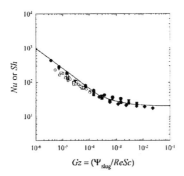

Fig. 27.8. Liquid to solid mass or heat transfer in a cylinder with a moving shaft. The symbols refer to CFD calculations of heat transfer. The line was calculated using the method described by Kreutzer *et al.* [20]. The open symbols refer to experiments by Bercic and Pintar [57].

$$Sh = 3.66 \left[1 + 0.0095 \frac{D_h}{L} Pe_m \right]^{0.45} \tag{27.22}$$

$$Nu = 3.66 \left[1 + 0.0095 \frac{D_h}{L} Pe_h \right]^{0.45} \tag{27.23}$$

These correlations have been frequently applied although they are only valid for a fixed length and do not represent the local value at $x = L$. Therefore, Groppi *et al.* [24] have refined the correlation for square channels as well as for constant wall temperature and heat flux, respectively:

$$Nu_T = 2.97 + 6584 \left[\frac{1000}{Gz} \right]^{-0.5174} exp \left[\frac{42.49}{Gz} \right] \tag{27.24}$$

$$Nu_H = 3.095 + 8.933 \left[\frac{1000}{Gz} \right]^{-0.5386} exp \left[-\frac{6.7275}{Gz} \right] \tag{27.25}$$

A two-dimensional model based on the axisymmetric momentum, mass and energy balance for the first order reaction occurring at the wall was developed by Hayes and Kolaczkowski [25]. The values of Nu and Sh do not correlate uniquely with Gr, although Kreutzer *et al.* observed in reality a good correlation shown in Fig. 27.8 [20]. Progress in computation and novel algorithms allow prediction of both mass and heat transfers in monolithic reactors. The prediction of mass transfer in a single capillary is representative for the whole monolith, while prediction of heat transfer is more complex since thermal coupling of adjacent capillaries affects the heat transfer [26].

27.4 ENZYME REACTORS

Monoliths used for enzyme reactors differ from the ceramic honeycomb materials. They rather resemble or are completely identical with materials typically used in bioseparations [27,28]. These materials are also considered as monolithic supports and, together with membranes, they are characterized by substantially reduced diffusional resistance [29]. They feature tortuous channels with an average pore size smaller than 1.5 μm [30]. Therefore they represent almost ideal supports for the immobilization of enzymes intended for conversion of substrates. Preliminary studies concerning enzymatic conversion within these supports [31] were carried out shortly after the methacrylate-based monoliths were introduced [32-34]. However until recently, the pace of further development in this area was rather slow. However, the progress achieved in the last 2 years has shown that the enzymes immobilized on the monolith can be used advantageously for both analytical and preparative applications. The advancement of monoliths towards miniaturization makes them also suitable for in-process control in flow injection systems [35,36]. Monoliths can be also produced on larger scale [37,38] and as thin sheets [39] thus making possible industrial application in large-scale reactor formats.

Turková *et al.* [40] and Drobník *et al.* [41] demonstrated that enzymes can be immobilized to poly(glycidyl methacrylate)-based particles without substantial loss of activity. These beads contained epoxy groups, which could be used directly for immobilization of ligands. They were also suitable for insertion of a spacer bearing a functional group, such as β-alanine or diethylamine that can be used for further reactions. These studies indicated that enzymes can also be readily immobilized on poly(glycidyl methacrylate) monoliths. Abou-Rebyeh *et al.* [31] carried out a preliminary study of enzymatic conversion using a fixed bed reactor based on monolithic support. The enzyme, carbonic anhydrase from human erythrocytes, was immobilized on a disk cast by radical copolymerization of glycidyl methacrylate and ethylene dimethacrylate in the presence of porogenic solvents. The epoxy groups were then utilized for immobilization of the enzyme. The disk was placed into a small cartridge and the substrate was drawn through the reactor by a peristaltic pump. Both 2-chloro-4-nitrophenyl acetate and 4-nitrophenylacetate were used as substrates for determination of the enzymatic activity. Although at that time no optimized hardware was available for this application, kinetic experiments under dynamic conditions involving low molecular weights substrates were carried out at rather low flow-rates of only up to 1.2 ml/min. However, the authors also demonstrated that an increase in flow rate

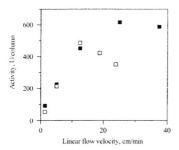

Fig. 27.9. Effect of linear flow velocity of N-benzoyl-L-arginine ethyl ester solution on enzymatic activity of trypsin immobilized on poly(glycidyl methacrylate-*co*-ethylene dimethacrylate) monolith (■) and beads (□) (Reprinted from ref. [42]. Copyright 1996 Wiley). Column: 50 × 8 mm i.d., temperature 25°C, substrate concentration 200 mmol/L.

enhanced the enzymatic conversion. This was an early indication that the reaction was not limited by diffusion. Unfortunately, the immobilized enzyme had only a limited stability and lost a considerable part of activity upon immobilization. Petro *et al*. [42] used trypsin as a model enzyme immobilized on a monolithic column to study enzymatic conversion of synthetic substrates. The free epoxy groups on the poly(glycidyl methacrylate-*co*-ethylene dimethacrylate) monolith were reacted first with ethylene diamine and then the enzyme was attached via the bifunctional reagent – glutaraldehyde. For comparison, the enzyme was also immobilized on macroporous beads using identical chemistry [40,41]. As shown in Fig. 27.9, an increase in the flow velocity through the monolithic reactor resulted in a higher substrate conversion until a maximum conversion was reached indicating that the reaction was not limited by diffusion within a broad range of flow rates. Only at a certain relatively high flow velocity, the reaction was either mass transfer limited or kinetically limited. In contrast, serious limitations occured for beds packed with particles already at low flow rates. This indicates that the kinetics in particle-based reactor was mass transfer limited at much lower velocity compared to the monolithic reactor. Both support and immobilization chemistry have been later improved. Xie *et al*. [30] developed a monolithic support based on poly(2-vinyl-4,4-dimethylazlactone-*co*-acrylamide-*co*-ethylene dimethacrylate). The immobilization of trypsin on this support was achieved in a single reaction step via the azlactone functionalities. This reactor could be operated at five times higher speed than that based on a poly(glycidyl methacrylate-*co*-ethylene dimethacrylate) monolith. The latter that had a higher porosity as well as pore size (5 μm), appeared to be more fragile, and could not be run at that high flow rates although a similar efficiency in terms of conversion of substrate per unit of time and reactor

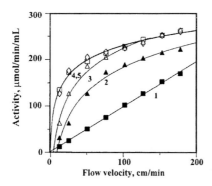

Fig. 27.10. Effect of linear flow velocity and concentration of N-benzoyl-L-arginine ethyl ester solution on enzymatic activity of trypsin immobilized on poly(2-vinyl-4,4-dimethyl-azlactone-*co*-arylamide-*co*-ethylene dimethacrylate) monolith. (Reprinted from ref. [30]. Copyright 1999 Wiley). Column: 20 × 1 mm i.d., temperature 25°C, substrate concentration 2 (curve 1), 5 (curve 2), 10 (curve 3), 20 (curve 4), and 30 mmol/L (curve 5).

volume was achieved (Fig. 27.10). However, a direct comparison of both systems is not possible, since different substrate concentrations were used.

Following theoretical calculation demonstrate the performance of an enzyme reactor involving a monolithic support. The actual reaction rates were taken from Xie *et al.* [30]. We have assumed a first order reaction rate. Replacing the surface area by the reactor volume yields

$$- r''' = \frac{1}{V} \cdot \frac{dN}{dt} = k'''C_A \tag{27.26}$$

where r''' is the reaction rate expressed in moles of substrate converted per unit of reactor volume and time, and k''' is the reaction rate constant with dimension s^{-1}. For calculation of the Thiele modulus, values of the effective diffusion coefficient D_{eff}, the reaction rate constant k''', and the effective pore length L are required. Reaction rate constant was determined from equation 27.26. Xie *et al.* [30] tested five different substrate concentrations (C_A = 2, 5, 10, 20, and 30 mmol/l). Value of D_{eff} was estimated as 1/10 of the molecular diffusivity D, which may represent a worst-case scenario for the monoliths. The diffusion coefficient was calculated according to Tyn and Gusek [43] by $D = 2.7\ 10^{-10}/M_R^{-1/3}$. Scanning electron micrographs revealed that poly 2-vinyl-4,4-dimethyl azlactone-*co*-ethylene dimethacrylate just as the other similar monoliths are composed of globules with a diameter of approximately 1 μm [44,45]. The efficiency and Thiele moduli at a flow velocity of 100 cm/min (residence

TABLE 27.4

EFFICIENCY OF REACTOR COMPRISING TRYPSIN IMMOBILIZED ON MONO-
LITHIC SUPPORT AS CALCULATED FROM DATA OF REF. [30] FOR CONVERSION
OF A SYNTHETIC SUBSTRATE

	Values for different substrate concentrations				
C_A (mol/m^3)	2	5	10	20	30
r''' (μmol/min.ml)	100	180	230	230	230
r''' (mol.m^{-3}.s^{-1})	1.67	3	3.83	3.83	3.83
k (s^{-1})	0.83	0.6	0.38	0.19	0.127
Thiele modulus (ϕ)	0.1469	0.1247	0.0996	0.0705	0.0575
Efficiency (η)	0.99287	0.99485	0.99670	0.99835	0.99890

time t_r = 0.02 min) are summarized in Table 27.4. Since we have to make a simplifi-
cation assuming a first order reaction kinetic and since most of the enzymatic
reactions follow the Michaelis-Menten kinetics also in the immobilized state, the
values presented in Table 27.4 have to be understood as only rough estimates. For
sufficiently small ϕ ($\phi \leq 0.3$), diffusion of substrate is fast relative to its consumption
and the overall substrate consumption rate is limited by the reaction rate on the
interior/exterior surface of the support. If this is the case, the surface substrate concentration is practically equal to that of the bulk phase (Fig. 27.1) and model studies
leading to correct immobilized-enzyme kinetics can be reasonably carried out using
the bulk concentration of substrate.

Ayhan *et al.* [46] immobilized urease on poly(2-hydroxyethyl methacrylate-*co*-eth-
ylene dimethacrylate) beads. They also observed that the value of ϕ was very low in
the range of 10^{-5} and the reaction rate was not limited by diffusion since the beads
were nonporous.

Horvath *et al.* [47] have studied open tubular enzyme reactors using capillaries
coated with a porous layer with immobilized trypsin already in 1972. Since this
technique also avoids diffusion controlled reaction rate, a perfect enzyme reactor in
the honeycomb monolithic format could be envisioned.

In the first experiments using monoliths as supports, the enzymes were immobilized via a variety of immobilization chemistries that recovered a sufficient activity
after the immobilization procedure has been completed [30,31,42]. Both carbonic
anhydrase and trypsin are rather small proteins. Therefore, these enzymes might be

readily immobilized without damaging their activity or affecting the enzyme-substrate interaction. In contrast, Josic *et al.* [48] in their preliminary study reported immobilization of more complex enzymes, such as invertase. Their experiments have demonstrated that the conversion of the low molecular weight substrate – saccharose (sucrose) is incomplete at higher concentration thus indicating a loss of activity already during immobilization. Therefore, substrate solutions with higher concentrations had to be recycled to achieve complete conversion. Since yeast invertase is a highly glycosilated protein, it was possible that immobilization of the enzyme led to its partial inactivation or a partial loss of its enzymatic activity due to multipoint binding.

The effect of a spacer used for immobilization on enzymatic activity has been investigated on a model system involving yeast glucose oxidase [48]. This enzyme is also a glycoprotein with a high degree of glycosilation. Its activity was low if the enzyme was immobilized directly via the epoxy groups of the support. An increase in activity was observed after using spacers such as ethylenediamine and glutaraldehyde inserted between the enzyme and the surface functionality of the support. This approach was previously described by Petro *et al.* [42]. However, only after the enzyme was attached through its oligosaccharide part to lectin Concanavalin A immobilized on the monolith, the desired distance between the enzyme and the support was achieved and the active center was fully accessible to the substrate. Applying this optimized immobilization technique, activities comparable to that of the commercially available detectors were achieved.

Vodopivec *et al.* [35] immobilized glucose oxidase on monolithic CIM disks and used them for on-line monitoring of glucose in fermentations involving *Saccharomyces cerevisiae* and *Aspergillus niger*. A disk shaped enzyme reactor having a diameter of 12 mm and thickness of 3 mm with immobilized glucose oxidase was built in a flow-injection analysis (FIA) system and filtered fermentation effluent was monitored. This CIM glucose oxidase disk FIA system exhibited a good signal reproducibility and linear response in the range of 10–200 mg/l. Although less stable than the commercial packed bed glucose oxidase reactor, it featured better sensitivity and could be applied even at high flow rates exceeding 3 ml/min. Both the conventional FIA and CIM glucose oxidase disk FIA systems afforded identical results. The values obtained with the CIM glucose oxidase disk FIA were also confirmed by off-line glucose determination using a HPLC reference method. The slight loss of enzyme activity during the fermentation could be compensated by autocorrelation measurements. The authors also assumed that non-specific adsorption might reduce the response. However, a time-response analysis has not been carried out.

A superporous agarose monolith has also been used to fabricate a minireactor for flow injections systems [36]. This type of monolith has different properties compared to materials designed for bioseparation [38]. A 15×5 mm i.d. gel plug of superporous

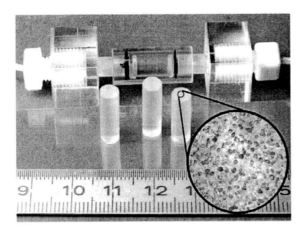

Fig. 27.11. Superporous agarose gel plugs used as mini-reactors for flow injections systems (Reprinted from ref. [36]. Copyright 2000 Kluwer). The plugs were inserted into a tight fitting glass tube provided with flow adapters. The flow pore structure is demonstrated by the inserted micrograph of a thin slice. The dark areas constitute the large pores, typically 50 μm in diameter. The light areas represent the agarose phase containing diffusion pores with a diameter of 30 nm, which are not visible. The units shown on the ruler are cm.

agarose was cast as shown in Fig. 27.11. This monolith contained both very large pores with a diameter of about 50 μm and "normal" 30 nm pores providing ample surface for covalent attachment of proteins such as enzymes and antibodies. To test the practical performance of the superporous monolithic support, two applications were studied: (i) on-line determination of glucose in culture broth with immobilized glucose oxidase and (ii) immunological quantification of intracellular β-galactosidase in *Escherichia coli*. For the latter, lysozyme was immobilized on the first plug to afford cell lysis while antibodies against β-galactosidase required for quantification of the enzyme were immobilized on the second plug. Lysozyme and the antibody against β-galactosidase were immobilized using the common cyanogen bromide activation [49]. Glucose oxidase was coupled after the tresyl chloride activation following the protocol of Nilsson and Mosbach [50].

Since monolithic supports with rather large channels were used, the distribution in residence times was investigated. Small tracer pulses were injected and the residence time t_r as well as the peak variance σ^2 were determined. For comparison of the plug flow characteristics at various flow rates, the Bodenstein number Bo, which is a measure of deviation from plug flow in chemical reactors and relates to the axial dispersion and the linear velocity [1], was calculated.

$$\frac{\sigma^2}{t_r^2} = \frac{2}{Bo} - \frac{2}{Bo^2} \cdot \left(1 - e^{-Bo}\right) \tag{27.27}$$

Reasonably high Bodenstein numbers of more than 14 for an acetone pulse and more than 6 for yeast cells indicated plug flow profile despite the large size of the channels. The reactor enabled the on-line lysis of *E.coli* and its performance exceeded that of an expanded bed [51]. Up to 87% of model cells could be lysed at a flow rate of 0.4 ml/min representing a t_r of 0.7 min. A strong dependence of conversion rate on the flow rate, which results from the diffusion in pores, was observed for all enzyme rectors. The agarose minireactor was stable for 500 min and online monitoring of glucose and intracellular β-galactosidase could be readily accomplished.

Another interesting enzyme reactor based on a superporous agarose monolith facilitated enzymatic conversion of lactose to glucose and galactose. Processing of surplus lactose is an important issue in dairy industry since this disaccharide is a byproduct in cheese making and usually is wasted. Another problem concerning lactose is the intolerance of a large part of the human population. Conversion of lactose into glucose and galactose enables persons with this deficiency to consume milk and milk derived products. Gustavsson and Larsson [38] have demonstrated the production of lactose-free milk by enzymatic conversion of lactose using their superporous agarose monolith with immobilized β-galactosidase from *Kluyveromyces* with the feed streams applied in radial direction. One liter of defatted milk was recirculated through the 61 mm i.d., 80 mm o.d., and 100 mm long reactor at a flow rate of 86 ml/min (Fig. 27.12). All lactose was converted after 90 min, which corresponds to an activity of 1400 U. Based on the amount of immobilized enzyme, 24 % of the immobilized protein was active. This example further demonstrates the universal applicability of monoliths.

Only limited number of studies concerned the use of monoliths with immobilized enzymes for conversion of high molecular weight substrates. So far, almost all experimental work has been carried out with immobilized trypsin. The results showed that for the high molecular weight substrates, the conversion depended on residence time in the reactor. Therefore, the situation was fundamentally different from the enzyme reactors converting low molecular weight substrates.

However, this was not observed with the elastase immobilized on methacrylate-based disks used by Lim *et al.* [52] for preparative hydrolysis of the inter-α-trypsin inhibitor from human plasma. Controlled enzymatic hydrolysis of proteins is a very popular biochemical tool for peptide mapping. The high molecular weight substrate had to be denatured by adding SDS and mercaptoethanol to exclude effects such as inaccessibility of the cleavable site.

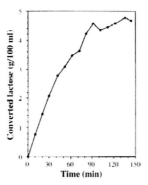

Fig. 27.12. Production of lactose-free milk using a superporous agarose monolith reactor (Reprinted from ref. [38] (Copyright 2001 Elsevier). After breakthrough, fractions were collected and analyzed for lactose content. The original milk contained 4.8 g lactose/l and 0.1 % fat.

Xie *et al.* [30] used a poly(2-vinyl-4.4-dimethylazlactone-*co*-acrylamide-*co*-ethylene dimethacrylate) monolith with immobilized trypsin as enzymatic reactor for the conversion of both high and low molecular weight substrates. The catalytic activity of the monolithic reactor for a low molecular weight substrate L-benzoyl arginine ethyl ester was maintained even at a high flow velocity of 180 cm/min representing a residence time of merely 0.1 min. However, this beneficial flow rate/activity behavior was not observed in comparative experiments with high molecular weight substrate casein. The authors explained this by pointing to the high viscosity of casein solutions and their colloidal character. This significantly contributes to increased flow resistance within the monolithic reactor and does not permit the use of the high flow rates possible for low molecular weight substrates. In a long-term experiment using a monolithic reactor with immobilized trypsin and casein as a substrate, the initial enzymatic activity did not change during the first 24-h period of continuous operation. However, the caseinolytic activity decreased significantly thereafter. This loss of activity is likely due to accumulation of colloidal particles within the reactor and plugging of the pores.

27.5 CONCLUSION

Enzyme reactors based on monolithic supports appeared promising in preliminary studies summarized in Table 27.5 [53]. They are well suited for the construction of enzyme reactors processing both small and high molecular weight substrates. How-

TABLE 27.5

OVERVIEW OF ENZYME REACTORS BASED ON MONOLITHIC SUPPORTS

Support	Functional group for coupling and/or method	Enzyme	Substrate	Ref.
Poly(glycidyl methacrylate-*co*-ethylene dimethacrylate)	Epoxy	Carbonic anhydrase	2-chloro-4-nitrophenyl acetate, 4-nitrophenylacetat	[31]
Poly(glycidyl methacrylate-*co*-ethylene dimethacrylate)	Epoxy with ethylene diamine spacer and glutaraldehyde	Trypsin	N-benzoyl-L-arginine ethyl ester	[42]
Poly(glycidyl methacrylate-*co*-ethylene dimethacrylate)	Epoxy	Glucose oxidase	Fermentation broth of *S. cerevisiae* and *A. niger*	[35]
Poly(glycidyl methacrylate-*co*-ethylene dimethacrylate)	Epoxy with ethylene diamine spacer and glutaraldehyde	Glucose oxidase	Fermentation broth	[48]
Poly(glycidyl methacrylate-*co*-ethylene dimethacrylate)	Epoxy	Glucose oxidase	Not reported	[54]

continued on the next page

TABLE 27.5 (continued)

Support	Functional group for coupling and/or method	Enzyme	Substrate	Ref.
Poly(glycidyl methacrylate-*co*-ethylene dimethacrylate)	Epoxy with ethylene diamine spacer and glutaraldehyde	Elastase	Inter-α-trypsin inhibitor	[52]
Poly(2-vinyl-4,4-dimethylazlactone-*co*-acrylamide-*co*-ethylene dimethacrylate)	Azlactone	Trypsin	N-benzoyl-L-arginine ethyl ester	[30]
Superporous agarose	Cyanogen bromide activation	Lysozyme	*E. coli* broth	[36]
Superporous agarose	Cyanogen bromide activation	β-galactosidase	*E. coli* homogenate	[36]
Superporous agarose	Tresyl chloride activation	Glucose oxidase	*E. coli* broth	[36]
Superporous agarose	Reductive amination	β-galactosidase	Defatted milk	[38]

ever, detailed studies of performance, stability, and scale up parameters remain to be done. Similarly, effectiveness factors and Thiele moduli have also to be worked out. However, the ability of monolithic enzyme reactors to be part of systems involving both fast and slow reactions is intriguing since preliminary experiments indicate that mass transfer does not appear to be a limitation.

27.6 REFERENCES

1   O. Levenspiel, Chemical Reaction Engineering, Wiley, New York, 1999.

2   D.K. Sang, K. Yong, Chem. Eng. Sci. 52 (1997) 3639.

3   K. Yong, J.W. Kwang, H.K. Myung, D.K. Sang, Chem. Eng. Sci. 52 (1997) 3723.

4   H.C. IJselendoorn, H.P.A. Calis, C.M. van den Bleek, Chem. Eng. Sci. 56 (2001) 841.

5   F. Kapteijn, T.A. Nijhuis, J.J. Heiszwolf, J.A. Moulijn, Catal. Today 66 (2001) 133.

6   J.L. Williams, Catal. Today 69 (2001) 3.

7   T.A. Nijhuis, M.T. Kreutzer, A.C.J. Romijn, F. Kapteijn, J.A. Moulijn, Catal. Today 66 (2001) 157.

8   J.C. Jansen, J.H. Koegler, H. van Bekkum, H.P.A. Calis, C.M. van den Bleek, F. Kapteijn, J.A. Moulijn, E.R. Geus, N. van der Puil, Microporous and Mesoporous Materials 21 (1998) 213.

9   A. Cybulski, J.A. Moulijn, Catal. Rev.-Sci. Eng. 36 (1994) 179.

10   S.T. Gulati, in: A. Cybulski, J.A. Moulijn (Eds.), Structured catalysts and reactors, Marcel Dekker, New york, 1998, p. 15.

11   E. Garcia-Bordeje, F. Kapteijn, J.A. Moulijn, Catal. Today 69 (2001) 357.

12   T. Vergunst, F. Kapteijn, J.A. Moulijn, Carbon, In Press, (2002).

13   F. Kapteijn, J.J. Heiszwolf, T.A. Nijhuis, J.A. Moulijn, CATTECH 3 (1999) 24.

14   M.V. Twigg, D.E. Webster in: A. Cybulski, J.A. Moulijn (Eds.), Structured catalysts and reactors, Marcel Dekker, New York, 1998.

15   P.J.M. Lebens, M.M. Stork, F. Kapteijn, S.T. Sie, J.A. Moulijn, Chem. Eng. Sci. 54 (1999) 2381.

16   S.T. Sie, P.J.M. Lebens in: A. Cybulski, J.A. Moulijn (Eds.), Structured catalyst and reactors, Marcel Dekker, New York, 1998.

17   A. Cybulski, A. Stankiewicz, R.K. Edvinsson Albers, J.A. Moulijn, Chem. Eng. Sci. 54 (1999) 2351.

18   D. Mewes, T. Loser, M. Millies, Chem. Eng. Sci. 54 (1999) 4729.

19   M.T. Kreutzer, J.J. Heiszwolf, T.A. Nijhuis, J.A. Moulijn, CATTECH 3 (1999) 24.

20   M.T. Kreutzer, P. Du, J.J. Heiszwolf, F. Kapteijn, J.A. Moulijn, Chem. Eng. Sci. 56 (2001) 6015.

21   I.V. Koptyug, L.Y. Ilyina, A.V. Matveev, R.Z. Sagdeev, V.N. Parmon, S.A. Altobelli, Catal. Today 69 (2001) 385.

22   R.B. Bird, W.E. Stewart, E.N. Lightfoot, Transport phenomena, Wiley, New York, 2002.

23   R.D. Hawthorn, AIChE Symp. Ser. 701 (1974) 428.

24   G. Groppi, A. Belloli, E. Tronconi, P. Forzatti, Chem. Eng. Sci. 50 (1995) 2705.

25   R.E. Hayes, S.T. Kolaczkowski, Catal. Today 47 (1999) 295.

26   S.T. Kolaczkowski, Catal. Today 47 (1999) 209.

27   D. Josic, A. Štrancar, Ind. Eng. Chem. Res. 38 (1999) 333.

28   D. Josic, A. Buchacher, A. Jungbauer, J. Chromatogr. B 752 (2001) 191.

29   G. Iberer, R. Hahn, A. Jungbauer, LC-GC 17 (1999) 998.

30   S. Xie, F. Svec, J.M. Fréchet, Biotech. Bioeng. 62 (1999) 30.

31   H. Abou-Rebyeh, F. Körber, K. Schubert-Rehberg, J. Reusch, D. Josic, J. Chromatogr. 566 (1991) 341.

32   T.B. Tennikova, D. Horák, F. Svec, J. Kolář, J. Čoupek, S.A. Trushin, V.G. Maltzev, B.G. Belenkii, J. Chromatogr. 435 (1988) 357.

33   T. Tennikova, F. Svec, B.G. Belenkii, J. Liq. Chromatogr. 13 (1990) 63.

34   T.B. Tennikova, M. Bleha, F. Svec, T.V. Almazova, B.G. Belenkii, J. Chromatogr. 555 (1991) 97.

35   M. Vodopivec, M. Berovic, J. Jancar, A. Podgornik, A. Štrancar, Anal. Chim. Acta 407 (2000) 105.

36   M.P. Nandakumar, E. Palsson, P.E. Gustavsson, P.O. Larsson, B. Mattiasson, Biosep. 9 (2000) 193.

37   A. Podgornik, M. Barut, A. Štrancar, D. Josic, T. Koloini, Anal Chem 72 (2000) 5693.

38   P.-E. Gustavsson, P.-O. Larsson, J. Chromatogr. A 925 (2001) 69.

39   M.A. Teeters, T.W. Root, E.N. Lightfoot, J. Chromatogr. A 944 (2002) 129.

40   J. Turková, K. Bláha, M. Malaníková, D. Vančurová, F. Svec, J. Kálal, Biochim. Biophys. Acta 524 (1978) 162.

41   J. Drobník, V. Saudek, F. Svec, J. Kálal, V. Vojtíšek, M. Bárta, Biotech. Bioeng. 21 (1979) 1317.

42   M. Petro, F. Svec, M.J. Frechet, Biotech. Bioeng. 49 (1996) 355.

43   M.T. Tyn, T.W. Gusek, Biotech. Bioeng. 35 (1990) 327.

44   R. Hahn, A. Jungbauer, Anal. Chem. 72 (2000) 4853.

45   R. Hahn, E. Berger, K. Pflegerl, A. Jungbauer, Angew. Chem. Int. Ed. Submitted (2002).

46   F. Ayhan, H. Ayhan, E. Piskin, A. Tanyolac, Biores. Techn. 81 (2002) 131.

47   C. Horvath, B.A. Solomon, Biotech. Bioeng. 14 (1972) 885.

48   D. Josic, H. Schwinn, A. Štrancar, A. Podgornik, M. Barut, Y.P. Lim, M. Vodopivec, J. Chromatogr. A 803 (1998) 61.

49   S.C. March, I. Parikh, P. Cuatrecasas, Anal. Biochem. 60 (1974) 149.

50   K. Nilsson, K. Mosbach, Meth. Enzymol. 104 (1984) 56.

51   M.P. Nandakumar, A.M. Lali, B. Mattiasson, Biosep. 8 (1999) 237.

52   Y.P. Lim, H. Callanan, D. Hixon, M. Barut, A. Podgornik, D. Josic, A. Štrancar in: D. Josic, A. Štrancar, A. Jungbauer (Eds.), ISPPP 2000, Ljubljana, Slovenia, 2000.

53   F. Svec, J.M. Fréchet, Science 273 (1996) 205.

54   A. Štrancar, P. Koselj, Anal. Chem. 68 (1996) 3483.

55   P.J.M. Lebens, R.K. Edvinsson, S. Tiong Sie, J.A. Moulijn, Ind. Eng. Chem. Res. 37 (1998) 3722.

56   J.J. Heiszwolf, M.T. Kreutzer, M.G. van den Eijnden, F. Kapteijn, J.A. Moulijn, Catal. Today 69 (2001) 51.

57   G. Beri, A. Pintar, Chem. Eng. Sci. 52 (1997) 3709.

F. Švec, T.B. Tennikova and Z. Deyl (Editors)
Monolithic Materials
Journal of Chromatography Library, Vol. 67

Chapter 28

Solid Phase Synthesis and Auxiliaries for Combinatorial Chemistry

Alois JUNGBAUER and Karin PFLEGERL

Institute for Applied Microbiology, University of Agricultural Sciences, Muthgasse 18, A-1190 Vienna, Austria

CONTENTS

28.1 SOLID PHASE SYNTHESIS

28.1.1 Rational of solid phase synthesis

Combinatorial chemistry including both solid phase and parallel synthesis is progressing at a rapid pace. The progress has been comprehensively reviewed [1-5]. Large numbers of combinatorial libraries are prepared on solid phases. Solid phase synthesis is preferably applied to ease the removal of side-reaction-products, scavengers, and unconverted substrates. This method is applied for sequential reactions, since it makes little sense for single step reactions. The solid phase can be a pellicular material, filter, or a monolith. Solid-phase synthesis is typically performed on beads with a diameter of 50–500 μm (swollen size) and a loading of 0.1–1.0 mmol/g corresponding to 30 pmol–150 nmol of attached compound per bead. However, for some purposes it may be desirable to have larger amounts of material anchored to each bead, enough to perform NMR, biological testing, immunization, or for affinity chromatography. A bead with a diameter of 2–3 mm in swollen state and a loading in the around 1 mmol/g possesses about 2–8 mol of reaction sites, corresponding to about 1-4 mg of compound with a relative molecular mass of 500. However, the use of beads of this size requires a long time for a reagent to diffuse into the center of a bead. This consequently contributes to an increase in the overall reaction time. The reaction becomes diffusion controlled. When a reagent A is added to the system containing beads, two processes occur: diffusion and reaction. The flow of A through the bead surface is proportional to the flux of A multiplied by the surface area. Since no reaction occurs until reactant A and a substitution site B come into contact, the total flow through the bead surface is constant and independent of the bead radius [6].

Direct measurements of diffusion rates are technically demanding [7,8] and special equipment is required when diffusion is rapid. Groth *et al.* [9] described a simple method based on image analysis to determine the apparent diffusion kinetic in a bead. The partial acetylation of an amino-functionalized resin was used as a model reaction system and successive staining with acetaldehyde and chloranil was employed to visualize the progress of reaction. The data obtained from the image analyzer were transformed into concentration as a function of time and could be approximated by a pseudo-first-order reaction kinetic which is expressed by

$$y = a_0 \left(1 - e^{-k_{obs}t} \right) \tag{28.1}$$

where y is the measured percentage of reacted sites, a_0 is the conversion after the reaction is complete, k_{obs} is the observed rate constant, and t is the time. In this case, a_0 always equals 100%. For typical commercial available resins, k_{obs} values ranging from 3.200 to 0.018 min^{-1} were observed. These values correspond to a $t_{1/2}$, a time after which the regent has diffused half distance into the particle, of 0.22 to 38.5 min. In contrast to agitation and sonication, temperature had a significant effect on the reaction rate. Other important effects are the degree of crosslinking, the bead diameter, and the polarity of the solvent.

In the case of monoliths and filters, the reactions take place in the pores while the reactants are transported through the material by convection. Thus, the residence time of the reactants is controlled by the convective flow through the monolith or filter. For monoliths the diffusional resistance was not yet measured. However, it can be assumed to be lower by orders of magnitude, since the diffusional path in monoliths is very short [10]. This is also an important advantage in a number of other applications such as enzyme reactors [11-13] and separations [14,15].

28.1.2 Advantages of monolithic supports

Polymer supports in the format of functionalized styrene-divinylbenzene beads have been the principal part of solid phase peptide synthesis since it was introduced by Merrifield in 1963 [16]. Not surprisingly, these beads have also become a key component in the more recent developments involving solid phase synthesis. Despite numerous ingenious alternatives [3], they remain the support of choice. However continuing developments in automation have now reached such a level that the weighing and dosing of resin beads into small packed bed reactors, small batch stirred reactors, microtiter well plates, "tea bags", etc. has become to some extent limiting the scale and speed of parallel synthesis. To fully employ the advantage of automation, it would be convenient to use a single polymer particulate or monolith of a convenient size. This material should provide sufficient capacity to obtain enough individual compounds for comprehensive structural characterization, initial biological screening of pharmaceutically and agrochemically active compounds, or screening of engineering properties of affinity ligands [17,18]. It would also be advantageous if such a support is inexpensive, easily prepared, and applicable in current automated systems. In principle, a single large bead might fulfill these requirements, but the maximum particle size readily available using suspension polymerization is ~1–2 mm. Methods for production of larger size particles are not simple and potentially very costly. In addition, such large beads would have serious mass transfer limitations [9].

28.1.2.1 Polystyrene monolithic disks

Monolithic polymer rods or cylinders with a diameter of 8–10 mm and a length of 50 mm were prepared by polymerizing appropriate monomer mixtures in glass tubes [19]. These rods were recovered by carefully breaking the tubes, and disks were cut from them using conventional machining tools. A special protocol had to be developed to produce defect free rods, which were soft enough to cut. For example, the free radical initiator – azobisisobutyronitrile – generated bubbles if polymerization was carried out at 80°C, whereas 2,3′-azobis (2,4-dimethyl(valeronitrile)) used at 60°C has led to defect-free monoliths. To get soft monoliths, swelling solvent must be present in the polymerization mixture. Thus, toluene was found very suitable for styrene/divinylbenzene mixtures. With no solvent present, hard and glassy monoliths were obtained, while the use of a precipitating porogen such as alcohols, as previously described by Svec and Fréchet [20], resulted in friable rods. Toluene was also useful for styrene/PEG 400 diacrylate mixture, while dimethyl formamide had to be used for styrene/PEG 1000 diacrylate polymerization mixtures. The styrene/PEG 1000 monoliths exhibited very robust swelling and deswelling properties[19].

28.1.2.2 Poly(chloromethylstyrene) monolithic disks

Formulations similar to those described above were selected for further investigation with vinylbenzyl chloride as a functional monomer [19]. Table 28.1 shows the characteristics of the monoliths. The last polymer was prepared from PEG1000 di-4-vinylbenzyl ether as the crosslinker to mimic the solvation behavior offered by PEG1000 diacrylate, but to eliminate the ester groups as the possible sites for the undesired side-reactions. Two levels of chloromethylstyrene (10 and 50 wt%) were investigated. The elemental analysis and corresponding loading with chloromethyl groups demonstrated reasonably good correlation with the level of the functional monomer used in polymerization (Table 28.1). The disks were mechanically stable and responded to solvents in a way predicted from the styrene-based model studies. Then the disks have been tested using a multi-step synthesis of 2-hydroxy-7-methyl-biphenylcarboxylic acid methyl ester which has been previously prepared on conventional and developmental resins [19]. The first reaction conditions were chosen to represent the worst-case scenario. The disks were simply immersed in the solutions. Inspection of the cross-section of the disk revealed clear "sandwich" morphology. The reaction in the 2.5 mm thick disk progressed to a depth of only 0.8 mm. This effect is similar to the shrinking core phenomenon observed for some reactions on beads and is assumed to result from a substantial effect of solvent/reagent diffusion limitation.

Using monoliths, this effect can be circumvented by performing the synthesis in a flow through mode. The monolith is fitted in a small column and the reactants and solvents must be transported through the column by applying a positive or negative pressure. It can be anticipated that the transfer of known reaction schemes proceeding very efficiently on beads to a disk format should be relatively straightforward. Fur-

TABLE 28.1

PREPARATION OF MONOLITHIC POLYMERS USING CHLOROMETHYLSTYRENE
AND VARIOUS CROSSLINKERS[a] [19]

Polymerization mixture

Crosslinker type [b]	wt %	Styrene wt %	VBC wt %	Porogen[c] type	wt %	Elemental analysis C	H	Cl	$-CH_2Cl$ mmol g^{-1}
VB	2.4	1.65	50	Tol	27	79.5	6.4	15.3	4.32
DVB	2.4	3.05	10	Tol	27	88.5	7.4	2.45	0.69
PEGdiac	50	0	5	DMF	25	65.1	6.65	12.6	3.55
PEGdiac	50	1.76	10	DMF	25	76.1	7.4	3.1	0.87
PEGdst	50	0	50	DMF	27	65.4	6.8	13.9	3.91

[a]Reaction conditions: initiator, azobis[2,4-dimethyl(valeronitrile)] 1% (with respect to monomers), temperature 50°C, time 24 h

[b]DVB – 80% divinylbenzene (balance is ethylstyrene)
PEGdiac – poly(ethyleneglycol)1000 diacrylate
PEGdst – poly(ethyleneglycol)1000 divinylbenzylether

[c]Tol – toluene, DMF- dimethylformamide

thermore, a single monolithic disk with a weight of 250 mg can be easily prepared and
handled and may allow the synthesis of up to 0.5 mmol of a single compound.

28.2 REACTIVE FILTRATION

Functionalized polymers are also being applied to facilitate the liquid phase synthesis. For example, excessive reagents that have been involved in the synthesis can
be separated by reaction with surfaces or the surface functionalities. The flow-through
application of functionalized disks for scavenging is called reactive filtration [21]. The
disks with different chemistries can be stacked together and scavenging reactions
performed sequentially while the reaction mixture passes through the stack of monoliths.

Polyethylene encased porous poly(chloromethylstyrene-*co*-divinylbenzene) disks
have been prepared by Tripp *et al.* [22]. The pore size was controlled by addition of
porogens to the polymerization mixture. A mean pore diameter of 1.0 μm and a
surface area of about 8.5 m^2/g were optimal. This surface area was considered as

Fig. 28.1. Reaction used to demonstrate the utility of reactive filtration. Electrophilic scavenger react with nucleophiles such as amines (Reprinted from ref. [22]. Copyright 2001 American Chemical Society).

sufficient for grafting and the porosity high enough to generate only a low back pressure. Although the monoliths were sufficiently mechanically stable, they were reinforced by a polyethylene ring. This measure also prevented the disks from fraying of their edges. An additional benefit is that the flat face of the ring enables the disk to be firmly sealed between the bottom and top face of the cartridge without exercising excessive force on the porous polymer monolith itself. The disks were grafted with 2-vinyl-4,4-dimethylazlactone (VAZ) and used as electrophilic scavenger that reacts with nucleophiles such as amines according to the reaction shown in Fig. 28.1. The advantage of using monoliths instead of beads is faster mass transfer. Usually, such reactions are performed with beads in a stirred vessel. Rather large beads are used and according to the nature of the material and to the reactor design, washing steps require large quantities of solvent. Beads with a smaller diameter can be packed into a column to work in a flow through mode but as a result a very high back pressure may occur. The monoliths are the ultimate solutions. The high porosity affords a low back pressure while the monolithic structure is responsible for the low mass transport resistance. Overall, work with such materials is very economic.

Similar porous poly(chloromethylstyrene-*co*-divinylbenzene) disks encased with a polyethylene ring were used as a solid-phase acylating reagent [23]. The preparation of this polymeric acylation reagent is shown in Fig. 28.2. The benzylic chloride group of poly(chloromethylstyrene-*co*-divinylbenzene) (CMS-DVB) monolith 1 serves as a handle for further functionalization. First, the CMS-DVB monolith reacts with 4,4'-azobis(4-cyanovaleric acid) (ACVA), an azo initiator containing reactive carboxylic acid functionalities via displacement of the halogen of the benzyl chloride groups located at the pore surface. When this reaction is completed, a nitrophenyl ester bond is formed. As an alternative, a solid phase acylating reagent was developed with chemistry based on 2,3,5,6-tetrachlorophenol. The functionalized disks were tested in conversion of benzylamine to N-benzylacetamide. A large portion of benzylamine (74%) was converted to the desired product after a residence time of less than 1 min.

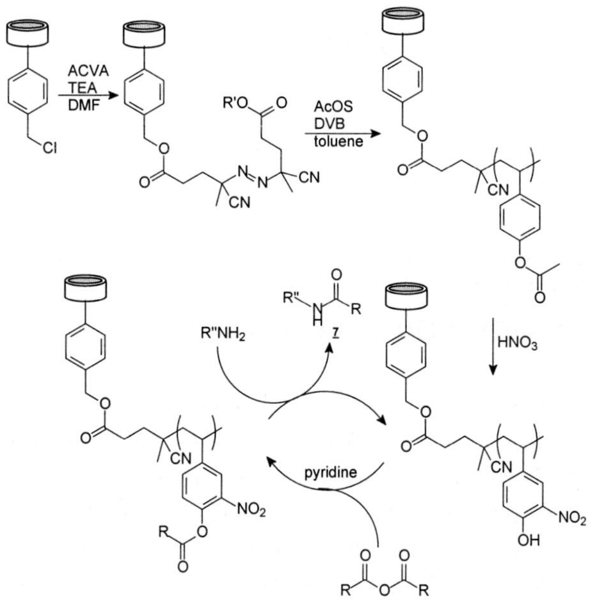

Fig. 28.2. Reaction scheme showing the preparation and use of monolith-supported nitrophenyl-esters acylating resin (Reprinted from ref. [23]. Copyright 2001 American Chemical Society).

An increase of the residence time to 8 minutes resulted in a 99% conversion. The flow-through implementation enhances reaction rate and leads to high conversions at shorter reaction times compared to more common bead-supported reagents. Typical reaction mixtures involving these beads must be maintained for several hours to achieve an extent of reaction comparable to that obtained with the monolith in less than 1 hour. In addition, the flow-though mode allows an easy connection of the cartridge containing the disk to a pump and use in an automated system. Furthermore, it was also demonstrated that the reagent could be regenerated.

The most exciting feature of monolith-supported reagents is that different reactions can be implemented in a sequence by stacking disks with various chemistries. To demonstrate this feature, an acylation and a scavenger disk were stacked. A single acetylating disk containing nitrophenyl ester functionalities converted benzylamine to an amide under conditions at which 86% conversion was achieved. Residual 14% of the original amine remains in solution together with the amide. The unreacted benzylamine was removed by the second disk grafted with poly(2-vinyl-4,4-dimethylazlactone). It scavenged all the benzylamine and a pure N-benzylacetamide was obtained.

This purification is readily achieved, because the chemistries of the two disks are strictly segregated and do not interfere with each other. It should be emphasized that this single path binary reaction cannot be accomplished with beads functionalized with relevant chemistries admixed to an amine solution, because both types of beads (reagent and scavenger) would compete for the amine at the same time and the yield as well as purity of the desired amide would be compromised. However, segregation of the chemistries of these disks enables to achieve several consecutive chemical transformations in a single path.

28.3 SOLID PHASE SYNTHESIS OF PEPTIDES

Curtius described in 1881 the first formation of a peptide bond in vitro [24]. He synthesized the dipeptide N-α-benzoyl-glycyl-glycine. One of his developments was also the introduction of the azide method for the coupling reaction that is practically free of racemization. In addition, he introduced the theoretical concept of protecting reactive amino acid side groups during synthesis. Emil Fischer later developed a series of theoretical and practical concepts for peptide synthesis, which have been valid ever since. Until 1919, he had synthesized more than 70 dipeptides, but he could not achieve the synthesis of oligopeptides. Within the following decades most important work has been done in the area of developing protection schemes for the N-terminus of the amino acids as well as for their side chains.

Merrifield introduced the concept of solid phase peptide synthesis (SPPS) in 1963, which is still used [16]. Synthesis of peptides with as many as 100 or more amino acid residue can be carried out using a solid polymer resin on which the growing peptide is attached while all chemicals are added in solution. Growth of the peptide is achieved via the C-terminus to the N-terminus. First, a cleavable linker is attached to the functionalized resin. In the following step, the N-protected first amino acid with activated carboxylic acid functionality is added. This carboxyl reacts with the functional group of the linker. After completion of the reaction, both remaining amino acid and side products of the coupling reaction are washed out. The resin with the attached amino acid remains in the reaction vessel. Before starting the next coupling step, the protected N-terminus is deprotected, and the resin is washed again. By repeating this sequence including reaction of the terminal free amino group with N-protected amino acid followed by deprotection, the peptide is synthesized step by step. After the complete peptide is synthesized, the covalent bond between the linker and the peptide is cleaved, and the peptide liberated. The most striking advantage of the SPPS concept is its amenability to automation. The simplicity of peptide synthesis using an entirely automated peptide synthesizer was very important for the breakthrough in peptide

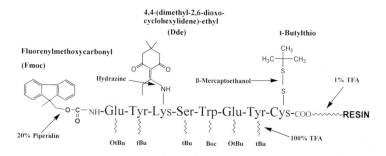

Fig. 28.3. Example of orthogonal protection during synthesis of a peptide directed against FVIII . Protecting groups: t-butyloxycarbonyl (Boc), t-butyl (tBu), and t-butoxy (OtBu).

synthesis and made available sufficient amounts of different peptides for various purposes.

The concept of protection of both side chains of amino acids and N-terminal reactive groups was first brought up by Curtius and developed by various chemists [24]. Nowadays, a large panel of cleavable linkers is available [25-32]. A variety of schemes were also developed for the protection of the terminal amino groups. For example, tert-butyloxycarbonyl (Boc) group for protection of lateral amino groups of lysine and histidine or the N-terminus of the peptide has been introduced by Barany and Merrifield [33]. The Boc group is acid labile and cleaves in the presence of trifluoroacetic acid (TFA). In 1970, Carpino and Han [34] developed the fluorenyl-9-methoxy carbonyl (Fmoc) group for protection of amino groups. This can be cleaved with bases such as piperidin. This enables orthogonal protection that can include combination of the base sensitive Fmoc-protection of the N-terminus with protection of the side groups using acid labile groups as shown as example in Fig. 28.3. This strategy can be also exploited for side chain directed labeling, modification, and immobilization of the peptide. During the cleavage of the Fmoc group after each synthesis cycle the side group protection is not affected. The linker between the peptide and the resin can also be acid labile. Thus, the deprotection of the side chain protection groups and cleavage of the linker can be accomplished in a simple step using TFA. The Boc strategy involves linkers, which are stable in acids and withstand the deprotection of the Boc groups during synthesis. Finally, hydrofluoric acid (HF) must be used to cleave the peptide from the resin. The Boc chemistry is considerably cheaper than Fmoc chemistry.

Independently of the protection of the N-terminus, the carboxyl of the amino acid must be activated. Several chemistries have been developed. According to Bodanszky [26] they can be classified in six different groups: azides, anhydrides, active esters, various coupling reagents, auxiliary nucleophiles, and enzymes. Based on these developments, several groups developed concepts and methods for multiple peptide synthesis (MPS) or simultaneous multiple peptide synthesis (SMPS) [35-38].

28.3.1 Styrene containing supports

Hird *et al.* [19] and Korol'kov *et al.* [39] introduced polymeric monoliths as supports for solid phase synthesis. Since the physical and chemical properties of macroporous monoliths are well characterized and the epoxy group of the polymerized glycidyl methacrylate can be utilized for a variety of chemical reactions, the peptide synthesis using these materials can be readily implemented. Korol'kov *et al.* [39] used porous poly(glycidyl methacrylate-*co*-styrene-*co*-ethylene dimethacrylate) monolithic disks and beads (GMA-ST-EDMA). These disks as well as beads had a capacity of 4 mmol/g of epoxy groups. Both the monolithic discs and the beads were first subjected to acid hydrolysis using dilute aqueous sulfuric acid to open the epoxyde group and form the vicinal diol functionality. Alternatively, these materials were treated with a three-fold volume excess of ethylenediamine to introduce the anchoring amino groups. For demonstration of the capability of this method, the nonapeptide Bradykinin was synthesized using these supports. The capacity of the beads was 0.3 mmol/g for the GMA-ST-EDMA and 0.1 mmol/g for GMA-EDMA. The amount of peptide could not be determined for the disks with styrene co-monomer, although it could be demonstrated that peptide with exact amino acid composition was grown. Only a much lower ligand density in the range between 0.0015 to 0.0022 mmol/g was achieved using GMA-EDMA disks. It seems likely that the reaction was performed by immersion of the disks. This may explain the low ligand density. The functionality of this material was tested using affinity chromatography. Antibodies have been isolated by passing pre-purified rabbit serum containing antibodies against bradykinin. The incorporation of styrene resulted into nonspecific adsorption of proteins. Although this was explained by the high ligand density, the hydrophobic nature of the styrene co-monomer appears to be responsible for these nonspecific interaction. Non-specific adsorption were not observed when the same peptide was immobilized on GMA-EDMA monoliths.

28.3.2 Methacrylate based disks

The idea to develop a combinatorial peptide synthesis protocol originates from the excellent performance of monoliths in affinity chromatography [40-46]. Peptides could be screened directly on cellulose sheets on which they have been prepared using spot synthesis. This approach could be readily transferred to CIM-disks since it was obvious to investigate peptide synthesis on chromatography carriers, which bear chemistries compatible with feed and regeneration solutions common in biotechnol-

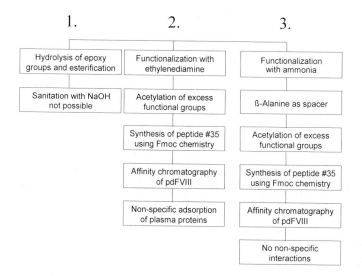

Fig. 28.4. Peptide synthesis on CIM-monoliths: strategy for the anchoring the peptide.

ogy [46]. These are typically aqueous buffers containing proteins and other bio-molecules. The alternative method developed by Korol'kov *et al.* [39] for peptide synthesis on monoliths does not meet the requirements for affinity chromatography since the support includes styrene, which is hydrophobic. Therefore proteins may irreversibly bind to the resin. The incorporation of styrene was motivated by the intention to prevent swelling and shrinking during synthesis.

In contrast, we aimed at the development of synthetic conditions using regular hydrophilic monoliths. The goal was to synthesize a combinatorial library of peptides on CIM-disk monoliths and to develop a screening procedure for selection of the best ligands. Pflegerl *et al.* [17,18] investigated three straightforward strategies for anchoring peptides: (i) Hydrolysis of the epoxy groups followed by esterification with the peptide, (ii) functionalization with ethylenediamine applying Fmoc-chemistry, and (iii) functionalization with ammonia and introduction of a β-alanine spacer applying Fmoc chemistry.

The individual steps of these synthetic strategies are shown in Fig. 28.4. In contrast to Korol'kov *et al.* [39], all reactants and washing solutions were drawn through the monolith. CIM columns were placed in a cartridge originally designed for chromatography and washed with dimethylformamide prior to the peptide synthesis. This was performed by injecting solutions of Fmoc N-terminal protected amino acid pentafluorophenyl esters in N-methylpyrolidone. The synthesis was carried out using a protocol similar to that of Frank, differing in use of CIM-disks used as the support.

The solutions can be simply pushed through the support by a syringe or connected to peptide synthesizer, which delivered all solutions in an automated fashion.

The first approach was found not well-suited for biotechnological applications since the ester bond does not resist alkaline conditions of sanitation with sodium hydroxide that is almost mandatory. A similar effect was observed by Korol'kov *et al.* [39]. Therefore this strategy was not pursued any further. Next, the peptides were synthesized using the second strategy. The function of the peptides was tested in affinity chromatography of FVIII. For that purpose, a series of peptides directed to FVIII was synthesized [46] and crude FVIII was loaded directly on monoliths. Non-specific adsorption of plasma proteins was observed. In order to avoid this undesired effect, β-alanine spacer was introduced instead of ethylenediamine and the third strategy used. The peptides were again synthesized with the help of Fmoc chemistry. In that case no non-specific adsorption occurred. Therefore, this strategy was applied in all following experiments.

In order to demonstrate the suitability of the support for the desired application, we also measured the swelling and shrinking during the peptide synthesis. The results are summarized in Table 28.2. The volume changes found in these experiments were considered acceptable for application of this method in parallel synthesis of peptides. For further demonstration of the reliability and reproducibility of this procedure, seven peptides were grown on CIM-disks using this protocol. The quality of synthesis was checked by amino acid analysis. A composition close to the theoretical was found in all cases. The different yields were due to variations in coupling times. Rigorous peptide characterization besides amino acid analysis was not possible, because the peptides were attached to the support without a cleavable linker to prevent leakage

TABLE 28.2

SWELLING PROPERTIES OF A CIM-DISK™ DURING PEPTIDE SYNTHESIS

Step	Volume, ml
Original CIM disk	0.34
DMF, 12 h	0.40
Cleavage solution 1[a], 30 min	0.43
Cleavage solution 2[b], 120 min	0.52
H_2O, 30 min	0.48
H_2O, 12 h	0.44

[a] 90% TFA, 3% TIBS, 2% H_2O, 1% phenol, 4% dichloromethane

[b] 50% TFA, 3% TIBS, 2% H_2O, 1% phenol, 44% dichloromethane

during consecutive chromatographic applications. Further characterization will require the development of novel analytical methods.

28.3.2.1 Development of a format for parallel synthesis

Conventional CIM disks™ had to be miniaturized for parallel synthesis in order to fit into a Microelute™ microtiter plate, which is equipped with a filter at the bottom. To ensure low consumption of solvent and amino acid during synthesis and low sample consumption during the screening procedure, a down-scaling of the commercially available 0.34 ml CIM-disks to so-called minidisks with a volume of 10 µl was performed. Additionally, the small monoliths were provided with a solvent resistant polypropylene ring and a sealing gasket. This gasket allowed tight fitting and avoided bypassing when the minidisks were placed in the 96-well microtiter plate conventionally used for solid phase extraction. The plate material was also tested for compatibility with the organic solvents used in the Fmoc-peptide synthesis. The synthesis station shown in Fig. 28.5 was adapted to this format. The individual steps of the parallel

A

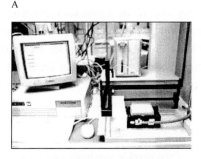

B

Fig. 28.5. Semi-automated spot robot with microlute™ plate (A) and microlute™ plate with a vacuum manifold (B).

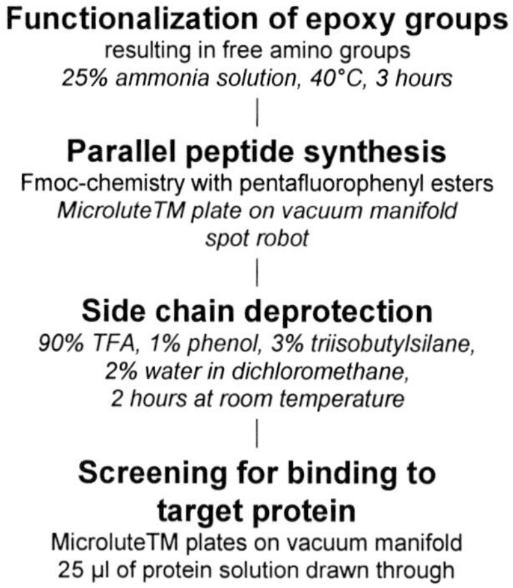

Functionalization of epoxy groups
resulting in free amino groups
25% ammonia solution, 40°C, 3 hours

|

Parallel peptide synthesis
Fmoc-chemistry with pentafluorophenyl esters
MicroluteTM plate on vacuum manifold
spot robot

|

Side chain deprotection
90% TFA, 1% phenol, 3% triisobutylsilane,
2% water in dichloromethane,
2 hours at room temperature

|

Screening for binding to
target protein
MicroluteTM plates on vacuum manifold
25 µl of protein solution drawn through

Fig. 28.6. Steps included in parallel peptide synthesis and screening on CIM-minidisks™.

synthesis are shown in Fig. 28.6. The porosity of the minidisks appears to be the critical issue in parallel peptide synthesis. All disks should have a similar porosity [17]. If this condition was not met, the flow resistance would vary for each microtiter well. Since all liquids are driven through the disk by vacuum, this effect would lead to different ligand densities and/or peptide quality.

28.3.2.2 Screening

Simultaneously with the development of the peptide synthesis on CIM-minidisks, a screening procedure shown in Fig. 28.7 has also been developed. In contrast to conventional parallel peptide synthesis systems, the monoliths offer two possibilities for screening functional ligands: The target protein is detected on the disk or, alternatively, the eluent from the liquid flown through the minidisks is screened for target protein. Using this screening procedure, the best affinant can be readily selected. In the first case, a positive signal indicates bound protein, while in the second, the protein must not be detected in the eluent. Fig. 28.7 also shows the different analytical methods we used to detect the bound proteins. Obviously, all analytical methods capable to detect the target protein are suited for screening of the functional ligand. Although the use of radiolabelled target protein would be also possible, the current trends prefer immuno-conjugates to radioactivity. However, not all enzymatic sub-

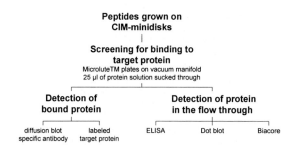

Fig. 28.7. Screening procedures for functional peptide ligands on CIM-minidisks™.

strates are suitable for immuno detection due to possible non-specific interaction with the polymethacrylate support, and radiolabeling may remain a useful alternative.

28.3.2.3 Combinatorial library

The preparation of a complete combinatorial library of selected octapeptides would require 1600 minidisks. Since this amount of disks was not available, we demonstrated the concept using mutational synthesis of a peptide directed against FVIII [17,47]. The amino acid residue at position three was varied to include all proteinogenic amino acids. Three different screening strategies were used to evaluate the library and compared to the conventional spot synthesis. The results of this screening are shown in Fig. 28.8. They document that our novel technology is more sensitive than that on spots and enables even cross checks since the bound material and the eluent can be analyzed independently. Our results also demonstrate that the direct synthesis of peptides on CIM-minidisks is possible and that functional ligands can be rapidly screened.

28.4 EUCLIDEAN SHAPE-ENCODED LIBRARIES

Euclidean shape-encoded libraries were developed to simplify decoding of chemical libraries by using monolithic supports. Vaino and Janda [48] exemplified the utility of Euclidean shape-encoded libraries on synthesis of a library of substituted ureas. Five Euclidean shapes shown in Fig. 28.9, circles, triangles, squares, pentagons, and hexagons, were selected, and 2 mm thick monoliths were prepared. By tailoring their dimensions (circle diameter of 5.0 mm; triangle of 5.9 mm/side; square of 3.3 mm/side; pentagon of 4.7 mm/side and hexagon of 3.3 mm/side), monoliths having equal surface areas and volumes but different minimum sizes were obtained. For a

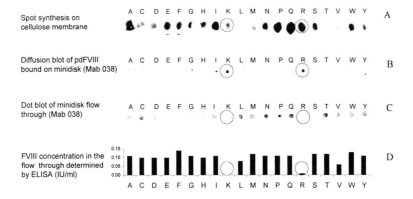

Fig. 28.8. Comparison of spot synthesis (A), diffusion blot from minidisks (B), dot-blot of an eluent flown through the mini disks (C), and measurement of FVIII concentrations in the minidisk eluent by ELISA (D). The peptide against blood coagulation factor VIII with the sequence EYKSWEYC was synthesized on minidisks and the randomized position 3 substituted using all proteinogenic amino acids.

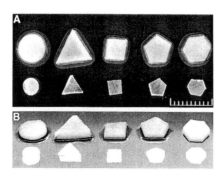

Fig. 28.9. Euclidean monolithic shapes in dry and swollen state. Top view of the monoliths swollen in 1,4 dioxane (A), view of dry monoliths at 45° angle (B) (Reprinted from ref. [48]. Copyright 2000 American Academy of Sciences). The scale is in millimeters.

successful application of these monoliths, the crosslinked polymer must be resilient to agitation, such as magnetic stirring, to manipulation with tweezers, and sieving. In addition, this supports have to swell sufficiently so that reactions could progress within the matrix while maintaining their shape over the course of several reactions. Therefore the monoliths were prepared from polystyrene incorporating a flexible crosslinking agent, 1,4-bis(vinylphenoxy)butane [49]. The reaction sequence used to demonstrate the utility of the concept is shown in Fig. 28.10. The split-and-mix

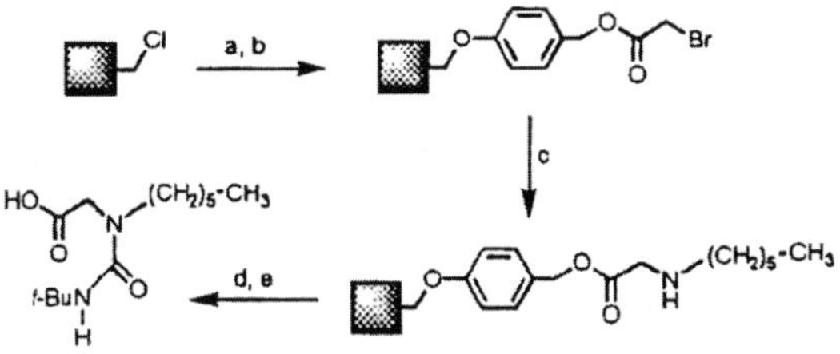

Fig. 28.10. Reaction sequence used for demonstration of the function of Euclidean shape-encoded monoliths (Reprinted from ref. [48]. Copyright 2000 American Academy of Sciences).

sequence leading to a library of substituted ureas was used as an example. A total of 24 different compounds were prepared using only 11 reaction steps. Preparing the same library by conventional parallel synthesis would have required 30 steps. Furthermore, the loading capacity of each of these monoliths is approximately 4,000 times higher than the amount of a compound required for the conventional spectroscopic characterization. It could be envisioned that this concept could be applied for the generation of much larger libraries. However, this would require an increase in sophistication of the system by adding further shapes. Also new methods of tagging must be involved.

28.5 CONCLUSION

Combinatorial and parallel syntheses are progressing at a very fast pace. A large number of reactions were performed using solid phases that act either as a support, as functional scavenger resins and extractants, or to ease the product work-up, purification, and isolation. The current state of the art solid phase syntheses are performed on beads with large diameters. However, beads exhibit a strong diffusional resistance and the reactions are mass transport controlled wit their diameter and reaction temperature playing a pivotal role. In order to overcome the diffusional limitations of the beads, monoliths have been used as supports. Both polystyrene and poly(vinylbenzylchloride)-based monoliths have been studied as a support. Poly(methacrylate)-based monoliths have been used for reactive filtration as scavengers and as supports for peptide synthesis. Peptides have been grown on poly(methacrylate) disks and the conjugate used for affinity chromatography. These supports offer desirable properties to be used in both peptide synthesis and affinity chromatography without significant unspecific adsorption of proteins. Although monolithic supports exhibit some distinct advantages over the conventional beads in combinatorial syntheses, they are currently less popular due to their novelty.

28.6 REFERENCES

1   H.M.I. Osborn, T.H. Khan, Tetrahedron 55 (1999) 1807.

2   P. Bravo, L. Bruche, C. Pesenti, F. Viani, A. Volonterio, M. Zanda, J. Flour. Chem. 112 (2001) 153.

3   R.E. Dolle, J. Comb. Chem. 3 (2001) 477.

4   S.W. Gerritz, Curr. Opin. Chem. Biol. 5 (2001) 264.

5   A. Ganesan, Drug Discovery Today 7 (2002) 47.

6   P.W. Atkins, in Physical Chemistry, Oxford University Press, Oxford, 1990, p. 840.

7   M.E. Wilson, K. Paech, W.J. Zhou, J. Org. Chem. 63 (1998) 5094.

8   M. Grotli, C.H. Gotfredsen, J. Rademann, J. Buchhardt, A.J. Clark, J. Duus, M. Meldal, J. Combinat. Chem. 2 (2000) 108.

9   T. Groth, M. Grotli, M. Meldal, J. Comb. Chem. 3 (2001) 461.

10   R. Hahn, A. Jungbauer, Anal. Chem. 72 (2000) 4853.

11   F. Svec, J.M. Fréchet, Science 273 (1996) 205.

12   F. Svec, P. Gemeiner, Biotech. Gen. Eng. Rev. 13 (1996) 217.

13   S. Xie, F. Svec, J.M.J. Fréchet, Biotech. Bioeng. 62 (1999) 30.

14   D. Josic, A. Buchacher, J. Biochem. Biophys. Meth. 49 (2001) 153.

15   D. Josic, A. Buchacher, A. Jungbauer, J. Chromatogr. B 752 (2001) 191.

16   R.B. Merrifield, J. Am. Chem. Soc. 85 (1963) 2149.

17   K. Pflegerl, A. Podgornik, E. Berger, A. Jungbauer, Biotech. Bioeng. In press (2002).

18   K. Pflegerl, A. Podgornik, E. Schallaun, A. Jungbauer, J. Comb. Chem. 4 (2002) 33.

19   N. Hird, I. Hughes, D. Hunter, M.G.J.T. Morrison, D.C. Sherrington, L. Stevenson, Tetrahedron 55 (1999) 9575.

20   F. Svec, J.M.J. Fréchet, Chem. Mater. (1995) 707.

21   J.A. Tripp, J.A. Stein, F. Svec, J.M. Fréchet, Org. Lett. 2 (2000) 195.

22   J.A. Tripp, F. Svec, J.M. Fréchet, J. Comb. Chem. 3 (2001) 216.

23   J.A. Tripp, F. Svec, J.M. Fréchet, J. Comb. Chem. 3 (2001) 604.

24   H.D. Jakubke, Peptide; Chemie und Biologie, Springer, Berlin, 1996.

25   M. Bodanszky, M.A. Bednarek, J. Prot. Chem. 8 (1989) 461.

26   M. Bodanszky, Principles of peptide synthesis, Springer, New York, 1993.

27   M.F. Songster, G. Barany, Methods Enzymol. 289 (1997) 126.

28   J.M. Stewart, Methods Enzymol. 289 (1997) 29.

29   O. Seitz, Angew. Chem. Int. Ed. 37 (1998) 3109.

30   K. Barlos, D. Gatos, Biopol. Pep. Sci. Section 51 (1999) 266.

31   F. Albericio, Biopol. Pep. Sci. Sect. 55 (2000) 123.

32   G.A. Grant, Synthetic petides, W.H. Freeman, New York, 1992.

33   G. Barany, R.B. Merrifield, Solid phase peptide synthesis, Academic Press, New York, 1979.

34   L.A. Carpino, G.Y. Han, J. Org. Chem. 37 (1972) 3404.

35   H.M. Geysen, R.H. Meloen, S.J. Barteling, Proc. Natl. Acad. Sci. USA 81 (1984) 3998.

36   R.A. Houghten, Proc. Natl. Acad. Sci. USA 82 (1985) 5131.

37   R. Frank, Tetrahedron 48 (1992) 9217.

38   S.P.A. Fodor, J.L. Read, M.C. Pirrung, L. Stryer, A.T. Lu, D. Solas, Science 251 (1991) 767.

39   V.I. Korol'kov, G.A. Platanova, V.V. Azanova, T.B. Tennikova, G.P. Vlasov, Letters Pept. Sci. 7 (2000) 53.

40   H. Abou-Rebyeh, F. Korber, K. Schubert-Rehberg, J. Reusch, D. Josic, J. Chromatogr 566 (1991) 341.

41   D. Josic, H. Schwinn, A. Strancar, A. Podgornik, M. Barut, Y.P. Lim, M. Vodopivec, J. Chromatogr A 803 (1998) 61.

42   X. Sun, Z. Chai, J. Chromatogr. A 943 (2002) 209.

43   L.G. Berruex, R. Freitag, T.B. Tennikova, J. Pharm. Biomed. Anal. 24 (2000) 95.

44   M. Schuster, E. Wasserbauer, A. Neubauer, A. Jungbauer, Bioseparation 9 (2000) 259.

45   Q. Luo, H. Zou, X. Xiao, Z. Guo, L. Kong, X. Mao, J. Chromatogr. A 926 (2001) 255.

46   K. Amatschek, R. Necina, R. Hahn, E. Schallaun, H. Schwinn, D. Josic, A. Jungbauer, J. High Resol. Chromatogr. 23 (2000) 47.

47   K. Pflegerl, R. Hahn, E. Berger, A. Jungbauer, J. Pept. Res. 59 (2002) 1.

48   A.R. Vaino, K.D. Janda, Proc. Natl. Acad. Sci. USA 97 (2000) 7692.

49   P.H. Toy, K.D. Janda, Tetrahedron Lett. 40 (1999) 6329.

Index of Monolithic Materials Used

Index of Compounds Separated

JOURNAL OF CHROMATOGRAPHY LIBRARY

A Series of Books Devoted to Chromatographic and Electrophoretic Techniques and their Applications

Although complementary to the Journal of Chromatography, each volume in the library series is an important and independent contribution in the field of chromatography and electrophoresis. The library contains no material reprinted from the journal itself.

Other volumes in this series